W9-AEJ-597

ADVANCED MECHANICS OF MATERIALS

ARTHUR P. BORESI
Professor and Head of Civil Engineering
University of Wyoming at Laramie

OMAR M. SIDEBOTTOM
Professor Emeritus of Theoretical and Applied Mechanics
University of Illinois at Urbana–Champaign

JOHN WILEY & SONS
New York Chichester Brisbane Toronto Singapore

Library of Congress Cataloging in Publication Data:

Boresi, Arthur P. (Arthur Peter), 1924–
Advanced mechanics of materials.

Includes bibliographical references and indexes.
1. Strength of materials. I. Sidebottom, Omar M.
(Omar Marion) II. Title.
TA405.B66 1985 620.1′12 84-17246
ISBN 0-471-88392-1

Printed in the United States of America

10 9 8 7 6 5 4 3 2 1

PREFACE

The present edition is an integration of both traditional methods and innovations in the field of engineering. The sometimes gradual, sometimes explosive developments in engineering are reflected in the focus and treatment of the material in this book. For example, the finite element method is introduced, a method that has permeated all of modern engineering with its approach to problem solving by means of carefully structured digital computer programs. At the same time, the book retains the essence of the method of advanced mechanics of materials, which approaches problem solving by carefully blending analysis, qualified approximations, and judgments based upon practical experience. In recognition and support of the role that computers play in modern engineering, several special-purpose computer programs are presented; however, this element is not overemphasized in an effort to avoid obscuring the power of the advanced mechanics method.

The overall structure of the book is as follows: Chapters 1, 2, 3, and 4 present a unified treatment of fundamentals that serves as a basis for the remainder of the text; important engineering problems are investigated in Chapters 5, 6, 7, 8, 9, and 11; and analyses of plates, stress concentrations, contact stresses, and finite elements are set forth in Chapters 10, 12, 13, 14, and 15.

As in previous editions, this edition discusses many types of load-carrying structural members and systems. The treatment of individual type members is based upon fundamental concepts of mechanics and is given in sufficient detail so that the resulting solutions are applicable to meaningful engineering problems. Numerous new illustrative problems are worked out for each type member as employed in practice, and a large number of new significant problems are included for the reader to solve (answers to many of the problems are also included). A limited number of digital computer programs are listed and a new chapter on finite element methods has been added.

In response to suggestions offered by users of previous editions, Chapter 10, "Flat Plates," has been shortened and in particular a large section of Part B, "Summary of Certain Plate Solutions," has been deleted, since there are several books available that treat flat plates as a separate subject. Also, much of the previous discussion on structural stability and buckling has been omitted. However, part of this topic has been moved to Chapter 3, "Failure Criteria," and is discussed from the viewpoint of failure criteria of buckling.

The book contains topics sufficient for two semesters or three quarters, thus giving the instructor an opportunity to choose topics of special interest to students in a one-semester course or a one- or two-quarter course. Depending on reader background, some of the fundamental chapters (Chapters 1 and 2, for example) may be bypassed; however, parts of the early chapters include material that is used throughout the remainder of the text. The book is not only suitable for senior and graduate engineering students but is also extremely relevant to the practicing engineer.

At the end of each chapter, a section called Additional References has been added to aid the serious student and the practitioner to seek out the latest reference materials. The book could also be used as a reference for courses in structural design, elastic plates, finite elements, and advanced reading, as well as for consulting engineers.

The authors hope that the user will find the text instructive, and they welcome reader comments, for it has been from such comments that the book has gradually changed over its long history to meet the requirements of modern engineering practice.

ARTHUR P. BORESI

May 1984 **OMAR M. SIDEBOTTOM**

PREFACE
to Third Edition

In acknowledgment of the widespread acceptance of the metric system, SI units are used throughout this major revision.

In modernizing and reorganizing the third edition, we have maintained the chief ingredients that made earlier editions so useful. In particular, we have followed the tradition of (1) the treatment of many important types of load carrying structural members, (2) the treatment of each type of member in sufficient detail so that the resulting solutions are applicable to meaningful engineering problems, (3) the presentation of numerous illustrative examples for each type member as employed in practice, and (4) the inclusion of a large number of problems for reader solution.

In recognition of the advances in engineering education that have occurred since the appearance of the second edition, in the third edition much of the general theory upon which the study of advanced mechanics of materials is based has been transferred from appendix to full chapter status. Therefore in Chapters 1 and 2, general theories of stress and strain and of stress-strain-temperature relations are presented. In Chapter 3, broad ideas of failure criteria are developed, and in Chapter 4 energy methods are considered in a unified manner, in contrast to the segmented approach used in the second edition.

The general concepts and theory developed in Chapters 1, 2, 3, and 4 are then employed to study in-depth individual types of structural members and specific concepts. Thus, we treat torsion of bars (Chapter 5), unsymmetrical bending of straight beams (Chapter 6), shear center for thin-wall beam cross sections (Chapter 7), curved beams (Chapter 8), beams on elastic foundations (Chapter 9), flat plates (Chapter 10), and the thick-wall cylinder (Chapter 11). Each of these topics is examined in depth with emphasis upon utility in the treatment of engineering problems. For problems that have exact solutions, only minor alterations of the earlier edition, mainly in presentation, have been made. However, problems that are solved at best by approximate methods, have been carefully reexamined and, where appropriate, modifications of the solutions presented in the second edition have been made. For example, the approximate solution for curved beams has been modified to make it readily applicable to complicated practical cross sections without the numerical integration required in the earlier editions. To note a second example, an approximate solution for plates is presented in Chapter 10, which can be applied directly to a broad class of plate problems, whereas results presented in earlier editions are practical for only a few simple plate shapes and a few simple loadings.

In addition, a number of topics not treated in earlier editions have been added. A few examples are the computation of upper bounds to the fully plastic loads of rectangular and circular plates, fully plastic torsion of noncircular sections, shear center of box sections, method of solution for ring loading of thin-wall cylinders, the use of the concept of orthogonal shearing stresses in the study of fatigue failure in contact stress problems, the presentation of failure criteria for high cycle fatigue including effects of range of stress and stress concentrations, presentation of a failure criterion of brittle fracture under plain strain conditions, and a discussion of basic concepts of fracture mechanics including the tabulation of toughness factors for various loading conditions.

Furthermore, several general topics of interest are discussed, namely, basic concepts of stress concentrations (Chapter 12); applications of effective stress concentration factors (Chapter 13); contact stresses (Chapter 14); and structural instability and buckling (Chapter 15). Incorporated in these latter chapters are discussions of significant engineering advancements and appropriate references that have appeared since the second edition.

Finally, Part V, "Influence of Small Inelastic Strains in Load Carrying Members" of the second edition has been deleted. However, many of those topics that remain important to modern engineering have been inserted in the articles of various chapters dealing with fully plastic loads.

Throughout the book, numerous sample problems have been worked in detail and a large number of problems have been added that are suitable for solution with a pocket calculator. Answers to many of the problems have been included.

Urbana-Champaign, Illinois — **ARTHUR P. BORESI**
March 1978 — **OMAR M. SIDEBOTTOM**

PREFACE to Second Edition

Many important contributions have been made to the subject of mechanics of materials since the first edition of this book appeared twenty years ago. Likewise, more students and engineers are adequately prepared for the study of advanced topics in this field of knowledge, and a greater need exists in engineering analysis and design for an understanding of such topics. As a consequence, the treatment in the second edition of the book has been made more penetrating and comprehensive, and much new material has been added. In fact, the second edition is essentially a new book, although the main objectives are the same as in the first edition.

Although methods of analysis are given careful attention throughout the book, equal emphasis is given to the engineering evaluation and interpretation of the analysis as influenced by the assumptions made and principles used.

The book was prepared primarily for advanced undergraduate and first-year graduate students in engineering, although in selecting the topics and methods of presentation, the needs of design and research engineers were kept in mind.

Where differential equations are involved in the analysis, their solutions are obtained, and the results, for a rather wide range of conditions, frequently are presented in the form of tables or curves. Numerical methods which are usually effective in the solution of differential equations for some combinations of physical conditions encountered in the analyses of this subject are not emphasized in this book.

The book consists of six parts, two new parts having been added: The Influence of Small Inelastic Strains on the Load-Carrying Capacity of Members, and Introduction to Instability—Buckling Loads. Useful material not previously published has also been presented in a number of the topics throughout the book. Likewise, two new appendixes have been added.

The purpose of the first appendix is to give the reader who is not familiar with the method employed in the theory of elasticity an opportunity to compare this method of analysis with that used in the so-called method of mechanics of materials as emphasized in this book. In treating certain topics in this edition the method of analysis and results of the mathematical theory of elasticity supplement the method of mechanics of materials. And, since it is assumed that students and engineers who have not had a formal course in the theory of elasticity may wish to make use of the treatment of these topics, Appendix I should serve a useful purpose. In like manner the addition of Appendix II, The Elastic Membrane (Soap-film) Analogy for Torsion, which makes use of the method discussed in Appendix I, should be helpful to the reader who wishes to understand the mathematical basis of the analogy.

Any part of the book after Part I is essentially independent of the other parts. Likewise the chapters in Part II, Special Topics on the Strength and Stiffness of Members Subjected to Static Loads, are not dependent on each other and hence may be studied in any order. Thus the book can readily be made to fit courses of different lengths, and also of different content and objectives. The complete book was prepared with the purpose of offering sufficient material for a course covering one academic year. Selected references at the end of each chapter suggest desirable sources of information for the reader who desires to pursue the subject further.

A detailed explanation is given in Chapter I of the main steps in the general procedure or method of analysis used in mechanics of materials. This general procedure is repeatedly illustrated throughout the book, especially in the chapters of Part II.

In Part II two new chapters have been introduced. These chapters deal with beams on elastic supports and with contact stresses. In these and a number of other topics the more complex results are given in readily useful form by means of tables and graphs. It is hoped that this feature will help make the book valuable in engineering design offices. Many illustrative problems are also given which emphasize applications of theory to design.

It is the aim in Part II to give a thorough treatment of a limited number of important topics rather than a briefer consideration of a relatively large number of topics. Thus, the student may be given the valuable experience of analyzing thoroughly by means of the methods and tools of mechanics of materials various types of engineering problems involving load-resisting members.

In Part III an attempt has been made to give a rational explanation of the significance of stress concentration in members of engineering machines and structures, thereby avoiding some of the confusion and difficulties which students frequently experience in the treatment of this topic.

Part IV on energy methods for determining the relationship between loads and deflections has been rewritten completely with a different approach, in order to present a more penetrating treatment. Two general methods of attack are used: work and energy and so-called complementary work and complementary energy. Emphasis is placed on the meaning, advantages, and disadvantages of these two methods of approach and on the limitations of less general but widely used procedures which are obtained from the more general methods, such as Castigliano's theorem and the dummy-load method.

The treatment in Part V of the inelastic behavior of load-resisting members contains a new, convenient approximate method of determining the load corresponding to a small specified amount of inelastic strain. The fact is emphasized in this part, as well as elsewhere in the book, that the load-carrying capacity of many members lies between two limiting values, namely: the load at which inelastic strain begins in the most-stressed fibers, and that at which a section or sections become fully plastic. The results of the method have been presented in convenient graphical form by means of interaction curves.

In Part VI a brief treatment is given of buckling of so-called thin-walled or slender members. Both elastic and inelastic (plastic) buckling are considered, primarily in relation to columns subjected to axial loads and to thin-walled cylinders subjected to uniform external pressure.

In this second edition many illustrative problems have been added as a means of introducing new methods or principles and not merely to show how to apply theories and methods previously explained.

Throughout the book there is considerable repetition of detailed statements of ideas, principles, and methods. This is done partly to make the discussion of a given topic less dependent on previously discussed topics than would otherwise be possible. The main purpose of the repetition, however, is to give emphasis to the ideas. The authors have found this procedure to be essential in the classroom and have introduced it to a limited extent in the book for the benefit of the student.

In the preparation of the second edition the authors were aided greatly by the constructive criticism of several of their colleagues who read considerable portions of the manuscript and by many students who studied various chapters in a graduate course. The authors are especially grateful to Professors Alfred M. Freudenthal and Winston E. Black for their careful examination of the manuscript and for helpful suggestions on many of the topics treated, and to Dr. C. K. Liu for contributions to the analysis of certain topics and problems and for preparing many of the drawings. Likewise, the authors are indebted to their colleagues Professors M. C. Steele and O. M. Sidebottom and to their former colleagues Professor M. C. Stippes, Dr. V. P. Jensen, and Mr. G. L. Armstrong for valuable suggestions in the treatment of a number of the topics.

The development of the topics in various parts of the second edition reflects also the many suggestions and comments made during the past twenty years by teachers, students, and practicing engineers who used the first edition. In addition, the authors, through oral discussions and correspondence, have obtained highly desirable information from a number of persons who were especially qualified by engineering experience to discuss recent developments in different phases of the subject; these generous contributions have been of great value to the book.

Urbana, Illinois **FRED B. SEELY**
October 1952 **JAMES O. SMITH**

PREFACE
to First Edition

The main title of this book might well have been "A Second Course in Mechanics of Materials." The topics considered in the book lie just beyond those usually included in a first course in strength of materials as given in most engineering schools in the United States.

The book is an outgrowth of notes used by the author during the past few years in a course for advanced undergraduate students and first year graduate students. It is well adapted for a course that either precedes or accompanies the study of the mathematical theory of elasticity.

The increasing use of analytical methods, in contrast with empirical rules, in solving the engineering problems that are continually arising in engineering industry has, in recent years, created a need for further training in the analysis of stresses and strains in various members of engineering structures and machines. There is likewise a need for a better understanding of the significance of calculated stresses in relation to the usable resistance of a member subjected to different types of loading.

It is hoped that this book may be a contribution towards filling this need both for teachers and students in institutions where a second course in strength of materials is given, and for some of the younger graduate engineers who may wish to make a further study of the subject on their own initiative and direction.

It is also hoped that the material herein presented may be found helpful in engineering offices where problems involving the analysis and significance of stresses in members are of importance.

In preparing the book the following objects have been kept in mind:

1. To review and make more useful the methods and results presented in the first course in strength of materials.
2. To show the limitations of the ordinary formulas of strength of materials, to consider the conditions under which these limitations are significant, and to extend the subject to include a variety of

important topics more complex than those usually considered in a first course.

3. To present a more comprehensive and useful view of the fundamental concepts and methods used in the analysis of stresses in structural and machine members.
4. To acquaint the student with various sources of information, largely through references, and thus give him an opportunity to appreciate how knowledge of this subject has grown.
5. To change the usual attitude of the student from one of dogmatic confidence in the methods employed and results obtained to one in which the methods and results are viewed as merely approximate but such that under certain conditions they become reliable and useful.

The book is divided into four parts as follows:

Part I. Preliminary considerations, consisting mainly of a discussion of the fundamental concepts involved in the subject, and a review of some of the more important methods used and results obtained in the usual first course in strength of materials.

Part II. Special topics, consisting of the analysis of stresses in a number of types of members not included, as a rule, in a first course in the subject.

Part III. Discussion of stress concentration and localized stress, in which non-mathematical methods of stress determination are emphasized.

Part IV. An introduction to the analysis of statically indeterminate stresses, in which methods involving elastic strain energy are used.

Throughout the book the engineering significance of the methods and results is strongly emphasized. Illustrative problems are frequently given, and many problems are offered for solution. References for further study are given at the end of each chapter.

In organizing the material herein presented, the results of experimental and analytical investigations from many sources have been used. The author wishes to acknowledge his indebtedness to those who have made this material available. Acknowledgment of the material used is given throughout the book where the material is presented.

Urbana, Illinois
August 1931

FRED B. SEELY

CONTENTS

CHAPTER 1

THEORIES OF STRESS AND STRAIN

Many topics in advanced mechanics of materials are discussed in this book. Load-stress and load-deflection relations for important structural members are derived. In addition, criteria are specified for determining the failure loads for specific members. Required prerequisite material is presented in Chapter 1, "Theories of Stress and Strain," Chapter 2, "Stress-Strain-Temperature Relations," and Chapter 3, "Failure Criteria." Depending upon student background, part or all of these chapters may be omitted.

Load-Stress and Load-Deflection Relations / For most of the members considered in this book we derive relations, in terms of known loads and known dimensions of the member, for either the distributions of normal and shearing stresses on a cross section of the member or for stress components that act at a point in the member. For a given member subjected to prescribed loads, the derivation of load-stress relations depends on the following conditions:

1. The equations of equilibrium (or equations of motion for bodies not in equilibrium).
2. The compatibility conditions that require deformed volume elements in the member to fit together without overlap or tearing (continuity conditions).
3. The stress-strain relations.
4. The material response.

Two different methods are used to satisfy Conditions (1) and (2): the method of mechanics of materials and the method of general continuum mechanics. Often load-stress and load-deflection relations have not been derived in this book by general continuum mechanics methods, either because the beginning student does not have the necessary background or because of the complexity of the general solutions. Instead, the method of mechanics of materials is used to obtain either exact solutions or reliable approximate solutions. In the method of mechanics of materials, the load-stress relations are derived first. They are then used to obtain load-deflection relations for the member.

A simple member such as a circular shaft of uniform cross section may be subjected to complex states of load that produce a multiaxial state of stress in the shaft. However, such states of load can be reduced to several simple types of load, such as axial centric, bending, and torsion. Each type of load, when acting alone, produces mainly one stress component, which is distributed over the cross section of the shaft. The method of mechanics of materials can be used to obtain load-stress relations for each type of load. If the deformations of the shaft that result from one type of load do not influence the magnitudes of the other types of loads and if the material remains linearly elastic for the combined loads, the stress components due to each type of load can be added together appropriately (that is, the method of superposition may be used). In a complex member, each load may have a significant influence on each component of the state of stress. Then, the method of mechanics of materials becomes cumbersome, and the use of the method of continuum mechanics may be more appropriate.

Method of Mechanics of Materials/The method of mechanics of materials is based on simplified assumptions related to the geometry of deformation (Condition 2) so that strain distributions for a cross section of the member can be determined. A basic assumption is that plane sections before loading remain plane after loading. The assumption can be shown to be exact for axially loaded members of uniform cross sections, for straight torsion members having uniform circular cross sections, and for straight beams of uniform cross sections subjected to pure bending. The assumption is approximate for other beam problems. The method of mechanics of materials is used in this book to treat all advanced beam topics (Chapters 6 to 9). In a similar way we assume that lines normal to the middle surface of an undeformed plate remain straight and normal to the middle surface after the load is applied. This assumption is used to simplify the plate problem in Chapter 10.

We review the steps used in the derivation of the flexure formula to illustrate the method of mechanics of materials and to show how the four conditions listed above are used. Consider a symmetrically loaded straight beam of uniform cross section subjected to a moment M that produces *pure bending*. (Note that the plane of loads lies in a plane of symmetry of every cross section of the beam.) It is required that we determine the normal stress distribution σ for a specified cross section of the beam. We assume that σ is the major stress component and, hence, ignore other effects. Pass a section through the beam, at the specified cross section, that cuts the beam into two lengths. Consider a free body diagram of one part. The applied moment M for this part of the beam is in equilibrium with internal forces represented by the sum of the forces that result from the normal stress σ that acts over the area of the cut section. Equations of equilibrium (Condition 1) relate the applied moment to internal forces. Since no axial external force acts, two integrals are obtained as follows: $\int \sigma \, dA = 0$ and $\int \sigma y \, dA = M$, where M is the applied external moment and y is the perpendicular distance from the neutral axis to the element of area dA.

Before the two integrals can be evaluated, we must know the distribution of σ over the cross section. Since the stress distribution is not known, it is determined indirectly through a strain distribution obtained by Condition 2. The continuity (Condition 2) is examined by consideration of two cross sections of the undeformed beam separated by an infinitesimal distance. Under the assumption that plane sections remain plane, the cross sections must rotate with respect to each other as the moment M is applied. There is a straight line in each cross section called the neutral axis along which the strains remain zero. Since plane sections remain plane, the strain distribution must vary linearly with the distance y as measured from this neutral axis.

Conditions 3 and 4 are now employed to obtain the relation between the assumed strain distribution and the stress distribution. Tension and compression stress-strain diagrams (Chapter 2) represent the material response (Condition 4) for the material in the beam. For sufficiently small strains, these diagrams indicate that the stresses and strains are related linearly. Their constant ratio, $\sigma/\epsilon = E$, is the modulus of elasticity for the material. The modulus of elasticity is found to be the same for tension and for compression for most engineering materials. Since other stress components in the beam are neglected, σ is the only stress component in the beam. Hence, by the stress-strain relations from Chapter 2 (Condition 3), $\sigma = E\epsilon$ for the beam. Therefore, both the stress σ and the strain ϵ vary linearly with the distance y as measured from the neutral axis of the beam. Hence, the equations of equilibrium can be integrated to obtain the flexure formula $\sigma = My/I$, where M is the applied moment at the given cross section of the beam and I is the moment of inertia of the beam cross section.

The method of mechanics of materials is used in Chapter 6 to treat unsymmetrical bending, in Chapter 7 to treat shear center, in Chapter 8 to treat curved beams, and in Chapter 9 to treat beams on elastic foundations.

Method of Continuum Mechanics, Theory of Elasticity/Many of the problems treated in this book—noncircular torsion (Chapter 5), plates (Chapter 10), thick walled cylinders (Chapter 11), contact stresses (Chapter 14), and stress concentrations (Chapters 12 and 13)—have multiaxial states of stress of such complexity that the method of mechanics of materials cannot be employed to derive load-stress and load-deflection relations simply as in the above example. Therefore, in such cases, the method of continuum mechanics is used. When we consider small displacements and when we deal with linear elastic material behavior only, the general method of continuum mechanics reduces to the method of the theory of linear elasticity.

In the derivation of load-stress and load-deflection relations by the theory of linear elasticity, an infinitesimal volume element at a point in a body with faces perpendicular to the coordinate axes is often employed. Condition 1 is represented by the differential equations of equilibrium (Chapter 1). Condition 2 is represented by the differential equations of compatibility (Chapter 1). The material response (Condition 4) for linearly elastic behavior is determined by one or more experimental tests that define the required elastic coefficients for the material. In this book we consider mainly isotropic materials for which two elastic coefficients are needed (Chapter 2). These coefficients can be obtained from a tension specimen if both axial and lateral strains are measured for every load applied to the specimen. Condition 3 is represented by the isotropic stress-strain relations developed in Chapter 2. If the differential equations of equilibrium and the differential equations of compatibility can be solved subject to specified stress-strain relations and specified boundary conditions, the states of stress and displacements for every point in the member are obtained.

Fully Plastic Loads/In addition to linear elastic behavior of members considered in this book, we derive fully plastic loads for many of the members. The major change in the analysis occurs in the material response. For fully plastic conditions, it is assumed that the material has a flat topped stress-strain diagram at the yield point stress (Chapter 2). Since stress is independent of strain at the fully plastic load, deflections of a member at the fully plastic load are not determinable.

Deflections by Energy Methods/Certain structures are made up of members whose cross sections remain essentially plane during the deflection of the structures. The deflected position of a cross section of a member of the structure is defined by three orthogonal displacement components of the centroid of the cross section and by three orthogonal rotation components of the cross section. These three components of displacement and of rotation of a cross section of a member of a structure are readily calculated by energy methods. For small displacements of the centroid, for small rotations, and for linearly elastic material behavior, Castigliano's theorem is recommended as a method for the computation of the displacements and rotations. The method is employed in Chapter 4 for structures made up of axially loaded members, beams, and torsion members, and in Chapter 8 for curved beams.

Failure Loads / Each member of a structure or a machine must be designed not only to resist a given design load, but also to include a factor of safety against overloads. Failure criteria and factors of safety are discussed in Chapter 3. Three classes of failure are treated. They are excessive elastic deflection, general yielding, and fracture. For members with elastic deflections proportional to loads, deflections are calculated by means of Castigliano's theorem (Chapter 4). Failure loads for members that fail by buckling, in which deflections are not proportional to loads, are also considered in Chapter 3.

The computation of exact failure loads for members that fail by general yielding is extremely difficult. Therefore, rather than attempt to compute exact failure loads, upper and lower bounds of the failure loads are computed. The lower bound to the failure load is calculated by the linear elastic method to be the load that initiates yielding in the member at locations other than at stress concentrations. The upper bound load for failure by general yielding is considered to be the load at which the cross section is fully plastic. The material property that is required to calculate the lower bound load for failure by general yielding is usually obtained from a tension test. However, often the state of stress in a member may be multiaxial. A criterion of failure is needed in order to use failure properties obtained from a specimen subjected to an axial state of stress to predict failure for a member subjected to multiaxial states of stress. To predict the lower bound failure by general yielding, two criteria of failure are generally employed: (1) the maximum shearing stress criterion and (2) the maximum octahedral shearing stress criterion. The stress components that specify the state of stress at a point in a member are calculated by load-stress formulas; generally these stress components do not directly represent the maximum shearing stress for the point, and they never represent directly the octahedral shearing stress for the point. Equations in Chapter 1 relate the maximum shearing stress and the octahedral shearing stress to these stress components.

Failure by fracture may be divided into three types: (1) the sudden fracture of members made of brittle materials, (2) the fracture of cracked or flawed members, and (3) the progressive fracture of members (fatigue). In this book the maximum principal stress failure criterion is used to predict failure loads of members made of brittle materials. Fracture mechanics techniques are presented to determine fracture loads of cracked members. It is assumed that yielding at the leading edge of the crack is minimal and that the opening mode of crack growth dominates the process. It is observed that yielding at the leading edge of the crack is minimal only if plane strain conditions are present over a large portion of the leading edge of the crack. Brittle fracture of cracked members are treated in Chapter 3 and in Chapter 12. Computation of failure loads for members subjected to fatigue loading are restricted to high cycle fatigue for which the number of cycles are greater than 10^6 cycles; yielding is assumed to be minimal for fatigue loading of these members. Fatigue of unnotched members are discussed in Chapter 3, and fatigue of notched members in Chapters 12 and 13.

1-1 DEFINITION OF STRESS AT A POINT

When loads are applied to a member, the loads are transmitted through the member as internal forces. To examine the internal forces, we pass a cutting plane through the member, which divides the member into two parts. Either part can be considered a free body diagram. The internal forces at the cut section are represented as a distributed load **F**. In general both the magnitude and direction of **F** vary over the cut section. If failure of the member (either because of general yielding or fracture) occurs on the cut section, the failure will be initiated at some point on the cut section. We find that the failure not only is influenced by the intensity of **F** at that point but also depends on the direction of **F** at that point. The intensity of **F** at that point is called the stress $\boldsymbol{\sigma}$ at that point; $\boldsymbol{\sigma}$ has the same direction as **F**. Since the direction of $\boldsymbol{\sigma}$ is important in design, we determine the normal and tangential components of $\boldsymbol{\sigma}$ which we designate $\boldsymbol{\sigma}_N$ and $\boldsymbol{\sigma}_S$. These stress components are defined in the following paragraphs.

Consider a general body subjected to forces acting on its surface (Fig. 1-1.1). Pass a fictitious plane Q through the body, cutting the body along surface A (Fig. 1-1.2). Designate one side of plane Q as positive and the other side as negative. The portion of the body on the positive side of Q exerts a force upon the portion of the body on the negative side. This force is transmitted through the plane Q by direct contact of the parts of the body on the two sides of Q. Let the force that is transmitted through an incremental area ΔA of A by the part on the positive side Q be denoted by **F**. In accordance with Newton's law of reaction, the portion of the body on the negative side of Q transmits through area ΔA a force $-\mathbf{F}$.

The force **F** may be resolved into components $\mathbf{F}_N$ and $\mathbf{F}_S$, along unit normal **N** and unit tangent **S**, respectively, to the plane Q. The force $\mathbf{F}_N$

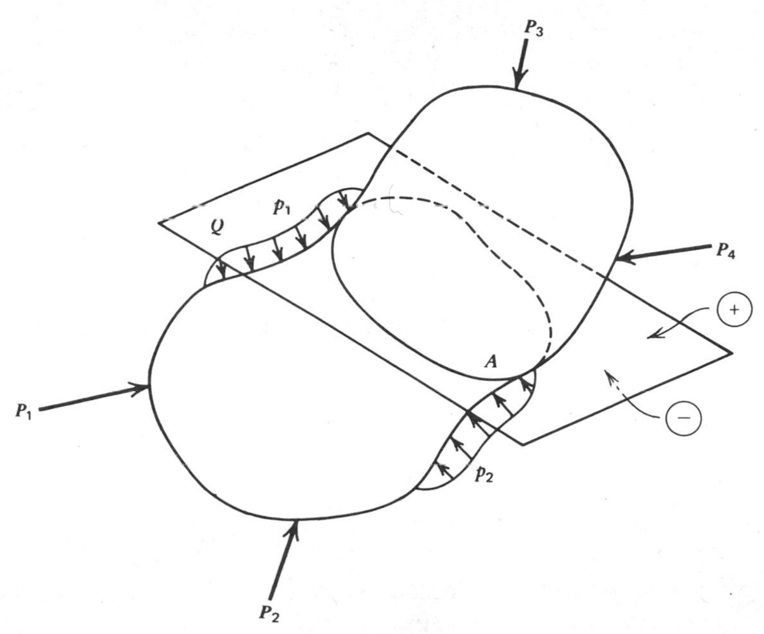

Fig. 1-1.1/A general loaded body cut by plane Q.

is called the *normal* (*perpendicular*) *force* on area ΔA and $\mathbf{F}_S$ is called the *shearing* (*tangential*) *force* on ΔA. The forces $\mathbf{F}$, $\mathbf{F}_N$, and $\mathbf{F}_S$ depend on the area ΔA and the orientation of plane Q. The magnitudes of the average forces per unit area are $F/\Delta A$, $F_N/\Delta A$, and $F_S/\Delta A$. These ratios are called the average stress, the average normal stress, and the average shearing stress, respectively, acting on area ΔA. The concept of

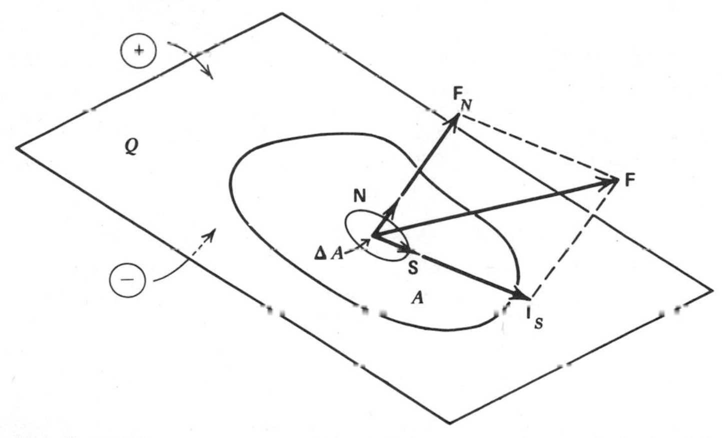

Fig. 1-1.2/Force transmitted through incremental area ΔA of cut body.

stress at a point is obtained by letting ΔA become an infinitesimal. Then the forces $\mathbf{F}$, $\mathbf{F}_N$, and $\mathbf{F}_S$ approach zero, but usually the ratios $\mathbf{F}/\Delta A$, $\mathbf{F}_N/\Delta A$, and $\mathbf{F}_S/\Delta A$ approach limits different from zero. The limiting ratio of $\mathbf{F}/\Delta A$ as ΔA goes to zero defines the stress vector $\boldsymbol{\sigma}$. Thus, the stress vector $\boldsymbol{\sigma}$ is given by

$$\boldsymbol{\sigma} = \lim_{\Delta A \to 0} \frac{\mathbf{F}}{\Delta A} \tag{1-1.1}$$

The stress vector $\boldsymbol{\sigma}$ always lies along the direction of the force vector $\mathbf{F}$, which in general is neither perpendicular nor tangent to the plane Q.

Similarly, the limiting ratios of $\mathbf{F}_N/\Delta A$ and $\mathbf{F}_S/\Delta A$ define the *normal stress vector* $\boldsymbol{\sigma}_N$ and the *shearing stress vector* $\boldsymbol{\sigma}_S$ that act at a point in the plane Q. In general we work with the magnitudes of these stress vectors defined by the relations

$$\sigma_N = \lim_{\Delta A \to 0} \frac{\mathbf{F}_N}{\Delta A} \qquad \sigma_S = \lim_{\Delta A \to 0} \frac{\mathbf{F}_S}{\Delta A} \tag{1-1.2}$$

The unit vectors associated with σ_N and σ_S are perpendicular and tangent, respectively, to the plane Q.

1-2 STRESS NOTATION

We use free body diagrams to specify the state of stress at a point and to obtain relations between various stress components. In general a free body diagram may be a diagram of a complete member, a portion of the member obtained by passing a cutting plane through the member, or a boxlike volume element of the member. The loads that act on any of these free body diagrams can be divided into two types as follows:

1. Surface forces, which are forces that act on the lateral surface of the free body diagram.
2. Body forces, which are forces that act throughout the volume of that portion of the member considered in the free body diagram.

Examples of surface forces are contact forces and distributed loads. Concentrated loads and reactions at a point are considered contact forces. Distributed loads may be either line loads with dimensions of force per unit length or surface loads with dimensions of force per unit area (dimensions of stress). Distributed loads on beams are often indicated as loads per unit length. Examples of surface loads are pressure exerted by a fluid in contact with the body and normal and shearing stresses that act on a cut section of the body.

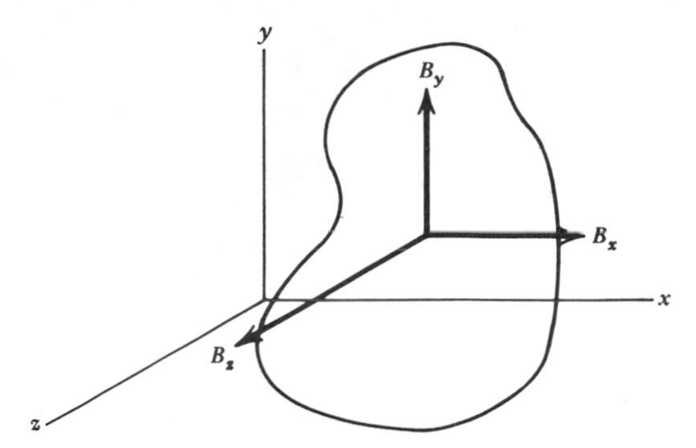

Fig. 1-2.1/Body forces.

Examples of body forces are gravitational forces, magnetic forces, and inertia forces. Since the body force is distributed throughout the volume of the free body diagram, it is convenient to define body force per unit volume. We use the notation **B** or (B_x, B_y, B_z) for body force per unit volume where **B** stands for body and subscripts (x, y, z) denote components in the (x, y, z)-directions, respectively, of the rectangular coordinate system (x, y, z) (see Fig. 1-2.1).

Consider now a free body diagram of a box-shaped volume element at a point 0 in a member, with sides parallel to the (x, y, z)-axes (Fig. 1-2.2). For simplicity, we show the volume element with one corner at point 0 and assume that the stress components are uniform (constant) throughout the volume element. The surface forces are represented by the product of the stress components (finite quantities shown in Fig. 1-2.2) and the areas† on which they act (product of two infinitesimal lengths of the sides of the volume element). Body forces, represented by the product of the components (B_x, B_y, B_z) and the volume of the element (product of the three infinitesimal lengths of the sides of the element), are higher order terms and are not shown on the free body diagram in Fig. 1-2.2. Consider the two faces perpendicular to the x-axis. The face from which the positive x-axis is extended is taken to be the positive face; the other face perpendicular to the x-axis is taken to be the negative face. The stress components σ_{xx}, σ_{xy}, and σ_{xz} acting on the positive face are taken to be in the positive sense as shown when they are directed in the positive x, y, and z senses. By Newton's law of reaction, the positive stress components σ_{xx}, σ_{xy}, and σ_{xz} shown acting on the negative face in Fig. 1-2.2 are in the negative (x, y, z) senses, respectively. In effect, a positive stress component σ_{xx} exerts a tension (pull) parallel to the x-axis. Equivalent sign conventions hold for the planes perpendicular to the y-

†In order to simplify Fig. 1-2.2 and others to follow, the areas on which the stress components act have been deleted. The reader must multiply each stress component by an appropriate area before applying equations of force equilibrium.

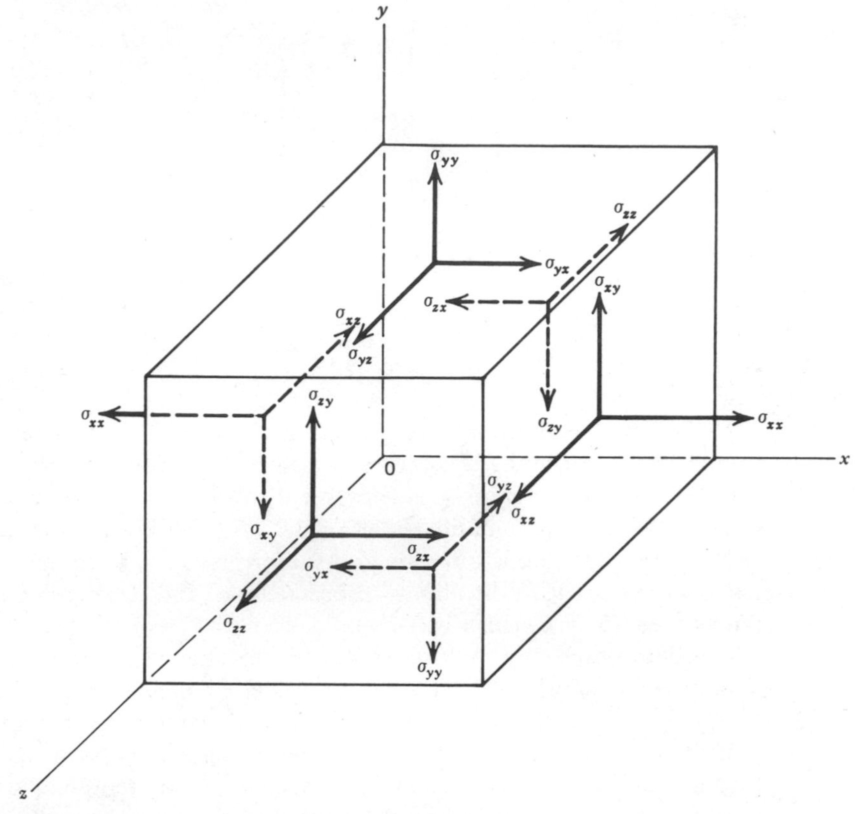

Fig. 1-2.2/Stress components at a point in loaded body.

and z-axis. Hence, associated with the concept of the state of stress at a point 0, nine components of stress exist:

$$(\sigma_{xx}, \sigma_{xy}, \sigma_{xz}), (\sigma_{yy}, \sigma_{yx}, \sigma_{yz}), (\sigma_{zz}, \sigma_{zx}, \sigma_{zy})$$

In the next article we show that the nine stress components may be reduced to six for most practical problems.

1-3 SYMMETRY OF THE STRESS ARRAY AND STRESS ON AN ARBITRARILY ORIENTED PLANE

Symmetry of Stress Components/The nine stress components relative to rectangular coordinate axes (x, y, z) may be tabulated in array form as follows:

$$\mathbf{T} = \begin{pmatrix} \sigma_{xx} & \sigma_{xy} & \sigma_{xz} \\ \sigma_{yx} & \sigma_{yy} & \sigma_{yz} \\ \sigma_{zx} & \sigma_{zy} & \sigma_{zz} \end{pmatrix} \tag{1-3.1}$$

Table 1-3.1

Summary of Stress Notations (Symmetric Stress Components)

I	σ_{xx}	σ_{yy}	σ_{zz}	$\sigma_{xy} = \sigma_{yx}$	$\sigma_{xz} = \sigma_{zx}$	$\sigma_{yz} = \sigma_{zy}$
II	σ_x	σ_y	σ_z	$\tau_{xy} = \tau_{yx}$	$\tau_{xz} = \tau_{zx}$	$\tau_{yz} = \tau_{zy}$
III	X_x	Y_y	Z_z	$X_y = Y_x$	$X_z = Z_x$	$Y_z = Z_y$
IV	$-X_x$	$-Y_y$	$-Z_z$	$-X_y = -Y_x$	$-X_z = -Z_x$	$-Y_z = -Z_y$
V	P	Q	R	S	T	U
VI	σ_{11}	σ_{22}	σ_{33}	$\sigma_{12} = \sigma_{21}$	$\sigma_{13} = \sigma_{31}$	$\sigma_{23} = \sigma_{32}$

where **T** symbolically represents the stress array called the stress tensor. In this array, the stress components in the first, second, and third rows act on planes perpendicular to the (x, y, z)-axes, respectively. Seemingly, nine stress components are required to describe the state of stress at a point in a member. However, if the only forces that act on the free body diagram in Fig. 1-2.2 are surface forces and body forces discussed in Art. 1-2, we can demonstrate from the equilibrium of the volume element in Fig. 1-2.2 that the three pairs of the shearing stresses are equal. Summation of force moments about the axes (x, y, z) leads to the result

$$\sigma_{yz} = \sigma_{zy}, \qquad \sigma_{zx} = \sigma_{xz}, \qquad \sigma_{xy} = \sigma_{yx} \tag{1-3.2}$$

Thus, with Eqs. (1-3.2), Eq. (1-3.1) may be written in the symmetrical form

$$\mathbf{T} = \begin{pmatrix} \sigma_{xx} & \sigma_{xy} & \sigma_{xz} \\ \sigma_{xy} & \sigma_{yy} & \sigma_{yz} \\ \sigma_{xz} & \sigma_{yz} & \sigma_{zz} \end{pmatrix} \tag{1-3.3}$$

Hence, for this type of stress theory, only six components of stress are required to describe the state of stress at a point in a member.

Although we do not consider body couples or surface couples in this book,[1†] it is possible for them to be acting in the free body diagram in Fig. 1-2.2. This means that Eqs. (1-3.2) are no longer true and that nine stress components are required to represent the unsymmetrical state of stress.

The stress notation described above is fairly widely used in engineering practice. The symmetric stress components used in this book are listed in Row I of Table 1-3.1. Several other stress notations are used in the technical literature. Frequently used symmetric stress notations are listed in Table 1-3.1. The symbolism indicated in Row VI is employed where index notation is used.

[†]Superior numbers are used to cite references at the end of the chapter.

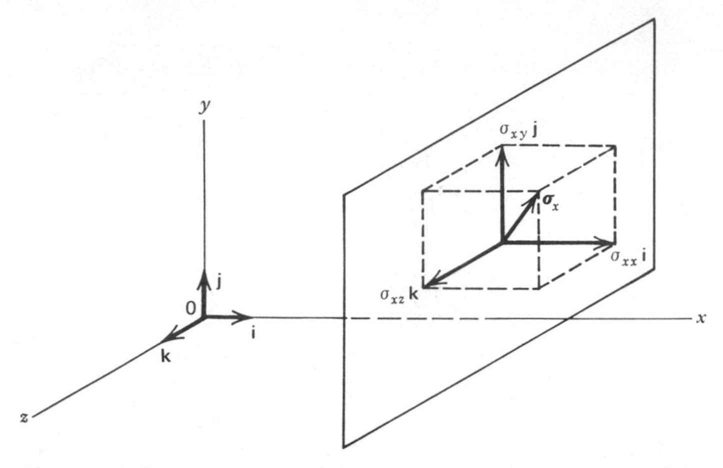

Fig. 1-3.1/Stress vector and its components acting on a plane perpendicular to the x-axis.

Stresses Acting on Arbitrary Planes/The stress vectors $\boldsymbol{\sigma}_x$, $\boldsymbol{\sigma}_y$, and $\boldsymbol{\sigma}_z$ on planes that are perpendicular, respectively, to the x, y, and z-axes are

$$\begin{aligned} \boldsymbol{\sigma}_x &= \sigma_{xx}\mathbf{i} + \sigma_{xy}\mathbf{j} + \sigma_{xz}\mathbf{k} \\ \boldsymbol{\sigma}_y &= \sigma_{yx}\mathbf{i} + \sigma_{yy}\mathbf{j} + \sigma_{yz}\mathbf{k} \\ \boldsymbol{\sigma}_z &= \sigma_{zx}\mathbf{i} + \sigma_{zy}\mathbf{j} + \sigma_{zz}\mathbf{k} \end{aligned} \tag{1-3.4}$$

where **i**, **j**, and **k** are unit vectors relative to (x, y, z)-axes (see Fig. 1-3.1 for $\boldsymbol{\sigma}_x$). Now consider the stress vector $\boldsymbol{\sigma}_P$ on an arbitrary oblique plane P through point 0 of a member (Fig. 1-3.2). For clarity, the plane P is

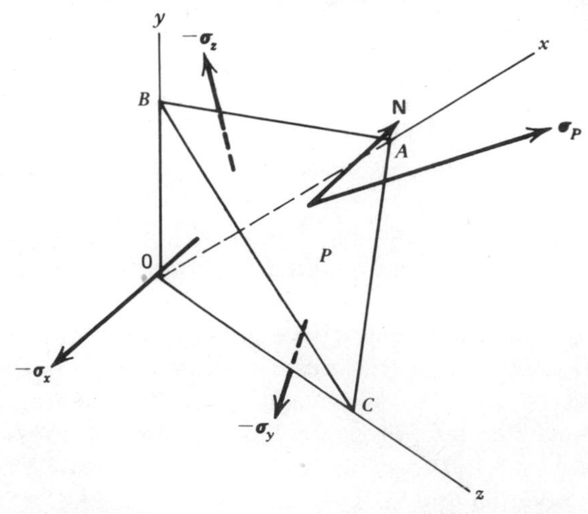

Fig. 1-3.2/Stress vector on arbitrary plane having a normal **N**.

shown removed from point 0. The unit normal vector to plane P is

$$\mathbf{N} = l\mathbf{i} + m\mathbf{j} + n\mathbf{k} \tag{1-3.5}$$

where (l, m, n) are the direction cosines of unit vector $\mathbf{N}$. Therefore, vectorial summation of forces acting on the tetrahedral element $0ABC$ yields the following (note that the ratios of areas $0BC$, $0AC$, $0AB$ to area ABC are equal to l, m, and n, respectively):

$$\boldsymbol{\sigma}_P = l\boldsymbol{\sigma}_x + m\boldsymbol{\sigma}_y + n\boldsymbol{\sigma}_z \tag{1-3.6}$$

Also, in terms of the projections $(\sigma_{Px}, \sigma_{Py}, \sigma_{Pz})$ of the stress vector $\boldsymbol{\sigma}_P$ along axes (x, y, z), we may write

$$\boldsymbol{\sigma}_P = \sigma_{Px}\mathbf{i} + \sigma_{Py}\mathbf{j} + \sigma_{Pz}\mathbf{k} \tag{1-3.7}$$

Comparison of Eqs. (1-3.6) and (1-3.7) yields, with Eqs. (1-3.4),

$$\begin{aligned} \sigma_{Px} &= l\sigma_{xx} + m\sigma_{yx} + n\sigma_{zx} \\ \sigma_{Py} &= l\sigma_{xy} + m\sigma_{yy} + n\sigma_{zy} \\ \sigma_{Pz} &= l\sigma_{xz} + m\sigma_{yz} + n\sigma_{zz} \end{aligned} \tag{1-3.8}$$

Equations (1-3.8) allow the computation of the components of stress on any oblique plane defined by unit normal $\mathbf{N}:(l, m, n)$, provided that the six components of stress

$$\sigma_{xx}, \sigma_{yy}, \sigma_{zz}, \sigma_{xy} = \sigma_{yx}, \sigma_{xz} = \sigma_{zx}, \sigma_{yz} = \sigma_{zy}$$

at point 0 are known. When point 0 lies on the surface of the member where the surface forces are represented by distributions of normal and shearing stresses, Eqs. (1-3.8) represent the *stress boundary conditions at point* 0.

Normal Stress and Shearing Stress on an Oblique Plane / The normal stress σ_{PN} on the plane P is the projection of the vector $\boldsymbol{\sigma}_P$ in the direction of $\mathbf{N}$; that is, $\sigma_{PN} = \boldsymbol{\sigma}_P \cdot \mathbf{N}$. Hence, by Eqs. (1-3.5), (1-3.7), and (1-3.8),

$$\begin{aligned} \sigma_{PN} &= l^2\sigma_{xx} + m^2\sigma_{yy} + n^2\sigma_{zz} + mn(\sigma_{yz} + \sigma_{zy}) + nl(\sigma_{xz} + \sigma_{zx}) \\ &\quad + lm(\sigma_{xy} + \sigma_{yx}) \\ &= l^2\sigma_{xx} + m^2\sigma_{yy} + n^2\sigma_{zz} + 2mn\sigma_{yz} + 2ln\sigma_{xz} + 2lm\sigma_{xy} \end{aligned} \tag{1-3.9}$$

By Eq. (1-3.9), the normal stress σ_{PN} on an oblique plane with unit normal $\mathbf{N}:(l, m, n)$ is expressed in terms of the six stress components $(\sigma_{xx}, \sigma_{yy}, \sigma_{zz}, \sigma_{xy}, \sigma_{xz}, \sigma_{zy})$. Often the maximum value of σ_{PN} at a point is of importance in design (see Art. 3-2). Of the infinite number of planes through point 0, σ_{PN} attains a maximum value called the maximum principal stress on one of these planes. The method of determining this stress and the orientation of the plane on which it acts is given in Art. 1-4.

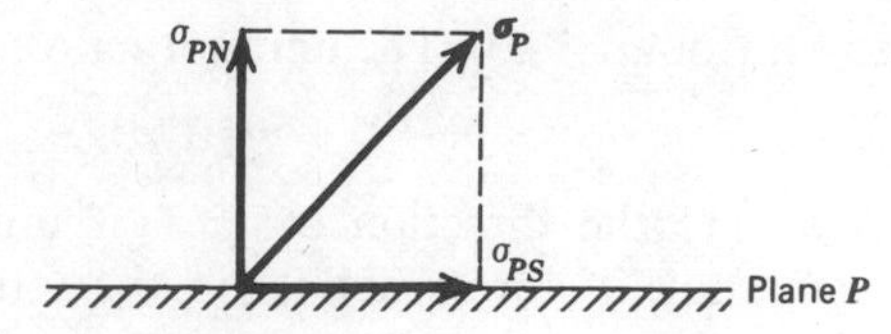

Fig. 1-3.3/Normal and shearing stress components of stress vector on an arbitrary plane.

To compute the magnitude of the shearing stress σ_{PS} on plane P, we note by geometry (Fig. 1-3.3) that

$$\sigma_{PS} = \sqrt{\sigma_P^2 - \sigma_{PN}^2} = \sqrt{\sigma_{Px}^2 + \sigma_{Py}^2 + \sigma_{Pz}^2 - \sigma_{PN}^2} \qquad (1\text{-}3.10)$$

Substitution of Eqs. (1-3.8) and (1-3.9) into Eq. (1-3.10) yields σ_{PS} in terms of $(\sigma_{xx}, \sigma_{yy}, \sigma_{zz}, \sigma_{xy}, \sigma_{xz}, \sigma_{zy})$ and (l, m, n). In certain criteria of failure the maximum value of σ_{PS} at a point in the body plays an important role (see Art. 3-2). The maximum value of σ_{PS} is equal to one half the difference of the maximum and minimum principal stresses (see Art. 1-4).

1-4
TRANSFORMATION OF STRESS. PRINCIPAL STRESSES. OTHER PROPERTIES

Transformation of Stress/Let (x, y, z) and (X, Y, Z) denote two rectangular coordinate systems with a common origin (Fig. 1-4.1). The cosines of the

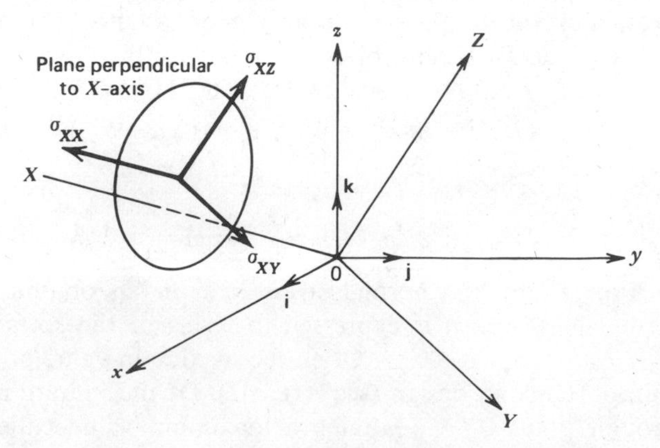

Fig. 1-4.1/Stress components on plane perpendicular to transformed X-axes.

Table 1-4.1

Direction Cosines

	x	y	z
X	l_1	m_1	n_1
Y	l_2	m_2	n_2
Z	l_3	m_3	n_3

angles between the coordinate axes (x, y, z) and the coordinate axes (X, Y, Z) are listed in Table 1-4.1. Each entry in Table 1-4.1 is the cosine of the angle between the two coordinate axes designated at the top of its column and to the left of its row.

The stress components $\sigma_{XX}, \sigma_{XY}, \sigma_{XZ}, \ldots$ are defined with reference to (X, Y, Z)-axes in the same manner as $\sigma_{xx}, \sigma_{xy}, \sigma_{xz}, \ldots$ are defined relative to the axes (x, y, z). Hence, σ_{XX} is the normal stress component on a plane perpendicular to axis X and σ_{XY}, σ_{XZ} are shearing stress components on this same plane (Fig. 1-4.1), and so on. Hence, by Eq. (1-3.9),

$$
\begin{aligned}
\sigma_{XX} &= l_1^2\sigma_{xx} + m_1^2\sigma_{yy} + n_1^2\sigma_{zz} \\
&\quad + 2m_1n_1\sigma_{yz} + 2n_1l_1\sigma_{zx} + 2l_1m_1\sigma_{xy} \\
\sigma_{YY} &= l_2^2\sigma_{xx} + m_2^2\sigma_{yy} + n_2^2\sigma_{zz} \\
&\quad + 2m_2n_2\sigma_{yz} + 2n_2l_2\sigma_{zx} + 2l_2m_2\sigma_{xy} \\
\sigma_{ZZ} &= l_3^2\sigma_{xx} + m_3^2\sigma_{yy} + n_3^2\sigma_{zz} \\
&\quad + 2m_3n_3\sigma_{yz} + 2n_3l_3\sigma_{zx} + 2l_3m_3\sigma_{xy}
\end{aligned}
\tag{1-4.1}
$$

The shearing stress component σ_{XY} is the component in the Y-direction of the stress vector on a plane perpendicular to the X-axis; that is, it is the Y-component of the stress vector $\boldsymbol{\sigma}_X$ acting on the plane perpendicular to the X-axis. Thus, σ_{XY} may be evaluated by forming the scalar product of the vector $\boldsymbol{\sigma}_X$ [determined by Eqs. (1-3.7) and (1-3.8) with $l_1 = l$, $m_1 = m$, $n_1 = n$] with a unit vector parallel to the Y-axis; that is, with the unit vector

$$\mathbf{N}_2 = l_2\mathbf{i} + m_2\mathbf{j} + n_2\mathbf{k} \tag{1-4.2}$$

Hence, by Eqs. (1-3.7), (1-3.8) and (1-4.2), σ_{XY} is determined; similar procedures also determine σ_{XZ} and σ_{YZ}. Hence,

$$
\begin{aligned}
\sigma_{XY} &= \boldsymbol{\sigma}_X \cdot \mathbf{N}_2 = \boldsymbol{\sigma}_Y \cdot \mathbf{N}_1 \\
&= l_1 l_2 \sigma_{xx} + m_1 m_2 \sigma_{yy} + n_1 n_2 \sigma_{zz} \\
&\quad + (m_1 n_2 + m_2 n_1)\sigma_{yz} + (l_1 n_2 + l_2 n_1)\sigma_{zx} \\
&\quad + (l_1 m_2 + l_2 m_1)\sigma_{xy} \\
\sigma_{XZ} &= \boldsymbol{\sigma}_X \cdot \mathbf{N}_3 = l_1 l_3 \sigma_{xx} + m_1 m_3 \sigma_{yy} + n_1 n_3 \sigma_{zz} \\
&\quad + (m_1 n_3 + m_3 n_1)\sigma_{yz} + (l_1 n_3 + l_3 n_1)\sigma_{zx} \\
&\quad + (l_1 m_3 + l_3 m_1)\sigma_{xy} \\
\sigma_{YZ} &= \boldsymbol{\sigma}_Y \cdot \mathbf{N}_3 = l_2 l_3 \sigma_{xx} + m_2 m_3 \sigma_{yy} + n_2 n_3 \sigma_{zz} \\
&\quad + (m_2 n_3 + m_3 n_2)\sigma_{yz} + (l_2 n_3 + l_3 n_2)\sigma_{zx} \\
&\quad + (l_2 m_3 + l_3 m_2)\sigma_{xy}
\end{aligned}
\tag{1-4.3}
$$

Equations (1-4.1) and (1-4.3) determine the stress components relative to rectangular axes (X, Y, Z) in terms of the stress components relative to rectangular axes (x, y, z); that is, they determine how the stress components transform under a rotation of rectangular axes. A set of quantities that transform according to this rule is called a second-order symmetrical tensor. Later it will be shown that strain components (see Art. 1-7) and moments and products of inertia (see Art. A-3) also transform under rotation of axes by similar relationships; hence, they too are second-order symmetrical tensors.

Principal Stresses/It may be shown that for any general state of stress at any point 0 in a body there exist three mutually perpendicular planes at point 0 on which the shearing stresses vanish. The remaining normal stress components on these three planes are called *principal stresses*. Correspondingly, the three planes are called principal planes, and the three mutually perpendicular axes that are normal to the three planes (hence, that coincide with the three principal stress directions) are called principal axes. Thus, by definition, principal stresses are directed along principal axes that are perpendicular to the principal planes. A cubic element subjected to principal stresses is easily visualized, since the forces on the surface of the cube are normal to the faces of the cube. More complete discussions of principal stress theory are present elsewhere.[1] Here we merely sketch the main results.

Principal Values and Directions/Since the shearing stresses vanish on principal planes, the stress vector on principal planes is given by $\boldsymbol{\sigma}_P = \sigma \mathbf{N}$, where σ is the magnitude of the stress vector $\boldsymbol{\sigma}_P$ and $\mathbf{N}$ is the unit normal to a principal plane. Let $\mathbf{N} = l\mathbf{i} + m\mathbf{j} + n\mathbf{k}$ relative to rectangular axes (x, y, z) with associate unit vectors $\mathbf{i}, \mathbf{j}, \mathbf{k}$. Thus, (l, m, n) are the direction cosines of unit normal $\mathbf{N}$. Thus, projections of $\boldsymbol{\sigma}_P$ along (x, y, z)-axes are $\sigma_{Px} = \sigma l$, $\sigma_{Py} = \sigma m$, $\sigma_{pz} = \sigma n$. Hence, by Eq. (1-3.8), we obtain

$$
\begin{aligned}
&l(\sigma_{xx} - \sigma) + m\sigma_{xy} + n\sigma_{xz} = 0 \\
&l\sigma_{xy} + m(\sigma_{yy} - \sigma) + n\sigma_{yz} = 0 \\
&l\sigma_{xz} + m\sigma_{yz} + n(\sigma_{zz} - \sigma) = 0
\end{aligned}
\tag{1-4.4}
$$

Since Eqs. (1-4.4) are linear homogeneous equations in (l, m, n) and since the trivial solution $l = m = n = 0$ is impossible because $l^2 + m^2 + n^2 = 1$ (law of direction cosines), it follows from the theory of linear algebraic equations that Eqs. (1-4.4) are consistent if and only if the determinant of the coefficients of (l, m, n) vanishes identically. Thus, we have

$$
\begin{vmatrix}
\sigma_{xx} - \sigma & \sigma_{xy} & \sigma_{xz} \\
\sigma_{xy} & \sigma_{yy} - \sigma & \sigma_{yz} \\
\sigma_{xz} & \sigma_{yz} & \sigma_{zz} - \sigma
\end{vmatrix} = 0
\tag{1-4.5}
$$

or, expanding the determinant, we obtain

$$
\sigma^3 - I_1\sigma^2 + I_2\sigma - I_3 = 0
\tag{1-4.6}
$$

where

$$
\begin{aligned}
I_1 &= \sigma_{xx} + \sigma_{yy} + \sigma_{zz} \\
I_2 &= \begin{vmatrix} \sigma_{xx} & \sigma_{xy} \\ \sigma_{xy} & \sigma_{yy} \end{vmatrix} + \begin{vmatrix} \sigma_{xx} & \sigma_{xz} \\ \sigma_{xz} & \sigma_{zz} \end{vmatrix} + \begin{vmatrix} \sigma_{yy} & \sigma_{yz} \\ \sigma_{yz} & \sigma_{zz} \end{vmatrix} \\
&= \sigma_{xx}\sigma_{yy} + \sigma_{xx}\sigma_{zz} + \sigma_{yy}\sigma_{zz} - \sigma_{xy}^2 - \sigma_{xz}^2 - \sigma_{yz}^2 \\
I_3 &= \begin{vmatrix} \sigma_{xx} & \sigma_{xy} & \sigma_{xz} \\ \sigma_{xy} & \sigma_{yy} & \sigma_{yz} \\ \sigma_{xz} & \sigma_{yz} & \sigma_{zz} \end{vmatrix}
\end{aligned}
\tag{1-4.7}
$$

The three roots $(\sigma_1, \sigma_2, \sigma_3)$ of Eq. (1-4.6) are the three principal stresses at point 0. The magnitudes and directions of σ_1, σ_2, and σ_3 for a given member depend only on the loads being applied to the member and cannot be influenced by the choice of coordinate axes (x, y, z) used to specify the state of stress at point 0. This means that I_1, I_2, and I_3 given by Eqs. (1-4.7) are *invariants of stress* and must have the same magnitudes for all choices of coordinate axes (x, y, z).

When $(\sigma_1, \sigma_2, \sigma_3)$ have been determined, the direction cosines of the three principal axes are obtained from Eqs. (1-4.4) by setting σ in turn equal to $(\sigma_1, \sigma_2, \sigma_3)$, respectively, and observing the direction cosine condition $l^2 + m^2 + n^2 = 1$ for each of the three values of σ.

In special cases, two principal stresses may be numerically equal. Then, Eqs. (1-4.4) show that the associated principal directions are not unique.

In particular, all planes through the unique principal axis for which the principal stress has a different value from the other two principal stresses are principal planes. Then, any two mutually perpendicular axes that are perpendicular to the unique principal axis will serve as principal axes. If all three principal stresses are equal, $\sigma_1 = \sigma_2 = \sigma_3$ at point 0, and all planes passing through point 0 are principal planes. Then any set of three mutually perpendicular axes at point 0 will serve as principal axes. A simple FORTRAN program for computation of principal stresses is listed in Table 1-4.2.

Octahedral Stress / Let (x, y, z) be principal axes. Consider the family of planes whose unit normals satisfy the relation $l^2 = m^2 = n^2 = \frac{1}{3}$ with respect to the principal axes (x, y, z). There are eight such planes (the octahedral planes) that make equal angles with respect to the (x, y, z)-directions. Therefore, the normal and shearing stress components associated with these planes are called the *octahedral normal stress* σ_{oct} and the *octahedral shearing stress* τ_{oct}. By Eqs. (1-3.8) and (1-3.9) and (1-3.10), we obtain

$$\begin{aligned} \sigma_{oct} &= \tfrac{1}{3}(\sigma_1 + \sigma_2 + \sigma_3) = \tfrac{1}{3} I_1 \\ 9\tau_{oct}^2 &= (\sigma_1 - \sigma_2)^2 + (\sigma_1 - \sigma_3)^2 + (\sigma_2 - \sigma_3)^2 = 2I_1^2 - 6I_2 \end{aligned} \tag{1-4.8}$$

since for the principal axes $\sigma_{xx} = \sigma_1$, $\sigma_{yy} = \sigma_2$, $\sigma_{zz} = \sigma_3$ and $\sigma_{xy} = \sigma_{yz} = \sigma_{zx} = 0$. (See Eqs. 1-4.7.) It follows that since (I_1, I_2, I_3) are invariants under rotation of axes, we may refer Eqs. (1-4.8) to arbitrary (x, y, z)-axes by replacing I_1, I_2, I_3 by their general forms as given by Eqs. (1-4.7). Thus for arbitrary (x, y, z)-axes,

$$\begin{aligned} \sigma_{oct} &= \tfrac{1}{3}(\sigma_{xx} + \sigma_{yy} + \sigma_{zz}) \\ 9\tau_{oct}^2 &= (\sigma_{xx} - \sigma_{yy})^2 + (\sigma_{xx} - \sigma_{zz})^2 + (\sigma_{yy} - \sigma_{zz})^2 \\ &\quad + 6\sigma_{xy}^2 + 6\sigma_{xz}^2 + 6\sigma_{yz}^2 \end{aligned} \tag{1-4.9}$$

The normal and shearing octahedral stresses play a role in certain failure criteria (Art. 3-2).

Mean and Deviator Stress / Experiments indicate that yielding and plastic deformation of many metals are essentially independent of the applied normal stress σ_m, where by definition

$$\sigma_m = \frac{\sigma_{xx} + \sigma_{yy} + \sigma_{zz}}{3} = \frac{\sigma_1 + \sigma_2 + \sigma_3}{3} = \frac{1}{3} I_1 \tag{1-4.10}$$

Comparing Eqs. (1-4.8), (1-4.9) and (1-4.10), we note that the mean normal stress σ_m is equal to σ_{oct}. Hence, most plasticity theories postulate that plastic behavior of materials is related primarily to that part of the stress tensor that is independent of σ_m. Therefore, the stress array (Eq. 1-3.3) is rewritten in the following form:

$$\mathbf{T} = \mathbf{T}_m + \mathbf{T}_d \tag{1-4.11}$$

Table 1-4.2
Fortran Program for Principal Stresses

```
TYPE,  PRINC
       PROGRAM PRINST(INPUT,OUTPUT, PRIN,PROUT, TAPE5
      A = PRIN,TAPE6 = PROUT)
       DIMENSION SIG(3)
       READ (5,*) NN
       WRITE (6,22)
22     FORMAT (" SIGXX SIGYY SIGZZ SIGXY SIGYZ SIGXZ SIG(1)
      ASIG(2)  SIG(3)")
       DO 23 N = 1,NN
     C NN IS THE NUMBER OF STATES OF STRESS TO BE SOLVED.
       READ (5,*)  SIGXX,SIGYY,SIGZZ,SIGXY,SIGYZ,SIGXZ
       P = (SIGXX + SIGYY + SIGZZ)/3.
       Q = SIGXX*SIGYY + SIGYY*SIGZZ + SIGXX*SIGZZ - SIGXY**2 -
      ASIGYZ**2 - SIGXZ**2
       R = SIGXX*SIGYY*SIGZZ + 2.*SIGXY*SIGYZ*SIGXZ -
      ASIGYY*SIGXZ**2 - SIGXX*SIGYZ**2 - SIGZZ*SIGXY**2
       QQ = ABS(Q/3. - P**2)
       RR = R/2. - P*Q/2. + P**3
       A = QQ**3
       B = 2.*SQRT(QQ)*RR/ABS(RR)
       THETA = ATAN(SQRT(A - RR**2)/RR)
       C = 2.*3.14159/3.
       CC = THETA/3.
       DO 25 I = 1,3
25     SIG(I) = B*COS(CC + (I - 1)*C) + P
       WRITE (6,26) SIGXX,SIGYY,SIGZZ,SIGXY,SIGYZ,SIGXZ,SIG(1)
      ASIG(2),SIG(3)
26     FORMAT (9F7.2)
23     CONTINUE
       STOP
       END
```

INPUT-OUTPUT DATA

```
8
120.,- 55.,- 85.,- 55.,33.,- 75.
-120.,40.,66.,45.,- 65.,25.
0.,100.,0.,60.,60.,45.
130.,- 70.,60.,50.,0.,0.
150.,- 90.,0.,60.,0.,0.
- 90.,- 60.,40.,70.,- 40.,- 55.
-150.,0.,80.,- 40.,0.,50.
0.,0.,0.,- 75.,65.,- 55.
/RWF
RWF FINISHED.
/FTN,I = PRINC,B = PRINCB,L = 0
     .100 CP SECONDS COMPILATION TIME
/PRINCB
     .022 CP SECONDS EXECUTION TIME
/TYPE,PROUT
```

SIGXX	SIGYY	SIGZZ	SIGXY	SIGYZ	SIGXZ	SIG(1)	SIG(2)	SIG(3)
120.00	-55.00	-85.00	-55.00	33.00	-75.00	162.54	-114.14	-68.40
-120.00	40.00	66.00	45.00	-65.00	25.00	-140.49	6.82	119.67
0.00	100.00	0.00	60.00	60.00	45.00	161.70	-45.00	-16.70
130.00	-70.00	60.00	50.00	0.00	0.00	-81.80	60.00	141.80
150.00	-90.00	0.00	60.00	0.00	0.00	164.16	-104.16	-.00
-90.00	-60.00	40.00	70.00	-40.00	-55.00	88.34	-148.54	-49.80
-150.00	0.00	80.00	-40.00	0.00	50.00	-169.46	8.28	91.18
0.00	0.00	0.00	-75.00	65.00	-55.00	130.34	-76.71	-53.63

where **T** symbolically represents the stress array and where

$$\mathbf{T}_m = \begin{pmatrix} \sigma_m & 0 & 0 \\ 0 & \sigma_m & 0 \\ 0 & 0 & \sigma_m \end{pmatrix} \tag{1-4.12}$$

and

$$\mathbf{T}_d = \begin{pmatrix} \dfrac{2\sigma_{xx} - \sigma_{yy} - \sigma_{zz}}{3} & \sigma_{xy} & \sigma_{xz} \\ \sigma_{xy} & \dfrac{2\sigma_{yy} - \sigma_{xx} - \sigma_{zz}}{3} & \sigma_{yz} \\ \sigma_{xz} & \sigma_{yz} & \dfrac{2\sigma_{zz} - \sigma_{yy} - \sigma_{xx}}{3} \end{pmatrix} \tag{1-4.13}$$

The array $\mathbf{T}_m$ is called the mean stress tensor. The array $\mathbf{T}_d$ is called the *deviator* stress tensor, since it is in a certain sense a measure of the deviation of the state of stress from a spherically symmetric state, that is, from the state of stress that exists in an ideal (frictionless) fluid.

Let (X, Y, Z) be the transformed axes that are in the principal stress directions. Then,

$$\sigma_{XX} = \sigma_1 \qquad \sigma_{YY} = \sigma_2 \qquad \sigma_{ZZ} = \sigma_3 \qquad \sigma_{XY} = \sigma_{YZ} = \sigma_{ZX} = 0$$

and Eq. (1-4.11) is simplified accordingly. Application of Eqs. (1-4.7) to Eqs. (1-4.12) and (1-4.13) yields the following stress invariants for $\mathbf{T}_m$ and $\mathbf{T}_d$:

$$\begin{aligned} \text{For } \mathbf{T}_m:\ & I_{1m} = I_1 = 3\sigma_m \\ & I_{2m} = \tfrac{1}{3}I_1^2 = 3\sigma_m^2 \\ & I_{3m} = \tfrac{1}{27}I_1^3 = \sigma_m^3 \end{aligned} \tag{1-4.14}$$

$$\begin{aligned} \text{For } \mathbf{T}_d:\ & I_{1d} = 0 \\ & I_{2d} = I_2 - \tfrac{1}{3}I_1^2 = -\tfrac{1}{6}\left[(\sigma_1 - \sigma_2)^2 + (\sigma_2 - \sigma_3)^2 + (\sigma_3 - \sigma_1)^2\right] \\ & I_{3d} = I_3 - \tfrac{1}{3}I_1 I_2 + \tfrac{2}{27}I_1^3 \\ & \quad = \tfrac{1}{27}(2\sigma_1 - \sigma_2 - \sigma_3)(2\sigma_2 - \sigma_3 - \sigma_1)(2\sigma_3 - \sigma_1 - \sigma_2) \end{aligned} \tag{1-4.15}$$

The principal values of the deviator tensor T_d are

$$\begin{aligned} S_1 &= \sigma_1 - \sigma_m = \frac{2\sigma_1 - \sigma_2 - \sigma_3}{3} = \frac{(\sigma_1 - \sigma_3) + (\sigma_1 - \sigma_2)}{3} \\ S_2 &= \sigma_2 - \sigma_m = \frac{(\sigma_2 - \sigma_3) + (\sigma_2 - \sigma_1)}{3} = \frac{(\sigma_2 - \sigma_3) - (\sigma_1 - \sigma_2)}{3} \\ S_3 &= \sigma_3 - \sigma_m = \frac{(\sigma_3 - \sigma_1) + (\sigma_3 - \sigma_2)}{3} = -\frac{(\sigma_1 - \sigma_3) + (\sigma_2 - \sigma_3)}{3} \end{aligned} \tag{1-4.16}$$

Accordingly, since $S_1 + S_2 + S_3 = 0$, only two of the principal stresses (values) of $\mathbf{T}_d$ are independent. Many of the formulas of the mathematical theory of plasticity are often written in terms of the stress invariants of the deviator stress tensor $\mathbf{T}_d$.

Plane Stress / In a large class of important problems, certain approximations may be applied to simplify the three-dimensional stress array (see Eq. 1-3.1). For example, simplifying approximations can be made in analyzing the deformations that occur in a thin flat plate subjected to forces applied along its edge and directed so that they lie in the middle surface of the plate. We define a thin plate to be a prismatic member (for example, a cylinder) of a very small length or thickness h. Accordingly, the middle surface of the plate, located halfway between its ends (faces) and parallel to them, may be taken as the (x, y)-plane. The thickness direction is then coincident with the direction of the z-axis. Since the plate is not loaded on its faces, $\sigma_{zz} = \sigma_{zx} = \sigma_{zy} = 0$ on its lateral surfaces ($z = \pm h/2$). Consequently, since the plate is thin, as a first approximation, it may be assumed that

$$\sigma_{zz} = \sigma_{zx} = \sigma_{zy} = 0 \tag{1-4.17}$$

throughout the plate thickness.

Furthermore, it may be assumed that the remaining stress components σ_{xx}, σ_{yy}, and σ_{xy} are independent of z. With these approximations, the stress array reduces to a function of the two variables (x, y): then it is called a *plane stress array* or the *tensor of plane stress*.

Consider a transformation from the (x, y, z) coordinate axes to the (X, Y, Z) coordinate axes for the condition that the z-axis and the Z-axis remain coincident under the transformation. Then, for a state of plane stress in the (x, y)-plane, Table 1-4.3 gives the direction cosines between the axes in a transformation from the (x, y) coordinate axes to the (X, Y) coordinate axes (Fig. 1-4.2). Hence, with Table (1-4.3) and Fig. (1-4.2), Eqs. (1-4.1) and (1-4.3) yield

$$\begin{aligned}
\sigma_{XX} &= \sigma_{xx}\cos^2\theta + \sigma_{yy}\sin^2\theta + 2\sigma_{xy}\sin\theta\cos\theta \\
\sigma_{YY} &= \sigma_{xx}\sin^2\theta + \sigma_{yy}\cos^2\theta - 2\sigma_{xy}\sin\theta\cos\theta \\
\sigma_{XY} &= -(\sigma_{xx} - \sigma_{yy})\sin\theta\cos\theta + \sigma_{xy}(\cos^2\theta - \sin^2\theta)
\end{aligned} \tag{1-4.18}$$

By means of trigonometric double angle formulas, Eq. (1-4.18) may be written in the form

$$\begin{aligned}
\sigma_{XX} &= \tfrac{1}{2}(\sigma_{xx} + \sigma_{yy}) + \tfrac{1}{2}(\sigma_{xx} - \sigma_{yy})\cos 2\theta + \sigma_{xy}\sin 2\theta \\
\sigma_{YY} &= \tfrac{1}{2}(\sigma_{xx} + \sigma_{yy}) - \tfrac{1}{2}(\sigma_{xx} - \sigma_{yy})\cos 2\theta - \sigma_{xy}\sin 2\theta \\
\sigma_{XY} &= -\tfrac{1}{2}(\sigma_{xx} - \sigma_{yy})\sin 2\theta + \sigma_{xy}\cos 2\theta
\end{aligned} \tag{1-4.19}$$

Table 1-4.3

	x	y	z
X	$l_1 = \cos\theta$	$m_1 = \sin\theta$	$n_1 = 0$
Y	$l_2 = -\sin\theta$	$m_2 = \cos\theta$	$n_2 = 0$
Z	$l_3 = 0$	$m_3 = 0$	$n_3 = 1$

Equations (1-4.18) or (1-4.19) express the stress components σ_{XX}, σ_{YY}, and σ_{XY} in the (X, Y) coordinate system in terms of the corresponding stress components σ_{xx}, σ_{yy}, and σ_{xy} in the (x, y) coordinate system for the plane transformation defined by Fig. (1-4.2) and Table 1-4.3.

Graphical Interpretation of Plane Stress. Mohr's Circle in Two Dimensions/In the form of Eq. (1-4.19), the plane transformation of stress components is particularly suited for graphical interpretation. Stress components σ_{XX} and σ_{XY} act on face BE in Fig. (1-4.3) that is located at a positive angle θ (counterclockwise) from face BC on which stress components σ_{xx} and σ_{xy} act. The variation of the stress components σ_{XX} and σ_{XY} with θ may be depicted graphically by constructing a diagram in which σ_{XX} and σ_{XY} are coordinates. For each plane BE, there is a point on the diagram whose coordinates correspond to values of σ_{XX} and σ_{XY}.

Rewriting the first of Eqs. (1-4.19) by moving the first term on the right side to the left side and squaring both sides of the resulting equation, squaring both sides of the last of Eq. (1-4.19), and adding, we obtain

$$\left[\sigma_{XX} - \tfrac{1}{2}(\sigma_{xx} + \sigma_{yy})\right]^2 + [\sigma_{XY} - 0]^2 = \tfrac{1}{4}(\sigma_{xx} - \sigma_{yy})^2 + \sigma_{xy}^2 \qquad (1\text{-}4.20)$$

Equation (1-4.20) is the equation of a circle in the σ_{XX}, σ_{XY} plane whose center C has coordinates

$$\left[\tfrac{1}{2}(\sigma_{xx} + \sigma_{yy}), 0\right] \qquad (1\text{-}4.21)$$

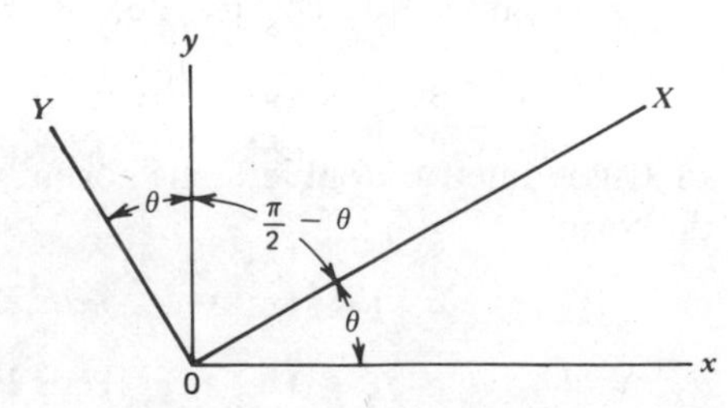

Fig. 1-4.2/Location of transformed axes for plane stress.

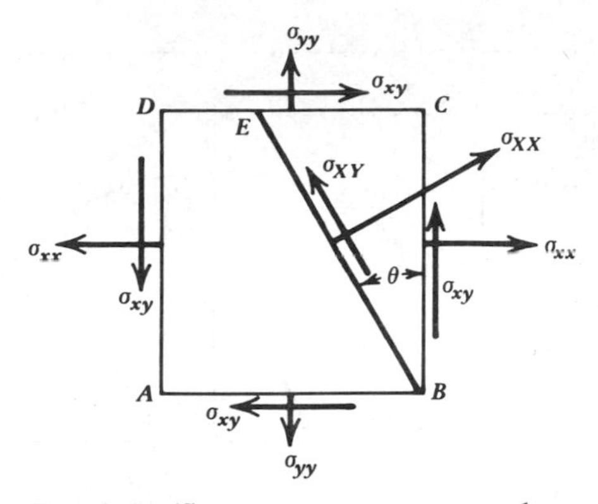

Fig. 1-4.3/Stress components on plane perpendicular to transformed X-axis for plane stress.

and whose radius R is given by the relation

$$R = \sqrt{\tfrac{1}{4}(\sigma_{xx} - \sigma_{yy})^2 + \sigma_{xy}^2} \tag{1-4.22}$$

Consequently, the geometrical representation of the first and third of Eqs. (1-4.19) is a circle (Fig. 1-4.4). This stress circle is frequently called *Mohr's circle* in honor of O. Mohr, who first employed it to study plane stress problems. It is necessary to take the positive direction of the σ_{XY}-axis downward so that the positive direction of θ in both Figs. 1-4.3 and 1-4.4 is counterclockwise.

Since σ_{xx}, σ_{yy}, and σ_{xy} are known quantities, the circle in Fig. 1-4.4 can be constructed using Eqs. (1-4.21) and (1-4.22). The interpretation of Mohr's circle of stress requires that one known point be located on the circle. When $\theta = 0$ (Fig. 1-4.3), the first and third of Eqs. (1-4.19) give

$$\sigma_{XX} = \sigma_{xx} \quad \text{and} \quad \sigma_{XY} = \sigma_{xy} \tag{1.4-23}$$

which are coordinates of point P in Fig. 1-4.4.

Principal stresses σ_1 and σ_2 are located at points Q and Q' in Fig. 1-4.4 and occur when $\theta = \theta_1$ and $\theta_1 + \pi/2$, measured counterclockwise from line CP. The two magnitudes of θ are given by the third of Eqs. (1-4.19) since $\sigma_{XY} = 0$ when $\theta = \theta_1$ and $\theta_1 + \pi/2$. Note that in Fig. 1-4.4, we must rotate through angle 2θ from line CP, which corresponds to a rotation of θ from plane BC in Fig. 1-4.3. (See also Eqs. 1-4.19.) Thus, by Eqs. (1-4.19), for $\sigma_{XY} = 0$, we obtain (see also Fig. 1-4.4)

$$\tan 2\theta = \frac{2\sigma_{xy}}{\sigma_{xx} - \sigma_{yy}} \tag{1-4.24}$$

Solution of Eq. (1-4.24) yields the values $\theta = \theta_1$ and $\theta_1 + \pi/2$.

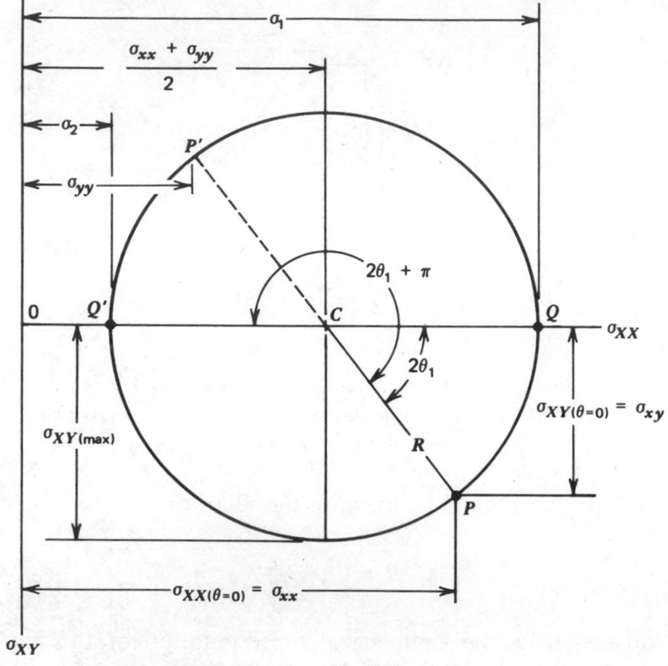

Fig. 1-4.4/Mohr's Circle for plane stress.

The magnitudes of the principal stresses from Mohr's circle of stress are

$$\begin{aligned} \sigma_1 &= \frac{\sigma_{xx}+\sigma_{yy}}{2} + \sqrt{\tfrac{1}{4}(\sigma_{xx}-\sigma_{yy})^2 + \sigma_{xy}^2} \\ \sigma_2 &= \frac{\sigma_{xx}+\sigma_{yy}}{2} - \sqrt{\tfrac{1}{4}(\sigma_{xx}-\sigma_{yy})^2 + \sigma_{xy}^2} \end{aligned} \tag{1-4.25}$$

and are in agreement with the values predicted by the procedure outlined earlier in this article.

Another known point on Mohr's circle of stress can be located although it is not needed for the interpretation of the circle. When $\theta = \pi/2$, the first and third of Eqs. (1-4.19) give

$$\sigma_{XX} = \sigma_{yy} \quad \text{and} \quad \sigma_{XY} = -\sigma_{xy} \tag{1-4.26}$$

These coordinates locate point P' in Fig. 1-4.4, which is on the opposite end of the diameter from point P.

EXAMPLE 1-4.1
Mohr's Circle in Two Dimensions

A piece of chalk is subjected to combined loading consisting of a tensile load P and a torque T (Fig. E1-4.1*a*). The chalk has an ultimate strength

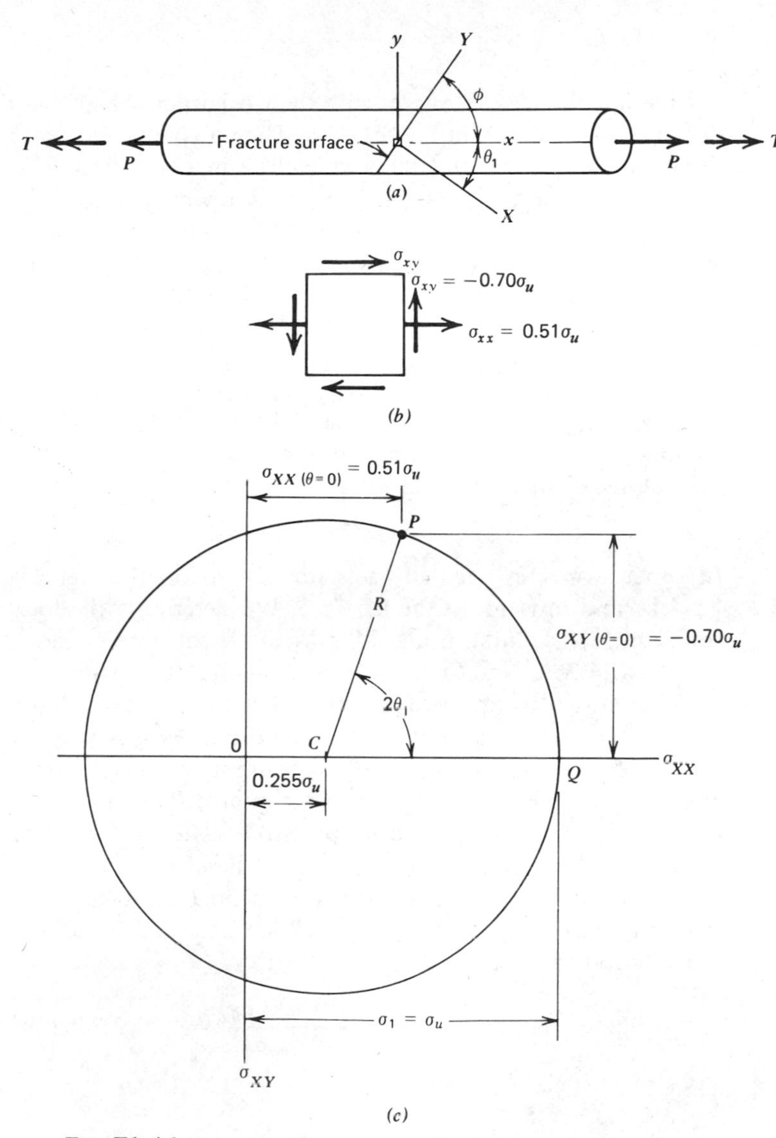

Fig. E1-4.1

σ_u as determined in a simple tensile test. The load P remains constant at such a value that it produces a tensile stress 0.51 σ_u on any cross section. The torque T is increased gradually until fracture occurs on some inclined surface.

Assuming that fracture takes place when the maximum principal stress σ_1 reaches the ultimate strength σ_u, determine the magnitude of the torsional shearing stress produced by torque T at fracture and determine the orientation of the fracture surface.

SOLUTION

Let us take the x- and y-axes with their origin at a point on the surface of the chalk as shown in Fig. E1-4.1a. Then a volume element taken from the chalk at the origin of the axes will be in plane stress (Fig. E1-4.1b) with $\sigma_{xx} = 0.51\sigma_u$, $\sigma_{yy} = 0$, and σ_{xy} unknown. The magnitude of the shearing stress σ_{xy} can be determined from the condition that the maximum principal stress σ_1 [given by Eq. (1-4.25)] is equal to σ_u; thus,

$$\sigma_u = 0.255\,\sigma_u + \sqrt{(0.255\,\sigma_u)^2 + \sigma_{xy}^2}$$

$$\sigma_{xy} = 0.700\,\sigma_u$$

Since the torque acting on the right end of the piece of chalk is counterclockwise, the shearing stress σ_{xy} acts down on the front face of the volume element (Fig. E1-4.1b) and is therefore negative. Thus,

$$\sigma_{xy} = -0.700\,\sigma_u$$

In other words σ_{xy} actually acts downward on the right face of Fig. E1-4.1b and upward on the left face. We determine the location of the fracture surface first using Mohr's circle of stress and then using Eq. (1-4.24). As indicated in Fig. E1-4.1c, the center C of Mohr's circle of stress lies on the σ_{XX}-axis at distance $0.255\sigma_u$ from the origin 0 [see Eq. (1-4.21)]. The radius R of the circle is given by Eq. (1-4.22); $R = 0.745\,\sigma_u$. When $\theta = 0$, the stress components $\sigma_{XX(\theta=0)} = \sigma_{xx} = 0.51\,\sigma_u$ and $\sigma_{XY(\theta=0)} = \sigma_{xy} = -0.700\,\sigma_u$ locate point P on the circle. Point Q representing the maximum principal stress is located by rotating clockwise through angle $2\theta_1$ from point P; therefore, the fracture plane is perpendicular to the X-axis, which is located at an angle θ_1 clockwise from the x-axis. The angle θ_1 can also be obtained from Eq. (1-4.24), as the solution of

$$\tan\theta_1 = \frac{2\sigma_{xy}}{\sigma_{xx}} = -\frac{2(0.700\,\sigma_u)}{0.51\,\sigma_u} = --2.7452$$

Thus,

$$\theta_1 = -0.6107 \text{ rad}$$

Since θ_1 is negative, the X-axis is located clockwise through angle θ_1 from the x-axis. The fracture plane is at angle ϕ from the x-axis. It is given as

$$\phi = \frac{\pi}{2} - |\theta_1| = 0.9601 \text{ rad}$$

The magnitude of ϕ depends on the magnitude of P. If $P = 0$, the chalk is subjected to pure torsion and $\phi = \pi/4$. If $P/A = \sigma_u$ (A is the cross-sectional area), the chalk is subjected to pure tension ($T = 0$) and $\phi = \pi/2$.

Extreme Values of Normal and Shearing Stresses/As discussed in Chapter 3, failure of load carrying members is often associated with either the maximum normal stress or the maximum shearing stress at the point in the member where failure is initiated. The maximum normal stress is equal to the maximum of the three principal stresses σ_1, σ_2, and σ_3. Procedures have been presented for determining the values of the principal stresses for either the general state of stress or for plane stress. For plane stress states, two of the principal stresses are given by Eqs. (1-4.25), the third being $\sigma_3 = \sigma_{zz} = 0$.

Even though the construction of Mohr's circle of stress was presented for plane stress ($\sigma_{zz} = 0$), the transformation equations given by either Eqs. (1-4.18) or (1-4.19) are not influenced by the magnitude of σ_{zz} but require only that $\sigma_{zx} = \sigma_{zy} = 0$ (Problem 1-4.2). Therefore, in terms of the principal stresses, Mohr's circle of stress can be constructed by using any two of the principal stresses, thus giving three Mohr's circles for any given state of stress. Consider any point in a stressed body for which values of σ_1, σ_2, and σ_3 are known. For any plane through the point, let the N-axis be normal to the plane and let the S-axis coincide with the shearing component of the stress for the plane. If we choose σ_{NN} and σ_{NS} as coordinate axes in Fig. 1-4.5, three Mohr's circles of stress can be constructed. As will be shown later, the stress components σ_{NN} and σ_{NS}

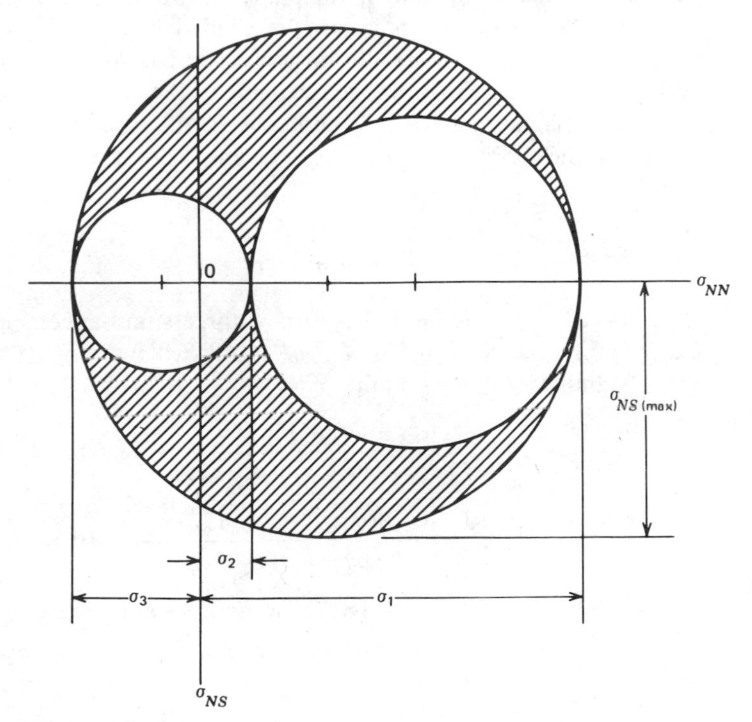

Fig. 1-4.5/Mohr's Circle in three dimensions.

for any plane passing through the point locates a point either on one of the three circles in Fig. 1-4.5 or in one of the two shaded areas. The maximum shearing stress τ_{max} for the point is equal to the maximum value of σ_{NS} and is equal in magnitude to the radius of the largest of the three Mohr's circles of stress. Hence,

$$\tau_{max} = \sigma_{NS(max)} = \frac{\sigma_{max} - \sigma_{min}}{2} \qquad (1\text{-}4.27)$$

where $\sigma_{max} = \sigma_1$ and $\sigma_{min} = \sigma_3$ (Fig. 1-4.5).

Mohr's Circles in Three Dimensions /[†]Once the state of stress at a point is expressed in terms of the principal stresses, three Mohr's circles of stress can be constructed as indicated in Fig. 1-4.5. Consider plane P whose normal relative to principal axes has direction cosines l, m, and n. The normal stress σ_{NN} on plane P is by Eq. (1-3.9)

$$\sigma_{NN} = l^2\sigma_1 + m^2\sigma_2 + n^2\sigma_3 \qquad (1\text{-}4.28)$$

Similarly, the square of the shearing stress σ_{NS} on plane P is, by Eqs. (1-3.8) and (1-3.10),

$$\sigma_{NS}^2 = l^2\sigma_1^2 + m^2\sigma_2^2 + n^2\sigma_3^2 - \sigma_{NN}^2 \qquad (1\text{-}4.29)$$

For known values of the principal stresses σ_1, σ_2, and σ_3 and of the direction cosines l, m, and n for plane P, graphical techniques can be developed to locate the point in the shaded area of Fig. 1-4.5 whose coordinates $(\sigma_{NN}, \sigma_{NS})$ are the normal and shearing stress components acting on plane P. However, we recommend the procedure in Art. 1-3 to determine magnitudes for σ_{NN} and σ_{NS}. In the discussion to follow, we show that the coordinates $(\sigma_{NN}, \sigma_{NS})$ locates a point in the shaded area of Fig. 1-4.5.

Since

$$l^2 + m^2 + n^2 = 1 \qquad (1\text{-}4.30)$$

Eqs. (1-4.28), (1-4.29), and (1-4.30) are three simultaneous equations in l^2, m^2, and n^2. Solving for l^2, m^2, and n^2 and noting that $l^2 \geq 0$, $m^2 \geq 0$, and $n^2 \geq 0$, we obtain

$$\begin{aligned} l^2 &= \frac{\sigma_{NS}^2 + (\sigma_{NN} - \sigma_2)(\sigma_{NN} - \sigma_3)}{(\sigma_1 - \sigma_2)(\sigma_1 - \sigma_3)} \geq 0 \\ m^2 &= \frac{\sigma_{NS}^2 + (\sigma_{NN} - \sigma_1)(\sigma_{NN} - \sigma_3)}{(\sigma_2 - \sigma_3)(\sigma_2 - \sigma_1)} \geq 0 \qquad (1\text{-}4.31) \\ n^2 &= \frac{\sigma_{NS}^2 + (\sigma_{NN} - \sigma_1)(\sigma_{NN} - \sigma_2)}{(\sigma_3 - \sigma_1)(\sigma_3 - \sigma_2)} \geq 0 \end{aligned}$$

[†]In the early history of stress analysis, Mohr's circles in three dimensions were used extensively. However, today, they are used principally as a heuristic device.

Ordering the principal stresses such that $\sigma_1 > \sigma_2 > \sigma_3$, we may write Eqs. (1-4.31) in the form

$$\sigma_{NS}^2 + (\sigma_{NN} - \sigma_2)(\sigma_{NN} - \sigma_3) \geq 0$$
$$\sigma_{NS}^2 + (\sigma_{NN} - \sigma_3)(\sigma_{NN} - \sigma_1) \leq 0$$
$$\sigma_{NS}^2 + (\sigma_{NN} - \sigma_1)(\sigma_{NN} - \sigma_2) \geq 0$$

These inequalities may be rewritten in the form

$$\begin{aligned} \sigma_{NS}^2 + \left[\sigma_{NN} - \frac{\sigma_2 + \sigma_3}{2}\right]^2 &\geq \frac{1}{4}(\sigma_2 - \sigma_3)^2 \\ \sigma_{NS}^2 + \left[\sigma_{NN} - \frac{\sigma_1 + \sigma_3}{2}\right]^2 &\leq \frac{1}{4}(\sigma_3 - \sigma_1)^2 \qquad (1\text{-}4.32) \\ \sigma_{NS}^2 + \left[\sigma_{NN} - \frac{\sigma_1 + \sigma_2}{2}\right]^2 &\geq \frac{1}{4}(\sigma_1 - \sigma_2)^2 \end{aligned}$$

The inequalities of Eqs. (1-4.32) may be interpreted graphically as follows: Let $(\sigma_{NN}, \sigma_{NS})$ denote abscissa and ordinate, respectively, on a graph (Fig. 1-4.5). Then, an admissible state of stress must lie within a region bounded by three circles obtained from Eqs. (1-4.32) where the equalities are taken (the shaded region in Fig. 1-4.5).

EXAMPLE 1-4.2

Three-Dimensional State of Stress

Let the state of stress at a point be given by $\sigma_{xx} = 120$ MPa, $\sigma_{yy} = 55$ MPa, $\sigma_{zz} = -85$ MPa, $\sigma_{xy} = -55$ MPa, $\sigma_{yz} = 33$ MPa, and $\sigma_{zx} = -75$ MPa. Determine the three principal stresses and the directions associated with the three principal stresses.

SOLUTION

Substituting the given stress components into Eq. (1-4.6) we obtain

$$\sigma^3 - 90\sigma^2 - 18014\sigma + 471{,}680 = 0$$

The three principal stresses are the three roots of this equation. They are

$$\sigma_1 = 176.80 \text{ MPa} \qquad \sigma_2 = -110.86 \text{ MPa} \qquad \sigma_3 = 24.06 \text{ MPa}$$

The direction cosines for any one of the principal stress directions are given by substituting the given principal stress into Eq. (1-4.4). Substitution of σ_1 into Eqs. (1-4.4) gives

$$\begin{aligned} (120 - 176.80)l_1 - 55m_1 - 75n_1 &= 0 \\ -55l_1 + (55 - 176.80)m_1 + 33n_1 &= 0 \qquad (1) \\ -75l_1 + 33m_1 + (-85 - 176.80)n_1 &= 0 \end{aligned}$$

where l_1, m_1, and n_1 are the direction cosines for the σ_1-direction. Only two of these equations are independent; in addition the direction cosines must satisfy the equation

$$l_i^2 + m_i^2 + n_i^2 = 1, \qquad i = 1, 2, \text{ or } 3 \tag{2}$$

The simultaneous solution of any two of Eqs. (1) along with Eq. (2) gives

$$l_1 = 0.8372 \qquad m_1 = -0.4587 \qquad n_1 = -0.2977$$

In a similar manner, we obtain sets of direction cosines for σ_2 and σ_3.

$$l_2 = 0.2872 \qquad m_3 = -0.0944 \qquad n_2 = 0.9532$$
$$l_3 = 0.4657 \qquad m_3 = 0.8834 \qquad n_3 = -0.0521$$

PROBLEM SET 1-4

1. Let the state of stress at a point be specified by the following stress components: $\sigma_{xx} = \sigma_{yy} = \sigma_{zz} = 0$, $\sigma_{xy} = -75$ MPa, $\sigma_{yz} = 65$ MPa, and $\sigma_{zx} = -55$ MPa. Determine the principal stresses, the direction cosines for the three principal stress directions, and the maximum shearing stress.

2. Consider a state of stress in which the nonzero stress components are σ_{xx}, σ_{yy}, σ_{zz}, and σ_{xy}. Note that this is not a state of plane stress since $\sigma_{zz} \neq 0$. Consider another set of coordinate axes (X, Y, Z) with the Z-axis coinciding with the z-axis and with the X-axis located counterclockwise through angle θ from the x-axis. Show that the transformation equations for this state of stress are identical with Eq. (1-4.18) or Eq. (1-4.19) for plane stress.

3. Let the state of stress at a point be specified by the following stress components: $\sigma_{xx} = 110$ MPa, $\sigma_{yy} = -86$ MPa, $\sigma_{zz} = 55$ MPa, $\sigma_{xy} = 60$ MPa, and $\sigma_{yz} = \sigma_{zx} = 0$. Determine the principal stresses, the direction cosines of the principal stress directions, and the maximum shearing stress.

   Ans. $\sigma_1 = 126.9$ MPa, $\sigma_2 = -102.9$ MPa, $\sigma_3 = 55.0$ MPa,
   $l_1 = 0.9625$, $m_1 = 0.2717$, $n_1 = 0$,
   $l_2 = 0.2717$, $m_2 = -0.9625$, $n_2 = 0$,
   $l_3 = 0$, $m_3 = 0$, $n_3 = 1$,
   $\sigma_{NS(\max)} = 114.9$ MPa

4. Solve Problem 3 using the results of Problem 2.

5. Let the state of plane stress be specified by the following stress components: $\sigma_{xx} = 90$ MPa, $\sigma_{yy} = -10$ MPa, $\sigma_{xy} = 40$ MPa. Let the X-axis lie in the (x, y)-plane and be located at $\theta = \pi/6$ clockwise from the x-axis. The direction cosines for the X-axis are $l = \cos(-\pi/6) = 0.8660$, $m = \sin(-\pi/6) = -0.5000$, $n = 0$. Determine

the normal and shearing stresses on a plane perpendicular to the X-axis; use Eqs. (1-3.8), (1-3.9), and (1-3.10).

Ans. $\sigma_{XX} = 30.36$ MPa, $\sigma_{XY} = 63.30$ MPa

In Problems 6 through 9, the Z-axis for the transformed axes coincides with the z-axis for the volume element on which the known stress components act.

6. The nonzero stress components are $\sigma_{xx} = 200$ MPa, $\sigma_{yy} = 100$ MPa, and $\sigma_{xy} = -50$ MPa. Determine the principal stresses and the maximum shearing stress. Determine the angle between the X-axis and the x-axis when the X-axis is in the direction of the principal stress with largest absolute magnitude.

7. The nonzero stress components are $\sigma_{xx} = -90$ MPa, $\sigma_{yy} = 50$ MPa, and $\sigma_{xy} = 60$ MPa. Determine the principal stresses and the maximum shearing stress. Determine the angle between the X-axis and the x-axis when the X-axis is in the direction of the principal stress with largest absolute magnitude.

   Ans. $\sigma_1 = 72.2$ MPa, $\sigma_2 = -112.2$ MPa, $\sigma_3 = 0$, $\tau_{max} = 92.2$ MPa X-axis located 0.3543 rad clockwise from x-axis.

8. The nonzero stress components are $\sigma_{xx} = 80$ MPa, $\sigma_{zz} = -60$ MPa, and $\sigma_{xy} = 30$ MPa. Determine the principal stresses and the maximum shearing stress. Determine the angle between the X-axis and the x-axis when the X-axis is in the direction of the principal stress with largest absolute magnitude.

9. The nonzero stress components are $\sigma_{xx} = 150$ MPa, $\sigma_{yy} = 70$ MPa, $\sigma_{zz} = -80$ MPa, and $\sigma_{xy} = -45$ MPa. Determine the principal stresses and the maximum shearing stress. Determine the angle between the X-axis and the x-axis when the X-axis is in the direction of the principal stress with largest absolute magnitude.

   Ans. $\sigma_1 = 170.2$ MPa, $\sigma_2 = 49.8$ MPa, $\sigma_3 = -80$ MPa, $\tau_{max} = 125.1$ MPa X-axis located 0.4221 rad clockwise from the x-axis.

10. Using transformation equations of plane stress, determine σ_{XX} and σ_{XY} for the X-axis located 0.5000 rad clockwise from the x-axis. The nonzero stress components are given in Problem 6.

11. Using transformation equations of plane stress, determine σ_{XX} and σ_{XY} for the X-axis located 0.1500 rad counterclockwise from the x-axis. The nonzero stress components are given in Problem 7.

    Ans. $\sigma_{XX} = -69.1$ MPa, $\sigma_{XY} = 78.0$ MPa

12. Using transformation equations of stress (see Problem 2), determine σ_{XX} and σ_{XY} for the X-axis located 1.0000 rad clockwise from the x-axis. The nonzero stress components are given in Problem 8.

13. Using transformation equations of stress (see Problem 2), determine σ_{XX} and σ_{XY} for the X-axis located 0.7000 rad counterclockwise from the x-axis. The nonzero stress components are given in Problem 9.

    Ans. $\sigma_{XX} = 72.5$ MPa, $\sigma_{XY} = -47.1$ MPa

14. Solve Problem 10 using Mohr's circle of stress.

15. Solve Problem 11 using Mohr's circle of stress.

16. Solve Problem 12 using Mohr's circle of stress.

17. Solve Problem 13 using Mohr's circle of stress.

18. A volume element at the free surface is shown in Fig. P1-4.18. The state of stress is plane stress with $\sigma_{xx} = 100$ MPa. Determine the other stress components.

19. Determine the unknown stress components for the volume element in Fig. P1-4.19.

    Ans. $\sigma_{xx} = 26.67$ MPa, $\sigma_{yy} = 172.50$ MPa

20. Determine the unknown stress components for the volume element in Fig. P1-4.20.

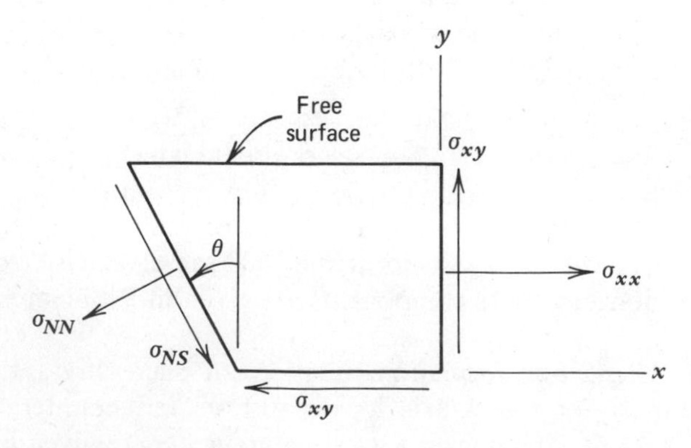

Fig. P1-4.18

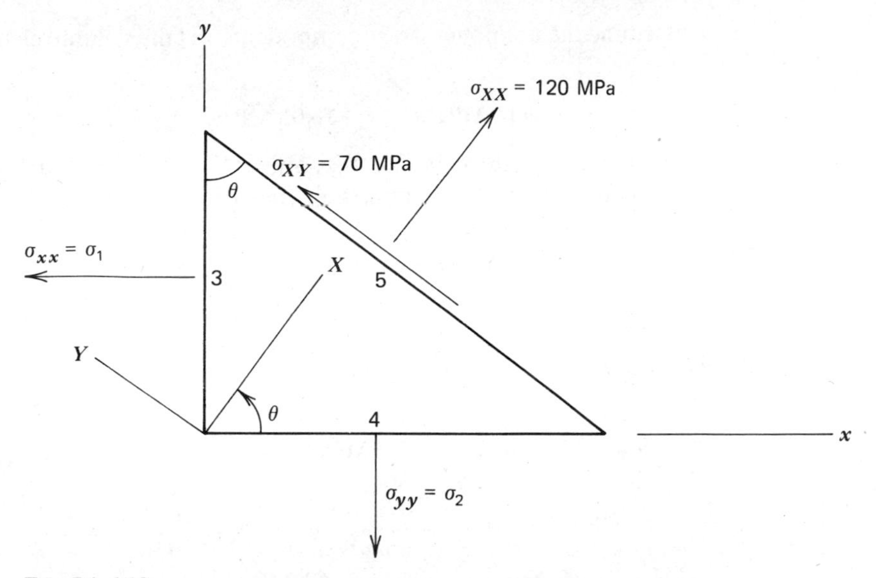

Fig. P1-4.19

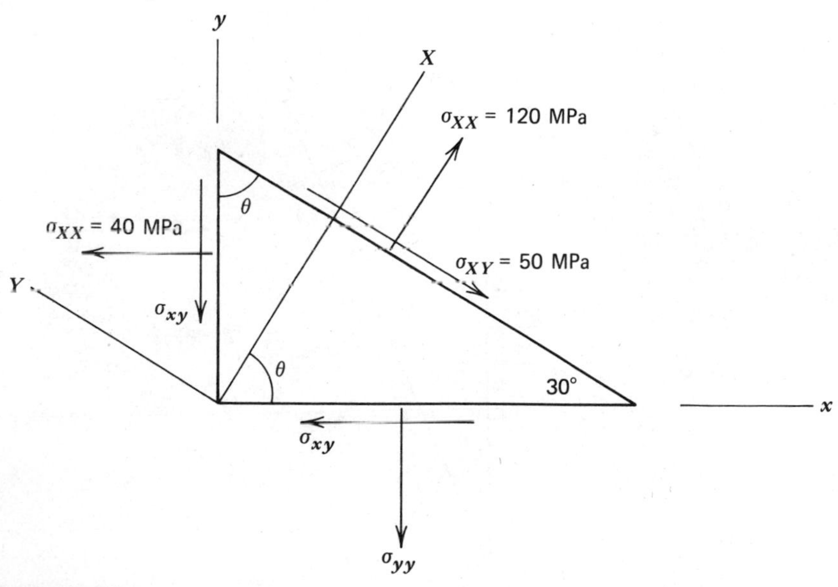

Fig. P1-4.20

21. Determine the unknown stress components for the volume element in Fig. P1-4.21.

    Ans. $\sigma_{xx} = -109.18$ MPa, $\sigma_{xy} = -10.01$ MPa

In Problems 22 through 26, determine the principal stresses, the maximum shearing stress, and the octahedral shearing stress.

22. The nonzero stress components are $\sigma_{xx} = -100$ MPa, $\sigma_{yy} = 60$ MPa, and $\sigma_{xy} = -50$ MPa.

23. The nonzero stress components are $\sigma_{xx} = 180$ MPa, $\sigma_{yy} = 90$ MPa, and $\sigma_{xy} = 50$ MPa.

    Ans. $\sigma_1 = 202.3$ MPa, $\sigma_2 = 67.7$ MPa, $\sigma_3 = 0$, $\tau_{max} = 101.1$ MPa, $\tau_{oct} = 84.1$ MPa

24. The nonzero stress components are $\sigma_{xx} = -150$ MPa, $\sigma_{yy} = -70$ MPa, $\sigma_{zz} = 40$ MPa, and $\sigma_{xy} = -60$ MPa.

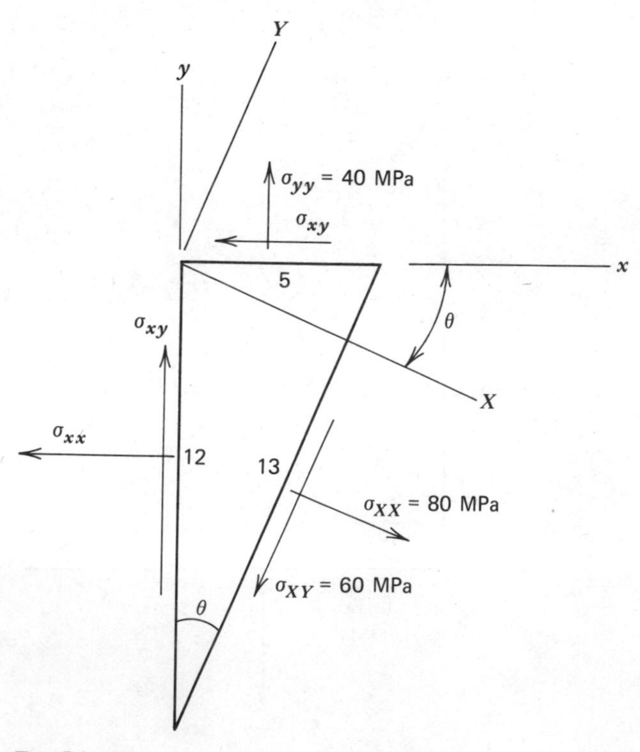

Fig. P1-4.21

25. The nonzero stress components are $\sigma_{xx} = 80$ MPa, $\sigma_{yy} = -35$ MPa, $\sigma_{zz} = -50$ MPa, and $\sigma_{xy} = 45$ MPa.

    Ans. $\sigma_1 = 95.5$ MPa, $\sigma_2 = -50.5$ MPa, $\sigma_3 = -50$ MPa, $\tau_{max} = 73.0$ MPa, $\tau_{oct} = 68.7$ MPa

26. The nonzero stress components are $\sigma_{xx} = 95$ MPa, $\sigma_{yy} = 0$, $\sigma_{zz} = 60$ MPa, and $\sigma_{xy} = -55$ MPa.

27. Let the state of stress at a point be given by $\sigma_{xx} = -120$ MPa, $\sigma_{yy} = 140$ MPa, $\sigma_{zz} = 66$ MPa, $\sigma_{xy} = 45$ MPa, $\sigma_{yz} = -65$ MPa, and $\sigma_{zx} = 25$ MPa. Determine the three principal stresses and the directions associated with the three principal stresses.

    Ans. $\sigma_1 = 180.2$ MPa, $\sigma_2 = 40.1$ MPa, $\sigma_3 = -134.3$ MPa,
    $l_1 = 0.0913$, $m_1 = 0.8740$, $n_1 = -0.4773$,
    $l_2 = 0.2584$, $m_2 = 0.4422$, $n_2 = 0.8589$,
    $l_3 = 0.9598$, $m_3 = -0.2062$, $n_3 = -0.1904$

28. Let the state of stress at a point be given by $\sigma_{xx} = 0$, $\sigma_{yy} = 100$ MPa, $\sigma_{zz} = 0$, $\sigma_{xy} = -60$ MPa, $\sigma_{yz} = 35$ MPa, and $\sigma_{zx} = 50$ MPa. Determine the three principal stresses.

29. Let the state of stress at a point be given by $\sigma_{xx} = 120$ MPa, $\sigma_{yy} = -55$ MPa, $\sigma_{zz} = -85$ MPa, $\sigma_{xy} = -55$ MPa, $\sigma_{yz} = 33$ MPa, and $\sigma_{zx} = -75$ MPa. Determine the three principal stresses and the maximum shearing stress.

    Ans. $\sigma_1 = 162.5$ MPa, $\sigma_2 = -114.1$ MPa, $\sigma_3 = -68.4$ MPa, $\tau_{max} = 138.3$ MPa

30. Let the state of stress at a point be given by $\sigma_{xx} = -90$ MPa, $\sigma_{yy} = -60$ MPa, $\sigma_{zz} = 40$ MPa, $\sigma_{xy} = 70$ MPa, $\sigma_{yz} = -40$ MPa, and $\sigma_{zx} = -55$ MPa. Determine the three principal stresses and the maximum shearing stress.

31. Let the state of stress at a point be given by $\sigma_{xx} = -150$ MPa, $\sigma_{yy} = 0$, $\sigma_{zz} = 80$ MPa, $\sigma_{xy} = -40$ MPa, $\sigma_{yz} = 0$, and $\sigma_{zx} = 50$ MPa. Determine the three principal stresses and the maximum shearing stress.

    Ans. $\sigma_1 = 91.2$ MPa, $\sigma_2 = 8.28$ MPa, $\sigma_3 = -169.5$ MPa, $\tau_{max} = 130.3$ MPa

1-5 DIFFERENTIAL EQUATIONS OF MOTION OF A DEFORMABLE BODY

In previous articles, we determined the stress components needed to specify the state of stress at a *point* 0 in a deformed body for a given set of orthogonal coordinate axes (x, y, z). We derived transformation equations that define the state of stress at point 0 for any other set of orthogonal axes (X, Y, Z) rotated with respect to (x, y, z). We derived relations that give at point 0 the principal stresses and their directions, the maximum shearing stress, the octahedral shearing stress, and the hydrostatic and deviatoric states of stress.

In this article, we derive differential equations of motion of a deformable solid body (differential equations of equilibrium if the deformed body has zero acceleration). These equations are needed when the theory of elasticity is used to derive load-stress and load-deflection relations for a member. We consider a general deformed body and choose a volume element at point 0 in the body as indicated in Fig. 1-5.1. The form of the differential equations of motion depend on the type of orthogonal coordinate axes employed. Hence we choose rectangular coordinate axes (x, y, z) whose directions are parallel to the edges of the deformed volume element. In this book, we restrict our consideration mainly to small displacements and, therefore, do not distinguish between coordinate axes in the deformed state and in the undeformed state.[1] Six cutting planes bound the volume element shown as a free body diagram in Fig. 1-5.2. In general, the state of stress changes with the location of point 0. In particular, the stress components undergo changes from one face of the volume element to another face. Body forces (B_x, B_y, B_z) are included in the free body diagram. Note that each stress component must be

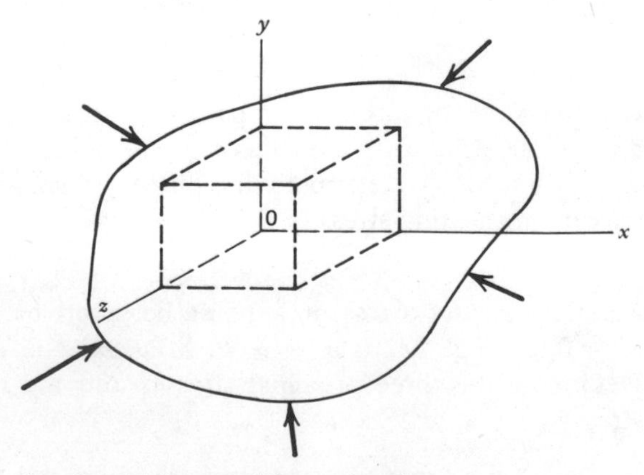

Fig. 1-5.1/General deformed body.

multiplied by the area on which it acts and each body force must be multiplied by the volume of the element since (B_x, B_y, B_z) have dimensions of force per unit volume.

The equations of motion for the volume element in Fig. 1-5.2 are obtained by summation of forces and summation of moments. In Art. 1-3 we have already used summation of moments to obtain the stress symmetry conditions (Eqs. 1-3.2). Summation of forces in the x-direction gives

$$\frac{\partial \sigma_{xx}}{\partial x} + \frac{\partial \sigma_{xy}}{\partial y} + \frac{\partial \sigma_{xz}}{\partial z} + B_x = 0$$

where σ_{xx}, $\sigma_{yx} = \sigma_{xy}$, and $\sigma_{xz} = \sigma_{zx}$ are stress components in the x-direction and B_x is the body force per unit volume in the x-direction including inertial (acceleration) forces. Summation of forces in the y- and z-directions yields similar results. The three equations of motion are thus

$$\begin{aligned}
\frac{\partial \sigma_{xx}}{\partial x} + \frac{\partial \sigma_{xy}}{\partial y} + \frac{\partial \sigma_{xz}}{\partial z} + B_x &= 0 \\
\frac{\partial \sigma_{yx}}{\partial x} + \frac{\partial \sigma_{yy}}{\partial y} + \frac{\partial \sigma_{yz}}{\partial z} + B_y &= 0 \\
\frac{\partial \sigma_{zx}}{\partial x} + \frac{\partial \sigma_{zy}}{\partial y} + \frac{\partial \sigma_{zz}}{\partial z} + B_z &= 0
\end{aligned} \tag{1-5.1}$$

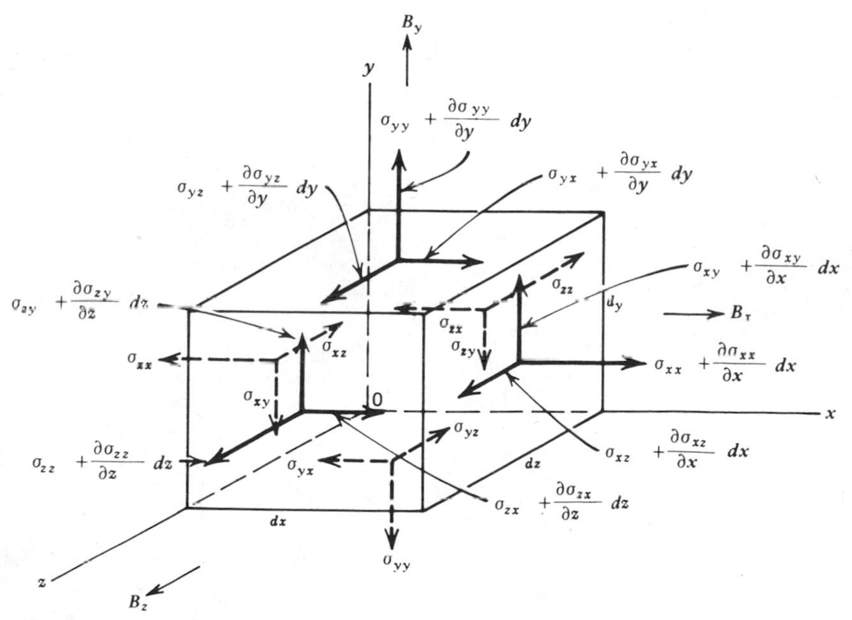

Fig. 1-5.2/Stress components showing changes from face to face along with body force per unit volume including inertial forces.

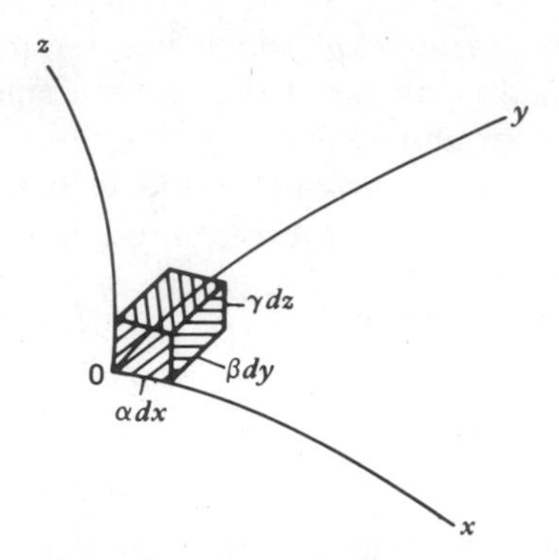

Fig. 1-5.3/Orthogonal curvilinear coordinates.

where (B_x, B_y, B_z) are the components of body force per unit volume including inertial forces. We use Eqs. (1-5.1) in the treatment of torsion of noncircular sections (Chapter 5).

As noted earlier, the form of the differential equations of motion depend on the coordinate axes; Eqs. (1-5.1) were derived for rectangular coordinate axes. In this book we need differential equations of motion in terms of cylindrical coordinates and plane polar coordinates. These are not derived here; instead, we present the most general form from the literature† and show how the general form can be reduced to desired forms. The equations of motion relative to orthogonal curvilinear coordinates (x, y, z) (see Fig. 1-5.3), are

$$\frac{\partial(\beta\gamma\sigma_{xx})}{\partial x} + \frac{\partial(\gamma\alpha\sigma_{xy})}{\partial y} + \frac{\partial(\alpha\beta\sigma_{xz})}{\partial z} + \gamma\sigma_{xy}\frac{\partial\alpha}{\partial y}$$

$$+ \beta\sigma_{xz}\frac{\partial\alpha}{\partial z} - \gamma\sigma_{yy}\frac{\partial\beta}{\partial x} - \beta\sigma_{zz}\frac{\partial\gamma}{\partial x} + \alpha\beta\gamma B_x = 0$$

$$\frac{\partial(\beta\gamma\sigma_{yx})}{\partial x} + \frac{\partial(\gamma\alpha\sigma_{yy})}{\partial y} + \frac{\partial(\alpha\beta\sigma_{yz})}{\partial z} + \alpha\sigma_{yz}\frac{\partial\beta}{\partial z}$$

$$+ \gamma\sigma_{yx}\frac{\partial\beta}{\partial x} - \alpha\sigma_{zz}\frac{\partial\gamma}{\partial y} - \gamma\sigma_{xx}\frac{\partial\alpha}{\partial y} + \alpha\beta\gamma B_y = 0$$

$$\frac{\partial(\beta\gamma\sigma_{zx})}{\partial x} + \frac{\partial(\gamma\alpha\sigma_{zy})}{\partial y} + \frac{\partial(\alpha\beta\sigma_{zz})}{\partial z} + \beta\sigma_{zx}\frac{\partial\gamma}{\partial x}$$

$$+ \alpha\sigma_{zy}\frac{\partial\gamma}{\partial y} - \beta\sigma_{xx}\frac{\partial\alpha}{\partial z} - \alpha\sigma_{yy}\frac{\partial\beta}{\partial z} + \alpha\beta\gamma B_z = 0 \qquad (1\text{-}5.2)$$

where (α, β, γ) are metric coefficients which are functions of the coordi-

† Pages 129 to 134 of reference 1.

nates (x, y, z) and are defined by

$$ds^2 = \alpha^2 dx^2 + \beta^2 dy^2 + \gamma^2 dz^2 \tag{1-5.3}$$

where ds is the differential arc length representing the diagonal of a volume element (Fig. 1-5.3) with edge lengths $\alpha\, dx$, $\beta\, dy$, and $\gamma\, dz$, and where (B_x, B_y, B_z) are the components of body force per unit volume including inertial forces. For rectangular coordinates, $\alpha = \beta = \gamma = 1$ and Eqs. (1-5.2) reduce to Eqs. (1-5.1).

Specialization of Equations (1-5.2)/Commonly employed orthogonal curvilinear systems in three-dimensional problems are the cylindrical coordinate system (r, θ, z) and the spherical coordinate system (r, θ, ϕ); in plane problems, the plane coordinate system (r, θ) is frequently used. We will now specialize Eqs. (1-5.2) for these systems:

(a) Cylindrical Coordinate System (r, θ, z)/In Eqs. (1-5.2), we let $x = r$, $y = \theta$, $z = z$. Then the differential length ds is defined by the relation

$$ds^2 = dr^2 + r^2 d\theta^2 + dz^2 \tag{1-5.4}$$

Comparison of Eqs. (1-5.3) and (1-5.4) yields

$$\alpha = 1 \qquad \beta = r \qquad \gamma = 1 \tag{1-5.5}$$

Substituting Eq. (1-5.5) into Eqs. (1-5.2), we obtain the differential equations of motion

$$\begin{aligned}
&\frac{\partial \sigma_{rr}}{\partial r} + \frac{1}{r}\frac{\partial \sigma_{r\theta}}{\partial \theta} + \frac{\partial \sigma_{rz}}{\partial z} + \frac{\sigma_{rr} - \sigma_{\theta\theta}}{r} + B_r = 0 \\
&\frac{\partial \sigma_{r\theta}}{\partial r} + \frac{1}{r}\frac{\partial \sigma_{\theta\theta}}{\partial \theta} + \frac{\partial \sigma_{\theta z}}{\partial z} + \frac{2\sigma_{r\theta}}{r} + B_\theta = 0 \\
&\frac{\partial \sigma_{rz}}{\partial r} + \frac{1}{r}\frac{\partial \sigma_{\theta z}}{\partial \theta} + \frac{\partial \sigma_{zz}}{\partial z} + \frac{\sigma_{rz}}{r} + B_z = 0
\end{aligned} \tag{1-5.6}$$

where $(\sigma_{rr}, \sigma_{\theta\theta}, \sigma_{zz}, \sigma_{r\theta}, \sigma_{rz}, \sigma_{\theta z})$ represent stress components defined relative to cylindrical coordinates (r, θ, z). We use Eqs. (1-5.6) in Chapter 11 to derive load-stress and load-deflection relations for thick-walled cylinders.

(b) Spherical Coordinate System (r, θ, ϕ)/In Eqs. (1-5.2), we let $x = r$, $y = \theta$, $z = \phi$, where r is the radial coordinate, θ is the colatitude, and ϕ is the longitude. Since the differential length ds is defined by

$$ds^2 = dr^2 + r^2 d\theta^2 + r^2 \sin^2\theta\, d\phi^2 \tag{1-5.7}$$

comparison of Eqs. (1-5.3) and (1-5.7) yields

$$\alpha = 1 \qquad \beta = r \qquad \gamma = r \sin\theta \tag{1-5.8}$$

Substituting Eq. (1-5.8) into Eqs. (1-5.2), we obtain the differential equations of motion

$$\frac{\partial \sigma_{rr}}{\partial r} + \frac{1}{r}\frac{\partial \sigma_{r\theta}}{\partial \theta} + \frac{1}{r\sin\theta}\frac{\partial \sigma_{r\phi}}{\partial \phi} + \frac{1}{r}(2\sigma_{rr} - \sigma_{\theta\theta} - \sigma_{\phi\phi} + \sigma_{r\theta}\cot\theta) + B_r = 0$$

$$\frac{\partial \sigma_{r\theta}}{\partial r} + \frac{1}{r}\frac{\partial \sigma_{\theta\theta}}{\partial \theta} + \frac{1}{r\sin\theta}\frac{\partial \sigma_{\theta\phi}}{\partial \phi} + \frac{1}{r}\left[(\sigma_{\theta\theta} - \sigma_{\phi\phi})\cot\theta + 3\sigma_{r\theta}\right] + B_\theta = 0$$

$$\frac{\partial \sigma_{r\phi}}{\partial r} + \frac{1}{r}\frac{\partial \sigma_{\theta\phi}}{\partial \theta} + \frac{1}{r\sin\theta}\frac{\partial \sigma_{\phi\phi}}{\partial \phi} + \frac{1}{r}(3\sigma_{r\phi} + 2\sigma_{\theta\phi}\cot\theta) + B_\phi = 0 \tag{1-5.9}$$

where $(\sigma_{rr}, \sigma_{\theta\theta}, \sigma_{\phi\phi}, \sigma_{r\theta}, \sigma_{r\phi}, \sigma_{\theta\phi})$ are defined relative to spherical coordinates (r, θ, ϕ).

(c) Plane Polar Coordinate System (r, θ)/In plane-stress problems relative to (x, y) coordinates, $\sigma_{zz} = \sigma_{xz} = \sigma_{yz} = 0$, and the remaining stress components are functions of (x, y) only (Art. 1-4). Letting $x = r$, $y = \theta$, $z = z$ in Eqs. (1-5.6) and noting that $\sigma_{zz} = \sigma_{rz} = \sigma_{\theta z} = (\partial/\partial z) = 0$, we obtain from Eq. (1-5.6)

$$\frac{\partial \sigma_{rr}}{\partial r} + \frac{1}{r}\frac{\partial \sigma_{r\theta}}{\partial \theta} + \frac{\sigma_{rr} - \sigma_{\theta\theta}}{r} + B_r = 0$$

$$\frac{\partial \sigma_{r\theta}}{\partial r} + \frac{1}{r}\frac{\partial \sigma_{\theta\theta}}{\partial \theta} + 2\frac{\sigma_{r\theta}}{r} + B_\theta = 0 \tag{1-5.10}$$

1-6 DEFORMATION OF A DEFORMABLE BODY

In the first four articles of this chapter, we examined the six stress components that define the state of stress at a point in a loaded member, derived the transformation equations of stress, and derived expressions for the maximum principal stress, the maximum shearing stress, and the maximum octahedral shearing stress at a point. These relations are of interest throughout the most of the book. Differential equations of equilibrium (differential equations of motion for members being accelerated) were derived in Art. 1-5. These are needed in chapters in which the method of theory of elasticity is used to derive load-stress and load-deflection relations. Additional differential equations of compatibility, needed for the method of theory of elasticity, are derived in Art. 1-9; the derivation employs small displacement approximations and the associated strain-displacement relations. Although small displacements are considered in most applications of this book, the more general finite strain-displacement relations are derived in this chapter so that the reader may better understand the approximations that lead to the strain-displacement relations of small-displacement theory.

In the derivation of strain-displacement relations for a member, we consider the member first to be unloaded (undeformed and unstressed) and next to be loaded (stressed and deformed). We let R represent the closed region occupied by the undeformed member and R^* represent the closed region occupied by the deformed member. Asterisks are used to designate quantities associated with the deformed state of members throughout the book.

Let (x, y, z) be rectangular coordinates (Fig. 1-6.1). A particle P located at the general coordinate point (x, y, z) in the undeformed body is then defined by the equations

$$\begin{aligned} x^* &= x^*(x, y, z) \\ y^* &= y^*(x, y, z) \\ z^* &= z^*(x, y, z) \end{aligned} \tag{1-6.1}$$

where the values of (x, y, z) are restricted to region R and (x^*, y^*, z^*) are restricted to region R^*. Equations (1-6.1) define the final location of a particle P that lies at a given point (x, y, z) in the undeformed member. It is assumed that the functions (x^*, y^*, z^*) are continuous and differentiable in the independent variables (x, y, z), since a discontinuity of these functions would imply a rupture of the member. Mathematically this means that Eqs. (1-6.1) may be solved for single-valued solutions of

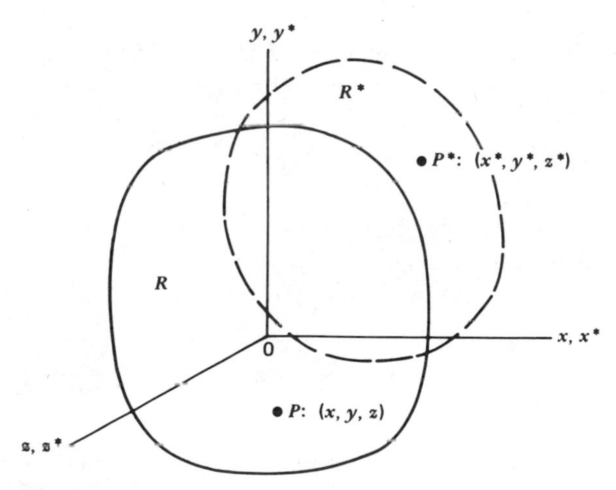

Fig. 1-6.1/Location of general point P in undeformed and deformed body.

(x, y, z); that is,

$$\begin{aligned} x &= x(x^*, y^*, z^*) \\ y &= y(x^*, y^*, z^*) \\ z &= z(x^*, y^*, z^*) \end{aligned} \tag{1-6.2}$$

Equations (1-6.2) define the initial location of a particle P that lies at point (x^*, y^*, z^*) in the deformed member. Functions (x, y, z) are continuous and differentiable in the independent variables (x^*, y^*, z^*).

When (x^*, y^*, z^*) are used as independent variables (Eq. 1-6.2), the point of view is that of the *Eulerian or spatial coordinate method*. When (x, y, z) are used as independent variables, the point of view is that of the *Lagrangian or material coordinate method*. It may be shown that for classical, small-displacement theories of elasticity and plasticity, it is not necessary to distinguish between the variables (x^*, y^*, z^*) and (x, y, z). We employ material coordinates in this book.

1-7 STRAIN THEORY. PRINCIPAL STRAINS†

Strain of a Line Element/When a member is deformed, the particle at point $P:(x, y, z)$ passes to the point $P^*:(x^*, y^*, z^*)$ (Fig. 1-7.1). Also, the particle at point $Q:(x+dx, y+dy, z+dz)$ passes to the point $Q^*:(x^* + dx^*, y^* + dy^*, z^* + dz^*)$, and the infinitesimal line element $PQ = ds$ passes into the line element $P^*Q^* = ds^*$. We define the *engineering strain* ϵ_E of the line element $PQ = ds$ as

$$\epsilon_E = \frac{ds^* - ds}{ds} \tag{1-7.1}$$

Therefore, by the definition, $\epsilon_E > -1$. Equation (1-7.1) is employed widely in engineering.

By Eqs. (1-6.1), we obtain the total differential

$$dx^* = \frac{\partial x^*}{\partial x}\, dx + \frac{\partial x^*}{\partial y}\, dy + \frac{\partial x^*}{\partial z}\, dz \tag{1-7.2}$$

† The theory presented in this article includes quadratic terms in the displacement components (u, v, w) and in the engineering strain ϵ_E. One may, if one wishes, discard all quadratic terms in u, v, w, and ϵ_E, and obtain directly the theory of strain for small deformations. (See Art. 1-9.)

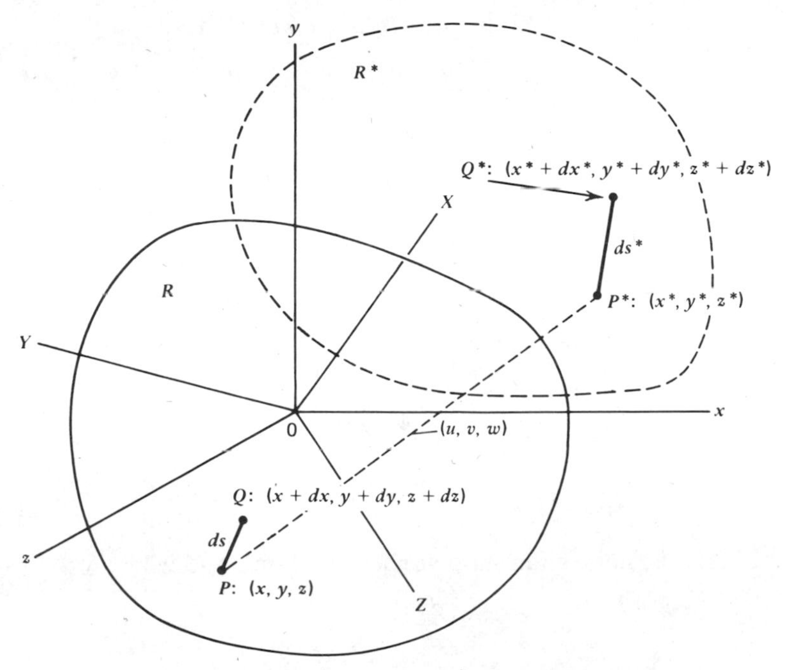

Fig. 1-7.1/Line segment PQ in undeformed and deformed body.

with similar expressions for dy^*, dz^*. Noting that

$$
\begin{aligned}
x^* &= x + u \\
y^* &= y + v \qquad\qquad (1\text{-}7.3) \\
z^* &= z + w
\end{aligned}
$$

where (u, v, w) denote the (x, y, z)-components of the displacement of P to P^*, and also noting that

$$
\begin{aligned}
(ds)^2 &= (dx)^2 + (dy)^2 + (dz)^2 \\
(ds^*)^2 &= (dx^*)^2 + (dy^*)^2 + (dz^*)^2 \qquad (1\text{-}7.4)
\end{aligned}
$$

we find† (retaining quadratic terms in derivatives of (u, v, w))

$$
\begin{aligned}
M = \frac{1}{2}\left[\left(\frac{ds^*}{ds}\right)^2 - 1\right] &= \epsilon_E + \tfrac{1}{2}\epsilon_E^2 = l^2\epsilon_{xx} + lm\epsilon_{xy} + ln\epsilon_{xz} \\
&\quad + ml\epsilon_{yx} + m^2\epsilon_{yy} + mn\epsilon_{yz} + nl\epsilon_{zx} + nm\epsilon_{zy} + n^2\epsilon_{zz} \\
&= l^2\epsilon_{xx} + m^2\epsilon_{yy} + n^2\epsilon_{zz} + 2lm\epsilon_{xy} + 2ln\epsilon_{xz} + 2mn\epsilon_{yz} \qquad (1\text{-}7.5)
\end{aligned}
$$

†Although one may compute ϵ_E directly from Eq. (1-7.1), it is mathematically simpler to form the quantity $M = \frac{1}{2}[(ds^*/ds)^2 - 1] = \frac{1}{2}[(1+\epsilon_E)^2 - 1] = \epsilon_E + \frac{1}{2}\epsilon_E^2$. Then one may compute ϵ_E from Eq. (1-7.5). For small ϵ_E (Art. 1-9), $\epsilon_E \cong M$.

where M is called the *magnification factor* and

$$\epsilon_{xx} = \frac{\partial u}{\partial x} + \frac{1}{2}\left[\left(\frac{\partial u}{\partial x}\right)^2 + \left(\frac{\partial v}{\partial x}\right)^2 + \left(\frac{\partial w}{\partial x}\right)^2\right]$$

$$\epsilon_{yy} = \frac{\partial v}{\partial y} + \frac{1}{2}\left[\left(\frac{\partial u}{\partial y}\right)^2 + \left(\frac{\partial v}{\partial y}\right)^2 + \left(\frac{\partial w}{\partial y}\right)^2\right]$$

$$\epsilon_{zz} = \frac{\partial w}{\partial z} + \frac{1}{2}\left[\left(\frac{\partial u}{\partial z}\right)^2 + \left(\frac{\partial v}{\partial z}\right)^2 + \left(\frac{\partial w}{\partial z}\right)^2\right] \tag{1-7.6}$$

$$\epsilon_{xy} = \epsilon_{yx} = \frac{1}{2}\left[\frac{\partial v}{\partial x} + \frac{\partial u}{\partial y} + \frac{\partial u}{\partial x}\frac{\partial u}{\partial y} + \frac{\partial v}{\partial x}\frac{\partial v}{\partial y} + \frac{\partial w}{\partial x}\frac{\partial w}{\partial y}\right]$$

$$\epsilon_{xz} = \epsilon_{zx} = \frac{1}{2}\left[\frac{\partial w}{\partial x} + \frac{\partial u}{\partial z} + \frac{\partial u}{\partial x}\frac{\partial u}{\partial z} + \frac{\partial v}{\partial x}\frac{\partial v}{\partial z} + \frac{\partial w}{\partial x}\frac{\partial w}{\partial z}\right]$$

$$\epsilon_{yz} = \epsilon_{zy} = \frac{1}{2}\left[\frac{\partial w}{\partial y} + \frac{\partial v}{\partial z} + \frac{\partial u}{\partial y}\frac{\partial u}{\partial z} + \frac{\partial v}{\partial y}\frac{\partial v}{\partial z} + \frac{\partial w}{\partial y}\frac{\partial w}{\partial z}\right]$$

are the finite strain-displacement relations† and where

$$l = \frac{dx}{ds}, \qquad m = \frac{dy}{ds}, \qquad n = \frac{dz}{ds} \tag{1-7.7}$$

are the direction cosines of line element ds.

We may interpret the quantities $\epsilon_{xx}, \epsilon_{yy}, \epsilon_{zz}$ physically, by considering line elements ds that lie parallel to the (x, y, z)-axes, respectively. For example, let the line element ds (Fig. 1-7.1) lie parallel to the x-axis. Then $l = 1$, $m = n = 0$, and Eq. (1-7.5) yields

$$M_x = \epsilon_{Ex} + \tfrac{1}{2}\epsilon_{Ex}^2 = \epsilon_{xx} \tag{1-7.5a}$$

where M_x and ϵ_{Ex} denote the magnification factor and the engineering strain of the element ds (parallel to the x-direction). Hence, ϵ_{xx}, physically, is the magnification factor of the line element at P that lies initially in the x-direction. In particular, if the engineering strain is small ($\epsilon_{Ex} \ll 1$), we obtain the result $\epsilon_{xx} \approx \epsilon_{Ex}$: namely that ϵ_{xx} is approximately equal to the engineering strain for small strains. Similarly, for the cases where initially ds lies parallel to the y-axis and then the z-axis, we obtain

$$\begin{aligned} M_y &= \epsilon_{Ey} + \tfrac{1}{2}\epsilon_{Ey}^2 = \epsilon_{yy} \\ M_z &= \epsilon_{Ez} + \tfrac{1}{2}\epsilon_{Ez}^2 = \epsilon_{zz} \end{aligned} \tag{1-7.5b}$$

Thus, $(\epsilon_{xx}, \epsilon_{yy}, \epsilon_{zz})$, physically represent the magnification factors for line elements that initially lie parallel to the (x, y, z)-axes, respectively.

To obtain a physical interpretation of the components $\epsilon_{xy}, \epsilon_{xz}, \epsilon_{yz}$, it is necessary to determine the rotation between two line elements initially

† In small displacement theory, the quadratic terms in Eqs. (1-7.6) are neglected. Then, Eqs. (1-7.6) reduce to Eqs. (1-9.1).

parallel to (x, y)-axes, (x, z)-axes and (y, z)-axes, respectively. To do this, we first determine the final direction of a single line element under the deformation. Then, we use this result to determine the rotation between two line elements.

Final Direction of Line Element / As a result of the deformation, the line element ds: (dx, dy, dz) deforms into the line element ds^*: (dx^*, dy^*, dz^*). By definition, the direction cosines of ds and ds^* are

$$l = \frac{dx}{ds} \qquad m = \frac{dy}{ds} \qquad n = \frac{dz}{ds}$$
$$l^* = \frac{dx^*}{ds^*} \qquad m^* = \frac{dy^*}{ds^*} \qquad n^* = \frac{dz^*}{ds^*} \tag{1-7.8}$$

Alternatively, we may write

$$l^* = \frac{dx^*}{ds}\frac{ds}{ds^*} \qquad m^* = \frac{dy^*}{ds}\frac{ds}{ds^*} \qquad n^* = \frac{dz^*}{ds}\frac{ds}{ds^*} \tag{1-7.9}$$

By Eqs. (1-7.2) and (1-7.3), we find

$$\frac{dx^*}{ds} = \left(1 + \frac{\partial u}{\partial x}\right)l + \frac{\partial u}{\partial y}m + \frac{\partial u}{\partial z}n$$
$$\frac{dy^*}{ds} = \frac{\partial v}{\partial x}l + \left(1 + \frac{\partial v}{\partial y}\right)m + \frac{\partial v}{\partial z}n \tag{1-7.10}$$
$$\frac{dz^*}{ds} = \frac{\partial w}{\partial x}l + \frac{\partial w}{\partial y}m + \left(1 + \frac{\partial w}{\partial z}\right)n$$

and by Eq. (1-7.1)

$$\frac{ds}{ds^*} = \frac{1}{1 + \epsilon_E} \tag{1-7.11}$$

Hence, Eqs. (1-7.9), (1-7.10) and (1-7.11) yield

$$(1 + \epsilon_E)l^* = \left(1 + \frac{\partial u}{\partial x}\right)l + \frac{\partial u}{\partial y}m + \frac{\partial u}{\partial z}n$$
$$(1 + \epsilon_E)m^* = \frac{\partial v}{\partial x}l + \left(1 + \frac{\partial v}{\partial y}\right)m + \frac{\partial v}{\partial z}n \tag{1-7.12}$$
$$(1 + \epsilon_E)n^* = \frac{\partial w}{\partial x}l + \frac{\partial w}{\partial y}m + \left(1 + \frac{\partial w}{\partial z}\right)n$$

Equations (1-7.12) represent the final direction cosines of line element ds when it passes into the line element ds^* under the deformation.

Rotation between Two Line Elements (Definition of Shearing Strain) / Next, let us consider two infinitesimal line elements PA and PB of

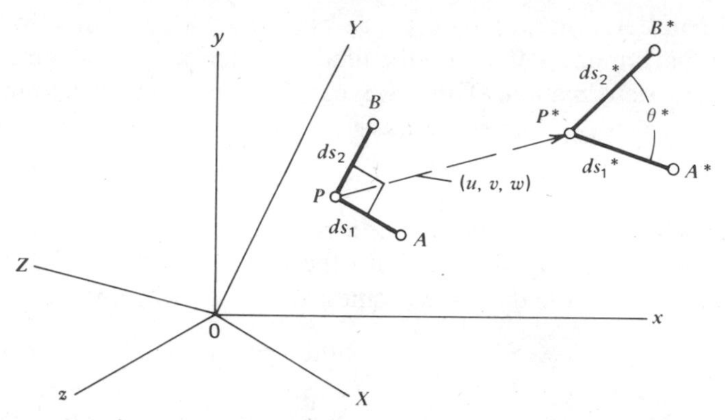

Fig. 1-7.2/Line segments PA and PB before and after deformation.

lengths ds_1 and ds_2 emanating from point P. For simplicity, let PA be perpendicular to PB† (Fig. 1-7.2). Let the direction cosines of lines PA and PB be (l_1, m_1, n_1) and (l_2, m_2, n_2), respectively. By the deformation, line elements PA, PB are transformed into line elements P^*A^*, P^*B^*, with direction cosines (l_1^*, m_1^*, n_1^*) and (l_2^*, m_2^*, n_2^*), respectively. Since PA is perpendicular to PB, by the definition of scalar product of vectors

$$\cos\frac{\pi}{2} = l_1 l_2 + m_1 m_2 + n_1 n_2 = 0 \tag{1-7.13}$$

Similarly, the angle θ^* between P^*A^* and P^*B^* is defined by

$$\cos\theta^* = l_1^* l_2^* + m_1^* m_2^* + n_1^* n_2^* \tag{1-7.14}$$

In turn, (l_1^*, m_1^*, n_1^*) and (l_2^*, m_2^*, n_2^*) are expressed in terms of (l_1, m_1, n_1) and (l_2, m_2, n_2), respectively, by means of Eq. (1-7.12). Hence, by Eqs. (1-7.12), (1-7.13), and (1-7.14), we may write with Eqs. (1-7.6)

$$\begin{aligned}\gamma_{12} &= (1+\epsilon_{E1})(1+\epsilon_{E2})\cos\theta^* \\ &= 2l_1 l_2 \epsilon_{xx} + 2m_1 m_2 \epsilon_{yy} + 2n_1 n_2 \epsilon_{zz} + 2(l_1 m_2 + l_2 m_1)\epsilon_{xy} \\ &\quad + 2(m_1 n_2 + m_2 n_1)\epsilon_{yz} + 2(l_1 n_2 + l_2 n_1)\epsilon_{xz}\end{aligned} \tag{1-7.15}$$

where γ_{12} is defined to be the *engineering shearing strain* between line elements PA and PB as they are deformed into P^*A^* and P^*B^* (Fig. 1-7.2).

To obtain a physical interpretation of ϵ_{xy}, we now let PA and PB be oriented initially parallel to axes (x, y), respectively. Hence, $l_1 = 1$, $m_1 = n_1 = 0$ and $l_2 = n_2 = 0$, $m_2 = 1$. Then Eq. (1-7.15) yields the result

$$\gamma_{12} = \gamma_{xy} = 2\epsilon_{xy} \tag{1-7.16}$$

† This restriction is not necessary but is used for simplicity. See Reference 1.

In other words, $2\epsilon_{xy}$ represents the engineering shearing strain between two line elements initially parallel to the (x, y)-axes, respectively. Similarly, we may consider PA and PB to be oriented initially parallel to (y, z)-axes and then to (x, z)-axes to obtain similar interpretations for $\epsilon_{yz}, \epsilon_{xz}$. Thus,

$$\gamma_{xy} = 2\epsilon_{xy} \qquad \gamma_{yz} = 2\epsilon_{yz} \qquad \gamma_{xz} = 2\epsilon_{xz} \tag{1-7.17}$$

represent the engineering shearing strains between two line elements initially parallel to (x, y)-, (y, z)-, and (x, z)-axes, respectively.

If the strains $\epsilon_{E1}, \epsilon_{E2}$ are small and if the rotations are small (e.g., $\theta^* \approx \pi/2$), Eq. (1-7.15) yields the approximation

$$\gamma_{12} = (1 + \epsilon_{E1})(1 + \epsilon_{E2}) \cos\theta^* \approx \frac{\pi}{2} - \theta^* \tag{1-7.18}$$

and the engineering shearing strain becomes approximately equal to the change in angle between line elements PA and PB.

Other results analogous to those of stress theory (Art. 1-4) also hold. For example, the symmetric array

$$\begin{pmatrix} \epsilon_{xx} & \epsilon_{xy} & \epsilon_{xz} \\ \epsilon_{xy} & \epsilon_{yy} & \epsilon_{yz} \\ \epsilon_{xz} & \epsilon_{yz} & \epsilon_{zz} \end{pmatrix} \tag{1-7.19}$$

is the *strain tensor*. Under a rotation of axis, the components of the strain tensor $(\epsilon_{xx}, \epsilon_{xy}, \epsilon_{xz}, \ldots)$ transform in exactly the same way as the stress tensor (Eqs. 1-4.1 and 1-4.3). (Compare Eqs. 1-3.3 and 1-7.19. Also compare Eqs. 1-3.9 and 1-7.5.) To show this transformation, consider again axes (x, y, z) and (X, Y, Z), as in Art. 1-4, Fig. 1-4.1 (also Fig. 1-7.1), and Table 1-4.1. The strain components $\epsilon_{XX}, \epsilon_{XY}, \epsilon_{XZ}, \ldots,$ are defined with reference to axes (X, Y, Z) in the same manner as $\epsilon_{xx}, \epsilon_{xy}, \epsilon_{xz}, \ldots,$ are defined relative to axes (x, y, z). Hence, ϵ_{XX} is the extensional strain of a line element at point P (Fig. 1-7.1) that lies in the direction of the X-axis, and ϵ_{XY} and ϵ_{XZ} are shearing components between line elements that are parallel to axes (X, Y) and (X, Z), respectively, and so on for $\epsilon_{YY}, \epsilon_{ZZ}, \epsilon_{YZ}$. Hence, if we let element ds lie parallel to the X-axis, Eq. (1-7.5), with Table 1-4.1, yields

$$\begin{aligned} \epsilon_{XX} = {} & l_1^2\epsilon_{xx} + m_1^2\epsilon_{yy} + n_1^2\epsilon_{zz} + 2l_1m_1\epsilon_{xy} \\ & + 2l_1n_1\epsilon_{xz} + 2m_1n_1\epsilon_{yz} \end{aligned} \tag{1-7.20a}$$

Similarly for the line elements that lie parallel to axes Y and Z, respectively, we have

$$\begin{aligned} \epsilon_{YY} = {} & l_2^2\epsilon_{xx} + m_2^2\epsilon_{yy} + n_2^2\epsilon_{zz} + 2l_2m_2\epsilon_{xy} \\ & + 2l_2n_2\epsilon_{xz} + 2m_2n_2\epsilon_{yz} \end{aligned} \tag{1-7.20b}$$

$$\begin{aligned} \epsilon_{ZZ} = {} & l_3^2\epsilon_{xx} + m_3^2\epsilon_{yy} + n_3^2\epsilon_{zz} + 2l_3m_3\epsilon_{xy} \\ & + 2l_3n_3\epsilon_{xz} + 2m_3n_3\epsilon_{yz} \end{aligned} \tag{1-7.20c}$$

Similarly, if we take line elements PA and PB parallel, respectively to axes X and Y (Fig. 1-7.2), Eqs. (1-7.15) and (1-7.17) yield the result

$$\begin{aligned}\tfrac{1}{2}\gamma_{XY}=\epsilon_{XY}=l_1l_2\epsilon_{xx}+m_1m_2\epsilon_{yy}+n_1n_2\epsilon_{zz}\\ +(l_1m_2+l_2m_1)\epsilon_{xy}+(m_1n_2+m_2n_1)\epsilon_{yz}\\ +(l_1n_2+l_2n_1)\epsilon_{xz}\end{aligned} \tag{1-7.20d}$$

In a similar manner, we find

$$\begin{aligned}\tfrac{1}{2}\gamma_{YZ}=\epsilon_{YZ}=l_2l_3\epsilon_{xx}+m_2m_3\epsilon_{yy}+n_2n_3\epsilon_{zz}\\ +(l_2m_3+l_3m_2)\epsilon_{xy}\\ +(m_2n_3+m_3n_2)\epsilon_{yz}\\ +(l_2n_3+l_3n_2)\epsilon_{xz}\end{aligned} \tag{1-7.20e}$$

$$\begin{aligned}\tfrac{1}{2}\gamma_{XZ}=\epsilon_{XZ}=l_1l_3\epsilon_{xx}+m_1m_3\epsilon_{yy}+n_1n_3\epsilon_{zz}\\ +(l_1m_3+l_3m_1)\epsilon_{xy}\\ +(m_1n_3+m_3n_1)\epsilon_{yz}\\ +(l_1n_3+l_3n_1)\epsilon_{xz}\end{aligned} \tag{1-7.20f}$$

where (l_1, m_1, n_1), (l_2, m_2, n_2), and (l_3, m_3, n_3) are the direction cosines of axes X, Y, and Z, respectively.

Equations (1-7.20) represent the transformation of the strain tensor $(\epsilon_{xx}, \epsilon_{yy}, \ldots, \epsilon_{yz})$ under a rotation from axes (x, y, z) to axes (X, Y, Z). (See Figs. 1-7.1 and 1-7.2 and also Fig. 1-4.1.)

Principal Strains/Under a deformation of the member (Art. 1-6), any infinitesimal sphere in the member is deformed into an ellipsoid, called the *strain ellipsoid*. The principal axes of the strain ellipsoid have the directions of the principal axes of strain (see below) at the center of the ellipsoid in the deformed member. The radii of the infinitesimal sphere that pass into the principal axes of the strain ellipsoid are initially perpendicular to each other, and they coincide with the principal axes of strain in the undeformed member. Hence, through any point in an undeformed member, there exist three mutually perpendicular line elements that remain perpendicular under the deformation. The strains of these three line elements are called the *principal strains* at the point. We denote them by $(\epsilon_{E1}, \epsilon_{E2}, \epsilon_{E3})$ and the corresponding principal values of the magnification factor $M=\epsilon_E+\frac{1}{2}\epsilon_E^2$ are denoted by (M_1, M_2, M_3). By analogy with stress theory (Art. 1-4), the principal values of the magnification factor are the three roots of the determinantal equation

$$\begin{vmatrix}\epsilon_{xx}-M & \epsilon_{xy} & \epsilon_{xz}\\ \epsilon_{xy} & \epsilon_{yy}-M & \epsilon_{yz}\\ \epsilon_{xz} & \epsilon_{yz} & \epsilon_{xx}-M\end{vmatrix}=0 \tag{1-7.21a}$$

or

$$M^3 - J_1 M^2 + J_2 M - J_3 = 0$$
$$M = \epsilon_E + \tfrac{1}{2}\epsilon_E^2 \tag{1-7.21b}$$

where

$$J_1 = \epsilon_{xx} + \epsilon_{yy} + \epsilon_{zz}$$
$$J_2 = \begin{vmatrix} \epsilon_{xx} & \epsilon_{xy} \\ \epsilon_{xy} & \epsilon_{yy} \end{vmatrix} + \begin{vmatrix} \epsilon_{xx} & \epsilon_{xz} \\ \epsilon_{xz} & \epsilon_{zz} \end{vmatrix} + \begin{vmatrix} \epsilon_{yy} & \epsilon_{yz} \\ \epsilon_{yz} & \epsilon_{zz} \end{vmatrix}$$
$$= \epsilon_{xx}\epsilon_{yy} + \epsilon_{xx}\epsilon_{zz} + \epsilon_{yy}\epsilon_{zz} - \epsilon_{xy}^2 - \epsilon_{xz}^2 - \epsilon_{yz}^2 \tag{1-7.22}$$
$$J_3 = \begin{vmatrix} \epsilon_{xx} & \epsilon_{xy} & \epsilon_{xz} \\ \epsilon_{xy} & \epsilon_{yy} & \epsilon_{yz} \\ \epsilon_{xz} & \epsilon_{yz} & \epsilon_{zz} \end{vmatrix}$$

are the *strain invariants* (see Eqs. 1-4.5, 1-4.6, 1-4.7). Because of the symmetry of the determinant of Eq. (1-7.21a), the roots $M_i : i = 1, 2, 3$ are always real. Also since $\epsilon_{Ei} > -1$, $M_i > -1$.

The three principal strain directions associated with the three principal strains $(\epsilon_{E1}, \epsilon_{E2}, \epsilon_{E3})$, Eq. 1-7.21b, are obtained as the solution for (l, m, n) of the equations

$$\begin{aligned} l(\epsilon_{xx} - M) + m\epsilon_{xy} + n\epsilon_{xz} &= 0 \\ l\epsilon_{xy} + m(\epsilon_{yy} - M) + n\epsilon_{yz} &= 0 \\ l\epsilon_{xz} + m\epsilon_{yz} + n(\epsilon_{zz} - M) &= 0 \end{aligned} \tag{1-7.23}$$
$$l^2 + m^2 + n^2 = 1$$

The solution $M = M_1$ yields the direction cosines for $\epsilon_E = \epsilon_{E1}$ and so on for $M = M_2(\epsilon_E = \epsilon_{E2})$, $M = M_3(\epsilon_E = \epsilon_{E3})$.

If (x, y, z) axes are principal strain axes, $\epsilon_{xx} = M_1$, $\epsilon_{yy} = M_2$, $\epsilon_{zz} = M_3$, $\epsilon_{xy} = \epsilon_{xz} = \epsilon_{yz} = 0$ and the expressions for the strain invariants J_1, J_2, J_3 reduce to

$$\begin{aligned} J_1 &= M_1 + M_2 + M_3 \\ J_2 &= M_1 M_2 + M_1 M_3 + M_2 M_3 \\ J_3 &= M_1 M_2 M_3 \end{aligned} \tag{1-7.24}$$

1-8 STRAIN OF A VOLUME ELEMENT†

Analogous to the strain and rotation (change in direction) of a line element, one may also define strain and rotation of a volume element. Let a volume element of a deformed member have the initial volume dV and

† This article is included for completeness but is not essential to the remainder of the text.

the final volume dV^*. Then the volumetric strain (also called the cubical strain or dilation in contrast to the lineal strain of a line element) is defined by

$$e = \frac{dV^* - dV}{dV} = \frac{dV^*}{dV} - 1 \tag{1-8.1}$$

The value of e is computed simply if one refers to principal axes of strain at the point, such that the volume element ($dV = ds_1\, ds_2\, ds_3$) is in the form of an infinitesimal rectangular parallelepiped with edges of length ds_1, ds_2, and ds_3 along the principal directions. Then since the edges remain perpendicular, the volume element remains a rectangular parallelepiped after the deformation and $dV^* = ds_1^*\, ds_2^*\, ds_3^*$ where ds_1^*, ds_2^*, and ds_3^* are the lengths of the sides of the volume element after deformation. Hence, by Eq. (1-7.1)

$$dV^* = ds_1^*\, ds_2^*\, ds_3^* = (1 + \epsilon_{E1})(1 + \epsilon_{E2})(1 + \epsilon_{E3})\, dV \tag{1-8.2}$$

Using Eqs. (1-8.1) and (1-8.2), the engineering volumetric strain is

$$e = \epsilon_{E1} + \epsilon_{E2} + \epsilon_{E3} + \epsilon_{E1}\epsilon_{E2} + \epsilon_{E1}\epsilon_{E3} + \epsilon_{E2}\epsilon_{E3} + \epsilon_{E1}\epsilon_{E2}\epsilon_{E3} \tag{1-8.3}$$

Analogous to the magnification factor of a line element (Eq. 1-7.5), we have

$$M_V = \frac{1}{2}\left[\left(\frac{dV^*}{dV}\right)^2 - 1\right] \tag{1-8.4}$$

where M_V is the magnification factor for the volume element. Now by Eq. (1-7.5)

$$\begin{aligned} M_1 &= \frac{1}{2}\left[\left(\frac{ds_1^*}{ds_1}\right)^2 - 1\right] \\ M_2 &= \frac{1}{2}\left[\left(\frac{ds_2^*}{ds_2}\right)^2 - 1\right] \\ M_3 &= \frac{1}{2}\left[\left(\frac{ds_3^*}{ds_3}\right)^2 - 1\right] \end{aligned} \tag{1-8.5}$$

where M_1, M_2, and M_3 are the principal strains. Solving Eqs. (1-8.5) for $(ds_1^*/ds_1)^2$, $(ds_2^*/ds_2)^2$, and $(ds_3^*/ds_3)^2$ and substituting into Eq. (1-8.4) for $(dV^*/dV)^2$, we find (see Eq. 1-7.5)

$$M_V = e + \tfrac{1}{2}e^2 = J_1 + 2J_2 + 4J_3 \tag{1-8.6}$$

where J_1, J_2, and J_3 are given in terms of M_1, M_2, and M_3 by Eqs. (1-7.24). If $e \ll 1$, Eq. (1-8.6) yields

$$M_V \approx e \approx J_1 \tag{1-8.7}$$

where $J_1 = M_1 + M_2 + M_3$ and by Eq. (1-8.3)

$$e \approx \epsilon_{E1} + \epsilon_{E2} + \epsilon_{E3} \tag{1-8.8}$$

1-9 SMALL-DISPLACEMENT THEORY

The deformation theory developed in Arts. 1-6, 1-7, and 1-8 is purely geometrical and the associated equations are exact. In the small-displacement theory, the quadratic terms in Eqs. (1-7.6) are discarded. Then,

$$\epsilon_{xx} \cong \frac{\partial u}{\partial x} \qquad \epsilon_{yy} \cong \frac{\partial v}{\partial y} \qquad \epsilon_{zz} \cong \frac{\partial w}{\partial z} \tag{1-9.1}$$
$$\epsilon_{xy} \cong \frac{1}{2}\left(\frac{\partial v}{\partial x} + \frac{\partial u}{\partial y}\right) \qquad \epsilon_{xz} \cong \frac{1}{2}\left(\frac{\partial w}{\partial x} + \frac{\partial u}{\partial z}\right) \qquad \epsilon_{yz} \cong \frac{1}{2}\left(\frac{\partial w}{\partial y} + \frac{\partial v}{\partial z}\right)$$

are the strain-displacement relations for small-displacement theory. Also, as noted in Art. 1-8, then

$$e \cong J_1 \tag{1-9.2}$$

Furthermore, the magnification factor reduces to

$$M \cong \epsilon_E \tag{1-9.3}$$

The above approximations, which are the basis for small-displacement theory, imply that the strains and the rotations (excluding rigid-body rotations) are small compared to unity. The latter condition is not necessarily satisfied in deformation of thin flexible bodies, such as rods, plates, and shells. For these bodies the rotations may be large. Consequently, the small-displacement theory must be used with caution: it is usually applicable for massive (thick) bodies, but it may give results that are seriously in error when applied to thin flexible bodies.

Strain Compatibility Relations / Elimination of the displacement components (u, v, w) from Eqs. (1-9.1) yields

$$\begin{aligned}
&\frac{\partial^2 \epsilon_{yy}}{\partial x^2} + \frac{\partial^2 \epsilon_{xx}}{\partial y^2} = 2\frac{\partial^2 \epsilon_{xy}}{\partial x\,\partial y}\\
&\frac{\partial^2 \epsilon_{zz}}{\partial x^2} + \frac{\partial^2 \epsilon_{xx}}{\partial z^2} = 2\frac{\partial^2 \epsilon_{xz}}{\partial x\,\partial z}\\
&\frac{\partial^2 \epsilon_{zz}}{\partial y^2} + \frac{\partial^2 \epsilon_{yy}}{\partial z^2} = 2\frac{\partial^2 \epsilon_{yz}}{\partial y\,\partial z}\\
&\frac{\partial^2 \epsilon_{zz}}{\partial x\,\partial y} + \frac{\partial^2 \epsilon_{xy}}{\partial z^2} = \frac{\partial^2 \epsilon_{yz}}{\partial z\,\partial x} + \frac{\partial^2 \epsilon_{zx}}{\partial y\,\partial z}\\
&\frac{\partial^2 \epsilon_{yy}}{\partial x\,\partial z} + \frac{\partial^2 \epsilon_{xz}}{\partial y^2} = \frac{\partial^2 \epsilon_{xy}}{\partial y\,\partial z} + \frac{\partial^2 \epsilon_{yz}}{\partial x\,\partial y}\\
&\frac{\partial^2 \epsilon_{xx}}{\partial y\,\partial z} + \frac{\partial^2 \epsilon_{yz}}{\partial x^2} = \frac{\partial^2 \epsilon_{xz}}{\partial x\,\partial y} + \frac{\partial^2 \epsilon_{xy}}{\partial x\,\partial z}
\end{aligned} \tag{1-9.4}$$

Equations (1-9.4) are known as the *strain compatibility equations of small-displacement theory*. It may be shown that if the strain components $(\epsilon_{xx}, \epsilon_{yy}, \epsilon_{zz}, \epsilon_{xy}, \epsilon_{xz}, \epsilon_{yz})$ satisfy Eqs. (1-9.4), there exist displacement components (u, v, w) that are solutions of Eqs. (1-9.1). More fully, in the small-displacement theory, the functions $(\epsilon_{xx}, \epsilon_{yy}, \epsilon_{zz}, \epsilon_{xy}, \epsilon_{xz}, \epsilon_{yz})$ are possible components of strain if, and only if, they satisfy Eqs. (1-9.4). For large displacement theory the equivalent results are very complicated.[2]

Strain-Displacement Relations for Orthogonal Curvilinear Coordinates/More generally, the strain-displacement relations (Eqs. 1-7.6) may be written for orthogonal curvilinear coordinates (Fig. 1-5.3). The derivation of the expressions for $(\epsilon_{xx}, \epsilon_{yy}, \epsilon_{zz}, \epsilon_{xy}, \epsilon_{xz}, \epsilon_{yz})$, is a routine problem.[1] For small-displacement theory the results are

$$
\begin{aligned}
\epsilon_{xx} &= \frac{1}{\alpha}\left[\frac{\partial u}{\partial x} + \frac{v}{\beta}\frac{\partial \alpha}{\partial y} + \frac{w}{\gamma}\frac{\partial \alpha}{\partial z}\right] \\
\epsilon_{yy} &= \frac{1}{\beta}\left[\frac{\partial v}{\partial y} + \frac{w}{\gamma}\frac{\partial \beta}{\partial z} + \frac{u}{\alpha}\frac{\partial \beta}{\partial x}\right] \\
\epsilon_{zz} &= \frac{1}{\gamma}\left[\frac{\partial w}{\partial z} + \frac{u}{\alpha}\frac{\partial \gamma}{\partial x} + \frac{v}{\beta}\frac{\partial \gamma}{\partial y}\right] \\
\epsilon_{xy} &= \frac{1}{2}\left[\frac{1}{\beta}\frac{\partial u}{\partial y} + \frac{1}{\alpha}\frac{\partial v}{\partial x} - \frac{v}{\alpha\beta}\frac{\partial \beta}{\partial x} - \frac{u}{\alpha\beta}\frac{\partial \alpha}{\partial y}\right] \\
\epsilon_{xz} &= \frac{1}{2}\left[\frac{1}{\alpha}\frac{\partial w}{\partial x} + \frac{1}{\gamma}\frac{\partial u}{\partial z} - \frac{u}{\alpha\gamma}\frac{\partial \alpha}{\partial z} - \frac{w}{\alpha\gamma}\frac{\partial \gamma}{\partial x}\right] \\
\epsilon_{yz} &= \frac{1}{2}\left[\frac{1}{\beta}\frac{\partial w}{\partial y} + \frac{1}{\gamma}\frac{\partial v}{\partial z} - \frac{w}{\beta\gamma}\frac{\partial \gamma}{\partial y} - \frac{v}{\beta\gamma}\frac{\partial \beta}{\partial z}\right]
\end{aligned}
\tag{1-9.5}
$$

where (u, v, w) are the projections of the displacement vector of point (x, y, z) on the tangents to the respective coordinate lines at that point and (α, β, γ) are the metric coefficients of the coordinate system (Eq. 1-5.3). Equations (1-9.5) are easily specialized for particular coordinates. For cylindrical coordinates $x = r$, $y = \theta$, $z = z$ and then $\alpha = 1$, $\beta = r$, $\gamma = 1$; for spherical coordinates, $x = r$, $y = \theta =$ colatitude, $z = \phi =$ longitude and then $\alpha = 1$, $\beta = r$, $\gamma = r \sin\theta$ (see Art. 1-5), etc.

Thus, we obtain for

Cylindrical Coordinates:

$$
\begin{aligned}
&\epsilon_{rr} = \frac{\partial u}{\partial r}, \qquad \epsilon_{\theta\theta} = \frac{u}{r} + \frac{1}{r}\frac{\partial v}{\partial \theta}, \qquad \epsilon_{zz} = \frac{\partial w}{\partial z} \\
&\gamma_{r\theta} = 2\epsilon_{r\theta} = \frac{1}{r}\frac{\partial u}{\partial \theta} + \frac{\partial v}{\partial r} - \frac{v}{r}, \qquad \gamma_{rz} = 2\epsilon_{rz} = \frac{\partial u}{\partial z} + \frac{\partial w}{\partial r} \\
&\gamma_{\theta z} = 2\epsilon_{\theta z} = \frac{\partial v}{\partial z} + \frac{1}{r}\frac{\partial w}{\partial \theta}
\end{aligned}
\tag{1-9.6}
$$

Spherical Coordinates:

$$\epsilon_{rr} = \frac{\partial u}{\partial r}, \qquad \epsilon_{\theta\theta} = \frac{u}{r} + \frac{1}{r}\frac{\partial v}{\partial \theta}, \qquad \epsilon_{\phi\phi} = \frac{u}{r} + \frac{v}{r}\cot\theta + \frac{1}{r\sin\theta}\frac{\partial w}{\partial \phi}$$

$$\gamma_{r\theta} = 2\epsilon_{r\theta} = \frac{1}{r}\frac{\partial u}{\partial \theta} + \frac{\partial v}{\partial r} - \frac{v}{r}, \qquad \gamma_{r\phi} = 2\epsilon_{r\phi} = \frac{1}{r\sin\theta}\frac{\partial u}{\partial \phi} + \frac{\partial w}{\partial r} - \frac{w}{r}$$

$$\gamma_{\theta\phi} = 2\epsilon_{\theta\phi} = \frac{1}{r}\left(\frac{\partial w}{\partial \theta} - w\cot\theta\right) + \frac{1}{r\sin\theta}\frac{\partial v}{\partial \phi} \tag{1-9.7}$$

Polar Coordinates:

$$\epsilon_{rr} = \frac{\partial u}{\partial r}, \qquad \epsilon_{\theta\theta} = \frac{u}{r} + \frac{1}{r}\frac{\partial v}{\partial \theta}, \qquad \gamma_{r\theta} = 2\epsilon_{r\theta} = \frac{1}{r}\frac{\partial u}{\partial \theta} + \frac{\partial v}{\partial r} - \frac{v}{r} \tag{1-9.8}$$

EXAMPLE 1-9.1

Three-Dimensional State of Strain

The parallelopiped in Fig. E1-9.1 is deformed into the shape indicated by the dashed straight lines (small displacements). The displacements are given by the following relations: $u = C_1xyz$, $v = C_2xyz$, and $w = C_3xyz$.

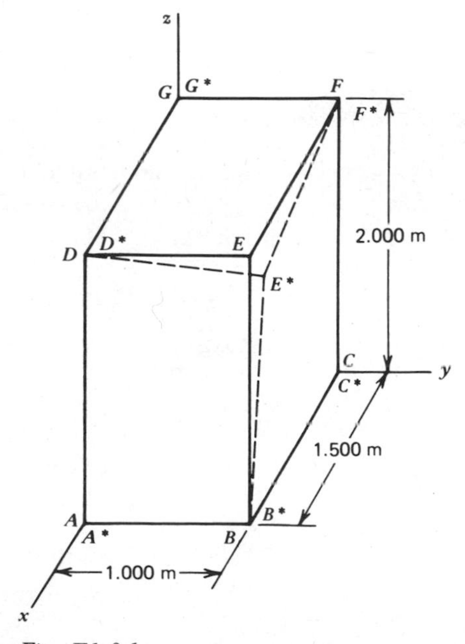

Fig. E1-9.1

(a) Determine the state of strain at point E when the coordinates of point E^* for the deformed body are (1.504, 1.002, 1,996). (b) Determine the normal strain at E in the direction of line EA. (c) Determine the shearing strain at E for the undeformed orthogonal lines EA and EF.

SOLUTION

The magnitudes of C_1, C_2, and C_3 are obtained from the fact that the displacements of point E are known as follows: $u_E = 0.004$ m, $v_E = 0.002$ m, and $w_E = -0.004$ m. Thus,

$$u = \frac{0.004}{3} xyz$$

$$v = \frac{0.002}{3} xyz$$

$$w = -\frac{0.004}{3} xyz$$

(a) The strain components for the state of strain at point E are given by Eqs. (1-9.1). At point E,

$$\epsilon_{xx} = \frac{\partial u}{\partial x} = \frac{0.004}{3} yz = 0.00267; \qquad \epsilon_{yy} = 0.00200; \qquad \epsilon_{zz} = -0.00200$$

$$\epsilon_{xy} = \frac{1}{2}\left(\frac{\partial v}{\partial x} + \frac{\partial u}{\partial y}\right) = \frac{1}{2}\left(\frac{0.002}{3} yz + \frac{0.004}{3} xz\right) = 0.00267$$

$$\gamma_{xy} = 2\epsilon_{xy} = 0.00583, \qquad \gamma_{xz} = 2\epsilon_{xz} = -0.00007,$$

$$\gamma_{yz} = 2\epsilon_{yz} = -0.00300$$

(b) Let the X-axis lie along the line from E to A. The direction cosines of EA are $l_1 = 0$, $m_1 = -1/\sqrt{5}$, and $n_1 = -2/\sqrt{5}$. Equations (1-7.5) and (1-9.3) give the magnitude for ϵ_{XX}. Thus,

$$\epsilon_{XX} = \epsilon_{yy} m_1^2 + \epsilon_{zz} n_1^2 + 2\epsilon_{yz} m_1 n_2$$

$$= \frac{0.00200}{5} - \frac{0.00200(4)}{5} - \frac{0.00300(2)}{5} = -0.00240$$

(c) Let the Y-axis lie along the line from E to F. The direction cosines of EF are $l_2 = -1$, $m_2 = 0$, and $n_2 = 0$. The shearing strain $\gamma_{XY} = 2\epsilon_{XY}$ is given by Eq. (1-7.20d). Thus,

$$\gamma_{XY} = 2\epsilon_{XY} = 2\epsilon_{xy} l_2 m_1 + 2\epsilon_{xz} l_2 n_1$$

$$= \frac{(0.00533)}{\sqrt{5}} + \frac{(-0.00007)(2)}{\sqrt{5}} = 0.00232$$

EXAMPLE 1-9.2

State of Strain in Torsion-Tension Member

A straight torsion-tension member with solid circular cross section has a length $L = 6$ m and a radius $R = 10$ mm. The member is subjected to tension and torsion loads that produce an elongation $\Delta L = 10$ mm and a rotation of one end of the member with respect to the other end of $\pi/3$ rad. Let the origin of the (r, θ, z) cylindrical coordinate axes lie at the centroid of one end of the member with the z-axis extending along the centroid axis of the member. The deformations of the member are assumed to occur under conditions of constant volume. The end $z = 0$ is constrained so that only radial displacements are possible there. (a) Determine the displacements for any point in the member and the state of strain for a point on the outer surface. (b) Determine the principal strains for the point where the state of strain was determined.

SOLUTIONS

The change in radius ΔR for the member is obtained from the condition of constant volume. Thus,

$$\pi R^2 L = \pi(R + \Delta R)^2(L + \Delta L)$$

$$10^2(6 \times 10^3) = (10 + \Delta R)^2(6010)$$

$$\Delta R = -0.00832 \text{ mm}$$

(a) The displacements components

$$u = -0.000832r \text{ (mm)}$$

$$v = 0.001745rz \text{ (mm)}$$

$$w = 0.001664z \text{ (mm)}$$

satisfy the displacement boundary conditions at $z = 0$. The strain components at the outer radius are given by Eqs. (1-9.6). They are

$$\epsilon_{rr} = \frac{\partial u}{\partial r} = -0.000832 \qquad \epsilon_{\theta\theta} = \frac{u}{r} + \frac{1}{r}\frac{\partial v}{\partial \theta} = -0.000832$$

$$\epsilon_{zz} = \frac{\partial w}{\partial z} = 0.001664 \qquad \gamma_{r\theta} = 2\epsilon_{r\theta} = \frac{1}{r}\frac{\partial u}{\partial \theta} + \frac{\partial v}{\partial r} - \frac{v}{r} = 0$$

$$\gamma_{rz} = 2\epsilon_{rz} = \frac{\partial u}{\partial z} + \frac{\partial w}{\partial r} = 0 \qquad \gamma_{\theta z} = 2\epsilon_{\theta z} = \frac{\partial v}{\partial z} + \frac{1}{r}\frac{\partial w}{\partial \theta} = 0.001745$$

(b) The three principal strains are the three roots of a cubic equation, Eq. (1-7.21b), where the three invariants of strain are defined by Eqs. (1-7.22). Choose the (x, y, z) coordinate axes at the point on the outer surface of the member where the strain components have been determined in part (a). Let $x = r$, $y = \theta$, and $z = z$. From Eqs. (1-7.22),

$$J_1 = \epsilon_{rr} + \epsilon_{\theta\theta} + \epsilon_{zz} = -0.000832 - 0.000832 + 0.001664 = 0$$

$$\begin{aligned} J_2 &= \epsilon_{rr}\epsilon_{\theta\theta} + \epsilon_{rr}\epsilon_{zz} + \epsilon_{\theta\theta}\epsilon_{zz} - \epsilon_{r\theta}^2 - \epsilon_{rz}^2 - \epsilon_{\theta z}^2 \\ &= -0.000832(-0.000832) + (-0.000832)(0.001664) \\ &\quad + (-0.000832)(0.001664) - \left(\frac{0.001745}{2}\right)^2 \\ &= -2.838 \times 10^{-6} \end{aligned}$$

$$J_3 = \begin{vmatrix} \epsilon_{rr} & \epsilon_{r\theta} & \epsilon_{rz} \\ \epsilon_{\theta r} & \epsilon_{\theta\theta} & \epsilon_{\theta z} \\ \epsilon_{zr} & \epsilon_{z\theta} & \epsilon_{zz} \end{vmatrix} = \begin{vmatrix} -0.000832 & 0 & 0 \\ 0 & -0.000832 & \dfrac{0.001745}{2} \\ 0 & \dfrac{0.001745}{2} & 0.001664 \end{vmatrix}$$

$$= 1.785 \times 10^{-9}$$

Substitution of these results into Eq. (1-7.21b) gives the following cubic equation in $\epsilon (= M)$.

$$\epsilon^3 - 2.838 \times 10^{-6}\epsilon - 1.785 \times 10^{-9} = 0$$

One principal strain, $\epsilon_1 = \epsilon_{rr} = -0.000832$, is known. Factoring out this root, we find

$$\epsilon^2 - 0.000832\epsilon - 2.146 \times 10^{-6} = 0$$

Solution of this quadratic equation yields the remaining two principal strains. Thus, the three principal strains are

$$\epsilon_1 = -0.00832$$

$$\epsilon_2 = 0.001939$$

$$\epsilon_3 = -0.001107$$

PROBLEM SET 1-9

1. The tension member in Fig. P1-9.1 has the following dimensions: $L = 5$ m, $b = 100$ mm, and $h = 200$ mm. The (x, y, z) coordinate axes are parallel to the edges of the member with origin 0 located at the centroid of the left end. Under the deformation produced by load P, the origin 0 remains located at the centroid of the left end and the coordinate axes remain parallel to the edges of the deformed mem-

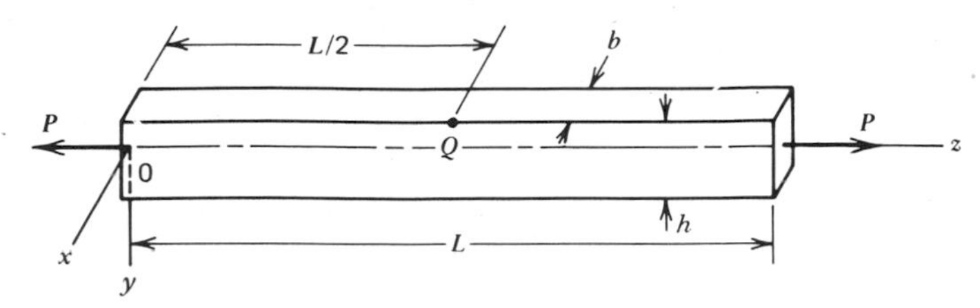

Fig. P1-9.1

ber. Under the action of load P, the bar elongates 20 mm. Assume that the volume of the bar remains constant with $\epsilon_{xx} = \epsilon_{yy}$. (a) Determine the displacements for the member and the state of strain at point Q assuming that the small-displacement theory holds. (b) Determine ϵ_{zz} at point Q based on the assumption that displacements are not small.

2. In many practical engineering problems, the state of strain is approximated by the condition that the normal and shearing strains for some direction, say the z-direction, are zero, that is, $\epsilon_{zz} = \epsilon_{zx} = \epsilon_{zy} = 0$ (plane strain). In Chapter 2, it is shown that analogously, $\epsilon_{zx} = \epsilon_{zy} = 0$, but $\epsilon_{zz} \neq 0$ for members made of isotropic materials and loaded such that the state of stress may be approximated by the condition $\sigma_{zz} = \sigma_{zx} = \sigma_{zy} = 0$ (plane stress). Assume that ϵ_{xx}, ϵ_{yy}, ϵ_{zz}, and ϵ_{xy} for the (x, y) coordinate axes shown in Fig. P1-9.2 are known. Let the (X, Y) coordinate axes be defined by a counterclockwise rotation through angle θ as indicated in Fig. P1-9.2. Analogous to the transformation for plane stress, show that the transformation equations of plane strain are $\epsilon_{XX} = \epsilon_{xx} \cos^2\theta + \epsilon_{yy} \sin^2\theta + 2\epsilon_{xy} \sin\theta \cos\theta$ and $\epsilon_{XY} = -\epsilon_{xx} \sin\theta \cos\theta + \epsilon_{yy} \sin\theta \cos\theta + \epsilon_{xy}(\cos^2\theta - \sin^2\theta)$.

3. The square plate in Fig. P1-9.3 is loaded so that the plate is in a state of plane strain ($\epsilon_{zz} = \epsilon_{zx} = \epsilon_{zy} = 0$). (a) Determine the displacements for the plate given the deformations shown and the strain compo-

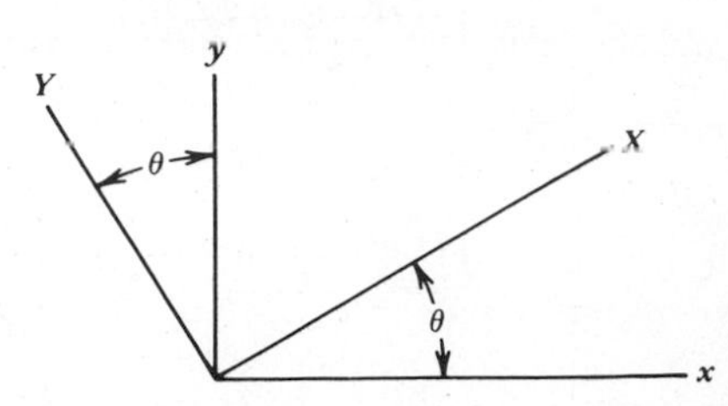

Fig. P1-9.2

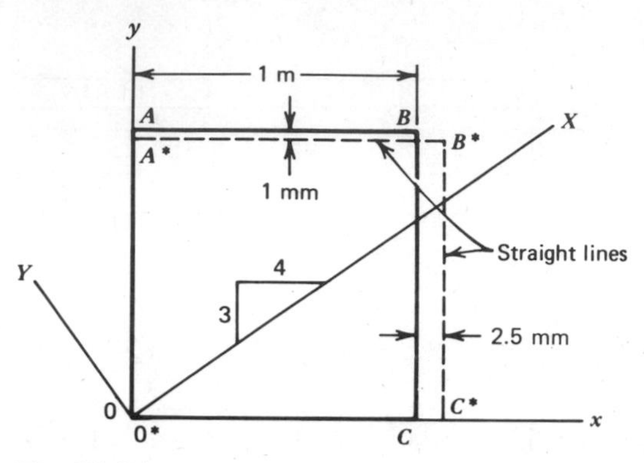

Fig. P1-9.3

nents for the (x, y) coordinate axes. (b) Determine the strain components for the (X, Y)-axes.

4. The square plate in Fig. P1-9.4 is loaded so that the plate is in a state of plane strain ($\epsilon_{zz} = \epsilon_{zx} = \epsilon_{zy} = 0$). (a) Determine the displacements for the plate for the deformations shown and the strain components for the (x, y) coordinate axes. (b) Determine the strain components for the (X, Y)-axes.

   Ans. (a) $u = -0.0020x - 0.0030y$, $v = 0.0010x + 0.0025y$, $\epsilon_{xx} = -0.0020$, $\epsilon_y = 0.0025$, $\gamma_{xy} = 2\epsilon_{xy} = -0.0020$; (b) $\epsilon_{XX} = -0.00174$, $\epsilon_{YY} = 0.00224$, $\gamma_{XY} = 2\epsilon_{XY} = 0.00290$.

5. Determine the orientation of the (X, Y) coordinate axes for principal directions in Problem 4. What are the principal strains?

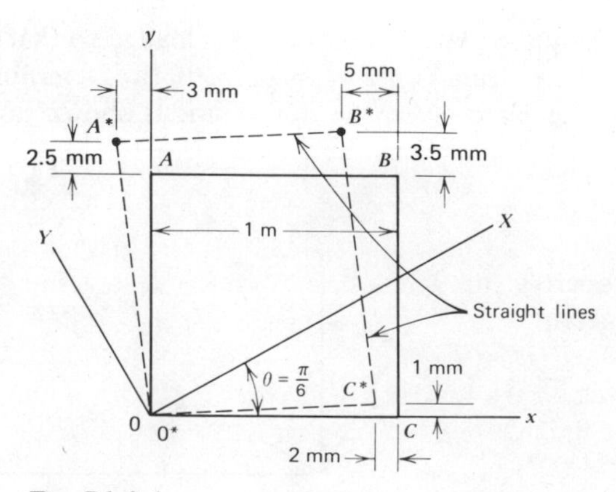

Fig. P1-9.4

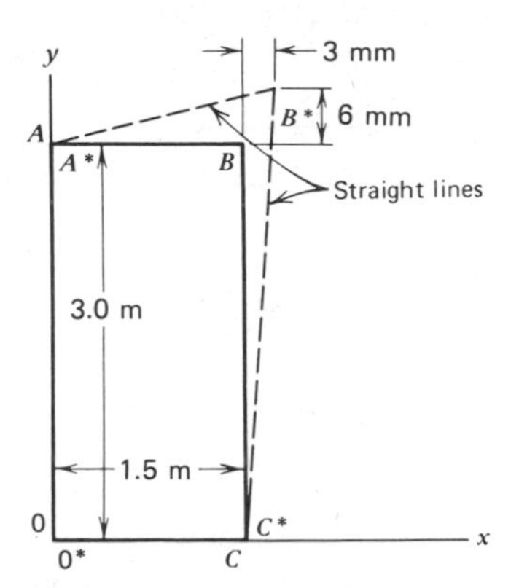

Fig. P1-9.6

6. The plate in Fig. P1-9.6 is loaded so that a state of plane strain ($\epsilon_{zz} = \epsilon_{zx} = \epsilon_{zy} = 0$) exists. (a) Determine the displacements for the plate for the deformations shown and the strain components at point B. (b) Let the X-axis extend from point 0 through point B. Determine ϵ_{XX} at point B.

   Ans. (a) (dimensions in m) $u = 0.000667xy$, $v = 0.001333xy$, $\epsilon_{xx} = 0.00200$, $\epsilon_{yy} = 0.00200$, $\gamma_{xy} = 2\epsilon_{xy} = 0.00500$; (b) $\epsilon_{XX} = 0.00400$.

7. The nonzero strain components at a point in a loaded member are $\epsilon_{xx} = 0.00180$, $\epsilon_{yy} = -0.00108$, and $\gamma_{xy} = 2\epsilon_{xy} = -0.00220$. Using the results of Problem 2, determine the principal strain directions and the principal strains.

8. Solve for the principal strains in Problem 7 by using Eqs. (1-7.21b) and (1-7.22).

   Ans. $\epsilon_1 = 0.00217$, $\epsilon_2 = -0.00145$, $\epsilon_3 = 0$

9. Determine the principal strains at point E for the deformed parallelopiped in Problem E1-9.1.

10. When solid circular torsion members are used to obtain material properties for finite strain applications, an expression for the engineering shearing strain γ_{zx} is needed, where the (x, z)-plane is a tangent plane and the z-axis is parallel to the axis of the member as indicated in Fig. P1-9.10. Consider an element $ABCD$ in Fig. P1-9.10 for the undeformed member. Assume that the member deforms such that the volume remains constant and the diameter remains unchanged. (This is an approximation to real behavior of many metals.)

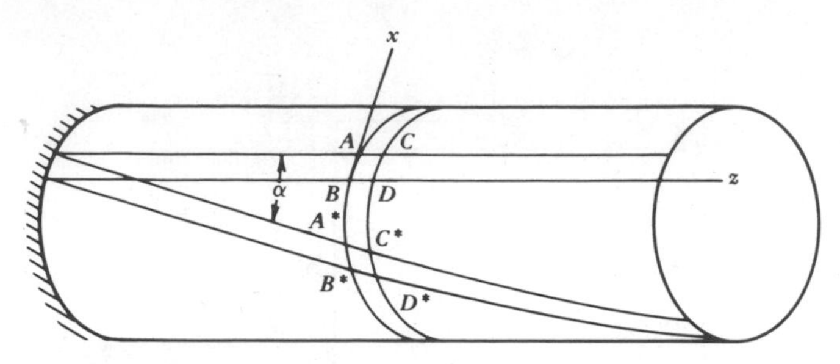

Fig. P1-9.10

Thus, for the deformed element $A^*B^*C^*D^*$, $A^*B^* = AB$, $C^*D^* = CD$, and the distance along the z-axis of the member between the parallel curved lines A^*B^* and C^*D^* remains unchanged. Show that Eq. (1-7.15) gives the result $\gamma_{zx} = \tan\alpha$, where α is the angle between AC and A^*C^*, where $\gamma_{zx} = 2\epsilon_{zx}$ is defined to be the engineering shearing strain.

REFERENCES

1. A. P. Boresi and P. P. Lynn, *Elasticity in Engineering Mechanics*, 2nd Ed., Prentice-Hall, Englewood Cliffs, New Jersey, 1974.
2. F. D. Murnahan, *Finite Deformation of an Elastic Solid*, Wiley, New York, 1951.

Additional References

1. F. Borg, *Fundamentals of Engineering Elasticity*, Van Nostrand, Princeton, New Jersey, 1962.
2. A. Cemal Eringen, *Mechanics of Continua*, Wiley, New York, 1967.
3. Y. C. Fung, *Continuum Mechanics*, 2nd. Ed., Prentice-Hall, Englewood Cliffs, New Jersey, 1977.
4. P. G. Hodge, Jr., *Continuum Mechanics*, McGraw-Hill, New York, 1970.
5. A. E. H. Love, *A Treatise on the Mathematical Theory of Elasticity*, 4th Ed., Dover, New York, 1944.
6. J. E. Marsden and T. J. R. Hughes, *Mathematical Foundations of Elasticity*, Prentice-Hall, Englewood Cliffs, New Jersey, 1983.
7. N. O. Myklestad, *Statics of Deformable Bodies*, Macmillan, New York, 1966.
8. G. Sines, *Elasticity and Strength*, Allyn & Bacon, Boston, 1969.

CHAPTER 2

STRESS-STRAIN-TEMPERATURE RELATIONS

In Chapter 1, the state of stress at a point was defined according to the six stress components of the symmetric stress tensor. The transformation of the stress components under a rotation of coordinate axes was developed, and equations of equilibrium (or equations of motion for accelerated bodies) were derived. The analogous theory of deformation, based upon geometrical concepts, was presented and strain-displacement relations, transformation of the strain components under a rotation of coordinate axes, and strain compatibility relations were derived.

To derive load-stress and load-deflection relations for specified structural members, the stress components must be related to the strain components. Consequently, in this chapter we discuss linear stress-strain-temperature relations. These relations may be employed in the study of linearly elastic material behavior. In addition, they may be employed in plasticity theories to describe the linearly elastic part of the plastic response of materials. More generally, nonlinear (inelastic) stress-strain relations are required for the plastic part of material behavior. Unfortunately, these relations take on different forms depending on the material behavior during plastic response. In this book, we consider only the limiting case of fully plastic loads. At the fully plastic load, stress components are assumed to be independent of the strain components and to remain constant with increasing strain.

Since empirical studies are required to determine material properties (for example, elastic coefficients for linearly elastic materials), the study of stress-strain relations is in part empirical. To obtain needed isotropic elastic material properties, we employ a tension specimen. If lateral as well as longitudinal strains are measured for linearly elastic behavior of the tension specimen, the resulting stress-strain-temperature data represent the material response for obtaining the needed elastic constants for the material. The main structure of the stress-strain-temperature relations, however, is studied theoretically by means of the first law of thermodynamics.

The stress-strain-temperature relations presented in this chapter are limited mainly to small strains and small rotations. The reader interested in large strains and large rotations may refer to the works of Green and Adkins.[1]

2-1 ELASTIC AND NONELASTIC RESPONSE OF A SOLID

Initially, we review the results of the simple tension test of a circular cylindrical material bar, which is clamped at one end, and is subjected to an axially directed tensile load (pull) P at the other end. It is assumed that the load is increased slowly from its initial value of zero, since the material response depends not only on the magnitude of the load, but on the rate of loading as well. It is customary to plot the tensile stress σ in the bar with increasing values of P as a function of the strain ϵ of the bar. In engineering practice, the tensile stress σ is usually approximated by $\sigma \approx P/A_0$, where A_0 is the original cross-sectional area of the bar. Then σ is proportional to load P. However, strictly speaking, according to the definition of stress (Art. 1-1), the stress is P/A, where A is the actual cross-sectional area of the bar when the load P acts. (The bar undergoes lateral contraction everywhere as it is loaded.) The difference between P/A_0 and P/A is usually negligible, provided that the strain is small (Art. 1-9).

For most structural metals, the stress-strain curve of a tension specimen takes the form shown in Fig. 2-1.1*a*. An important structural metal, mild steel, has a distinctive stress-strain curve as shown in Fig. 2-1.1*b*. Two scales are generally used for strains in order to better demonstrate small strain behavior of the metal. For a load P, which produces sufficiently small strains, the strain disappears upon removal of the load. Then the tension specimen (or more generally any member) is said to be strained within the limit of *perfect elasticity*. If the strain is proportional to the load, the tension specimen (or any member) is said to be strained within the limit of *linear elasticity*. The limit of perfect elasticity is frequently referred to simply as the *elastic limit* σ_{EL}, whereas the limit of linear elasticity is referred to as the *proportional limit* σ_{PL}. These limits may be different for metals (Fig. 2-1.1*a*). However, generally for materials that behave as shown in Fig. 2-1.1*a*, neither σ_{EL} nor σ_{PL} are obtained from tensile tests. Rather the *yield stress* Y is used in design.

For sufficiently small strains (the small scale strains shown in Fig. 2-1.1), the stress-strain curve differs little whether the original cross-sectional area A_0 or the instantaneous area A is used. When the load produces a stress σ that exceeds the elastic limit, the strain does not disappear upon unloading (Fig. 2-1.1*a*). The strain ϵ_s that remains after

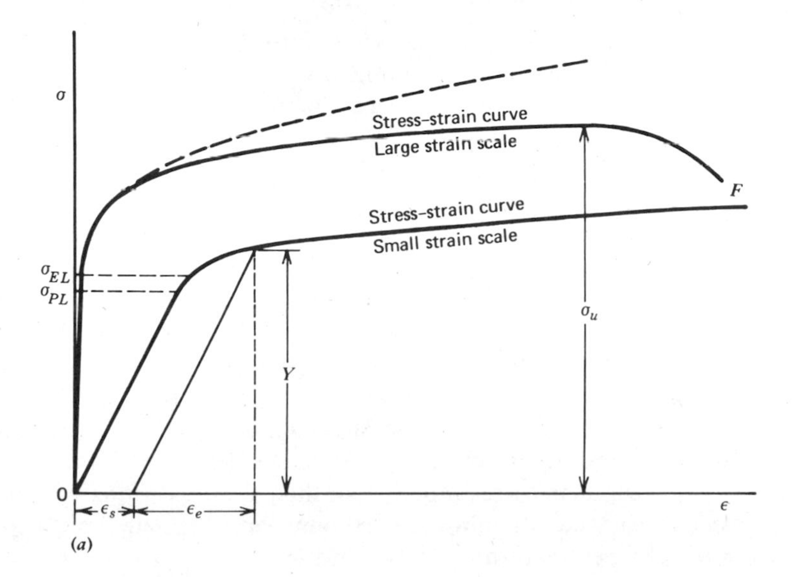

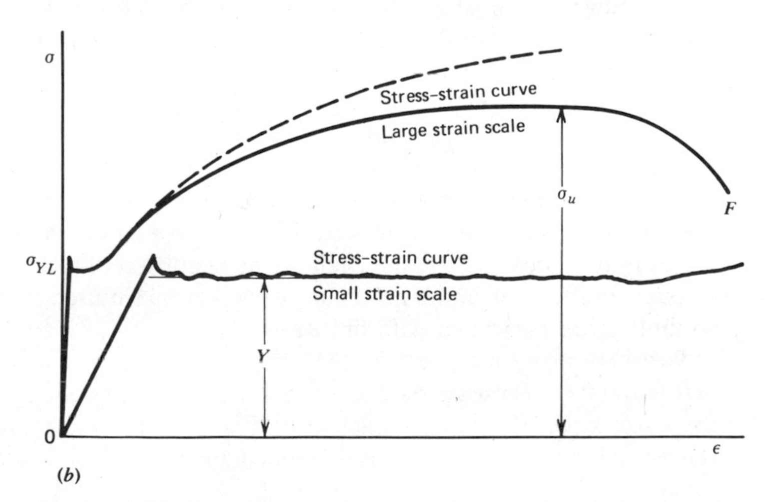

Fig. 2-1.1/Typical stress-strain curves for metals. (a) Stress-strain curve for strain-hardening metal. (b) Stress-strain curve for mild steel.

unloading is called "set" (Fig. 2-1.1*a*). The strain ϵ_e that is recovered when the load is removed is called the "elastic strain." Hence, beyond the elastic limit, the strain ϵ is a sum of the "set" ϵ_s and the "elastic strain" ϵ_e, or $\epsilon = \epsilon_s + \epsilon_e$. For materials that behave in the manner indicated in Fig. 2-1.1*a*, the yield stress Y is defined as the stress corresponding to a specified "set," usually $\epsilon_s = 0.002$ or 0.2 percent strain. As discussed in Art. 3-3, failure loads based on general yielding as the mode of failure are calculated assuming that the material behavior is linearly elastic up to the yield stress. The working load is obtained by dividing the failure load by a factor of safety. The resulting working load causes a linearly elastic material response unless the factor of safety approaches a magnitude of 1.00.

For mild steel, Fig. 2-1.1*b*, the stress-strain curve reaches a local maximum called the upper yield or plastic limit σ_{YL} after which it drops to a local minimum (the lower yield point), and runs approximately parallel (in a wavy fashion) to the strain axis for a short distance. For mild steel, the lower yield point stress is assumed to be the yield stress Y.

Consider the stress-strain diagrams in Fig. 2-1.1 for the large strain scale. In the plastic or flow region, a relatively small change in load causes a large change in strain. In this region, as the strain becomes large, considerable difference in the stress-strain curve occurs, depending on whether area A_0 or area A is used in the definition of stress. With area A_0, the curve first rises rapidly and then slowly, turning with its concave side down and attaining a maximum value σ_u (the so-called *ultimate strength*) before turning downward to fracture (point F). Physically, after σ_u is reached, the so-called "neckingdown" of the bar occurs. This neckingdown is a drastic reduction of the cross-sectional area of the bar in the neighborhood of the region in the bar where the fracture ultimately occurs. If load P is referred to the cross sectional area A, and hence $\sigma = P/A$, the stress-strain curve obtained in the plastic region (dashed line in Fig. 2-1.1) differs considerably from the stress-strain curve relative to area A_0. We have discontinued the dashed curve at the ultimate strength. The shape of the stress-strain curve in the region of fracture is greatly influenced by the strain definition. Since we limit strains to small values in this book, we use the engineering definition of strain (Eq. 1-7.1) throughout the book and do not consider other, often used, definitions of strain that are associated with finite strains.

There are many materials whose tensile specimens do not neck down before fracture. These materials are called *brittle materials*. For example, the stress-strain diagram of cast iron (Fig. 2-1.2) exhibits little plastic range, and fracture occurs almost immediately at the end of the elastic range.

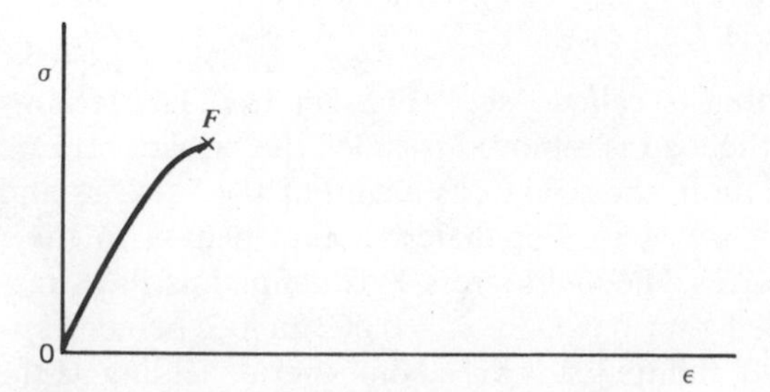

Fig. 2-1.2/Stress-strain curve for cast iron.

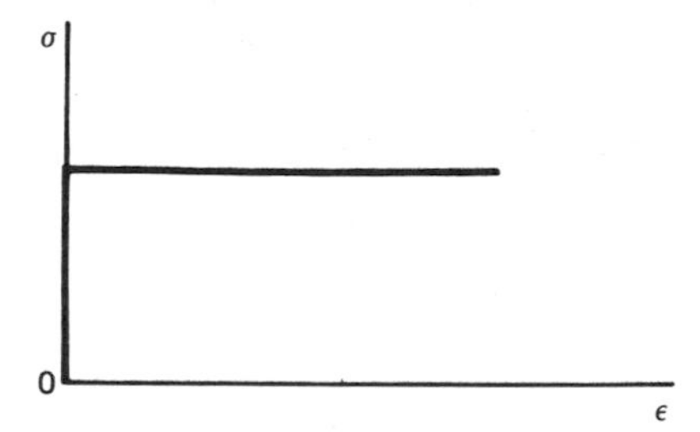

Fig. 2-1.3/Stress-strain curve for clay.

Some materials, lead and clay, for example (Fig. 2-1.3), respond to a tensile load almost entirely in a plastic manner. This response is referred to as *perfectly plastic.*

Load Carrying Members/In this article we have been concerned with the tension specimen used to obtain material properties. Stress-strain data obtained from such tension specimens are used to represent material responses for most members considered in this book. As indicated in the introduction to Chapter 1, the material response, the stress-strain relations (derived later in this chapter), the equations of equilibrium (equations of motion for accelerated members), and the compatibility conditions are used to derive load-stress and load-deflection relations for members. We restrict our study to the behavior of solid members (that is, members composed of materials that possess large cohesive forces, in contrast to fluids that can sustain only relatively small tension forces) and that have the ability to instantly recover their original size and shape when the forces producing the deformations are removed. This property of instant recovery of initial size and shape upon removal of load has been defined earlier in this article as perfect elasticity. In most of our discussion, we limit our consideration to linear perfect elasticity. We assume that the magnitudes of the stress components at any point P in a member depend at all times solely on the simultaneous deformation in the immediate neighborhood of the point P. In general, the state of stress at point P depends not only on the forces acting on the member at any instant, but also on the previous history of deformation of the member. For example, the state of stress at point P may depend on residual stresses due to previous history of cold work or cold forming of the member. The stress components at point P obtained from the load-stress relations derived later in this book must be added to the residual stresses at point P to obtain the actual state of stress at point P. All of the problems in this book assume that residual stresses are negligibly small, or that they may be added to nonresidual stresses.

Generally, a structural member is acted on continuously by forces. For example, in the vicinity of the earth a member is acted on by the earth's gravitational force, even in the absence of other forces. Only in interstellar space does a member approach being free of the action of forces, although even there it is acted on by the gravitational attractions of the distant stars. Therefore, the *zero state* or the *zero configuration* from which the deformations of the member are measured is arbitrary. However, once the zero configuration is specified, the strains of the member measured from the zero state determine the member's internal configuration.

Whenever a member exhibits the phenomenon of *hysteresis*—that is, of returning to its original size and shape only slowly or not at all—its behavior is not perfectly elastic. The study of members that recover their sizes and shapes only gradually after a load is removed is discussed in the theory of viscoelasticity. The study of members that do not return to their original sizes after removal of load is generally considered in the theory of plasticity.

Finally, the complete description of the initial state of a member requires that the temperature at every point in the member, as well as its initial configuration be specified. This is because, in general, a change in temperature produces a change in configuration. In turn, a change in configuration may or may not be accompanied by a change in temperature.

2-2 FIRST LAW OF THERMODYNAMICS, INTERNAL-ENERGY DENSITY, COMPLEMENTARY INTERNAL-ENERGY DENSITY

The derivation of load-stress and load-deflection relations by either the method of mechanics of materials or the method of elasticity requires stress-strain relations that relate the components of the strain tensor to components of the stress tensor. The form of these relations depends on material behavior. In this book, the materials are assumed to be isotropic, that is, at any point they have the same properties in all directions. Stress-strain relations for linearly elastic isotropic materials are well known and are presented in Art. 2-4. Stress-strain relations may be treated theoretically by the use of the first law of thermodynamics.[2] However, the elastic coefficients that enter into these relations are obtained experimentally.

To apply the first law of thermodynamics, we consider a loaded member in equilibrium in its deflected configuration. The deflections are assumed to be known. They are specified by known displacement components (u, v, w) for each point in the deflected member; positive u, v, and w are components of displacement of a point in the positive direction of rectangular orthogonal coordinate axes (x, y, z), respectively. We allow each point to undergo infinitesimal increments (variations) in the displacement components (u, v, w) indicated by $(\delta u, \delta v, \delta w)$. The stress components at every point of the member are considered to be unchanged under variations of the displacements. These displacement variations are arbitrary except that two or more particles cannot occupy the same point in space, nor can a single particle occupy more than one position (the member does not tear). In addition, displacements of certain points in the member may be specified (for instance, at a fixed support); such specified displacements are referred to as forced boundary conditions.[2] The variations of the strain components resulting from variations $(\delta u, \delta v, \delta w)$ are, by Eq. (1-9.1),

$$
\begin{aligned}
\delta\epsilon_{xx} &= \frac{\partial(\delta u)}{\partial x} & \delta\epsilon_{xy} &= \frac{1}{2}\left[\frac{\partial(\delta v)}{\partial x} + \frac{\partial(\delta u)}{\partial y}\right] \\
\delta\epsilon_{yy} &= \frac{\partial(\delta v)}{\partial y} & \delta\epsilon_{yz} &= \frac{1}{2}\left[\frac{\partial(\delta w)}{\partial y} + \frac{\partial(\delta v)}{\partial z}\right] \qquad (2\text{-}2.1) \\
\delta\epsilon_{zz} &= \frac{\partial(\delta w)}{\partial z} & \delta\epsilon_{zx} &= \frac{1}{2}\left[\frac{\partial(\delta w)}{\partial x} + \frac{\partial(\delta u)}{\partial z}\right]
\end{aligned}
$$

These equations are used later in the analysis.

To introduce force quantities, consider now a free body diagram of an arbitrary volume V^* of the deformed member enclosed by a closed surface S^*. We assume that the member is in static equilibrium during the displacement variations $(\delta u, \delta v, \delta w)$. Therefore, the part of the member considered in volume V^* is in equilibrium under the action of surface forces (represented by stress distributions on surface S^*) and by body forces (represented by distributions of body forces per unit volume B_x, B_y, and B_z in volume V^*).

For adiabatic conditions (no net heat flow into V^*), the first law of thermodynamics states that, during the displacement variations $(\delta u, \delta v, \delta w)$, the variation in work of the external forces δW_e is equal to the variation of internal energy δU for each volume element. Hence, for V^*, we have

$$\delta W_e = \delta U \qquad (2\text{-}2.2)$$

It is convenient to divide δW_e into two parts: the work of the surface forces δW_S and the work of the body forces δW_B. At point P of surface S^*, consider an increment of area dS. The stress vector $\boldsymbol{\sigma}_P$ acting on dS has components σ_{Px}, σ_{Py}, and σ_{Pz} defined by Eqs. (1-3.8). The forces are equal to the product of these stress components and dS. The work δW_S is equal to the work of these forces. Thus,

$$\begin{aligned}\delta W_S &= \iint_{S^*} \sigma_{Px}\,\delta u\,dS + \iint_{S^*} \sigma_{Py}\,\delta v\,dS + \iint_{S^*} \sigma_{Pz}\,\delta w\,dS \\ &= \iint_{S^*} \big[(\sigma_{xx} l + \sigma_{yx} m + \sigma_{zx} n)\,\delta u + (\sigma_{xy} l + \sigma_{yy} m + \sigma_{zy} n)\,\delta v \\ &\qquad + (\sigma_{xz} l + \sigma_{yz} m + \sigma_{zz} n)\,\delta w\big]\,dS\end{aligned} \tag{2-2.3}$$

For a volume element dV of volume V^*, the body forces are given by products of dV and the body force components per unit volume (B_x, B_y, B_z). The work δW_B of the body forces that act throughout V^* is

$$\delta W_B = \iiint_{V^*} (B_x\,\delta u + B_y\,\delta v + B_z\,\delta w)\,dV \tag{2-2.4}$$

The variation of work δW_e of the external forces that act on volume V^* with surface S^* is equal to the sum of δW_S and δW_B. The surface integral in Eq. (2-2.3) may be converted into a volume integral by use of the divergence theorem.[3] Thus,

$$\begin{aligned}\delta W_e = \delta W_S + \delta W_B = \iiint_{V^*} \Big[&\frac{\partial}{\partial x}(\sigma_{xx}\,\delta u + \sigma_{xy}\,\delta v + \sigma_{xz}\,\delta w) \\ &+ \frac{\partial}{\partial y}(\sigma_{yx}\,\delta u + \sigma_{yy}\,\delta v + \sigma_{yz}\,\delta w) \\ &+ \frac{\partial}{\partial z}(\sigma_{zx}\,\delta u + \sigma_{zy}\,\delta v + \sigma_{zz}\,\delta w) \\ &+ B_x\,\delta u + B_y\,\delta v + B_z\,\delta w \Big]\,dV\end{aligned} \tag{2-2.5}$$

With Eqs. (2-2.1) and (1-5.1), Eq. (2-2.5) reduces to

$$\begin{aligned}\delta W_e = \iiint_{V^*} (&\sigma_{xx}\delta\epsilon_{xx} + \sigma_{yy}\delta\epsilon_{yy} + \sigma_{zz}\delta\epsilon_{zz} + 2\sigma_{xy}\delta\epsilon_{xy} \\ &+ 2\sigma_{yz}\delta\epsilon_{yz} + 2\sigma_{zx}\delta\epsilon_{zx})\,dV\end{aligned} \tag{2-2.6}$$

which is the desired form for substitution into Eq. (2-2.2).

The internal energy U for volume V^* is expressed in terms of the internal energy per unit volume, that is, in terms of the *internal-energy density* U_0. Thus,

$$U = \iiint_{V^*} U_0\,dV$$

and the variation of internal energy becomes

$$\delta U = \iiint_{V^*} \delta U_0\,dV \tag{2-2.7}$$

Substitution of Eqs. (2-2.6) and (2-2.7) into Eq. (2-2.2) gives the variation of the internal-energy density δU_0 in terms of the stress components and the variation in strain components. Thus,

$$\delta U_0 = \sigma_{xx}\delta\epsilon_{xx} + \sigma_{yy}\delta\epsilon_{yy} + \sigma_{zz}\delta\epsilon_{zz} + 2\sigma_{xy}\delta\epsilon_{xy} + 2\sigma_{yz}\delta\epsilon_{yz} + 2\sigma_{zx}\delta\epsilon_{zx} \tag{2-2.8}$$

This equation is used later in the derivation of expressions that relate the stress components to the strain energy density U_0 (see Eqs. 2-2.11).

Elasticity and Internal-Energy Density / The strain-energy density U_0 is a function of certain variables; we need to determine these variables. For linearly elastic material behavior, the total internal energy U in a loaded member is equal to the potential energy of the internal forces (called the *elastic strain energy*). For small displacements, each stress component is assumed to be linearly related to strain components; therefore, the internal-energy density U_0 at a given point in the member can be expressed in terms of the six strain components of the strain tensor. If the material is nonhomogeneous (has different properties at different points in the member), the function U_0 depends on location (x, y, z) in the member as well. The possibility arises that the zero state of deformation (strain) may not coincide with a zero state of stress. Hence, in a zero state of strain nonzero stresses (residual stresses) may exist. This condition may easily arise in cases of prior inelastic deformation. Cases also arise in which we may wish to consider a prescribed nonzero state of deformation as the zero configuration of potential energy. This situation arises in the analysis of buckling problems where the zero configuration of potential energy is often taken as the unbuckled state that exists just prior to buckling.

Generally, the strain-energy density U_0 depends on the temperature T. However, usually small elastic deformations do not cause large changes in temperature. Consequently, thermal stress problems may be treated approximately with the assumption that the time rate of change of temperature is sufficiently slow so that transient inertial effects may be ignored. Then, the stress distribution at any instant is the same as if the temperature distribution at that instant were maintained constant (Art. 2-5).

Since the strain-energy density function U_0 generally depends on the strain components, the coordinates, and the temperature, we may express it as a function of these variables. Thus,

$$U_0 = U_0(\epsilon_{xx}, \epsilon_{yy}, \epsilon_{zz}, \epsilon_{yz}, \epsilon_{zx}, \epsilon_{xy}, x, y, z, T) \tag{2-2.9}$$

Then, if the displacements (u, v, w) undergo a variation $(\delta u, \delta v, \delta w)$, the strain components take variations $\delta\epsilon_{xx}$, $\delta\epsilon_{yy}$, $\delta\epsilon_{zz}$, $\delta\epsilon_{yz}$, $\delta\epsilon_{zx}$, and $\delta\epsilon_{xy}$, and the function U_0 takes on the variation

$$\delta U_0 = \frac{\partial U_0}{\partial \epsilon_{xx}} \delta\epsilon_{xx} + \frac{\partial U_0}{\partial \epsilon_{yy}} \delta\epsilon_{yy} + \frac{\partial U_0}{\partial \epsilon_{zz}} \delta\epsilon_{zz}$$
$$+ \frac{\partial U_0}{\partial \epsilon_{yz}} \delta\epsilon_{yz} + \frac{\partial U_0}{\partial \epsilon_{zx}} \delta\epsilon_{zx} + \frac{\partial U_0}{\partial \epsilon_{xy}} \delta\epsilon_{xy} \tag{2-2.10}$$

Therefore, since Eqs. (2-2.8) and (2-2.10) are valid for arbitrary variations $(\delta u, \delta v, \delta w)$, comparison yields

$$\sigma_{xx} = \frac{\partial U_0}{\partial \epsilon_{xx}} \qquad \sigma_{yy} = \frac{\partial U_0}{\partial \epsilon_{yy}} \qquad \sigma_{zz} = \frac{\partial U_0}{\partial \epsilon_{zz}}$$
$$\sigma_{yz} = \frac{1}{2}\frac{\partial U_0}{\partial \epsilon_{yz}} \qquad \sigma_{zx} = \frac{1}{2}\frac{\partial U_0}{\partial \epsilon_{zx}} \qquad \sigma_{xy} = \frac{1}{2}\frac{\partial U_0}{\partial \epsilon_{xy}} \tag{2-2.11}$$

Although Eqs. (2-2.11) were derived for rectangular coordinate axes (x, y, z), they also are valid for other orthogonal axes such as cylindrical and spherical coordinates.

Elasticity and Complementary Internal-Energy Density / In many members of engineering structures there may be one dominant component of stress, call it σ. This situation may arise in axially loaded members, simple columns, beams, or torsional members. Then the strain-energy density U_0 (Eq. 2-2.9) depends mainly on the associated strain component ϵ; consequently, for a given temperature T, σ depends mainly on ϵ. A graph of the relationship between σ and ϵ for a tension specimen is the stress-strain diagram (Fig. 2-2.1).

By Eq. (2-2.11), $\sigma = dU_0/d\epsilon$ and, therefore, $U_0 = \int \sigma\, d\epsilon$. It follows therefore that U_0 is represented by the area under the stress-strain diagram (Fig. 2-2.1). Although this area may be measured from an arbitrary abscissa (ϵ value), for simplicity $\epsilon = 0$ for $\sigma = 0$ in Fig. 2-2.1. The rectangular area $(0,0), (0,\epsilon), (\sigma,\epsilon), (\sigma,0)$ is represented by the prod-

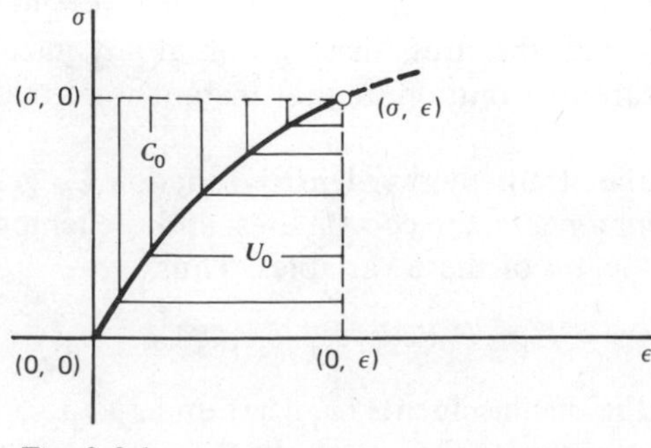

Fig. 2-2.1

uct $\sigma\epsilon$. Hence, this area is given by

$$\sigma\epsilon = U_0 + C_0 \tag{2-2.12}$$

where C_0 is called the *complementary internal-energy density or complementary strain energy density.* C_0 is represented by the area above the stress-strain curve and below the horizontal line from $(\sigma, 0)$ to (σ, ϵ). Hence by Fig. 2-2.1,

$$C_0 = \int \epsilon \, d\sigma \tag{2-2.13}$$

or

$$\epsilon = \frac{dC_0}{d\sigma} \tag{2-2.14}$$

The above graphical interpretation of the complementary strain energy is applicable only for the case of a single nonzero component of stress. However, an analytical generalization for several nonzero components of stress has been given by A. M. Legendre.[3] To achieve this generalization, we assume that Eqs. (2-2.11) may be integrated to obtain the strain components as functions of the stress components. Thus we obtain

$$\begin{aligned} \epsilon_{xx} &= f_1(\sigma_{xx}, \sigma_{yy}, \sigma_{zz}, \sigma_{yz}, \sigma_{zx}, \sigma_{xy}) \\ \epsilon_{yy} &= f_2(\sigma_{xx}, \sigma_{yy}, \sigma_{zz}, \sigma_{yz}, \sigma_{zx}, \sigma_{xy}) \\ &\vdots \\ \epsilon_{xy} &= f_6(\sigma_{xx}, \sigma_{yy}, \sigma_{zz}, \sigma_{yz}, \sigma_{zx}, \sigma_{xy}) \end{aligned} \tag{2-2.15}$$

where $f_1, f_2, \ldots, f_6$ denote functions of the stress components. Substitution of Eqs. (2-2.15) into Eqs. (2-2.9) yields U_0 as a function of the six stress components. Then direct extension of Eq. (2-2.12) yields

$$C_0 = -U_0 + \sigma_{xx}\epsilon_{xx} + \sigma_{yy}\epsilon_{yy} + \sigma_{zz}\epsilon_{zz} + 2\sigma_{yz}\epsilon_{yz} + 2\sigma_{zx}\epsilon_{zx} + 2\sigma_{xy}\epsilon_{xy} \tag{2-2.16}$$

By Eqs. (2-2.15) and (2-2.16), the complementary energy density C_0 may be expressed in terms of the six stress components. Hence, differentiating Eq. (2-2.16) with respect to σ_{xx}, noting by the chain rule of differentiation that

$$\begin{aligned} \frac{\partial U_0}{\partial \sigma_{xx}} &= \frac{\partial U_0}{\partial \epsilon_{xx}}\frac{\partial \epsilon_{xx}}{\partial \sigma_{xx}} + \frac{\partial U_0}{\partial \epsilon_{yy}}\frac{\partial \epsilon_{yy}}{\partial \sigma_{xx}} + \frac{\partial U_0}{\partial \epsilon_{zz}}\frac{\partial \epsilon_{zz}}{\partial \sigma_{xx}} \\ &\quad + \frac{\partial U_0}{\partial \epsilon_{yz}}\frac{\partial \epsilon_{yz}}{\partial \sigma_{xx}} + \frac{\partial U_0}{\partial \epsilon_{zx}}\frac{\partial \epsilon_{zx}}{\partial \sigma_{xx}} + \frac{\partial U_0}{\partial \epsilon_{xy}}\frac{\partial \epsilon_{xy}}{\partial \sigma_{xx}} \end{aligned} \tag{2-2.17}$$

and employing Eq. (2-2.11), we find

$$\epsilon_{xx} = \frac{\partial C_0}{\partial \sigma_{xx}} \tag{2-2.18}$$

Similarly, taking derivatives of Eq. (2-2.16) with respect of the other stress components ($\sigma_{yy}, \sigma_{zz}, \sigma_{yz}, \sigma_{zx}, \sigma_{xy}$), we obtain the generalization of Eq. (2-2.14).

$$
\begin{aligned}
\epsilon_{xx} &= \frac{\partial C_0}{\partial \sigma_{xx}} \qquad \epsilon_{yy} = \frac{\partial C_0}{\partial \sigma_{yy}} \qquad \epsilon_{zz} = \frac{\partial C_0}{\partial \sigma_{zz}} \\
\epsilon_{yz} &= \frac{1}{2}\frac{\partial C_0}{\partial \sigma_{yz}} \qquad \epsilon_{zx} = \frac{1}{2}\frac{\partial C_0}{\partial \sigma_{zx}} \qquad \epsilon_{xy} = \frac{1}{2}\frac{\partial C_0}{\partial \sigma_{xy}}
\end{aligned}
\tag{2-2.19}
$$

Because of their relationship to Eqs. (2-2.11), Eqs. (2-2.19) are said to be conjugate to Eqs. (2-2.11). Equations (2-2.19) are known also as the *Legendre transform* of Eqs. (2-2.11). In certain theories of plasticity, it is assumed that the increase of internal energy during adiabatic deformation depends only on the final state of stress, and not on the path by which the state of stress is reached, provided that the stresses (loads) increase monotonically during the deformation process. These theories do not distinguish between inelasticity and nonlinear elasticity, unless unloading occurs. Hence, in these theories the concept of complementary strain is often employed.[4] For a linearly elastic material $C_0 = U_0$.

2-3 HOOKE'S LAW. ANISOTROPIC ELASTICITY

In the most general form, Hooke's law asserts that each of the stress components is a linear function of the components of the strain tensor; that is,

$$
\begin{aligned}
\sigma_{xx} &= C_{11}\epsilon_{xx} + C_{12}\epsilon_{yy} + C_{13}\epsilon_{zz} + C_{14}\epsilon_{xy} + C_{15}\epsilon_{xz} + C_{16}\epsilon_{yz} \\
\sigma_{yy} &= C_{21}\epsilon_{xx} + C_{22}\epsilon_{yy} + C_{23}\epsilon_{zz} + C_{24}\epsilon_{xy} + C_{25}\epsilon_{xz} + C_{26}\epsilon_{yz} \\
\sigma_{zz} &= C_{31}\epsilon_{xx} + C_{32}\epsilon_{yy} + C_{33}\epsilon_{zz} + C_{34}\epsilon_{xy} + C_{35}\epsilon_{xz} + C_{36}\epsilon_{yz} \\
\sigma_{xy} &= C_{41}\epsilon_{xx} + C_{42}\epsilon_{yy} + C_{43}\epsilon_{zz} + C_{44}\epsilon_{xy} + C_{45}\epsilon_{xz} + C_{46}\epsilon_{yz} \\
\sigma_{xz} &= C_{51}\epsilon_{xx} + C_{52}\epsilon_{yy} + C_{53}\epsilon_{zz} + C_{54}\epsilon_{xy} + C_{55}\epsilon_{xz} + C_{56}\epsilon_{yz} \\
\sigma_{yz} &= C_{61}\epsilon_{xx} + C_{62}\epsilon_{yy} + C_{63}\epsilon_{zz} + C_{64}\epsilon_{xy} + C_{65}\epsilon_{xz} + C_{66}\epsilon_{yz}
\end{aligned}
\tag{2-3.1}
$$

where the 36 coefficients, $C_{11}, \ldots, C_{66}$, are called elastic coefficients (stiffness).

In general, the coefficients C_{ij}, $i, j = 1, 2, 3, 4, 5, 6$ are not constants, but may depend upon the location in the body as well as on time and temperature. Ordinarily the C_{ij} decrease with increasing temperature.

In reality, Eq. (2-3.1) is no law, but merely an approximation that is valid for small strains, since any continuous function is approximately linear in a sufficiently small range of the variables. For a given temperature, time, and location in the body, the coefficients C_{ij} are constants that are characteristics of the material.

Equations (2-2.11) and (2-3.1) yield

$$
\begin{aligned}
\frac{\partial U_0}{\partial \epsilon_{xx}} &= \sigma_{xx} = C_{11}\epsilon_{xx} + C_{12}\epsilon_{yy} + C_{13}\epsilon_{zz} + C_{14}\epsilon_{xy} + C_{15}\epsilon_{xz} + C_{16}\epsilon_{yz} \\
&\cdots\cdots\cdots\cdots\cdots\cdots\cdots\cdots\cdots\cdots \\
&\cdots\cdots\cdots\cdots\cdots\cdots\cdots\cdots\cdots\cdots \\
\frac{\partial U_0}{\partial \epsilon_{yz}} &= \sigma_{yz} = C_{61}\epsilon_{xx} + C_{62}\epsilon_{yy} + C_{63}\epsilon_{zz} + C_{64}\epsilon_{xy} + C_{65}\epsilon_{xz} + C_{66}\epsilon_{yz}
\end{aligned}
\tag{2-3.2}
$$

Hence, the appropriate differentiations of Eqs. (2-3.2) yield

$$
\begin{aligned}
&\frac{\partial^2 U_0}{\partial \epsilon_{xx}\partial \epsilon_{yy}} = C_{12} = C_{21}, \frac{\partial^2 U_0}{\partial \epsilon_{xx}\partial \epsilon_{zz}} = C_{13} = C_{31}, \ldots, \\
&\ldots, \frac{\partial^2 U_0}{\partial \epsilon_{yz}\partial \epsilon_{xy}} = C_{46} = C_{64}, \frac{\partial^2 U_0}{\partial \epsilon_{yz}\partial \epsilon_{xz}} = C_{56} = C_{65}
\end{aligned}
\tag{2-3.3}
$$

These equations show that $C_{12} = C_{21}, C_{13} = C_{31}, \ldots, C_{ik} = C_{ki}, \ldots, C_{56} = C_{65}$; that is, the elastic coefficients C_{ij} are symmetrical. Therefore, there are only *twenty-one distinct* C's. In other words, the general anisotropic linear elastic material has 21 elastic coefficients. In view of the preceding relation, the strain energy density of a general anisotropic material is (by integration† of Eqs. 2-3.2)

$$
\begin{aligned}
U_0 = &\tfrac{1}{2}C_{11}\epsilon_{xx}^2 + \tfrac{1}{2}C_{12}\epsilon_{xx}\epsilon_{yy} + \cdots + \tfrac{1}{2}C_{16}\epsilon_{xx}\epsilon_{yz} \\
&+ \tfrac{1}{2}C_{12}\epsilon_{xx}\epsilon_{yy} + \tfrac{1}{2}C_{22}\epsilon_{yy}^2 + \cdots + \tfrac{1}{2}C_{26}\epsilon_{yy}\epsilon_{yz} \\
&+ \tfrac{1}{2}C_{13}\epsilon_{xx}\epsilon_{zz} + \tfrac{1}{2}C_{23}\epsilon_{yy}\epsilon_{zz} + \cdots + \tfrac{1}{2}C_{36}\epsilon_{zz}\epsilon_{yz} \\
&+ \cdots\cdots\cdots\cdots\cdots\cdots\cdots\cdots \\
&+ \tfrac{1}{2}C_{16}\epsilon_{xx}\epsilon_{yz} + \tfrac{1}{2}C_{26}\epsilon_{yy}\epsilon_{yz} + \cdots + \tfrac{1}{2}C_{66}\epsilon_{yz}^2
\end{aligned}
\tag{2-3.4}
$$

In this general form, Eq. (2-3.4) is important in the study of crystals.[5] We also note that Eq. (2-3.4) holds for orthogonal curvilinear coordinates (see the remark after Eq. 2-2.11.)

† Here, we discard an arbitrary function of (x, y, z), since we are interested in derivatives of U_0 with respect to $\epsilon_{xx}, \epsilon_{yy}, \ldots, \epsilon_{yz}$. Furthermore, in agreement with Eq. (2-3.1), we assume that the stress components $\sigma_{xx}, \sigma_{yy}, \ldots, \sigma_{yz}$ vanish identically with the strain components. Accordingly, linear terms in $\epsilon_{xx}, \epsilon_{yy}, \ldots, \epsilon_{yz}$ are discarded from Eq. (2-3.4). If the stress components do not vanish with the strain components (for example, in residual stresses), arbitrary functions of (x, y, z) must be added to Eq. (2-3.1). In turn, these functions lead to linear terms in $\epsilon_{xx}, \epsilon_{yy}, \ldots, \epsilon_{yz}$ in Eq. (2-3.4).

2-4
HOOKE'S LAW. ISOTROPIC ELASTICITY

Isotropic Materials. Homogeneous Materials / If the orientations of crystals and grains constituting the material of a solid member are distributed sufficiently randomly, any part of the member will display essentially the same material properties in all directions. If a solid member is composed of such randomly oriented crystals and grains, it is said to be *isotropic*. Thus, isotropy may be considered as a directional property of the material. Accordingly, if a material member is isotropic, its physical properties at a point P in the member are invariant under a rotation with respect to axes with origin at P. A material is said to be *elastically isotropic* if its characteristic elastic constants are invariant under any rotation of coordinates.

If the material properties are identical for every point in a member, the member is said to be *homogeneous*. In other words, homogeneity implies that the physical properties of a member are invariant under a translation. Alternatively, a member whose material properties change from point to point is said to be nonhomogeneous. For example, since in general the elastic constants are functions of temperature, a member subjected to a nonuniform temperature distribution is nonhomogeneous. Accordingly, the property of nonhomogeneity is a scalar property; that is, it depends only on the location of a point in the member, not on any direction at the point. Consequently, the material in a member may be nonhomogeneous, but isotropic. For example, consider a flat sandwich plate formed by a layer of aluminum bounded by layers of steel. If the point considered is in a steel layer, the material properties have certain values that are generally independent of direction. That is, the steel is essentially isotropic. Furthermore, if the temperature is approximately constant throughout the plate, the material properties do not change greatly from point to point. If the point considered is in the aluminum, the material properties differ from those of steel. Therefore, taken as a complete body, the sandwich plate exhibits nonhomogeneity. However, at any point in the sandwich plate, the properties are essentially independent of direction.†

Analogously, a member may be nonisotropic, but homogeneous. For example, the physical properties of a crystal depend on direction in the crystal, but the properties vary little from one point to another.[5]

If an elastic member is composed of isotropic materials, the strain energy density depends only on the principal strains (which are in-

†An exception occurs at the boundaries between the aluminum layer and the steel layers. Here, the sandwich plate is nonisotropic in nature.

variants), since for isotropic materials the elastic constants are invariants under arbitrary rotations (see Eq. 2-4.2).

Strain Energy Density of Isotropic Elastic Materials / The strain energy density of an elastic isotropic material depends only on the principal strains $(\epsilon_1, \epsilon_2, \epsilon_3)$. Accordingly, if the elasticity is linear, Eq. (2-3.4) yields

$$U_0 = \tfrac{1}{2}C_{11}\epsilon_1^2 + \tfrac{1}{2}C_{12}\epsilon_1\epsilon_2 + \tfrac{1}{2}C_{13}\epsilon_1\epsilon_3 + \tfrac{1}{2}C_{12}\epsilon_1\epsilon_2 + \tfrac{1}{2}C_{22}\epsilon_2^2 + \tfrac{1}{2}C_{23}\epsilon_2\epsilon_3 + \tfrac{1}{2}C_{13}\epsilon_1\epsilon_3 + \tfrac{1}{2}C_{23}\epsilon_2\epsilon_3 + \tfrac{1}{2}C_{33}\epsilon_3^2 \qquad (2\text{-}4.1)$$

We note that a strain energy density function U_0 exists for either adiabatic or isothermal (constant temperature) deformations. However, the numerical values of the elastic coefficients C_{ij} differ in these two cases.[5]

By symmetry, the naming of the principal axes is arbitrary. Hence, $C_{11} = C_{22} = C_{33} = C_1$, and $C_{12} = C_{23} = C_{13} = C_2$. Consequently, Eq. (2-4.1) contains only two distinct coefficients. Hence, for linear elastic isotropic material, the strain energy density may be expresses in the form

$$U_0 = \tfrac{1}{2}\lambda(\epsilon_1 + \epsilon_2 + \epsilon_3)^2 + G(\epsilon_1^2 + \epsilon_2^2 + \epsilon_3^2) \qquad (2\text{-}4.2)$$

where $\lambda = C_2$ and $G = (C_1 - C_2)/2$ are elastic coefficients called Lame's elastic coefficients. If the material is homogeneous and temperature is constant everywhere, λ and G are constants at all points. In terms of the strain invariants (see Eq. 1-7.22 and Eq. 1-9.3), Eq. (2-4.2) may be written in the following form:

$$U_0 = (\tfrac{1}{2}\lambda + G)J_1^2 - 2GJ_2 \qquad (2\text{-}4.3)$$

Returning to orthogonal curvilinear coordinates (x, y, z) and introducing the general definitions of J_1 and J_2 from Eq. (1-7.22), we obtain

$$U_0 = \tfrac{1}{2}\lambda(\epsilon_{xx} + \epsilon_{yy} + \epsilon_{zz})^2 + G(\epsilon_{xx}^2 + \epsilon_{yy}^2 + \epsilon_{zz}^2 + 2\epsilon_{xy}^2 + 2\epsilon_{xz}^2 + 2\epsilon_{yz}^2) \qquad (2\text{-}4.4)$$

where $\epsilon_{xx}, \epsilon_{yy}, \epsilon_{zz}, \epsilon_{xy}, \epsilon_{xz}, \epsilon_{yz})$ are strain components relative to orthogonal coordinates (x, y, z); see Eqs. (1-9.5) and the remark after Eq. (2-3.4). Equations (2-2.11) and (2-4.4) now yield Hooke's law for a linear elastic isotropic material in the form (for orthogonal curvilinear coordinates x, y, z):

$$\begin{array}{lll} \sigma_{xx} = \lambda e + 2G\epsilon_{xx} & \sigma_{yy} = \lambda e + 2G\epsilon_{yy} & \sigma_{zz} = \lambda e + 2G\epsilon_{zz} \\ \sigma_{xy} = 2G\epsilon_{xy} & \sigma_{xz} = 2G\epsilon_{xz} & \sigma_{yz} = 2G\epsilon_{yz} \end{array} \qquad (2\text{-}4.5)$$

where $e \cong \epsilon_{xx} + \epsilon_{yy} + \epsilon_{zz} = J_1$ is the classical small-displacement cubical strain (see Eq. 1-9.2). Thus, we have shown that for isotropic linear elastic materials, the stress-strain relations involves only two elastic constants. An analytic proof of the fact that no further reduction is possible on a theoretical basis can be constructed.[6]

By means of Eqs. (2-4.5) we find (with Eqs. 1-4.7 and 1-7.22)

$$\begin{aligned} I_1 &= (3\lambda + 2G)J_1 \\ I_2 &= \lambda(3\lambda + 4G)J_1^2 + 4G^2J_2 \\ I_3 &= \lambda^2(\lambda + 2G)J_1^3 + 4\lambda G^2 J_1 J_2 + 8G^3 J_3 \end{aligned} \tag{2-4.6}$$

which relate the stress invariants, I_1, I_2, I_3 to the strain invariants J_1, J_2, J_3.

Inverting Eqs. (2-4.5), we obtain

$$\begin{aligned} \epsilon_{xx} &= \frac{1}{E}\left[(1+\nu)\sigma_{xx} - \nu I_1\right] = \frac{1}{E}(\sigma_{xx} - \nu\sigma_{yy} - \nu\sigma_{zz}) \\ \epsilon_{yy} &= \frac{1}{E}\left[(1+\nu)\sigma_{yy} - \nu I_1\right] = \frac{1}{E}(\sigma_{yy} - \nu\sigma_{xx} - \nu\sigma_{zz}) \\ \epsilon_{zz} &= \frac{1}{E}\left[(1+\nu)\sigma_{zz} - \nu I_1\right] = \frac{1}{E}(\sigma_{zz} - \nu\sigma_{xx} - \nu\sigma_{yy}) \\ \epsilon_{xy} &= \frac{1}{2G}\sigma_{xy} = \frac{1+\nu}{E}\sigma_{xy} \\ \epsilon_{xz} &= \frac{1}{2G}\sigma_{xz} = \frac{1+\nu}{E}\sigma_{xz} \\ \epsilon_{yz} &= \frac{1}{2G}\sigma_{yz} = \frac{1+\nu}{E}\sigma_{yz} \end{aligned} \tag{2-4.7}$$

where

$$E = \frac{G(3\lambda + 2G)}{\lambda + G} \qquad \nu = \frac{\lambda}{2(\lambda + G)} \tag{2-4.8}$$

are elastic coefficients called Young's modulus and Poisson's ratio, respectively. Alternatively, Eqs. (2-4.5) may be written in terms of E and ν as follows:

$$\begin{aligned} \sigma_{xx} &= \frac{E}{(1+\nu)(1-2\nu)}\left[(1-2\nu)\epsilon_{xx} + \nu J_1\right] \\ \sigma_{yy} &= \frac{E}{(1+\nu)(1-2\nu)}\left[(1-2\nu)\epsilon_{yy} + \nu J_1\right] \\ \sigma_{zz} &= \frac{E}{(1+\nu)(1-2\nu)}\left[(1-2\nu)\epsilon_{zz} + \nu J_1\right] \\ \sigma_{xy} &= \frac{E}{1+\nu}\epsilon_{xy} \qquad \sigma_{xz} = \frac{E}{1+\nu}\epsilon_{xz} \qquad \sigma_{yz} = \frac{E}{1+\nu}\epsilon_{yz} \end{aligned} \tag{2-4.9}$$

Substitution of Eqs. (2-4.7) into Eq. (2-4.4) yields the strain-energy density U_0 in terms of stress quantities. Thus, we obtain

$$U_0 = \frac{1}{2E}\left[\sigma_{xx}^2 + \sigma_{yy}^2 + \sigma_{zz}^2 - 2\nu\left(\sigma_{xx}\sigma_{yy} + \sigma_{xx}\sigma_{zz} + \sigma_{yy}\sigma_{zz}\right) + 2(1+\nu)\left(\sigma_{xy}^2 + \sigma_{xz}^2 + \sigma_{yz}^2\right)\right] \qquad (2\text{-}4.10)$$

$$= \frac{1}{2E}\left[I_1^2 - 2(1+\nu)I_2\right]$$

If the axes (x, y, z) are directed along the principal axes of strain, then $\epsilon_{xy} = \epsilon_{xz} = \epsilon_{yz} = 0$. Hence, by Eq. (2-4.9), $\sigma_{xy} = \sigma_{xz} = \sigma_{yz} = 0$. Therefore, the axes (x, y, z) must lie along the principal axes of stress. Consequently, for an isotropic material, the principal axes of stress are coincident with the principal axes of strain. *Hence, when we deal with isotropic materials, no distinction need be made between principal axes of stress and principal axes of strain. Such axes are called simply principal axes.*

EXAMPLE 2-4.1

Flat Plate Bent Around a Circular Cylinder

A flat rectangular plate lies in the (x, y) plane, (Fig. E2-4.1*a*). The plate, of uniform thickness $h = 2.00$ mm, is bent around a circular cylinder (Fig. E2-4.1*b*) with the y-axis parallel to the axis of the cylinder. The plate is made of an isotropic aluminum alloy ($E = 72.0$ GPa and $\nu = 0.33$). The radius of the cylinder is 600 mm. (a) Assuming that plane sections, x = constant for the undeformed plate, remain plane after deformation, determine the maximum circumferential stress $\sigma_{\theta\theta(\max)}$ in the plate for linearly elastic behavior. (b) The reciprocal of the radius of curvature R for a beam subject to pure bending is the curvature $\kappa = 1/R = M/EI$. For the plate, derive a formula for the curvature $\kappa = 1/R$ in terms of the

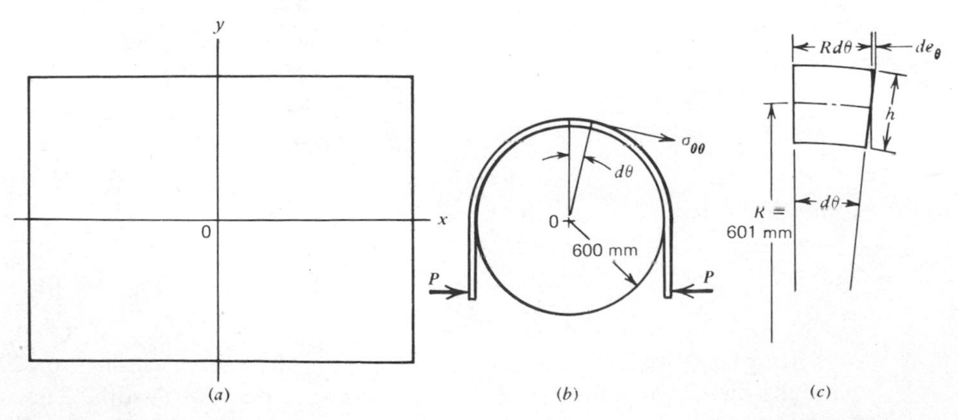

Fig. E2-4.1

applied moment M per unit width and the flexural rigidity $D = Eh^3/[12(1-\nu^2)]$ of the plate.

SOLUTION

(a) We assume that the middle surface of the plate remains unstressed and that the stress through the thickness is negligible. Hence, the flexure formula is valid for the bending of the plate. Therefore, $\sigma_{\theta\theta} = \sigma_{yy} = 0$ for the middle surface and $\sigma_{rr} = 0$ throughout the plate thickness h. Equations (2-4.7) yield results $\epsilon_{rr} = \epsilon_{\theta\theta} = \epsilon_{yy} = 0$ in the middle surface of the plate. Since the length of the plate in the y-direction is large compared to the thickness h, the plate deforms approximately under conditions of plane strain; that is, $\epsilon_{yy} \approx 0$ throughout the plate thickness. Equations (2-4.7) give

$$\epsilon_{yy} = 0 = \frac{1}{E}\sigma_{yy} - \frac{\nu}{E}\sigma_{\theta\theta}$$

throughout the plate thickness. Thus, for plane strain relative to the r, θ plane

$$\sigma_{yy} = \nu\sigma_{\theta\theta} \tag{1}$$

With Eqs. (2-4.7), Eq. (1) yields

$$\epsilon_{\theta\theta} = \frac{1}{E}\sigma_{\theta\theta} - \frac{\nu}{E}\sigma_{yy} = \frac{(1-\nu^2)}{E}\sigma_{\theta\theta} \tag{2}$$

The relation between the radius of curvature R of the deformed plate and $\epsilon_{\theta\theta}$ may be determined by the geometry of deformation of a plate segment (Fig. E2-4.1*c*). By similar triangles, we find from Fig. E2-4.1*c* that

$$\frac{R\,d\theta}{R} = \frac{2\,de_\theta}{h} = \frac{2\epsilon_{\theta\theta(\max)}R\,d\theta}{h}$$

or

$$\epsilon_{\theta\theta(\max)} = \frac{h}{2R} \tag{3}$$

Equations (2) and (3) yield the result

$$\sigma_{\theta\theta(\max)} = \frac{Eh}{2(1-\nu^2)R} = \frac{72.0 \times 10^3(2)}{2(1-0.33^2)(601)} = 134 \text{ MPa} \tag{4}$$

(b) In plate problems, it is convenient to consider a unit width of the plate (y-direction) and let M be the moment per unit width. The moment of inertia for this unit width is $I = bh^3/12 = h^3/12$. Since

$\sigma_{\theta\theta(\max)} = M(h/2)/I$, this relation may be used with Eq. (4) to give

$$\frac{1}{R} = \frac{\sigma_{\theta\theta(\max)}(2)(1-\nu^2)}{EH} = \frac{Mh(12)}{2h^3}\frac{2(1-\nu^2)}{Eh} = \frac{M}{D} \tag{5}$$

where

$$D = \frac{Eh^3}{12(1-\nu^2)} \tag{6}$$

is called the flexure rigidity of the plate.

PROBLEM SET 2-4

1. A square plate with 800 mm sides parallel to the x- and y-axes has a uniform thickness $h = 10$ mm and is made of an isotropic steel ($E = 200$ GPa and $\nu = 0.29$). The plate is subjected to a uniform state of stress. If $\sigma_{zz} = \sigma_{zx} = \sigma_{zy} = 0$ (plane stress), $\sigma_{xx} = \sigma_1 = 500$ MPa, and $\epsilon_{yy} = 0$ for the plate, determine $\sigma_{yy} = \sigma_2$ and the final dimensions of the plate assuming linearly elastic conditions.

2. The plate in Problem 1 is subjected to plane strain ($\epsilon_{zz} = \epsilon_{zx} = \epsilon_{zy} = 0$). If $\sigma_{xx} = \sigma_1 = 500$ MPa and $\epsilon_{xx} = 2\epsilon_{yy}$, determine the magnitude of $\sigma_{yy} = \sigma_2$ and $\sigma_{zz} = \sigma_3$, assuming linearly elastic conditions.

   Ans. $\sigma_{yy} = 377.2$ MPa, $\sigma_{zz} = 254.4$ MPa

3. A triaxial state of principal stress acts on the faces of a unit cube. Show that these stresses will not produce a volume change if $\nu = 1/2$. The material is a linearly elastic isotropic material. If $\nu \neq 1/2$, show that the condition necessary for the volume to remain unchanged is for $\sigma_1 + \sigma_2 + \sigma_3 = 0$.

4. A member is made of an isotropic linearly elastic aluminum alloy ($E = 72.0$ GPa and $\nu = 0.33$). Consider a point in the free surface that is tangent to the (x, y)-plane. If $\sigma_{xx} = 250$ MPa, $\sigma_{yy} = -50$ MPa, and $\sigma_{xy} = -150$ MPa, determine the directions for strain gauges at that point to measure two of the principal strains. What are the magnitudes of these principal strains?

   Ans. $\epsilon_1 = 0.00485$ at 0.3927 rad clockwise from x-axis,
   $\epsilon_2 = -0.00299$

5. A member made of isotropic bronze ($E = 82.6$ GPa and $\nu = 0.35$) is subjected to a state of plane strain ($\epsilon_{zz} = \epsilon_{zx} = \epsilon_{zy} = 0$). Determine σ_{zz}, ϵ_{xx}, ϵ_{yy}, and $\gamma_{xy} = 2\epsilon_{xy}$, if $\sigma_{xx} = 90$ MPa, $\sigma_{yy} = -50$ MPa, and $\sigma_{xy} = 70$ MPa.

6. Solve Problem 1 for the condition that $\epsilon_{xx} = 2\epsilon_{yy}$.

Ans. $\sigma_{yy} = 345.0$ MPa, $\epsilon_{xx} = 0.00200$, $\epsilon_{yy} = 0.00100$, $\epsilon_{zz} = -0.00123$, $L_x = 801.60$ mm, $L_y = 800.80$ mm, $L_z = 9.99$ mm

2-5 EQUATIONS OF THERMOELASTICITY FOR ISOTROPIC MATERIALS

Consider a member made of an isotropic elastic material in an arbitrary zero configuration. Let the temperature of the member be increased by a small amount T. Since the member is isotropic, all infinitesimal line elements in the volume undergo equal expansions. Furthermore, all line elements maintain their initial directions. Therefore, the strain components due to the temperature change T are, with respect to rectangular Cartesian coordinates (x, y, z),

$$\epsilon'_{xx} = \epsilon'_{yy} = \epsilon'_{zz} = kT \qquad \epsilon'_{xy} = \epsilon'_{xz} = \epsilon'_{zy} = 0 \tag{2-5.1}$$

where k denotes the linear coefficient of thermal expansion of the material. For a nonhomogeneous member, k may be a function of coordinates and of temperature; that is, $k = k(x, y, z, T)$.

Now let the member be subjected to forces that induce stresses $\sigma_{xx}, \sigma_{yy}, \ldots, \sigma_{yz}$ at point 0 in the member. Accordingly, if $\epsilon_{xx}, \epsilon_{yy}, \ldots, \epsilon_{yz}$ denote the strain components at point 0 after the application of the forces, the new change in strain produced by the forces is represented by the equations

$$\begin{aligned} &\epsilon''_{xx} = \epsilon_{xx} - kT \qquad \epsilon''_{yy} = \epsilon_{yy} - kT \qquad \epsilon''_{zz} = \epsilon_{zz} - kT \\ &\epsilon''_{xy} = \epsilon_{xy} \qquad \epsilon''_{xz} = \epsilon_{xz} \qquad \epsilon''_{yz} = \epsilon_{yz} \end{aligned} \tag{2-5.2}$$

In general, T may depend on the location of point 0 and on time t. Hence $T = T(x, y, z, t)$. Substitution of Eq. (2-5.2) into Eqs. (2-4.5) yields

$$\begin{aligned} &\sigma_{xx} = \lambda e + 2G\epsilon_{xx} - cT \qquad \sigma_{yy} = \lambda e + 2G\epsilon_{yy} - cT \\ &\sigma_{zz} = \lambda e + 2G\epsilon_{zz} - cT \\ &\sigma_{xy} = 2G\epsilon_{xy} \qquad \sigma_{xz} = 2G\epsilon_{xz} \qquad \sigma_{yz} = 2G\epsilon_{yz} \end{aligned} \tag{2-5.3}$$

where

$$c = (3\lambda + 2G)k = Ek/(1 - 2\nu) \tag{2-5.4}$$

Similarly, substitution of Eqs. (2-5.3) into Eqs. (2-4.7) yields

$$
\begin{aligned}
\epsilon_{xx} &= \frac{1}{E}\left[\sigma_{xx} - \nu(\sigma_{yy} + \sigma_{zz})\right] + kT \\
\epsilon_{yy} &= \frac{1}{E}\left[\sigma_{yy} - \nu(\sigma_{xx} + \sigma_{zz})\right] + kT \\
\epsilon_{zz} &= \frac{1}{E}\left[\sigma_{zz} - \nu(\sigma_{xx} + \sigma_{yy})\right] + kT \\
\epsilon_{xy} &= \frac{(1+\nu)}{E}\sigma_{xy} \qquad \epsilon_{xz} = \frac{(1+\nu)}{E}\sigma_{xz} \qquad \epsilon_{yz} = \frac{(1+\nu)}{E}\sigma_{yz}
\end{aligned}
\tag{2-5.5}
$$

Finally, substituting Eqs. (2-5.5) into Eqs. (2-4.3) or (2-4.4), we find

$$U_0 = (\tfrac{1}{2}\lambda + G)J_1^2 - 2GJ_2 - cJ_1T + \tfrac{3}{2}ckT^2 \tag{2-5.6}$$

In terms of the strain components (see Eqs. (1-7.22)), we obtain

$$
\begin{aligned}
U_0 = {} & \tfrac{1}{2}\lambda(\epsilon_{xx} + \epsilon_{yy} + \epsilon_{zz})^2 + G(\epsilon_{xx}^2 + \epsilon_{yy}^2 + \epsilon_{zz}^2 + 2\epsilon_{xy}^2 + 2\epsilon_{xz}^2 + 2\epsilon_{yz}^2) \\
& - c(\epsilon_{xx} + \epsilon_{yy} + \epsilon_{zz})T + \tfrac{3}{2}ckT^2
\end{aligned}
\tag{2-5.7}
$$

Equations (2-5.3) and (2-5.5) are the basic stress-strain relations of classical thermoelasticity for isotropic materials. For temperature changes T the strain energy density is modified by a temperature-dependent term that is proportional to the volumetric strain $e = J_1 = \epsilon_{xx} + \epsilon_{yy} + \epsilon_{zz}$ and by a term proportional to T^2 (Eqs. 2-5.6 and 2-5.7).

We find by Eqs. (2-5.5) and (2-5.7)

$$U_0 = \frac{1}{2E}\left[I_1^2 - 2(1+\nu)I_2\right] \tag{2-5.8}$$

and

$$
\begin{aligned}
U_0 = \frac{1}{2E}\Big[& \sigma_{xx}^2 + \sigma_{yy}^2 + \sigma_{zz}^2 - 2\nu(\sigma_{xx}\sigma_{yy} + \sigma_{xx}\sigma_{zz} + \sigma_{yy}\sigma_{zz}) \\
& + 2(1+\nu)(\sigma_{xy}^2 + \sigma_{xz}^2 + \sigma_{yz}^2)\Big]
\end{aligned}
\tag{2-5.9}
$$

in terms of stress components. Equation (2-5.9) does not contain T explicitly. However, the temperature distribution may affect the stresses.

2-6 INITIATION OF YIELD. YIELD CRITERIA

Introduction/One of the primary purposes of the theory of plasticity is the explanation of the deformational behavior of materials stressed beyond the elastic limit (see Art. 2-1). A second purpose is to determine the stress at which plastic behavior initially begins (that is, at which behavior departs from pure elastic). In general, the deformation of a material subjected to static loads may be characterized as occurring in four stages. In the first stage, the deformation is such that the displacements are small (infinitesimal) with resulting small strains. In the region of infinitesimal displacements and linear elastic material behavior, either the method of mechanics of materials or the method of elasticity are employed to describe the material response to load. In the elastic region of deformation, the material returns to its initial (unstrained) state when the loads are removed. In the second stage of deformation, the loads are sufficiently large to exceed the proportional limit stress (strain) value. However, the loads are not so large that the strains become excessive (that is, so large that noticeable changes in the member's dimensions or even fracture may occur). In this second region of deformation, most members do not completely regain their initial unstrained state when loads are removed. Therefore, we say that the member is deformed inelastically or plastically. The portion of strain (deformation) remaining after the removal of load is called the *permanent strain*. In the third stage of deformation, the member undergoes sufficiently large inelastic strains so that noticeable changes in dimensions occur. For example, metal forming processes such as cold rolling and deep drawing fall in the third category of deformation. Finally, all deformations that are sufficiently large to produce fracture of the member may be considered to form a fourth stage of deformation, the *fracture stage*.

In this book we consider mainly the first stage and the initiation of the second stage of deformation. In this article we are primarily concerned with the phenomenon of the transition from the elastic state to the inelastic state (initiation of yield). Several hypotheses have been proposed for the prediction of this transition. These hypotheses are called *yield criteria*. Since initiation of yielding is one of the modes of failure considered in Chapter 3, yield criteria are also called *failure criteria* (Art. 3-3).

As in the theory of elasticity, several conditions are of importance in the theory of plasticity. For example, the shape of the member, the temperature distribution in the member, and the nature of the loads acting on the member are important. In addition, the magnitude of the temperature, the rate of load application, the time history of the load, as well as other effects, are additional factors that complicate the nature of the inelastic response of a member. However, ordinarily, theories of plasticity do not encompass time effects. Furthermore, the large majority of the theories of plasticity are restricted to members made of homogeneous materials, and only limited treatments of members made of anisotropic materials have appeared, although some attempts have been made to include the anisotropy caused by inelastic deformations. There are two main theories of plasticity: the deformational theory and the incremental theory. In essence, the deformational theory is a nonlinear theory of elasticity, whereas the incremental theory places importance on the incremental inelastic strains measured from a given deformed state of a member. In general, these theories require certain knowledge of the physical response of a material to complex loading situations. However, because of inherent difficulties in the theory and in the experimental measurement of the response of a member to loading, the theories are based on certain assumptions. For example, the stress-strain relationships for a member subjected to a three-dimensional state of inelastic strain are ordinarily deduced (postulated) on the basis of a simple uniaxial tension or compression test. Accordingly, one must employ with caution results derived for more complex states of strain, when the material properties are extrapolated from the uniaxial state to the three-dimensional state.

We shall employ a continuum mechanics phenomenological approach to the observation of inelastic strains. That is, the experimentally measured inelastic strains are considered to be average strains determined over gauge lengths that are of finite length. The experimental values so determined are thus only approximations to the inelastic strain at a point. However, the theory will treat pointwise values of strain (since the theory is based on a continuous medium model). Accordingly, our concept of inelastic strain differs from that of the experimental metallurgist who treats the inelastic strain as a microscopic quantity associated with the slip of crystals. Nevertheless, certain microscopic concepts are employed to arrive at hypotheses to predict the inception of yielding (that is, the inception of inelastic strains). For example, it is well known that crystals have slip planes along which the resistance to shear force is relatively small. Consequently, one hypothesis of yielding for certain ductile metals is that the material yields when the maximum shearing stress (see Art. 1-4, Eq. 1-4.27) at any point in the member reaches a critical value that causes slip (yield) of the crystal along its slip planes. Another hypothesis of yield is based upon the maximum octahedral shearing stress (see Art. 1-4, Eq. 1-4.8 or Eq. 1-4.9) that attains a critical value at any point in the member.

Model Behavior. One-Dimensional Case / As noted in Art. 2-1, the shape of the tensile stress-strain curve depends on the material being tested. However, under slowly applied loads certain features of the stress-strain curve are similar for all structural materials. For example, if the load is sufficiently small, the relation between stress and strain is essentially linearly elastic; that is, the stress-strain curve is a straight line, and loading and unloading proceed along this straight line (we say the process is reversible). As the load is increased to a sufficiently large value, the stress-strain curve becomes nonlinear (departs from a straight line). Depending on the material, the loading-unloading process may be elastic (reversible) or plastic (nonreversible). If the path of unloading (on the curve of stress-

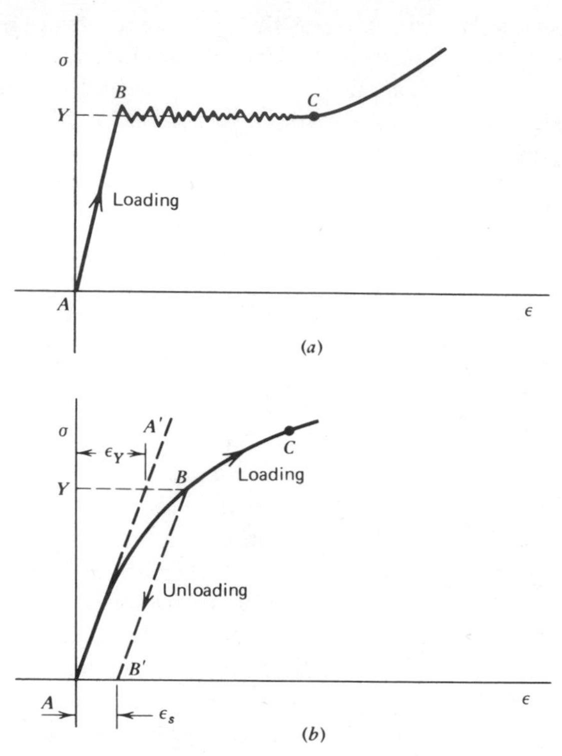

Fig. 2-6.1/In (*b*) line BB' is parallel to line AA' which is tangent to curve ABC at A.

strain) coincides with the path of loading (reversible) we say the material is nonlinearly elastic. If the unloading path does not follow the loading path, we say that the behavior is nonelastic or plastic. The change between linearly elastic and plastic (or nonlinear elastic) response may be abrupt (Fig. 2-1.1*b*) or gradual (Figs. 2-1.1*a* and 2-1.2). In the former case, at the initiation of plastic behavior (characterized by the yield point stress) a noticeable departure from linearity occurs. In the latter case, the yield stress Y is defined (conventionally) as the stress (read from the stress-strain curve) for which a definite permanent strain ϵ_S (say 0.002) remains after complete unloading (see Fig. 2-6.1*b*).

Since actual stress-strain curves, such as ABC of Fig. 2-6.1, are difficult to use in mathematical solutions of complex problems, idealizations of stress-strain curves (i.e., idealized models of material behavior) are usually used in analysis. For example the curve of Fig. 2-6.1*a* may be modeled as shown in Fig. 2-6.2*a*. Since part BC of the stress-strain curve is parallel to the strain axis (Fig. 2-6.2*a*), the material is said to behave in an *elastic-perfectly* (*ideally*) *plastic manner*, or the material is said to be *elastic-perfectly plastic*. For materials that strain harden (Fig. 2-6.1*b*), the stress-strain may be idealized as indicated in Fig. 2-6.2*b*; the material is said to be an *elastic-linear strain hardening plastic material*. For the

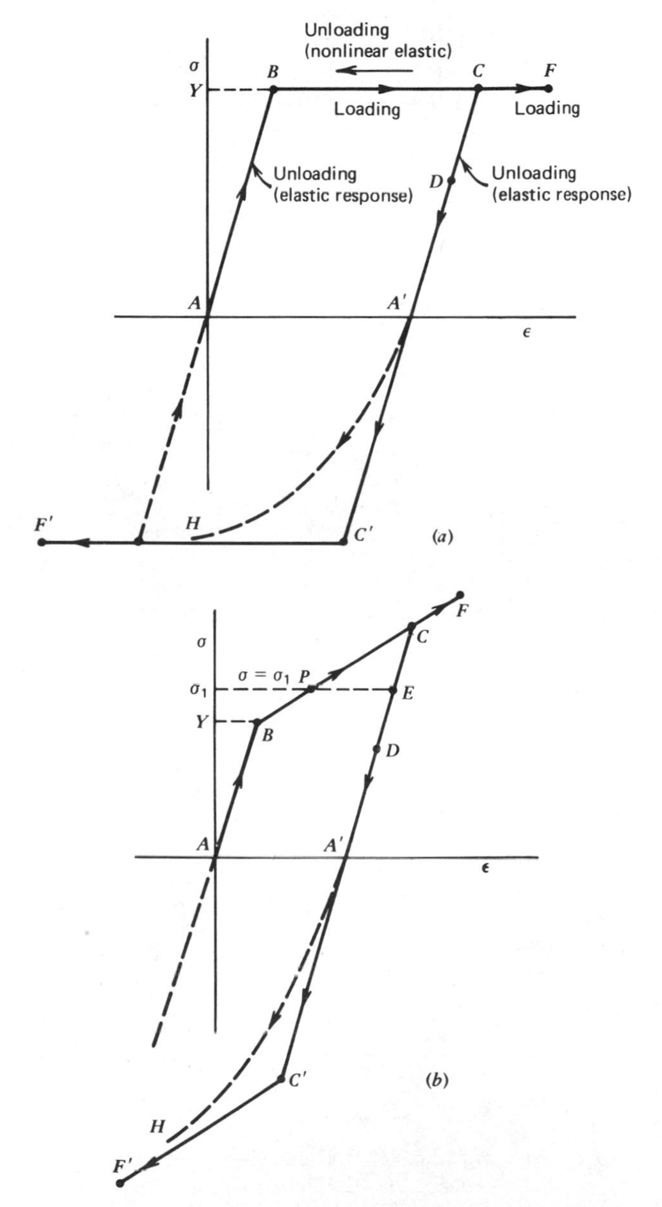

Fig. 2-6.2/(*a*) Elastic-perfectly plastic response. (*b*) Elastic-strain hardening plastic response.

idealized diagram, the yield stress Y is taken to be the intersection of the two straight lines and not the stress for a specified set ϵ_s.

In some problems, the elastic strain (associated with point B in Fig. 2-6.2) is very small compared to the total strain (say associated with point C). Then a further idealization of the stress-strain behavior is modeled as shown in Fig. 2-6.3, and the materials are said to be rigid

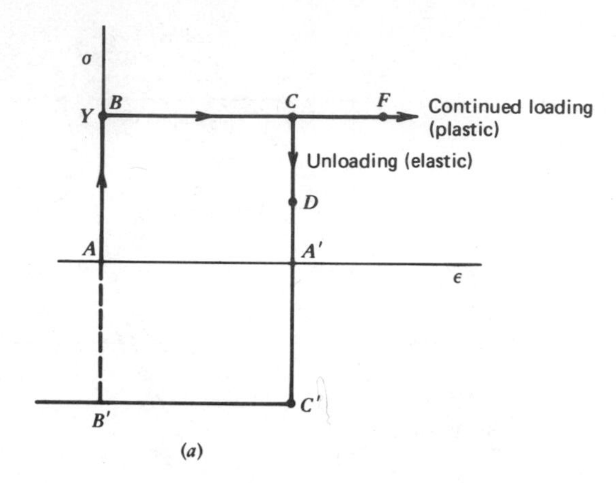

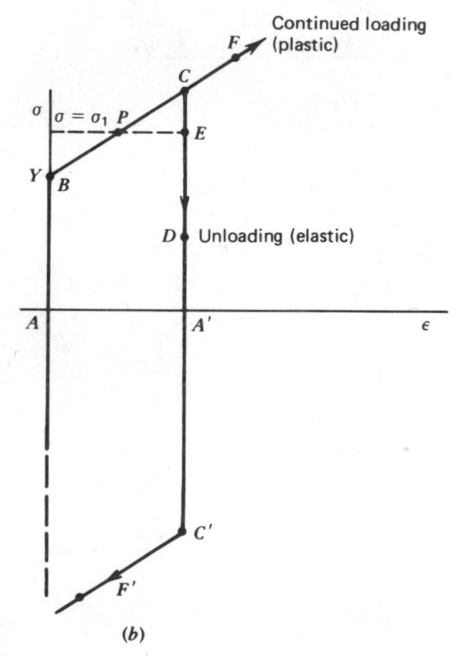

Fig. 2-6.3/(a) Rigid elastic-perfectly plastic response. (*b*) Rigidly elastic-strain hardening plastic response.

elastic-perfect plastic (Fig. 2-6.3*a*) and rigid elastic-strain hardening plastic (Fig. 2-6.3*b*). Briefly, the material (plastic) behavior modeled by the stress-strain curves of Fig. 2-6.3 is said to be *rigid-plastic.*

As noted above, the distinction between nonlinear elastic and plastic behavior occurs upon unloading; the nonlinear elastic unloading, being reversible, merely follows the stress-strain curve *CBA* back to a state of zero stress and strain at *A*. However, when the first increment of load is removed from a plastic material, experiments show (approximately) that the corresponding increment of strain is elastic so that the initial slope of this unloading curve will be the same as the initial slope of the loading stage (see Fig. 2-6.2, which is Fig. 2-6.1 idealized). Thus, the initial slopes of lines *CD* will be approximately the same as those of lines *AB* in Figs. 2-6.2*a* and 2-6.2*b*, respectively.

Eventually, under the idealized model behavior, the load will decrease to zero and a permanent strain *AA′* remains (Figs. 2-6.2 and 2-6.3). If unloading is continued, that is, if increasing compressive load is applied, eventually yielding in compression will occur at *C′*. For the elastic-perfectly plastic material, the magnitude of *A′C′* is approximately the same as *AB*. In other words, the yield stress in compression remains equal to *Y*, the same as in tension. However, for a strain-hardening material, the compressive yield stress may depend on the amount of tensile hardening that has taken place (Fig. 2-6.2*b*); that is, *AB* and *A′C′* have different magnitudes in general.

Stress-strain diagrams for unloading and reversed loading are often idealized as indicated by *CDC′F′* in Fig. 2-6.2. However, the response of real materials may deviate from the idealized curve. For some materials, experimental data show that the curve for unloading and for reversed loading may follow a path *CDH* in Fig. 2-6.2.

As noted above, the essential differences between nonlinear elasticity and plasticity occur during the unloading phase. In particular, for the nonlinear elastic response during loading or unloading, there is always a unique relation between stress and strain; that is, for each value of stress there is one and only one value of strain. However, in a plastic response, the correspondence of stress and strain upon unloading is different from the correspondence during loading. Specifically, in a plastic response a given value of stress (strain) may correspond to two values of strain (stress), and the uniqueness of relation between stress and strain is lost.

Two values of strain may be specified in Fig. 2-6.2*b* for one value of stress σ_1 for a specimen loaded to point *C*. If $\sigma_1 > Y$ and the specimen is being loaded, the strain at point *P* corresponds to stress σ_1. If the specimen is being unloaded from point *C*, the strain at point *E* also corresponds to stress σ_1. Since the response of the material to load is quite different at points *P* and *E*, extreme care must be taken to distinguish between them. To do so, two conditions must be considered. First, a material will possess, in general, an initial yield stress Y_T in tension and an initial yield stress Y_C in compression. The initial material response will be elastic so long as the stress remains between these two yield values. For the initial stressing of most materials, the yield stresses in tension and compression are equal in magnitude ($Y_T = Y_C = Y$). For a perfectly plastic material, the current yield stresses maintain their initial values throughout the stress history. However, for strain hardening materials, the current yield stresses depend on the deformation history. For example, after loading to point *C* and unloading to point *D*, we note that at point D in Figs. 2-6.2 and 2-6.3 the current tensile and compressive yield stresses are the values associated with points C and C', respectively. The initial tensile and compressive yield stresses (before loading) are plus and minus the yield stress Y associated with point B.

Thus, the response is elastic if the stress lies between the current tensile and compressive yield stresses. However, if the stress level reaches the yield stress value, a second condition must be established to ascertain whether or not the subsequent behavior is to be elastic or plastic. This condition may be established by noting (see Figs. 2-6.2 and 2-6.3) that the behavior will be plastic if the stress rate is nondecreasing at the current tensile yield stress and nonincreasing at the current compressive yield stress. Thus, at point C, the behavior can be either plastic along CF or elastic along CD, with similar conditions prevailing at point C'. Generalizations of these concepts for multiaxial states of stresses are available.[7] In this book, we shall be concerned primarily with the initiation of yield at a given point in a solid. We shall not consider general plastic stress-strain relations.

Yield Criteria / Since plasticity theories are concerned with the behavior of a solid loaded beyond the yield stress, a hypothesis of yielding is required. Experimentally, the yield stress is determined, ordinarily, from a simple tension test as the stress at which the idealized stress-strain relation ceases to be linear as indicated in Fig. 2-6.2. Theoretically, the yield stress is assumed to be determined on the basis of a given yield hypothesis. Common yield hypotheses are associated with critical values of quantities such as maximum normal stress, maximum extensional strain, maximum shearing stress, total energy of deformation, energy of distortion, or maximum octahedral shearing stress. Experiments indicate that no one of these quantities may be used to accurately predict yielding for all materials. However, for ductile or semiductile metals, the experimentally determined yield stress is reasonably accurately predicted by conditions based on the critical values of either the maximum shearing stress or the maximum octahedral shearing stress. Therefore, plasticity theories often employ yield hypotheses based on either a *maximum shearing stress condition* or *a maximum octahedral shearing stress condition*.

The maximum shearing stress yield condition, originally proposed by Tresca and later employed by Saint-Venant, implies that yielding begins when the maximum shearing stress $\tau_{max} = \frac{1}{2}(\sigma_{max} - \sigma_{min})$ attains a critical value, where σ_{max} and σ_{min} denote the maximum and minimum principal stress components. In a uniaxial test (tension or compression) $\sigma_1 = \sigma$, $\sigma_2 = \sigma_3 = 0$. Then $\tau_{max} = \frac{1}{2}\sigma$. Inherently in the maximum shearing stress yield criterion it is assumed that the critical value of τ_{max} is the same in tension and in compression. Thus, when σ attains a critical value, say Y ($\tau_{max} = \frac{1}{2}Y$), it is assumed that yielding begins. The critical value Y is called the yield stress. Its value is determined from the simple tension (or compression) test. Furthermore, it is assumed that the value of Y is not affected by the magnitude of the mean stress. Thus, the maximum shearing stress yield criterion assumes that yielding occurs when the shearing stress τ attains the maximum value $\frac{1}{2}Y$. That is, at yield

$$\tau_{\max} = \tfrac{1}{2}Y \tag{2-6.1}$$

The energy-of-distortion yield condition, first proposed by von Mises and later interpreted by Hencky, is based on the assumption that yielding first begins when the strain energy density (after discarding the portion of strain energy density associated with the mean stress) attains a critical value. Accordingly, noting that the strain energy density may be expressed in terms of principal stress components in the form (Eq. 2-4.10 with $\sigma_{xz} = \sigma_{xy} = \sigma_{yz} = 0$ and $\sigma_{xx} = \sigma_1$, $\sigma_{yy} = \sigma_2$, $\sigma_{zz} = \sigma_3$)

$$U_0 = \frac{(\sigma_1 - \sigma_2)^2 + (\sigma_2 - \sigma_3)^2 + (\sigma_3 - \sigma_1)^2}{12G} + \frac{(\sigma_1 + \sigma_2 + \sigma_3)^2}{18K} \tag{2-6.2}$$

where $G, 2G = E/(1+\nu)$, is the shearing modulus of the material and $K, 3K = E/(1-2\nu)$, is the elastic bulk modulus, and noting that $(\sigma_1 + \sigma_2 + \sigma_3)/3$ is the mean stress, we obtain, after discarding the mean stress effect,

$$U_{D0} = \frac{(\sigma_1 - \sigma_2)^2 + (\sigma_2 - \sigma_3)^2 + (\sigma_3 - \sigma_1)^2}{12G} \tag{2-6.3}$$

Hence, the energy of distortion yield condition states that yielding occurs when the energy of distortion U_{D0} (given by Eq. 2-6.3) attains a critical value. The critical value of U_{D0} is determined on the basis of a simple tension test. Thus, for a uniaxial test, $\sigma_1 = \sigma$, $\sigma_2 = \sigma_3 = 0$, and Eq. (2-6.3) yields

$$U_{D0} = \frac{\sigma^2}{6G} \tag{2-6.4}$$

When $\sigma = Y$, the specimen yields and

$$U_{DY} = \frac{Y^2}{6G} \tag{2-6.5}$$

where U_{DY} is the value of U_{D0} at $\sigma = Y$. Accordingly, by Eqs. (2-6.3) and (2-6.5), the yield condition is taken as

$$\frac{1}{12G}\left[(\sigma_1 - \sigma_2)^2 + (\sigma_2 - \sigma_3)^2 + (\sigma_3 - \sigma_1)^2\right] = \frac{Y^2}{6G} \tag{2-6.6}$$

Furthermore, recalling that

$$9\tau_{\text{oct(max)}}^2 = (\sigma_1 - \sigma_2)^2 + (\sigma_2 - \sigma_3)^2 + (\sigma_3 - \sigma_1)^2 \tag{1-4.8}$$

we note that Eqs. (2-6.6) and (1-4.8) determine the value of $\tau_{\text{oct(max)}}$ at yield; namely,

$$\tau_{\text{oct(max)}} = \frac{\sqrt{2}}{3}Y = 0.471Y \tag{2-6.6a}$$

In other words, the energy of distortion yield condition may be expressed either by Eq. (2-6.6) or by Eq. (2-6.6a). For this reason, the energy of distortion yield criterion is commonly referred to as the maximum octahedral shearing stress yield criterion. Since U_{D0} is independent of the sign of the uniaxial load (tension or compression) the maximum octahedral shearing stress yield condition is assumed valid for tension or compression loading.

Although the physical basis for the von Mises-Hencky theory is vague, its use leads to a *workable* theory of plasticity. Physically, the von Mises-Hencky theory denotes that the octahedral shearing stress reaches the critical value $0.471Y$ at yield.

Graphical Interpretation of Yield Criteria. Yield Surface/Since the maximum shearing stress condition and the maximum octahedral shearing stress condition are representable in terms of principal stresses $(\sigma_1, \sigma_2, \sigma_3)$, it is logical to consider graphical representations of the yield criteria relative to principal stress axes. More explicitly, we consider a *stress space* $(\sigma_1, \sigma_2, \sigma_3)$ with rectangular axes $\sigma_1, \sigma_2, \sigma_3$ (Fig. 2-6.4). Since the yield criteria (Eqs. 2-6.1 and 2-6.6) are expressible in terms of $(\sigma_1, \sigma_2, \sigma_3)$, the yield condition for a material can be expressed by an equation of the type

$$f(\sigma_1, \sigma_2, \sigma_3, Y) = 0 \qquad (2\text{-}6.7)$$

where Y is a constant for the material.

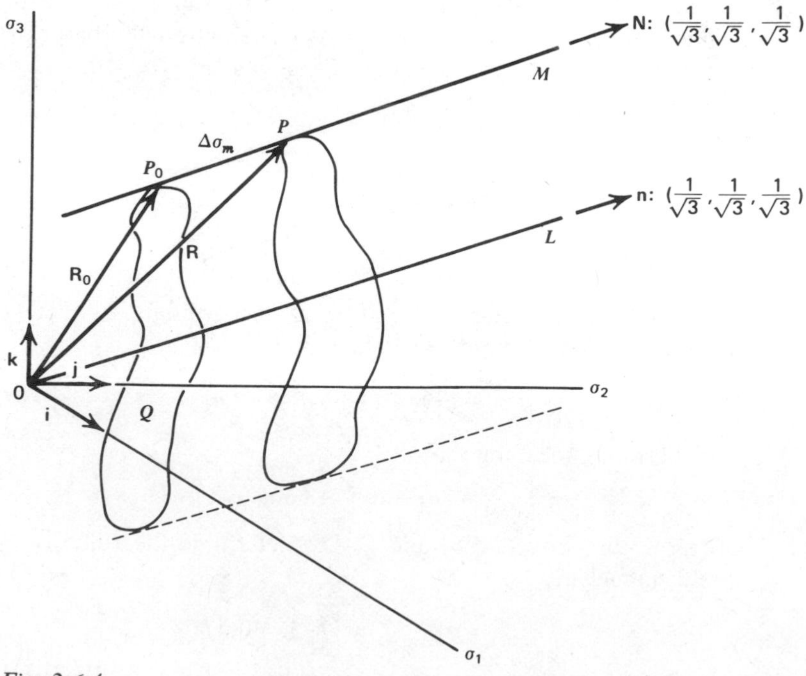

Fig. 2-6.4

Geometrically, Eq. (2-6.7) represents a surface in the stress space, the so-called *yield surface*. Since the yield criteria incorporate the assumption that yield is independent of a hydrostatic state of stress $\sigma_1 = \sigma_2 = \sigma_3 = \sigma$, we note that this condition is represented by the line L, through the origin ($\sigma_1 = \sigma_2 = \sigma_3 = 0$), with direction cosines given by the unit vector $\mathbf{n}: (1/\sqrt{3}, 1/\sqrt{3}, 1/\sqrt{3})$. Alternatively, we may think of line L as representing points of mean stress $\sigma_m = (\sigma_1 + \sigma_2 + \sigma_3)/3 = \sigma$. Hence along line L, the principal values of the stress deviator tensor T_d (Eq. 1-4.16) are $S_1 = S_2 = S_3 = 0$, and yielding cannot occur for values of $(\sigma_1, \sigma_2, \sigma_3)$ that lie on line L. Therefore, for yielding, the stress components $(\sigma_1, \sigma_2, \sigma_3)$ must lie off the line L. Next, consider a point P_0, with radius vector $\mathbf{R}_0$, that lies on line M parallel to L. The direction cosines of M are therefore given by $\mathbf{N}: (1/\sqrt{3}, 1/\sqrt{3}, 1/\sqrt{3})$. The point P_0 lies in a plane Q that is perpendicular to lines L and M and that is defined by the equation

$$\mathbf{R}_0 \cdot \mathbf{N} = \left(\mathbf{i}\sigma_1^0 + \mathbf{j}\sigma_2^0 + \mathbf{k}\sigma_3^0\right) \cdot \left(\frac{\mathbf{i} + \mathbf{j} + \mathbf{k}}{\sqrt{3}}\right) = C$$

where $(\sigma_1^0, \sigma_2^0, \sigma_3^0)$ are the coordinates of point P_0 and

$$C = \frac{1}{\sqrt{3}}\left(\sigma_1^0 + \sigma_2^0 + \sigma_3^0\right) = \sqrt{3}\,\sigma_m^0 \tag{2-6.8}$$

where σ_m^0 is the mean stress at P_0. Hence, Eq. (2-6.8) represents planes in the stress space for which all points in the plane represent states of stress with the same mean stress. If $C = 0$, Eq. (2-6.8) represents a state of zero mean stress and the plane passes through the origin ($\sigma_1 = \sigma_2 = \sigma_3 = 0$) of the stress space.

Any other point $P: (\sigma_1, \sigma_2, \sigma_3)$ on line M lies on a plane at a distance $\Delta\sigma_m$ from point P_0. Hence,

$$\mathbf{R} = \mathbf{R}_0 + \mathbf{N}(\Delta\sigma_m) \tag{2-6.9}$$

where $\mathbf{R}$ is the radius vector to P. In scalar notation, Eq. (2-6.9) yields

$$\sigma_1 = \sigma_1^0 + \frac{1}{\sqrt{3}}\Delta\sigma_m, \qquad \sigma_2 = \sigma_2^0 + \frac{1}{\sqrt{3}}\Delta\sigma_m, \qquad \sigma_3 = \sigma_3^0 + \frac{1}{\sqrt{3}}\Delta\sigma_m \tag{2-6.10}$$

Thus, the stress state $(\sigma_1, \sigma_2, \sigma_3)$ differs from the stress state $(\sigma_1^0, \sigma_2^0, \sigma_3^0)$ only by a factor involving the mean stress increment $\Delta\sigma_m$. Since the yield hypotheses under consideration assume that the mean stress does not affect yielding of a solid, the surface generated by a set of lines M parallel to line L, with each of lines M at a critical distance from L (for example, at a distance equivalent to a critical value of maximum shearing stress) is called a *yield surface*. Since its generators are straight lines (lines M), the yield surface is a cylinder in the stress space $(\sigma_1, \sigma_2, \sigma_3)$.

The generators of the yield surface cut out a closed curve called a *yield locus* in planes Q perpendicular to line L. Since the yield surface is a

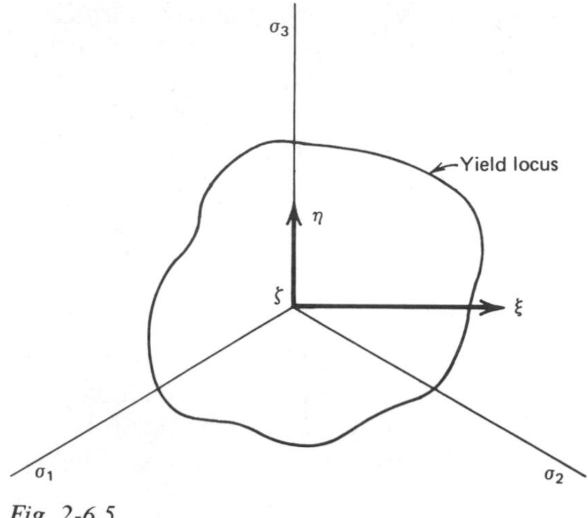

Fig. 2-6.5

cylinder, the shape of the yield locus is the same for all planes Q. Accordingly, it suffices to discuss the yield surface in terms of right-handed rectangular coordinates (ξ, η, ζ) with origin at $\sigma_1 = \sigma_2 = \sigma_3 = 0$ and with coordinates ξ, η in plane Q and ζ perpendicular to plane Q (Fig. 2-6.5). Since generators of the yield surface cut a closed curve (the yield locus) in planes perpendicular to the ζ-axes, they cut the yield locus in the (ξ, η)-plane. When $\sigma_3 = 0$, we note that line L coincides with the σ_3-axis, and the yield locus is traced in the (σ_1, σ_2)-plane (Figs. 2-6.4 and 2-6.5). Consequently, biaxial states of stress will suffice for testing of yield hypotheses. The yield locus will be different for different hypotheses.

Tresca-Saint Venant Yield Hypothesis. The Maximum Shearing-Stress Yield Criterion/For the maximum shearing-stress criterion, the yield locus is a regular hexagon in the (ξ, η)-plane. To prove this result, we define the principal shearing stresses

$$\begin{aligned}\tau_1 &= \tfrac{1}{2}(\sigma_2 - \sigma_3)\\ \tau_2 &= \tfrac{1}{2}(\sigma_3 - \sigma_1)\\ \tau_3 &= \tfrac{1}{2}(\sigma_1 - \sigma_2)\end{aligned} \tag{2-6.11}$$

We note that if Eq. (2-6.1) is valid, Eqs. (2-6.11) give

$$\begin{aligned}\sigma_2 - \sigma_3 &= \pm Y\\ \sigma_3 - \sigma_1 &= \pm Y\\ \sigma_1 - \sigma_2 &= \pm Y\end{aligned} \tag{2-6.12}$$

where Y is the magnitude of the uniaxial yield stress, yield occurs, and the stress state is on the yield surface. Hence, Eqs. (2-6.12) are the equation of the yield locus in $(\sigma_1, \sigma_2, \sigma_3)$ space. To determine the equa-

Table 2-6.1

	σ_1	σ_2	σ_3
ξ	$-\frac{1}{\sqrt{2}}$	$\frac{1}{\sqrt{2}}$	0
η	$-\frac{1}{\sqrt{6}}$	$-\frac{1}{\sqrt{6}}$	$\frac{2}{\sqrt{6}}$
ζ	$\frac{1}{\sqrt{3}}$	$\frac{1}{\sqrt{3}}$	$\frac{1}{\sqrt{3}}$

tion of the yield locus in terms of (ξ, η), we note that

$$\begin{aligned} \sigma_1 &= -\frac{1}{\sqrt{2}}\xi - \frac{1}{\sqrt{6}}\eta + \frac{1}{\sqrt{3}}\zeta \\ \sigma_2 &= \frac{1}{\sqrt{2}}\xi - \frac{1}{\sqrt{6}}\eta + \frac{1}{\sqrt{3}}\zeta \\ \sigma_3 &= \qquad\quad \frac{2}{\sqrt{6}}\eta + \frac{1}{\sqrt{3}}\zeta \end{aligned} \tag{2-6.13}$$

since the direction cosine table is given by Table 2-6.1. Hence, Eqs. (2-6.12) and (2-6.13) yield

$$\begin{aligned} \xi - \sqrt{3}\,\eta &= \pm\sqrt{2}\,Y \\ \xi + \sqrt{3}\,\eta &= \pm\sqrt{2}\,Y \\ \xi \qquad &= \pm\frac{1}{\sqrt{2}}Y \end{aligned} \tag{2-6.14}$$

Equations (2-6.14) represent six straight lines in the (ξ, η)-plane that form a regular hexagon (Fig. 2-6.6).

von Mises-Hencky Yield Hypothesis. Maximum Energy of Distortion Yield Criterion. The Maximum Octahedral Shearing Stress Yield Criterion / For the maximum octahedral shearing stress criterion, the yield locus is a circle, in the (ξ, η)-plane, which circumscribes the regular hexagon yield locus of the maximum shearing stress criterion. To show this result, we note by Eqs. (1-4.8) and (2-6.6a) that

$$(\sigma_1 - \sigma_2)^2 + (\sigma_2 - \sigma_3)^2 + (\sigma_3 - \sigma_1)^2 = 2Y^2 \tag{2-6.15}$$

Then substitution of Eqs. (2-6.13) into Eq. (2-6.15) yields

$$\xi^2 + \eta^2 = \tfrac{2}{3}Y^2 \tag{2-6.16}$$

Equation (2-6.16) represents a circle in the (ξ, η)-plane with radius $a = \sqrt{\frac{2}{3}}\,Y$. This circle (the von Mises circle) circumscribes the hexagon yield

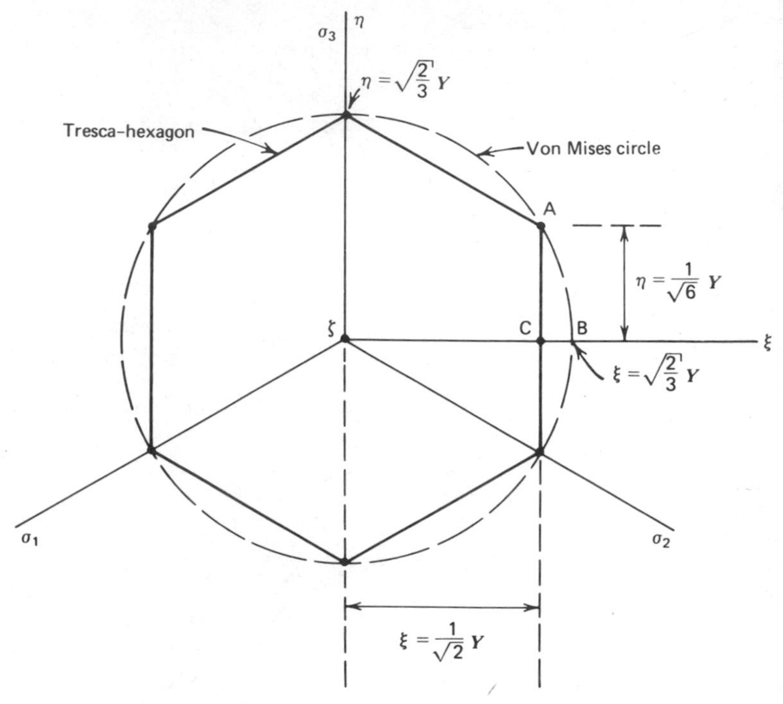

Fig. 2-6.6/Yield locus in (ξ, η) plane.

locus of the maximum shearing stress (Fig. 2-6.6). Relative to the maximum shearing stress yield locus, the largest difference between the two hypotheses is approximately 15.5 percent. To verify this statement, we note that by Eqs. (2-6.13),

$$
\begin{aligned}
\sigma_2 - \sigma_3 &= \frac{1}{\sqrt{2}}\xi - \frac{3}{\sqrt{6}}\eta \\
\sigma_3 - \sigma_1 &= \frac{1}{\sqrt{2}}\xi + \frac{3}{\sqrt{6}}\eta \qquad (2\text{-}6.17) \\
\sigma_1 - \sigma_2 &= -\sqrt{2}\,\xi
\end{aligned}
$$

For point A, Fig. 2-6.6, $\xi = 1/\sqrt{2}\,Y$, $\eta = 1/\sqrt{6}\,Y$. Hence, for point A, Eqs. (2-6.17) gives

$$\sigma_2 - \sigma_3 = 0 \qquad \sigma_3 - \sigma_1 = Y \qquad \sigma_1 - \sigma_2 = -Y \qquad (2\text{-}6.18)$$

Hence, for the maximum shearing stress criterion, Eqs. (2-6.18) yield

$$\tau_{\max} = \frac{1}{2}(\sigma_3 - \sigma_1) = \frac{Y}{2} \qquad (2\text{-}6.19)$$

For the maximum octahedral shearing stress criterion of yield, Eq. (2-6.15) must be satisfied. Substitution of Eqs. (2-6.18) into Eq. (2-6.15) yields an identity. Consequently, for the maximum octahedral shearing stress criterion, Eqs. (2-6.18) also yield $\tau_{\max} = Y/2$ for point A.

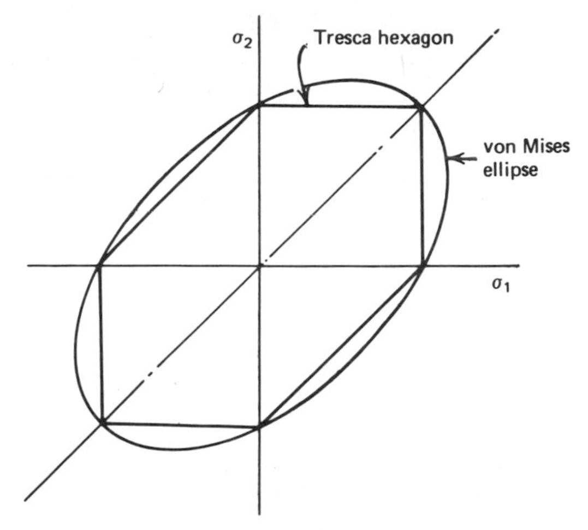

Fig. 2-6.7

For point B (Fig. 2-6.6) $\xi = \sqrt{\frac{2}{3}}\,Y$, $\eta = 0$, and Eqs. (2-6.17) yield

$$\sigma_2 - \sigma_3 = \frac{1}{\sqrt{3}}Y \qquad \sigma_3 - \sigma_1 = \frac{1}{\sqrt{3}}Y \qquad \sigma_1 - \sigma_2 = -\frac{2}{\sqrt{3}}Y \tag{2-6.20}$$

The maximum shearing stress criterion predicts that yield will occur at point C, before the stress level reaches point B, at a value $\tau_{max} = Y/2$; see Eqs. (2-6.17) with $\xi = 1/\sqrt{2}\,Y$, $\eta = 0$. On the other hand, the maximum octahedral shearing stress criterion predicts that yield occurs at point B [where $\tau_{oct(max)} = (\sqrt{2}/3)Y$]. Consequently, by Eqs. (2-6.20), the maximum octahedral shearing stress criterion predicts a corresponding maximum shearing stress at yield of

$$\tau_{max} = \frac{1}{2}(\sigma_1 - \sigma_2) = \frac{1}{\sqrt{3}}Y = 0.577Y \tag{2-6.21}$$

In summary, for points A and C the maximum shearing stress criterion predicts $\tau_{max} = \frac{1}{2}Y$ at yield; whereas, the von Mises criterion predicts $\tau_{max} = \frac{1}{2}Y$ for point A and $\tau_{max} = (1/\sqrt{3})Y = 0.577Y$ for point B.

For the case $\sigma_3 = 0$, the yield locus in the (σ_1, σ_2)-plane becomes a nonregular hexagon for the maximum shearing stress criterion and the corresponding circumscribing ellipse for the maximum octahedral shearing stress criterion (Fig. 2-6.7).

REFERENCES

1. A. E. Green and J. E. Adkins, *Large Elastic Deformations*, Oxford University Press, London, 1960.

2. H. L. Langhaar, *Energy Methods in Applied Mechanics*, Wiley, New York, 1965.
3. A. P. Boresi and P. P. Lynn, *Elasticity in Engineering Mechanics*, 2nd Ed., Prentice-Hall, Englewood Cliffs, New Jersey, 1974.
4. J. O. Smith and O. M. Sidebottom, *Inelastic Behavior of Load-Carrying Members*, Wiley, New York, 1965.
5. J. F. Nye, *Physical Properties of Crystals*, Oxford University Press, London, 1957.
6. H. Jeffreys, *Cartesian Tensors*, Cambridge University Press, London, 1957.
7. A. Mendelson, *Plasticity: Theory and Applications*, Macmillan, New York, 1968.

Additional References

1. P. P. Benham and R. Hoyle, Eds., *Thermal Stress*, Pitman, London, 1964.
2. B. A. Boley and J. H. Weiner, *Theory of Thermal Stresses*, Wiley, New York, 1962.
3. D. J. Johns, *Thermal Stress Analysis*, Pergamon Press, Oxford, England, 1965.
4. A. D. Kovalenko, *Thermoelasticity*, Wolters-Noordhoff, Groningen, The Netherlands, 1969.
5. S. S. Manson, *Thermal Stress and Low-Cycle Fatigue*, McGraw-Hill, New York, 1966.
6. H. Parkus, *Thermoelasticity*, Blaisdell, Waltham, Massachusetts, 1968.

CHAPTER 3

FAILURE CRITERIA

To design a structural part or system to perform a given function, the designer must have a clear understanding of the possible ways or modes by which the part or system may fail to perform the function. In other words, the designer must determine possible *modes of failure* of the system and then establish suitable *failure criteria* that accurately predict the various modes of failure. In general, the determination of modes of failure requires extensive knowledge of the response of a structural system to loads. In particular, it requires a comprehensive stress analysis of the system. Since the response of a structural system depends strongly on the material used, the mode of failure depends strongly on the type of material. In turn, the mode of failure of a given material also depends on the manner or the history of loading—for example, slowly, rapidly, repeatedly applied and removed, and repeatedly reversed (for instance cyclically repeated tension and compression), etc. Accordingly, suitable failure criteria must include effects for different materials, different loading procedures, as well as factors that influence the stress distribution (for example, supports and cracks) in the member.

A major part of this book is concerned with (1) stress analysis, (2) the behavior of various materials subjected to stress and strain, and (3) the theoretical determination of the relationship between appropriate values of stress, strain, displacement, loads, number of repetitions of load, etc. and the mode of failure. The specific mode of failure is related to a significant (critical) value of one of the quantities (for example, load, stress, and strain) associated with failure by an appropriate failure criterion. In addition, attention is devoted to the significance of relations between critical values, determined theoretically, and values determined by experiments on, or by experience with, the response of the structural part or system to loads. In particular, the establishment of *factors of safety* is examined. For example, let P_f be a theoretical critical (failure) load associated with a critical value of a significant quantity (say the maximum shearing stress in the member) for a specific mode of failure (say yielding). Let P_w be a safe working load determined on the basis of experiments or experience with similar members made of the same material under the same loading conditions. Then, the *factor of safety SF* is defined by

$$SF = \frac{P_f}{P_w} \tag{A}$$

In industrial applications, the magnitude of the factor of safety SF may range from 1 to 3. For example, in aircraft and space vehicle industries, where it is necessary to reduce the weight of the structures as much as possible, the SF may be as low as 1. In the nuclear reactor industries, where safety is of prime importance in the face of many unknown effects, SF may be as high as 5.

If P_f and P_w are each directly proportional to stress, the factor of safety SF may be expressed in terms of equivalent stresses. However, since the usual function of a structural part of a system is to safely support or transfer design loads, the factor of safety should be applied to the loads.

The prediction of working stresses (safe working loads) at complex points of ordinary structures can be attempted only within a rather broad range. If it is absolutely essential to establish working stresses within a narrow range, it is usually necessary to employ experimental stress analysis on the particular system. Ordinarily because of the complexity of full scaled testing, this analysis requires the building of a suitable model and making significant measurements on the model. Structural systems requiring the smallest range of working stress values are optimized usually by a combination of analytical and experimental analyses.

The terms working stresses (loads), allowable stresses (loads), or limit stresses (loads) represent calculated (or experimentally verified) maximum permissible working stresses (loads), agreed upon (and established as a working code) by a consensus of expert authorities in the field of the particular type of structural system. These working stresses are usually referred to simple tests of simple geometric shapes (for instance, tension test of a bar, torsion test of a cylinder, and shear test of a block). The limiting values of stress (working stress) are laid down in national codes and rules and as such are semilegal and binding upon the structural profession.

The concept of factors of safety and its *arbitrary use* may be considered a first rough approach (a so-called first generation design practice) to failure safe design. The use of a factor of safety *in conjunction with a set* of codes and rules has resulted in an enforced, more "strict quality" of design, with a "benefit" of decreasing the factor of safety to lie within a smaller range. As such, this use of factors of safety may be considered a second generation design practice. Unfortunately, dependence on such practices does not assure a safe design, since there are well-documented histories of failure with design safety factors of 5 or higher. Such high safety factors are not acceptable for a number of practical reasons (including economics). Clearly, a failure safe design requires a precise understanding of the actual failure mechanism and the ability to describe the failure mechanism in terms of its relation to structural loads, geometric shapes, and material response to load and to environment, including the effects of the degree of uncertainty in these relations.

In the light of public concern with questions of human safety and of potential environmental impact of failures of large oil tankers, bridges, airplanes, trains, and nuclear power plants, for example, the design engineer of today is forced more and more to work toward a truly failure safe design. At the same time, because of economics, the so-called factor of safety must be kept as small as possible while attaining the failure safe design.

In this chapter, we study modes of failure and their relations to associated failure criteria that hopefully will help the design engineer achieve a failure safe design.

3-1 MODES OF FAILURE

When a structural member is subjected to loads, its response depends not only on the type of material from which it is made, but also on the environmental conditions and the manner of loading. Depending on how the member is loaded, it may fail by *excessive displacement*, which results in the member being unable to perform its design function; it may fail by *plastic deformation* (*yielding*), which may cause a permanent undesirable change in shape; it may fail due to a *fracture* (break), which depending on the material and the nature of loading, may be of a *ductile type* preceded by appreciable plastic deformation or of a *brittle type* with little or no prior plastic deformation. Materials such as glasses, ceramics, rocks, plain concrete, and cast iron are examples of materials (brittle materials) that fracture in a brittle manner under normal environmental conditions and the slow application of tension load. In uniaxial compression, they also fracture in a brittle manner, but the nature of the fracture is quite different from that in tension. Depending on a number of conditions such as environment, rate of load, nature of loading, and presence of cracks or flaws, structural metals may exhibit ductile or brittle fracture.

One type of loading that may result in brittle fracture of ductile metals is that of repeated loads. For example, if a uniaxially loaded bar with smooth surface is subjected to repeated cycles of alternately applied tensile and compressive loads of equal magnitude, it may fail by fracture (usually in a brittle manner for high cycle fatigue) at a stress level considerably below the magnitude of stress that causes failure by fracture under a noncyclic static load. Fracture of a structural member under repeated loads is commonly called *fatigue fracture* or *failure*. Fracture by fatigue may start by the initiation of one or more small cracks, usually in the neighborhood of the maximum critical stress in the member. Repeated cycling of the load causes the crack or cracks to propagate until the structural member is no longer able to carry the load across the cracked region, and the member ruptures.

Another manner in which a structural member may fail is that of elastic or plastic instability. In this failure mode the structural member may undergo large displacements from its design configuration when the applied load reaches a critical value, the so-called *buckling load* (or *instability load*). This type of failure may result in excessive displacement or loss of ability (because of yielding or fracture) to carry the design load. In addition to the above failure modes, a structural member may fail because of environmental corrosion (chemical action).

To elaborate upon the modes of failure of structural members we discuss more fully the following categories of failure modes:

1. Failure by excessive deflection
2. Failure by yielding
   (a) Ordinary (room) temperatures
   (b) Elevated temperatures (creep)
3. Failure by fracture
   (a) Sudden fracture of brittle materials
   (b) Fracture of cracked or flawed members
   (c) Progressive fracture (fatigue)
   (d) Fracture with time at elevated temperatures

These failure modes and their associated failure criteria are most meaningful for simple structural members (for example, tension members, columns, beams, circular cross section torsion members). For more complicated two- and three-dimensional problems, the significance of such simple failure modes is open to question.

Many of these modes of failure for simple structural members are well-known to engineers. However, under unusual conditions of load or environment, new types of failure modes may occur. For example, in nuclear reactor systems, cracks in pipe loops have been attributed to stress-assisted corrosion cracking, with possible side effects attributable to residual welding stresses.[1]

The physical action in a structural member leading to failure is usually a complicated phenomenon, and in the following discussion the phenomena are necessarily over-simplified, but they nevertheless retain the essential features of the failures.

1. Failure by Elastic Deflection/The maximum load that may be applied to a member without causing it to cease to function properly may be limited by the permissible elastic strain or deflection of the member. Elastic deflection that may cause damage to a member can occur under these different conditions:

(a) Deflection under conditions of stable equilibrium, such as the stretch of a tension member, the angle of twist of a shaft, and the deflection of a beam, particularly under gradually applied (so-called static) loads. Elastic deflections, under conditions of equilibrium, are computed in Chapter 4. See also Arts. 3-2 and 3-6.

(b) Buckling, or the rather sudden deflection associated with unstable equilibrium and often resulting in total collapse of the member, which occurs, for example, when an axial load that is applied gradually to a very slender column exceeds slightly the Euler critical load, or when an external fluid pressure is applied to a cylindrical shell or thin-walled pipe that suddently collapses when the pressure reaches a critical value. See Arts. 3-2 and 3-6.

(c) Elastic deflections that are the amplitudes of the vibration of a member sometimes are associated with failure of the member resulting from objectionable noise, shaking forces, collision of moving parts with stationary parts, etc., which result from the vibrations.

When a member fails by elastic deformation, the significant equations for design are those that relate loads and elastic deflection. For example, the equations for the three members mentioned under (a) are: $e = PL/AE$, $\theta = TL/GJ$ and $\delta = \alpha(WL^3/EI)$ in which α is a constant for a given beam and given location in the beam depending on the type of supports and the type of loading. It will be noted that these equations contain the significant property of the material involved in the elastic deflection, namely the modulus of elasticity E (sometimes called the stiffness) or $G = E/[2(1 + \nu)]$. The stresses set up by the loads are not the significant quantities; that is, the stresses do not limit the loads that can be applied to the member without causing structural damage and, hence, the strength properties of the material (e.g., yield stress) are not of primary importance. In different words, if a member of given dimensions fails to perform its load resisting function because of excessive elastic deflection, its load carrying capacity is not increased by making the member of stronger material, but rather by making it stiffer by using a material with a higher modulus of elasticity or by changing the shape and dimensions of the member. As a rule, however, the most effective method of decreasing the deflection of a member is by changing the shape or increasing the dimensions of its cross section, rather than by making the member of a stiffer material; moreover, if a member is made of steel, its stiffness cannot be increased greatly by substituting another material, since steel is one or the stiffest of structural materials available.

2. Failure by Extensive Yielding/Another condition that may cause a member to fail is inelastic (plastic) deformation of a considerable portion of the member, denoted by extensive yielding to differentiate it from (localized) yielding of a very small portion of a member. Again, we note that these modes of failure are most significant with regard to simple structural members such as axially loaded members, beams, torsion members, columns, or possibly thin sheets or plates subject to inplane forces. These modes of failure may also be applied fairly directly to pressure vessels, and pipes, for example. Extensive yielding of a metal member may result from either one of two different conditions, depending on whether the existing temperature of the member is above or below the recrystallization temperature of the metal. Temperatures above the recrystallization temperature such as those to which steel may be subjected in steam turbines, oil-cracking apparatus, etc., are frequently referred to as *elevated* temperatures, and temperatures below the recrystallization temperature as *ordinary* temperatures or *room* temperatures.

(a) Extensive Yielding at Ordinary Temperatures/Polycrystalline metals are made up of extremely large numbers of very small units called crystals or grains. The crystals have slip planes on which the resistance to shearing stress is relatively small. During yielding in which crystal defects called dislocations[2] move in the slip planes, one part of the crystal moves relative to the other part. General or extensive yielding of a metal is the summation of these very minute shearing deformations or slips in an extremely large number of crystals. Slip planes in adjacent crystals are not likely to be parallel to one another, but rather are oriented randomly throughout the metal. Hence, the yield strength of a metal is a statistical value representing the strength and strengthening effect of a large number of crystals.

After yielding has occurred in some crystals at a given load, these crystals will not yield further in the same direction without an increase in load. During plastic straining, dislocation entanglements are formed that make the motion of dislocations more and more difficult. This is especially true in polycrystalline metals since intersecting slip planes are more likely to be produced in the individual grains, and dislocation entanglements are more likely to occur at the intersections of the slip planes. A higher and higher stress will be required to push new dislocations through these entanglements. This increased resistance that accompanies yielding is called *strain hardening*, *work hardening*, *cold working*, or *strain strengthening*. This strain hardening is permanent (expect for time effects such as ageing and recovery) provided that the temperature at which it occurs is below the recrystallization temperature for the metal. Hence, below the recrystallization temperature yielding does *not* continue under the same load. However, when the temperature is above the recrystallization temperature (discussed further under the subsection "Extensive Yielding at Elevated Temperatures"), yielding continues unabated.

There is considerable but not necessarily conclusive evidence, therefore, to support the assumption that when a metal member fails by general yielding at *ordinary temperatures*, the significant quantity associated with the failure is shearing stress.[†] This stress is the shearing yield point stress if the material possesses a yield point, or it is the shearing yield stress (stress corresponding to an arbitrary offset) if the material does not exhibit a yield point. In certain failure criteria, the *tensile* stress is considered to be the significant quantity, and its maximum allowable value is taken as the *tensile* yield point or *tensile* yield stress Y.

When failure occurs by extensive yielding, stress concentrations usually are *not* significant because of the interaction and adjustment that take place between crystals in the regions of stress concentrations. Slip in a few weak or poorly oriented crystals or highly stressed crystals does not limit the general load carrying capacity of the member, but merely causes readjustment of stresses that permit the stronger or less stressed crystals to take higher stresses and thereby causes the stress distribution to approach more closely the distribution that occurs in a member free from stress concentrations. Thus the member as a whole acts substantially as does an ideal homogeneous elastic member free from abrupt changes of section, until slip in a very large number of crystals occurs, resulting in failure by general yielding.

The action or behavior that leads to this type of failure is, therefore, a statistical action of a large number of the structural units (crystals) of which the member is made, as opposed to local or individual action or behavior. The laws governing this statistical action are substantially the same as are assumed for ideal homogeneous material. Failure criteria for general yielding failure, presented in Art. 3-3, require nominal stress values calculated from common elementary equations or formulas in mechanics of materials. These formulas assume that the material in the member is linearly elastic, isotropic, homogeneous, and ductile, and that the member is free of residual stresses and stress concentrations. Furthermore, the member is assumed to be subjected to static loads at ordinary uniform temperature.

It is important to observe, however, that if a member that fails by yielding, as discussed in the foregoing paragraph, is replaced by one

[†]Other criteria of failure, such as octahedral shearing stress, have also been propounded (see "Failure Criteria," Art. 3-3).

made of a stronger material (higher yield stress Y) with a resulting change of dimensions of the member, the mode of failure may change completely to that of elastic deflection or buckling or to failure by mechanical vibrations, etc. Hence, the entire basis of design may be changed when conditions are altered to prevent a given mode of failure.

(b) Extensive Yielding at Elevated Temperatures. Creep/If a metal is subjected to loads at a temperature *above* the recrystallization temperature, the increase in strength (strain hardening) that accompanies slip[†] in the crystals is not permanent. The recrystallization temperature is the temperature above which crystals that have slipped reform themselves into unstrained crystals. Therefore the increase in strength accompanying slip is soon offset or neutralized by the annealing action at the elevated temperature. Hence, *continuing* deformation or yielding, called creep, occurs at the same load at which yielding started.

The procedure by which strain hardening is modified by elevated temperatures is thought to occur mainly as follows: Along each slip plane in a crystal, crystal defects (for example, dislocation entanglements) are produced. When the temperature is above a certain value, these crystal defects are reduced. As a result, the crystal structure more closely resembles the atomic structure of the parent crystal on either side of the slip plane so that the crystal tends to assume the same condition or state that existed before slip took place. Then further slip under the same load occurs, etc. In time, sufficient creep at elevated temperature may occur to cause failure at stresses that would be considered very low and safe for the material to resist at ordinary temperature.

The quantity associated with the failure of a metal by creep at elevated temperature is usually considered to be stress. However, the relation between loads and stresses in members at elevated temperatures is more complex and difficult to express mathematically than at ordinary temperatures, except perhaps for members subjected to axial tensile loads. For example, the formula $\sigma = Mc/I$ does not give the stress in a beam subjected to loads that cause creep at elevated temperatures because the stress in the beam is not linearly proportional to the corresponding strain.

The maximum utilizable strength of the material at a given elevated temperature—called the creep strength— is usually considered to be the stress corresponding to a given amount of creep in a given time (such as for instance 1 percent strain in 10,000 hours).

[†] It is assumed that the stress is sufficiently large to cause deformation mainly by slip within the crystals. At lower stresses creep (continuing deformation at constant load) may be the result of viscous flow of the unordered, disorganized material in the crystal boundaries. For creep at low stress levels, the failure is usually by brittle fracture and not general yielding.

3. Failure by Fracture / Some members cease to function satisfactorily because they break (fracture) before either excessive elastic deflection or extensive yielding occurs. Four rather different modes or mechanisms of fracture occur especially in a metal as described briefly below.

(a) Sudden Fracture of Brittle Material / Some materials—so-called brittle materials—function satisfactorily in resisting loads under static conditions until the material breaks rather suddenly with little or no evidence of plastic deformation. Ordinarily, the tensile stress in members made of such materials is considered to be the significant quantity associated with the failure, and the tensile static ultimate strength σ_u is taken as the measure of the maximum utilizable strength of the material.

(b) Fracture of Cracked or Flawed Members / A member made of *ductile* material and subjected to uniaxial stress rarely fractures under static loads because structural damage (failure) usually occurs by general yielding before fracture takes place. However, at regions of abrupt changes in section and at edges of defects, for example where the distribution of stress is nonuniform and the state of stress is more generally triaxial, failure of the member sometimes occurs as a brittle fracture, especially when subzero temperatures are encountered, even though the material is classed as ductile and the member is subjected to static loads. If high stress concentrations are present, the tendency toward brittle fracture is greatly increased; the tendency is further increased under a combination of impact loads and subzero temperatures.

Recently, fracture criteria have been based on certain concepts introduced into fracture mechanics analysis (see Art. 3-4 and Chapter 12).

(c) Progressive Fracture / If a metal that ordinarily fails by general yielding under a static load is subjected to repeated cycles of completely reversed stress, it may fail by fracture without visual evidence of yielding, provided that the repeated stress is greater than a value called the *fatigue strength*. Under such conditions, minute cracks start at one or more points in the member, usually at points of high *localized* stress such as at abrupt changes in section, and gradually spread by fracture of the material at the edges of the cracks where the stress is highly concentrated. The *progressive fracture* continues until the member finally breaks. This mode of failure is usually called a *fatigue failure*, but it is better designated as *failure by progressive fracture* resulting from repeated loads.

The quantity that is usually considered most significant in failure by progressive fracture is *localized* tensile stress (although the fatigue crack sometimes occurs on the plane of maximum shearing stress), and the maximum utilizable strength of the material is considered to be the stress (*fatigue strength*) corresponding to a given "life" (number of repetitions of stress). If the material has an *endurance limit* (fatigue strength for infinite life) and a design for so-called infinite life is desired, then the endurance limit is the limiting resistance value or maximum utilizable strength of the material (Art. 3-5).

(d) Fracture with Time at Elevated Temperatures/A fourth type of fracture of materials (mostly metals) may occur at elevated temperatures under a static load that is applied for a long period of time. The material separates with very little evidence of yielding mainly because of viscous flow of the unordered material in the grain boundaries of the metal.

3-2
FAILURE CRITERIA. EXCESSIVE DEFLECTIONS

Small Deflections/Deflections of systems or members of systems for linearly elastic conditions and small displacements are considered in Chapter 4. Often these deflections are limited by design considerations (for instance, machine tools). The failure loads for such members are easily calculated. They are the loads required to produce the limiting deflections.

Large Deflections: Elastic Stability (Buckling) and Jamming/When an initially straight, slender column with pivot ends is subjected to a large compressive force, theoretically, failure may occur by elastic buckling when the load exceeds the critical (buckling) load (Figs. 3-2.1 and 3-2.2),

$$P_{cr} = \frac{\pi^2 EI}{L^2} \tag{3-2.1}$$

where E is the modulus of elasticity, I is the moment of inertia of the cross sectional area about the neutral axis of bending and L is the length of the pivot end column. The load-end deflection curve $0C$ for the column (ideal column) is shown in Fig. 3-2.2. When the load P reaches the value P_{cr}, two possibilities exist. If the column is perfectly straight, if the load P lies exactly along the central longitudinal axis of the column, and if the column is subjected to no lateral force, the load P may be increased along line $0C$. More realistically, however, in practice, the line of action of P does not lie *exactly* along the central axis of the column and the column is not exactly straight. Hence, in general the force P produces a bending moment in the column. In other words, most columns in practice perform their load carrying function as a beam-column.[3]

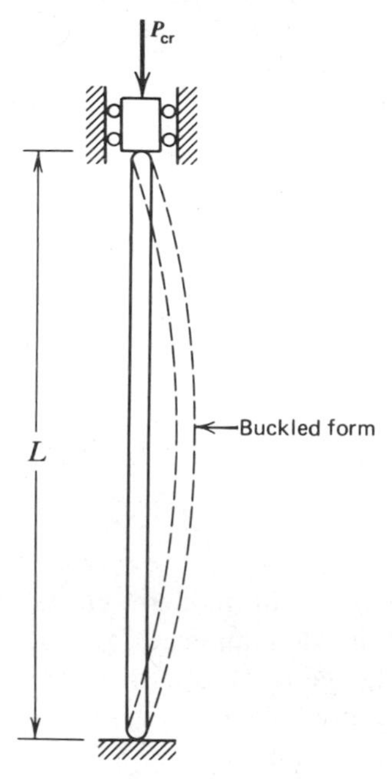

Fig. 3-2.1/Pivot end column.

For an ideal column, for sufficiently small values of P, the elastic restoring forces in the column are sufficient to restore the column to its equilibrium position (straight position), when given a slight lateral displacement. Thus, for sufficiently small values of P, the member is said to be in a stable equilibrium state. As the value of P increases, a load is reached for which the elastic restoring force is just large enough to maintain equilibrium in a displaced state, but it is not capable of returning the column to the original (straight) equilibrium position. (Then $P = P_{cr}$.) For values of P greater than P_{cr}, the slightest lateral movement of the column from its straight position results in a catastrophic (large) displacement that culminates in either yielding or fracture, or in excessive deflection that exceeds the clearance tolerances of the member.† As seen by Eq. (3-2.1), the critical value P_{cr} depends on the moment of inertia I of the cross section, on the elastic modulus E, and on the value of L. For structural materials, the value of E may vary by a factor of as great as 15 or more (e.g., between steel and concrete). However for

†If the column remains elastic, its ends may meet (pin-end column, Fig. 3-2.1) or the column may be inverted (fixed-free column, Art. 3-6). The problem of determining the deflections of such a column is called the elastica problem. More frequently, the member will exceed clearance tolerances with respect to other parts of the system of which it is a part; that is, it will *jam*. In general the problem of *jamming* may also occur in systems with small tolerances, such as required in electrical motors and generators, and internal combustion engines.

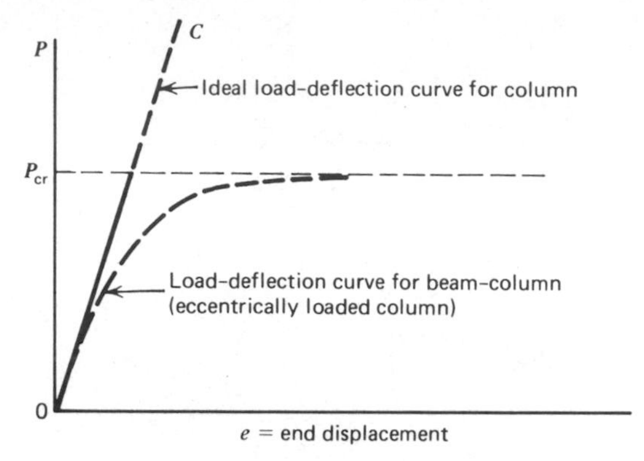

Fig. 3-2.2

structural steels used in modern steel construction E is essentially constant. Hence, in such application, control of the critical value of P is maintained by the value of L and the moment of inertia I of the cross section; the magnitude of P_{cr} is greatly influenced by the nature of the end supports for columns without pivot ends (Art. 3-6). For given end conditions, buckling control of slender columns is maintained most frequently by controlling L or by designing the column cross section to maximize I. To maximize I, box-girder cross sections are commonly employed in column design.[3]

If therefore a slender column is subjected to *any* lateral disturbance when the value of P exceeds P_{cr}, it undergoes large displacements. Hence, since most materials cannot withstand the associated strains without yielding or breaking, failure may eventually occur by complete *plastic collapse* resulting from extensive yielding or by fracture.

In summary, failure by deflection may be attributed to three causes: (1) Excessive stable elastic deflections. (2) Excessive unstable (buckling) deflections, which initially are elastic but may become inelastic with continued deflection under load. (3) Excessive deflections because of plastic collapse. In Chapter 4, we consider the calculations of stable elastic deflections for simple members and framework structures. In Art. 3-6, we treat the problem of instability (buckling).

3-3 FAILURE CRITERIA. YIELD INITIATION. EXTENSIVE YIELD

Over the years, at least six failure criteria have been proposed to predict failure of members; these criteria are mainly for modes of failure that involve one of the following: general yielding of members made of ductile materials, brittle fracture of members made of brittle materials, and brittle fatigue fracture of members that are made of ductile materials and that are subjected to high cycle fatigue loading (more than 10^6 cycles). The appropriate failure criterion to be used in a given design situation depends on the mode of failure. In this article, we evaluate six failure criteria for the general yielding mode of failure.

Before discussing specific failure criteria, we define what is meant by the general yielding mode of failure. We consider members made of ductile materials and subjected to static loads. In this book, we define the failure load for the general yielding mode of failure as the load for which the load-deflection curve for the member becomes nonlinear. Hence, since the effect of stress concentrations on the overall shape of the load-deflection curve is small,[3] we consider initiation of yielding to be caused by a load that produces a critical value of the important nominal stress associated with yielding, without regard to stress concentration effects.

Our definition of failure load for the general yielding mode of failure is considered, by some engineers, to be the lower bound load for the general yielding mode of failure. For members made of elastic-perfectly plastic materials (Fig. 2-6.2a), these engineers consider the upper bound load for the general yielding mode of failure to be the fully plastic load. For example, consider the nondimensional load-deflection curves for two simple structural members (Fig. 3-3.1).

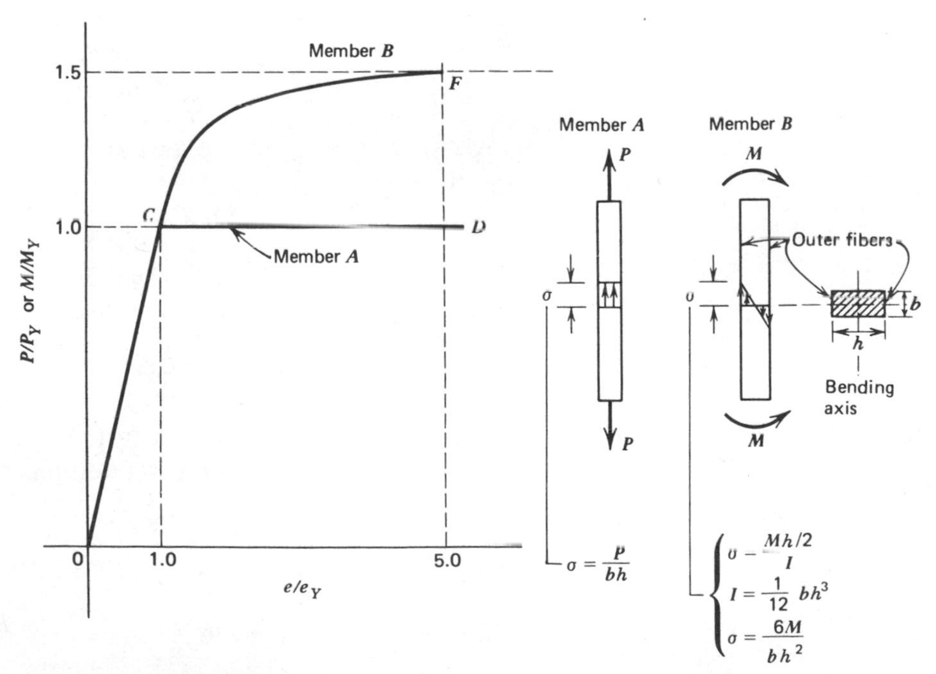

Fig. 3-3.1/Nondimensional load-strain curves for members A and B.

Member A is an axially loaded tension member and member B is a beam with rectangular cross section subjected to pure bending. The members are made of the same material which, for simplicity, we take to be elastic-perfectly plastic. For member A, let P_Y denote the maximum elastic load, that is, the load that causes the stress at some point in the member to reach the yield point stress Y. For member B, let M_Y denote the maximum elastic moment, that is, the moment for which a point on the outer fiber of the beam reaches the yield point stress Y. Thus, for members of rectangular cross section of width b and depth h, loads P_Y and M_Y are given by

$$P_Y = Ybh \quad \text{and} \quad M_Y = Y\frac{bh^2}{6} \tag{3-3.1}$$

These loads are defined in this book as the failure loads for the general yielding mode of failure. Now consider the fully plastic loads for these members, since some engineers use fully plastic loads as failure loads for the general yielding mode of failure.

Let e_Y be the elongation of member A when $P = P_Y$, and let e_Y be the elongation of the outer fibers of the beam (member B) when $M = M_Y$. Then the load-deflection curve for member A is $0CD$ and for member B is $0CF$. The load-deflection curves have been extended to $e/e_Y = 5$, although most structural members would have been considered to have failed prior to the occurrence of such large elongations.

Since member A has a uniform distribution of stress over its cross section, and since the material is elastic-perfectly plastic, member A cannot support any load beyond P_Y. Thus, the load P_Y that initiates yielding does not increase, and the fully plastic load P_P (the load at which the entire cross section has yielded) is equal to P_Y. That is, for member A, the fully plastic load P_P is

$$P_P = P_Y \tag{3-3.2}$$

The behavior of member B is considerably different from that of member A. Since $M = M_Y$ produces yielding only on the outer fibers of the beam, the beam is capable of supporting considerably larger values of M, since beam fibers in the interior have not yielded. Thus, the load-deflection curve $0CF$ for member B continues to rise (elastic-plastic beam[4]) until all fibers across the beam cross section have yielded (fully-plastic beam), at which instant, $M = M_P$. This value M_P is greater than the value M_Y. In particular,

$$M_P = Y\frac{bh^2}{4} = 1.5M_Y \tag{3-3.3}$$

Thus, we see that for member A, initiation of yielding ($P = P_Y$) and extensive yielding (fully plastic load $P = P_P$) occur at the same load $P = P_Y = P_P$. However, for member B, initiation of yielding occurs for

$M = M_Y$ and extensive yielding occurs for $M = M_P = 1.5M_Y$. For materials in a ductile state, the effect of stress concentrations have little influence on these values of load.

In the following failure criteria, we assume that failure occurs by yielding when yielding is initiated at some point in a simple structural member. Thus, for the simple members of Fig. 3-3.1, failure occurs by yielding for loads corresponding to point C on the load-deflection diagrams; that is, for $P = P_Y$ and $M = M_Y$. However, we note that current practices of plastic collapse design also employ the criterion of a fully plastic load. Then for member B, $M = M_P = 1.5M_Y$ defines the load at failure. Therefore, in this book, we also, upon occasion, calculate fully plastic loads for simple structural members. We remark, however, that the concept of fully plastic load loses its meaning when the material in the member is capable of strain-hardening (Fig. 2-6.2b). For materials that strain harden (Fig. 2-1.1a), a load corresponding to the fully plastic load is often calculated. Assume the stress-strain curve to be flat topped at the yield stress Y; the resulting fully plastic load will be less than the load required to cause $e/e_Y = 5$ (Fig. 3-3.1) as long as the offset strain ϵ_s is appreciably less than $5\epsilon_Y$.

We noted in Art. 2-6 that a large volume of experimental evidence indicates that the maximum shearing stress and maximum octahedral shearing stress criteria of failure are found to be valid for the general yielding mode of failure. (A general yielding failure criterion and corresponding yield condition (Art. 2-6) are synonymous when the general yielding failure criterion predicts loads to initiate yielding.) These two along with four other criteria of failure will now be evaluated first for a tension specimen (uniaxial state of stress) and then for members having other states of stress. A typical tensile stress-strain curve for a specimen of ductile steel as obtained from a tension test is shown in Fig. 3-3.2. When the specimen starts to yield, the following six quantities attain so-called critical values at the same load P_Y.

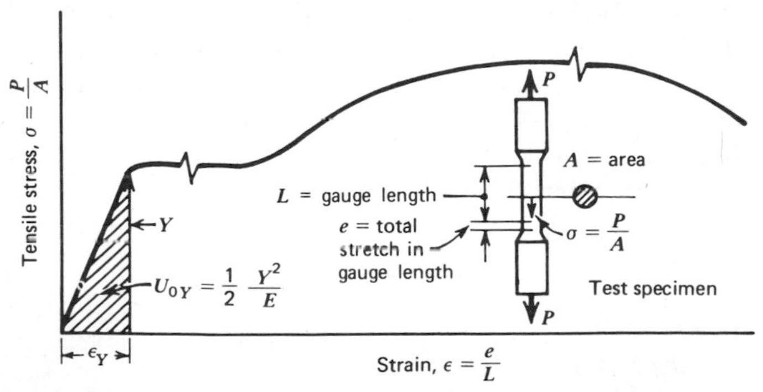

Fig. 3-3.2/Typical stress-strain diagram for ductile steel.

1. The maximum principal stress ($\sigma_{max} = P_Y/A$) reaches the tensile elastic strength (yield stress or yield point) Y of the material.
2. The maximum shearing stress ($\tau_{max} = P_Y/2A$) reaches the shearing yield stress or yield point τ_Y of the material, $\tau_Y = Y/2$ (see Eq. 2-6.1).
3. The maximum tensile strain ϵ_{max} reaches the value ϵ_Y.
4. The strain energy density U_0 absorbed by the material per unit volume reaches the value $U_{0Y} = Y^2/2E$ (see Eq. 2-4.10, with $\sigma_{xx} = Y$; all other stresses zero).
5. The strain energy density of distortion U_{D0} (energy accompanying a change in shape) absorbed by the material reaches a value $U_{DY} = Y^2/6G$ (see Eq. 2-6.5).
6. The maximum octahedral shearing stress reaches the value $\tau_{oct(max)} = \sqrt{2}\,Y/3 = 0.471Y$ (see Eq. 2-6.6a).

These six quantities are summarized in Table 3-3.1, as determined by a simple tension test.

The six critical values (Table 3-3.1) occur simultaneously in a tensile specimen, in which the state of stress is uniaxial. Hence, it is impossible to determine from a tension test which one of the quantities is the cause of the beginning of inelastic action. If, however, the state of stress is biaxial or triaxial, the six quantities do not occur simultaneously at a given load P_Y. It is a matter of considerable importance in design as to which one of the quantities is assumed to limit the loads that can be applied to a member without causing inelastic action. The six limiting quantities as given in Table 3-3.1 suggest six criteria of failure or six different methods for using data obtained in the tension test to predict inelastic action when the state of stress in the member is not uniaxial.

Failure Criteria for Yielding/The six quantities listed in Table 3-3.1 suggest the following six criteria of failure for a material that fails by yielding under static loading.

1. The *maximum principal stress* criterion, often called Rankine's criterion. It states that inelastic action at any point in a member at which *any state of stress* exists *begins* when the maximum principal stress at the point reaches a value equal to the tensile (or compressive) yield stress Y of the material as found in a simple tension (or compression) test, regardless of the normal or shearing stresses that occur on other planes through the point. Thus, according to this criterion, if the block in Fig. 3-3.3*a* reaches its yield stress when subjected to the stress σ_1, the yield stress will still be σ_1 even if the block is subjected to the stress σ_2 (Fig. 3-3.3*b*) in addition to σ_1.

Table 3-3.1

Quantity	*Critical Value in Terms of Tension Test*
1. Maximum principal stress	
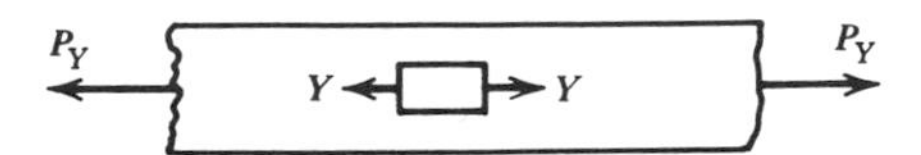	$Y = P_Y/A$
2. Maximum shearing stress	
	$\tau_Y = P_Y/2A = Y/2$
3. Maximum strain	
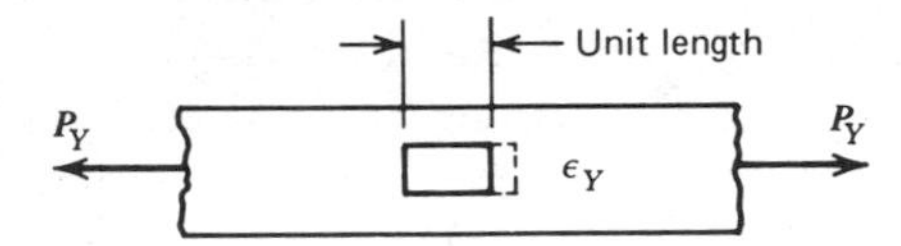	$\epsilon_Y = Y/E$
4. Strain energy density	
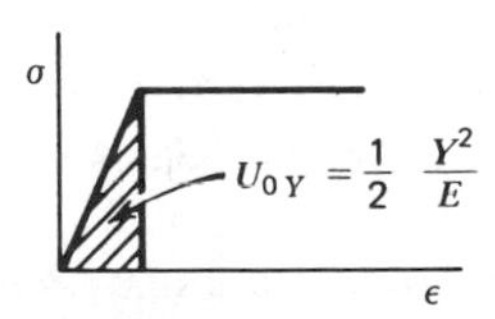	$U_{0Y} = \frac{1}{2}(Y^2/E)$
5. Strain energy density of distortion	$U_{DY} = \dfrac{Y^2}{6G}, \quad G = \dfrac{E}{2(1+\nu)}$
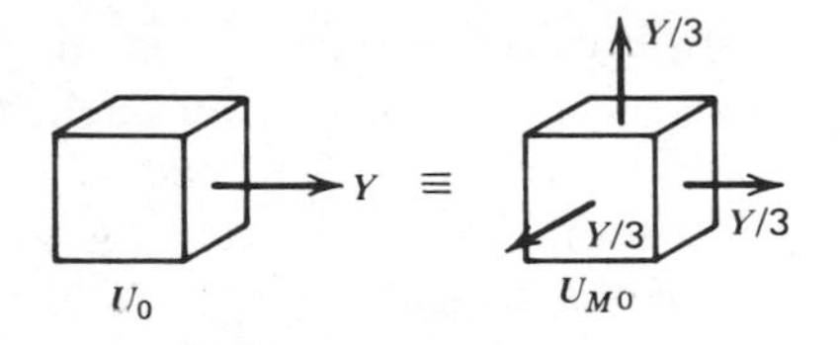 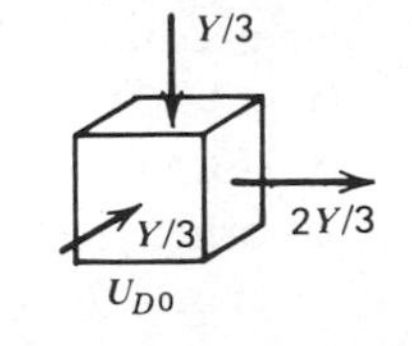	
6. Octahedral shearing stress	$\tau_{oct} = (\sqrt{2}/3)Y = 0.471Y$
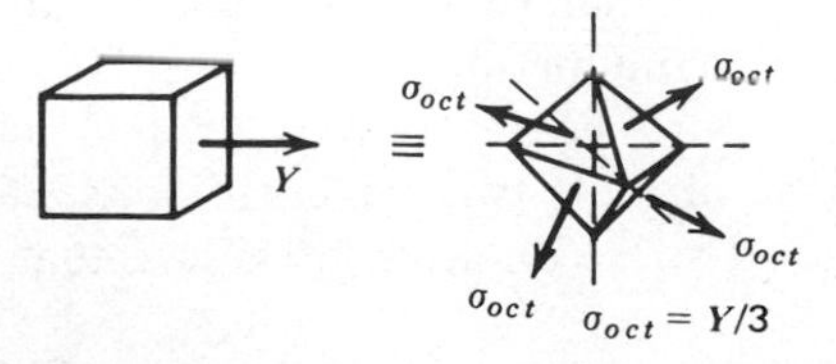 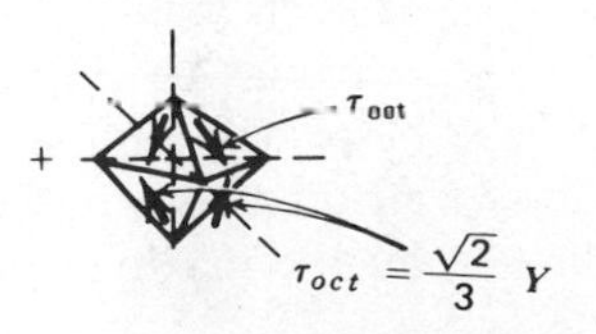	

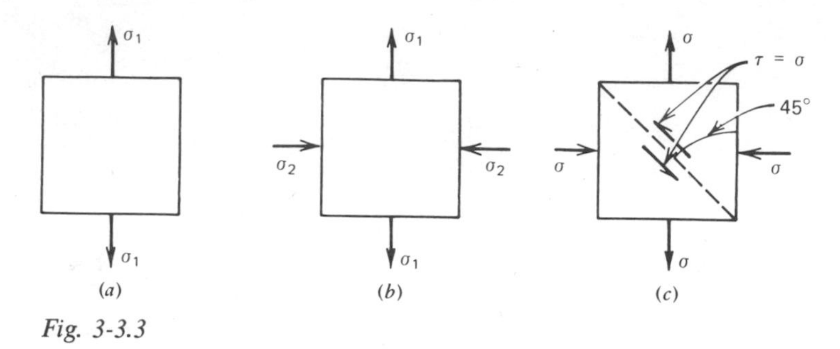

Fig. 3-3.3

If σ_1 and σ_2 are equal to σ in magnitude but are of opposite sign, shearing stress τ equal to σ in magnitude is developed on 45° diagonal planes as shown in Fig. 3-3.3*c*. A state of stress like that shown on the 45° plane in Fig. 3-3.3*c* occurs in a cylindrical bar subjected to pure torsion. Thus, if the maximum principal stress criterion is to hold for all kinds of loading, the shearing yield stress τ_Y of the material must be equal to the tensile yield stress Y. But for ductile metals the shearing yield stress τ_Y as found from the torsion test is much less than the tensile yield stress Y as found from the tension test. It is evident, therefore, that for ductile materials the maximum principal stress criterion of failure is limited. For brittle materials that *do not fail by yielding but fail by brittle fracture*, the maximum principal stress criterion may be satisfactory.

2. The *maximum shearing stress* criterion, sometimes called Tresca's criterion, Coulomb's criterion, or Guest's law. It states that inelastic action at any point in a member at which any state of stress exists *begins when* the maximum shearing stress at the point reaches a value equal to the maximum shearing stress in a tension specimen when yielding starts. This means that the shearing yield stress τ_Y must be no more than one-half the tensile yield stress Y, since the maximum shearing stress in a tension specimen (on a 45° oblique plane) is one-half the maximum tensile stress in the specimen.

The maximum shearing stress criterion seems to be fairly well justified for ductile material in which relatively large shearing stresses are developed. However, for a state of pure shear such as occurs in a torsion test, the shearing yield stress τ_Y of some ductile metals is found by experience to be approximately 0.577 of the tensile yield stress Y ($\tau_Y = 0.577Y$). Hence for such a state of stress the maximum shearing stress criterion errs (on the side of safety) by approximately 15 percent; that is, it predicts yielding at loads less than those that actually cause yielding.

It may be shown that maximum and minimum principal stresses may be resolved into a state of pure shear τ combined with equal tensions σ in all directions in the plane of these two principal stresses.† It is assumed by the maximum shearing stress criterion of failure that the maximum

†See Fig. 1-4.4, where $\sigma = (\sigma_1 + \sigma_2)/2$ and $\tau = \frac{1}{2}(\sigma_1 - \sigma_2)$.

shearing stresses alone produce inelastic action and that the equal tensile stresses have no influence on yield initiation. However, if the state of stress consists of triaxial tensile stresses of nearly equal magnitude, shearing stresses are very small and failure will probably occur by brittle fracture rather than by yielding.

3. The *maximum strain* criterion, often called St. Venant's criterion. It states that inelastic action at a point in a member at which any state of stress exists *begins when* the maximum strain at the point reaches a value equal to that which occurs when inelastic action begins in the material under a uniaxial state of stress such as occurs in the tension test. This value, ϵ_Y, occurs simultaneously with the tensile yield stress Y of the material. Thus $\epsilon_Y = Y/E$.

For example, according to this criterion, inelastic action in the block of Fig. 3-3.3*a begins when* σ_1 becomes equal to Y since $\epsilon_Y = Y/E$, but in Fig. 3-3.3*b* $\epsilon = (\sigma_1/E) + \nu(\sigma_2/E)$. Hence, inelastic action is predicted for a value of σ_1 less than Y, since the strain in the direction of σ_1 is increased by the amount $\nu(\sigma_2/E)$. Therefore, according to this criterion of failure, σ_1 is somewhat smaller than Y at yielding if the second normal stress σ_2 is a compressive stress. If σ_2 is a tensile stress, the maximum value of σ_1 that can be applied without causing yielding is somewhat larger than Y.

The maximum strain criterion of the breakdown of elastic action is an improvement over the maximum principal stress criterion but, like the latter, it does not reliably predict failure by yielding.

4. The *strain energy* density criterion, proposed by Beltrami and by Haigh. It states that inelastic action at any point in a member resulting from any state of stress *begins when* the strain energy per unit volume (strain energy density) absorbed at the point is equal to the strain energy absorbed per unit volume by the material when subjected to the yield stress Y under a uniaxial state of stress, as occurs in a simple tensile test. As given in Table 3-3.1, the value of this maximum strain energy density is $U_{0Y} = \frac{1}{2}(Y^2/E)$. Expressions for the strain energy absorbed per unit volume for various states of stress can be calculated by the results given in Art. 2-4. According to the strain energy density criterion none of these expressions can exceed the value $\frac{1}{2}(Y^2/E)$ without causing yielding.

5. The *strain energy density of distortion* criterion, which grew out of the analytical work of Huber, von Mises, and Hencky and out of the results of tests by Bridgman on various materials. The results show that many materials remain elastic when subjected to very high hydrostatic pressures. This criterion states that inelastic action at any point in a member under any combination of stresses begins when the strain energy density of distortion U_{D0} (Eq. 2-6.3) absorbed at the point is equal to the strain energy density of distortion U_{DY} in a tension specimen when the stress in the tension specimen is equal to the yield stress Y. As given in Table 3-3.1, $U_{DY} = Y^2/6G$.

The maximum strain energy of distortion criterion differs from the maximum strain energy density criterion as follows. In the maximum strain energy density criterion, it is assumed that the entire strain energy density is associated with the beginning of inelastic action. However, tests of various materials under very high hydrostatic stresses show that the materials may withstand, without inelastic action taking place, strain energy density values that are many times greater than those obtained in the simple axial load compression test. Hence, since in the hydrostatic tests the strain energy density is used in producing volume changes only, it was proposed that the energy absorbed in changing volume has no effect in causing failure by yielding, and that failure by inelastic action is associated only with energy absorbed in changing shape. It was further assumed that if it were possible to make tests of materials under a negative hydrostatic pressure, which would create three equal tensile principal stresses, the same results as found for three equal principal compressive stresses would be obtained; that is, no yielding would take place, although fracture eventually would occur. Since change of shape involves shearing stresses, the strain energy density of distortion criterion is sometimes called (somewhat erroneously) the shear energy criterion.

The expression for the energy of distortion per unit volume is given by Eq. (2-6.3). The strain energy density of distortion criterion of failure (yield condition in Art. 2-6) is given by Eq. (2-6.6).

6. The *octahedral shearing stress criterion*. It gives the same results as does the energy of distortion criterion. For example, by Eqs. (1-4.8) and (2-6.3), we obtain

$$\tau_{\text{oct}} = \frac{2}{\sqrt{3}}\sqrt{GU_{D0}} \tag{3-3.4}$$

where U_{D0} is the strain energy density of distortion. The criterion of failure according to the maximum energy of distortion criterion states that inelastic action *begins when* U_{D0} becomes equal to $U_{DY} = Y^2/6G$ (see Eq. 2-6.5). If we substitute this value of U_{DY} in Eq. (3-3.4), the octahedral shearing stress at yield is (Eq. 2-6.6a)

$$\tau_{\text{oct(max)}} = (\sqrt{2}/3)Y = 0.471Y \tag{3-3.5}$$

This value of $\tau_{\text{oct(max)}}$ is the same as the value of octahedral shearing stress that occurs in the standard tensile test at $\sigma = Y$. Thus, the maximum octahedral shearing stress criterion may be stated as follows: Inelastic action at any point in a member under any combination of stresses *begins when* the maximum octahedral shearing stress $\tau_{\text{oct(max)}}$ becomes equal to $0.471Y$, where Y is the tensile yield stress of the material as determined from the standard tension test. The octahedral shearing stress criterion of failure makes it possible to apply the energy of distortion criterion of failure by dealing only with stresses instead of dealing with energy directly; this procedure to some engineers seems desirable because stress is a more familiar quantity in engineering design than is energy.

Generally, it is found by experiments that initiation of yielding in most materials is predicted fairly well by either the maximum shearing stress criterion (criterion 2) or the octahedral shearing stress criterion (criteria 6 or 5). In recent years, fracture mechanics analysis has also been employed to establish general yield criteria (see Art. 3-4).

Significance of Failure Criteria / A rational procedure of design requires that a general mode of failure under assumed service conditions be determined or assumed (failure usually is by yielding or by fracture) and that a quantity (for example, stress, strain, or energy) be chosen that is considered to be the cause of the failure. This means that there is a maximum or critical value of the quantity selected which limits the loads that can be applied to the member. Furthermore, a *suitable* test of the material must be made for determining the critical value. This critical value is frequently referred to as the maximum utilizable "strength" of the material. It is important to understand how criteria of failure fit into the rational design.

For a given general mode of failure, each criterion of failure identifies a (significant) quantity that causes failure, when the value of the quantity reaches a critical value. A tension test is considered to be a suitable test for determining the critical or maximum value of this significant quantity.

It is important to note that if an appropriate or suitable test could always be selected so that the material would be subjected to the same conditions of stress to which it is subjected in the actual member, there would be no need for criteria of failure. For example, in Table 3-3.1, the maximum utilizable "strength" of a material as determined by each of the several criteria of failure is obtained from a tension test. Hence, in the design of members in which the state of stress is uniaxial, each criterion predicts the same dimensions for the member. Similarly, if the maximum or limiting values of the various quantities that are considered to be the cause of failure are obtained from a torsion test, each criterion of failure predicts that members made of the same material subjected to a state of stress of pure shear (torsion) be given the same dimensions, since all the quantities assumed to cause failure reach their limiting values simultaneously.

If a single criterion of failure applied to all conditions in which load-resisting members are used, it could be used to predict the nature of the quantity (for example, stress, strain, or energy) as well as a critical value (as obtained from a specified test) that would limit the load that could be applied to the member without causing structural damage. This generality, however, is too much to expect when we consider the radically different modes of failure (ranging from incipient yielding to brittle fracture) and the simplifying conditions that are necessary to impose in a practical test. In general, we are limited, because of practical considerations, to one or two test members to obtain material properties: the tension test or the torsion test.

Interpretation of Failure Criteria for General Yielding/Two of the criteria of failure, the maximum shearing stress and the maximum octahedral shearing stress criteria of failure, are interpreted graphically in Fig. 2-6.6. There, it is shown that the greatest difference between the two criteria is exhibited when material properties obtained from a tension test are used to predict failure loads for a torsion member made of the same material. In Table 3-3.2, we interpret five failure criteria by the use of a tension specimen to obtain the yield stress Y, by the use of a hollow torsion specimen to obtain the shearing yield stress τ_Y, and by the use of each of the five failure criteria to predict the relation between τ_Y and Y (see column 4). If all the criteria of failure are correct, they all should predict the same relation between τ_Y and Y in column 4 of Table 3-3.2.

Experimental data for metals indicate that $\tau_Y = 0.500Y$ for some metals (particularly some elastic-perfectly plastic metals), that $\tau_Y = 0.577Y$ for some metals, and that, for most of the remaining metals, the value of τ_Y falls between $0.500Y$ and $0.577Y$. Thus, experimental evidence indicates that either the maximum shearing stress or the maximum octahedral shearing stress criteria of failure can be used to predict failure loads for members that fail by general yielding. If tension specimens are used to obtain material properties (yield stress Y), the maximum shearing stress criterion of failure predicts failure loads that are conservative for most metals, with a maximum error of 15 percent. If hollow torsion specimens are used to obtain material properties (shearing yield stress τ_Y), the maximum octahedral shearing stress criterion of failure predicts failure loads that are conservative for most metals, with a maximum error of 15 percent.

The states of stress in the tension and torsion tests represents about as wide a range of stress conditions as occurs in most engineering members that fail by yielding under static loads. In the tension test $\sigma_{max}/\tau_{max} = 2$, and in the torsion test $\sigma_{max}/\tau_{max} = 1$. For some triaxial states of stress σ_{max}/τ_{max} is greater than 2, approaching infinity when the triaxial stresses are equal and of similar sign. However, in this case, failure occurs because of brittle fracture, if the stresses are tensile.

For states of stress in which σ_{max}/τ_{max} lies between 1 and 2 as, for example, in a cylindrical shaft subjected to a twisting moment T and to a bending moment M producing the state of stress shown in Fig. 3-3.4, the results predicted by the various criteria are given in Fig. 3-3.4.

Table 3-3.2

Comparison of Maximum Utilizable Values of a Material Quantity According to Various Criteria of Failure for States of Stress in the Tension and Torsion Tests

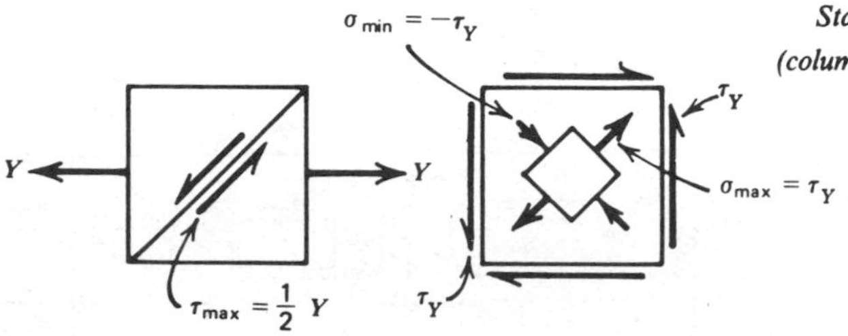

(1) *Failure Criterion*	*(2)* *Predicted Maximum Utilizable Value as Obtained from a Tensile Test*	*(3)* *Predicted Maximum Utilizable Value as Obtained from a Torsion Test*	*(4)* *Relation between Values of Y and τ_Y if the Criterion of Failure is Correct for Both States of Stress (column 2 = column 3)*
Maximum normal stress	$\sigma_{max} = Y$	$\sigma_{max} = \tau_Y$	$\tau_Y = Y$
Maximum strain $\nu = \frac{1}{4}$	$\epsilon_{max} = \frac{Y}{E}$	$\epsilon_{max} = \frac{5}{4}\frac{\tau_Y}{E}$	$\tau_Y = 0.80\,Y$
Maximum shearing stress	$\tau_{max} = \frac{1}{2}Y$	$\tau_{max} = \tau_Y$	$\tau_Y = 0.50\,Y$
Maximum octahedral stress	$\tau_{oct(max)} = \frac{\sqrt{2}}{3}Y$	$\tau_{oct(max)} = \frac{\sqrt{2}}{\sqrt{3}}\tau_Y$	$\tau_Y = 0.577\,Y$
Maximum strain energy density of distortion	$U_{DY} = \frac{Y^2}{6G}$	$U_{DY} = \frac{\tau_Y^2}{2G}$	$\tau_Y = 0.577\,Y$

The values of diameter d, which are just large enough to prevent inelastic action in the shaft, are computed by each failure criterion and these values of d are then compared by obtaining the ratios of the various values of d to the value d_s, computed by the maximum shearing stress criterion.

These ratios are obtained for combinations of T and M ranging from M acting alone ($T/M = 0$) to T acting alone. The case for which $T/M = \infty$, or T acting alone, is shown by the horizontal asymptote (right side of graph). The maximum shearing stress criterion predicts the largest required diameter and the maximum principal stress criterion the smallest required diameter for all values of T/M, except for $T/M = 0$, where all criteria predict equal diameters.

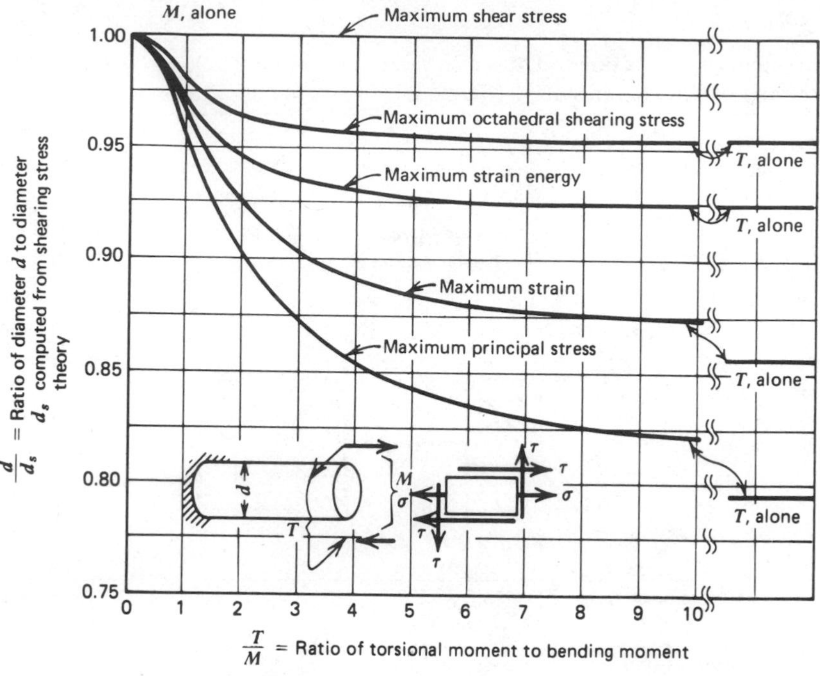

Fig. 3-3.4/Comparison of criteria of failure.

Figure 3-3.5 compares in another way the two most appropriate criteria of failure when the mode of failure is by general yielding and when the state of stress is the same as that considered in Fig. 3-3.4 for the range $\sigma_{max}/\tau_{max} = 2$ (bending alone) to $\sigma_{max}/\tau_{max} = 1$ (torsion alone). The equations represented by these curves are found as follows: For any combination of τ and σ, yielding starts according to the maximum shearing stress criterion when

$$\sqrt{(\sigma/2)^2 + \tau^2} = Y/2 \quad \text{or} \quad 4\left(\frac{\tau}{Y}\right)^2 + \left(\frac{\sigma}{Y}\right)^2 = 1 \tag{3-3.6}$$

Likewise, yielding starts according to the maximum octahedral shearing stress criterion when

$$\frac{\sqrt{2\sigma^2 + 6\tau^2}}{3} = \frac{\sqrt{2Y^2}}{3} \quad \text{or} \quad 3\left(\frac{\tau}{Y}\right)^2 + \left(\frac{\sigma}{Y}\right)^2 = 1 \tag{3-3.7}$$

For other types of loading the maximum shearing stress criterion of failure is satisfied by setting the maximum shearing stress

$$\tau_{max} = \frac{\sigma_{max} - \sigma_{min}}{2} \tag{3-3.8}$$

at failure, in the member to be designed, equal to the maximum shearing

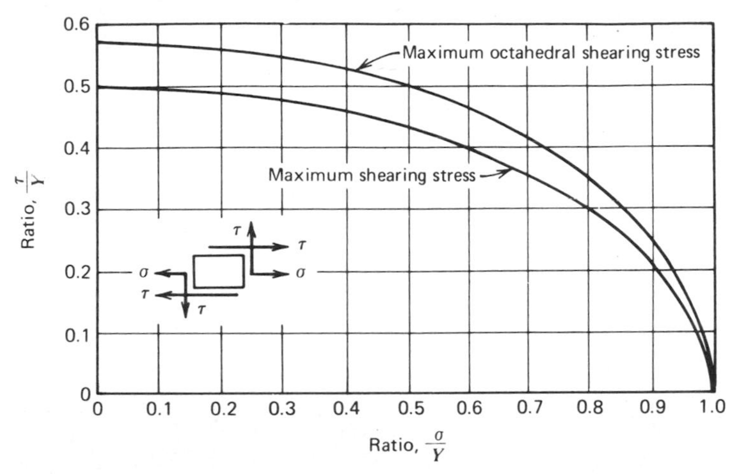

Fig. 3-3.5/Comparison of two of the criteria of failure.

stress at failure in the member used to obtain material properties. The maximum octahedral shearing stress criterion (or energy of distortion criterion) of failure is satisfied by setting the maximum octahedral shearing stress

$$\tau_{\text{oct(max)}} = \frac{1}{3}\sqrt{\begin{array}{l}(\sigma_{xx}-\sigma_{yy})^2+(\sigma_{yy}-\sigma_{zz})^2+(\sigma_{zz}-\sigma_{xx})^2\\ +6(\sigma_{xy}^2+\sigma_{yz}^2+\sigma_{zx}^2)\end{array}} \tag{3-3.9}$$

at failure, in the member to be designed, equal to the maximum octahedral shearing stress at failure in the member used to obtain material properties.

Other Factors to Be Considered/Failure criteria do not take account of all conditions that the engineer must consider in the problem of failure, even for failure by yielding of ductile materials subjected to static loads at ordinary temperatures. In many cases some inelastic strain (yielding) may occur without destroying the usefulness of a member, and these inelastic strains cause a readjustment of stresses that may permit an appreciable increase in the loads on the member.[2]

In conclusion, we remark that although the capacity of a material to work harden may permit higher applied loads to be incorporated into a design, present day codes do not usually take work hardening into account. Hence, the so-called plastic design concept is based on the idea of allowing design loads to exceed those that initiate yield (see Eq. 3-3.1 for pure bending of beams having rectangular cross sections), but not to exceed the fully plastic yield (plastic collapse) load. (See Eq. 3-3.3 for pure bending of beams having rectangular cross sections.) As we have seen it is relatively simple to calculate fully plastic loads for simple frame members like beams and shafts. However, for more complicated massive parts or members, the calculation of fully plastic loads becomes extremely difficult because of the effect upon general yielding of the more general (triaxial) states of stress. Also, as noted in Art. 3-1, the results obtained through the use of a factor of safety in arriving at safe working loads (failure-safe design) for complicated massive parts should be considered at best to be a crude approximation.

EXAMPLE 3-3.1

Cylindrical Steel Shaft Subjected to Torsion and Bending

A circular cylindrical shaft is made of steel with a yield stress $Y = 700 MN/m^2$ (MPa). The shaft is subjected to a static bending moment $M = 13.0$ kN · m and a static torsional moment $T = 30.0$ kN · m (Fig. 3-3.4). Also, for the steel the modulus of elasticity is $E = 200$ GPa and Poisson's ratio is $\nu = 0.29$. Employing a factor of safety of $SF = 2.60$, determine the minimum required safe diameter for the shaft.

SOLUTION

Assuming that the failure is by yield initiation, we note that either the maximum octahedral shearing stress (or equivalently the distortion strain energy) criterion or the maximum shearing stress criterion is applicable.

For the octahedral shearing stress criterion we obtain by Fig. 3-3.4 and Eq. (3-3.7) or Eq. (3-3.9)

$$\tau_{\text{oct(max)}} = \frac{\sqrt{2}}{3} Y = \frac{1}{3}\sqrt{2\sigma^2 + 6\tau^2} \tag{1}$$

which simplifies to

$$Y = \sqrt{\sigma^2 + 2\tau^2} \tag{2}$$

Yielding in the shaft is to be designed to occur when the loads M and T are increased by the safety factor. Thus, $\sigma = (SF)Mc/I = 32(SF)M/\pi d^3$ and $\tau = (SF)Tc/J = 16(SF)T/\pi d^3$. Substitution into Eq. (2) yields

$$Y = \sqrt{\sigma^2 + 3\tau^2} = \frac{16SF}{\pi d^3}\sqrt{4M^2 + 3T^2} \tag{3}$$

or

$$d_{\min} = \left[\frac{16SF}{\pi Y}\sqrt{4M^2 + 3T^2}\right]^{1/3}$$

Thus, with numerical values, we get

$$d_{\min} = 103 \text{ mm}$$

Hence, by the octahedral shearing stress criterion a minimum shaft diameter of 103 mm is required to prevent excessive inelastic action.

Alternatively, if the maximum shearing stress criterion is employed, the yield condition becomes (Eq. 3-3.6)

$$\tau_{max} = \frac{Y}{2} = \frac{1}{2}\sqrt{\sigma^2 + 4\tau^2}$$

or

$$d_{min} = \left[\frac{32SF}{\pi Y}\sqrt{M^2 + T^2}\right]^{1/3}$$

Thus, with numerical values, we get

$$d_{min} = 107 \text{ mm}$$

Hence, a diameter not less than 107 mm would probably be required to prevent initiation of yielding of the shaft. We observe that the properties E and ν of the steel do not enter into the computations.

PROBLEM SET 3-3

1. A shaft has a diameter of 20 mm and is made of an aluminum alloy with yield stress $Y = 330$ MPa. The shaft is subjected to an axial load $P = 50.0$ kN. (a) Determine the torque T that can be applied to the shaft to initiate yielding. (b) Determine the torque T that can be applied to the shaft if the shaft is designed with a factor of safety $SF = 1.75$ for both P and T against initiation of yielding. Use the maximum octahedral shearing stress criterion of failure.

2. A low carbon steel shaft is designed to have a diameter of 30 mm. It is to be subjected to an axial load $P = 30.0$ kN, a moment $M = 150$ N · m, and a torque $T = 250$ N · m. If the yield point for the steel is $Y = 280$ MPa, determine the factor of safety used in the design of the shaft based on the maximum shearing stress criterion of failure assuming that failure occurs at initiation of yielding.

   Ans. $SF = 2.05$

3. A closed end thin-walled cylinder of titanium alloy Ti-6AL-4B ($Y = 800$ MPa) has an inside diameter of 38 mm and a wall of thickness of 2 mm. The cylinder is subjected to an internal pressure $p = 22.0$ MPa and an axial load $P = 50.0$ kN. Determine the torque T that can be applied to the cylinder if the factor of safety for design is $SF = 1.90$. The design is based on the maximum shearing stress criterion of failure assuming that failure occurs at initiation of yielding.

4. A load $P = 30.0$ kN is applied to the crank pin of the crank shaft in Fig. P3-3.4 to rotate the shaft at constant speed. The crank shaft is made of a ductile steel with a yield stress $Y = 276$ MPa. Determine the diameter of the crank shaft if it is designed using the maximum shearing stress criterion for initiation of yielding and the factor of safety is $SF = 2.00$.

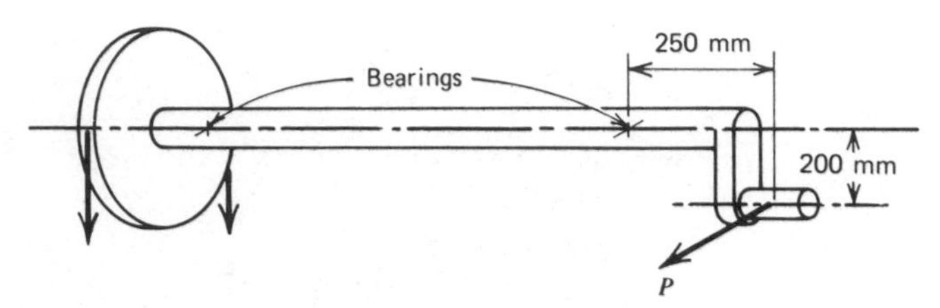

Fig. P3-3.4

Ans. $d = 89.2$ mm

5. Solve Problem 4 using the maximum octahedral shearing stress criterion of failure.

6. The shaft in Fig. P3-3.6 is supported in flexible bearings at A and D, and two gears B and C are attached to the shaft at the locations shown. The gears are acted upon by tangential forces as shown by the end view. The shaft is made of a ductile steel having a yield stress $Y = 290$ MPa. If the factor of safety for the design of the shaft is $SF = 1.85$, determine the diameter of the shaft using the maximum shearing stress criterion for the initiation of yielding failure.

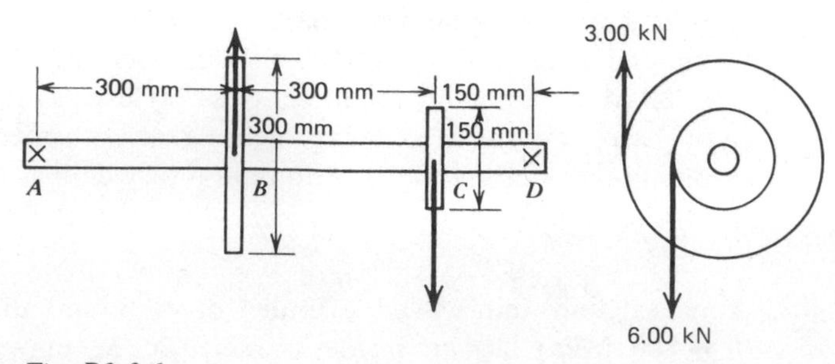

Fig. P3-3.6

Ans. $d = 33.5$ mm

7. Let the 6.00 kN load on the smaller gear of Problem 6 be horizontal instead of vertical. Determine the diameter of the shaft.

8. The 100 mm diameter bar shown in Fig. P3-3.8 is made of a ductile steel that has a yield stress $Y = 420$ MPa. The free end of the bar is subjected to a load P making equal angles with the positive directions of the three coordinate axes. Using the maximum octahedral shearing stress criterion of failure, determine the magnitude of P that will initiate yielding.

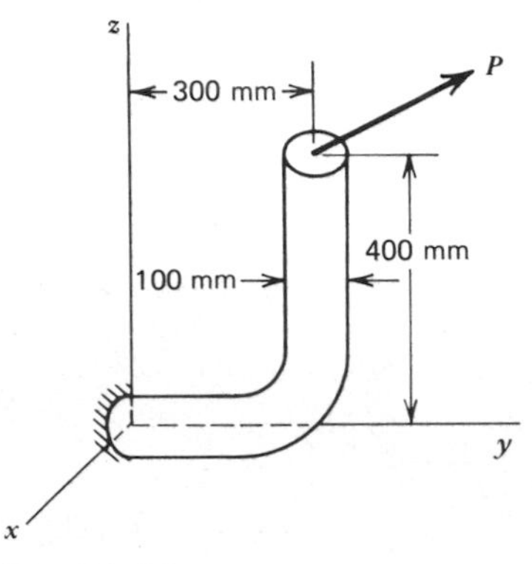

Fig. P3-3.8

Ans. $P = 149.5$ kN

9. The shaft in Fig. P3-3.9 has a diameter of 20 mm and is made of a ductile steel ($Y = 400$ MPa). It is subjected to a combination of static loads as follows: axial load $P = 25.0$ kN, bending moment $M_x = 50.0$ N · m, and torque $T = 120$ N · m. (a) Determine the factor of safety for design based on the maximum octahedral shearing stress criterion of failure. (b) Determine the factor of safety for design based on the maximum shearing stress criterion of failure. (c) Determine the maximum and minimum principal stresses and indicate the direction that they act at point B shown.

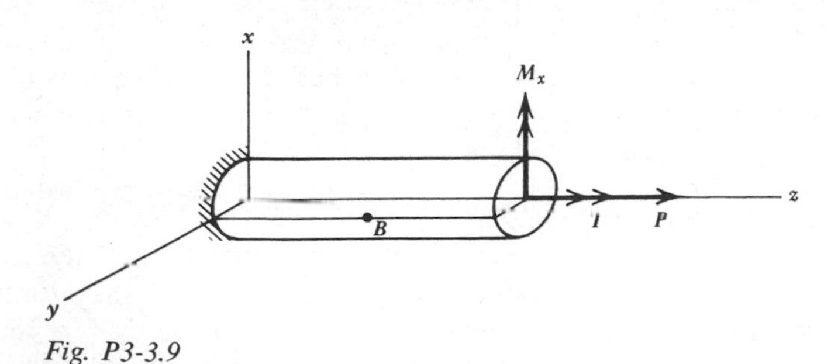

Fig. P3-3.9

Ans. (a) $SF = 2.05$, (b) $SF = 1.91$,
(c) $\sigma_{max} = 176.3$ MPa, $\sigma_{min} = -33.1$ MPa, $\theta = -0.4088$ rad.

10. Let the material properties for the shaft in Problem 9 be obtained by using a hollow torsion specimen. The shearing yield stress is found to be $\tau_Y = 200$ MPa. Resolve parts *a* and *b*.

11. The member in Fig. P3-3.11 has a diameter of 20 mm and is made of a ductile metal. Static loads P and Q are parallel to the y-axis and z-axis, respectively. Determine the magnitude of the yield stress Y of the material if yielding is impending.

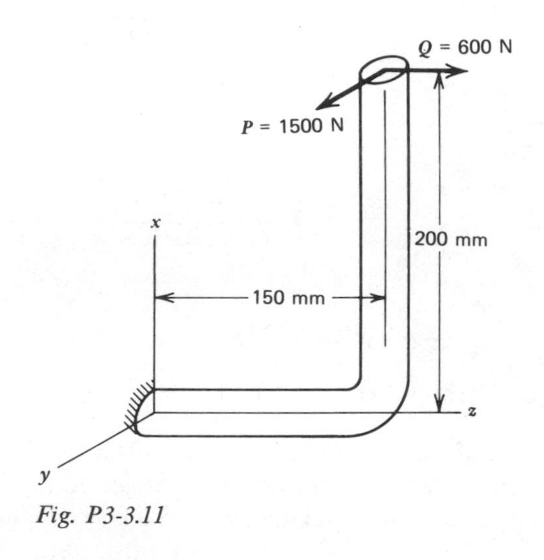

Fig. P3-3.11

Ans. $Y = 502.6$ MPa based on maximum shearing stress criterion of failure.

12. A 50 mm diameter structural steel shaft is subjected to a torque $T = 1.20$ kN · m and to an axial load P. A hollow torsion specimen made of the same steel indicated a shearing yield point $\tau_Y = 140$ MPa. If the shaft is designed for a factor of safety $SF = 2.00$, determine the magnitude of P based on (a) the maximum shearing stress criterion of failure and (b) the maximum octahedral shearing stress criterion of failure.

13. A solid aluminum alloy ($Y = 320$ MPa) shaft extends 200 mm from a bearing support to the center of a 400 mm diameter pully (Fig. P3-3.13). The belt tensions T_1 and T_2 vary in magnitude with time. Their maximum values are $T_1 = 1800$ N and $T_2 = 180$ N. If the maximum values of the belt tensions are applied only a few times during the life of the shaft, determine the required diameter of the shaft if the factor of safety is $SF = 2.20$.

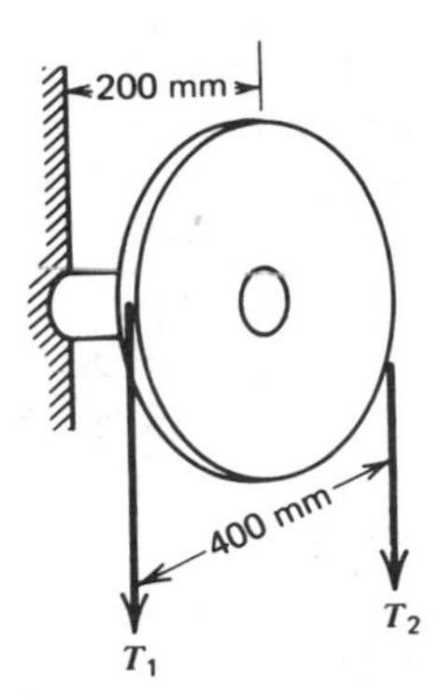

Fig. P3-3.13

Ans. $d = 32.97$ mm based on the maximum shearing stress criterion of failure.

14. A closed thin-wall tube has a mean radius of 40.0 mm and a wall thickness of 4.00 mm. It is subjected to an internal pressure of 11.0 MPa. The axis of the tube lies along the z-axis. In addition to internal pressure, the tube is subjected to an axial load $P = 80.0$ kN, bending moments $M_x = 660$ N · m and $M_y = 480$ N · m, and torque $T = 3.60$ kN · m. If yielding is impending in the tube, determine the yield stress Y of the material based on the maximum shearing stress criterion of failure.

15. Solve Problem 14 by using the maximum octahedral shearing stress criterion of failure.

    Ans. $Y = 216.6$ MPa

3-4 FAILURE CRITERIA. FRACTURE

As noted in Arts. 3-2 and 3-3, failure of a structural system may occur by excessive deflection, yield, or fracture. Unfortunately, these modes of failure do not occur in a singular fashion, since prior to failure, say, by fracture, yielding of a member may occur. Furthermore, a member may undergo considerable deflection before it fails, say by extensive yielding. Consequently, failure criteria are usually based on the dominant failure mode. Thus, for yield-dominant failures, the criteria of octahedral shearing stress (also distortional strain energy density) and maximum shearing stress seem most appropriate. For fracture-dominant failure, several types of fracture must be considered. For example, fracture may occur in a "sudden" manner (brittle materials at ordinary temperatures or structural steels at low temperatures), it may occur as brittle fracture of cracked or flawed members, it may occur in progressive stages (so-called fatiguing) at general levels of stress below yield and, finally, it may occur with time at elevated temperatures (creep rupture). In contrast to yield dominate failures, different types of failure criteria are applicable to different types of fracture dominate failures.

Material defects are of significance in all kinds of failures. However, different types of defects influence various modes of failure differently. For example, for initiation of yielding, the significant defects tend to distort and interrupt crystal lattice planes and interfere with easy glide of dislocations. These defects are of the nature of dislocation entanglements, interstitial atoms, out-of-size substitutional atoms, grain-boundary spacings, bonded precipitate particles, etc. In general, these defects provide resistance (to yielding) that is essential to the proper performance of high-strength metals. On the other hand, little resistance to yielding is provided by larger defects such as inclusions, porosity, surface scratches, and small cracks although such defects may alter the net load-bearing section.

For failure by fracture before extensive yielding of the section (fracture-dominant failure), the significant defects (size scale) depend principally on the toughness of the material. Unfortunately, there is no clear boundary between yielding (ductile-type material) failures and fracture-dominant (brittle-type material) failures. Indeed classification of many materials as ductile or brittle is meaningless unless physical factors such as temperature, state of stress, rate of loading, and chemical environment are specified. For example, many materials can be made to behave in a ductile manner for a given set of conditions and in a brittle manner for another set of conditions. To be more precise, one should speak of a material in a *brittle state* or a material in a *ductile state*. However, here too difficulties arise, since there is not always a clear demarcation between the brittle state and the ductile state. Nevertheless, it is fortuitous that for an important range of materials and conditions in either the ductile state or the brittle state, time effects, temperature, stress gradients, microstructural features, and size effects, for example, are of secondary importance. For the ductile state, it is possible to postulate failure criteria based on concepts of macroscopic states of stress that define critical values of quantities for which yielding begins (Art. 3-3). As noted in Art. 3-3, certain of these failure criteria (maximum octahedral shearing stress, distortion strain energy density) discard mean pressure effects; whereas, others of these criteria discard one or more of the principal stresses (e.g., maximum shearing stress criterion, which does not include influence of the intermediate principal stress, and the maximum principal stress criterion, which ignores the two lesser principal stresses).

Under similar circumstances, it is also possible to postulate reasonable failure criteria based on macroscopic stress concepts for the onset of brittle fracture. In general, in contrast to materials in the ductile state, failure (fracture) states for materials in the brittle state are sensitive to both the magnitude and sign of the mean stress. The fracture states for isotropic materials in the brittle state, are frequently (conveniently) represented by pyramidlike surfaces in principal stress space, which are cut (limited) by suitable tension (critical value) cutoffs.[5,6]

Brittle fracture problems that we consider are subdivided into three types as follows: (1) brittle fracture of members free of cracks and flaws under static loading conditions, (2) brittle fracture originating at cracks and flaws in members under static loading conditions, and (3) brittle fracture resulting from high cycle fatigue loading. Another type of brittle fracture, which we do not consider, occurs at elevated temperatures after long time creep in which small deformations occur as creep in grain boundaries (see reference 7).

Brittle Fracture of Members Free of Cracks and Flaws / So-called brittle materials (such as glass, gray cast iron, and chalk) exhibit nearly linear tensile stress-strain diagrams up to their ultimate strengths. If, at the fracture location in a member made of a brittle material, the principal stress of maximum absolute magnitude is tensile in nature, fracture will occur on the plane on which this principal stress acts. Then, the maximum principal stress criterion of failure is considered to be valid for design purposes. When mean stresses at the fracture locations in members are tension, the brittle materials in these members are considered to be loaded in a brittle state. It may be possible for the same materials in other members to be loaded in a ductile state if the mean stresses in these members are large compression. However, failure criteria for such loadings are not considered in this book.

Brittle Fracture of Cracked or Flawed Members / Cracks may be present in the material of members before loading, may be created (initiated) by flaws (high stress concentrations) at low nominal stress levels, or may be initiated and made to propagate with number of cycles because of fatigue loading. Failure by fracture (complete fracture) results when a crack propagates sufficiently far through the member so that the member is unable to support the load and hence fractures into two or more pieces.[†] In general, brittle fracture consists of at least two stages: initiation of crack (crack initiation) and propagation of crack (crack extension or crack propagation). Once a crack has been initiated, subsequent crack propagation may occur in several ways depending on the relative dis-

[†] In pipe systems, which carry fluids and other materials, failure may occur when a crack propagates through the pipe thickness, allowing fluid to escape from the pipe. Such failures may be extremely harmful to life and property if the liquid is a dangerous chemical or is contaminated, say by nuclear fission products such as occur in nuclear reactor pipe loop systems (see Reference 1).

placements of the particles in the two faces (surfaces) of the crack. Three basis modes of crack surface displacements are Mode I: the opening mode (Fig. 3-4.1*a*), Mode II: the (edge) sliding mode (Fig. 3-4.1*b*), and Mode III: the tearing mode (Fig. 3-4.1*c*). In Mode I, the opening mode, the crack surfaces move directly apart. In Mode II, the sliding mode, the crack surfaces move (slide) normal to the crack edge and remain in the plane of the crack. In the tearing mode, Mode III, the crack surfaces move parallel to the crack edge and again remain in the plane of the crack. The most general case of crack surface displacements are obtained by superposition of these basic three modes. We follow the convention of adding Roman numeral subscripts I, II, III to symbols associated with quantities that describe Modes I, II, III, respectively.

In isotropic materials, brittle fracture usually occurs in Mode I. Consequently, we confine our attention mainly to Mode I in establishing fracture criteria for sudden fracture of flawed members when the materials in these members is loaded in the brittle state. While fractures induced by sliding (Mode II) and tearing (Mode III) do occur, their

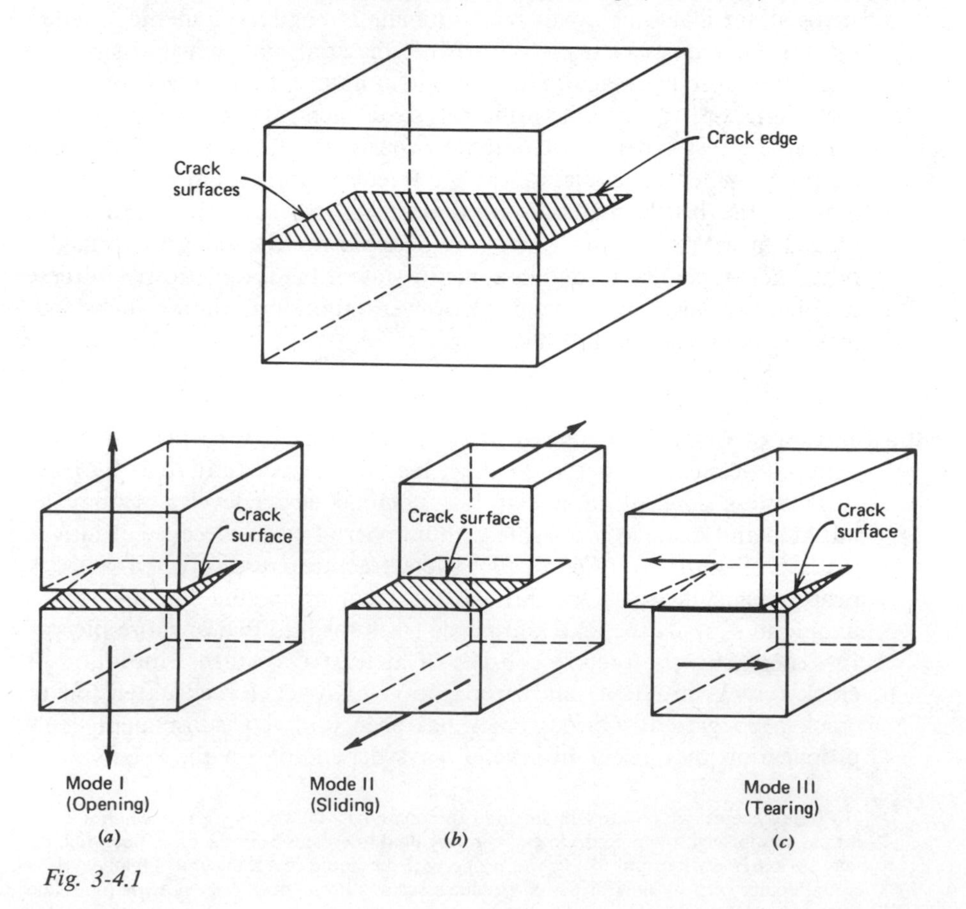

Fig. 3-4.1

frequency is much less than the opening mode fracture (Mode I). Although the combined influence of two or three modes of crack extension has been studied,[8] we do not consider such problems here. However, we do note that improvement of fracture resistance for Mode I usually results in improved resistance to combined mode crack extension.

The crack surfaces, which are stressfree boundaries in the neighborhood of the crack tip, strongly influence the distribution of stress around the crack. More remote boundaries and remote loading forces affect mainly the intensity of the stress field at the crack tip. Elastic stress analysis (Chapter 12) of cracks leads to the concept of *stress intensity factor* K, which is employed to describe the elastic stress field surrounding the crack tip. As noted above, the motion of crack surfaces can be divided into three types, with corresponding stress fields. Hence, three stress intensity factors K_{I}, K_{II}, and K_{III} are employed to characterize the stress fields for these three modes. The dimensions of stress intensity factor K is [stress] $\times$ [length]$^{1/2}$, and depends on specimen dimensions and loading conditions. In general, K is proportional to [average stress] $\times$ [crack length]$^{1/2}$ (Chapter 12). When K is known for a given mode (say K_{I}), stresses and displacements in the neighborhood of the crack tip can be calculated (Chapter 12). The stresses are inversely proportional to the square of the distance from the crack tip, becoming infinite at the tip. In general, fracture criteria for brittle fracture are based on critical values of the stress intensity factor, for which the crack rapidly propagates (leading to fracture).

In order to determine the load or loads required to cause brittle fracture of a cracked member or structure, it is necessary that relations be developed so that K can be determined for the member or structure and that the critical value of K be determined for the material. Test specimens have been developed to measure the critical value of K for the opening mode (Mode I); the critical value is designated as K_{IC}, and it is called fracture toughness. Fracture toughness, K_{IC}, is considered to be the material property measure of resistance to brittle fracture.[9]

The designs of test specimens currently recommended in ASTM and British Standards to determine values of K_{IC} are indicated in Fig. 3-4.2. The relative dimensions of the two test specimens in Fig. 3-4.2 are specified by the magnitude of W. The minimum magnitude of W depends on the values of both K_{IC} and the yield stress Y of the material. The derivation of K is based on a linear elastic solution. If the magnitude of $B = W/2$ for the specimens in Fig. 3-4.2 is small, appreciable inelastic deformation occurs at the tip of the crack and the measured value of K_{I} is larger than K_{IC}. As B is increased in magnitude, the measured value of K_{I} approaches a minimum value obtained when the inelastic deformation at the tip of the crack is held to a minimum as will occur when the state of stress at the tip of the crack over most of the width of the specimen is that associated with plane strain. The state of stress at

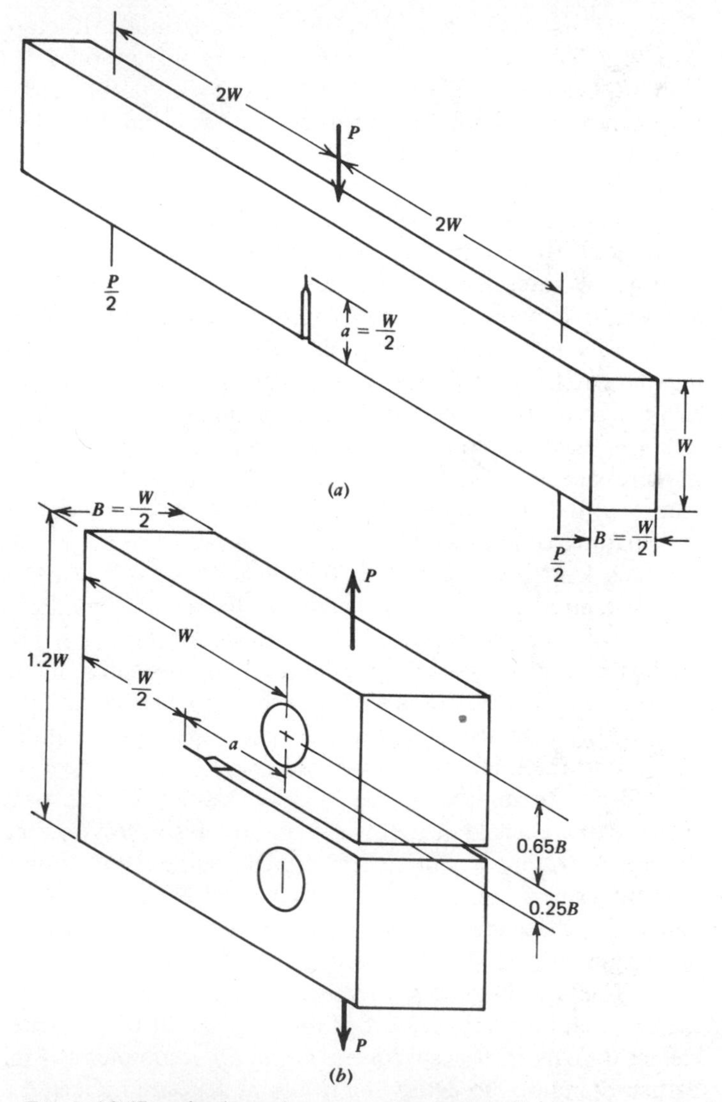

Fig. 3-4.2/Standard toughness specimens. (*a*) Single-edge cracked bend specimen. (*b*) Compact tension specimen.

the free surface ends of the crack is plane stress. In order to insure plane strain conditions it has been recommended[10] that the magnitude of B satisfy the relation

$$B \geq 2.5\left[\frac{K_{IC}}{Y}\right]^2 \tag{3-4.1}$$

Because of the large size of specimens required to satisfy Eq. (3-4.1) for some materials, the expense of specimen preparation and the expense of

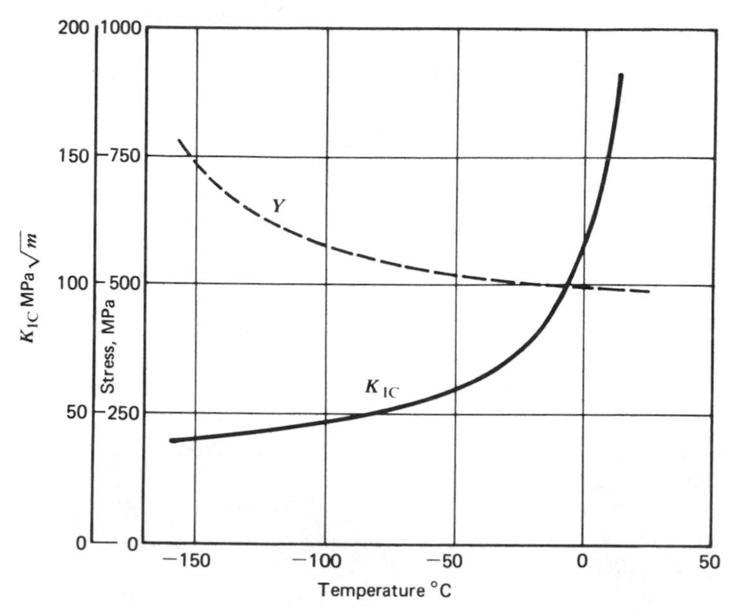

Fig. 3-4.3/Temperature dependence of K_{IC} for A533B steel.

testing may be large. Wessel[11] has obtained values of K_{IC} for an A533B steel at several temperatures as indicated in Fig. 3-4.3. At temperatures of −150°C and 10°C, Eq. (3-4.1) gives values for B of 7.4 mm and 242.0 mm, respectively. The large variation of K_{IC} with temperature indicated in Fig. 3-4.3 is typical of relatively low strength structural steels. Nonferrous alloys and very high strength steels show rather small variation of K_{IC} with temperature. Room temperature values of K_{IC} for several metals are listed in Table 3-4.1. Except for the data for A533B steel that were taken from Fig. 3-4.3, the other data in Table 3-4.1 were taken from published papers.†

In order to use values of K_{IC} from Table 3-4.1 in design, it is necessary that formulas be derived for typical load carrying members. A few formulas for K_{IC} for several geometric configurations and loads are given in Table 3-4.2. These formulas along with others may be found in reference 8. The formulas in Table 3-4.2 require values for K_I. We

†H. D. Greenberg, E. T. Wessel, and W. H. Pryle, "Fracture Toughness of Turbine Generator Rotor Forgings," *Engineering Fracture Mechanics*, Vol. 1, No. 4, 1970, pp. 653–674. E. A. Steigerwald, "Plane Strain Fracture Toughness of High Strength Materials," *Engineering Fracture Mechanics*, Vol. 1, No. 3, 1969, pp. 473–494. W. A. Logsdon, "An Evaluation of the Crack Growth and Fracture Properties of AISI 403 Modified 12 Cr. Stainless Steel," *Engineering Fracture Mechanics*, Vol. 7, No. 1, 1975, pp. 23–40. F. G. Nelson, P. E. Schilling, and J. G. Faufman, "The Effect of Specimen Size on the Results of Plane Strain Fracture Toughness Tests," *Engineering Fracture Mechanics*, Vol. 4, No. 1, 1972, pp. 33–50. C. N. Freed, "A Comparison of Fracture Toughness Parameters for Titanium Alloys," *Engineering Fracture Mechanics*, Vol. 1, No. 1, 1968, pp. 175–190.

Table 3-4.1

K_{IC} Critical Stress Intensity Factor (Fracture Toughness)
(Room Temperature Data)

Material	σ_u MPa	Y MPa	K_{IC} MPa$\sqrt{m}$	*Minimum Values for B, a, t* mm
	Alloy Steels			
A533B	—	500	175	306.0
2618 Ni Mo V	—	648	106	66.9
V1233 Ni Mo V	—	593	75	40.0
124 K 406 Cr Mo V	—	648	62	22.9
17-7PH	1289	1145	77	11.3
17-4PH	1331	1172	48	4.2
Ph 15-7Mo	1600	1413	50	3.1
AISI 4340	1827	1503	59	3.9
	Stainless Steel			
AISI 403	821	690	77	31.1
	Aluminum Alloys			
6061-T651	352	299	29	23.5
2219-T851	454	340	32	22.1
7075-T7351	470	392	31	15.6
7079-T651	569	502	26	6.7
2024-T851	488	444	23	6.7
	Titanium Alloys			
Ti-6Al-4Zr-2Sn-0.5Mo-0.5V	890	836	139	69.1
Ti-6Al-4V-2Sn	852	798	111	48.4
Ti-6.5Al-5Zr-1V	904	858	106	38.2
Ti-6Al-4Sn-1V	889	878	93	28.0
Ti-6Al-6V-2.5Sn	1176	1149	66	8.2

assume that the dimensions of each member are such that the state of stress at the crack tip over most of the thickness of the member is that associated with plane strain so that $K_I = K_{IC}$. In order to insure that the state of stress is plane strain for each of the cases in Table 3-4.2, it is assumed that the magnitudes of the crack half length a and the thickness t satisfy the relation

$$a, t \geq 2.5\left[\frac{K_{IC}}{Y}\right]^2 \tag{3-4.2}$$

Fracture mechanics analysis is also employed in establishing failure criteria for general yielding as well as for fracture criteria for materials loaded in the ductile state.[10] These topics are beyond the scope of this book.

Table 3-4.2

Stress Intensity Factors K_1

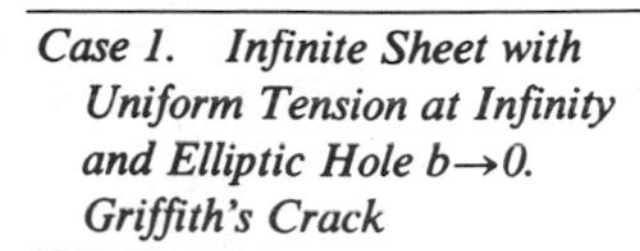

Case 1. Infinite Sheet with Uniform Tension at Infinity and Elliptic Hole b→0. Griffith's Crack

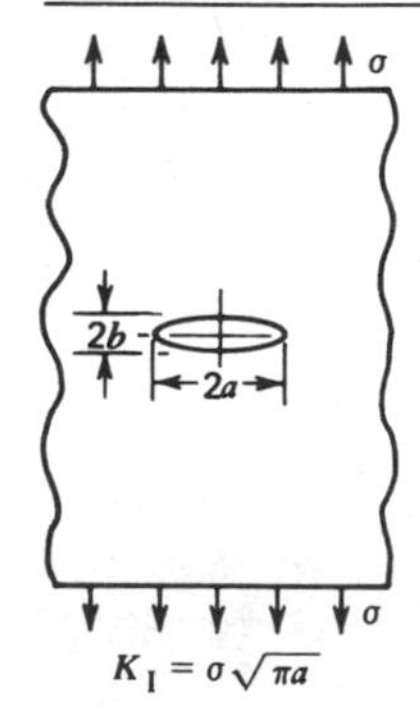

$K_I = \sigma\sqrt{\pi a}$

Case 2. Periodic Array of Elliptic Holes (b→0) in Infinite Sheet with Uniform Tension at Infinity.

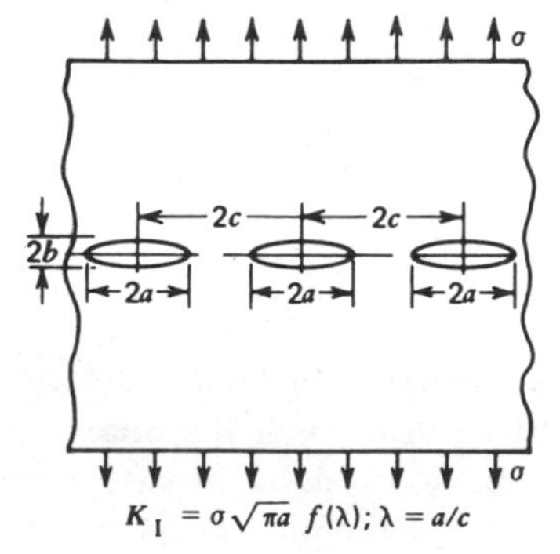

$K_I = \sigma\sqrt{\pi a}\, f(\lambda);\ \lambda = a/c$

λ	$f(\lambda)$
0.1	1.00
0.2	1.02
0.3	1.04
0.4	1.08
0.5	1.13
0.6	1.21

Case 3. Central Crack in Finite Width Strip Subject to Uniform Tension at Infinity b→0.

$K_I = \sigma\sqrt{\pi a}\, f(\lambda);\ \lambda = a/c$

λ	$f(\lambda)$
0.1	1.01
0.2	1.03
0.3	1.06
0.4	1.11
0.5	1.19
0.6	1.30

Case 4. Single Edge Crack in Finite-Width Sheet.

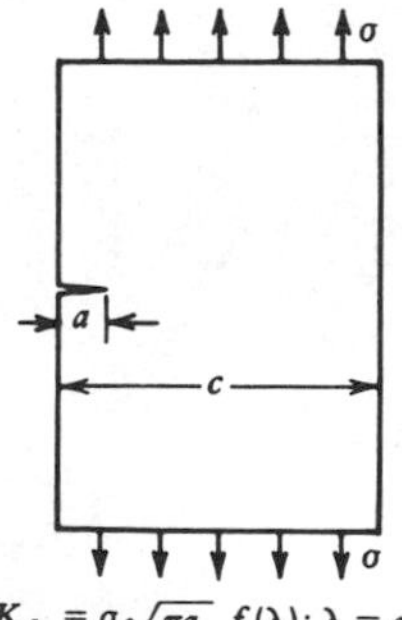

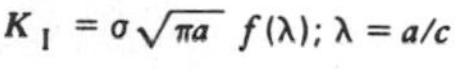

$K_I = \sigma\sqrt{\pi a}\, f(\lambda);\ \lambda = a/c$

λ	$f(\lambda)$
$0 (c \to \infty)$	1.12
0.2	1.37
0.4	2.11
0.5	2.83

Case 5. Double Edge Crack in Finite-Width Sheet.

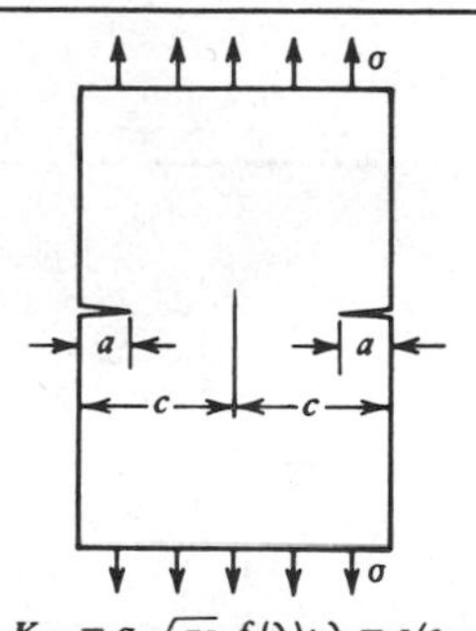

$K_I = \sigma\sqrt{\pi a}\, f(\lambda);\ \lambda = a/c$

λ	$f(\lambda)$
$0 (c \to \infty)$	1.12
0.2	1.12
0.4	1.14
0.5	1.15
0.6	1.22

Case 6. Edge Crack in Beam in Bending.

$K_I = \sigma\sqrt{\pi a}\, f(\lambda)$

$\lambda = a/2c$

$\sigma = \dfrac{3M}{2tc^2}$

λ	$f(\lambda)$
0.1	1.02
0.2	1.06
0.3	1.16
0.4	1.32
0.5	1.62
0.6	2.10

EXAMPLE 3-4.1

Longitudinal Cracks in Pressurized Pipes

General experience in nondestructive testing of pressurized pipes made of various materials indicate that longitudinal cracks of maximum length of 10 mm may be present. There is concern that the pipe will undergo sudden fracture. Hence, an estimate of the maximum allowable pressure is required. Consider two cases, one for which 17-4PH precipitation hardening steel heat treated to the properties in Table 3-4.1 is used and the other for which Ti-6Al-4Sn-1V titanium alloy heat treated to the properties in Table 3-4.1 is used.

SOLUTION

By fracture mechanics concepts, unstable crack growth (crack propagation) occurs at a load level for which the potential energy available for crack growth exceeds the work done in extending the crack (creating additional crack surface). For the pressurized pipe, the stress state of the crack corresponds to that of Case 1 of Table 3-4.2; see also Fig. E3-4.1. Thus, $K_1 = \sigma\sqrt{\pi a}$, where a is the half-crack length, $\sigma = pr/t$, where p is the internal pressure, r is the pipe inner radius, and t is the pipe thickness. By fracture mechanics concepts, $K_{\mathrm{I}} = K_{\mathrm{IC}}$ for unstable crack growth.

Case A/17-4PH precipitation hardening steel. By Table 3-4.1, $K_{\mathrm{IC}} = 48$ $\mathrm{MPa}\sqrt{\mathrm{m}}$. We find the maximum allowable pressure to be

$$p_{\max} = \frac{t}{r}\frac{K_{\mathrm{IC}}}{\sqrt{\pi a}} = \frac{48\sqrt{1000}}{\sqrt{5\pi}}\frac{t}{r} = 382\frac{t}{r} \tag{1}$$

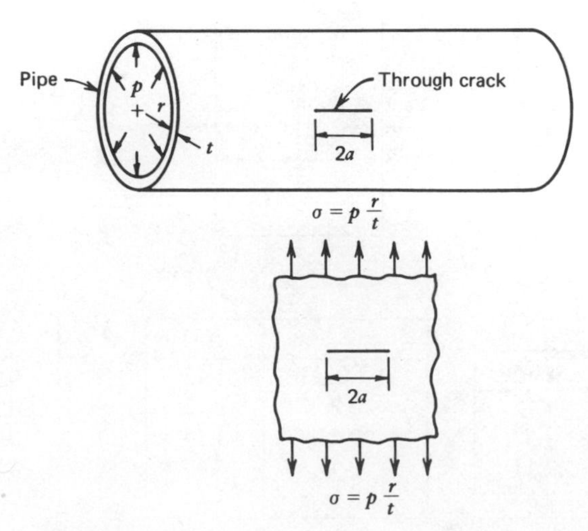

Fig. E3-4.1

If, on the other hand, the pressure p is fixed, the critical value of the ratio t/r is

$$\left(\frac{t}{r}\right)_{\text{critical}} = 0.00261p \tag{2}$$

These results assume that the thickness t is greater than 4.2 mm so that Eq. (3-4.2) is satisfied. Alternatively, by a conventional criterion of brittle fracture, say by the maximum principal stress criterion,

$$p_{\max} = \sigma_u \frac{t}{r} = 1331 \frac{t}{r} \text{ MPa} \tag{3}$$

where no account is taken of existing cracks in the material. Thus, by fracture mechanics analysis, the critical pressure that is predicted is about 71 percent less than that predicted by the maximum principal stress criterion. If the half crack length is decreased in magnitude, the fracture pressure given by Eq. (1) increases in magnitude; the computed value becomes conservative for $a < 4.2$ mm since Eq. (2) is no longer satisfied. For very small values of a, the failure pressure is not given by Eq. (3) since general yielding failure will occur at

$$p_{\max} = Y \frac{t}{r} = 1172 \frac{t}{r} \text{ MPa}$$

Case B / Ti-6Al-4Sn-1V titanium alloy. By Table 3-4.1, $K_{\text{IC}} = 93 \text{ MPa}\sqrt{m}$. As in Case A, by fracture mechanics

$$p_{\max} = \frac{t}{r} \frac{K_{\text{IC}}}{\sqrt{\pi a}} = \frac{93\sqrt{1000}}{\sqrt{5\pi}} \frac{t}{r} = 742 \frac{t}{r} \text{ MPa}$$

This pressure probably would not cause brittle fracture. Even if the thickness t was equal to or greater than 28 mm, the crack half length $a = 5$ mm is much less than that required (28 mm) to satisfy Eq. (3-4.2).

Another example problem of brittle fracture of cracked members is given in Chapter 12, Example 12-3.1. Additional problems are listed in Problem Set 3-4 and Problem Set 12-3.

Brittle Fracture Resulting from High Cycle Fatigue Loading / Many current studies of fatigue are devoted to problems of low cycle fatigue of members made of ductile materials. For such problems, large plastic strains occur at the section of the member where fracture finally occurs. Consequently, we consider the material in members that undergo low cycle fatigue to be in a ductile state. Failure resulting from low cycle fatigue are beyond the scope of this book, and the reader is referred to the literature.[12,13] Fatigue failures may occur with only small plastic strains. Such failures are called failures due to high cycle fatigue. For members made of ductile metals, high cycle fatigue failure occurs after about 10^6 cycles. An example problem and additional problems of high cycle fatigue are given at the end of Section 3-5.

PROBLEM SET 3-4

1. A circular shaft is made of gray cast iron, which may be considered to be linearly elastic up to its ultimate strength $\sigma_u = 145$ MPa. The shaft is subjected to a moment $M = 5.50$ kN · m and a torque $T = 5.00$ kN · m. Determine the diameter d of the shaft if the factor of safety against brittle fracture is $SF = 4.00$.

2. A piece of chalk of diameter d is subjected to an axial load P and to a torque T. Assume that the chalk remains linearly elastic up to the ultimate strength σ_u. The axial load $P = \sigma_u \pi d^2/12$. Determine the magnitude of the torque T that will cause brittle fracture.

   Ans. $T = \pi d^3 \sigma_u / 8\sqrt{6}$

3. A 50 mm diameter shaft is made of a brittle material. The shaft is subjected to a static torque $T = 1.20$ kN · m. A bending moment M is increased in magnitude until fracture. The fracture surface is found to make an angle of 1.000 rad with a longitudinal line drawn on the shaft. If the maximum principal stress criterion of failure is valid for this material and for this loading, determine the magnitude of M and the ultimate strength σ_u for the material.

4. A circular shaft is made of gray cast iron, which may be considered to be linearly elastic up to its ultimate strength $\sigma_u = 150$ MPa. The shaft has a diameter of 125 mm and is subjected to a bending moment $M = 7.50$ kN · m. Determine the maximum value for torque T that can be applied to the shaft if it has been designed with a factor of safety of 3.00 for both M and T.

   Ans. $T = 8.98$ kN · m

5. Let the bending moment for Problem 4 be applied by a dead load which is known to remain constant with time. The variation of torque T with time is unknown. Determine the limiting value for T if the factor of safety for M remains 3.00, while the factor of safety for T is increased to 5.00.

6. A long strip of aluminum alloy 2024-T851 has a width of 150 mm and a thickness of 8.00 mm. An edge crack of length $a = 9.00$ mm (Case 4 of Table 3-4.2) is located at one edge of the strip near the

center of its length. Determine the magnitude of the axial load P that will cause brittle fracture.

Ans. $P = 137.3$ kN

7. The long strip in Problem 6 has a double edge crack with $a = 9.00$ mm (Case 5 in Table 3-4.2). Determine the magnitude of the axial load P that will cause brittle fracture.

8. A 2024-T851 aluminum alloy pipe is used as a tension member. The pipe has an outside diameter of 100 mm and a wall thickness of 8.00 mm. An inspection of the pipe locates a circumferential through thickness crack having a length of 15.0 mm (Case 3 of Table 3-4.2). If the axial load P is increased to failure, will the failure be brittle fracture? What is the failure load?

   Ans. Yes. $P = 342.7$ kN

9. Let the tension member in Problem 8 be made of AISI 403 stainless steel. If the axial load P is increased to failure, will the failure be brittle fracture? What is the lower limit for the failure load?

10. A 60.0 mm square beam is made of AISI 4340 steel that has been heat treated to give the properties indicated in Table 3-4.1. On the tension side of the beam a transverse crack has a depth of 8.00 mm (Case 6 of Table 3-4.2). Determine the magnitude of the moment M that will cause brittle fracture.

    Ans. $M = 12.97 \text{ kN} \cdot \text{m}$

11. A simple beam has a span of 4.00 m, a depth of 250 mm, and a width of 100 mm. The beam is made of 6061-T651 aluminum alloy and is loaded by a concentrated load P at midspan. The design load for the beam has been calculated using a factor of safety of 3.00 assuming general yielding theory of failure. Determine the magnitude of P. An inspection of the beam located a transverse crack at a distance of 1.50 m from one end. The crack has a depth of 24.0 mm. What is the factor of safety for the beam against brittle fracture?

3-5 PROGRESSIVE FRACTURE (HIGH CYCLE FATIGUE FOR NUMBER OF CYCLES $N > 10^6$)

A basic concept in fracture predictions by fracture mechanics analysis is the existence of a critical crack size for a given geometry and load. In some practical applications, the size of the critical crack or defect is so large that the defect can usually be detected and corrected before the part is put into service. However, most parts contain subcritical cracks or flaws. These subcritical cracks may, during operation, grow to critical size and cause catastrophic failures. Several mechanisms of subcritical flaw growth exist. Of particular importance in practical problems are the mechanisms of fatigue and stress corrosion cracking. Here we consider briefly fatigue criteria associated with subcritical flaw (crack) growth by the mechanism of fatigue. The mechanism of stress corrosion cracking is left to more specialized works.[1] However, one should note that fatigue crack growth processes cannot be fully explained unless effects of environment (corrosion) are considered. Encouraged by the success of linear elastic fracture analysis in explaining sudden brittle fracture, several investigators have attempted to describe subcritical crack growth in terms of fracture mechanics parameters. The objective of the fracture mechanics approach is essentially to replace uncertainties (that is, the degree of ignorance) in conventional design factors by more reliable quantitative parameters that are more direct measures of the material fracture resistance. Early results seem to indicate that a material exhibiting high static fracture toughness also gives good resistance to subcritical crack propagation due to fatigue. By fatigue failure (progressive fracture), we mean one that occurs after a number of cycles under alternating stresses with peak stresses below the ultimate strength of the material in a simple tension test. We restrict our discussion to ordinary (room) temperatures. Fatigue fracture at high temperature (thermal fatigue) has been treated in the literature.[7,12] To simplify our discussion, we further divide our treatment of the fatigue growth of subcritical cracks into the initiation of cracks as microcracks and the propagation of cracks as macrocracks to fracture. For example, consider a smooth shaft rotating in bearings and subjected to loads that produce bending moments. As the shaft rotates, the maximum fiber stresses alternate between tension and compression. In turn these cyclic components at a surface point set up alternating shearing stresses, maximum on 45° planes with the tension-compression direction. If these stresses exceed locally the elastic limit, alternating plastic deformation (strain) is produced in the surface grains. Since the plastic deformation is not fully reversible, at least two effects result: (a) a general strain hardening of the surface grains that localizes the deformation along active slip-bands inclined roughly at 45° to the direction of the maximum principal stress and (b) a nonreversible flow at the surface producing *extrusions* that pile up material on the surface and associated *intrusions* that act as microcracks along the active slip-bands. An intrusion initially propagates along an active slip-band as a so-called *stage I* crack until it reaches a length sufficiently large with respect to the member for the crack tip stress field to become dominant. Under continued repeated loading, the intrusion then propagates as a *stage II* crack, normal to the maximum principal stress until the member breaks by a fast tensile fracture. During stage II propagation, *striations or ripples* occur on portions of the fatigue crack surface perpendicular to the tensile direction. The growth of the crack from intrusion to the stage II propagation is a rapidly accelerating process. Hence, the process is strongly controlled by the initiation of the intrusion. Fairly large amounts of alternating plastic deformations are required to form intrusions and extrusions on an initially smooth surface. Consequently, rather large alternating stresses are needed to precipitate fatigue fracture. Hence, it follows that once a crack has been initiated in any initially smooth surface, it propagates rapidly due to the high stress.

Conventional fatigue (endurance) testing has been concerned primarily with the testing of specimens with smooth surfaces, under conditions of rotating-bending or uniaxial tension-compression cycling. The results of these tests are presented in the form of plots of stress (applied alternating stress magnitude $\pm\sigma$) versus the number N of stress cycles (usually represented as $\log N$) required to cause fracture. These plots are called σ-N diagrams (often called S-N diagrams in the literature) (Fig. 3-5.1). Wöhler discovered that the steel in the railroad car axles he tested exhibited a behavior called an *endurance limit*: a stress level below which a material can undergo repeated cycling of stress indefinitely and show no evidence of fracture. However, later investigators found that many materials did not exhibit the endurance limit response, but rather continued to exhibit fracture, provided that the repetition of load was continued for a sufficiently large number of times (Fig. 3-5.1). Thus in general, under fatigue testing of smooth specimens, materials exhibit one of two types of responses. In mild steel or in certain other steels, an endurance limit is observed, below which the specimen seems to last indefinitely. On the other hand, many materials do not exhibit a clear-cut endurance limit, but the σ-N curves continues downward as N increases. For these materials (for example, most nonferrous materials), it is customary to *define* the stress to cause failure in a given number of cycles (say $N = 10^8$) as the endurance limit stress (Fig. 3-5.1).

The endurance limit σ_L is an important material property for members subjected to fatigue loading as long as the number of cycles of loading approaches the number associated with σ_L. It should be noted that other fatigue properties for each material can be obtained from the σ-N curve. Many members are subjected to fewer cycles than are associated with the endurance limit. For each value of N in Fig. 3-5.1 there is a stress σ_{am}, the fatigue strength; a specimen subjected to completely reversed cycles of stress at σ_{am} will fracture after N cycles. Note that $\sigma_{am} = \sigma_L$ at the endurance limit.

Typical σ-N curves for completely reversed loading of smooth specimens of a structural steel, a stainless steel, and an aluminum alloy are shown in Fig. 3-5.2. If a large number of specimens of one of the metals

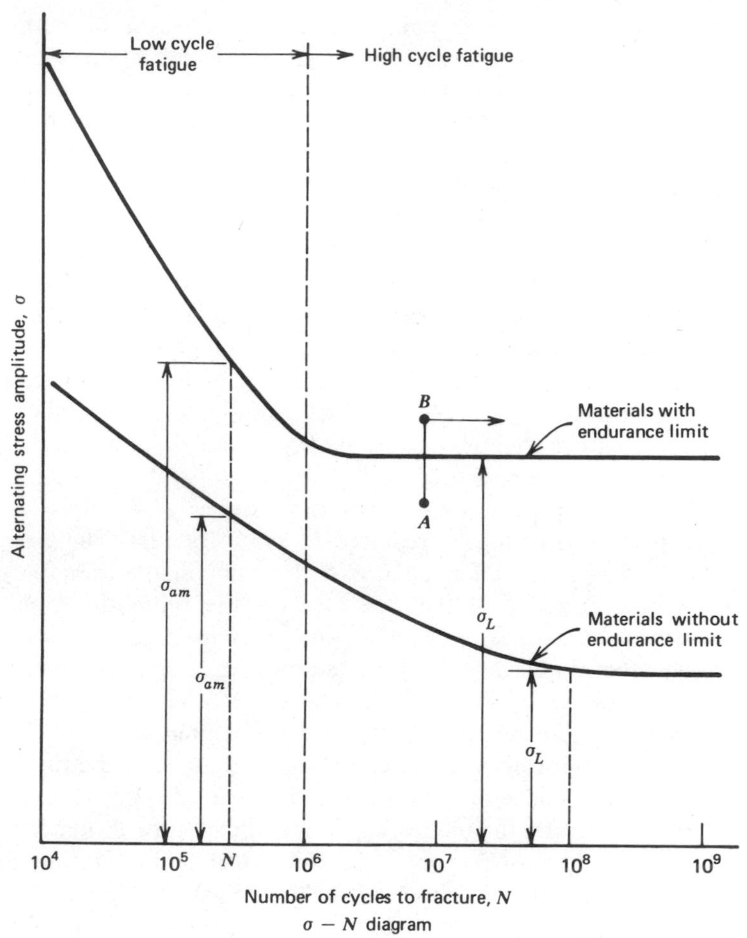

Fig. 3-5.1/σ-N diagram.

in Fig. 3-5.2 were tested at one stress level, the data would indicate appreciable scatter. The σ-N curve usually reported for a given metal (Fig. 3-5.2) is often taken to represent a 50 percent probability of failure curve. If a large number of fatigue specimens of one of the metals in Fig. 3-5.2 were tested at a given fatigue strength σ_{am}, approximately 50 percent of the specimens would be expected to fail prior to N cycles of load corresponding to the given σ_{am}. The statistical nature[14] of fatigue data may be represented either as a series of σ-N curves representing different probabilities of failure or as a σ-N band (Fig. 3-5.3). Because of the large expense involved, σ-N probability curves or σ-N bands (Fig. 3-5.3) are seldom obtained.

The experimental σ-N curves in Figs. 3-5.1 and 3-5.2 remain fairly valid for constant amplitude ($\pm\sigma_{am}$) tests. However, deviations from

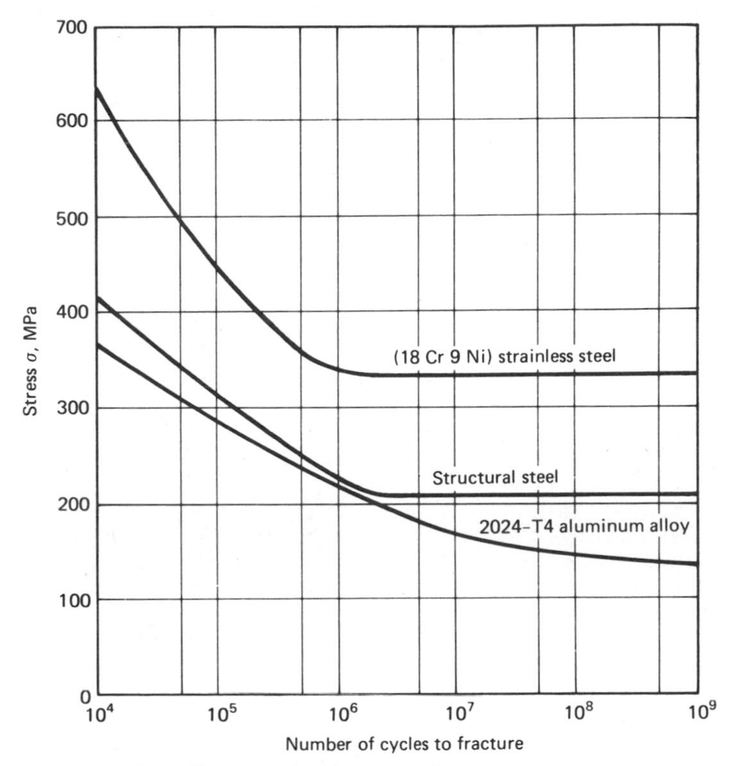

Fig. 3-5.2/σ-N diagrams for three metals.

constant amplitude alternating stress may alter the σ-N curve. For example, if a steel is subjected to cyclic stress of constant amplitude for a sufficiently long time below the endurance limit (point A in Fig. 3-5.1), its endurance limit may be increased (point B). This process, known as *coaxing*, is sometimes employed to improve resistance to fatigue fracture.

In addition to coaxing, various other factors affect the fatigue strength. For example, the fatigue strength of a material may be altered by such factors as frequency of cycling, cold working of the material, temperature, corrosion, residual stresses, surface finish, and mean stress.

As noted above, the σ-N curve gives the fatigue strength σ_{am} for specified N for members subjected to completely reversed loading (loading under the condition of zero mean stress). Nonzero mean stresses have a marked effect on the fatigue strength. There have been several relations proposed to describe the effects of mean stress. Three such relations are (for one dimensional testing)

(a) Soderberg relation

$$\frac{\sigma_a}{\sigma_{am}} + \frac{\sigma_m}{Y} = 1 \tag{3-5.1}$$

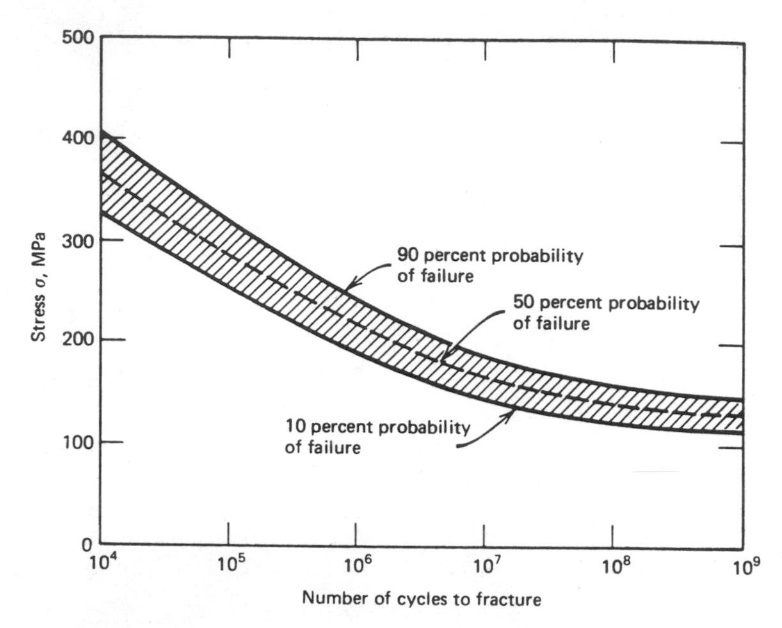

Fig. 3-5.3/σ-N band indicating scatter of fatigue data.

(b) Gerber relation

$$\frac{\sigma_a}{\sigma_{am}} + \left(\frac{\sigma_m}{\sigma_u}\right)^2 = 1 \qquad (3\text{-}5.2)$$

(c) Goodman relation

$$\frac{\sigma_a}{\sigma_{am}} + \frac{\sigma_m}{\sigma_u} = 1 \qquad (3\text{-}5.3)$$

where σ_a is the stress amplitude, σ_{am} is the fatigue strength for given N for zero mean stress, Y is the yield stress, σ_m is the mean stress, and σ_u is ultimate strength. The relation between σ_a and σ_m for cyclic loading with unequal stresses is indicated in Fig. 3-5.4. For most metals, the Soderberg relation yields conservative estimates of critical stress amplitude σ_a (or range of stress $2\sigma_a$). The Goodman relation gives reasonably good results for brittle materials, whereas it is conservative for ductile materials. The Gerber relation yields fairly good estimates for σ_a for ductile materials. The Soderberg relation, the Gerber relation, and the Goodman relation are interpreted in Fig. 3-5.5. For any mean stress σ_m, the ordinate to a particular curve gives the magnitude of σ_a for that relation. The dashed line CD is generally used along with the Gerber and Goodman relations since failure by general yielding is assumed to occur along CD.

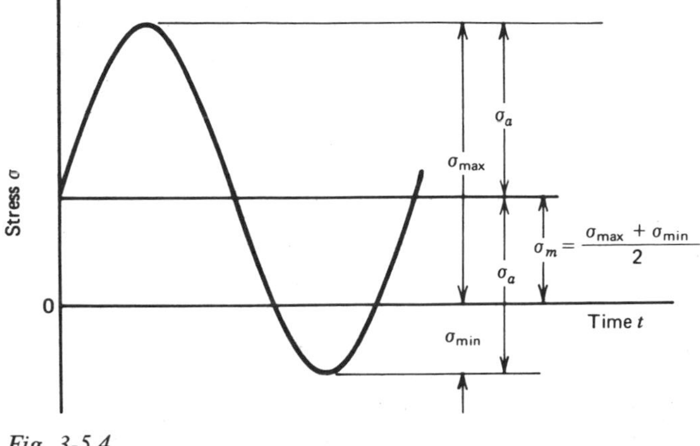

Fig. 3-5.4

In the case of fatigue loading, design applications in this book are limited to high cycle fatigue (Fig. 3-5.1) of members made of ductile metals and subjected to cyclic loading with constant mean stress and constant amplitude of alternating stress. The material property (fatigue strength) is assumed to be obtained from smooth specimens (free of stress concentrations) subject to completely reversed loading under a uniaxial state of stress (tension-compression specimens or reversed bending specimens). The fatigue strength for a specified number of cycles N, where $N > 10^6$, is specified by the magnitude of σ_{am}. The effect of mean

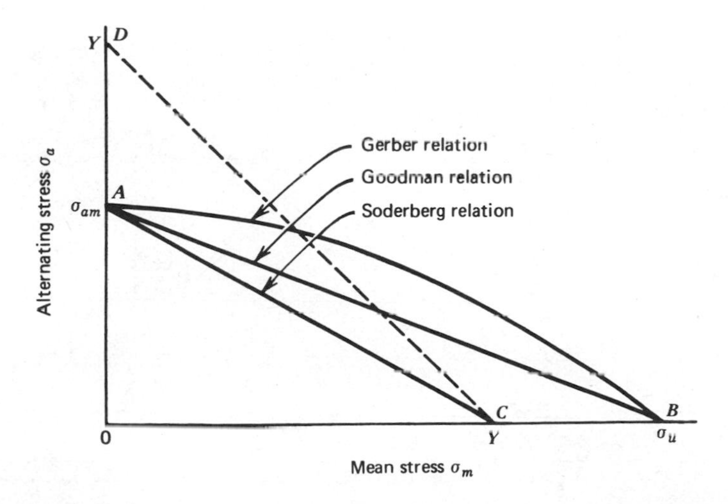

Fig. 3-5.5

stress σ_m is assumed to be given by either the Soderberg relation (Eq. 3-5.1), the Gerber relation (Eq. 3-5.2), or the Goodman relation (Eq. 3-5.3). The criteria of failure for members subject to multiaxial states of stress are assumed to be the same as for the general yielding failure. Both the maximum shearing stress criterion and the maximum octahedral shearing stress criterion of failure are widely used in the design for high cycle fatigue. In the case of low cycle fatigue, many criteria of failure have been proposed; however, none has been widely accepted at the present time.

Stress Concentrations/Stress concentrations (Chapter 12) greatly increase the stresses in the neighborhood of the stress concentrations and generally limit design loads when the member is subjected to fatigue loading. The effect of stress concentrations in fatigue loading is discussed in Art. 13-4 and is illustrated by example problems and by Problems 3 to 6 in Problem Set 13-4.

EXAMPLE 3-5.1

Fatigue of Torsion-Bending Member

The member in Fig. E3-5.1 is made of steel ($Y = 345$ MPa and $\sigma_u = 586$ MPa), has a diameter $d = 20$ mm, lies in the plane of the paper, and has a radius of curvature $R = 800$ mm. The member is simply supported at A and B and is subjected to a cyclic load P at C normal to the plane of the member. (a) The load varies from P_{max} to $P_{min} = -5P_{max}/6$. The en-

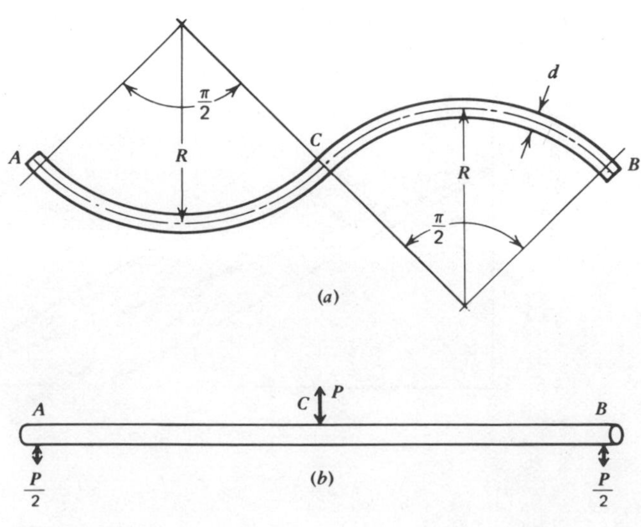

Fig. E3-5.1

durance limit for $N = 10^7$ for the steel is $\sigma_{am} = 290$ MPa. Determine the magnitude of P_{max} based on a factor of safety $SF = 1.80$ against failure at $N = 10^7$ cycles. Use the maximum octahedral shearing stress criterion of failure. Assume that the Gerber relation (Eq. 3-5.2) is valid. (b) Obtain the solution for $P_{min} = -P_{max}/2$.

SOLUTION

(a) The magnitude of the alternating component of stress σ_a is obtained by Eq. (3-5.2). For linearly elastic behavior, $\sigma_{min} = -5\sigma_{max}/6$. But $\sigma_{max} - \sigma_{min} = 2\sigma_a$ and $\sigma_{max} = \sigma_m + \sigma_a$. Hence, $\sigma_m = \sigma_a/11$. Substituting this value of σ_m and values of Y and σ_u in Eq. (3-5.2), we obtain $\sigma_a = 289$ MPa. Thus, $\sigma_{max} = \frac{12}{11}\sigma_a = 315$ MPa and $\sigma_{min} = -263$ MPa. This result indicates that a smooth fatigue specimen cycled between these stress levels would not fracture before 10^7 cycles. Since σ_{max} is less than Y, failure would be by fatigue and not by general yielding.

The load P on member ACB in Fig. E3-5.1 can be cycled from $P_{max}(SF)$ to $P_{min}(SF)$ through 10^7 cycles before fracture by fatigue. The reactions at A and B when $P_{max}(SF)$ is applied are equal to $P_{max}(SF)/2 = 0.90P_{max}$. The reaction $0.90P_{max}$ produces a moment and torque of equal magnitude at the critical section at C. Thus,

$$M = T = 0.90P_{max}R = 720P_{max} \text{ (N} \cdot \text{mm)}$$

The bending stress σ due to M and the shearing stress τ due to T at C are

$$\sigma = \frac{Mc}{I} = \frac{720P_{max}(10)(4)}{\pi(10)^4} = 0.917P_{max} \text{ (MPa)}$$

$$\tau = \frac{Tc}{J} = \frac{720P_{max}(10)(2)}{\pi(10)^4} - 0.458P_{max} \text{ (MPa)}$$

The magnitude of P_{max} is obtained by means of Eq. (3-3.7) when $\sigma_{max} = 315$ MPa is substituted for Y. Thus,

$$3\left[\frac{0.458P_{max}}{315}\right]^2 + \left[\frac{0.917P_{max}}{315}\right]^2 = 1$$

$$P_{max} = 260 \text{ N} \quad \text{and} \quad P_{min} = -216 \text{ N}$$

(b) For $\sigma_{min} = -\sigma_{max}/2$, we obtain $\sigma_m = \sigma_a/3$. Substitution of this value of σ_m along with values for Y and σ_u into Eq. (3-5.2) gives $\sigma_a = 282$ MPa and $\sigma_{max} = 376$ MPa. Since σ_{max} is greater than Y, failure of the member occurs by general yielding and not by fatigue. Substitution of values of σ, τ, and Y into Eqs. (3-3.7) gives $P_{max} = 285$ N and $P_{min} = -142$ N.

PROBLEM SET 3-5

1. A tension member is cycled an indefinitely large number of times from $P_{min} = -10.0$ kN to $P_{max} = 16.0$ kN. The member is made of a steel ($\sigma_u = 700$ MPa, $Y = 450$ MPa, and $\sigma_{am} = \sigma_L = 350$ MPa). Using the Gerber relation, determine the diameter of the rod for a factor of safety $SF = 2.20$.

2. Let the tension member in Problem 1 be cycled an indefinitely large number of times between $P_{min} = 0$ to $P_{max} = 16.0$ kN. Determine the diameter of the tension member for a factor of safety $SF = 2.20$. What is the mode of failure?

   Ans. $d = 9.98$ mm; general yielding mode of failure.

3. A cast iron I-beam has a depth of 150 mm, a width of 100 mm, and equal flange and web thicknesses of 20 mm. The beam is subjected to 10^6 cycles of loading from $M_{min} = 5.00$ kN · m to M_{max}. Consider the cast iron to be a brittle material ($\sigma_u = 200$ MPa and $\sigma_{am} = 90.0$ MPa for $N = 10^6$). Using the Goodman relation, determine M_{max} based on a factor of safety $SF = 2.50$.

4. A thin-wall cylinder is made of 2024-T4 aluminum alloy ($\sigma_u = 430$ MPa, $Y = 330$ MPa, $\sigma_{am} = 190$ MPa for $N = 10^6$). The cylinder has an inside diameter of 300 mm and a wall thickness of 8.00 mm. The ends are strengthened so that fatigue failure is assumed to occur in the cylinder away from the ends. The pressure in the cylinder is cycled 10^6 times between a vacuum of 2.00 MPa and $p_{max} = 7.00$ MPa. What is the factor of safety against fatigue failure if design is based on the Gerber relation?

   Ans. $SF = 2.13$

5. A shaking mechanism of a machine has a crank shown in Fig. P3-5.5. The crank is made of a stress relieved cold worked SAE 1040 steel ($\sigma_u = 830$ MPa, $Y = 660$ MPa, and $\sigma_{am} = \sigma_L = 380$ MPa). A completely reversed load $P = 500$ N is to be applied for up to 10^8 cycles to the crank pin, normal to the plane of the crank. Determine the diameter d of the shaft based on a factor of safety $SF = 1.75$ using the octahedral shearing stress criterion of failure.

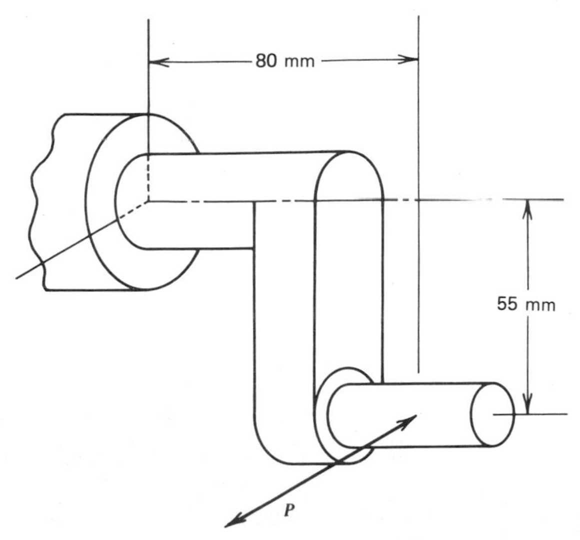

Fig. P3-5.5

6. The crank in Problem 5 has a diameter $d = 15.0$ mm and is made of 2024-T4 aluminum alloy ($\sigma_u = 430$ MPa, $Y = 330$ MPa, $\sigma_{am} = 160$ MPa for $N = 10^8$). Let the load vary between 0 and $P_{\max} = 500$ N for 10^8 cycles. Assume that the Gerber relation (Eq. 3-5.2) is valid. (a) Is the design governed by general yielding failure or fatigue failure? (b) Determine the magnitude of the safety factor SF used in the design based on the octahedral shearing stress criterion of failure.

   Ans. (a) Fatigue failure; (b) $SF = 2.03$

7. Let the shaft in Problem 3-3.6 be made of SAE 1040 steel ($\sigma_u = 830$ MPa, $Y = 660$ MPa, $\sigma_{am} = \sigma_L = 380$ MPa). If the shaft is rotated under constant load an indefinitely large number of times, determine the diameter of the shaft for a factor of safety $SF = 2.00$.

8. A 30.0 mm diameter shaft is subjected to cyclic combined bending and torsion loading such that $M = 200$ P N · mm and $T = 150$ P N · mm where the magnitude of P varies from $P_{\min} = -0.60\ P_{\max}$ to $P_{\max}$. The shaft is made of a stress-relieved cold-worked SAE 1060 steel ($\sigma_u = 810$ MPa, $Y = 620$ MPa, $\sigma_{am} = \sigma_L = 410$ MPa). Using a factor of safety $SF = 1.80$, determine $P_{\max}$ for 10^7 cycles of loading. Use the octahedral shearing stress criterion of failure and the Gerber relation.

   Ans. $P_{\max} = 3.12$ kN

3-6
FAILURE CRITERIA. BUCKLING

In Art. 3-2, we noted that the maximum load that can be applied to a member without causing it to cease to function satisfactorily in a structure or a machine (and hence to fail structurally) may be limited by the elastic deflection of the member. It was also noted that at least two different types of action or behavior of a member related to elastic deflection may limit the load carrying capacity.

One action consists of the deflection of a member which, as the loads increase, remains in stable equilibrium for all elastic deflections, as, for example, a simply supported beam. In this type of action the deflections and stresses are usually proportional to the loads and, furthermore, the elastic deflections are usually small enough so that second order terms may be neglected in mathematical expressions for the deflections. The relations between the loads and the elastic deflections of a member under conditions in which deflections are proportional to loads may be obtained by several methods (Chapter 4).

The second structural action in which elastic deflection may limit the maximum load that can be applied to the member without causing the member to fail structurally is denoted as *elastic buckling*. Such buckling action can occur in members that have certain relative proportions. Such members are frequently called thin-wall or lightweight members and include slender columns, thin-wall cylinders under axial compression, uniform external lateral (radial) pressure, or torsion, wide-flanged I-beams, and thin plates under edge compression or edge shear.

The significant fact or dominant characteristic of buckling, however, is the same in all cases, namely, that the elastic deflections and stresses in the member are *not* proportional to the loads after buckling takes place, even though the material acts elastically (stress is proportional to strain).

Elastic buckling action arises out of a condition of neutral equilibrium that develops when the applied load on the member reaches a so-called critical value. If the load is slightly below this critical load, the member is in equilibrium. But if the load is increased slightly above the critical load, the deflection of the member can increase abruptly (not in proportion to the load) and, unless it is extremely slender, the member can pass into a completely unstable condition, owing to inelastic action, and can undergo total collapse. Buckling, therefore, is frequently referred to as *structural instability*. Furthermore, in this type of failure the critical or buckling load usually represents the practical maximum or ultimate load that the member can be expected to resist, even though when the member is subjected to this load the stress in the material may be less than the compressive proportional limit. When elastic buckling is the cause of failure of a member, therefore, the problem is not that of determining the relationship between loads and deflection, but that of obtaining an expression for the critical or buckling load.

The first published analysis of elastic instability or buckling has been attributed to Leonhard Euler[16] (1707–1783), who in 1744 published a penetrating paper on the subject. The simplest illustration of elastic buckling is that of the failure of a straight, slender column subjected to a gradually applied axial load. A brief review of the elastic buckling of a slender column is given in the next article.

Caution/The influence on buckling of temperature, especially elevated temperature, and of time, such as in prolonged loading leading to creep, are *not* considered in the subsequent discussions. In practice, these effects may be extremely important.[17]

1. Elastic Buckling of an Ideal Slender Column/If a slender column is perfectly straight, the load truly axial, and the material perfectly homogeneous, the column will remain straight under any value of the load; it will not bend. If at a certain (critical) load, however, a small lateral force is applied, giving the column any small deflection, and the lateral force is then removed, the critical axial load is the load that will hold it in the slightly bent position (see Art. 3-2). The critical load, however, is nearly constant for a range of small elastic deflections in which the bent column is said to be in *neutral* equilibrium and, therefore, the deflections (and hence strains and stresses) in the bent positions are not proportional to the applied load, although the material still acts elastically. This behavior is

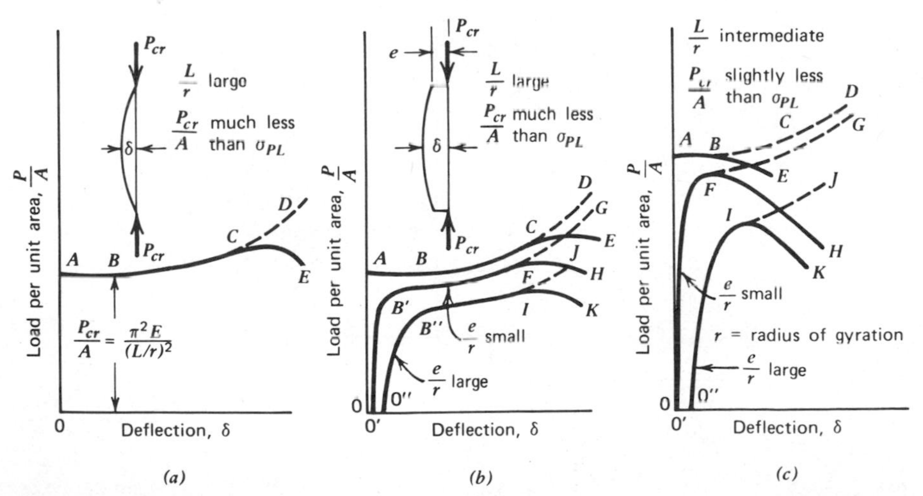

Fig. 3-6.1/Relation between load and deflection for columns.

represented by the line $0AB$ in Fig. 3-6.1*a*, where $0A$ represents the critical or buckling load for the ideal slender column. The length AB represents a relatively small deflection (larger deflections indicated by BCD are considered in the next article); the value of the deflection δ in the sketch of the deflected column is greatly exaggerated.

The expression for the critical load may be found by applying to the deflected column the elastic curve equation $EI(d^2y/dz^2) = M$ for a bar subjected to bending that causes small deflections. For the pivot-end column shown in Fig. 3-6.2*a* the bending moment M is equal to $-Py$ and hence the equation for the slightly deflected column is $EI(d^2y/dz^2) = -Py$. The solution[18] of this equation for the critical or buckling value for P, that is, the smallest value of P that will hold the column in a slightly deflected form, gives [see also Eq. (3-2.1)]

$$P_{cr} = \frac{\pi^2 EI}{L^2} \quad \text{or} \quad \sigma_{cr} = \frac{P_{cr}}{A} = \frac{\pi^2 E}{(L/r)^2} \tag{3-6.1}$$

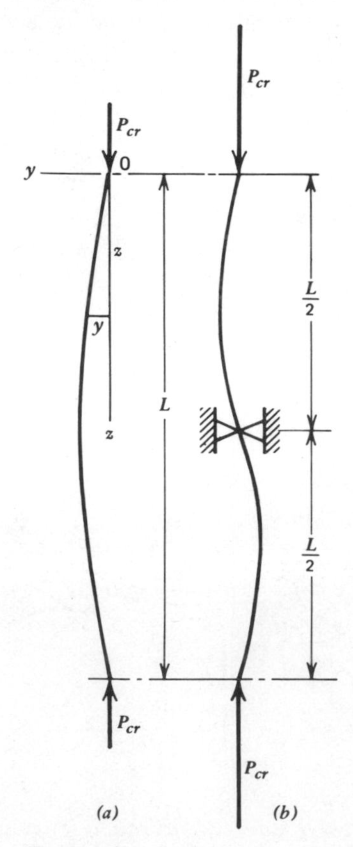

Fig. 3-6.2

in which E is the modulus of elasticity of the material, L is the length of the column, I is the moment of inertia of the cross section about the centroidal axis having the least moment of inertia, r is the least radius of gyration of the cross section ($I = Ar^2$), L/r is the slenderness ratio, and σ_{cr} is the uniform stress in the column when the load P_{cr} is acting on the column just before the lateral force is applied (σ_{cr} is called the buckling stress). It is important to note that the only property of the material that influences the elastic buckling load is the stiffness of the material as measured by the modulus of elasticity. The influence of the relative values of the dimensions on the buckling load (or load per unit area) is indicated by the slenderness ratio L/r. Equation (3-6.1) is called Euler's formula; it is discussed further later in this article.

Higher Buckling Loads / A slender column (one having a relatively large L/r ratio), and other members having relative dimensions such that they fail by buckling before the stress exceeds the proportional limit of the material, may have several elastic buckling or critical loads. If the column is restrained from buckling under the lowest buckling load, it may buckle at a higher value of the load, as the load is increased. For example, a pivot-end slender column (Fig. 3-6.2*a*) can buckle in the form of one lobe at the load given by Eq. (3-6.1). If, however, the member is prevented from bending in the form of one lobe by restraints at the middle as shown in Fig. 3-6.2*b*, the load can be increased above the first critical load until the column buckles in the form of two lobes shown in Fig. 3-6.2*b*. For this case, the critical load is given by the expression

$$P'_{cr} = \frac{\pi^2 EI}{\left(\frac{1}{2}L\right)^2} \quad \text{or} \quad \frac{P'_{cr}}{A} = 4\frac{\pi^2 E}{(L/r)^2}$$

and hence is four times the value of the critical load given by Eq. (3-6.1). In general, the critical elastic buckling load (or load per unit area) for an ideal slender column is given by the formula

$$\frac{P_{cr}}{A} = K\frac{\pi^2 E}{(L/r)^2}$$

where the value K depends on the restraints to which the column is subjected.

Similarly, if lateral restraints are applied at *two* equally spaced points, the elastic buckling load is increased further. However, then the column may fail structurally by direct yielding or by inelastic buckling rather than by elastic buckling. In other words, the critical elastic buckling load has little physical significance unless that critical value is reached before the proportional limit of the material is developed.

Local Buckling/A change in the relative dimensions of a member may also bring about a radical change in the type of elastic buckling from one involving the member as a whole to one involving only a part of the member. For example, aluminum columns[19] with equal-leg angle sections behave as shown in Fig. 3-6.3. When the ratio t/b for a leg of the angle section is relatively large, the column buckles as a unit in accordance with the Euler column formula (Eq. 3-6.1), but when the value of t/b is relatively small, buckling of one side or leg of the angle occurs first; this action is sometimes designated as sheet buckling, crimpling, or wrinkling and is, in general, a local action in contrast to general buckling of the member as a unit. Figure 3-6.4 shows similar wrinkling of the thin flanges of a channel section or one-half of an H section.[20] Under most condi-

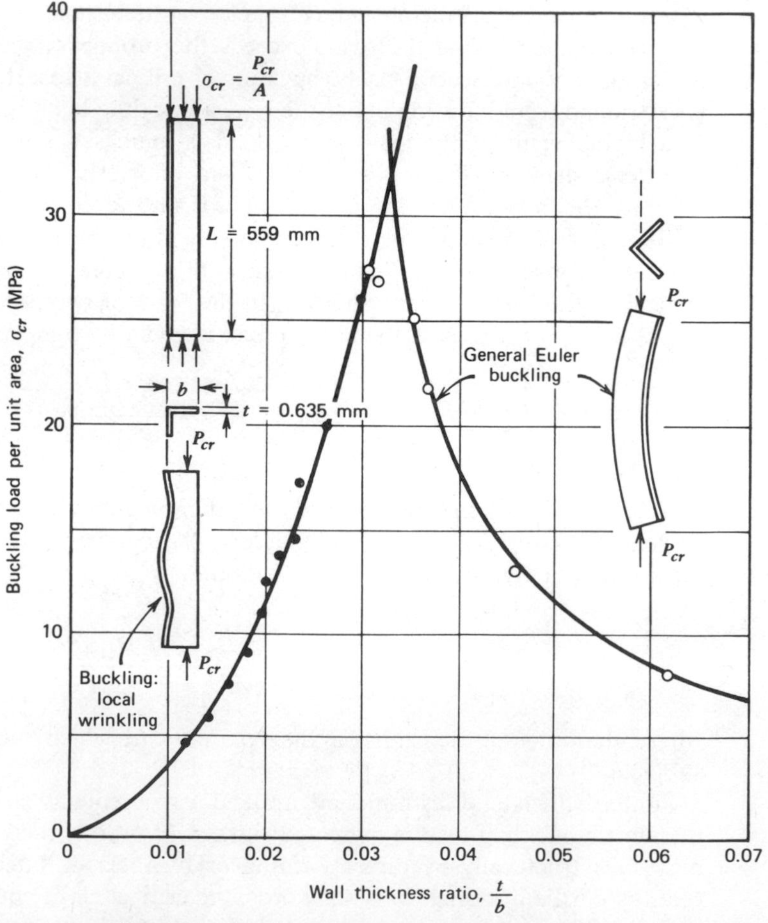

Fig. 3-6.3/Buckling loads for local buckling and Euler buckling for columns made of 245 TR aluminum alloy ($E = 74.5$ GPa).

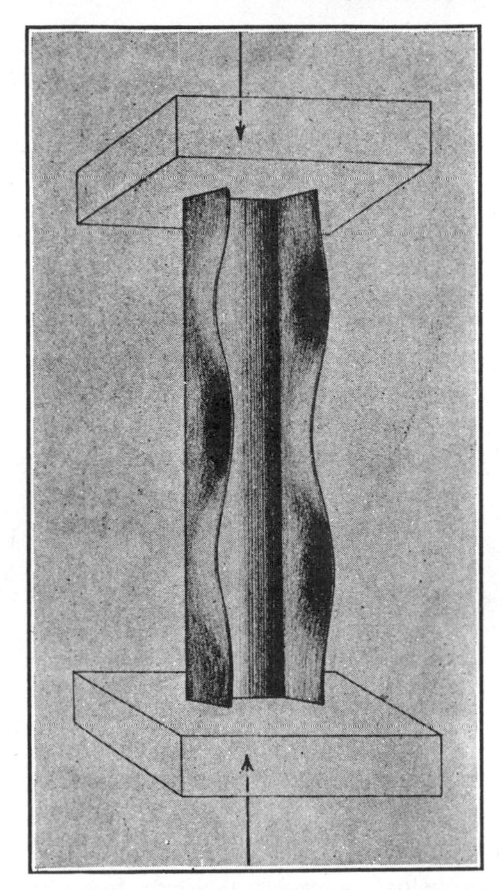

Fig. 3-6.4/Local buckling (wrinkling) of thin flanges of channel section (or half of H section; see reference 20).

tions local wrinkling reduces the buckling load for the member, but under certain conditions it may be taken advantage of to redistribute the load in the member so that additional load may be applied before the strength of the member as a whole is developed.

PROBLEM SET 3-6.1

1. An airplane compression member 2.00 m long has an elliptical cross section with major and minor diameters of $b = 150$ mm and $a = 50.0$ mm, respectively. Calculate the slenderness ratio for pivot ends. If the member is made of spruce wood with $E = 11.0$ GPa, determine the load P that can be carried by the column based on a factor of safety of $SF = 1.50$, $I = \pi ba^3/64$ and $A = \pi ab/4$.

2. Three pivot-end columns each have a cross-sectional area of 2000 mm^2 and a length of 750 mm. They are made of 7075-T5 aluminum alloy ($E = 72.0$ GPa and $\sigma_{PL} = 448$ MPa). One of the columns has a solid square cross section. A second column has a solid circular cross section. The third has a hollow circular cross section with an inside diameter of 30.0 mm. Determine the critical buckling load for each of the columns.

   Ans. $P_{cr(\text{square})} = 421$ kN; $P_{cr(\text{circle})} = 402$ kN; $P_{cr(\text{hollow circular})} = 686$ kN

3. Assume that the dimensions of the angle section of the pivot-end column shown in Fig. 3-6.3 are $t = 0.635$ mm, $b = 12.0$ mm, and $L = 559$ mm. Show by Fig. 3-6.3 that the column fails as a unit by elastic buckling (not by local buckling). Calculate the average stress in the column at the critical buckling load.

2. Imperfect Slender Column/Deviations from the ideal conditions assumed in Section 1 always exist in actual columns and also in other members that fail by elastic buckling. It is important, therefore, to consider the influence on the critical elastic buckling load of such imperfections, especially of slight eccentricity of loading or initial crookedness in the column. If these deviations from ideal conditions are small, the elastic deflection of the slender column starts with the first increment of load and increases slowly as the load increases until the load approaches the critical value for the ideal column, as given by Eq. (3-6.1). However, by the time this load is reached the deflection of a slender column may be very large and, hence, may no longer be approximated by the expression $1/R = -d^2y/dz^2$, used in deriving Eq. (3-6.1). Therefore, a discussion of the load-deflection relationship for large deflections of an ideal column is desirable to explain the effect of deviations from ideal conditions. It should be noted here, however, that many columns "fail" before such large deflections occur.

Large Deflections/Southwell[21] has shown that, when large deflections occur and the correct relationship between the curvature and deflection is used, the load found to keep a slender ideal elastic column in a bent position may be larger than P_{cr} as given by Eq. (3-6.1), especially if the column is very slender (L/r large) and if P_{cr}/A is much less than the proportional limit σ_{PL}. At larger deflections the load for such columns increases as shown by the ordinates to the curve BCD in Figs. 3-6.1a and b. However, if at some deflection, such as at C in Figs. 3-6.1a and b, the proportional limit of the material is exceeded (due to excessive bending and direct compression), the maximum resisting bending moment is developed after a slight additional deflection (as will be explained later), and the load rises slightly above the load at C and then drops as indicated by the curve CE instead of continuing to rise as indicated by CD.

A real column with its imperfections is frequently considered to be equivalent to an ideal column with a small eccentricity of loading e, as shown in Fig. 3-6.1b. The curve $0'B'FG$ represents the load-deflection relationship for this column. The curve approaches the curve $0ABCD$ if the material remains elastic; but if at some deflection, such as represented by F, the proportional limit is exceeded, the load rises slightly and then drops as shown by FH. Similar remarks apply to the curve $0''B''IJ$, which indicates the influence of a larger eccentricity e. It should be noted that the maximum load occurring between F and H for a small eccentricity e is very nearly equal to P_{cr} as obtained from Eq. (3-6.1), which is for small deflections of an ideal column; hence Euler's formula may in some cases give a satisfactory value for the buckling load on very slender real columns. However, for ideal columns that are less slender, in which P_{cr}/A is only slightly less than σ_{PL}, the actual load-deflection relationship is represented in Fig. 3-6.1c by $0ABE$ instead of $0ABCD$. This behavior is explained by the fact that only a small amount of lateral deflection at point B is necessary to raise the maximum stress in the column to the proportional limit.

The curves $0'FH$ and $0''IK$ in Fig. 3-6.1c show that in a column having an intermediate value of L/r the relative effect of eccentricity of loading is much more pronounced than in the very slender column of Figs. 3-6.1b; hence Euler's formula may not give a satisfactory value of the buckling load for real columns with intermediate value of slenderness ratios.

Failure of Columns. Slender Columns/The question arises as to whether it is appropriate to use the term buckling (or instability) in describing the failure of a slender column that is *loaded eccentrically*. If the columns whose load-deflection relationships are represented by the curves $0ABCE$, $0'B'FH$ and $0''B''IK$ in Fig. 3-6.1b are used as machine or structural parts, their failure, in most instances, is considered to take place by excessive deflection at a load and deflection represented approximately by the points B, B', and B''. But the loads at B, B', and B'' are smaller than the loads that are required to cause a condition of instability or total collapse as indicated by the ordinates to C, F, and I, which represent the load and deflection where inelastic strain begins, for, as noted previously, the type of instability that is associated with total collapse is initiated by inelastic strains, although the total collapse doesn't occur until some (relatively small) additional load is applied. Hence, failure of eccentrically loaded slender members (that is, the condition that limits the maximum utilizable load) is not a condition of instability in the usual meaning of the term.

Columns of Intermediate Slenderness / Such eccentrically loaded columns are represented by curves in Fig. 3-6.1*c* in which it is shown that after relatively small lateral deflections occur the column reaches a condition of instability associated with total collapse, points B, F, and I. In other words, inelastic strain occurs and is followed after some increase in load by instability and collapse at relatively small lateral deflections—not greatly different from those indicated in Fig. 3-6.1*b* by the points B, B', and B''.

Which Type of Failure Occurs? / Although these two types of failure—failure by excessive deflection before the collapse load occurs and failure by collapsing at or before excessive deflection occurs—are usually associated with two ranges of values of L/r, it is not easy to determine which type of failure takes place for a given value of L/r except, perhaps, when L/r is very large, as may occur in certain lightweight construction. In other words, for L/r values in the usual range used in columns for heavy machines and structures, the type of failure is not easily determined. Furthermore, for curves $0'FH$ and $0''IK$ in Fig. 3-6.1*c* it is not known how much the load increases after inelastic strains have started, that is, how far to the left and below the points F and I the inelastic strains begin that lead finally to collapse, for the amount of increase of load depends on the shape of the cross section, on the shape of the stress-strain diagram of the material, as discussed in Section 4 below and on the value of L/r. Thus a wholly rational method or formula is difficult to devise and, hence, empirical methods are frequently used in design specifications, etc.

3. Inelastic Buckling / Buckling, as has been noted, is not restricted solely to elastic action. If the value of L/r is relatively small so that the compressive stress in the column reaches the compressive proportional limit of the material before the load reaches the value of the elastic buckling load, the behavior of the column may be very similar to that of elastic buckling, namely, rather abrupt increases in deflection at a fairly well-defined load. This behavior is called *inelastic* or *plastic buckling*.

The problem of determining the buckling load for a column that does not buckle until it is strained inelastically may be considered for two kinds of material behavior. First, the material in the column is assumed to have a flat top stress-strain curve, that is, the slope of the compressive stress-strain curve changes abruptly to zero when the proportional limit is reached; the stress-strain curve suddenly becomes horizontal and remains so until relatively large inelastic strains have developed. For such material the proportional limit and the yield point are equal. Mild steel usually exhibits this type of stress-strain curve.

A column having a relatively small L/r (less than approximately 100 for structural steel) and made of such material usually buckles when the average stress in the column is slightly less than the proportional limit (and yield point) of the material. When the average stress in the column is equal to the yield point, it is impossible for the column to develop a resisting moment (see Section 4 below) until the column has attained sufficiently large deflection so that the most stressed material in the column strain hardens. The column usually fails by excessive deflection before this happens.

Under the second kind of material behavior, it is assumed that the compressive stress-strain curve changes in slope gradually as the stress is increased above the proportional limit of the material; aluminum alloys and some heat-treated steels, for example, exhibit this type of behavior. Inelastic buckling loads for columns made of material having this type of stress-strain curves are determined in the following articles.

Significance of Inelastic Buckling for Material Not Having a Yield Point / In certain applications, especially in aircraft structures where the strength-weight ratio of a member is of great importance, columns made of material having this second type of stress-strain diagram frequently have relative dimensions such that small amounts of inelastic strains occur in the member before it buckles. These inelastic strains, however, are not sufficient to cause damage to the column. In fact, the maximum inelastic strain in the column at the impending buckling load is of the same order of magnitude as that of the elastic strain at the proportional limit of the material and is usually much less than the strain corresponding to the yield stress based on a 0.2 percent offset. By permitting this small amount of inelastic strain a larger design load may be justified than if only strains within the elastic range were permitted. It is important to note, however, that when the buckling load is finally reached the deflection of the column may increase suddenly, leading to a catastrophic collapse.

Similar considerations apply to some so-called heavy structures although to a lesser degree since the strength-weight ratio is usually of less importance in heavy structures such as building frames and bridges. For a further brief discussion of this topic, see Examples 3-6.1 and 3-6.2.

4. Two Formulas for Inelastic Buckling of an Ideal Column / As previously noted, the buckling load for a column is considered to be the axial load that holds the column in a slightly deflected position. Hence, since an ideal column will not bend under any axial load, a small lateral force must be applied to produce the initial deflection. This procedure, however, may be carried out in either of two ways: (a) the lateral force may be assumed to be applied first and then the axial load required to hold the column in the slightly bent position is applied and the lateral force removed; or (b) the unknown buckling critical load may be assumed to be applied first, and then the lateral force is applied to cause the deflection and is then removed. For *elastic* behavior of the column the solution for the (Euler) buckling load will be the same for the two procedures, since the physical process constitutes a reversible system and hence does not depend on the strain history in arriving at a given physical state.

If the physical process is *nonreversible*, such as occurs in the inelastic behavior of materials, the order of applying the forces to the column in the two procedures leads to different values for the buckling loads. The main condition involved in the process may be emphasized by stating that, for inelastic behavior, a single-valued relationship between loads and deflections (or between stress and strain) does not usually exist. However, a single-valued relationship between stress and strain will exist not only for elastic strains but also for inelastic strains provided that all strains increase as the load increases (no fiber in the member is allowed to unload).

Accordingly two formulas for the buckling load may be obtained depending upon which of the following assumptions are employed: (a) the lateral force and the last increment of the axial load are assumed to be applied simultaneously so that the strains in all the fibers at any cross section increase although they are not uniformly distributed on the section after the lateral force is applied, or (b) an axial load equal to the buckling load is assumed to be applied first, followed by the application of a small lateral force that deflects the column; the bending in this case causes the strains in the fibers on the convex side to decrease (unload) and on the concave side to increase.

The essential difference in the two assumptions lies in the fact that under the second assumption the strains in some of the fibers on the convex side decrease elastically and hence the change in stress $\Delta\sigma$ accompanying the decrease in strain $\Delta\epsilon$ is given by $\Delta\sigma = E\,\Delta\epsilon$, in which E is the elastic modulus of elasticity, whereas under the first assumption $\Delta\sigma = E_T\,\Delta\epsilon$ for each fiber, in which E_T is the tangent modulus (slope of the tangent to the stress-strain curve) corresponding to the inelastic stress σ ($\sigma = P/A$ in which P is the buckling load).

The buckling load found with assumption (a) is called the tangent-modulus load and is given by the expression $P_T = \pi^2 E_T I/L^2$. For assumption (b), the buckling load is called the double-modulus load and is given by the expression $P_D = \pi^2 E_D I/L^2$, in which E_D is called the effective modulus or double modulus since it is expressed in terms of E and E_T.

The double-modulus theory was considered to be the more accurate theory of inelastic column buckling until 1946, when Shanley[22] showed that it represented a paradox requiring physically unattainable conditions. A development of the double-modulus theory may be found in reference 23, and a comparison of predicted values of buckling loads as given by the tangent modulus theory and the double-modulus theory is given in reference 24. It should be noted that neither of these theories applies to a column made of a material having a yield point (a flat top stress-strain curve; see Figs. 2-6.1*a* and 2-6.2*a*), since for such materials both E_T and E_D are zero and, hence, $P_T = P_D = 0$.

A development of the tangent-modulus formula follows. We recommend the use of the tangent-modulus equation, since it generally leads to a good estimate of the maximum (buckling) load that a real column having slight imperfections can safely be expected to resist.

5. Tangent-Modulus Formula for an Inelastic Buckling Load. Load at Which Inelastic Bending of Ideal Column Begins/Let Fig. 3-6.5*a* represent a column subjected to a gradually increasing axial load P. For convenience, let the column have a rectangular cross section. It is also assumed that the slenderness ratio L/r is sufficiently small to preclude elastic buckling. The load P may therefore attain a value that causes a uniform stress σ on the cross section of the column that is greater than the proportional limit σ_{PL} of the material. The stress-strain diagram for the material is shown in Fig. 3-6.5*d*. In this diagram σ represents the uniformly distributed compressive stress on each cross section of the column and ϵ the corresponding strain; it is assumed that the value of

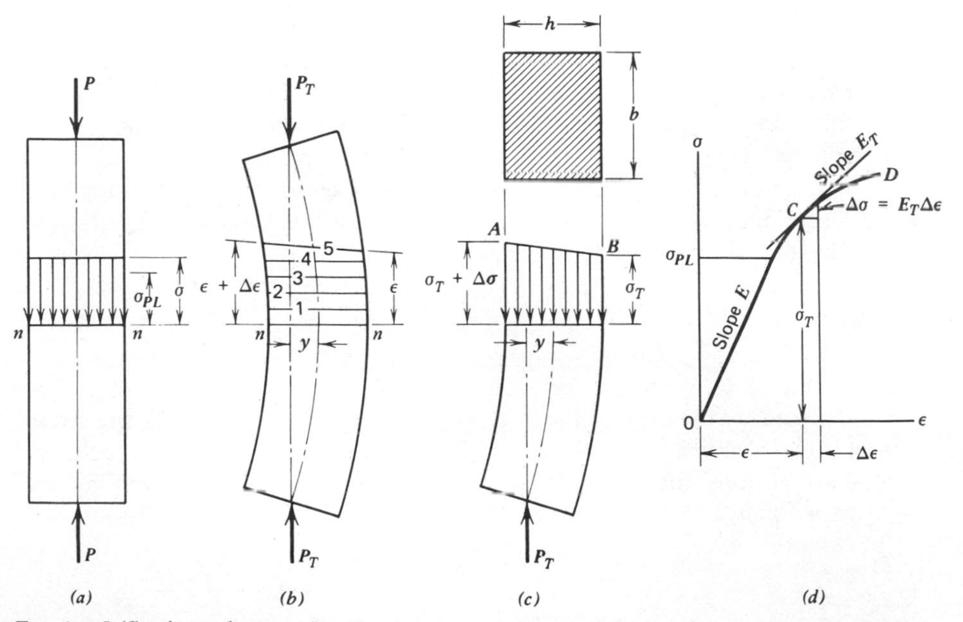

Fig. 3-6.5/Strain and stress distribution for tangent-modulus load.

the buckling load P_T (or buckling stress $\sigma_T = P_T/A$) and the corresponding inelastic strain are represented by a point in the neighborhood of C on the stress-strain curve. Hence, as previously noted, inelastic buckling involves only small inelastic strains.

The problem is to find the smallest load $P_T = A\sigma_T$ that will cause the ideal column to remain in a slightly bent position when a small lateral force is applied (simultaneously with the last increment of axial load) and then is removed. As increments of the axial load P are applied to the ideal column, the longitudinal strain at section nn increases but remains uniformly distributed as shown by the lines marked 1, 2, 3, and 4 in Fig. 15-5.1b. As P approaches the value P_T, which we wish to determine, let a small lateral force be applied *simultaneously* with the last increment of load as P attains the value P_T. The resulting distribution of strain is as shown by the line marked 5 in Fig. 3-6.5b in which the lateral bending is greatly exaggerated.

The resulting stress distribution on section nn is shown in Fig. 3-6.5c by the sloping line AB and is obtained from Fig. 3-6.5d by taking the stresses corresponding to the strains. The assumption that line AB is straight is equivalent to the assumption that the slope of tangent lines to the stress-strain curve, such as that shown at C in Fig. 3-6.5d, is constant during the change in strain from ϵ to $\epsilon + \Delta\epsilon$. This assumption is justified because the increment $\Delta\epsilon$ is small for the small lateral bending imposed on the column. The slope of the tangent line at any point such as C is called the *tangent modulus* at point C and is denoted by E_T; the increment of stress corresponding to $\Delta\epsilon$ is therefore $\Delta\sigma = E_T\Delta\epsilon$. The desired value P_T of the axial load P may now be found in the same manner that the Euler load for the beginning of *elastic* buckling is usually obtained. Thus, let Fig. 3-6.5c be a free-body diagram showing the forces acting on the lower half of the column. Equilibrium of the column requires that the external bending moment $P_T y$ for any cross section shall be equal and opposite to the resisting moment about the centroidal axis of the cross section of the internal forces on the section. This fact is expressed by the equation

$$P_T y = \frac{(\Delta\sigma/2)I}{h/2} \tag{3-6.2}$$

In Eq. (3-6.2), let $\Delta\sigma$ be replaced by $E_T\Delta\epsilon$ and, in turn, let $\Delta\epsilon$ be replaced by the expression h/R, which is obtained by relating the strain in the extreme fiber to the radius of curvature R of the column. Furthermore, for small deflections the curvature $1/R$ is given by the expression $1/R = -d^2y/dz^2$. With these substitutions, Eq. (3-6.2) becomes

$$E_T I \frac{d^2y}{dz^2} = -P_T y \tag{3-6.3}$$

The solution of this differential equation as given by Eq. (3-6.1) is

$$P_T = \frac{\pi^2 E_T I}{L^2} \quad \text{or} \quad \frac{P_T}{A} = \frac{\pi^2 E_T}{(L/r)^2} \tag{3-6.4}$$

in which P_T may be considered to be either the smallest load that will hold the ideal column in a slightly bent form or the largest load under which the ideal column will not bend. This formula is called the *tangent-modulus formula* or *Engesser's formula*.† It is generally regarded as the maximum (buckling) load that a real column having slight imperfects can safely be expected to resist.

The solution of the tangent-modulus equation for a column of given material and dimensions involves a trial-and-error process for the reason that a value of E_T cannot be selected unless P_T is known. Furthermore, a stress-strain diagram for the given material must be available. The method of solution is illustrated in the following examples.

EXAMPLE 3-6.1

Tangent-Modulus Load P_T

A straight column having a square cross section $b = 25.0$ mm on a side and a length of $L = 250$ mm is loaded axially through special bearing blocks that allow free rotation when bending of the column starts. The member is made of material for which the compressive stress-strain curve is shown by $0BC$ in Fig. E3-6.1. The tangent modulus for this stress-strain curve is shown by abscissas to the curve $DEFG$ on the upper scale. Compute the load P_T.

SOLUTION

We must by trial and error select from the curves in Fig. E3-6.1 a set of corresponding values of stress σ_T and tangent modulus E_T that will satisfy Eq. (3-6.4). As a first trial value, select $E_T = 31.0$ GPa, which from curve DEF corresponds to $\sigma_T = P_T/A = 262$ MPa. For rectangular sections $r = b/\sqrt{12} = 7.217$ mm and $L/r = 34.64$. The right side of Eq. (3-6.4) is

$$\frac{\pi^2 E_T}{(L/r)^2} = \frac{\pi^2(31 \times 10^3)}{(34.64)^2} = 255 \text{ MPa}$$

Since the left side is 262 MPa, a new trial is necessary. For the second trial assume that $E_T = 31.6$ GPa, which from curve DEF corresponds to $\sigma_T = 261$ MPa. The right side of Eq. (3-6.4) is now 260 MPa, which is

† This theory is due in part to F. R. Shanley. See references 25 and 26.

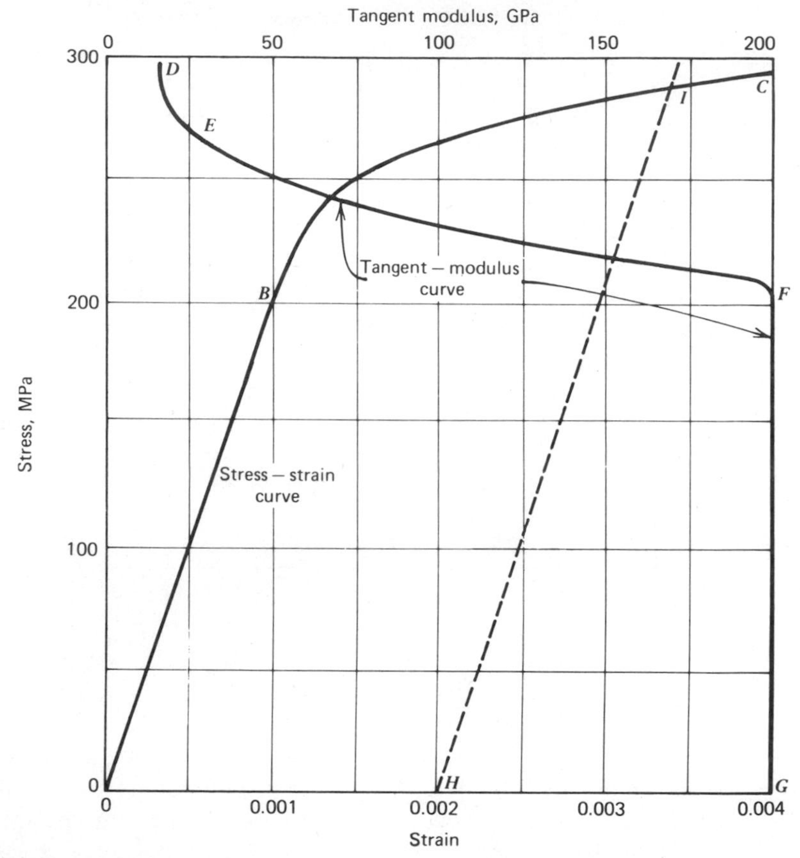

Fig. E3-6.1/Values of tangent modulus for a material not having a yield point.

very close to the assumed value. Hence the answer is $P_T \cong 260(25)^2 =$ 163,000 N = 163 kN.

EXAMPLE 3-6.2

In Example 3-6.1 the stress at the tangent-modulus load was found to be 260 MPa which from curve $0BC$ corresponds to a strain of 0.00180. If it is specified that the strain (or stress) in the column must not exceed the value corresponding to the yield stress of the material, based on 0.2 percent offset, will this stress and strain at the tangent modulus load be within the required limit?

SOLUTION

The yield stress based on 0.2 percent offset, as shown by the line HI in Fig. E3-6.1, is 288 MPa, which corresponds to a strain of 0.0034.

Therefore the stress and strain at the tangent-modulus load are less than values at the yield stress. It should be pointed out that the stress and strain at which the tangent-modulus load occurs in nearly all inelastic columns are smaller than the stress and strain values corresponding to the yield stress. This fact shows that, although the tangent-modulus formula is obtained on the assumption that some inelastic strain occurs, the inelastic strains that correspond to this load are smaller than the inelastic strains (the offset) that are usually assumed to be permissible without causing damage to the load resisting behavior of the material or structure.

PROBLEM SET 3-6.2

1. Solve Example 3-6.1 for a column length of $L = 300$ mm.

2. A column made of the steel whose compressive stress-strain diagram is shown in Fig. E3-6.1 has a length of 1.00 m and a solid circular cross section. The column must carry a design load of 810 kN with a factor of safety of $SF = 2.00$. Determine the required diameter of the column.

   Ans. $D = 90.4$ mm

3. An aluminum alloy has a modulus of $E = 72.0$ GPa and a proportional limit of $\sigma_{PL} = 310$ MPa. The equation $\sigma = A\epsilon^n$ where A and n are material constants accurately approximates the compressive stress-strain diagram for small inelastic strains when it is made to coincide with the stress-strain diagram at the proportional limit and to pass through another test point whose coordinates are $\sigma = 370$ MPa and $\epsilon = 0.00600$. (a) Determine values for A and n. (b) A rectangular section column with dimensions 40 mm by 60 mm has fixed ends and is found to buckle at an average stress of $P_T/A = 345$ MPa; determine the length L of the column.

4. For the aluminum alloy in Problem 3, determine the range of the slenderness ratios such that $P_T/A = \sigma_{PL}$.

   Ans. $(L/r)_{max} = 47.9$ and $(L/r)_{min} = 35.0$

REFERENCES

1. W. L. Clark and G. M. Gordon, "Investigation of Stress Corrosion Cracking Susceptibility of Fe-Ni-Ci alloys in Nuclear Reactor Water Environments," *Corrosion*, Vol 29, Part 1, 1973.

2. M. M. Eisenstadt, *Introduction to Mechanical Properties of Metals*, Macmillan, New York, 1971.
3. J. O. Smith and O. M. Sidebottom, *Elementary Mechanics of Deformable Bodies*, Macmillan, New York, 1969.
4. J. O. Smith and O. M. Sidebottom, *Inelastic Behavior of Load-Carrying Members*, Wiley, New York, 1965.
5. H. Liebowitz, *Fracture: An Advanced Treatise*, Chapter 4, Vol. II, Academic Press, New York, 1968.
6. B. Paul, "Generalized Pyramidal Fracture and Yield Criteria," *International Journal of Solids and Structures*, Vol. 4, 1968, pp. 175–196.
7. A. I. Smith and A. M. Nicolson, *Advances in Creep Design*, Halsted Press Division of Wiley, New York, 1971.
8. P. G. Paris and G. C. Sih, "Stress Analysis of Cracks," *Fracture Toughness and Its Applications*, American Society of Testing Materials, STP 381, 1965, p. 30.
9. J. E. Snawley and W. F. Brown, Jr., "Fracture Toughness Testing Methods," *Fracture Toughness Testing and Its Applications*, American Society of Testing Materials, STP 381, 1965, p. 133.
10. J. F. Knott, *Fundamentals of Fracture Mechanics*, Wiley, New York, 1973, p. 136. See also, D. Broek, *Elementary Engineering Fracture Mechanics*, 3rd Ed., Martinus Nijhoff, Boston, 1982.
11. E. T. Wessel, "Practical Fracture Mechanics for Structural Steel," Paper H, UKAEA/Chapman and Hall, 1969.
12. S. S. Manson, *Thermal Stress and Low-Cycle Fatigue*, McGraw-Hill, New York, 1966.
13. B. I. Sandor, *Fundamentals of Cyclic Stress and Strain*, The University of Wisconsin Press, Madison, Wisconsin, 1972.
14. G. Sines and J. L. Waisman, *Metal Fatigue—University of California Engineering Series*, McGraw-Hill, New York, 1959.
15. M. W. Brown and K. J. Miller, "A Theory for Fatigue Failure Under Multiaxial Stress-Strain Condition," *The Institute of Mechanical Engineers*, Proceedings, Vol. 187, No. 65, 1973, pp. 745–755.
16. Leonard Euler, *Elastic Curves* (*Des Curvie Elasticis*, Lausanne and Geneva, 1744), translated and annotated by W. A. Oldfather, C. A. Ellis, and D. M. Brown, ISIS, No. 58, Vol. XX, 1, November 1933.
17. A. M. Freudenthal, B. A. Boley, and H. Liebowitz, *High Temperature Structures and Materials*, Proceedings of the Third Symposium on Naval Structural Mechanics, Pergamon Press, New York, 1963.
18. J. O. Smith and O. M. Sidebottom, *Elementary Mechanics of Deformable Bodies*, Macmillan, New York, 1969.
19. F. J. Bridget, C. C. Jerome, and A. B. Vosseller, "Some New Experiments On Buckling of Thin-Walled Construction," *Transactions of the American Society of Mechanical Engineers*, Vol. 56, 1934, pp. 569–578.

20. E. Z. Stowell, et al., "Buckling Stresses for Flat Plates and Sections," *Proceedings of the American Society of Civil Engineers*, Vol. 77, July 1951.
21. R. V. Southwell, *Introduction to the Theory of Elasticity*, 2nd Ed., Oxford University Press, England, 1941, p. 434.
22. F. R. Shanley, "The Column Paradox," *Journal of Aeronautic Sciences*, Vol. 13, No. 12, December 1946, p. 492.
23. F. Bleich, *Buckling Strength of Metal Structures*, McGraw-Hill, New York, 1952, pp. 9–14.
24. A. P. Boresi, O. M. Sidebottom, et al., *Advanced Mechanics of Materials*, 3rd Ed., Wiley, New York, 1978, pp. 649–653.
25. F. R. Shanley, "Inelastic Column Theory," *Journal of Aeronautic Sciences*, Vol. 14, No. 5, May 1947, p. 261.
26. F. R. Shanley, "Applied Column Theory," *Transactions of the American Society of Civil Engineers*, Vol. 115, 1950, pp. 698–727.
27. W. R. Osgood, "The Double-Modulus Theory of Column Action," *Civil Engineering*, March 1935, p. 173.
28. S. P. Timoshenko and J. M. Gere, *Theory of Elastic Stability*, 2nd Ed., McGraw-Hill, New York, 1961.
29. R. G. Strum, "A Study of the Collapsing Pressure of Thin-Walled Cylinders," *Bulletin* 329, *Engineering Experiment Station*, University of Illinois, Urbana, 1941.

Additional References

1. D. O. Brush and Bo O. Almroth, *Buckling of Bars, Plates, and Shells*, McGraw-Hill, New York, 1975.
2. R. M. Caddell, *Deformation and Fracture of Solids*, Prentice-Hall, Englewood Cliffs, New Jersey, 1974.
3. A. Chajes, *Principles of Structural Stability Theory*, Prentice-Hall, Englewood Cliffs, New Jersey, 1974.
4. S. W. Freiman and E. R. Fuller, Jr., Ed., *Fracture Mechanics for Ceramics, Rocks, and Concrete*, American Society for Testing Materials, Philadelphia, 1981.
5. R. W. Hertzberg, *Deformation and Fracture of Engineering Materials*, 2nd Ed., Wiley, New York, 1983.
6. R. Narayanan, Ed., *Axially Compressed Structures: Stability and Strength*, Applied Science Publishers, New York, 1982.

CHAPTER 4

APPLICATION OF ENERGY METHODS: ELASTIC DEFLECTIONS AND STATICALLY INDETERMINATE MEMBERS AND STRUCTURES

Energy methods are widely used to obtain solutions to elasticity problems and to determine deflections of structures and machines. Since energy is a scalar quantity, energy methods are sometimes called scalar methods. In this chapter, energy methods are employed to obtain elastic deflections of statically determinate members and structures and to determine redundant reactions and deflections of statically indeterminate members and structures. The applications of energy methods in this book are limited mainly to linearly elastic material behavior and to small displacements. However, in Arts. 4-1 and 4-2, energy methods are applied to two nonlinear problems to demonstrate their generality.

Castigliano's theorem on deflections is restricted to small displacements and is used to obtain elastic deflections and to determine redundant reactions. In applications to linearly elastic material behavior, the theorem is generally expressed in terms of the total strain energy of the structure. However, Castigliano's theorem on deflections is based on the concept of complementary energy and not on the concept of strain energy. For the determination of the deflections of structures, two energy principles are presented: (1) the principle of stationary potential energy and (2) Castigliano's theorem on deflections. The general proofs of these principles are not presented in this book. Instead, the reader is referred to proofs given by H. L. Langhaar.[1]

In the application of the principle of stationary potential energy and of Castigliano's theorem on deflection to problems of members and structures, it is assumed that every plane section of each member before deformation remains plane after deformation. Therefore, the displacement of a given cross section of a member is specified by three components of the displacement of the centroid of its cross section, and by three angles, which define the rotation of the plane of the cross section. In this chapter, it is assumed that each member of the structure is straight (curved members are treated in Chapter 8). Rectangular coordinate axes (x, y, z) are chosen for each member with the z-axis directed along the axis of the member and with (x, y)-axes taken as principle axes of the cross section (see the Appendix). Principle axes (x, y) are assumed to maintain the same direction for every cross section of each straight member of the structure.

4-1 PRINCIPLE OF STATIONARY POTENTIAL ENERGY

We employ the concept of generalized coordinates $(x_1, x_2, \ldots, x_n)$ to describe the configuration (shape) of a member or structure (system) in equilibrium (see Art. 1-2 of reference 1). Since plane cross sections of the members of the structure are assumed to remain plane, the generalized coordinates denote displacements and rotations (changes in angles) that occur in the structure. If a finite number of coordinates suffices to specify the configuration of the system, the system is said to possess a finite number of degrees of freedom. If a finite number of coordinates cannot specify the system configuration, the system is said to have infinitely many degrees of freedom. We shall consider applications in this book in which the number of degrees of freedom equals the number of generalized coordinates required to specify the configuration of the system.

For a system with a finite number of generalized coordinates $(x_1, x_2, \ldots, x_n)$, the virtual work δW that corresponds to a virtual (imagined) displacement from the configuration $(x_1, x_2, \ldots, x_n)$ to the configuration $(x_1 + \delta x_1, x_2 + \delta x_2, \ldots, x_n + \delta x_n)$, where $(\delta x_1, \delta x_2, \ldots, \delta x_n)$ denote virtual (imagined) displacements, is given by

$$\delta W = Q_1 \delta x_1 + Q_2 \delta x_2 + \cdots + Q_n \delta x_n \tag{4-1.1}$$

where $(Q_1, Q_2, \ldots, Q_n)$ are functions of the generalized coordinates $(x_1, x_2, \ldots, x_n)$. The functions $(Q_1, Q_2, \ldots, Q_n)$ are called components of generalized load by analogy to the fact that $F_x\,dx + F_y\,dy + F_z\,dz$ represents the work performed by a force (F_x, F_y, F_z) when the particle on which it acts is displaced by the amount (dx, dy, dz). Let Q_i be defined for a given cross section of the structure (system); Q_i is a force if x_i is a displacement component for the centroid of the cross section, and Q_i is a moment if x_i is a change in angle component for the cross section.

The virtual work δW corresponding to an arbitrary displacement of a mechanical system may be separated into the sum

$$\delta W = \delta W_e + \delta W_i \tag{4-1.2}$$

where δW_e is the virtual work of the external forces and δW_i is the virtual work of the internal forces.

Analogous to the expression for δW (Eq. (4-1.1)), under a virtual displacement $(\delta x_1, \delta x_2, \ldots, \delta x_n)$, we have

$$\delta W_e = P_i \delta x_1 + P_2 \delta x_2 + \cdots + P_n \delta x_n \tag{4-1.3}$$

where $(P_1, P_2, \ldots, P_n)$ are functions of the generalized coordinates $(x_1, x_2, \ldots, x_n)$. By analogy to the Q_i in Eq. (4-1.1), the functions $(P_1, P_2, \ldots, P_n)$ are called the components of generalized *external* load. If the generalized coordinates $(x_1, x_2, \ldots, x_n)$ denote displacements and rotations (changes in angles) that occur in a member or structure (system), the variables $(P_1, P_2, \ldots, P_n)$ may usually be identified as the components of the prescribed external forces and couples that act on the member or structure (system).

Now let us imagine that the virtual displacement takes the system completely around any closed path. At the end of the closed path we have $\delta x_1 = \delta x_2 = \cdots = \delta x_n = 0$. Hence, by Eq. (4-1.3), $\delta W_e = 0$. In our applications, we consider only members or structures that undergo elastic behavior. Then the virtual work δW_i of the internal forces is equal to the negative of the virtual change in the elastic strain energy δU, that is,

$$\delta W_i = -\delta U \tag{4-1.4}$$

where $U = U(x_1, x_2, \ldots, x_n)$ is the total strain energy of the member or structure (system). Since the system travels around a closed path, it returns to its initial state and, hence, $\delta U = 0$. Consequently, by Eq. (4-1.4), $\delta W_i = 0$. Accordingly, the total virtual work δW (Eq. 4-1.2) also vanishes around a closed path. The condition $\delta W = 0$ for virtual displacements that carry the system around a closed path indicates that the system is conservative.

For a conservative system (e.g., elastic structure loaded by conservative external forces) the virtual change δU in strain energy of the structure under the virtual displacement $(\delta x_1, \delta x_2, \ldots, \delta x_n)$ is

$$\delta U = \frac{\partial U}{\partial x_1}\delta x_1 + \frac{\partial U}{\partial x_2}\delta x_2 + \cdots + \frac{\partial U}{\partial x_n}\delta x_n \tag{4-1.5}$$

Then, Eqs. (4-1.1), (4-1.2), (4-13), (4-1.4), and (4-1.5) yield the result

$$Q_1\,\delta x_1 + Q_2\,\delta x_2 + \cdots + Q_n\,\delta x_n = P_1\,\delta x_1 + P_2\,\delta x_2 + \cdots + P_n\,\delta x_n - \frac{\partial U}{\partial x_1}\delta x_1 - \frac{\partial U}{\partial x_2}\delta x_2 - \cdots - \frac{\partial U}{\partial x_n}\delta x_n$$

or

$$Q_i = P_i - \frac{\partial U}{\partial x_i}, \qquad i = 1, 2, \ldots, n \tag{4-1.6}$$

For any stationary (equilibrium) system with finite degrees of freedom, the vanishing of the components Q_i of the generalized force is sufficient for equilibrium.[1] Therefore, by Eq. (4-1.6), an elastic system with n degrees of freedom is in equilibrium if (see Art. 1-9 of reference 1)

$$P_i = \frac{\partial U}{\partial x_i}, \qquad i = 1, 2, \ldots, n \tag{4-1.7}$$

In the case of a structure, the strain energy U is obtained as the sum of the strain energies of the members of the structure. Note the similarity between Eqs. (4-1.7) and Eqs. (2-2.11).

As a simple example, consider a uniform bar loaded at its ends by an axial load P. Let the bar be made of a nonlinear elastic material with the load-elongation curve indicated in Fig. 4-1.1. The area below the curve

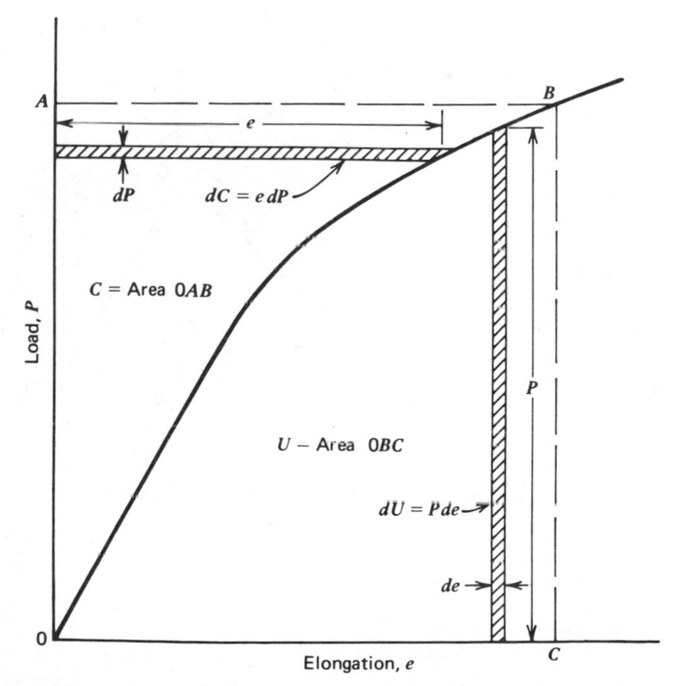

Fig. 4-1.1/Nonlinear elastic load-elongation curve.

represents the total strain energy U stored in the bar, that is $U = \int P\, de$, then by Eq. (4-1.7), $P = \partial U/\partial e$, where P is the generalized external force and e is the generalized coordinate. If the load-elongation data for the bar are plotted as a stress-strain curve (see Fig. 2-2.1), the area below the curve is the strain energy density U_0 stored in the bar. Then, $U_0 = \int \sigma\, d\epsilon$ and, by Eqs. (2-2.11), $\sigma = \partial U_0/\partial \epsilon$.

Equation (4-1.7) is valid for nonlinear problems in which the nonlinearity is due either to finite geometry changes, or to nonlinear elastic material behavior, or to both. The following example problem indicates the application of Eq. (4-1.7) for finite geometry changes.

EXAMPLE 4-1.1

Equilibrium of Simple Linear Elastic Pin-Joined Truss

Two members AB and CB of lengths L_1 and L_2, respectively, of a pin-joined truss are attached to a rigid foundation at points A and C, as shown in Fig. E4-1.1. The cross sectional area of member AB is A_1 and that of member CB is A_2. The corresponding modulii of elasticity are E_1

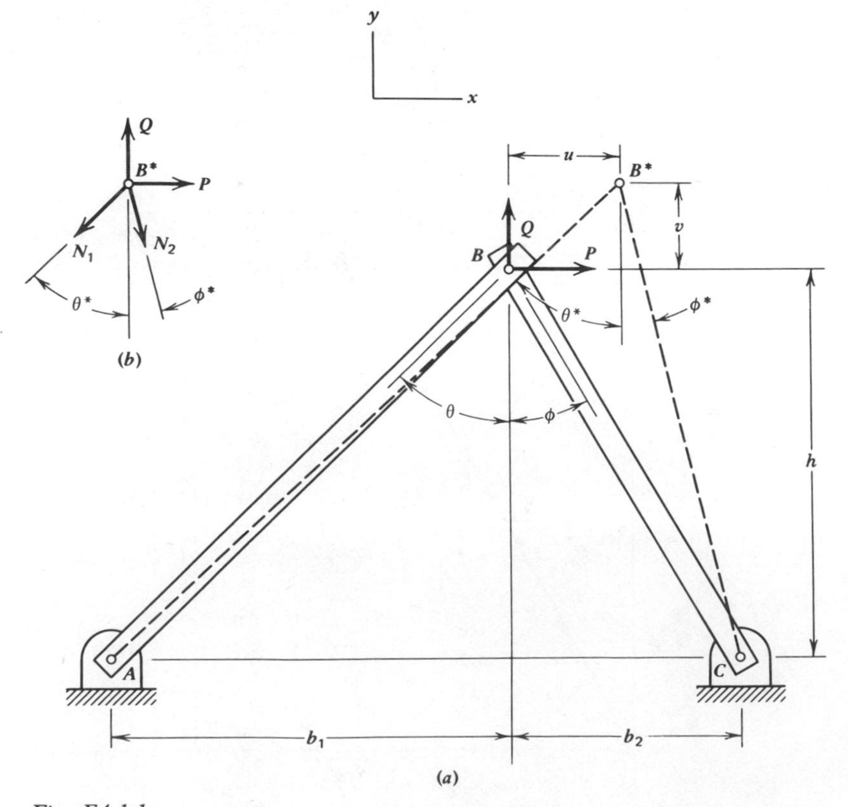

Fig. E4-1.1

and E_2. Under the action of horizontal and vertical forces P and Q, pin B undergoes finite horizontal and vertical displacement components u and v, respectively (Fig. E4-1.1). Under the displacements u and v, the bars AB and CB are assumed to remain linearly elastic. (a) Derive formulas for P and Q in terms of u and v. (b) Let $E_1A_1/L_1 = K_1 = 2.00$ N/mm and $E_2A_2/L_2 = K_2 = 3.00$ N/mm, and let $b_1 = h = 400$ mm and $b_2 = 300$ mm. For $u = 30$ mm and $v = 40$ mm, determine the values of P and Q using the formulas derived in Part (a). (c) Consider the equilibrium of the pin B in the displaced position B^* and verify the results of Part (b). (d) For small displacement components u and v ($u, v \ll L_1, L_2$), linearize the formulas for P and Q derived in Part (a).

SOLUTION

(a) For this problem the generalized external forces are $P_1 = P$ and $P_2 = Q$ and the generalized coordinates are $x_1 = u$ and $x_2 = v$. For the geometry of Fig. E4-1.1, the elongations e_1 and e_2 of members 1 (member AB with length L_1) and 2 (member CB with length L_2) can be obtained in terms of u and v.

$$\begin{aligned} (L_1 + e_1)^2 &= (b_1 + u)^2 + (h + v)^2; \qquad L_1^2 = b_1^2 + h^2 \\ (L_2 + e_2)^2 &= (b_2 - u)^2 + (h + v)^2; \qquad L_2^2 = b_2^2 + h^2 \end{aligned} \tag{1}$$

Hence, solving for (e_1, e_2), we obtain

$$\begin{aligned} e_1 &= \sqrt{(b_1 + u)^2 + (h + v)^2} - L_1 \\ e_2 &= \sqrt{(b_2 - u)^2 + (h + v)^2} - L_2 \end{aligned} \tag{2}$$

Each member is assumed to remain linearly elastic. Consequently, the strain energies U_1 and U_2 of members AB and CB are

$$\begin{aligned} U_1 &= \frac{1}{2} N_1 e_1 = \frac{E_1 A_1}{2L_1} e_1^2 \\ U_2 &= \frac{1}{2} N_2 e_2 = \frac{E_2 A_2}{2L_2} e_2^2 \end{aligned} \tag{3}$$

where N_1 and N_2 are the tensions in the two members. The elongations of the two members are given by the relation $e_i = N_i L_i / E_i A_i$. The total strain energy U for the structure is equal to the sum $U_1 + U_2$ of the strain energies of the two members; therefore by Eqs. (3),

$$U = \frac{E_1 A_1}{2L_i} e_1^2 + \frac{E_2 A_2}{2L_2} e_2^2 \tag{4}$$

The magnitudes of P and Q are obtained by substitution of Eq. (4) into

Eq. (4-1.7). Thus,

$$
\begin{aligned}
P &= \frac{\partial U}{\partial u} = \frac{E_1 A_1 e_1}{L_1} \frac{\partial e_1}{\partial u} + \frac{E_2 A_2 e_2}{L_2} \frac{\partial e_2}{\partial u} \\
Q &= \frac{\partial U}{\partial v} = \frac{E_1 A_1 e_1}{L_1} \frac{\partial e_1}{\partial v} + \frac{E_2 A_2 e_2}{L_2} \frac{\partial e_2}{\partial v}
\end{aligned}
\tag{5}
$$

The partial derivatives of e_1 and e_2 with respect to u and v are obtained from Eqs. (2). Taking the derivatives and substituting in Eqs. (5), we find

$$
\begin{aligned}
P = {} & \frac{E_1 A_1 (b_1 + u)}{L_1} \frac{\sqrt{(b_1 + u)^2 + (h + v)^2} - L_1}{\sqrt{(b_1 + u)^2 + (h + v)^2}} \\
& - \frac{E_2 A_2 (b_2 - u)}{L_2} \frac{\sqrt{(b_2 - u)^2 + (h + v)^2} - L_2}{\sqrt{(b_2 - u)^2 + (h + v)^2}} \\
Q = {} & \frac{E_1 A_1 (h + v)}{L_1} \frac{\sqrt{(b_1 + u)^2 + (h + v)^2} - L_1}{\sqrt{(b_1 + u)^2 + (h + v)^2}} \\
& + \frac{E_2 A_2 (h + v)}{L_2} \frac{\sqrt{(b_2 - u)^2 + (h + v)^2} - L_2}{\sqrt{(b_2 - u)^2 + (h + v)^2}}
\end{aligned}
\tag{6}
$$

These equations for loads P and Q even for this simple structure are complicated.

(b) Substitution of the values K_1, K_2, b_1, b_2, h, L_1, L_2, u, and v into Eqs. (6) gives

$$
\begin{aligned}
P &= 43.8 \text{ N} \\
Q &= 112.4 \text{ N}
\end{aligned}
\tag{7}
$$

(c) The values of P and Q may be verified by determining the tensions N_1 and N_2 in the two members, by determining directions of the axes of the two members for the deformed configuration, and by applying equations of equilibrium to a free body diagram of pin B^*. Elongations $e_1 = 49.54$ mm and $e_2 = 16.24$ mm are given by Eqs. (2). The tensions N_1 and N_2 are

$$
\begin{aligned}
N_1 &= e_1 K_1 = 99.08 \text{ N} \\
N_2 &= e_2 K_2 = 48.72 \text{ N}
\end{aligned}
$$

Angles θ^* and ϕ^* for the directions of the axes of the two members for the deformed configurations are found to be 0.7739 rad and 0.5504 rad, respectively. The free body diagram of pin B^* is shown in Fig. E4-1.1*b*.

The equations of equilibrium are

$$\sum F_x = 0 = P - N_1 \sin\theta^* + N_2 \sin\phi^*; \qquad \text{hence, } P = 43.8 \text{ N}$$

$$\sum F_y = 0 = Q - N_1 \cos\theta^* - N_2 \cos\phi^*; \qquad \text{hence, } Q = 112.4 \text{ N}$$

These values of P and Q agree with those of Eqs. (7).

(d) If displacements u and v are very small compared to b_1 and b_2, and, hence, with respect to L_1 and L_2, simple approximate expressions for P and Q may be obtained. For example, we find by the binomial expansion to linear terms in u and v that

$$\sqrt{(b_1 + u)^2 + (h + v)^2} = L_1 + \frac{b_1 u}{L_1} + \frac{hv}{L_1}$$

$$\sqrt{(b_2 - u)^2 + (h + v)^2} = L_2 - \frac{b_2 u}{L_2} + \frac{hv}{L_2}$$

With these approximations, Eqs. (6) yield the linear relations

$$P = \frac{E_1 A_1 b_1}{L_1^3}(b_1 u + hv) + \frac{E_2 A_2 b_2}{L_2^3}(b_2 u - hv)$$

$$Q = \frac{E_1 A_1 h}{L_1^3}(b_1 u + hv) + \frac{E_2 A_2 h}{L_2^3}(-b_2 u + hv)$$

If these equations are solved for the displacements u and v, the resulting relations are identical to those derived by means of Castigliano's theorem on deflections for linearly elastic materials (Arts. 4-3 and 4-4).

4-2 CASTIGLIANO'S THEOREM ON DEFLECTIONS

The derivation of Castigliano's theorem on deflections is based on the concept of complementary energy C of the member or structure (system). Consequently, the theorem is sometimes called the "principle of complementary energy." As noted in Art. 2-2, the complementary energy C is equal to the strain energy U in the case of linear material response. However, for the nonlinear material response, complementary energy and strain energy are not equal. For example, the complementary energy C of a nonlinear elastic tension member subject to an axial load is equal in magnitude to the area $0AB$ above the load-elongation curve $0B$ (see Fig. 4-1.1 and also Fig. 2-2.1 and Arts. 2-2), whereas the strain energy U is equal to the area $0BC$ below the curve $0B$. Hence, for this case $C \neq U$.

In the derivation of Castigliano's theorem the complementary energy C is regarded as a function of concentrated forces $(F_1, F_2, \ldots, F_p)$ that act

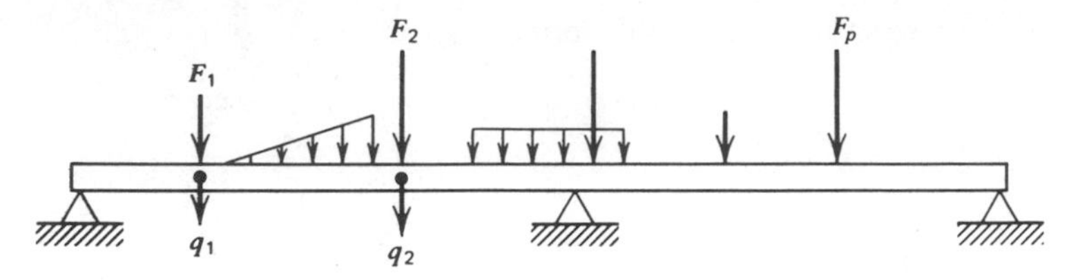

Fig. 4-2.1

on the system that is mounted on rigid supports (say a beam, Fig. 4-2.1). The complementary energy C depends also upon distributed loads that act on the beam (Fig. 4-2.1), as well as upon the weight of the beam. However, these distributed forces do not enter explicitly into consideration in the derivation. In addition, the beam may be subjected to temperature effects (e.g., thermal strains[1]).

Castigliano's theorem may be stated generally as follows (see reference 1, Art. 4-10):

> If an elastic system is mounted so that rigid-body displacements of the entire system are prevented, and if certain concentrated forces of magnitudes $F_1, F_2, \ldots, F_p$ act on the system, in addition to distributed loads and thermal strains, the displacement component q_i of the point of application of the force F_i, in the direction of F_i, is determined by the equation
>
> $$q_i = \frac{\partial C}{\partial F_i} \qquad i = 1, 2, \ldots, p \tag{4-2.1}$$

Thus, with reference to Fig. 4-2.1, the displacement q_1 at the location of F_1 in the direction of F_1 is given by the relation $q_1 = \partial C / \partial F_1$.

The derivation of Eq. (4-2.1) is based on the assumption of small displacements; therefore, Castigliano's theorem is restricted to small displacements of the system (beam). The complementary energy C of a structure composed of m members may be expressed by the relation

$$C = \sum_{i=1}^{m} C_i \tag{4-2.2}$$

where C_i denotes the complementary energy of the ith member, provided that the deflections produced by the loads are not so great as to cause the geometry of the loaded structure to be significantly different from that of the unloaded structure. In other words, the displacements of the structure must be everywhere "small" in order that Eq. (4-2.2) and hence Eq. (4-2.1), apply.†

Castigliano's theorem on deflections may be extended to compute the rotation of line elements, in a system, subjected to couples. For example,

†See Art. 4-10 of reference 1 and Chapter V of reference 2.

consider again a beam that is mounted on rigid supports and that is subjected to external concentrated forces of magnitudes $F_1, F_2, \ldots, F_p$ (Fig. 4-2.2). Let two of the concentrated forces (F_1, F_2) be parallel, lie in a principal plane of the cross section, have opposite senses, and act perpendicular to the ends of a line element of length b in the beam (Fig. 4-2.2a). Then, Eq. (4-2.1) shows that the rotation θ (Fig. 4-2.2b) of the line segment due to the deformations is given by the relation

$$\theta = \frac{1}{b}\frac{\partial C}{\partial F_1} + \frac{1}{b}\frac{\partial C}{\partial F_2} \tag{a}$$

where we have employed the condition of small displacements. To interpret this result, we employ the chain rule of partial differentiation of the complementary function C with respect to a scalar variable S, considering the magnitudes of F_1 and F_2 to be functions of S. Thus, we have

$$\frac{\partial C}{\partial S} = \frac{\partial C}{\partial F_1}\frac{\partial F_1}{\partial S} + \frac{\partial C}{\partial F_2}\frac{\partial F_2}{\partial S} \tag{b}$$

In particular, if we take the variable S equal to F_1, F_2, that is, $S = F_1 = F_2 = F$, where F denotes the magnitudes F_1, F_2, $\partial F_1/\partial S = \partial F_2/\partial S = 1$, and we obtain by Eq. (b)

$$\frac{\partial C}{\partial F} = \frac{\partial C}{\partial F_1} + \frac{\partial C}{\partial F_2} \tag{c}$$

Consequently, Eqs. (a) and (c) yield

$$\theta = \frac{1}{b}\frac{\partial C}{\partial F} \tag{d}$$

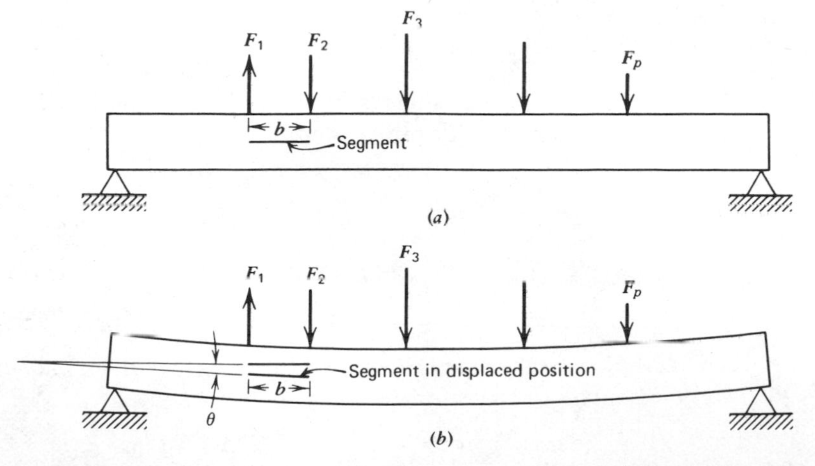

Fig. 4-2.2/(a) Beam before deformation. (b) Beam after deformation.

and since the equal and opposite forces F_1, F_2 constitute a couple of magnitude $M = bF$, Eq. (d) may be written in the form $\theta = \partial C/\partial M$. More generally, for couples M_i and rotations θ_i, we may write

$$\theta_i = \frac{\partial C}{\partial M_i} \qquad i = 1, 2, \ldots, s \tag{4-2.3}$$

Hence, Eq. (4-2.3) determines the angular displacement θ_i of the arm of a couple of magnitude M_i that acts on an elastic structure. The sense of θ_i is the same as that of the couple M_i.

Equations (4-2.1) and (4-2.3) are restricted to small displacements. However, they may be applied to structures that possess nonlinear material behavior.[1,2]

The following example problem indicates the application of Eq. (4-2.1) for nonlinear elastic material behavior.

EXAMPLE 4-2.1

Equilibrium of Simple Nonlinear Elastic Pin-Joined Truss

Let the two members of the pin-joined truss in Fig. E4-1.1 be made of a nonlinear elastic material whose stress-strain diagram is approximated by the relation $\epsilon = \epsilon_0 \sinh(\sigma/\sigma_0)$, where ϵ_0 and σ_0 are material constants. The truss is subjected to known loads P and Q. By means of Castigliano's theorem on deflections, determine the small displacement components u and v. Let $P = 10.0$ kN, $Q = 30.0$ kN, $\sigma_0 = 70.0$ MPa, $\epsilon_0 = 0.00100$, $b_1 = h = 400$ mm, $b_2 = 300$ mm, and $A_1 = A_2 = 300$ mm^2. Show that the values for u and v so obtained agree with those obtained by a direct application of equations of equilibrium and the consideration of the geometry of the deformed truss.

SOLUTION

Let N_1 and N_2 be the tensions in members 1 (AB) and 2 (CB). From the equilibrium conditions for pin B, we find

$$N_1 = \frac{L_1(Qb_2 + Ph)}{h(b_1 + b_2)}$$
$$N_2 = \frac{L_2(Qb_1 - Ph)}{h(b_1 + b_2)} \tag{1}$$

The complementary energy C for the truss is equal to the sum of the complementary energies for the two members. Thus,

$$C = C_1 + C_2 = \int_0^{N_1} e_1\, dN_1 + \int_0^{N_2} e_2\, dN_2 \tag{2}$$

With $e_1 = \epsilon_1 L_1$ and $e_2 = \epsilon_2 L_2$, Eq. (2) becomes

$$C = \int_0^{N_1} L_1 \epsilon_0 \sinh \frac{N}{A_1 \sigma_0}\, dN_1 + \int_0^{N_2} L_2 \epsilon_0 \sinh \frac{N_2}{A_2 \sigma_0}\, dN_2 \tag{3}$$

The displacement components u and v are obtained by substitution of Eq. (3) into Eq. (4-2.1). Thus, we find

$$\begin{aligned} u = q_P = \frac{\partial C}{\partial P} &= L_1 \epsilon_0 \left[\sinh \frac{N_1}{A_1 \sigma_0}\right] \frac{\partial N_1}{\partial P} + L_2 \epsilon_0 \left[\sinh \frac{N_2}{A_2 \sigma_0}\right] \frac{\partial N_2}{\partial P} \\ v = q_Q = \frac{\partial C}{\partial Q} &= L_1 \epsilon_0 \left[\sinh \frac{N_1}{A_1 \epsilon_0}\right] \frac{\partial N_1}{\partial Q} + L_2 \epsilon_0 \left[\sinh \frac{N_2}{A_2 \sigma_0}\right] \frac{\partial N_2}{\partial Q} \end{aligned} \tag{4}$$

The partial derivatives of N_1 and N_2 with respect to P and Q are obtained by means of Eqs. (1). Taking derivatives and substituting into Eqs. (4), we obtain

$$\begin{aligned} u &= \frac{L_1^2 \epsilon_0}{b_1 + b_2} \sinh \frac{L_1(Qb_2 + Ph)}{A_1 \sigma_0 h(b_1 + b_2)} - \frac{L_2^2 \epsilon_0}{b_1 + b_2} \sinh \frac{L_2(Qb_1 - Ph)}{A_2 \sigma_0 h(b_1 + b_2)} \\ v &= \frac{L_1 b_2 \epsilon_0}{h(b_1 + b_2)} \sinh \frac{L_1(Qb_2 + Ph)}{A_1 \sigma_0 h(b_1 + b_2)} + \frac{L_2 b_1 \epsilon_0}{h(b_1 + b_2)} \sinh \frac{L_2(Qb_1 - Ph)}{A_2 \sigma_0 h(b_1 + b_2)} \end{aligned} \tag{5}$$

Substitution of values for P, Q, σ_0, ϵ_0, b_1, b_2, h, A_1, and A_2, gives

$$\begin{aligned} u &= 0.4709 \text{ mm} \\ v &= 0.8119 \text{ mm} \end{aligned} \tag{6}$$

An alternate method of calculating u and v is as follows: determine tensions N_1 and N_2 in the two members by Eqs. (1); next, determine elongations e_1 and e_2 for the two members, and use these values of e_1 and e_2 along with geometric relations to calculate values for u and v. Equations (1) give $N_1 = 26.268$ kN and $N_2 = 14.286$ kN. Elongations e_1 and e_2 are given by the relations

$$e_1 = L_1 \epsilon_0 \sinh \frac{N_1}{A_1 \sigma_0} = 565.68(0.00100) \sinh \frac{26{,}268}{300(70)} = 0.9071 \text{ mm}$$

$$e_2 = L_2 \epsilon_0 \sinh \frac{N_2}{A_2 \sigma_0} = 500.00(0.00100) \sinh \frac{14{,}286}{300(70)} = 0.3670 \text{ mm}$$

With e_1 and e_2 known, values of u and v are given by the following geometric relations:

$$u = \frac{e_1 \cos \phi - e_2 \cos \theta}{\sin \theta \cos \phi + \cos \theta \sin \phi} = 0.4709 \text{ mm}$$

$$v = \frac{e_1 \sin \phi + e_2 \cos \theta}{\sin \theta \cos \phi + \cos \theta \sin \phi} = 0.8119 \text{ mm}$$

These values of u and v agree with those of Eqs. (6). Thus, Eq. (4-2.1) gave the correct values of u and v for this problem of nonlinear material behavior.

4-3
CASTIGLIANO'S THEOREM ON DEFLECTIONS FOR LINEAR LOAD-DEFLECTION RELATIONS

In the remainder of this chapter and in Chapter 8, we limit our consideration to linear elastic material behavior and to small displacements. The resulting load-deflection relation for either a member or a structure is linear, and the strain energy U is equal to the complementary energy C. Then Eqs. (4-2.1) and (4-2.3) may be written

$$q_i = \frac{\partial U}{\partial F_i} \qquad i = 1, 2, \ldots, p \tag{4-3.1}$$

$$\theta_i = \frac{\partial U}{\partial M_i} \qquad i = 1, 2, \ldots, s \tag{4-3.2}$$

where $U = U(F_1, F_2, \ldots, F_p, M_1, M_2, \ldots, M_s)$

The strain energy U is, by Eq. (2-2.7)

$$U = \iiint U_0 \, dV \tag{4-3.3}$$

where U_0 is the strain energy density. In this chapter we restrict ourselves to linear elastic isotropic homogeneous materials for which the strain energy density is (See Eq. 2-4.10)

$$U_0 = \frac{1}{2E}\left(\sigma_{xx}^2 + \sigma_{yy}^2 + \sigma_{zz}^2\right) - \frac{\nu}{E}\left(\sigma_{xx}\sigma_{yy} + \sigma_{yy}\sigma_{zz} + \sigma_{zz}\sigma_{xx}\right) + \frac{1}{2G}\left(\sigma_{yz}^2 + \sigma_{zx}^2 + \sigma_{xy}^2\right) \tag{4-3.4}$$

With load-stress formulas derived for the members of the structure, U_0 may be expressed in terms of the loads that act on the structure. Then, Eq. (4-3.3) gives U as a function of the loads. Equations (4-3.1) and (4-3.2) can then be used to obtain displacements at the points of applications of the concentrated forces or the rotations in the direction of the concentrated moments. Three types of loads are considered in this chapter for the various members of a structure as follows: (1) axial loading, (2) bending of straight beams, and (3) torsion. In practice, it is convenient to obtain the strain energy for each type of load acting alone and to add these strain energies together to obtain the total strain energy U, instead of using load-stress formulas and Eqs. (4-3.3) and (4-3.4) to obtain U.

Strain Energy U_N for Axial Loading of Members/The equation for strain energy U_N for axial loading is derived for the tension members shown in Fig. 4-3.1*a* and in Fig. 4-3.1*d*. In general, the cross-sectional area A of the tension member may vary "slowly" with axial coordinate z. The line of action of the loads (the z-axis) passes through the centroid of every cross section of the tension member. Consider two sections BC and DF of the tension member in Fig. 4-3.1*a* at distance dz apart. After the loads are applied, these sections are displaced to B^*C^* and D^*F^* (shown by the enlarged free body diagram in Fig. 4-3.1*b*) and the original length dz has elongated an amount de_z. For linear elastic material behavior, de_z varies linearly with N as indicated in Fig. 4-3.1*c*. The shaded area below the straight line is equal to the strain energy dU_N for the segment dz of the tension member. The strain energy U_N for the tension member becomes

$$U_N = \int dU_N = \int \tfrac{1}{2} N \, de_z \tag{4-3.5}$$

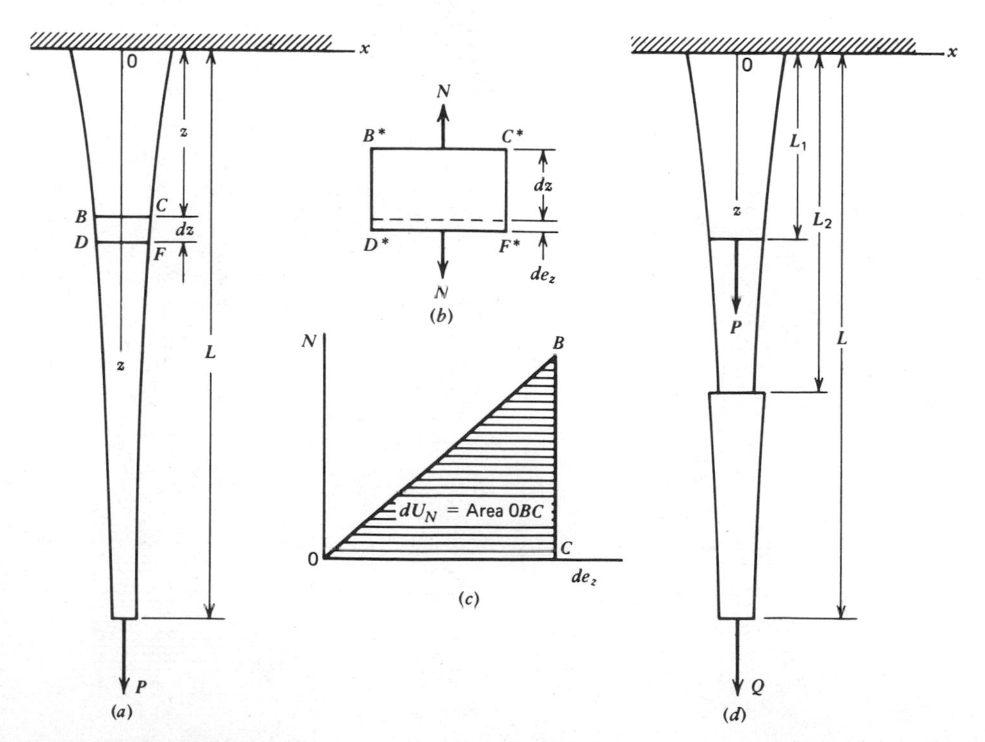

Fig. 4-3.1/Strain energy due to axial loading of member.

Noting that $de_z = \epsilon_{zz}\,dz$, and assuming that the cross-sectional area varies sufficiently slowly, we have $\epsilon_{zz} = \sigma_{zz}/E$ (see Eq. 2-4.7), and $\sigma_{zz} = N/A$, where A is the cross-sectional area of the member at section z. Then, we write Eq. (4-3.5) in the form

$$U_N = \int_0^L \frac{N^2}{2EA}\,dz \tag{4-3.6}$$

At abrupt changes in load or cross section, the values of N or A change abruptly. Then, we must account approximately for these changes by writing U_N in the form (see Fig. 4-3.1d)

$$U_N = \int_0^{L_1} \frac{N^2}{2EA}\,dz + \int_{L_1}^{L_2} \frac{N^2}{2EA}\,dz + \int_{L_2}^{L} \frac{N^2}{2EA}\,dz \tag{4-3.6a}$$

where an abrupt change in load occurs at L_1 and an abrupt change in cross-sectional area occurs at L_2.

Strain Energies U_M and U_S for Beams / Consider a beam of uniform cross section, as in Fig. 4-3.2a (or a beam with sufficiently slowly varying cross section). We take the z-axis along the axis of the beam with the y-axis down and the x-axis normal to the plane of the paper. The y-axis is assumed to be a principal axis for each cross section of the beam (see the Appendix), the (y, z)-plane is assumed to be a principal plane for the beam, and the loads P, Q, and R are assumed to lie in the principal

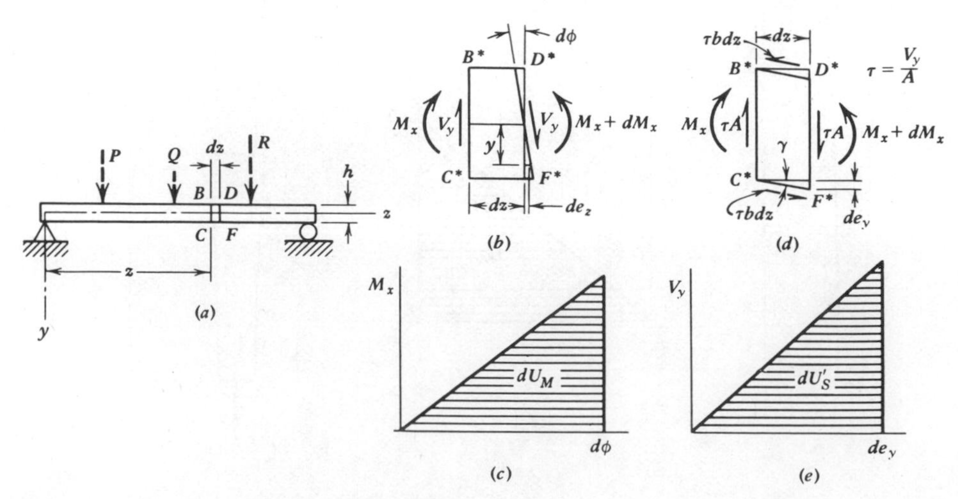

Fig. 4-3.2/Strain energy due to bending.

(y, z)-plane (the loading is said to be symmetrical). The flexure formula

$$\sigma_{zz} = \frac{M_x y}{I_x} \tag{4-3.7}$$

is assumed to hold, where M_x is the moment with respect to the centroidal x-axis, I_x is the moment of inertia of the cross section at z about the x-axis, and y is measured from the (x, z)-plane. Before applying the loads P, Q, and R to the beam, consider two sections BC and DF of the beam at distance dz apart. After the loads are applied to the beam, plane sections BC and DF are displaced to B^*C^* and D^*F^* and are assumed to remain plane. A free body diagram of the deformed beam segment is shown enlarged in Fig. 4-3.2b, where plane D^*F^* is rotated through angle $d\phi$ with respect to B^*C^*. For linear elastic material behavior, $d\phi$ varies linearly with M_x as indicated in Fig. 4-3.2c. The shaded area below the straight line is equal to the strain energy dU_M for the beam segment dz. An additional strain energy dU_S due to the shear V_y is considered later. The strain energy U_M for the beam due to M_x becomes

$$U_M = \int dU_M = \int \frac{1}{2} M_x \, d\phi \tag{4-3.8}$$

Noting that $d\phi = de_z/y$, $de_z = \epsilon_{zz}\, dz$, and assuming that $\epsilon_{zz} = \sigma_{zz}/E$ and $\sigma_{zz} = M_x y/I_x$, we may write Eq. (4-3.8) in the form

$$U_M = \int \frac{M_x^2}{2EI_x} \, dz \tag{4-3.9}$$

where in general M_x is a function of z. Equation (4-3.9) represents the strain energy due to bending about the x-axis. A similar relation is valid for bending about the y-axis for loads lying in the (x, z)-plane. For abrupt changes in moment M_x or cross section, the value of U_M may be computed following the same procedure as for U_N (Eq. 4-3.6a).

Equation (4-3.9) is exact for pure bending but is only approximate for shear loading as indicated in Fig. 4-3.2d. However, more exact solutions and experimental data indicate that Eq. (4-3.9) is fairly accurate, except for relatively short beams. Since an exact expression for the strain energy U_S due to shear loading of a beam is difficult to obtain, an approximate expression for U_S is also often used. When corrected by an appropriate coefficient, the use of this approximate expression often leads to fairly reliable results. The correction coefficients for given beam cross sections are discussed later.

Because of the shear V_y (Fig. 4-3.2b) shearing stresses σ_{zy} are developed in each cross section; the magnitude of σ_{zy} is zero at both the top and the bottom of the beam since the beam is not subjected to shear loads on the top or bottom surfaces. We define an average value of σ_{zy} as $\tau = V_y/A$. We assume that this average shearing stress acts on the beam segment (Fig. 4-3.2*d*) and, for convenience, assume that the beam cross section is rectangular with width b. Because of the shear, the displacement of face D^*F^* with respect to face B^*C^* is de_y. For linear elastic material behavior, de_y varies linearly with V_y, as indicated in Fig. 4-3.2*e*. The shaded area below the straight line is equal to the strain energy dU_S' for the beam segment dz. A correction coefficient k is now defined such that the exact expression for the shearing strain energy dU_S of the element is equal to $k\,dU_S'$. Then, the shearing strain energy U_S for the beam due to shear V_y is

$$U_S = \int k\,dU_S' = \int \frac{k}{2} V_y\,de_y \tag{4-3.10}$$

Noting that $de_y = \gamma\,dz$, and assuming that $\gamma = \tau/G$ and $\tau = V_y/A$, we may write Eq. (4-3.10) in the form

$$U_S = \int \frac{kV_y^2}{2GA}\,dz \tag{4-3.11}$$

Equation (4-3.11) represents the strain energy for shear loading of a beam. The value of V_y is generally a function of z. Also, the cross-sectional area A may vary slowly with z. For abrupt changes in shear V_y or cross-sectional area A, the value of U_S may be computed following the same procedure as for U_N (Eq. 4-3.6a).

An exact expression of U_S may be obtained provided the exact shearing stress distribution σ_{zy} due to the shear V_y is known. Then substitution into Eq. (4-3.4) to obtain U_0 (for σ_{zy} the only nonzero stress component) and then substitution into Eq. (4-3.3) yields U_S. However, the exact distribution of σ_{zy} is often difficult to obtain, and approximate distributions are used. For example, consider a segment dy of width b of a beam cross section. In the engineering theory of beams, the stress component σ_{zy} is assumed to be uniform over width b. With this assumption[3]

$$\sigma_{zy} = \frac{V_y Q}{I_x b} \tag{4-3.12}$$

where Q is the moment about the x-axis of the area above the line of length b with ordinate y. Generally, σ_{zy} is not uniform over width b. Nevertheless, if for a beam of rectangular cross section, one assumes that σ_{zy} is uniform over width b, it may be shown that $k = 1.20$.

Exact values of k are not generally available. Fortunately, in practical problems, the shearing strain energy U_S is often small compared to U_M. Hence, for practical problems, the need for exact values of U_S is not critical. Consequently, as an expedient approximation, we recommend that the correction coefficient k in Eq. (4-3.11) be obtained as the ratio of the shearing stress at the neutral surface of the beam calculated using Eq. (4-3.12) to the average shearing stress V_y/A. For example, by this

Table 4-3.1

Correction Coefficients for Strain Energy due to Shear

Beam Cross Section	$k^{\dagger}$
Rectangle	1.50
Circle	1.33
Thin-wall circular	2.00
I-section, channels, box-section‡	1.00

† Calculated by Eq. (4-3.13).

‡ The area A for the I-section, channel, or box-section is the area of the web ht where h is the beam depth and t is the web thickness.

procedure, the magnitude of k for the rectangular cross section is

$$k = \frac{V_y Q}{I_x b}\frac{A}{V_y} = \frac{Qbh}{I_x b} = \frac{\left(\frac{bh}{2}\right)\left(\frac{h}{4}\right)h}{\frac{1}{12}bh^3} = 1.50 \tag{4-3.13}$$

This value is larger and hence more conservative than the more exact value 1.20. Approximate values of k, calculated by this method are listed in Table 4-3.1 for several beam cross sections. For I-sections, channels, and box-sections, $k = 1.00$ provided that the area A in Eq. (4-3.11) be taken as the area of the web for these cross sections.

Strain Energy U_T for Torsion/The strain energy U_T for a torsion member with circular cross section (Fig. 4-3.3a) may be derived as follows. Let the z-axis lie along the centroidal axis of the torsion member. Before torsional loads T_1 and T_2 are applied, sections BC and DF are a distance dz apart. After the torsional loads are applied, these sections become sections B^*C^* and D^*F^*, with section D^*F^* rotated relative to section B^*C^* through the angle $d\beta$, as shown in the enlarged free body diagram of the element of length dz (Fig. 4-3.3b). For linear elastic material behavior, $d\beta$ varies linearly with T (Fig. 4-3.3c). The shaded area below the inclined straight line is equal to the torsional strain energy dU_T for the segment dz of the torsion member. Hence, the total torsional strain energy U_T for the torsional member becomes

$$U_T = \int dU_T = \int \frac{1}{2} T\, d\beta \tag{4-3.14}$$

Noting that $b\,d\beta = \gamma\, dz$, and assuming that $\gamma = \tau/G$ and $\tau = Tb/J$ (b is the radius of the torsion member and J is the polar moment of inertia of

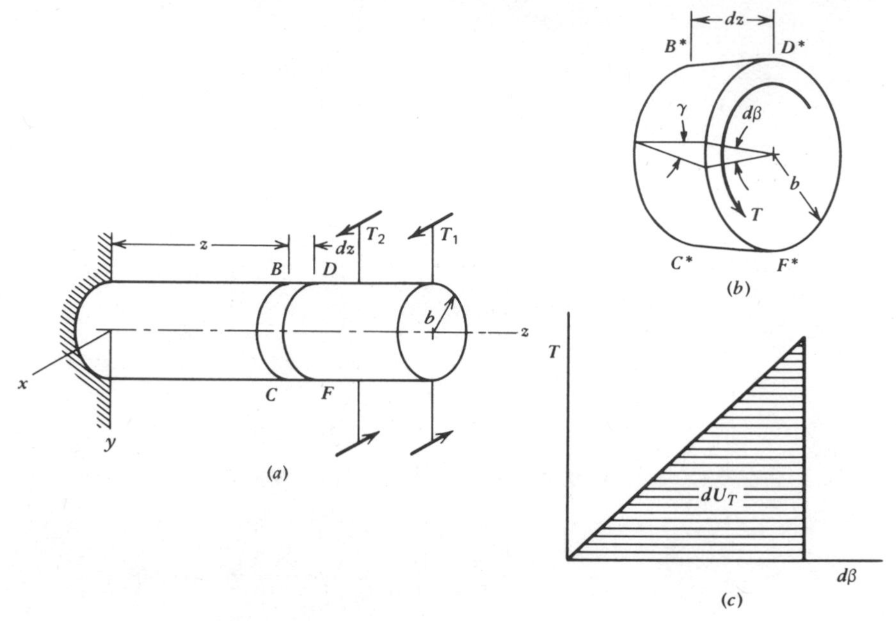

Fig. 4-3.3/Strain energy due to torsion.

the cross section), we may write Eq. (4-3.14) in the form

$$U_T = \int \frac{T^2}{2GJ}\, dz \tag{4-3.15}$$

Equation (4-3.15) represents the torsional strain energy for a torsion member with circular cross section. The unit angle of twist θ for a torsion member of circular cross section is given by $\theta = T/GJ$. Torsion of noncircular cross sections is treated in Chapter 5. Equation (4-3.15) is valid for other cross sections if the unit angle of twist θ for a given cross section replaces T/GJ in Eq. (4-3.15). For abrupt changes in torsional load T or cross sectional area A, the value of U_T may be computed following the same procedure as for U_N (Eq. 4-3.6a).

4-4 DEFLECTIONS OF STATICALLY DETERMINATE STRUCTURES

In the analysis of many engineering structures, the equations of static equilibrium are both necessary and sufficient to solve for unknown reactions and for tensions, shears, bending moments, and torques in the members of the structure. For example, the simple structure shown in

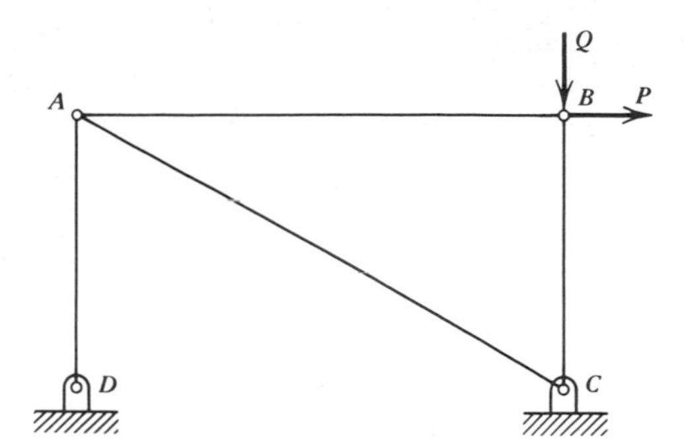

Fig. 4-4.1/Statically determinate pin-joined truss.

Fig. E4-1.1 is such a structure, since the equations of static equilibrium are sufficient to solve for the tensions N_1, and N_2 in members AB and CB, respectively. Structures for which the equations of static equilibrium are sufficient to determine the unknown tensions, shears, etc. uniquely are said to be *statically determinate structures*. Implied in the expression "statically determinate" is the condition that the deflections due to the loads are so small that the geometry of the initially unloaded structure remains essentially unchanged and the angles between members are essentially constant. If these conditions were not true, the internal tensions, etc., could not be determined without including the effects of the deformation and, hence, they could not be determined solely upon the basis of the equations of equilibrium.

The truss shown in Fig. 4-4.1 is a statically determinate truss. A physical characteristic of a statically determinate structure is that every member is essential for the proper functioning of the structure under the various loads to which it is subjected. For example, if member AC were to be removed from the truss of Fig. 4-4.1, the truss would be unable to support the loads and it would collapse.

Often additional members are added to structures in order to stiffen the structure (reduce deflections) or to strengthen the structure (increase its load carrying capacity). For example, for such purposes an additional diagonal member BD may be added to the truss of Fig. 4-4.1; see Fig. 4-4.2. Since the equations of static equilibrium are just sufficient for the analysis of the truss of Fig. 4-4.1, they are not adequate for the analysis of the truss of Fig. 4-4.2. Accordingly, the truss of Fig. 4-4.2 is said to be a *statically indeterminate structure*. The analysis of statically indeterminate structures requires additional information (additional equations) beyond that obtained from the equations of static equilibrium.

In this article, the analysis of statically determinate structures is discussed. The analysis of statically indeterminate structures is presented in Art. 4-5.

The strain energy U for a structure is equal to the sum of the strain energies of its members. The loading for the jth member of the structure is assumed to be such that the strain energy U_j for that member is

$$U_j = U_{Nj} + U_{Mj} + U_{Sj} + U_{Tj} \tag{4-4.1}$$

where U_{Nj}, U_{Mj}, U_{Sj}, and U_{Tj} for the jth member are given by Eqs. (4-3.6), (4-3.9), (4-3.11), and (4-3.15), respectively. In the remainder of this chapter each beam is assumed to undergo bending about a principal axis of the beam cross section (see the Appendix and Chapter 6); Eqs. (4-3.9) and (4-3.11) are valid only for bending about a principal axis. For simplicity, we consider bending about the x-axis (taken to be a principal axis) and let $M_x = M$ and $V_y = V$.

With the total strain energy U of the structure known, the deflection q_i of the structure at the location of a concentrated force F_i in the direction of F_i is (see Eq. 4-3.1)

$$\begin{aligned} q_i &= \frac{\partial U}{\partial F_i} \\ &= \sum_{j=1}^{m} \left[\int \frac{N_j}{E_j A_j} \frac{\partial N_j}{\partial F_i} dz + \int \frac{k_j V_j}{G_j A_j} \frac{\partial V_j}{\partial F_i} dz \right. \\ &\qquad \left. + \int \frac{M_j}{E_j I_j} \frac{\partial M_j}{\partial F_i} dz + \int \frac{T_j}{G_j J_j} \frac{\partial T_j}{\partial F_i} dz \right] \end{aligned} \tag{4-4.2}$$

and the angle (slope) change θ_i of the structure at the location of a concentrated moment M_i in the direction of M_i is (see Eq. 4-3.2)

$$\begin{aligned} \theta_i &= \frac{\partial U}{\partial M_i} \\ &= \sum_{j=1}^{m} \left[\int \frac{N_j}{E_j A_j} \frac{\partial N_j}{\partial M_i} dz + \int \frac{k_j V_j}{G_j A_j} \frac{\partial V_j}{\partial M_i} dz \right. \\ &\qquad \left. + \int \frac{M_j}{E_j I_j} \frac{\partial M_j}{\partial M_i} dz + \int \frac{T_j}{G_j J_j} \frac{\partial T_j}{\partial M_i} dz \right] \end{aligned} \tag{4-4.3}$$

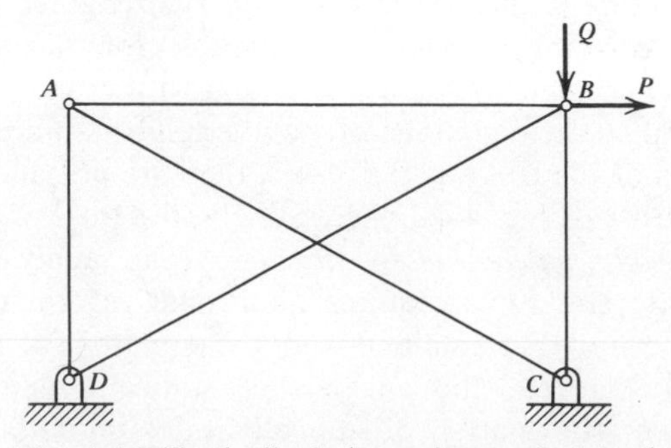

Fig. 4-4.2/Statically indeterminate pin-joined truss.

where m is the number of members in the structure. The similarity between Eqs. (4-4.2) and (4-4.3) and similar relations derived using the unit dummy load method should be noted. For example, consider the effect of a unit concentrated force acting on the structure at the same location and with the same sense as F_i. Then, we may interpret the quantities

$$n_{ji} = \frac{\partial N_j}{\partial F_i} \qquad v_{ji} = \frac{\partial V_j}{\partial F_i} \qquad m_{ji} = \frac{\partial M_j}{\partial F_i} \qquad t_{ji} = \frac{\partial T_j}{\partial F_i} \tag{4-4.4}$$

as the normal load, shear, moment, and torque, respectively, at a given location in the structure due to this unit force. The quantities n_{ji}, v_{ji}, m_{ji}, and t_{ji} due to the unit load correspond to the quantities N_j, V_j, M_j, and T_j due to the actual loads acting on the structure. Similarly, if a unit concentrated moment is considered to act on the structure at the same location and in the same sense as M_i, the quantities

$$n_{ji} = \frac{\partial N_j}{\partial M_i} \qquad v_{ji} = \frac{\partial V_j}{\partial M_i} \qquad m_{ji} = \frac{\partial M_j}{\partial M_i} \qquad t_{ji} = \frac{\partial T_j}{\partial M_i} \tag{4-4.5}$$

are the normal load, shear, moment, and torque, respectively, due to this unit moment. The results indicated in Eqs. (4-4.4) and (4-4.5) may be summarized in the following theorem (the Unit-Dummy Load Theorem):

> To compute the displacement q_i (or rotation θ_i of a given line element) of a structure, place a unit load at the point in the structure in the direction of q_i (or let the line element of the structure be subjected to a unit moment in the given sense of rotation). Denote the normal loads, shears, moments, and torques due to the unit load (or due to the unit moment) by n_{ji}, v_{ji}, m_{ji}, and t_{ji}, respectively. Denote the tensions, shears, moments, and torques due to the actual loads on the structure by N_j, V_j, M_j, and T_j. Then the displacement q_i is given by Eq. (4-4.2) in which the partial derivatives are replaced by n_{ji}, v_{ji}, m_{ji}, or t_{ji} as indicated by Eqs. (4-4.4). In like manner, Eq. (4-4.3) can be used to determine rotation θ_i with Eq. (4-4.5).

This theorem is limited to small deflections of linear elastic structures that consist of tension members, columns, beams, and torsion members. The structure may contain curved members (see the subsection below), provided that the curvatures are not so great that the elementary formulas for straight members cease to be applicable. Many applications of the above theorem are given in the book by Van Den Broek.[4]

Curved Beams Treated as Straight Beams / The strain energy due to bending (see Eq. 4-3.9) was derived assuming that the beam is straight. The magnitude of U_M for curves beams is derived in Chapter 8 where it is shown that the error in using Eq. (4-3.9) to determine U_M is negligible as long as the

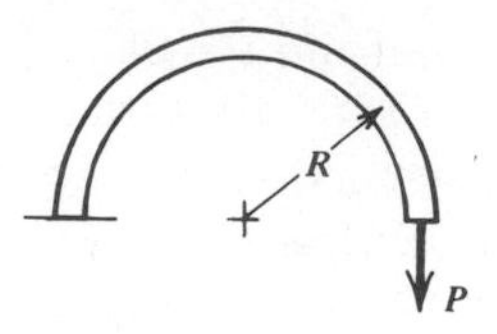

Fig. 4-4.3

radius of curvature of the curved beam is more than twice its depth. Consider the curved beam in Fig. 4-4.3 whose strain energy is the sum of U_N, U_S, and U_M each of which is caused by the same load P. If the radius of curvature R of the curved beam is large compared to the beam depth, the magnitudes of U_N and U_S will be small compared to U_M and can be neglected. We assume that U_N and U_S can be neglected when the ratio of length to depth is greater than 10. The resulting error is often less than 1 percent and will seldom exceed 5 percent. A numerical result is obtained in Example 4-4.1.

EXAMPLE 4-4.1

Cantilever Beam Loaded in its Plane

The cantilever beam in Fig. E4-4.1 has a rectangular cross section and is subjected to equal loads P at the free end and at the center as shown. (a) Determine the deflection of the free end of the beam. (b) What is the error in neglecting the strain energy due to shear if beam length L is five times the beam depth h? Assume that the beam is made of steel ($E = 200$ GPa and $G = 77.5$ GPa).

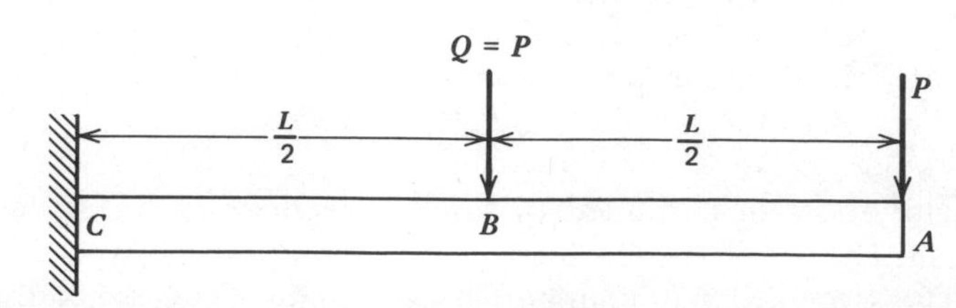

Fig. E4-4.1

SOLUTION

(a) To determine the dependencies of the shear V and moment M on the end load P, it is necessary to distinguish between the loads at A and B. Let the load at B, the center of the beam, be designated by Q. The moment and shear functions are continuous from A to B and from B to C. From A to B, we have

$$V = P, \qquad \therefore \ \frac{\partial V}{\partial P} = 1$$

$$M = Pz, \qquad \therefore \ \frac{\partial M}{\partial P} = z$$

From B to C, we have

$$V = P + Q, \qquad \therefore \ \frac{\partial V}{\partial P} = 1$$

$$M = P\left(z + \frac{L}{2}\right) + Qz, \qquad \therefore \ \frac{\partial M}{\partial P} = z + \frac{L}{2}$$

where we have chosen point B as the origin of coordinate z for the length from B to C. Equation (4-4.2) gives (with $Q = P$)

$$q_P = \int_0^{L/2} \frac{1.2P}{GA}(1)\, dz + \int_0^{L/2} \frac{Pz}{EI}(z)\, dz + \int_0^{L/2} \frac{2.4P}{GA}(1)\, dz$$

$$+ \int_0^{L/2} \frac{P(2z + L/2)}{EI}\left(z + \frac{L}{2}\right) dz = \frac{1.8PL}{GA} + \frac{7PL^3}{16EI} \tag{1}$$

(b) Since the beam has a rectangular section, $A = bh$ and $I = bh^3/12$. Equation (1) can be rewritten as follows:

$$\frac{Ebq_P}{P} = \frac{1.8LE}{Gh} + \frac{7(12)L^3}{16h^3}$$

$$= \frac{1.8(5)(200)}{77.5} + \frac{7(12)(5^3)}{16}$$

$$= 23.23 + 656.25$$

$$= 679.48$$

$$\text{Error in neglecting shear term} = \frac{23.23(100)}{679.48} = 3.42 \text{ percent.}$$

Alternatively, one could have used the value $k = 1.50$ (Table 4-3.1). Then the estimate of shear contribution would have been increased by the ratio $1.50/1.20 = 1.25$. Overall the shear contribution would still remain small.

EXAMPLE 4-4.2

Curved Beam Loaded in Its Plane

The curved beam in Fig. E4-4.2 has a 30 mm square cross section and a radius of curvature $R = 65$ mm. The beam is made of a steel for which

$E = 200$ GPa and $\nu = 0.29$. (a) If $P = 6.00$ kN, determine the component of deflection of the free end of the curved beam in the direction of P. (b) What is the error in the deflection if U_N and U_S are neglected? (c) Show that the same result would have been obtained using the dummy load method.

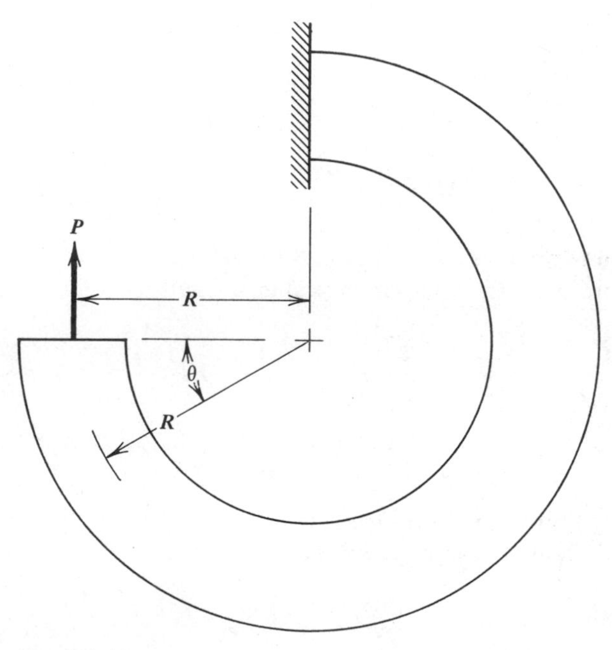

Fig. E4-4.2

SOLUTION

The shearing modulus for the steel is $G = E/(2 + 2v) = 77.5$ GPa.

(a) It is convenient to use polar coordinates. For a cross section of the curved beam located at angle θ from the section on which P is applied (Fig. E4-4.2),

$$
\begin{aligned}
N &= P\cos\theta & \frac{\partial N}{\partial P} &= \cos\theta \\
V &= P\sin\theta & \frac{\partial V}{\partial P} &= \sin\theta \qquad (1) \\
M &= PR(1-\cos\theta) & \frac{\partial M}{\partial P} &= R(1-\cos\theta)
\end{aligned}
$$

Substitution into Eq. (4-4.2) gives

$$q_P = \int_0^{3\pi/2} \frac{P\cos\theta}{EA}(\cos\theta)R\,d\theta + \int_0^{3\pi/2} \frac{kP\sin\theta}{GA}(\sin\theta)R\,d\theta + \int_0^{3\pi/2} \frac{PR(1-\cos\theta)}{EI}R(1-\cos\theta)R\,d\theta \quad (2)$$

Using the trigonometric identities $\cos^2\theta = \frac{1}{2} + \frac{1}{2}\cos 2\theta$ and $\sin^2\theta = \frac{1}{2} - \frac{1}{2}\cos 2\theta$, we find that

$$q_P = \frac{PR}{EA}\int_0^{3\pi/2}\left(\tfrac{1}{2} + \tfrac{1}{2}\cos 2\theta\right)d\theta + \frac{1.2PR}{GA}\int_0^{3\pi/2}\left(\tfrac{1}{2} - \tfrac{1}{2}\cos 2\theta\right)d\theta + \frac{PR^3}{EI}\int_0^{3\pi/2}\left(1 - 2\cos\theta + \tfrac{1}{2} + \tfrac{1}{2}\cos 2\theta\right)d\theta$$

$$q_P = \frac{3\pi PR}{4EA} + \frac{1.2(3\pi)PR}{4GA} + \left(\frac{9\pi}{4} + 2\right)\frac{PR^3}{EI}$$

$$= \frac{3\pi(65)(6000)}{4(200\times10^3)(30)^2} + \frac{1.2(3\pi)(65)(6000)}{4(77{,}500)(30)^2} + \left(\frac{9\pi}{4}+2\right)\frac{(65)^3(6000)(12)}{(200\times10^3)(30)^4}$$

$$= 0.0051 + 0.0158 + 1.1069 = 1.1278$$

Again, as noted in Example 4-4.1, we could have used the value $k = 1.50$, with a resulting slight overall change in the shear contribution.

(b) In case U_N and U_S are neglected

$$q_P = 1.1069 \text{ mm}$$

and the percentage error in the deflection calculation is

$$\text{percent error} = \frac{(1.1278 - 1.1069)100}{1.1278} = 1.85 \text{ percent}$$

This error is small enough to be neglected for most engineering applications. The ratio of length to depth for this beam is $3\pi(65)/[2(30)] = 10.2$.

(c) Let a unit load of 1 N replace load P in Fig. E4-4.2. The normal load n, shear load v, and moment m for the section of the curved beam located at angle θ from the section where the unit load is applied are

$$n = \cos\theta$$
$$v = \sin\theta \quad (3)$$
$$m = R(1-\cos\theta)$$

The deflection q_P predicted by the dummy load method is given by the relation

$$q_P = \int_0^\pi \frac{N}{EA} nR\, d\theta + \int_0^\pi \frac{kV}{GA} vR\, d\theta + \int_0^\pi \frac{M}{EI} mR\, d\theta \tag{4}$$

When values of N, V, and M from Eqs. (1) and values of n, v, and m from Eqs. (3) are substituted into Eq. (4), the resulting equation is identical with Eq. (2).

EXAMPLE 4-4.3

Pin-Connected Structure

The pin-connected structure in Fig. E4-4.3 is made of an aluminum alloy for which $E = 72.0$ GPa. The magnitudes of the loads are $P = 10$ kN and $Q = 5$ kN. Members BC, CD, and DE are subjected to compression loads; they have box-sections and a cross sectional area of 900 mm^2. The remaining members have a cross sectional area of 150 mm^2. Determine the change of slope of member BE caused by the loads P and Q.

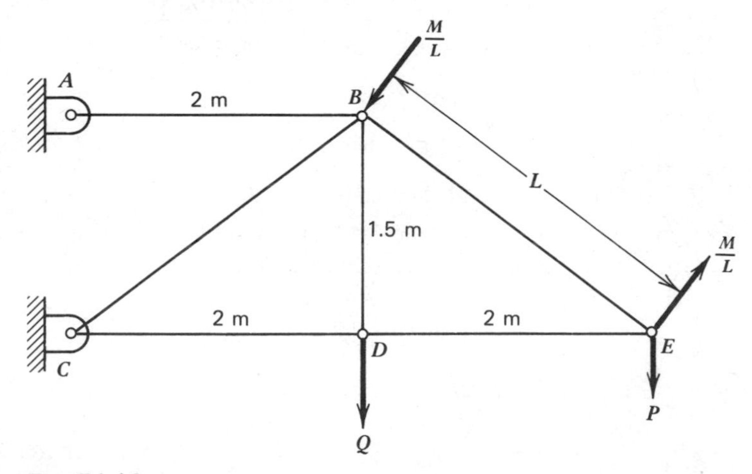

Fig. E4-4.3

SOLUTION

In order to determine the angle change of member BE by energy methods, a moment M must be acting on member BE. Hence, let M be an imaginary counterclockwise moment represented by a couple with equal and opposite forces M/L ($L = BE = 2.5 \times 10^3$ mm) applied perpendicular to BE at points B and E as indicated in Fig. E4-4.3. Equations of equilibrium give the following values for the normal loads in the members of the structure.

$$N_{AB} = \frac{4}{3}(Q + 2P) - \frac{5M}{3L} \qquad \frac{\partial N_{AB}}{\partial M} = -\frac{5}{3L}$$

$$N_{BC} = -\frac{5}{3}(Q + P) \qquad \frac{\partial N_{BC}}{\partial M} = 0$$

$$N_{BD} = Q \qquad \frac{\partial N_{BD}}{\partial M} = 0$$

$$N_{BE} = \frac{5P}{3} - \frac{4M}{3L} \qquad \frac{\partial N_{BE}}{\partial M} = -\frac{4}{3L}$$

$$N_{CD} = N_{DE} = -\frac{4P}{E} + \frac{5M}{3L} \qquad \frac{\partial N_{CD}}{\partial M} = \frac{5}{3L}$$

After the partial derivatives $\partial N_j/\partial M$ have been taken, the magnitude of M in the N_j is set equal to zero. The values of N_j and $\partial N_j/\partial M$ are then substituted into Eq. (4-4.3) to give

$$\theta_{BE} = \sum_{j=1}^{6} \frac{N_j L_j}{E_j A_j} \frac{\partial N_j}{\partial M} = \frac{N_{AB} L_{AB}}{EA_{AB}} \frac{\partial N_{AB}}{\partial M} + \frac{N_{BC} L_{BC}}{EA_{BC}} \frac{\partial N_{BC}}{\partial M}$$

$$+ \frac{N_{BD} L_{BD}}{EA_{BD}} \frac{\partial N_{BD}}{\partial M} + \frac{N_{BE} L_{BE}}{EA_{BE}} \frac{\partial N_{BE}}{\partial M}$$

$$+ 2\frac{N_{CD} L_{CD}}{EA_{CD}} \frac{\partial N_{CD}}{\partial M}$$

$$\theta_{BE} = \frac{4(25{,}000)(2000)}{3(72{,}000)(150)}\left[-\frac{5}{3(2500)}\right]$$

$$+ \frac{5(10{,}000)(2500)}{3(72{,}000)(150)}\left[-\frac{4}{3(2500)}\right]$$

$$- \frac{2(4)(10{,}000)(2000)}{3(72{,}000)(900)}\left[\frac{5}{3(2500)}\right]$$

$$= -0.004115 - 0.002058 - 0.000549$$

$$= -0.00672 \text{ rad}$$

The negative sign for θ_{BE} indicates that the angle change is clockwise; that is, the angle change has a sign opposite to that assumed for M.

EXAMPLE 4-4.4

Curved Beam Loaded Perpendicular to Its Plane

The semicircular curved beam of radius R in Fig. E4-4.4 has a circular cross section of radius r. The curved beam is fixed at 0 and lies in the (x, y)-plane with center C on the x-axis. Load P parallel to the z-axis acts at a section $\pi/2$ from the fixed end. Determine the z-component of the deflection of the free end. Assume that R/r is sufficiently large for U_S to be neglected.

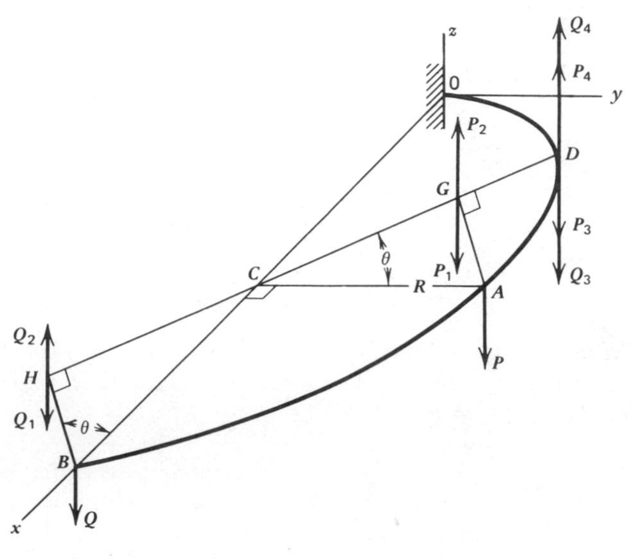

Fig. E4-4.4

SOLUTION

In order to determine the z-component of the deflection of the free end of the curved beam, an imaginary load Q parallel to the z-axis is applied at B as indicated in Fig. E4-4.4; the curved beam is indicated by its center line in order to simplify the figure. Consider a section D of the curved beam at angle θ measured from section A at the load P. Draw perpendicular lines AG and BH from points A and B to a line through points C and D. Apply equal and opposite forces P_1 and P_2 at G and equal and opposite forces P_3 and P_4 at D with magnitudes equal to P and action lines parallel to P. Also, apply equal and opposite forces Q_1 and Q_2 at H and equal and opposite forces Q_3 and Q_4 at D with magnitudes equal to Q and action lines parallel to Q. Note that the sum of couples formed by P and P_2 and by Q and Q_2 is the moment at section D; furthermore, the sum of the couples formed by P_1 and P_4 and by Q_1 and Q_4 is the torque at section D. The strain energy due to shear ($V = P_3 + Q_3$ at D) will not be considered.

$$M = PR\sin\theta + QR\cos\theta \qquad \frac{\partial M}{\partial Q} = R\cos\theta$$

$$T = PR(1-\cos\theta) + QR(1+\sin\theta) \qquad \frac{\partial M}{\partial Q} = R(1+\sin\theta)$$

These values are substituted into Eq. (4-4.2) to give with $Q = 0$

$$q_Q = \int_0^{\pi/2} \frac{PR\sin\theta}{EI} R(\cos\theta) R\,d\theta + \int_0^{\pi/2} \frac{PR(1-\cos\theta)}{GJ} R(1+\sin\theta) R\,d\theta$$

$$= \frac{PR^3}{EI}\frac{1}{2} + \frac{PR^3}{GJ}\left(\frac{\pi}{2} - \frac{1}{2}\right)$$

$$= \frac{2PR^3}{\pi E r^4}[1 + (1+\nu)(\pi - 1)]$$

PROBLEM SET 4-4

1. Determine the horizontal component of deflection of the free end of the curved beam described in Example 4-4.2. Assume that U_N and U_S are so small that they can be neglected.

2. For the pin-connected structure in Fig. E4-4.3, determine the component of the deflection of point E in the direction of force P.

   Ans. $q_P = 25.60$ mm

3. Find the vertical deflection of point C in the truss shown in Fig. P4-4.3. All members have the same cross section and are made of the same material.

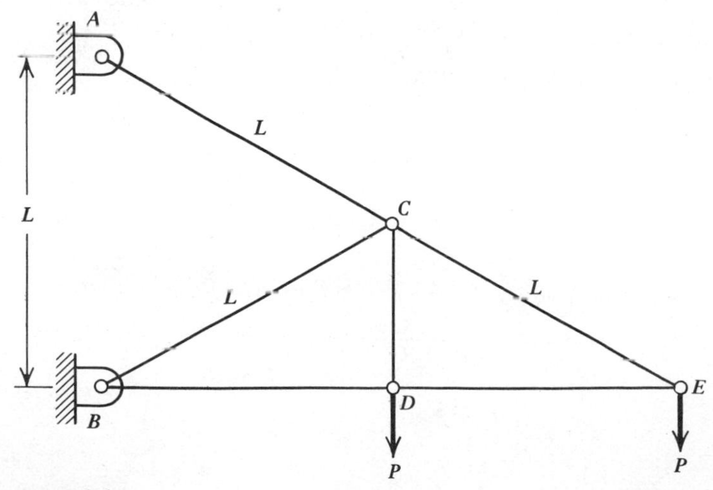

Fig. P4-4.3

4. The beam in Fig. P4-4.4 has the central half of the beam enlarged so that the moment of inertia I is twice the value for each end section. Determine the deflection of the beam at the center of the beam.

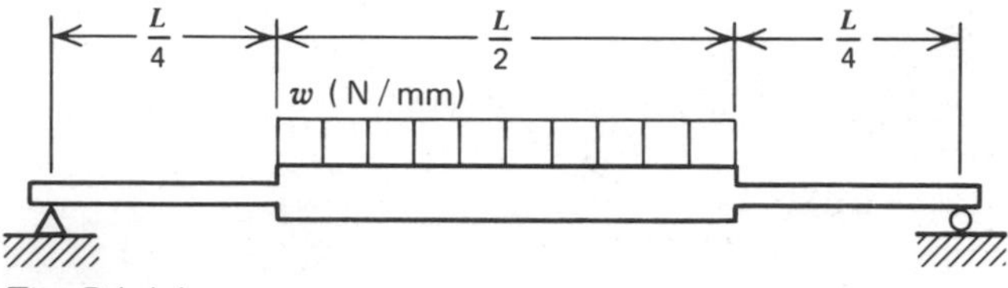

Fig. P4-4.4

Ans. $q_{max} = 65wL^4/(6144EI)$

5. Member ABC in Fig. P4-4.5 has a uniform symmetrical cross section and a depth that is small compared to L and R. Determine the component of the deflection of point C in the direction of load P.

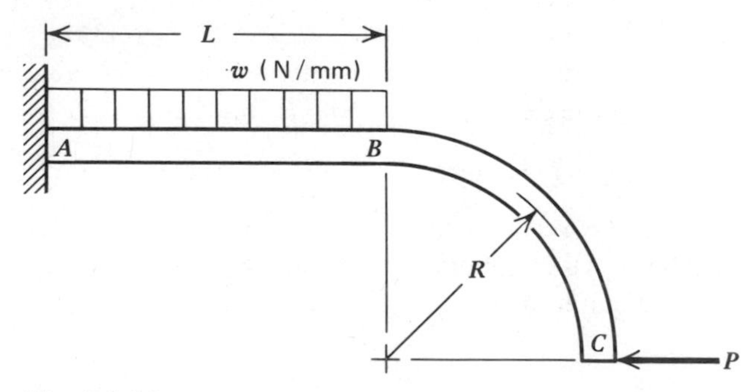

Fig. P4-4.5

6. Member $0AB$ in Fig. P4-4.6 lies in one plane and has the shape of two quadrants of a circle. Assuming that U_S and U_N can be ne-

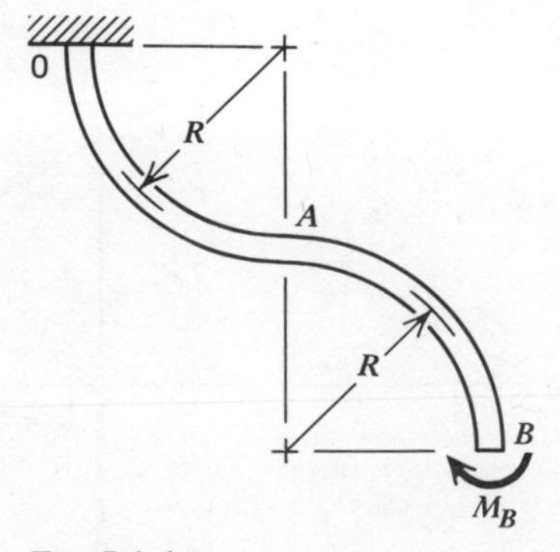

Fig. P4-4.6

glected, determine the vertical component of the deflection of point B.

Ans. Vertical $q_B = \dfrac{\pi R^2 M_B}{EI}$

7. Determine the horizontal deflection of point B for the member in Fig. P4-4.6.

8. Determine the change in slope of the cross section at point B for the member in Fig. P4-4.6.

   Ans. $\theta_B = \pi R M_B / EI$

9. Determine the x- and y-components of the deflection of point B of the semicircular beam in Fig. P4-4.9. The depth of the beam is small compared to R.

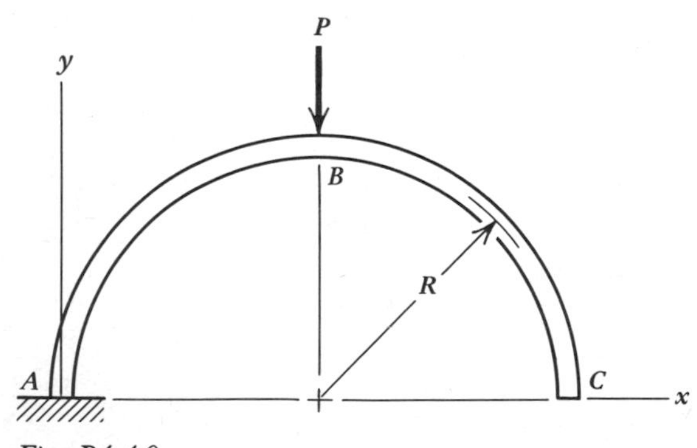

Fig. P4-4.9

10. Determine the vertical component C_y of the deflection of point C for the semicircular beam in Problem 9.

    Ans. $C_y = PR^3\left(1 + \frac{\pi}{4}\right)/EI$

11. The structure in Fig. P4-4.11 is made up of a cantilever beam $AB(E_1, I_1, A_1)$ and two identical rods BC and $CD(E_2, A_2)$. Let A_1 be large compared to A_2 and L_1 be large compared to the beam depth. (a) Determine the component of the deflection of point C in the direction of load P. (b) If $E_1 = E_2 = E$, the beam and rods have solid circular cross sections with radii r_1 and r_2, respectively, and $L_1 = L_2 = 25r_1$, determine the ratio of r_1 to r_2 such that the beam and rods contribute equally to q_P.

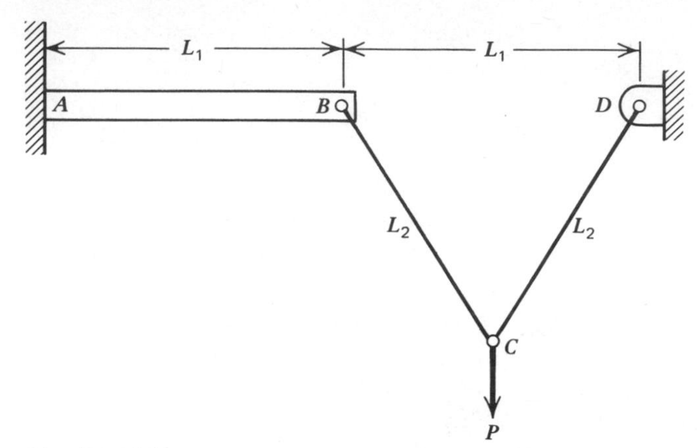

Fig. P4-4.11

12. Beam ABC in Fig. P4-4.12 is simply supported and subjected to a linearly varying distributed load as shown. Determine the deflection of the center of the beam.

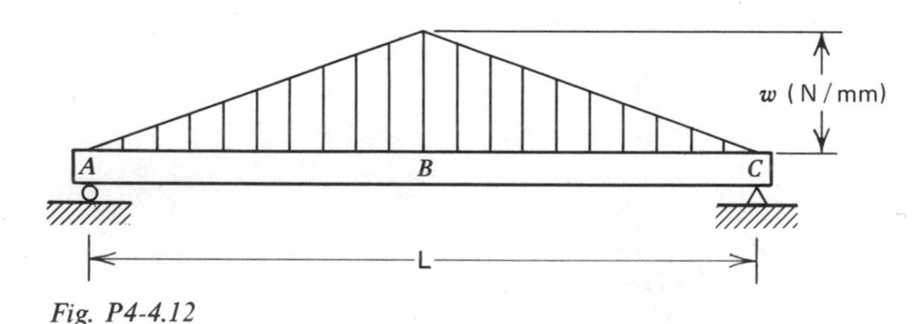

Fig. P4-4.12

Ans. $q_B = wL^4/120EI$

13. Member ABC in Fig. P4-4.13 has a circular cross section with radius r. It has a right angle bend at B and is loaded by a load P perpendicular to the plane of ABC. Determine the component of deflection of point C in the direction of P. Assume that L_1 and L_2 are each large compared to r.

14. Member ABC in Fig. P4-4.14 lies in the plane of the paper, has a uniform circular cross section, and is subjected to torque T_0, also in the plane of the paper at C as shown. Determine the displacement of point C. $G = E/2(1+\nu)$.

Ans. $q_C = T_0[4R^2 + 4R^2\nu - \pi R^2\nu + 4LR + 2L^2]/4EI$

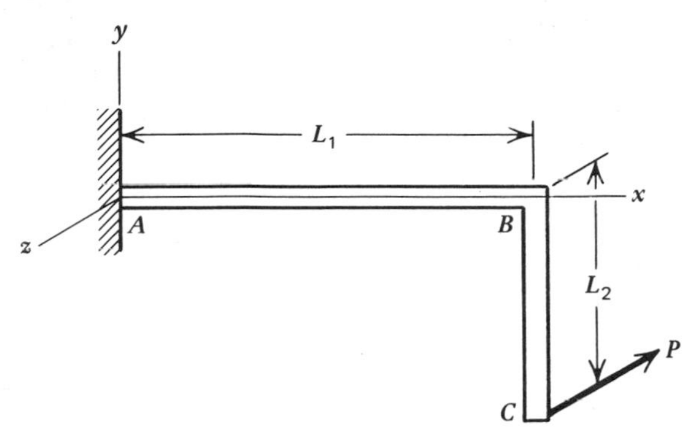

Fig. P4-4.13

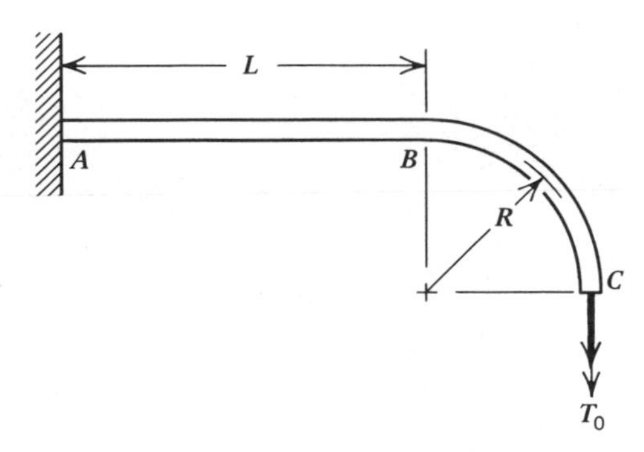

Fig. P4-4.14

15. For the member in Problem 14, determine the rotation of the section at C in the direction of T_0.

16. Member ABC in Fig. P4-4.16 lies in the plane of the paper, has a uniform circular cross section, and is subjected to a uniform load w

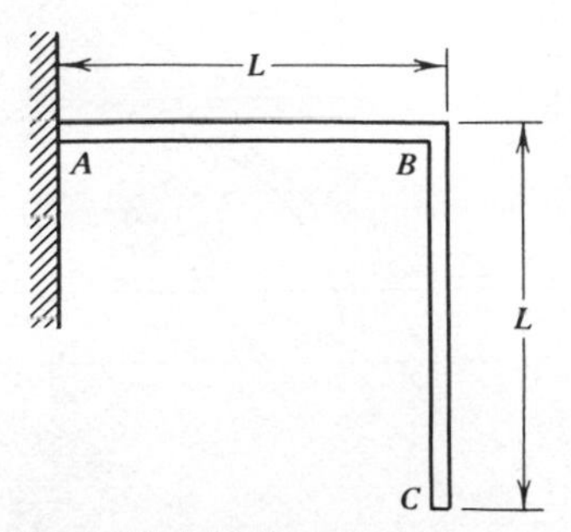

Fig. P4-4.16

(N/mm) that acts perpendicular to the plane of ABC. Determine the deflection of point C if length L is large compared to the diameter of the member. $G = E/2(1+\nu)$

Ans. $q_C = wL^4(13+6\nu)/12EI$

17. Member ABC in Fig. P4-4.17 lies in the plane of the paper, has a uniform circular cross section, and is subjected to a couple PL with loads P perpendicular to the plane of ABC. Determine the deflection of point C if R and L are large compared to the diameter of the member.

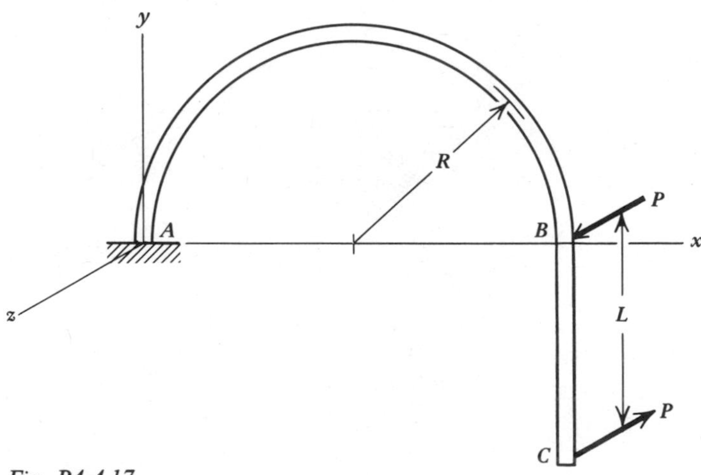

Fig. P4-4.17

18. The semicircular member in Fig. P4-4.18 lies in the (x, y)-plane and has a circular cross section with radius r. The member is fixed at A

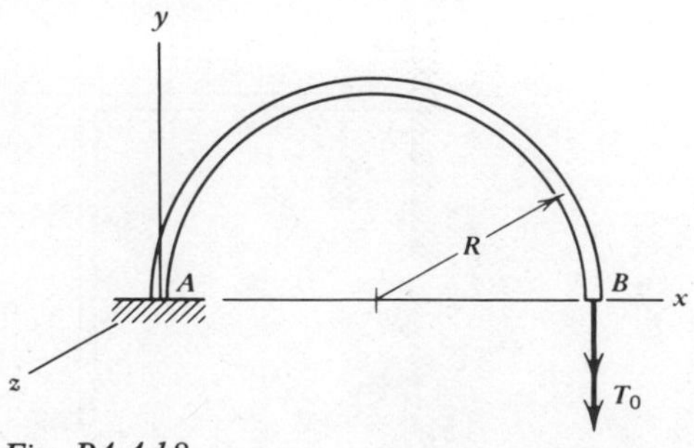

Fig. P4-4.18

and is subjected to torque T_0 at the free end at B. Determine the angle of twist of the member at B. $G = E/2(1+\nu)$.

Ans. $\theta_B = 2T_0R(2+\nu)/Er^4$

19. For the semicircular member in Problem 18, determine the x-, y-, and z-components of the deflection of point B.

20. A bar having a circular cross section is fixed at the origin 0 as shown in Fig. P4-4.20 and has right angle bends at points A and B. Length $0A$ lies along the z-axis; length AB is parallel to the x-axis; and length BC is parallel to the y-axis. Determine the x-, y-, and z-components of the deflection of point C. Moment M_C is a couple lying in a plane parallel to the (x, y)-plane. $G = E/2(1+\nu)$.

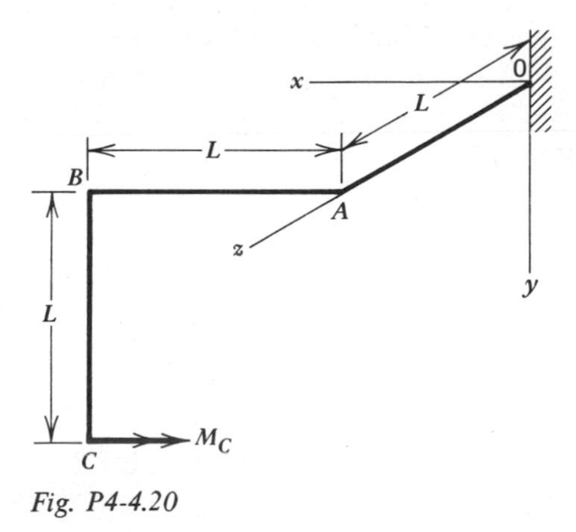

Fig. P4-4.20

Ans. $u = 0$; $v = M_CL^2/2EI$; $w = -M_CL^2(5+2\nu)/2EI$

21 through 40. Add 20 to the problem number from 1 through 20 and solve the resulting problem by the unit-dummy load method.

4-5 STATICALLY INDETERMINATE STRUCTURES

As observed in Art. 4-4, a statically determinate structure (Fig. 4-4.1) may be made statically indeterminate by the addition of a member (member BD in Fig. 4-4.2). Alternatively, a statically indeterminate structure is rendered statically determinate if certain members are removed. For example, the truss of Fig. 4-4.2 is rendered statically determinate if member BD (or equally well member AC) is removed. Such a member in a statically indeterminate structure is said to be *redundant* since after its removal the structure can still carry the loads without collapsing. In general, statically indeterminate structures contain one or more redundant members or supports. Removal of these redundant members or supports renders the structure statically determinate.

In the general analysis of structures, internal forces in each member of the structure are required so that the strain energy of the structure may be determined. In the case of statically indeterminate structures, these internal forces cannot be determined directly by the equations of equilibrium either because the number of external support reactions may exceed the number that can be determined by the equations of static equilibrium (Figs. 4-5.1a and b) or because the internal forces in redundant members cannot be determined from the equations of static equilibrium (Figs. 4-5.2a and b). For example, in Fig. 4-5.1a, the rigid stop at B is assumed to just make contact with the beam in the unloaded state. When the beam is loaded, the stop exerts an external reaction R (Fig. 4-5.1c) on the beam, which is not determinate by the equations of static equilibrium. If the stop at B were to be removed, the beam would function as a simple cantilever beam, although the end deflection of the beam may then be larger than desirable. Hence, the stop at B is redundant and, if it is removed, the beam is rendered statically determinate.

Since the reaction R in Fig. 4-5.1c is redundant, additional information is required for its determination. As we shall see, the fact that the

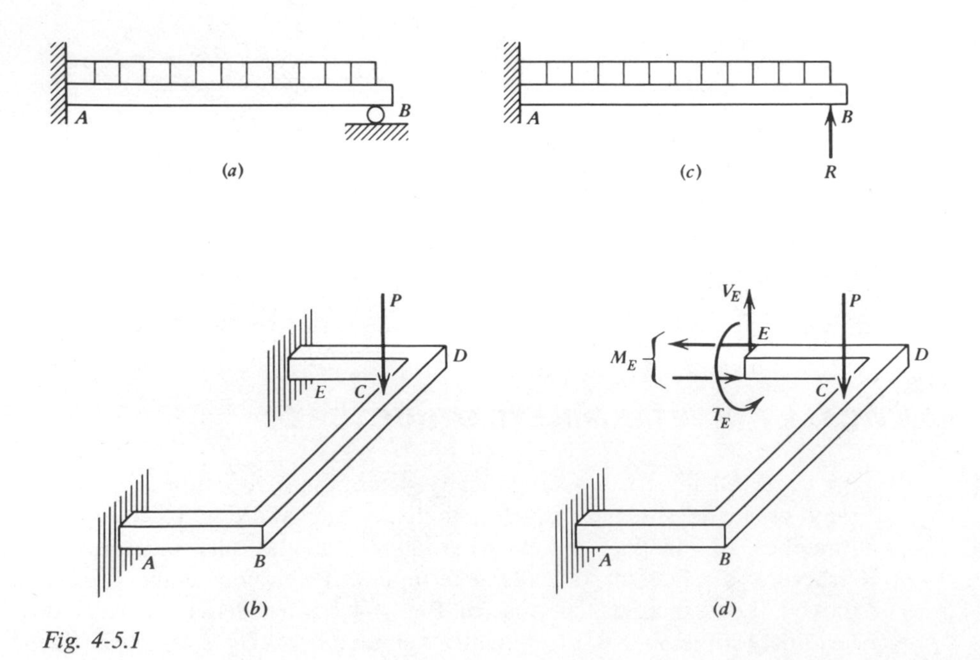

Fig. 4-5.1

rigid stop at B prevents the tip of the beam from displacing vertically may be used, in conjunction with Castigliano's theorem on deflections, to obtain the additional equation (in addition to the static equilibrium equations) needed to determine the external redundant reaction R.

Likewise, the support at A (or the support at E) for member $ABCDE$ in Fig. 4-5.1b is redundant. Hence, either the support at A or at E (but not both) may be removed to render the member statically determinate. If both supports at A and E are moved, of course, the member would be unsupported.

If member $ABCDE$ of Fig. 4-5.1b is assumed to be unstressed in the absence of load P, the supports at A and E produce no reactions on the member (assuming that the weight of the member is negligibly small and that thermal effects, expansions, and contractions are absent). However, when the load P is applied, displacements, bending rotations, and twisting rotations are prevented at the supports. For example, when the force P is applied, the support at E produces three internal reactions at section E (Fig. 4-5.1d): a vertical force V_E, which prevents vertical deflection at E, a bending moment M_E, which prevents bending rotation of the section at E, and a torque T_E, which prevents angular rotation of the section at E. The fact that the vertical deflection, the bending rotation, and the angular rotation are prevented at section E may be used, in conjunction with Castigliano's theorem on deflections, to obtain the additional equations needed to determine the internal reactions at E.

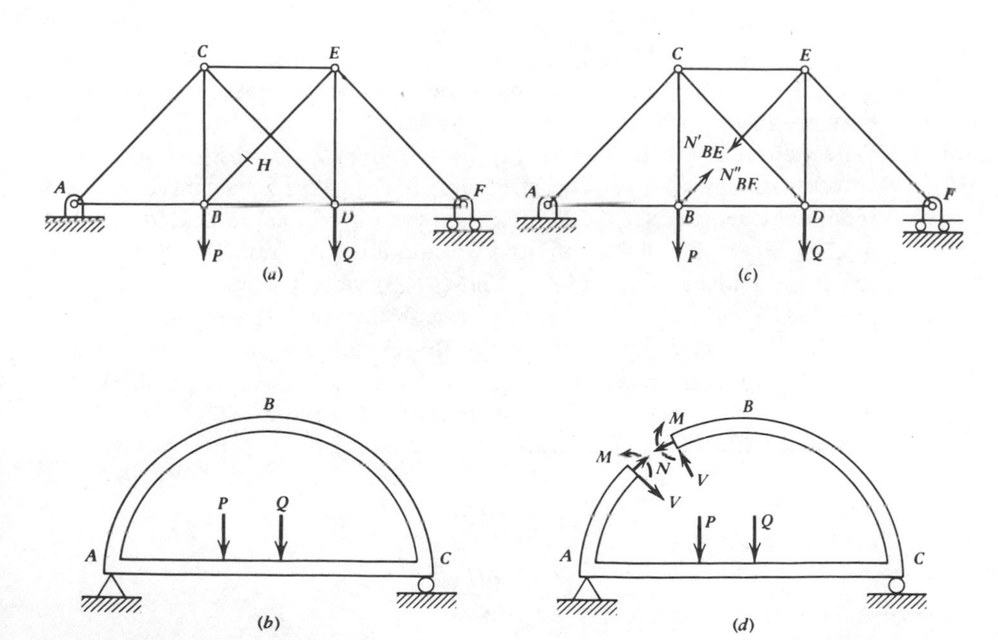

Fig. 4-5.2

The structures in Fig. 4-5.2 do not contain redundant reactions but do contain redundant members. In Fig. 4-5.2*a*, the member *BE* (or *CD*) of the truss is redundant. Hence, the truss is statically indeterminate. If either member *BE* or member *CD* is removed, the truss is rendered statically determinate. Likewise, the member *ABC* of the statically indeterminate structure in Fig. 4-5.2*b* is redundant. It may be removed to render the structure statically determinate.

Since the truss of Fig. 4-5.2*a* is pin-joined, the redundant member *BE* (or *CD*) is subject to an internal tension (or compression) only. Hence, the only redundant internal force for the truss is the tension in member *BE* (Fig. 4-5.2*c*). However, the redundant member *ABC* of the structure in Fig. 4-5.2*d* may support three internal reactions: the tension N, the shear V, and the moment M. The additional equations (in addition to the equations of static equilibrium) required to determine the additional unknowns (the redundant reactions caused by redundant members) in statically indeterminant structures may be obtained by the application of Castigliano's theorem on deflections.

In particular, we can show that

$$\frac{\partial U}{\partial F_i} = 0 \tag{4-5.1}$$

for every internal redundant force $(F_1, F_2, \ldots,)$ of the structure, and

$$\frac{\partial U}{\partial M_i} = 0 \tag{4-5.2}$$

for every internal and redundant moment $(M_1, M_2, \ldots,)$ of the structure. Equations (4-5.1) and (4-5.2) are readily verified for the structures in Fig. 4-5.1. The beam in Fig. 4-5.1*a* has a redundant external reaction R at B that produces an internal reaction at section B equal to R. Since the deflection at point B is zero, Eq. (4-4.2) gives $q_R = \partial U/\partial R = 0$, which agrees with Eq. (4-5.1). The structure in Fig. 4-5.1*b* has three internal redundant reactions (V_E, M_E, T_E) at section E, as indicated in Fig. 4-5.1*d*. Since the deflection and rotations at E remain zero as the structure is loaded, Eqs. (4-4.2) and (4-4.3) yield the results $\partial U/\partial V_E = \partial U/\partial M_E = \partial U/\partial T_E = 0$, which agree with Eqs. (4-5.1) and (4-5.2).

It is not directly apparent that Eqs. (4-5.1) and (4-5.2) are valid for the internal redundant reactions in the structures in Fig. 4-5.2. To show that they are valid, let N_{BE} be the redundant internal reaction for the pin-joined truss (Fig. 4-5.2*a*). Pass a section through some point H of member *BE* and apply equal and opposite tensions N'_{BE} and N''_{BE}, as indicated in Fig. 4-5.2*c*. Since the component of the deflection of point H along member *BE* is not zero, it is not obvious that

$$\frac{\partial U}{\partial N_{BE}} = 0 \tag{4-5.3}$$

In order to prove that Eq. (4-5.3) is valid, it is necessary to distinguish between tensions N'_{BE} and N''_{BE}. The displacement of point H in the direction of N'_{BE} is given by (see Eq. 4-4.2)

$$q_{N'_{BE}} = \frac{\partial U}{\partial N'_{BE}} \tag{4-5.4}$$

and in the direction of N''_{BE}, the displacement is given by

$$q_{N''_{BE}} = \frac{\partial U}{\partial N''_{BE}} \tag{4-5.5}$$

These displacements $q_{N'_{BE}}$ and $q_{N''_{BE}}$ are collinear, have equal magnitudes, but have opposite senses. Hence, by Eqs. (4-5.4) and (4-5.5) we have

$$\frac{\partial U}{\partial N'_{BE}} + \frac{\partial U}{\partial N''_{BE}} = 0 \tag{4-5.6}$$

The reduction of Eq. (4-5.6) to Eq. (4-5.3) then follows by the same technique employed in the reduction of Eq. (a) of Art. 4-2 to Eq. (d) of Art. 4-2, since $N'_{BE} = N''_{BE} = N_{BE}$. In a similar manner it may be shown for the structure in Fig. 4-5.2b that

$$\frac{\partial U}{\partial N} = \frac{\partial U}{\partial V} = \frac{\partial U}{\partial M} = 0 \tag{4-5.7}$$

where N, V, and M are the internal reactions for any given section of member ABC.

Note: In the application of Eqs. (4-5.1) and (4-5.2) to the system with redundant supports or redundant members, it is assumed that the unloaded system is stressfree. Consequently, redundant supports exert no force on the structure initially. However, in certain applications, these conditions do not hold. For example, consider the beam in Fig. 4-5.1. Initially, the right end of the beam may be lifted off the support, or the end support may exert a force on the beam either because of support settlement or because of thermal expansions or contractions. As a result, the end of the beam (in the absence of the redundant support) may be raised a distance q_1 above the location of the support before the beam is loaded (Fig. 4-5.3a) or it may be a distance q_2 below the support location (Fig. 4-5.3b).

If the displacement magnitudes q_1 or q_2 of the end of the beam (in the absence of the support) are known, we may compute the reaction R for the loaded beam (Figs. 4-5.3c and d) by the relations

$$q_1 = -\frac{\partial U}{\partial R} \quad \text{or} \quad q_2 = \frac{\partial U}{\partial R} \tag{4-5.8}$$

where the minus sign indicates that the displacement q_1 and the force R have opposite senses.

If known residual stresses (say due to fabrication processes) are present in a structure before it is loaded, the total stresses in the structure after it is loaded may be computed in two steps. First, we assume that the

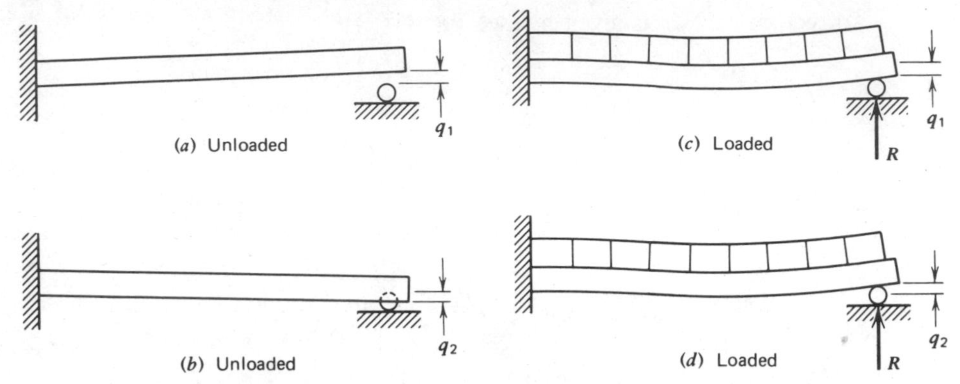

Fig. 4-5.3/Effect of support settlement or thermal expansion or contraction on redundant supports of loaded beams. (*a*) Unloaded. (*b*) Unloaded. (*c*) Loaded. (*d*) Loaded.

structure is stressfree when unloaded, and we calculate the redundant reactions by means of Eqs. (4-5.1) and (4-5.2). Stresses are calculated for the known loads and known reactions. Next, we superimpose these calculated stresses upon the residual stresses to determine the total stresses in the structure.

EXAMPLE 4-5.1

Statically Indeterminate Cantilever Beam

The beam in Fig. E4-5.1*a* is fixed at the left end, is simply supported at the right end, and is subjected to a concentrated load P at the center. (a) Determine the magnitude of the reaction R (Fig. E4-5.1*b*) at the right end. (b) Determine the deflection of the beam under load P. (c) If the simple support at the right end settles a vertical distance $PL^3/32EI$, determine the new magnitude of the reaction R.

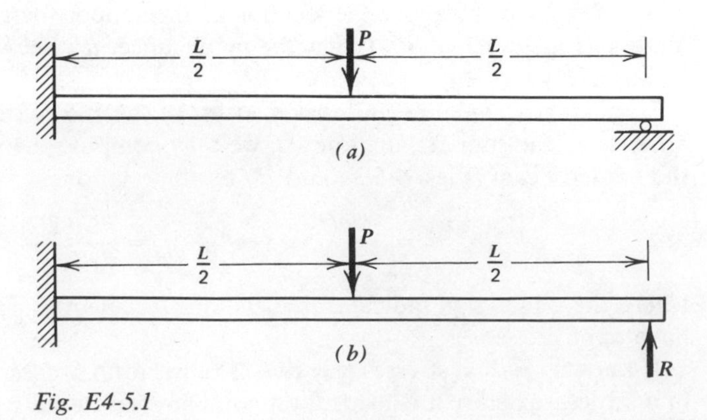

Fig. E4-5.1

SOLUTION

(a) Since the vertical displacement of the beam at the simple support is zero, Eq. (4-5.1) gives (with z measured to the left from R)

$$\frac{\partial U}{\partial R} = 0 = \int_0^{L/2} \frac{Rz}{EI}(z)\,dz + \int_{L/2}^{L} \frac{Rz - P(z - L/2)}{EI}(z)\,dz$$

$$0 = \frac{R}{3}(L/2)^3 + \frac{R}{3}\left[L^3 - (L/2)^3\right] - \frac{P}{3}\left[L^3 - (L/2)^3\right]$$

$$+ \frac{PL}{4}\left[L^2 - (L/2)^2\right]$$

$$R = \frac{5P}{16}$$

(b) Reaction R must be treated as independent of P in determining the deflection of the beam under load P. Therefore, the strain energy for the right half of the beam is independent of P. With z measured to the left from P, we have

$$q_P = \int_0^{L/2} \frac{Pz - R(z + L/2)}{EI}(z)\,dz = \frac{PL^3}{24EI} - \frac{5RL^3}{48EI} = \frac{7PL^3}{768EI}$$

(c) The vertical displacement at the simple support has a sense opposite to the sense of R; therefore, Eq. (4-5.1) gives (with z measured to the left from R)

$$\frac{\partial U}{\partial R} = -\frac{PL^3}{32EI} = \int_0^{L/2} \frac{Rz}{EI}(z)\,dz + \int_{L/2}^{L} \frac{Rz - P(z - L/2)}{EI}(z)\,dz$$

Hence,

$$-\frac{PL^3}{32} = \frac{RL^3}{3} - \frac{5PL^3}{48}, \qquad \text{or } R = \frac{7P}{32}$$

Alternatively, R could be assumed to be a function of P; i.e., $R = 5P/16$ in Part (b). Then, we would find $q_P = 7PL^3/(768EI) + Cq_R$, where C is an unknown nonzero constant and q_R is the displacement of the right end of the beam. Since the support at the right end of the beam prevents displacement, $q_R = 0$. Hence, both approaches lead to the same value for q_P. Note that this alternate method would not give the correct solution for q_P if $q_R \neq 0$.

EXAMPLE 4-5.2

Statically Indeterminate System

Determine the reactions at C for member ABC in Fig. E4-5.2a and the component of the deflection of point B in the direction of P. Assume U_N and U_S are so small that they can be neglected.

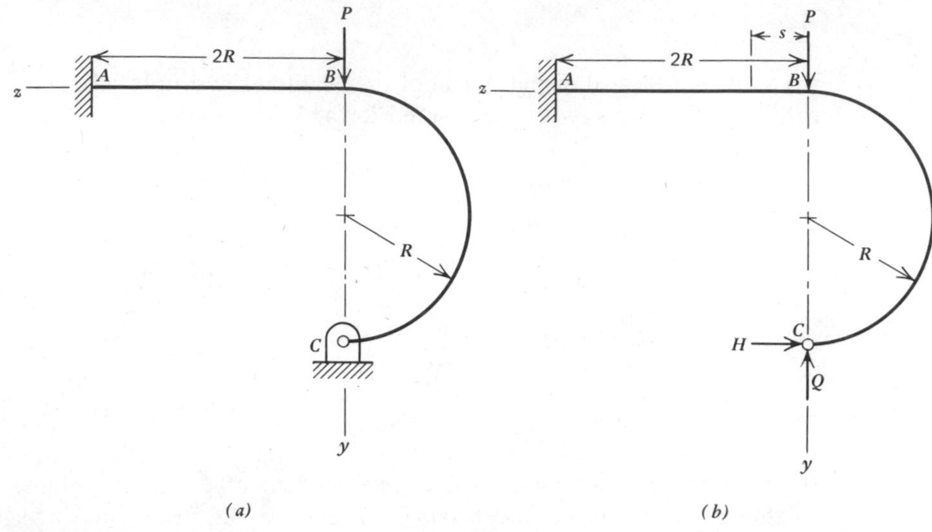

Fig. E4-5.2

SOLUTION

The support at C allows rotation but prevents displacement. Our first problem is to determine the redundant reactions Q and H (Fig. E4-5.2*b*) at C. Since the y-displacement at C is zero, Eq. (4-5.1) gives

$$\frac{\partial U}{\partial Q} = 0 = \int_0^{\pi} \frac{[QR\sin\theta - HR(1-\cos\theta)]}{EI} R(\sin\theta) R\, d\theta + \int_0^{2R} \frac{[(Q-P)s + 2HR]}{EI} s\, ds$$

or

$$Q\left(\frac{\pi}{2} + \frac{8}{3}\right) + 2H - \frac{8P}{3} = 0$$

or

$$4.2375Q + 2H - 2.6667P = 0 \tag{1}$$

Since the z-displacement at C is zero, Eq. (4-5.1) gives

$$\frac{\partial U}{\partial H} = 0 = \int_0^{\pi} \frac{[QR\sin\theta - HR(1-\cos\theta)}{EI} [-R(1-\cos\theta)] R\, d\theta + \int_0^{2R} \frac{[(Q-P)s + 2HR]}{EI} 2R\, ds$$

or

$$2Q + H\left(\frac{3\pi}{2} + 8\right) - 4P = 0$$

or

$$2Q + 12.7124H - 4P = 0 \tag{2}$$

The simultaneous solution of Eqs. (1) and (2) gives

$$Q = 0.5193P$$
$$H = 0.2329P$$

In the application of Castigliano's theorem to determine the deflection of the point B, the quantities H, Q, and P are considered to be independent. Then, since the moment in the curved part BC [namely, $QR\sin\theta - HR(1-\cos\theta)$] is independent of P, we need consider only the strain energy of the part AB. Thus,

$$q_P = \frac{\partial U}{\partial P} = \int_0^{2R} \frac{[(Q-P)s + 2HR]}{EI}(-s)\,ds$$
$$= \frac{1}{EI}\left(\frac{8}{3}PR^3 - \frac{8}{3}QR^3 - 4HR^3\right)$$

or

$$q_P = 0.3503\frac{PR^3}{EI}$$

Alternatively, Q and H may be assumed to be functions of P. Then, we find $q_P = 0.3503\ PR^3/EI + C_1 q_C^h + C_2 q_C^v$, where (C_1, C_2) are unknown nonzero constants and (q_C^h, q_C^v) are horizontal and vertical components of displacement at point C. Since the support at C prevents displacement, $q_C^h = q_C^v = 0$. Hence, both approaches lead to the same value for q_P.

EXAMPLE 4-5.3

Statically Indeterminate Truss

The king post truss in Fig. E4-5.3 is constructed of a 160 mm deep by 60 mm wide rectangular steel beam ABC ($E_{AC} = 200$ GPa and $Y_{AC} = 240$ MPa), a 15 mm diameter steel rod ADC ($E_{DC} = 200$ GPa and $Y_{DC} = 500$ MPa), and a 40 mm by 40 mm white oak compression member BD ($E_{BD} = 12.4$ GPa and $Y_{BD} = 29.6$ MPa). Determine the magnitude of the

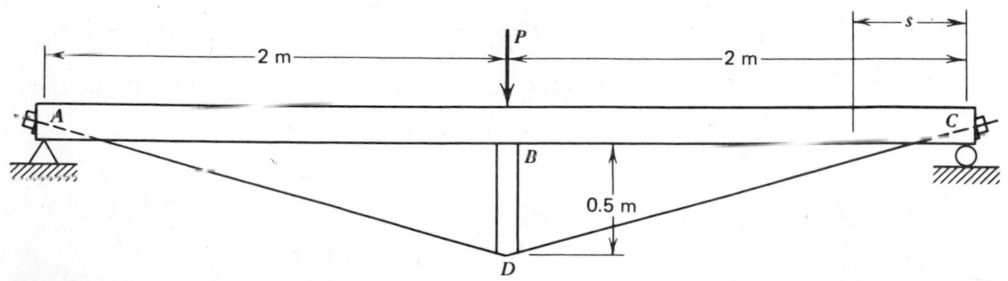

Fig. E4-5.3

load P that can be applied to the king post truss if all parts are designed using a factor of safety $SF = 2.00$. Neglect stress concentrations.

SOLUTION

Let member BD be the redundant member of the king post truss. We will include strain energy U_N for both member BD and member ADC; however, U_N and U_S for the beam are so small compared to U_M that they can be neglected. Let the compression load in member BD be N_{BD}. Equations of equilibrium at joint D give

$$N_{DC} = \sqrt{4.25}\, N_{BD}$$

The bending moment in the beam at distance s from either C or A is

$$M = \left[\frac{P}{2} - \frac{N_{BD}}{2}\right] s$$

Equation (4-5.1) gives

$$\frac{\partial U}{\partial N_{BD}} = 0 = \frac{N_{BD} L_{BD}}{E_{BD} A_{BD}} + 2 \frac{N_{DC} L_{DC}}{E_{DC} A_{DC}} \frac{\partial N_{DC}}{\partial N_{BD}} + 2 \int_0^{L_{BC}} \frac{M}{E_{AC} I_{AC}} \frac{\partial M}{\partial N_{BD}}\, ds$$

$$= \frac{500 N_{BD}}{E_{BD} A_{BD}} + \frac{2(\sqrt{4.25} \times 10^3)(4.25) N_{BD}}{E_{DC} A_{DC}} + 2 \int_0^{2000} \frac{\left[\frac{P}{2} - \frac{N_{BD}}{2}\right] s}{E_{AC} I_{AC}} \left(-\frac{s}{2}\right) ds$$

which can be simplified to give

$$P = N_{BD}\left[1 + \frac{500}{E_{BD} A_{BD}} \frac{3 E_{AC} I_{AC}}{4 \times 10^9} + \frac{\sqrt{4.25}\,(8.5 \times 10^3)}{E_{DC} A_{DC}} \frac{3 E_{AC} I_{AC}}{4 \times 10^9}\right] \tag{1}$$

But $A_{BD} = 40(40) = 1600$ mm^2, $A_{DC} = \pi(15)^2/4 = 176.7$ mm^2, and $I_{AC} = 60(160)^3/12 = 20.48 \times 10^6$ mm^4. These along with other given values when substituted in Eq. (1) give

$$P = 2.601 N_{BD}$$

The axial loads in members BD and ADC and the maximum moment in member ABC can now be written as functions of P.

$$N_{BD} = 0.384P \text{ (N)}$$

$$N_{DC} = 0.793P \text{ (N)}$$

$$M_{max} = 616P \text{ (N} \cdot \text{mm)}$$

Since the working stress for each member is half the yield stress for the member, a limiting value of P is obtained for each member. For compression member BD

$$\frac{Y_{BD}}{2} = \frac{29.6}{2} = \frac{N_{BD}}{A_{BD}} = \frac{0.384P}{1600}$$

$$P = 61{,}700 \text{ N}$$

For tension member ADC

$$\frac{Y_{DC}}{2} = \frac{500}{2} = \frac{N_{DC}}{A_{DC}} = \frac{0.797P}{176.7}$$

$$P = 55{,}700 \text{ N}$$

For beam ABC

$$\frac{Y_{AC}}{2} = \frac{240}{2} = \frac{M_{\max} c}{I_{AC}} = \frac{616P(80)}{20.48 \times 10^6}$$

$$P = 49{,}900 \text{ N}$$

Thus, the design load for the king post truss is 49.9 kN.

EXAMPLE 4-5.4

Spring-Supported I-Beam

An aluminum alloy I-beam (depth $= 100$ mm, $I = 2.45 \times 10^6$ mm^4, $E = 72.0$ GPa) has a length of 6.8 m and is supported by seven springs ($K = 110$ N/mm) spaced at distance $l = 1.10$ m center to center along the beam, (Fig. E4-54*a*). A load $P = 12.0$ kN is applied at the center of the beam over the center spring. Determine the load carried by each spring, the deflection of the beam under the load, the maximum bending moment, and the maximum bending stress in the beam.

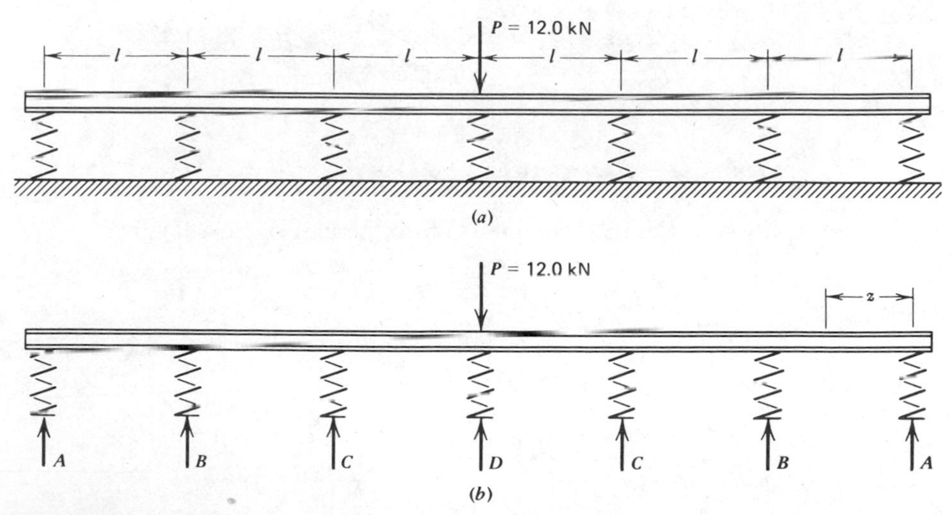

Fig. E4-5.4

SOLUTION

It is assumed that the springs are attached to the beam so that the springs can develop tensile as well as compressive forces. A free body diagram of the beam with springs attached is shown in Fig. E4-5.4*b*. Let the loads carried by springs B, C, and D be redundant reactions. The magnitudes of these redundants are obtained using Eq. (4-5.1)

$$\frac{\partial U}{\partial B} = \frac{\partial U}{\partial C} = \frac{\partial U}{\partial D} = 0 \tag{1}$$

The strain energy U for the beam and springs (neglecting U_S for the beam) is given by the relation

$$U = 2\int_0^l \frac{M^2}{2EI}\,dz + 2\int_l^{2l} \frac{M^2}{2EI}\,dz + 2\int_{2l}^{3l} \frac{M^2}{2EI}\,dz + 2\left(\frac{A^2}{2K} + \frac{B^2}{2K} + \frac{C^2}{2K}\right) + \frac{D^2}{2K} \tag{2}$$

The moments in the three integrals are functions of the reaction A, which can be eliminated from Eq. (2) by the equilibrium force equation for the y-direction.

$$A = \frac{P}{2} - B - C - \frac{D}{2} \tag{3}$$

The moments for the three segments of the beam are

$$\begin{aligned}
&0 \le z \le l \\
&M = Az \qquad\qquad\qquad\qquad\quad = \frac{P}{2}z - Bz - Cz - \frac{D}{2}z \\
&l \le z \le 2l \\
&M = Az + B(z-l) \qquad\qquad\ \ = \frac{P}{2}z - Bl - Cz - \frac{D}{2}z \\
&2l \le z \le 3l \\
&M = Az + B(z-l) + C(z-2l) = \frac{P}{2}z - Bl - 2Cl - \frac{D}{2}z
\end{aligned} \tag{4}$$

Substitution of Eqs. (2), (3), and (4) into the first of Eqs. (1) gives

$$\begin{aligned}
\frac{\partial U}{\partial B} = 0 = {} & \frac{2}{EI}\int_0^l \left(\frac{P}{2}z - Bz - Cz - \frac{D}{2}z\right)(-z)\,dz \\
& + \frac{2}{EI}\int_l^{2l}\left(\frac{P}{2}z - Bl - Cz - \frac{D}{2}z\right)(-l)\,dz \\
& + \frac{2}{EI}\int_{2l}^{3l}\left(\frac{P}{2}z - Bl - 2Cl - \frac{D}{2}z\right)(-l)\,dz \\
& + \frac{2}{K}\left(\frac{P}{2} - B - C - \frac{D}{2}\right)(-1) + \frac{2B}{K}
\end{aligned}$$

which can be simplified to give

$$0 = 12BEI + 6CEI + 3DEI - 3PEI - 13PKl^3 + 14BKl^3 + 23CKl^3 + 13DKl^3 \quad (5)$$

Substitution of Eqs. (2), (3), and (4) into the second and third of Eqs. (1) gives, after simplification,

$$0 = 6BEI + 12CEI + 3DEI - 3PEI - 23PKl^3 + 23BKl^3 + 40CKl^3 + 23DKl^3 \quad (6)$$

$$0 = 6BEI + 6CEI + 9DEI - 3PEI - 27PKl^3 + 26BKl^3 + 46CKl^3 + 27DKl^3 \quad (7)$$

Equations (5), (6), and (7) are three simultaneous equations in the three unknowns B, C, and D. Their magnitudes depend upon the magnitudes of E, I, and K. Using the values specified in the problem, we have by Eqs. (5), (6), and (7)

$$\begin{aligned} 0 &= B + 1.0622C + 0.5838D - 0.5838P \\ 0 &= B + 1.8015C + 0.8804D - 0.8804P \\ 0 &= B + 1.6019C + 1.1389D - 0.9213P \end{aligned} \quad (8)$$

The solutions of Eqs. (8) and (3) are

$$\begin{aligned} A &= -0.0379P = -455 \text{ N} \\ B &= 0.1014P = 1217 \text{ N} \\ C &= 0.2578P = 3094 \text{ N} \\ D &= 0.3573P = 4288 \text{ N} \end{aligned}$$

The maximum deflection of the beam is the deflection under the load P, which is equal to the deflection of the spring at D.

$$q_P = \frac{D}{K} = \frac{4288}{110} = 38.98 \text{ mm}$$

$$M_{\max} = 3lA + 2lB + lC = 4.58 \times 10^6 \text{ N} \cdot \text{mm}$$

$$\sigma_{\max} = \frac{M_{\max} c}{I} = 93.5 \text{ MPa}$$

Except for the simplifying assumptions that the shear introduced negligible error in the flexure formula and contributed negligible strain energy, the above solution is exact. A simple approximate solution of the same problem is presented in Chapter 9.

PROBLEM SET 4-5

1. Let tension member EF be added to the structure in Fig. E4-4.3 as indicated in Fig. P4-5.1. Member EF is made of the same material

and has the same cross sectional area as member AB. The loads, material and cross sectional dimensions are indicated in Example 4-4.3. Determine the load in member EF and the deflection of point E in the direction of force P.

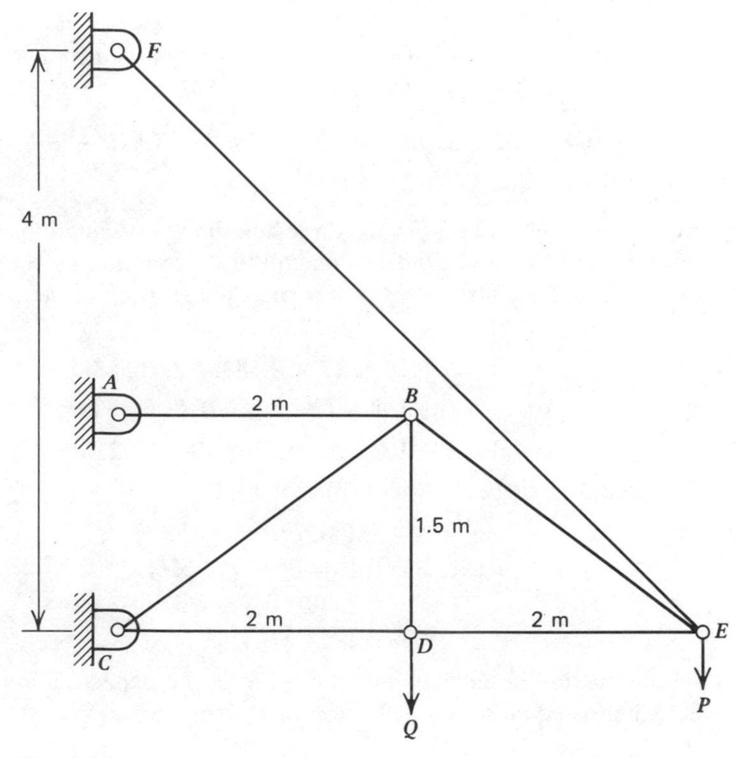

Fig. P4-5.1

2. The beam in Fig. P4-5.2 is fixed at the right end and simply supported at the left end. Determine the reaction R at the left end assuming that length L of the beam is large compared to its depth.

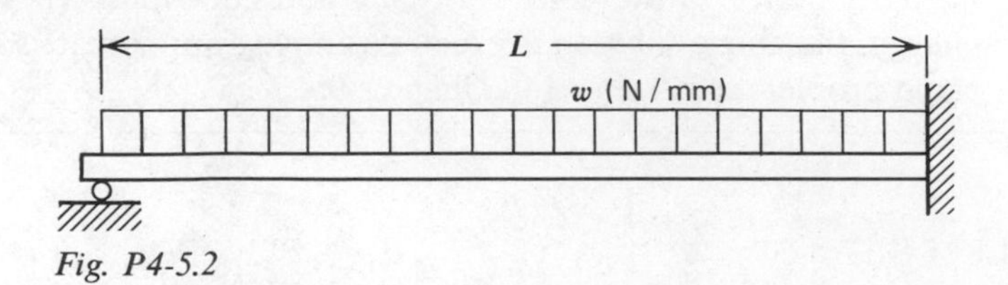

Fig. P4-5.2

Ans. $R = 3wL/8$

3. The beam in Problem 2 has a circular cross section with a diameter of 40 mm, has a length of 2.00 m, and is made of a steel ($E = 200$ GPa) having a working stress of 140 MPa. (a) Determine the magnitude of w. (b) How much would the stress in the beam be increased for the same value of w if the left end of the beam deflects 5.00 mm before making contact with the support?

4. The beam in Fig. P4-5.4 is subjected to two loads P and is supported at three locations A, B, and C as shown. Determine the reaction at B assuming that the beam length is large compared to its depth.

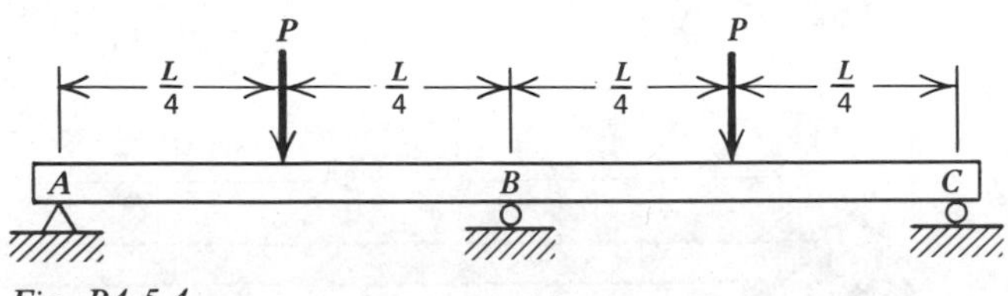

Fig. P4-5.4

Ans. Reaction at $B = 11P/8$

5. The beam in Fig. P4-5.5 is fixed at the left end and is simply supported at its center B. Assuming that the beam length is large compared to its depth, determine the reaction at B and the slope of the beam over the support at B.

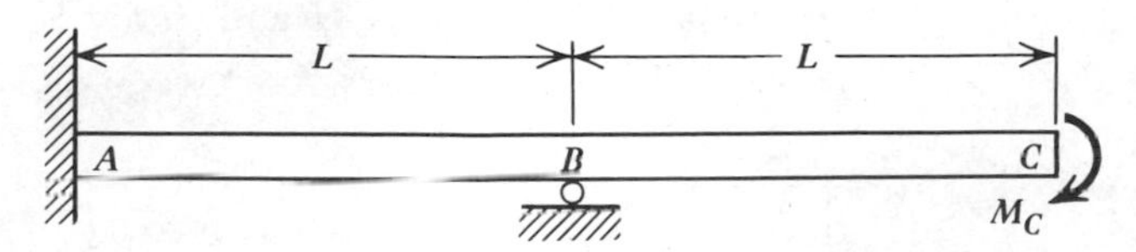

Fig. P4-5.5

6. Member ABC in Fig. P4-5.6 has a constant cross section. Assuming that length R is large compared to the depth of the member, determine the horizontal H and vertical V components of the pin reaction at C.

Ans. $V = 0.673\ wR$; $H = 0.608\ wR$

7. The beam in Fig. P4-5.7 is fixed at the right end and rests on a coil spring with spring constant K at the left end. Assuming that the beam length is large compared to its depth, determine the force R in the spring.

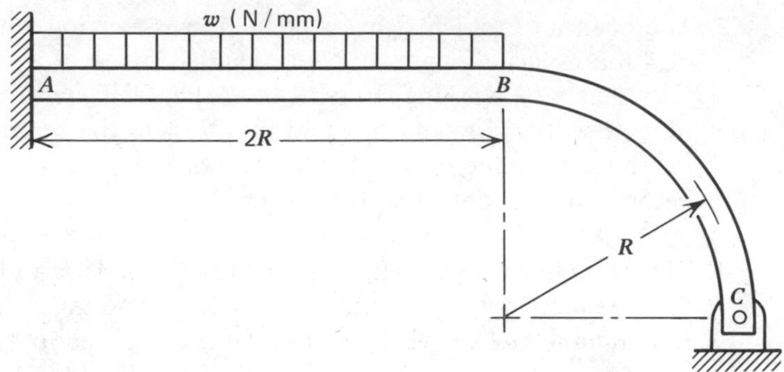

Fig. P4-5.6

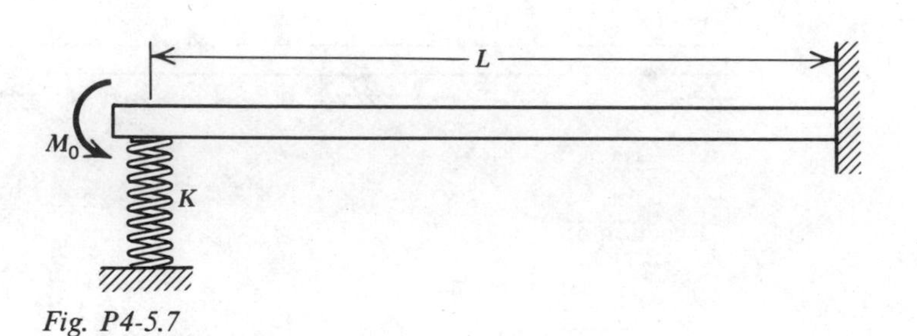

Fig. P4-5.7

8. The structure in Fig. P4-5.8 is constructed of two steel columns AB and CD with moment of inertia I_1 and steel beam BC with moment of inertia I_2. Assume that lengths H and L are large compared to

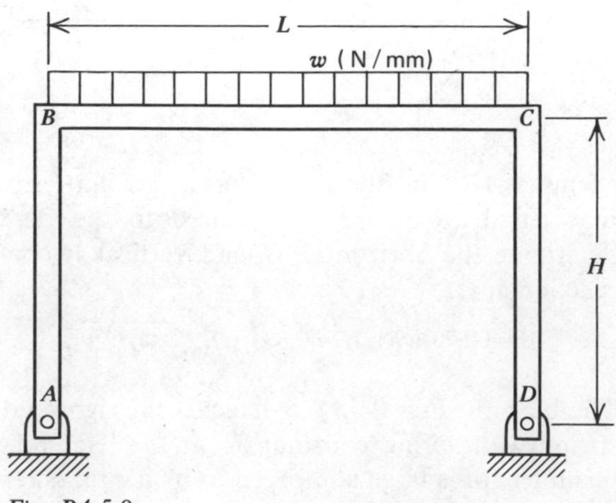

Fig. P4-5.8

the depths of the members, determine the horizontal component of the pin reaction at D.

Ans. Horizontal reaction at $D = wL^3I_1/(8H^2I_2 + 12HLI_1)$

9. Assuming that lengths R and L are large compared to the depth of the member, determine the maximum moment for the link shown in Fig. P4-5.9.

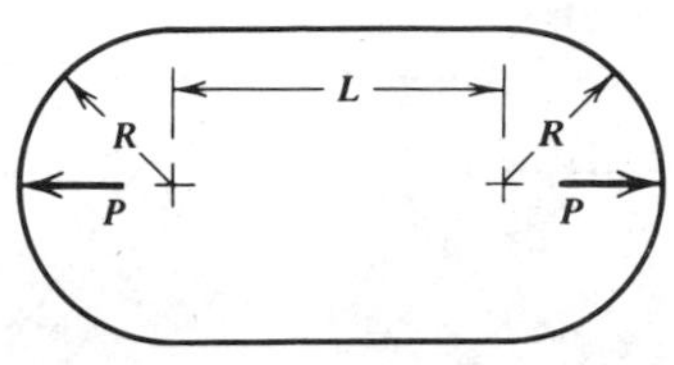

Fig. P4-5.9

10. Member $ABCD$ in Fig. P4-5.10 lies in the plane of the paper. If length L is large compared to the depth of the member, determine the pin reaction V at D and the horizontal displacement q_H of the pin at D.

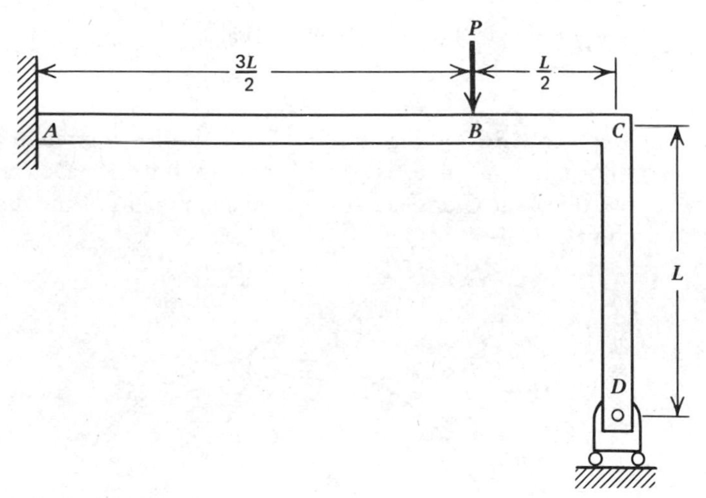

Fig. P4-5.10

Ans. $V = 81P/128$, $q_H - 9PL^3/64EI$

11. Let the pin at D for member $ABCD$ in Problem 10 be prevented from displacing as load P is applied. Determine pin reactions V and H at D.

12. The structure in Fig. P4-5.12 is made up of a steel ($E = 200$ GPa) rectangular beam ABC with depth $h = 40.0$ mm and width $b = 30.0$ mm and two wood ($E = 10.0$ GPa) pin-connected members BD and CD with 25.0 mm square cross sections. If load $P = 9.00$ kN is applied to the beam at C, determine the reaction V at support D and the maximum stresses in the steel beam and wood members.

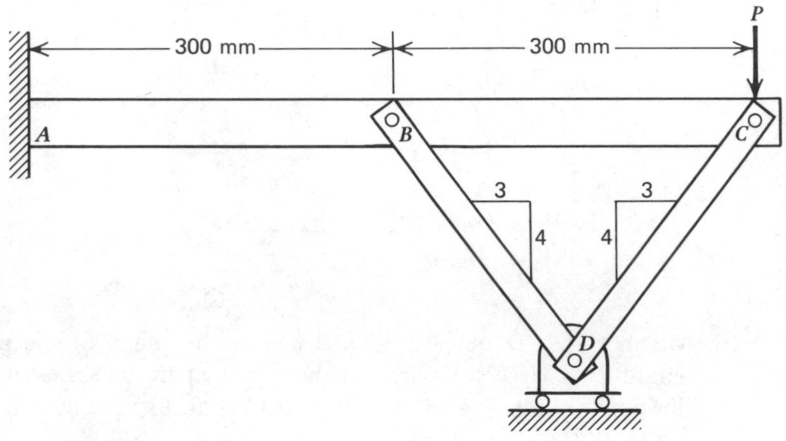

Fig. P4-5.12

Ans. $V = 13.09$ kN; $\sigma_{\text{beam}} = 92.1$ MPa; $\sigma_{\text{compression member}} = 13.1$ MPa

13. Member ABC in Fig. P4-5.13 has a uniform circular cross section with radius r that is small compared to R. Determine the pin reaction V at C and the horizontal component of the displacement of point B.

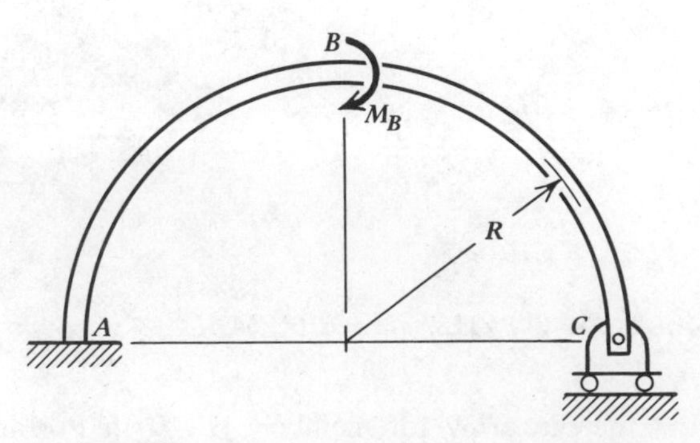

Fig. P4-5.13

14. Member ABC in Fig. P4-5.14 has a right angle bend at B, lies in the (x, z)-plane, and has a circular cross section with diameter d which is small compared to either length L_1 or length L_2. The reaction at C prevents deflection in the y-direction only. Determine the reaction V at C when the moment M_0 is applied at C. $G = E/2(1 + \nu)$.

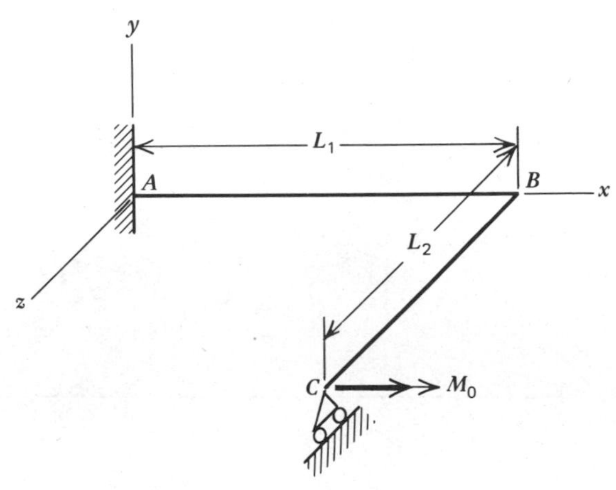

Fig. P4-5.14

Ans. $V = M_0[3L_2^2 + 6L_1L_2(1+\nu)]/[2L_1^3 + 2L_2^3 + 6L_1L_2^2(1+\nu)]$

15. Member AB in Fig. P4-5.15 is a quadrant of a circle lying in the (x, z)-plane, has a circular cross section of radius r, which is small compared to R, and is supported by a spring (spring constant K) at B, whose action line is parallel to the y-axis. Determine the reaction in the spring when torque T_0 is applied at B with action line parallel to the negative z-axis.

16. The structure in Fig. P4-5.16 has a uniform circular cross section with diameter d, which is small compared to either H or L. The structure is fixed at 0 and C and lies in the (x, z)-plane. The load P is parallel to the y-axis. Determine the magnitudes of the moment and torque at 0 and C. $G = E/2(1 + \nu)$.

Ans. $M_0 = M_C = PH/2$; $T_0 = T_C = PL^2/8[L + 2H(1+\nu)]$

17. Each of the three members of the structure in Fig. P4-5.17 is made of a ductile steel ($E = 200$ GPa, $\nu = 0.29$) with yield stress $Y = 420$ MPa. Member $0A$ has a diameter of 100 mm, is fixed at 0, and is welded to beam AB, which has a rectangular cross section with depth of 75.0 mm and a width of 50.0 mm. Tension member BC has

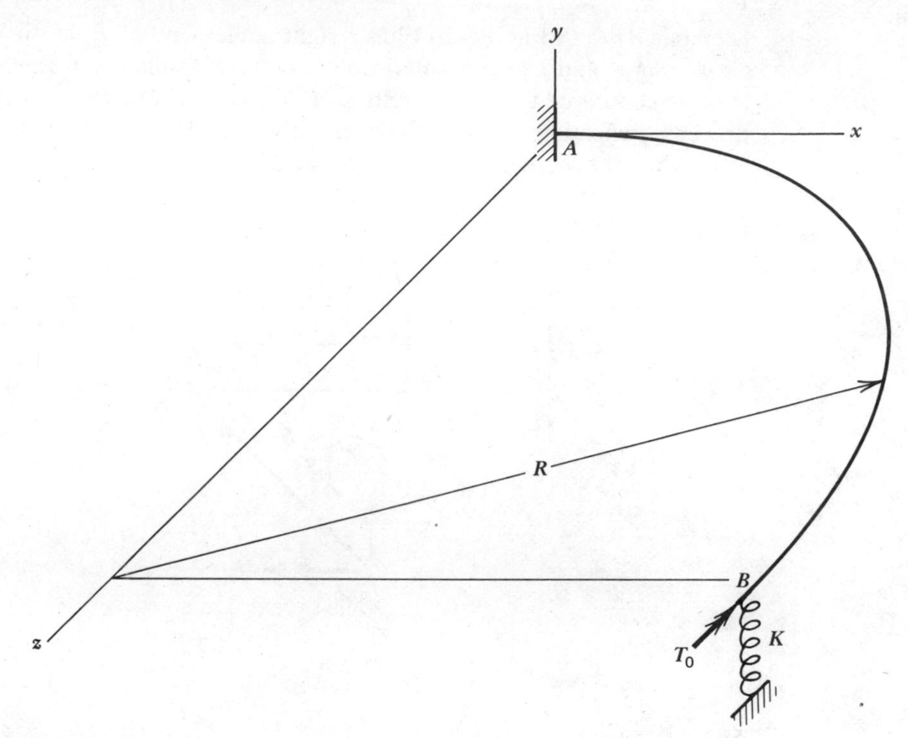

Fig. P4-5.15

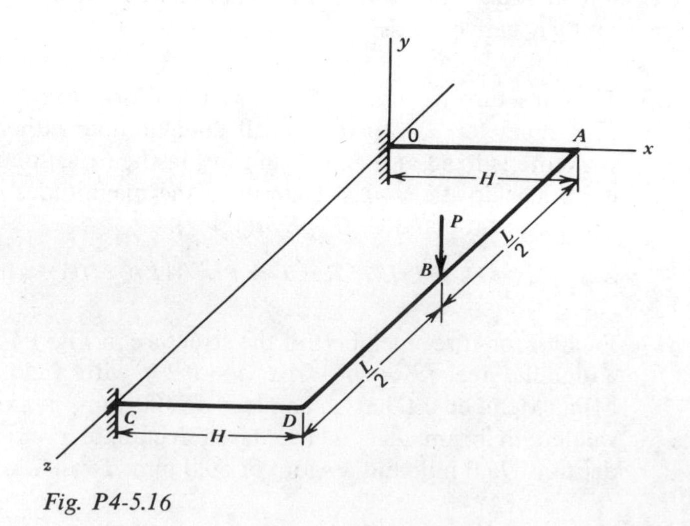

Fig. P4-5.16

a circular cross section with a diameter of 7.50 mm. All of the members are unstressed when $P = 0$. Determine the value of P based on a factor of safety of $SF = 2.00$ against initiation of yielding. Neglect stress concentrations. $G = E/2(1 + \nu)$.

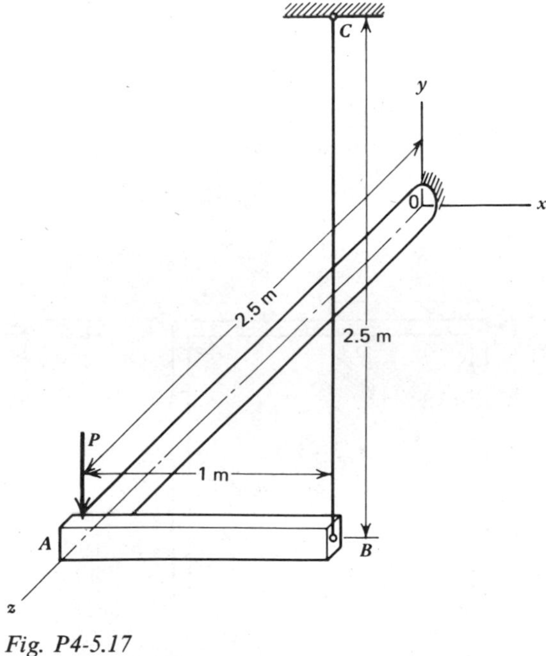

Fig. P4-5.17

18. Member BCD and tension member BD in Fig. P4-5.18 are made of materials having the same modulus of elasticity. Member BCD has a

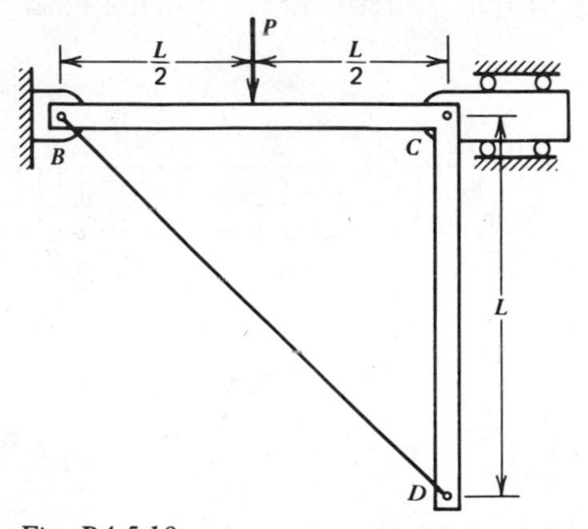

Fig. P4-5.18

constant moment of inertia I, has a cross-sectional area that is large compared to area A of tension member BD, and has a maximum cross-sectional dimension that is small compared to L. Determine the axial load N in member BD.

Ans. $N = 3PAL^2/(96I + 16\sqrt{2}\ AL^2)$

19. Member $BCDF$ in Fig. P4-5.19 has the same moment of inertia I at every section. Determine the internal reactions N_D, V_D, and M_D at section D. Length L is large compared to the maximum cross-sectional dimension.

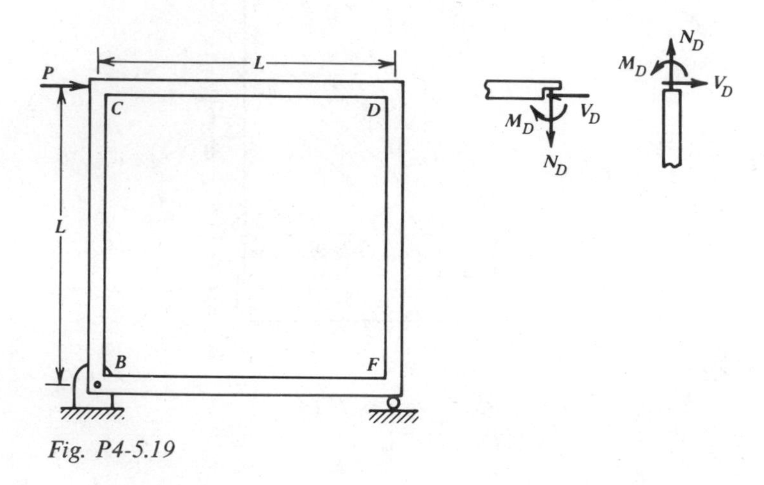

Fig. P4-5.19

20. Member BCD in Fig. P4-5.20 has the same moment of inertia I at every section. Determine the internal reactions N_B, V_B, and M_B at

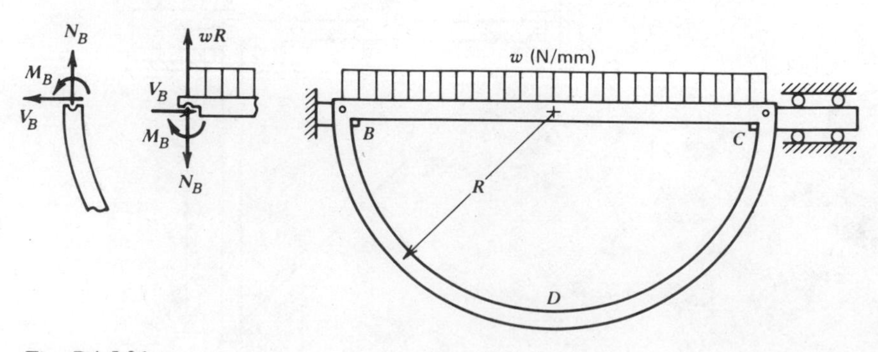

Fig. P4-5.20

section B. Length R is large compared to the maximum cross-sectional dimensions.

Ans. $N_B = 0$; $V_B = 8wR/(3\pi^2 + 6\pi - 24)$;
$M_B = -2\pi R^2 w/(3\pi^2 + 6\pi - 24)$

21 through 40. Add 20 to the problem number from 1 through 20 and solve the resulting problem by the unit-dummy load method.

REFERENCES

2. H. L. Langhaar, *Energy Methods in Applied Mechanics*, Wiley, New York, 1962.
2. T. M. Charlton, *Energy Principles in Applied Statics*, Blackie and Son, Limited, London, 1959.
3. J. O. Smith and O. M. Sidebottom, *Elementary Mechanics of Deformable Bodies*, Macmillan, London, 1969.
4. J. A. Van Den Broek, *Elastic Energy Theory*, 2nd Ed., Wiley, New York, 1942.

Additional References

1. B. A. Finlayson, *The Method of Weighted Residuals and Variational Principles*, Academic, New York, 1972.
2. S. G. Mikhlin, *Variational Methods in Mathematics and Physics*, Pergamon Press, New York, 1964.
3. K. Washizu, *Variational Methods in Elasticity and Plasticity*, 3rd Ed., Pergamon Press, New York, 1982.

CHAPTER 5

TORSION

In this chapter, we treat the problem of torsion of cylindrical bars with noncircular cross sections. First, we treat the case in which each torsion member is made of a linearly elastic isotropic material. The last article of the chapter treats fully plastic torsion. For cylindrical bars with circular cross sections, the torsion formulas are readily derived by the special methods of mechanics of materials. However, for noncircular cross sections, more general methods are required. In the following articles we treat noncircular cross sections by several methods, one of which is the semi-inverse method of Saint-Venant.[1] General relations are derived that are applicable for both the linear elastic torsion problem and for the fully plastic torsion problem. In order to aid in the solution of the resulting differential equation for some linear elastic torsion problems, the Saint-Venant solution is used in conjunction with the Prandtl elastic-membrane (soap-film) analogy.

In spirit, the semi-inverse method of Saint-Venant is in part comparable to the method of mechanics of materials in that certain assumptions based upon an understanding of the mechanics of the problem are introduced initially. However, these assumptions are not so specific as to attempt to meet all the requirements of the problem. Rather, sufficient freedom is allowed so that the equations that describe the boundary value problem of solids may be employed to determine the solution more completely. For the case of circular cross sections, the method of Saint-Venant leads to an exact solution (subject to appropriate boundary conditions) for the torsion problem. Because of its importance in engineering, the torsion problem of circular cross sections is discussed first.

5-1 TORSION OF A CYLINDRICAL BAR OF CIRCULAR CROSS SECTION

Consider a solid cylinder with circular cross section A and with length L. Let the cylinder be subjected to a twisting couple $\mathbf{T}$ applied at the right end (Fig. 5-1.1). An equilibrating torque $-\mathbf{T}$ acts on the left end. The vectors that represent **T** and $-$**T** are directed along the z-axis, the centroidal-axis of the shaft (Fig. 5-1.1). Under the action of the torque, an originally straight generator of the cylinder AB will deform into a helical curve A^*B^*. However, because of the radial symmetry of the circular cross section and because a deformed cross section must appear to be the same from both ends of the torsion member, plane cross sections of the torsion member normal to the z-axis remain plane after deformation and all radii remain straight. Furthermore, for small displacements, each radius remains inextensible. In other words, the torque **T** causes each cross section to rotate as a rigid body about the z-axis (axis of the couple); this axis is called the *axis of twist*. Furthermore, if we measure the rotation β of each section relative to the plane $z=0$, the rotation β of a given section will depend on its distance from the plane $z=0$. For small deformations, following Saint-Venant, we assume that the amount of rotation of a given section depends linearly on its distance z from the plane $z=0$. Thus, the rotation β of a section relative to the plane $z=0$ is

$$\beta = \theta z \tag{5-1.1}$$

where θ is the angle of twist per unit length of the shaft. Under the condition that plane sections remain plane and that Eq. (5-1.1) holds, we now seek to satisfy the equations of elasticity; that is, we employ the semi-inverse method of seeking the elasticity solution.

Since cross sections remain plane (do not displace in the z-direction), the displacement component w, parallel to the z-axis, is zero. To calculate the (x, y)-components of displacements u and v, consider a cross section at distance z from the plane $z=0$. Consider a point in the circular cross section (Fig. 5-1.2) with radial distance $0P$. Under the deformation, radius $0P$ rotates into the radius $0P^*$ ($0P^* = 0P$). In terms of the angular displacement β of the radius, the displacement compo-

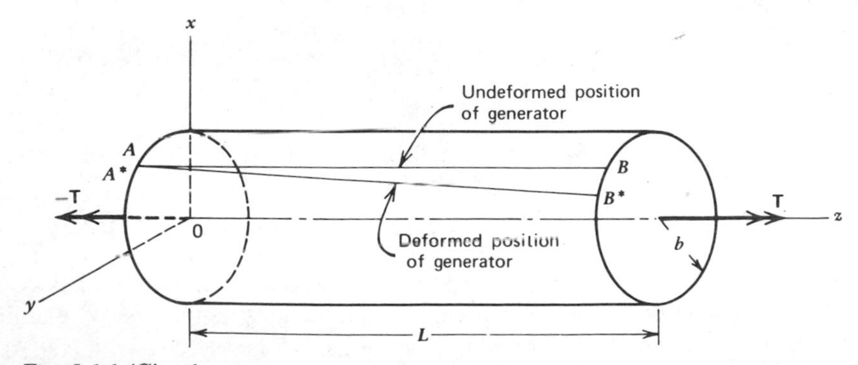

Fig. 5-1.1/Circular cross section torsion member.

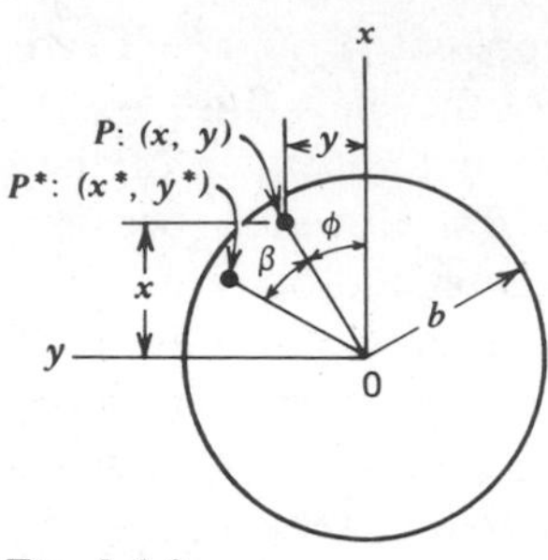

Fig. 5-1.2

nents (u, v) are

$$u = x^* - x = 0P[\cos(\beta + \phi) - \cos\phi]$$
$$v = y^* - y = 0P[\sin(\beta + \phi) - \sin\phi] \tag{5-1.2}$$

Expanding $\cos(\beta + \phi)$ and $\sin(\beta + \phi)$ and noting that $x = 0P\cos\phi$, $y = 0P\sin\phi$, we may write Eqs. (5-1.2) in the form

$$\begin{aligned} u &= x(\cos\beta - 1) - y\sin\beta \\ v &= x\sin\beta + y(\cos\beta - 1) \end{aligned} \tag{5-1.3}$$

Restricting the displacement to be small, we obtain (since then $\sin\beta \approx \beta$, $\cos\beta \approx 1$), with the assumption that $w = 0$,

$$u = -y\beta, \qquad v = x\beta, \qquad w = 0 \tag{5-1.4}$$

to first degree terms in β. Substitution of Eq. (5-1.1) into Eq. (5-1.4) yields

$$u = -\theta yz, \qquad v = \theta xz, \qquad w = 0 \tag{5-1.5}$$

On the basis of the foregoing assumptions, Eqs. (5-1.5) represent the displacement components of a circular shaft subjected to a torque **T**.

Substitution of Eqs. (5-1.5) into Eqs. (1-9.1) yields the strain components (ignoring temperature effects):

$$\epsilon_{xx} = \epsilon_{yy} = \epsilon_{zz} = \epsilon_{xy} = 0, \qquad 2\epsilon_{zx} = \gamma_{zx} = -\theta y, \qquad 2\epsilon_{zy} = \gamma_{zy} = \theta x \tag{5-1.6}$$

With Eqs. (5-1.6), Eqs. (2-4.7) yield the stress components for linear elasticity:

$$\sigma_{xx} = \sigma_{yy} = \sigma_{zz} = \sigma_{xy} = 0, \qquad \sigma_{zx} = -\theta G y, \qquad \sigma_{zy} = \theta G x \tag{5-1.7}$$

Since Eqs. (5-1.7) are linear in (x, y), they automatically satisfy compatibility, (Eqs. 1-9.4). Furthermore, they satisfy equations of equilibrium provided the body forces are zero (Eqs. 1-5.1).

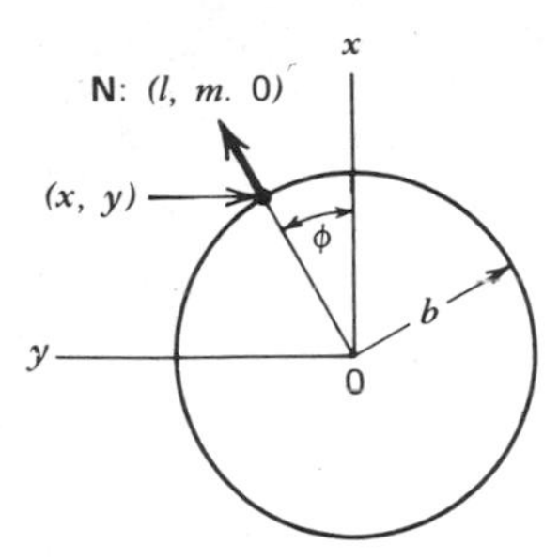

Fig. 5-1.3

To satisfy the boundary conditions, Eqs. (5-1.7) must yield no forces on the lateral boundary; on the ends, they must yield stresses such that the net moment is equal to **T** and the resultant force vanishes. Since the direction cosines of the unit normal to the lateral surface are $(l, m, 0)$ (see Fig. 5-1.3), the first two of Eqs. (1-3.8) are satisfied identically. The last of Eqs. (1-3.8) yields

$$l\sigma_{zx} + m\sigma_{zy} = 0 \tag{5-1.8}$$

By Fig. 5-1.3,

$$l = \cos\phi = \frac{x}{b}, \qquad m = \sin\phi = \frac{y}{b} \tag{5-1.9}$$

Substitution of Eqs. (5-1.7) and (5-1.9) into Eqs. (5-1.8) yields

$$-\frac{xy}{b} + \frac{xy}{b} = 0$$

Therefore, the boundary conditions on the lateral boundary are satisfied.

On the ends, the stresses must be distributed so that the net moment is **T**. Since all stress components except σ_{zx} and σ_{zy} vanish, $\Sigma F_z = \int \sigma_{zz}\, dA = 0$. Furthermore, with σ_{zx} and σ_{zy} given by Eqs. (5-1.7), $\Sigma F_x = \int \sigma_{zx}\, dA = -\theta G \int y\, dA = 0$ and $\Sigma F_y = \int \sigma_{zy}\, dA = \theta G \int x\, dA = 0$ since the first moment of an area for a centroidal axis is zero. Also summation of moments with respect to the z-axis yields (Fig. 5-1.4):

$$\sum M_z = T = \int_A (x\sigma_{zy} - y\sigma_{zx})\, dA \tag{5-1.10}$$

Substitution of Eqs. (5-1.7) into Eq. (5-1.10) yields

$$T = G\theta \int_A (x^2 + y^2)\, dA = G\theta \int_A r^2\, dA \tag{5-1.11}$$

Since the last integral is the polar moment of inertia ($J = \pi b^4/2$) of the circular cross section, Eq. (5-1.11) yields

$$\theta = \frac{T}{GJ} \tag{5-1.12}$$

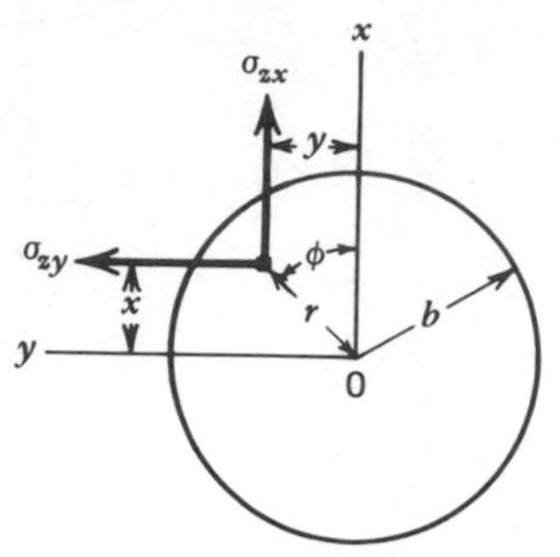

Fig. 5-1.4

which relates the angular twist θ per unit length of the shaft to the magnitude T of the applied torque.

Since compatibility and equilibrium are satisfied, Eqs. (5-1.7) represent the solution of the elasticity problem, provided the stress components σ_{zx} and σ_{zy} are distributed over the two end planes of the torsion member according to Eqs. (5-1.7). In applying torsional loads to most torsion members of circular cross section, the distributions of σ_{zx} and σ_{zy} on the ends probably do not satisfy Eqs. (5-1.7). In this case, it is assumed that σ_{zx} and σ_{zy} undergo a redistribution with distance from the ends of the bar until, at a small distance of several typical cross-sectional dimensions from the ends, the distributions are essentially given by Eqs. (5-1.7). This concept of redistribution of the applied end stresses with distance from the ends was employed by Saint-Venant. It is known as the Saint-Venant principle.[1]

Since the solution of Eqs. (5-1.7) indicates that σ_{zx} and σ_{zy} are independent of z, the stress distribution is the same for all cross sections. Thus, the stress vector $\boldsymbol{\tau}$ for any point P in a cross section is given by the relation

$$\boldsymbol{\tau} = -\theta G y\mathbf{i} + \theta G x\mathbf{j} \tag{5-1.13}$$

The stress vector $\boldsymbol{\tau}$ lies in the plane of the cross section, and it is perpendicular to the radius vector $\mathbf{r}$ joining point P to the origin 0. By Eq. (5-1.13), the magnitude of $\boldsymbol{\tau}$ is

$$\tau = \theta G\sqrt{x^2 + y^2} = \theta G r \tag{5-1.14}$$

Hence, τ is a maximum for $r = b$; that is, τ attains a maximum value of $\theta G b$.

Substitution of Eq. (5-1.12) into Eq. (5-1.14) yields the result

$$\tau = \frac{Tr}{J} \tag{5-1.15}$$

which relates the magnitude τ of the shearing stress to the magnitude T of the torque. The above results holds also for cylindrical bars with

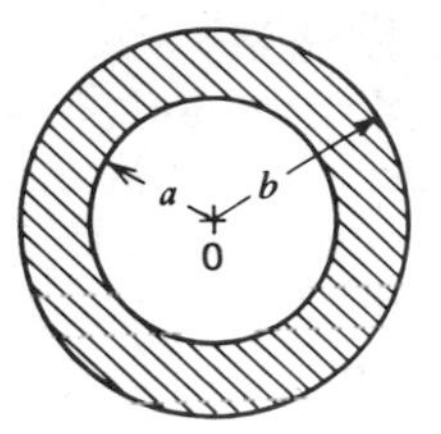

Fig. 5-1.5

hollow circular cross sections (Fig. 5-1.5), with an inner radius a and an outer radius b; for this cross section $J = \pi(b^4 - a^4)/2$ and $a \leq r \leq b$.

The analysis for the torsion of noncircular cross sections proceeds in much the same fashion as for circular cross sections except that a solution cannot be obtained if the displacement component w (Eqs. 5-1.5) is taken to be zero. In the case of noncircular cross sections, Saint-Venant assumed more generally that w is a function of (x, y), the cross section coordinates. Then, the cross section does not remain plane but "warps"; that is, different points in the cross section, in general, undergo different displacements in the z-direction.

5.2 SAINT-VENANT'S SEMI-INVERSE METHOD

Consider a torsion member with a uniform cross section of general shape as shown in Fig. 5-2.1. Axes (x, y, z) are taken as for the circular cross section (Fig. 5-1.1). The applied stress distribution on the ends $(\sigma_{zx}, \sigma_{zy})$ produces a torque **T**. In general, any number of stress distributions on

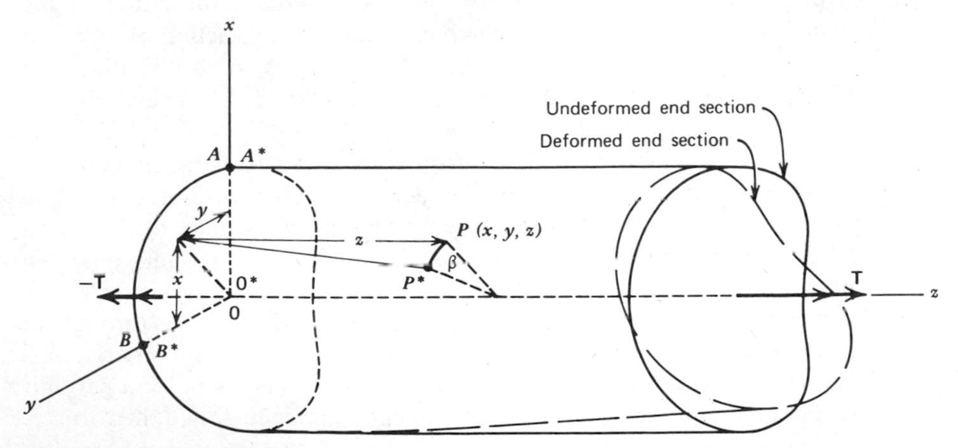

Fig. 5-2.1/Torsion member.

the end sections may produce a torque **T**. According to Saint-Venant's principle, the stress distribution on sections sufficiently far removed from the ends depends principally on the magnitude of T and not upon the stress distribution on the ends. Thus, for sufficiently long torsion members, the end stress distribution does not affect the stress distributions in a large part of the member.

Saint-Venant's semi-inverse method starts by an approximation of the displacement components due to torque **T**. This approximation is based upon observed geometric changes in the deformed torsion member.

Geometry of Deformation/ As with circular cross sections, Saint-Venant assumed that every straight torsion member with constant cross section (relative to axis z) has an axis of twist about which each cross section rotates approximately as a rigid body. Let the z-axis in Fig. 5-2.1 be the axis of twist.

For the torsion member in Fig. 5-2.1, let $0A$ and $0B$ be line segments in the cross section for $z = 0$, which coincide with the x- and y-axes, respectively. After deformation, by rigid body displacements, we may translate the new position of 0, that is, 0^* back to coincide with 0, align the axis of twist along the z-axis, and rotate the deformed torsion member until the projection of 0^*A^* on the (x, y)-plane coincides with the x-axis. Because of the displacement (z-displacement) of points in each cross section, 0^*A^* does not, in general, lie in the (x, y)-plane. However, the amount of warping is small for small displacements; therefore, line $0A$ and curved line 0^*A^* are shown as coinciding in Fig. 5-2.1. Experimental evidence indicates also that the distortion (warping) of each cross section is essentially the same. Furthermore, experimental evidence indicates that the in-plane cross-sectional dimensions of the torsion member are not changed significantly by the deformations, particularly for small displacements. In other words, deformation in the plane of the cross section is negligible. Hence, the projection of 0^*B^* on the (x, y)-plane coincides approximately with the y-axis, indicating that ϵ_{xy} ($\gamma_{xy} = 2\epsilon_{xy}$) is approximately zero (see Art. 1-7; particularly Eq. 1-7.18).

Consider a point P with coordinates (x, y, z) in the undeformed torsion member (Fig. 5-2.1). Under deformation, P goes into P^*. The point P, in general, is displaced by an amount w parallel to the z-axis because of the warping of the cross section and by amounts u and v parallel to the x- and y-axes, respectively. The displacements (u, v) are due principally to rotation of the cross section in which P lies through an angle β, with respect to the cross section at the origin. These observations led Saint-Venant to assume that $\beta = \theta z$, where θ is angle of twist per unit length and therefore that the displacement components take the form

$$u = -\theta z y, \qquad v = \theta z x, \qquad w = \theta \psi(x, y) \tag{5-2.1}$$

where ψ is the warping (out of plane distortion) function (compare Eqs. 5-2.1 for the general cross section with Eqs. 5-1.5 for the circular cross section). It may be shown that the function $\psi(x, y)$ may be determined such that the equations of elasticity are satisfied. Since we have assumed displacement components (u, v, w), the small-displacement compatibility conditions (Eqs. 1-9.4) are automatically satisfied.

The state of strain at a point in the torsion member is given by substitution of Eqs. (5-2.1) into Eqs. (1-9.1) to obtain

$$
\begin{aligned}
&\epsilon_{xx} = \epsilon_{yy} = \epsilon_{zz} = \epsilon_{xy} = 0 \\
&2\epsilon_{zx} = \gamma_{zx} = \theta\left(\frac{\partial \psi}{\partial x} - y\right) \\
&2\epsilon_{zy} = \gamma_{zy} = \theta\left(\frac{\partial \psi}{\partial y} + x\right)
\end{aligned}
\qquad (5\text{-}2.2)
$$

If the equation for γ_{zx} is differentiated partially with respect to y, the equation for γ_{zy} is differentiated partially with respect to x, and the second of these resulting equations is subtracted from the first, the warping function ψ may be eliminated to give the relation

$$
\frac{\partial \gamma_{zx}}{\partial y} - \frac{\partial \gamma_{zy}}{\partial x} = -2\theta \qquad (5\text{-}2.3)
$$

If the torsion problem is formulated in terms of $(\gamma_{zx}, \gamma_{zy})$, Eq. (5-2.3) is a geometrical condition (compatibility condition) to be satisfied for the torsion problem.

Stresses at a Point and Equations of Equilibrium / For torsion members made of isotropic materials, stress-strain relations for either elastic (the first of Eqs. 5-2.2 and Eqs. 2-4.7) or inelastic conditions indicate that

$$
\sigma_{xx} = \sigma_{yy} = \sigma_{zz} = \sigma_{xy} = 0 \qquad (5\text{-}2.4)
$$

The stress components $(\sigma_{zx}, \sigma_{zy})$ are nonzero. If body forces and acceleration terms are neglected, these stress components may be substituted into Eqs. (1-5.1) to obtain equations of equilibrium for the torsion member.†

$$
\frac{\partial \sigma_{zx}}{\partial z} = 0 \qquad (5\text{-}2.5)
$$

$$
\frac{\partial \sigma_{zy}}{\partial z} = 0 \qquad (5\text{-}2.6)
$$

$$
\frac{\partial \sigma_{zy}}{\partial y} + \frac{\partial \sigma_{zx}}{\partial x} = 0 \qquad (5\text{-}2.7)
$$

† This approach was taken by Prandtl. See Art. 7-3, reference 1.

Equations (5-2.5) and (5-2.6) indicate that σ_{zx} and σ_{zy} are independent of z. These stress components must satisfy Eq. (5-2.7), which expresses a necessary and sufficient condition for the existence of a stress function $\phi(x, y)$ (the so-called Prandtl stress function) such that

$$\begin{aligned} \sigma_{zx} &= \frac{\partial \phi}{\partial y} \\ \sigma_{zy} &= -\frac{\partial \phi}{\partial x} \end{aligned} \tag{5-2.8}$$

Thus, the torsion problem is transformed into the determination of the stress function ϕ. Boundary conditions put restrictions on ϕ.

Boundary Conditions/Since the lateral surface of a torsion member is free of applied stress, the resultant shearing stress τ in the cross section of the torsion member, at the boundary of the cross section, must be directed perpendicular to the normal to the boundary (Figs. 5-2.2*a* and *b*). The two shearing stress components σ_{zx} and σ_{zy} that act on the cross-sectional element with sides *dx*, *dy*, and *ds* may be written in terms of τ (Fig. 5-2.2*b*) in the form

$$\begin{aligned} \sigma_{zx} &= \tau \sin \alpha \\ \sigma_{zy} &= \tau \cos \alpha \end{aligned} \tag{5-2.9}$$

where according to Fig. 5-2.2*b*,

$$\sin \alpha = \frac{dx}{ds} \qquad \cos \alpha = \frac{dy}{ds} \tag{5-2.10}$$

Since the component of τ in the direction of the normal **n** to the boundary is zero, projections of σ_{zx} and σ_{zy} in the normal direction (Fig. 5-2.2*b*) yield, with Eq. (5-2.10)

$$\begin{aligned} \sigma_{zx} \cos \alpha - \sigma_{zy} \sin \alpha &= 0 \\ \sigma_{zx} \frac{dy}{ds} - \sigma_{zy} \frac{dx}{ds} &= 0 \end{aligned} \tag{5-2.11}$$

Substituting Eqs. (5-2.8) into Eq. (5-2.11), we find

$$\frac{\partial \phi}{\partial x}\frac{dx}{ds} + \frac{\partial \phi}{\partial y}\frac{dy}{ds} = \frac{d\phi}{ds} = 0$$

or

$$\phi = \text{constant on the boundary } S \tag{5-2.12}$$

Since the stresses are given by partial derivatives of ϕ (see Eqs. 5-2.8), it

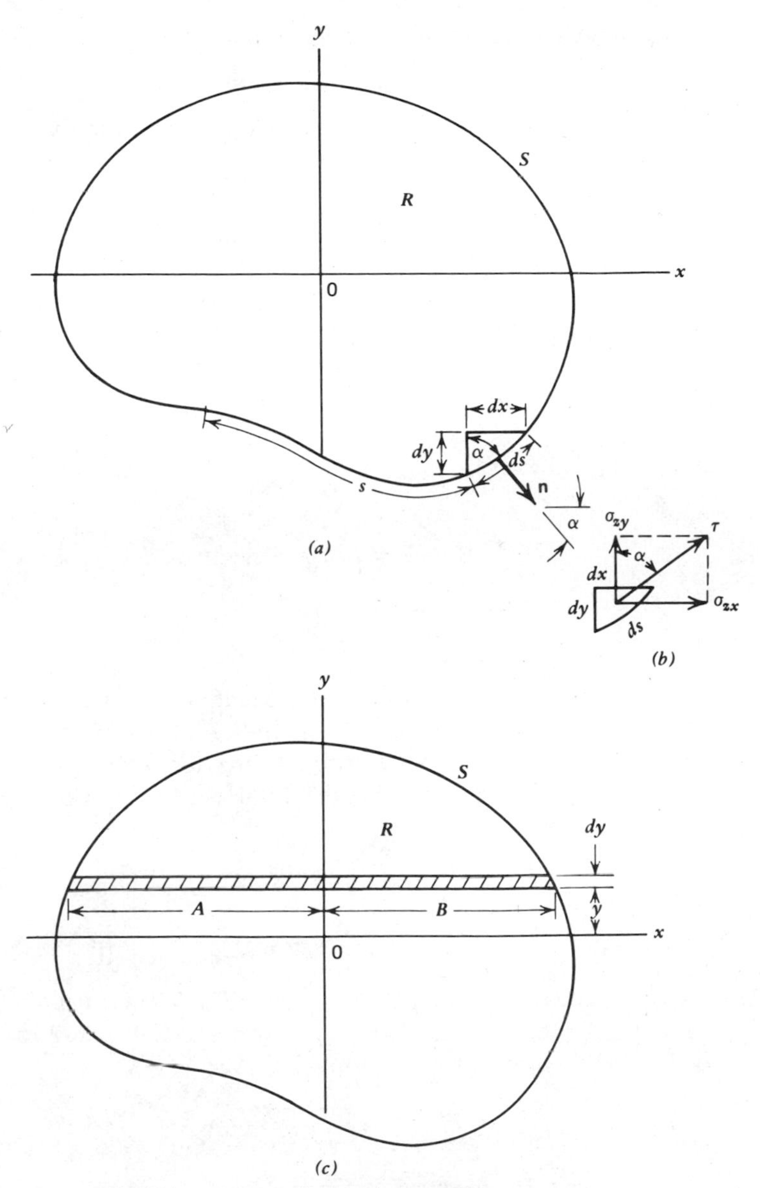

Fig. 5-2.2/Cross section of a torsion member.

is permissible to take this constant to be zero; thus

$$\phi = 0 \text{ on the boundary } S \tag{5-2.13}$$

The preceding argument can be used to show that the shearing stress

$$\tau = \sqrt{\sigma_{zx}^2 + \sigma_{zy}^2} \tag{5-2.14}$$

at any point in the cross section is directed tangent to the curve $\phi = \text{constant}$ through the point.

The distributions of σ_{zx} and σ_{zy} on a given cross section must satisfy the following equilibrium equations:

$$\sum F_x = 0 = \int \sigma_{zx}\, dx\, dy = \int \frac{\partial \phi}{\partial y}\, dx\, dy \tag{5-2.15}$$

$$\sum F_y = 0 = \int \sigma_{zy}\, dx\, dy = -\int \frac{\partial \phi}{\partial x}\, dx\, dy \tag{5-2.16}$$

$$\begin{aligned} \sum M_z = T &= \int (x\sigma_{zy} - y\sigma_{zx})\, dx\, dy \\ &= -\int \left(x\frac{\partial \phi}{\partial x} + y\frac{\partial \phi}{\partial y} \right) dx\, dy \end{aligned} \tag{5-2.17}$$

In satisfying the second equilibrium equation, consider the strip across the cross section of thickness dy as indicated in Fig. 5-2.2*c*. Since the stress function does not vary in the y-direction for this strip, the partial derivative can be replaced by the total derivative. For the strip, Eq. (5-2.16) becomes

$$\begin{aligned} dy \int \frac{\partial \phi}{\partial x}\, dx = dy \int \frac{d\phi}{dx}\, dx = dy \int_{\phi(A)}^{\phi(B)} d\phi \\ = dy\, [\phi(B) - \phi(A)] = 0 \end{aligned} \tag{5-2.18}$$

since ϕ is equal to zero on the boundary. The same is true for every strip so that $\Sigma F_y = 0$ is satisfied. In a similar manner, Eq. (5-2.15) is verified. In Eq. (5-2.17), consider the term

$$-\int\int x \frac{\partial \phi}{\partial x}\, dx\, dy$$

which becomes for the strip in Fig. 5-2.2*c*

$$-dy \int x \frac{d\phi}{dx}\, dx = -dy \int_{\phi(A)}^{\phi(B)} x\, d\phi \tag{5-2.19}$$

Integrating the latter integral by parts, we obtain, since $\phi(B) = \phi(A) = 0$,

$$-dy \int_{\phi(A)}^{\phi(B)} x\, d\phi = -dy \left[x\phi|_A^B - \int_{x_A}^{x_B} \phi\, dx \right] = dy \int_{x_A}^{x_B} \phi\, dx \tag{5-2.20}$$

Summing up for the other strips and repeating the process using strips of thickness dx for the other term in Eq. (5-2.17), we obtain the relation

$$T = 2\int\int \phi \, dx \, dy \tag{5-2.21}$$

The stress function ϕ can be considered to represent a surface over the cross section of the torsion member. This surface is in contact with the boundary of the cross section (see Eq. 5-2.13). Hence, Eq. (5-2.21) indicates that the torque is equal to twice the volume between the stress function and the plane of the cross section.

Note: Equations (5-2.3), (5-2.8), (5-2.13), and (5-2.21), as well as other equations in this article, have been derived for torsion members that have uniform cross sections that do not vary with z, that have simply connected cross sections, that are made of isotropic materials, and that are loaded so that deformations are small. These equations are used to obtain solutions for torsion members; they do not depend on any assumption as to material behavior except that the material is isotropic; therefore, they are valid for any specified material response (elastic or inelastic).

Two types of typical material response are considered in this chapter: a linearly elastic response and an elastic-perfectly plastic response (Fig. 2-6.2*a*). The linearly elastic response leads to the linearly elastic solution of torsion, whereas, the elastic-perfectly plastic response leads to the fully plastic solution of torsion of a bar for which the entire cross section yields (deforms plastically). The material properties associated with various material responses are determined by appropriate tests. Usually, we assume that the material properties are determined by either a tension test or by a torsion test of a cylinder with thin-wall annular cross section.

5-3 LINEAR ELASTIC SOLUTION

Stress-strain relations for linear elastic behavior of an isotropic material are given by Hooke's law (see Eqs. 2-4.7). By Eqs. (2-4.7) and (5-2.8), we obtain

$$\begin{aligned} \sigma_{zx} &= \frac{\partial \phi}{\partial y} = G\gamma_{zx} \\ \sigma_{zy} &= -\frac{\partial \phi}{\partial x} = G\gamma_{zy} \end{aligned} \tag{5-3.1}$$

Substitution of Eqs. (5-3.1) into Eq. (5-2.3) yields

$$\frac{\partial^2 \phi}{\partial x^2} + \frac{\partial^2 \phi}{\partial y^2} = -2G\theta \tag{5-3.2}$$

If the unit angle of twist θ is specified for a given torsion member, and if

ϕ satisfies the boundary condition indicated by Eq. (5-2.13), then Eq. (5-3.2) uniquely determines the stress function $\phi(x, y)$. Once ϕ has been determined, the stresses are given by Eqs. (5-2.8) and the torque is given by Eq. (5-2.21). The elastic solution of the torsion problem for many practical cross sections requires special methods[1] for determining the function ϕ and is beyond the scope of this book. As indicated in the following paragraphs, an indirect method may be used to obtain solutions for certain types of cross sections, although it is not a general method.

Let the boundary of the cross section for a given torsion member be specified by the relation

$$F(x, y) = 0 \tag{5-3.3}$$

Furthermore, let the torsion member be subjected to a specified unit angle of twist, and define the stress function by the relation

$$\phi = BF(x, y) \tag{5-3.4}$$

where B is a constant. The constant B is determined in terms of θ, if after substitution of Eq. (5-3.4) into Eq. (5-3.2), the left side of Eq. (5-3.2) is a constant. With B determined, the stress function ϕ for the torsion member is uniquely defined by Eq. (5-3.4). This indirect approach may, for example, be used to obtain the solutions for torsion members whose cross sections are in the form of a circle, an ellipse, or an equilateral triangle.

Elliptical Cross Section/Let the cross section of a torsion member be bounded by an ellipse (Fig. 5-3.1). The stress function ϕ for the elliptical cross section may be written in the form

$$\phi = B\left(\frac{x^2}{h^2} + \frac{y^2}{b^2} - 1\right) \tag{5-3.5}$$

since $F(x, y) = x^2/h^2 + y^2/b^2 - 1 = 0$ on the boundary (Eq. 5-3.3). Sub-

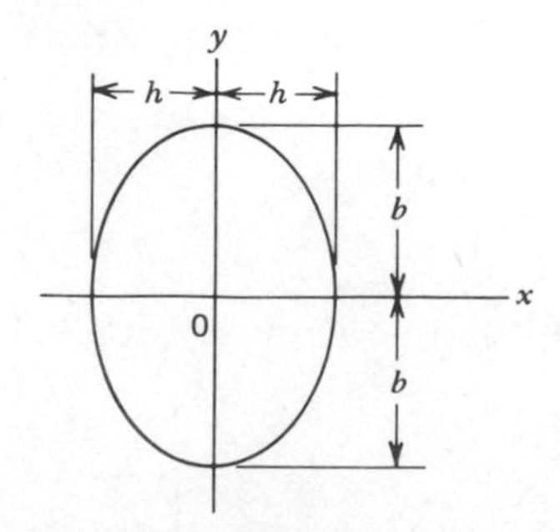

Fig. 5-3.1/Ellipse.

stituting Eq. (5-3.5) into Eq. (5-3.2), we obtain

$$B = -\frac{h^2 b^2 G\theta}{h^2 + b^2} \tag{5-3.6}$$

in terms of the geometrical parameters (h, b), the shear modulus G and the unit angle of twist θ. With ϕ determined, the shearing stress components for the elliptical cross section are, by Eqs. (5-2.8),

$$\sigma_{zx} = \frac{\partial \phi}{\partial y} = \frac{2By}{b^2} = -\frac{2h^2 G\theta y}{h^2 + b^2} \tag{5-3.7}$$

$$\sigma_{zy} = -\frac{\partial \phi}{\partial x} = -\frac{2Bx}{h^2} = \frac{2b^2 G\theta x}{h^2 + b^2} \tag{5-3.8}$$

It may be shown that τ_{max} occurs at the boundary nearest the centroid of the cross section. Its value is

$$\tau_{max} = \sigma_{zy(x=h)} = \frac{2b^2 hG\theta}{h^2 + b^2} \tag{5-3.9}$$

The torque T for the elliptical cross section torsion member is obtained by substituting Eq. (5-3.5) into Eq. (5-2.21). Thus, we obtain

$$T = \frac{2B}{h^2}\int x^2\,dA + \frac{2B}{b^2}\int y^2\,dA - 2B\int dA = \frac{2B}{h^2}I_y + \frac{2B}{b^2}I_x - 2BA$$

Determination of I_x, I_y in terms of (b, h) allows us to write

$$T = -\pi Bhb \tag{5-3.10}$$

The torque may be expressed in terms of either τ_{max} or θ by means of Eqs. (5-3.6), (5-3.9), and (5-3.10). Thus,

$$\tau_{max} = \frac{2T}{\pi bh^2} \qquad \theta = \frac{T(b^2 + h^2)}{G\pi b^3 h^3} \tag{5-3.11}$$

where $G\pi b^3 h^3/(b^2 + h^2)$ is called the torsional rigidity (stiffness) of the section.

Equilateral Triangle Cross Section / Let the boundary of a torsion member be an equilateral triangle (Fig. 5-3.2). The stress function can be shown to be given by the relation

$$\phi = \frac{G\theta}{2h}\left(x - \sqrt{3}\,y - \frac{2h}{3}\right)\left(x + \sqrt{3}\,y - \frac{2h}{3}\right)\left(x + \frac{h}{3}\right) \tag{5-3.12}$$

Proceeding as for the elliptical cross section, we find

$$\tau_{max} = \frac{15\sqrt{3}\,T}{2h^3} \qquad \theta = \frac{15\sqrt{3}\,T}{Gh^4} \tag{5-3.13}$$

where $Gh^4/15\sqrt{3}$ is called the torsional rigidity of the section.

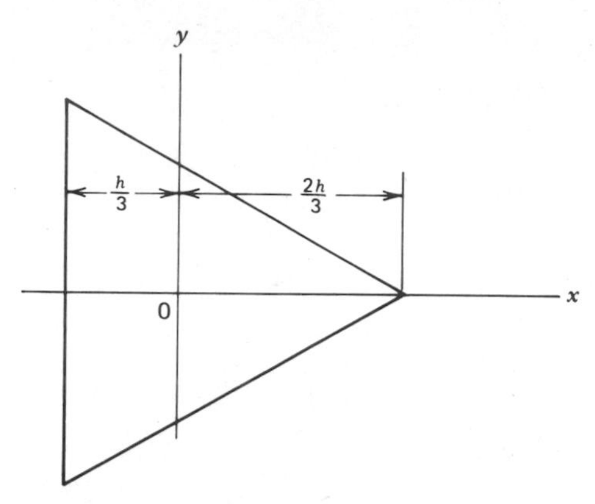

Fig. 5-3.2/Equilateral triangle.

Rectangular Cross Sections/The indirect method outlined above fails for rectangular cross sections. Special methods,[1] which are beyond the scope of this book, are required to obtain the torsion solution for rectangular cross sections. We merely summarize some of the results here. Consider the rectangular cross section shown in Fig. 5-3.3. Relations between the cross-sectional dimensions, T, τ_{max}, and θ take the form

$$\theta = \frac{T}{k_1 G(2b)(2h)^3} \qquad \tau_{max} = \frac{T}{k_2(2b)(2h)^2} \tag{5-3.14}$$

where τ_{max} is the maximum shearing stress at the center of the long side at the boundary. Values of the parameters k_1 and k_2 are tabulated in Table 5-3.1 for several values of the ratio b/h. In Eq. (5-3.14), the factor $k_1 G(2b)(2h)^3$ is the torsional rigidity of the section.

Other Cross Sections/There are many torsion members whose cross sections are so complex that exact analytical solutions are difficult to obtain. How-

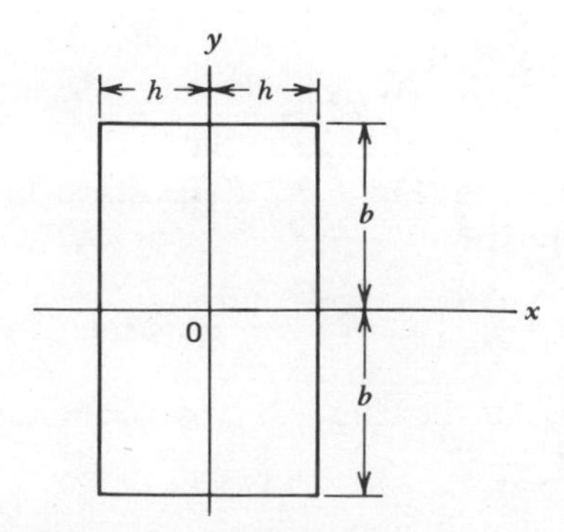

Fig. 5-3.3/Rectangle.

Table 5-3.1

b/h	1	1.5	2	2.5	3	4	6	10	∞
k_1	0.141	0.196	0.229	0.249	0.263	0.281	0.299	0.312	0.333
k_2	0.208	0.231	0.246	0.256	0.267	0.282	0.299	0.312	0.333

ever, approximate solutions may be obtained by Prandtl's membrane analogy (see Art 5-4). An important class of torsion members are those with thin walls. Included in the class of thin-walled torsion members are rolled sections and box sections. Approximate solutions for these types of section are obtained in Arts. 5-5 and 5-6 by means of the Prandtl membrane analogy.

EXAMPLE 5-3.1

Rectangular Section Torsion Member

The rectangular section torsion member in Fig. E5-3.1 has a width of 40 mm. The first 3.00 m length of the torsion member had a depth of 60 mm, and the remaining 1.50 m length has a depth of 30 mm. The torsion member is made of steel for which $G = 77.5$ GPa. If $T_1 = 750$ N · m and $T_2 = 400$ N · m, determine the maximum shearing stress in the torsion member. Determine the angle of twist of the free end. The support at the left end prevents rotation of this cross section but does not prevent warping of the cross section.

SOLUTION

For the left portion of the torsion member,

$$\frac{b}{h} = \frac{30}{20} = 1.5$$

From Table 5-3.1, we find $k_1 = 0.196$ and $k_2 = 0.231$. For the right

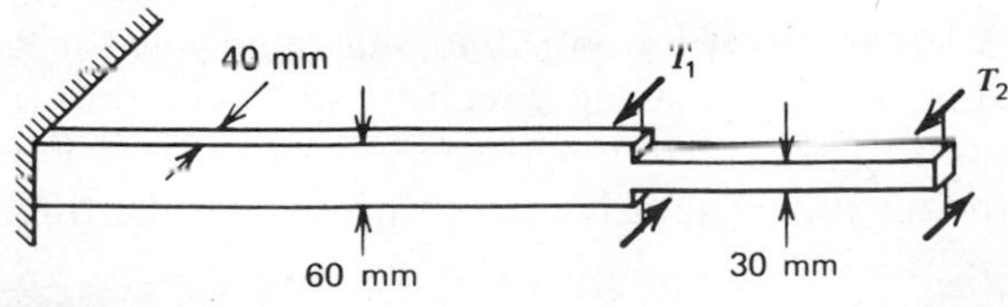

Fig. E5-3.1

portion of the torsion member,

$$\frac{b}{h} = \frac{20}{15} = 1.33$$

A linear interpolation between the values 1 and 1.5 in Table 5-3.1 gives $k_1 = 0.178$ and $k_2 = 0.223$. The torque in the left portion of the torsion member is $T = T_1 + T_2 = 1.15$ kN · m; the maximum shearing stress in this portion of the torsion member is

$$\tau_{max} = \frac{T}{k_2(2b)(2h)^2} = \frac{1{,}150{,}000}{0.231(60)(40)^2} = 51.9 \text{ MPa}$$

The torque in the right portion of the torsion member is equal to $T_2 = 400$ N · m; the maximum shearing stress in this portion of the torsion member is

$$\tau_{max} = \frac{400{,}000}{0.223(40)(30)^2} = 49.8 \text{ MPa}$$

Hence, the maximum shearing stress occurs in the left portion of the torsion member and is equal to 51.9 MPa.

The angle of twist is equal to the sum of the angles of twist for the left and right portions of the torsion member. Thus,

$$\text{Angle of Twist} = \frac{1{,}150{,}000(3000)}{0.196(77{,}500)(60)(40)^3} + \frac{400{,}000(1500)}{0.178(77{,}500)(40)(30)^3}$$
$$= 0.0994 \text{ rad}$$

PROBLEM SET 5-3

1. Derive the relation for the shearing stress distribution on the x-axis for the equilateral triangle in Fig. 5-3.2.

2. Derive Eqs. (5-3.13) for the equilateral triangle.

3. A square shaft may be used to transmit power from a farm tractor to farm implements. A 25.0 mm square shaft is made of a steel having a yield stress of $Y = 380$ MPa. Determine the torque that can be applied to the shaft based on a factor of safety of $SF = 2.00$ by using the octahedral shearing stress criterion of failure.

4. A square shaft has 42.0 mm sides and has the same cross-sectional area as shafts having circular and equilateral triangular cross sections. If each shaft is subjected to a torque of 1.00 kN · m, determine the maximum shearing stress for each of the three shafts.

   Ans. $\tau_{square} = 64.89$ MPa, $\tau_{circle} = 47.82$ MPa, $\tau_{triangle} = 76.86$ MPa

5. The shafts in Problem 4 are made of a steel for which $G = 77.5$ GPa. Determine the unit angle of twist for each shaft.

6. The left-hand section of the torsion member in Fig. E5-3.1 is 2.00 m long, and the right-hand section is 1.00 m long. It is made of an aluminum alloy for which $G = 27.1$ GPa. Determine the magnitude of T_2 if $T_1 = 350$ N · m and the maximum shearing stress is 45.0 MPa. Neglect stress concentrations at changes in section. Determine the angle of twist of the free end. The support at the left end prevents rotation of this cross section but does not prevent warping of the cross section.

   Ans. $T_2 = 247.8$ N · m; Angle of Twist = 0.1990 rad

7. A torsion member has an elliptical cross section with major and minor dimensions of 50.0 mm and 30.0 mm, respectively. The yield stress of the material in the torsion member is $Y = 400$ MPa. Determine the maximum torque that can be applied to the torsion member based on a factor of safety of $SF = 1.85$ using the maximum shearing stress criterion of failure.

5-4 THE PRANDTL ELASTIC-MEMBRANE (SOAP-FILM) ANALOGY

The equation that defines the small displacement of a plane elastic membrane subject to lateral pressures is identical in mathematical form to the stress function equation (Eq. 5-3.2). Hence, the displacement function of the membrane is mathematically equivalent to the stress function, provided that the boundary shape of the membrane is identical to the boundary shape of the cross section of the torsion member. In the following discussion, we outline the physical and mathematical procedures that lead to a complete analogy between the membrane problem and the torsion problem.

In the (x, y)-plane, cut an opening that has the same shape as the cross section of the torsion member to be investigated. Cover the opening with a homogeneous elastic membrane, such as a soap film, and apply a pressure to one side of the membrane. The pressure causes the membrane to bulge, forming a curved surface. If the slope of the membrane surface is sufficiently small, it may be shown that the lateral displacement of the membrane and the stress function $\phi(x, y)$ satisfy the same mathematical equation in (x, y).

As noted above, the Prandtl membrane analogy is based upon the equivalence of the torsion equation, (Eq. 5-3.2, repeated here for convenience)

$$\frac{\partial^2 \phi}{\partial x^2} + \frac{\partial^2 \phi}{\partial y^2} = -2G\theta \tag{5-4.1}$$

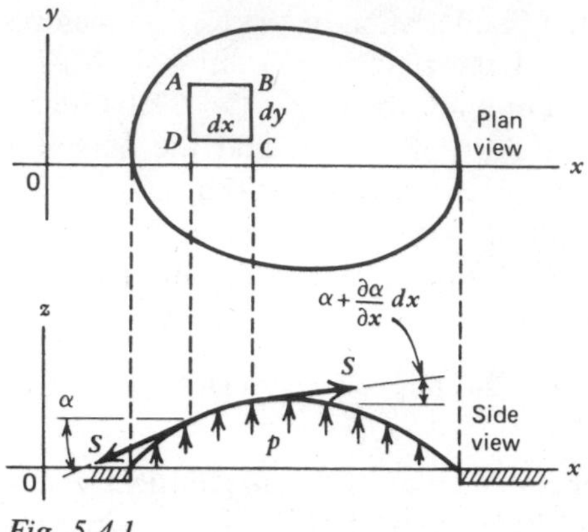

Fig. 5-4.1

and the membrane equation (to be derived in the next paragraph)

$$\frac{\partial^2 z}{\partial x^2} + \frac{\partial^2 z}{\partial y^2} = -\frac{p}{S} \tag{5-4.2}$$

where z denotes the lateral displacement of an elastic membrane subjected to a lateral pressure p in terms of force per unit area and an initial (large) tension S (Fig. 5-4.1) in terms of force per unit length.

For the derivation of Eq. (5-4.2), consider an element $ABCD$ of dimensions dx, dy of the elastic membrane shown in Fig. 5-4.1. The net vertical force due to the tension S acting along edge AD of the membrane is (assuming small displacements so that $\sin \alpha \approx \tan \alpha$)

$$-S\,dy \sin \alpha \approx -S\,dy \tan \alpha = -S\,dy \frac{\partial z}{\partial x}$$

and, similarly, the net vertical force due to the tension S (assumed to remain constant for sufficiently small values of p) acting along edge BC is

$$S\,dy \tan\left(\alpha + \frac{\partial \alpha}{\partial x} dx\right) = S\,dy \frac{\partial}{\partial x}\left(z + \frac{\partial z}{\partial x} dx\right)$$

Similarly for edges AB and DC we obtain

$$-S\,dx \frac{\partial z}{\partial y}, \qquad S\,dx \frac{\partial}{\partial y}\left(z + \frac{\partial z}{\partial y} dy\right)$$

Consequently, the summation of force in the vertical direction yields for the equilibrium of the membrane element $dx\,dy$

$$S \frac{\partial^2 z}{\partial x^2} dx\,dy + S \frac{\partial^2 z}{\partial y^2} dx\,dy + p\,dx\,dy = 0$$

or

$$\frac{\partial^2 z}{\partial x^2} + \frac{\partial^2 z}{\partial y^2} = -\frac{p}{S}$$

By comparison of Eqs. (5-4.1) and (5-4.2), we arrive at the following analogous quantities:

$$z = c\phi \qquad \frac{p}{S} = c2G\theta \tag{5-4.3}$$

where c is a constant of proportionality. Hence,

$$\frac{z}{p/S} = \frac{\phi}{2G\theta} \qquad \phi = \frac{2G\theta S}{p} z \tag{5-4.4}$$

Accordingly, the membrane displacement z is proportional to the Prandtl stress function ϕ and, since the shearing-stress components σ_{zx}, σ_{zy} are equal to the appropriate derivatives of ϕ with respect to x and y (see Eqs. 5-3.1), it follows that the stress components are proportional to the derivatives of the membrane displacement z with respect to the coordinates (x, y) in the flat plate to which the membrane is attached (Fig. 5-4.1). In other words, the stress components at a point (x, y) of the bar are proportional to the slopes of the membrane at the corresponding point (x, y) of the membrane. Consequently, the distribution of shearing stress components in the cross section of the bar is easily visualized by forming a mental image of the slope of the corresponding membrane. Furthermore, for simply connected cross sections, since z is proportional to ϕ, by Eqs. (5-2.21) and (5-4.4), we note that the twisting moment T is proportional to the volume enclosed by the membrane and the (x, y)-plane (Fig. 5-4.1). For the multiply connected cross section additional conditions arise.[1]

Recommended experimental techniques for the use of the membrane analogy are reported in a paper by Thoms and Masch.[2] The experimental technique requires that p/S be determined for the membrane. The procedure usually followed is to machine a circular hole in the plate in addition to the cross section of interest. If the same membrane is used for both openings, the values of p/S are the same for both openings, and the corresponding values of $G\theta$ are the same for the corresponding torsion members (see Eqs. 5-4.4). Since the relations between T, θ, and τ for a torsion member with a solid circular cross section are given by Eqs. (5-1.12) and (5-1.15), measurements obtained on the two elastic membranes may be used to compare torques and shearing stresses for the two corresponding torsion members for *the same unit angle of twist*. In particular, then the ratio of the two volumes between the membrane surfaces and the flat plate is equal to the ratio of the torques of the two corresponding torsion members. The ratio of the maximum slopes of the two membrane surfaces is equal to the ratio of the maximum shearing stresses in the two corresponding torsion members.

Another important aspect of the elastic membrane analogy is that, without performing experiments, valuable deductions can be made by merely visualizing the shape that the membrane must take. For example, if several membranes cover holes machined in a flat plate, the corresponding torsion members have equal values of $G\theta$; therefore, the stiffnesses (see Eqs. 5-3.11, 5-3.13, and 5-3.14) of torsion members made of materials having the same G are proportional to the volumes between the membranes and the flat plate. For cross sections with equal area, one can deduce that a long narrow rectangular section has the least stiffness and the circular section has the greatest stiffness.

Important conclusions may also be drawn in regard to the magnitude of the shearing stress and hence to the cross section for minimum shearing stress. Consider the angle section shown in Fig. 5-4.2*a*. At the external corners A, B, C, E, and F the membrane has zero slope and the shearing stress is zero; therefore, external corners do not constitute a design problem. However, at the reentrant corner at D (shown as right angle in Fig. 5-4.2*a*) the corresponding membrane would have an infinite slope which indicates an infinite shearing stress in the torsion member. Actually, an exact right angle is improbable in practice. Therefore, in practical problems, the magnitude of the shearing stress at D would be finite, but would be very large compared to that at other points in the cross section.

Remark on Reentrant Corners/If a torsion member with cross section shown in Fig. 5-4.2*a* is made of a ductile material, and if it is subjected to static loads, the material in the neighborhood of D yields and the load is distributed to adjacent material, so that the stress concentration at point D is not particularly important. If, on the other hand, the material is brittle or the torsion member is subjected to fatigue loading, the shearing stress at D limits the load carrying capacity of the member. In such a case, the maximum shearing stress in the torsion member may be decreased by removing some material as shown in Fig. 5-4.2*b*. However, preferably, the member should be redesigned to alter the cross section (Fig. 5-4.2*c*). The maximum shearing stress would then be about the

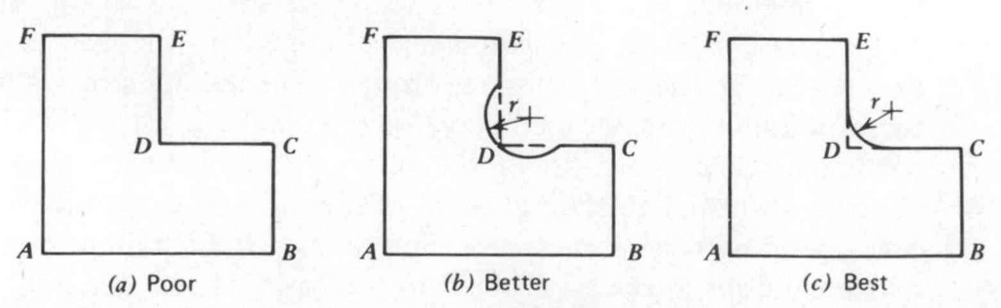

Fig. 5-4.2/Angle section of a torsion member.

same for the two cross sections shown in Figs. 5-4.2*b* and *c* for a given unit angle of twist; however, a torsion member with the cross section shown in Fig. 5-4.2*c* would be stiffer for a given unit angle of twist.

5-5
NARROW RECTANGULAR CROSS SECTION

The cross sections of many members of machines and structures are made up of narrow rectangular parts. These members are used mainly to carry tension, compression, and bending loads. However, they may be also required to carry secondary torsional loads. Since it is simple to use for this cross section, we use the elastic membrane analogy to obtain the solution of a torsion member whose cross section is in the shape of a narrow rectangle.

Consider a bar subjected to torsion. Let the cross section of the bar be a solid rectangle with width $2h$ and depth $2b$, where $b \gg h$ (Fig. 5-5.1). The associated membrane is shown in Fig. 5-5.2.

Except for the region near $x = \pm b$, the membrane deflection is approximately independent of x. Hence, if we assume that the deflected membrane is independent of x and that the deflection with respect to y is parabolic, the displacement equation of the membrane is approximately

$$z = z_0\left[1 - \left(\frac{y}{h}\right)^2\right] \tag{5-5.1}$$

where z_0 is the maximum deflection of the membrane. Note that Eq. (5-5.1) satisfies the condition $z = 0$ on the boundaries $y = \pm h$. Also, if p/S is a constant in Eq. (5-4.2), the parameter z_0 may be selected so that Eq. (5-5.1) represents a solution of Eq. (5-4.2). Consequently, Eq. (5-5.1) is an approximate solution of the membrane displacement. For example,

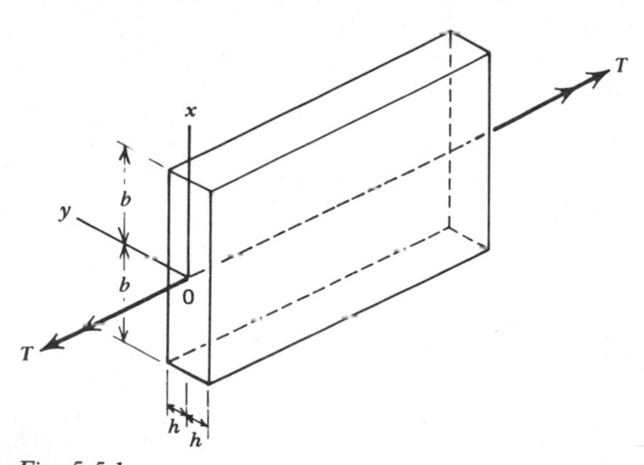

Fig. 5-5.1

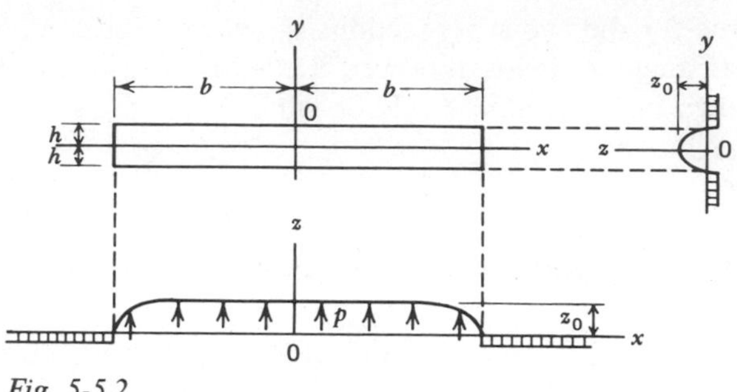

Fig. 5-5.2

by Eq. (5-5.1) we find

$$\frac{\partial^2 z}{\partial x^2} + \frac{\partial^2 z}{\partial y^2} = -\frac{2z_0}{h^2} \tag{5-5.2}$$

By Eqs. (5-5.2), (5-4.2), and (5-4.3), we may write $-2z_0/h^2 = -2cG\theta$ and Eq. (5-5.1) becomes

$$\phi = G\theta h^2\left[1 - \left(\frac{y}{h}\right)^2\right] \tag{5-5.3}$$

Consequently, Eqs. (5-2.8) yield

$$\sigma_{zx} = \frac{\partial\phi}{\partial y} = -2G\theta y \qquad \sigma_{zy} = -\frac{\partial\phi}{\partial x} = 0 \tag{5-5.4}$$

and we note that the maximum value of σ_{zx} is

$$\tau_{\max} = 2G\theta h \qquad \text{for } y = \pm h \tag{5-5.5}$$

Equations (5-2.21) and (5-5.3) yield

$$T = 2\int_{-b}^{b}\int_{-h}^{h}\phi\,dx\,dy = \frac{1}{3}G\theta(2b)(2h)^3 = GJ\theta \tag{5-5.6}$$

where

$$J = \tfrac{1}{3}(2b)(2h)^3 \tag{5-5.7}$$

is the torsional rigidity factor and GJ is the torsional rigidity.

In summary, we note that the solution is approximate and, in particular, the boundary condition for $x = \pm b$ is not satisfied. From Eqs. (5-5.5) and (5-5.6) we obtain

$$\tau_{\max} = \frac{3T}{(2b)(2h)^2} = \frac{2Th}{J} \qquad \theta = \frac{3T}{G(2b)(2h)^3} = \frac{T}{GJ} \tag{5-5.8}$$

Note that Eqs. (5-5.8) agree with Eqs. (5-3.14) since, from Table 5-3.1, $k_1 = k_2 = \frac{1}{3}$.

Cross Sections Made up of Long Narrow Rectangles/Many rolled composite sections are made up of joined long narrow rectangles. For these cross sections, it is convenient to define the torsional rigidity factor J by the relation

$$J = C\frac{1}{3}\sum_{i=1}^{n}(2b_i)(2h_i)^3 \tag{5-5.9}$$

where C is a correction coefficient. If $b_i > 10h_i$ for each rectangle part of the composite cross section (see Table 5-3.1), $C \approx 1$. For many rolled sections, b_i may be less than $10h_i$ for one or more of the rectangles making up the cross section. In this case it is recommended that $C = 0.91$. When $n = 1$ and $b > 10h$, $C = 1$ and Eq. (5-5.9) is identical with Eq. (5-5.7). For $n > 1$, Eqs. (5-5.8) take the form

$$\tau_{max} = \frac{2Th_{max}}{J} \qquad \theta = \frac{T}{GJ} \tag{5-5.10}$$

where h_{max} is the maximum value of the h_i.

PROBLEM SET 5-5

1. Find the maximum shearing stress and unit angle of twist of the bar having the cross section shown in Fig. P5-5.1 when subjected to a torque at its ends of 600 N · m. The bar is made of a steel for which $G = 77.5$ GPa.

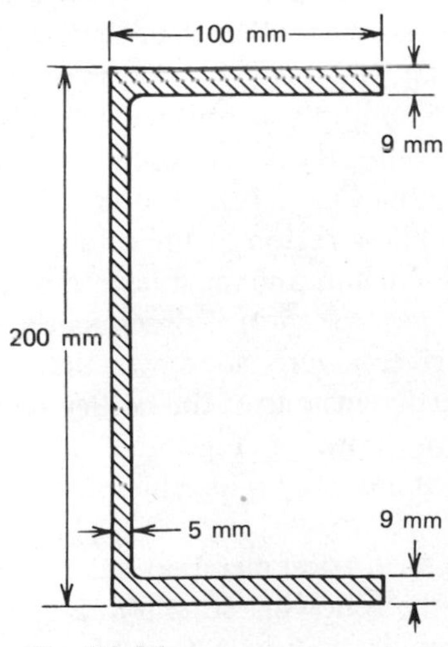

Fig. P5-5.1

2. An aluminum alloy extruded section (Fig. P5-5.2) is subjected to a torsional load. Determine the maximum torque that can be applied to the member if the maximum shearing stress is 75.0 MPa. Neglect stress concentrations at changes in section.

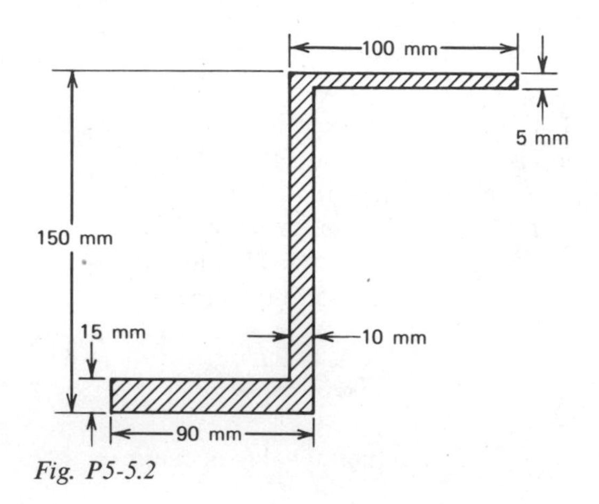

Fig. P5-5.2

Ans. $T = 665.4 \text{ N} \cdot \text{m}$

5-6 HOLLOW THIN-WALL TORSION MEMBERS. MULTIPLY CONNECTED CROSS SECTION

In general, the solution for a torsion member with a multiply connected cross section is more complex than that for the solid (simply connected cross section) torsion member. For simplicity, we refer to the torsion member with a multiply connected cross section as a hollow torsion member. The complexity of the solution can be illustrated for the hollow torsion member in Fig. 5-6.1. No shearing stresses act on the lateral surface of the hollow region of the torsion member; therefore, the stress function and the membrane must have zero slope over the hollow region (see Eqs. 5-2.8 and Art. 5-4). Consequently, the associated elastic membrane may be given a zero slope over the hollow region by machining a flat plate to the dimensions of the hollow region and displacing the plate a distance z_1, as shown in Fig. 5-6.1. However, the distance z_1 is not known. Furthermore, only one value of z_1 is valid for specified values of p and S.

The solution for torsion members having thin-wall noncircular sections is based upon the following simplifying assumption. Consider the thin-wall torsion member in Fig. 5-6.2a. The plateau (region of zero slope)

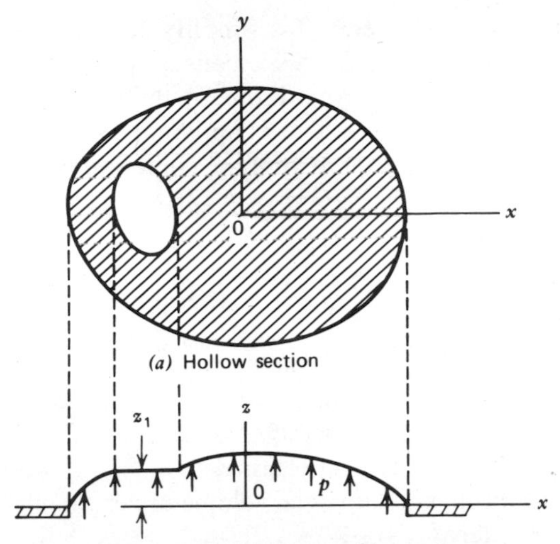

(a) Hollow section

(b) Intersection of (x, z)–plane with membrane

Fig. 5-6.1/Hollow torsion.

over the hollow area and the resulting membrane are shown in Fig. 5-6.2*b*. If the wall thickness is small compared to the other dimensions of the cross section, sections through the membrane, made by planes parallel to the z-axis and perpendicular to the outer boundary of the cross section, are approximately straight lines. It is *assumed that these intersections are straight lines*. Because the shearing stress is given by the

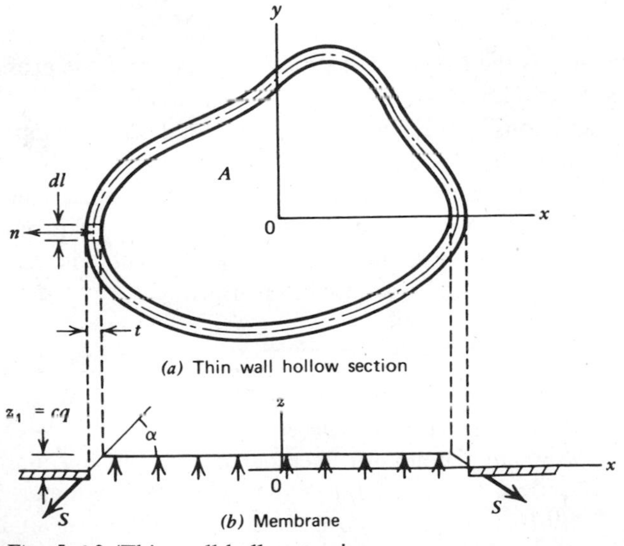

(a) Thin wall hollow section

(b) Membrane

Fig. 5-6.2/Thin-wall hollow torsion.

slope of the membrane, this simplifying assumption leads to the condition that the shearing stress is constant through the thickness. However, the shearing stress around the boundary is not constant, unless the thickness t is constant. This is apparent by Fig. 5-6.2b since $\tau = \partial\phi/\partial n$, where n is normal to a membrane contour curve z = constant. Hence, by Eqs. (5-4.4) and Fig. 5-6.2b, $\tau = (2G\theta S/p)\partial z/\partial n = (2G\theta S/p)\tan\alpha$. Finally, by Eq. (5-4.3),

$$\tau = \frac{1}{c}\tan\alpha = \frac{1}{c}\sin\alpha \qquad \text{(since } \alpha \text{ is assumed to be small)} \qquad (5\text{-}6.1)$$

The quantity $q = \tau t$, with dimensions N/mm, is commonly referred to in the literatures as the *shear flow*. As indicated in Fig. 5-6.2b, the shear flow is constant around the cross section of a thin-wall hollow torsion member. Since ϕ is proportional to z (Eq. (5-4.4)), by Eq. (5-2.21), the torque is proportional to the volume under the membrane. Thus, we have approximately ($z_1 = c\phi_1$)

$$T = 2A\phi_1 = 2Az_1/c = 2Aq = 2A\tau t \qquad (5\text{-}6.2)$$

in which A is the area enclosed by the mean perimeter of the cross section (see the area enclosed by the dot-dashed line in Fig. 5-6.2a). A relation between τ, G, θ, and the dimensions of the cross section may be derived from the equilibrium conditions in the z-direction. Thus,

$$\sum F_z = pA - \oint S\sin\alpha\, dl = 0,$$

and by Eqs. (5-6.1) and (5-4.3),

$$\frac{1}{A}\oint \tau\, dl = \frac{p}{cS} = 2G\theta \qquad (5\text{-}6.3)$$

where l is the length of the mean perimeter of the cross section and S is the tensile force per unit length of the membrane.

Equations (5-6.2) and (5-6.3) are based on the simplifying assumption that the wall thickness is sufficiently small so that the shearing stress may be assumed to be uniform through the wall thickness. For the cross section considered in the following illustrative problem, the resulting error is negligibly small when the wall thickness is less than one-tenth of the minimum cross-sectional dimension.

EXAMPLE 5-6.1

Hollow Thin-Wall Circular Torsion Member

A hollow circular torsion member has an outside diameter of 22.0 mm and inside diameter of 18.0 mm, with mean diameter $D = 20.0$ mm and $t/D = 0.10$. (a) Let the shearing stress at the mean diameter be $\tau = 70.0$ MPa. Determine T and θ using Eqs. (5-6.2) and (5-6.3) and compare these values with values obtained using the elasticity theory. $G = 77.5$ GPa. (b) Let a cut be made through the wall thickness along the entire length of the torsion member and let the maximum shearing stress in the resulting torsion member be 70.0 MPa. Determine T and θ.

SOLUTION

The area A enclosed by the mean perimeter is

$$A = \frac{\pi D^2}{4} = 100\pi \text{ mm}^2$$

The torque is given by Eq. (5-6.2):

$$T = 2A\tau t = 2(100\pi)(70)(2) = 87{,}960 \text{ N} \cdot \text{mm} = 87.96 \text{ N} \cdot \text{m}$$

Because the wall thickness is constant, Eq. (5-6.3) gives

$$\theta = \frac{\tau\pi D}{2GA} = \frac{70(\pi)(20)}{2(77{,}500)(100\pi)} = 0.0000903 \text{ rad/mm}$$

Elasticity values of T and θ are given by Eqs. (5-1.15) and (5-1.12). Thus, with

$$J = \frac{\pi}{32}(22^4 - 18^4) = 4040\pi \text{ mm}^4$$

we find that

$$T = \frac{\tau J}{r} = \frac{70(4040\pi)}{10} = 88{,}840 \text{ N} \cdot \text{mm} = 88.84 \text{ N} \cdot \text{m}$$

and

$$\theta = \frac{\tau}{Gr} = \frac{70}{77{,}500(10)} = 0.0000903 \text{ rad/mm}$$

The approximate solution agrees with the elasticity theory in the prediction of the unit angle of twist and yields torque that differs only by 1 percent. Note that the approximate solution assumes that the shearing stress was uniformly distributed while the elasticity solution indicates that the maximum shearing stress is 10 percent greater than the value at the mean diameter, since the elasticity solution indicates that τ is proportional to r.

(b) When a cut is made through the wall thickness along the entire length of the torsion member, the torsion member becomes a long narrow rectangle, for which the theory of Art. 5-5 applies. Thus, with $h = 1$ and $b = 10\pi$

$$\theta = \frac{\tau_{\max}}{2Gh} = \frac{70}{2(77{,}500)(1)} = 0.0004515 \text{ rad/mm}$$

$$T = \frac{8bh^2\tau_{\max}}{3} = \frac{8(10\pi)(1)^2(70)}{3} = 5{,}865 \text{ N}\cdot\text{mm} = 5.865 \text{ N}\cdot\text{m}$$

Hence, after the cut, the torque is $\frac{1}{15}$ of the torque for part (a) while the unit angle of twist is 5 times greater than that for part (a).

Hollow Thin-Wall Torsion Member Having Several Compartments/Thin-wall hollow torsion members may have two or more compartments. Consider the torsion member whose cross section is shown in Fig. 5-6.3*a*. Section *aa* through the membrane analogy is shown in Fig. 5-6.3*b*. The plateau over each compartment is assumed to have a different elevation z_i. If there are N compartments, there are $N + 1$ unknowns to be determined. For a specified torque T the unknowns are the N values for the q_i and the unit angle of twist θ, which is assumed to be the same for each compartment. The $N + 1$ equations are given by the equation

$$T = 2\sum_{i=1}^{N} A_i \frac{z_i}{c} = 2\sum_{i=1}^{N} A_i q_i \tag{5-6.4}$$

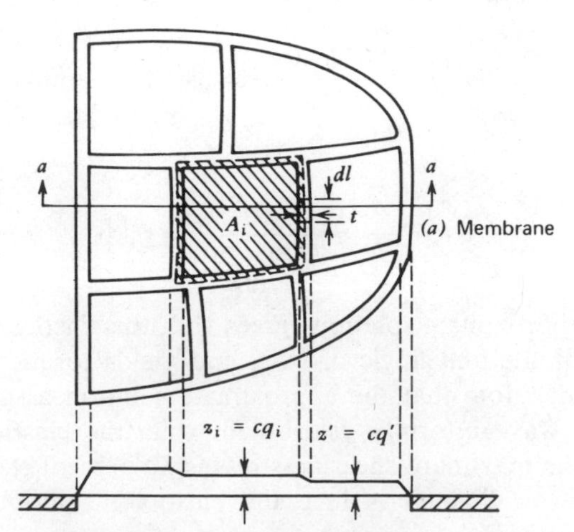

Fig. 5-6.3/Multicompartment hollow thin-wall torsion member.

and N additional equations similar to Eq. (5-6.3)

$$\theta = \frac{1}{2GA_i} \oint_{l_i} \frac{q_i - q'}{t} dl \qquad i = 1, 2, \ldots, N \tag{5-6.5}$$

where A_i is the area bounded by the mean perimeter for the ith compartment, q' is the shear flow for the compartment adjacent to the ith compartment where dl is located, t is the thickness where dl is located, and l_i is the length of the mean perimeter for the ith compartment. We note that q' is zero at the outer boundary. The maximum shearing stress occurs where the membrane has the greatest slope, that is, where $(q_i - q')/t$ takes on its maximum value for the N compartments.

EXAMPLE 5-6.2

Two Compartment Hollow Thin-Wall Torsion Member

A hollow thin-wall torsion member has two compartments with cross-sectional dimensions as indicated in Fig. E5-6.2. The material is an aluminum alloy for which $G = 26.0$ GPa. Determine the torque and unit angle of twist if the maximum shearing stress, at locations away from stress concentrations, is 40.0 MPa.

SOLUTION

Possible locations of the maximum shearing stress are in the outer wall of compartment 1 where $t_1 = 4.5$ mm, in the outer wall of compartment 2 where $t_2 = 3.0$ mm, and the wall between the two compartments where $t_3 = 1.5$ mm. The correct location requires that we determine the ratio of

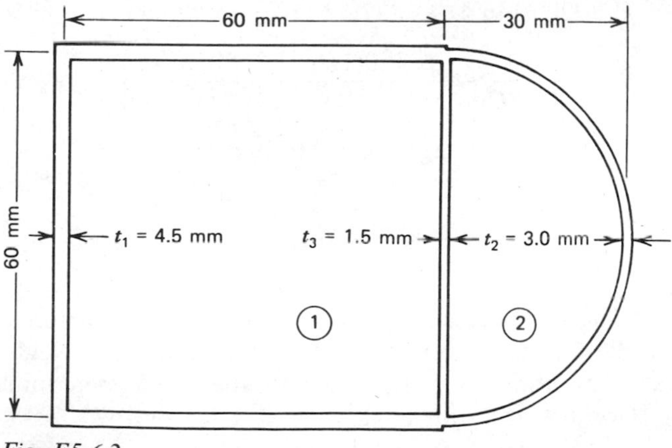

Fig. E5-6.2

q_1 to q_2. First, we will write down the three equations given by Eq. (5-6.4) and Eqs. (5-6.5).

$$T = 2(A_1 q_1 + A_2 q_2) = 7200 q_1 + 2827 q_2 \tag{1}$$

$$\theta = \frac{1}{2GA_1}\left[\frac{q_1 l_1}{t_1} + \frac{(q_1 - q_2) l_3}{t_3}\right] = \frac{1}{7200G}\left[\frac{180 q_1}{4.5} + \frac{60(q_1 - q_2)}{1.5}\right] \tag{2}$$

$$\theta = \frac{1}{2GA_2}\left[\frac{q_2 l_2}{t_2} + \frac{(q_2 - q_1) l_3}{t_3}\right] = \frac{1}{900\pi G}\left[\frac{30\pi q_2}{3.0} + \frac{60(q_2 - q_1)}{1.5}\right] \tag{3}$$

Since the unit angle of twist given by Eq. (2) is equal to that given by Eq. (3), the ratio of q_1 to q_2 is found to be

$$\frac{q_1}{q_2} = 1.220$$

This ratio is less than t_1/t_2; therefore, the maximum shearing stress does not occur in the walls with thickness $t_1 = 4.5$ mm. Let us assume that it occurs in the wall with thickness t_2.

$$q_2 = \tau_{max} t_2 = 40.0(3) = 120.0 \text{ N/mm} \tag{4}$$

$$q_1 = 1.220(120.0) = 146.4 \text{ N/mm} \tag{5}$$

$$q_1 - q_2 = 26.4 \text{ N/mm}$$

$$\tau = \frac{q_1}{t_1} = \frac{146.4}{4.5} = 32.5 \text{ MPa}, \quad \tau_2 = \frac{q_2}{t_2} = 40.0 \text{ MPa}, \quad \tau_3 = \frac{q_1 - q_2}{t_3}$$
$$= 17.6 \text{ MPa}$$

The magnitudes of q_1 and q_2 given by Eqs. (4) and (5) were based on the assumption that $\tau_2 = \tau_{max} = 40.0$ MPa; it is seen that the assumption is valid. These values for q_1 and q_2 may be substituted into Eqs. (1) and (2) to determine T and θ. Thus,

$$T = 7200(146.4) + 2827(120.0) = 1{,}393{,}000 \text{ N} \cdot \text{mm} = 1.393 \text{ kN} \cdot \text{m}$$

$$\theta = \frac{1}{7200(26{,}000)}\left[\frac{180(146.4)}{4.5} + \frac{60(26.4)}{1.5}\right]$$
$$= 0.0000369 \text{ rad/mm} = 0.0369 \text{ rad/m}$$

PROBLEM SET 5-6

1. The hollow circular and square thin-wall torsion members in Fig. P5-6.1 have identical values for b and t. Neglecting the stress concentrations at the corners of the square, determine the ratio of the torques and unit angle of twists for the two torsion members, for equal shearing stresses in each.

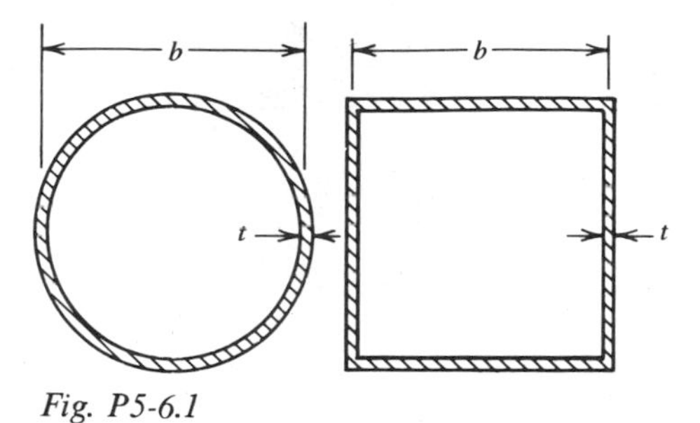

Fig. P5-6.1

2. A hollow thin-wall brass tube has an equilateral triangular cross section. The mean length of each side of the triangle is 40.0 mm. The wall thickness is 4.00 mm. Determine the torque and unit angle of twist for an average shearing stress of 20.0 MPa, and for $G = 31.1$ GPa.

   Ans. $T = 110.8\ \text{N} \cdot \text{m}$; $\theta = 0.0559\ \text{rad/m}$

3. A hollow rectangular thin-wall steel torsion member has the cross section shown in Fig. P5-6.3. The steel has a yield stress $Y = 360$ MPa and a shearing modulus of elasticity of $G = 77.5$ GPa. Determine the maximum torque that may be applied to the torsion member, based on a factor of safety of $SF = 2.00$ for the octahedral shearing stress criterion of failure. What is the unit angle of twist when the maximum torque is applied?

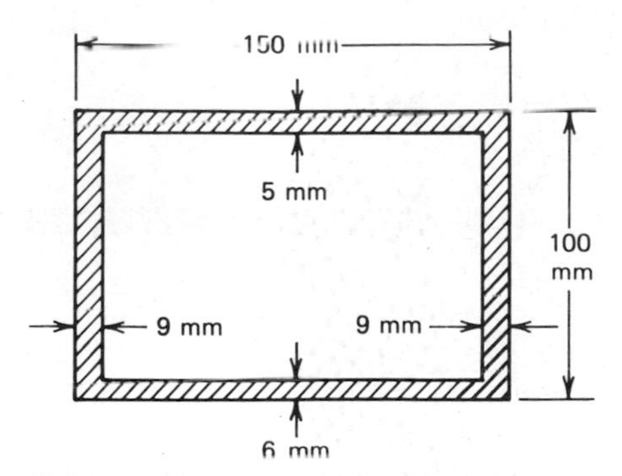

Fig. P5-6.3

4. The hollow thin-wall torsion member of Fig. P5-6.4 has uniform thickness walls. Show that walls, *BC*, *CD*, and *CF* are stressfree.

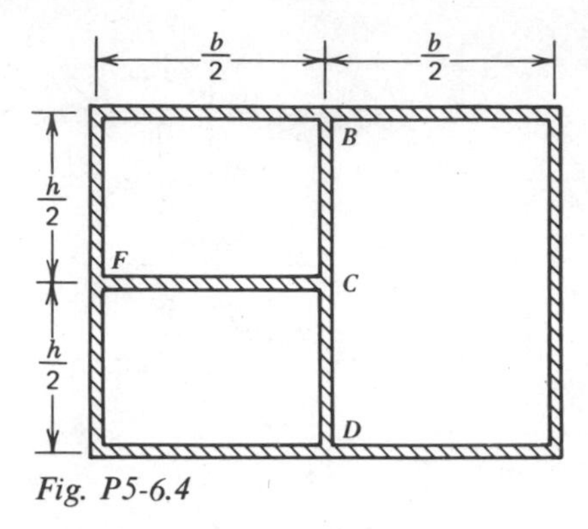

Fig. P5-6.4

5. The aluminum ($G = 27.1$ GPa) hollow thin-wall torsion member in Fig. P5-6.5 has the dimensions shown. Its length is 3.00 m. If the member is subjected to a torque $T = 11.0$ kN · m, determine the maximum shearing stress and the angle of twist.

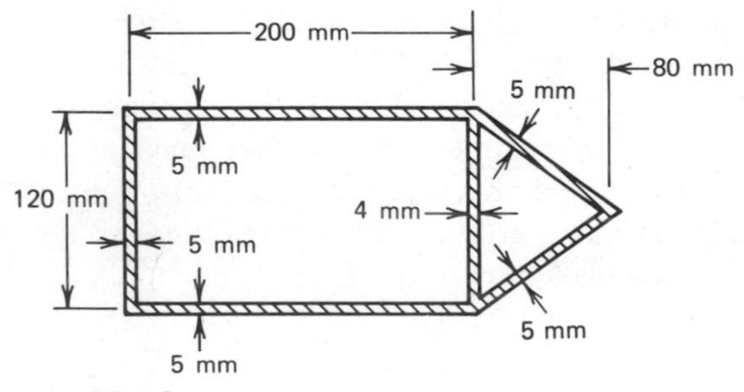

Fig. P5-6.5

5-7
THIN-WALL TORSION MEMBERS WITH RESTRAINED ENDS

Torsion members with noncircular cross sections warp when subjected to torsional loads. However, if a torsion member is welded at one end, or both ends, to heavy supports, warping at the end sections is prevented. Hence, for a torsion member with a noncircular cross section, a normal stress distribution, that prevents warping, occurs at the restrained end. In addition, a shearing stress distribution is developed at the restrained end to balance the torsional load.

Consider the I-section torsion member (Fig. 5-7.1) constrained against warping at the wall. At a section near the wall (Fig. 5-7.1*a*) the torsional load is transmitted mainly by lateral shearing force V in each flange. This shearing force produces lateral bending of each flange. As a result, on the basis of a linear bending theory for each flange, a linear normal stress distribution in each flange is produced at the wall. In addition, the shearing stress distribution in each flange, at the wall, is similar to that for shear loading of a rectangular beam. At small distances away from the wall (Fig. 5-7.1*b*), partial warping occurs, and the torsional load is transmitted partly by the shear $V' < V$ and partly by torsional shear. At greater distances from the wall (Fig. 5-7.1*c*), the effect of the restrained end diminishes rapidly, and the torque is transmitted mainly by torsional shearing stresses. These remarks are illustrated further by a solution for an I-section torsion member presented later in this article.

Another important class of thin-wall torsion members with restrained ends are thin-wall hollow torsion members. Contrary to torsion members with simply connected cross sections, noncircular thin-wall hollow torsion members may, under certain conditions, twist without warping. A

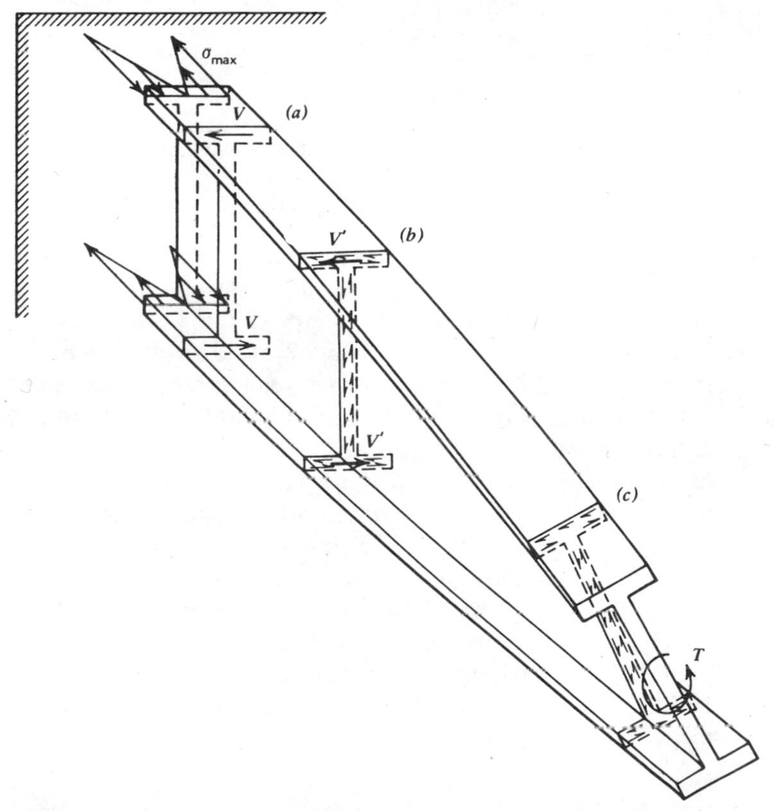

Fig. 5-7.1/General torsional effect of torsional load on T-section torsion member. (*a*) Lateral shear mainly. (*b*) Partly lateral shear and partly torsional shear. (*c*) Torsional shear mainly.

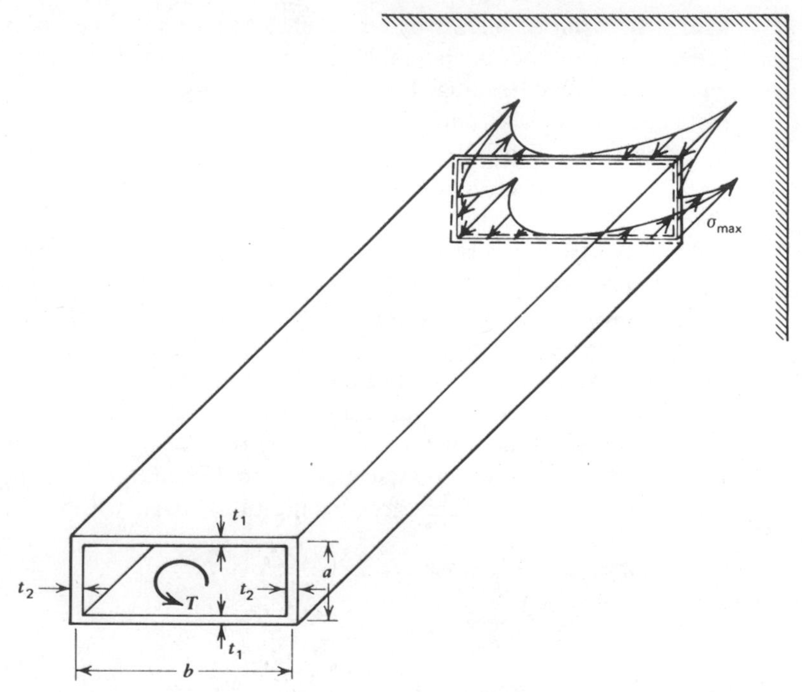

Fig. 5-7.2/Thin-wall rectangular section torsion member with a restrained end.

solution presented by von Karman and Chien[3] for constant thickness hollow torsion members indicates that a torsion member with equilateral polygon cross section does not warp. If $t_1 = t_2$, the rectangular section hollow torsion member in Fig. 5-7.2 tends to warp when subjected to torsion loads and, hence, tends to develop a normal stress distribution at a restrained end. However, a solution presented by Smith, Thomas, and Smith[4] indicates that the torsion member in Fig. 5-7.2 does not warp if $bt_1 = at_2$. They have presented a solution for the case when $a/b = 3/8$, $t_1 = b/32$, and $t_2 = b/16$. They found that the normal stress distribution at the end to be nonlinear as indicated in Fig. 5-7.2 with $\sigma_{max} = 0.0114T$, where σ_{max} has the dimensions of MPa and T has the dimensions of N · m. For hollow torsion members with rectangular sections of constant thickness, similar normal stress distributions at a restrained end are predicted in the papers by von Karmon and Chien[3] and by Smith, et al.[4] Readers are referred to these two papers for thin-wall hollow torsion members with restrained ends.

I-Section Torsion Member Having One Section Restrained from Warping/ Consider an I-section torsion member subjected to a twisting moment T (Fig. 5-7.3*a*). Let the section at the wall be restrained from warping. A small distance from the wall, say at section AB, partial warping takes place and the twisting moment T may be considered to be made up of two parts. One part is a twisting moment T_1 produced by the lateral shearing forces since these forces constitute a couple with moment arm h. Hence,

$$T_1 = V'h \tag{5-7.1}$$

The second part is twisting moment T_2, which produces warping of the section. Hence, T_2 is given by Eq. (5-5.10) as

$$T_2 = JG\theta \tag{5-7.2}$$

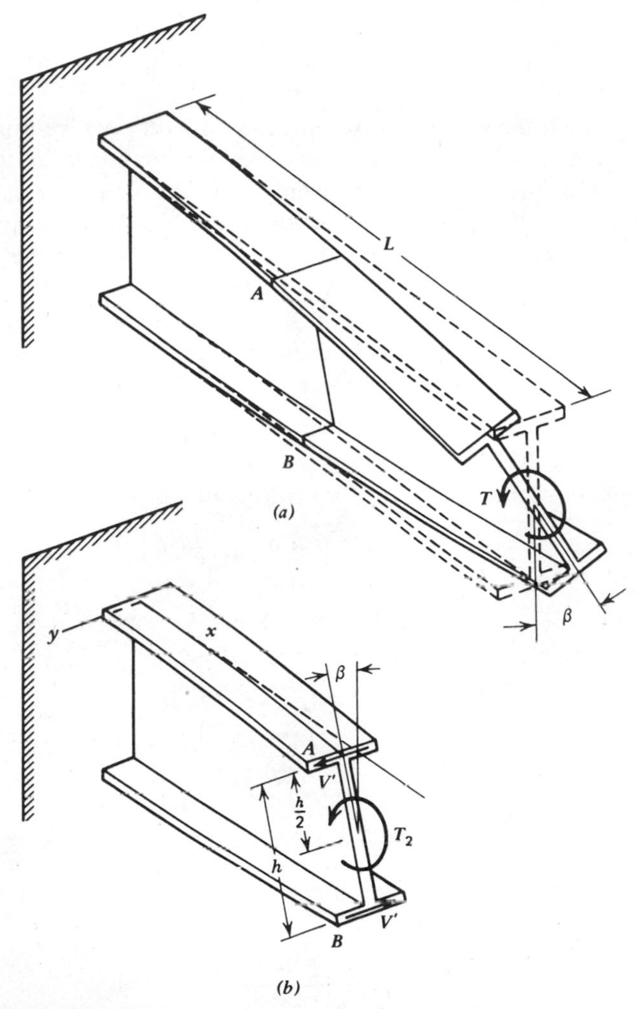

Fig. 5-7.3/Effect of twisting moment applied to an I-section torsion member with one end fixed.

The values of T_1 and T_2 are unknown since the values of V' and θ at any section are not known. Values of these quantities must be found before the lateral bending stresses in the flanges or the torsional shearing stresses in the I-section can be computed. For this purpose two equations are needed. From the condition of equilibrium, one of these equations is

$$T_1 + T_2 = T$$

which by Eqs. (5-7.1) and (5-7.2) may be written

$$V'h + JG\theta = T \tag{5-7.3}$$

For the additional equation, we may use the elastic curve equation for bending in the lateral direction of the upper flange in Fig. 5-7.3*b*, which is

$$\frac{EI}{2}\frac{d^2y}{dx^2} = -M \tag{5-7.4}$$

in which the x- and y-axes are chosen with positive directions as shown in Fig. 5-7.3; M is the lateral bending moment in the flange at any section, producing lateral bending in the flange; I is the moment of inertia of the entire cross section of the beam with respect to the axis of symmetry in the web, so that $\frac{1}{2}I$ closely approximates the value of the moment of inertia of a flange cross section. But Eq. (5-7.4) does not contain either of the desired quantities V' and θ. These quantities are introduced into Eq. (5-7.4) as follows: In Fig. 5-7.3*b* the deflection of the flange at section AB is

$$y = \frac{h}{2}\beta \tag{5-7.5}$$

Differentiation of Eq. (5-7.5) twice with respect to x gives

$$\frac{d^2y}{dx^2} = \frac{h}{2}\frac{d^2\beta}{dx^2} \tag{5-7.6}$$

and, since $d\beta/dx = \theta$, Eq. (5-7.6) may be written

$$\frac{d^2y}{dx^2} = \frac{h}{2}\frac{d\theta}{dx} \tag{5-7.7}$$

The substitution of this value of d^2y/dx^2 into Eq. (5-7.4) gives

$$\frac{EIh}{4}\frac{d\theta}{dx} = -M \tag{5-7.8}$$

In order to introduce V' into Eq. (5-7.8), use is made of the fact that $dM/dx = V'$. Thus, by differentiating both sides of Eq. (5-7.8) with respect to x, we obtain

$$\frac{EIh}{4}\frac{d^2\theta}{dx^2} = -V' \tag{5-7.9}$$

Equations (5-7.3) and (5-7.9) are simultaneous equations in V' and θ.

The value of V' obtained from Eq. (5-7.9) is substituted into Eq. (5-7.3), which then becomes

$$-\frac{EIh^2}{4JG}\frac{d^2\theta}{dx^2}+\theta=\frac{T}{JG} \tag{5-7.10}$$

For convenience let

$$\alpha=\frac{h}{2}\sqrt{\frac{EI}{JG}} \tag{5-7.11}$$

so that Eq. (5-7.10) may be written

$$-\alpha^2\frac{d^2\theta}{dx^2}+\theta=\frac{T}{JG} \tag{5-7.12}$$

The solution of this equation, as given in textbooks on differential equations, is

$$\theta=Ae^{x/\alpha}+Be^{-x/\alpha}+\frac{T}{JG} \tag{5-7.13}$$

Two known pairs of values of x and θ or its derivative are needed to determine values of the constants A and B in Eq. (5-7.13). At the fixed end where $x=0$, $\theta=d\beta/dx=(2/h)(dy/dx)=0$ (since the slope is zero). At the free end where $x=L$, $d\theta/dx=0$, (see Eq. 5-7.8) since at the free end the bending moment M in the flange is zero. The values of A and B are determined from these two conditions and are substituted in Eq. (5-7.13), which gives the angle of twist per unit length:

$$\theta=\frac{T}{JG}\left[1-\frac{\cosh(L-x)/\alpha}{\cosh(L/\alpha)}\right] \tag{5-7.14}$$

The total angle of twist at the free end is

$$\beta=\int_0^L\theta\,dx=\frac{T}{JG}\left(L-\alpha\tanh\frac{L}{\alpha}\right) \tag{5-7.15}$$

The twisting moment T_2 at any section of the beam is obtained by substitution of the value of θ from Eq. (5-7.14) into Eq. (5-7.2). Thus

$$T_2=T\left[1-\frac{\cosh(L-x)/\alpha}{\cosh(L/\alpha)}\right] \tag{5-7.16}$$

The torsional shearing stresses at any section are computed by substituting this value of T_2 into Eq. (5-5.10). The lateral bending moment M in the flanges of the beam at any section is obtained by substituting $d\theta/dx$ from Eq. (5-7.14) into Eq. (5-7.8), which gives

$$M=-\frac{T}{h}\alpha\frac{\sinh(L-x)/\alpha}{\cosh(L/\alpha)} \tag{5-7.17}$$

Note that Eq. (5-7.17) shows that the maximum value of M occurs at the fixed end, for $x = 0$, and is

$$M_{\max} = \frac{T}{h} \alpha \tanh \frac{L}{\alpha} \tag{5-7.18}$$

For all except relatively short beams the length L is large as compared with the value of α, and the value of $\tanh(L/\alpha)$ is approximately equal to 1 when $L/\alpha > 2.5$. In Eqs. (5-7.15) and (5-7.18) the substitution of $\tanh(L/\alpha) = 1$ gives

$$\beta = T(L - \alpha)/JG \tag{5-7.19}$$

$$M_{\max} = T\alpha/h \tag{5-7.20}$$

These approximate values of β and $M_{\max}$ obtained from Eqs. (5-7.19) and (5-7.20) lead to the following procedure for solving for the angle of twist and the maximum longitudinal stresses resulting from a twisting moment in an I-section torsion member with one section restrained from warping. Let Fig. 5-7.4*b* represent an I-section torsion member that is fixed at one end and loaded at the free end by the twisting moment T.

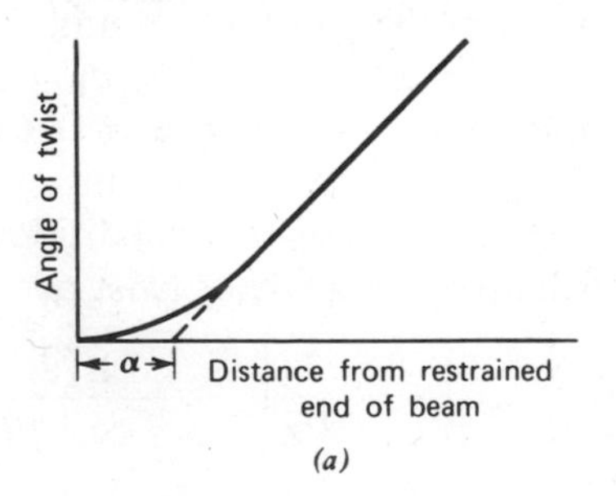

(a)

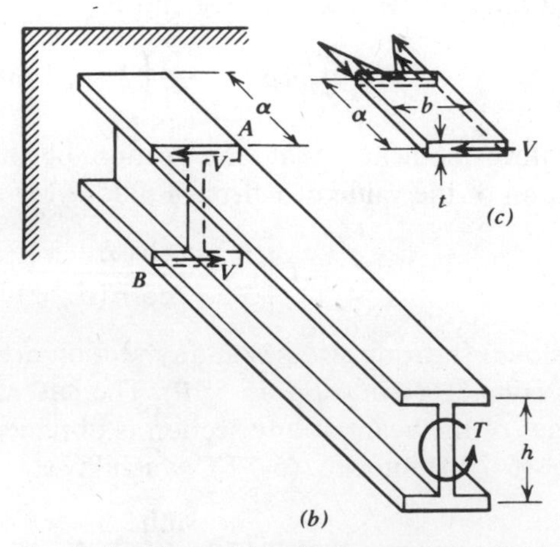

(b)

Fig. 5-7.4/Lateral bending stresses at fixed section in a flange of an I-section torsion member.

Figure 5-7.4*a* represents a typical curve showing the relation between the angle of twist of the I-section torsion member and the distance from the fixed section of the beam. In Fig. 5-7.4*a* the distance from the fixed section to the section AB at which the straight-line portion of the curve intersects the horizontal axis is very nearly equal to the distance α as given by Eq. (5-7.11). Thus, from this fact and from Eq. (5-7.19) the length $L-\alpha$ of the beam between the free end and section AB may be considered as being twisted under pure torsion for the purpose of computing the angle of twist. From Eq. (5-7.20) the sections of the beam within the length α from the fixed section to section AB may be considered as transmitting the entire twisting moment T by means of the lateral shears V in the flanges. Therefore

$$T = Vh \quad \text{or} \quad V = T/h \tag{5-7.21}$$

The force V causes each flange of length α to bend laterally, producing a longitudinal stress at each edge of the flange, tensile stress at one edge, and compressive stress at the other (Fig. 5-7.4*c*). Assuming that the lateral bending of each flange has a rectangular cross section, we have at the fixed end

$$\sigma = \frac{M\frac{1}{2}b}{I_f} = \frac{T}{h}\alpha\frac{\frac{1}{2}b}{\frac{1}{12}tb^3} = \frac{6T\alpha}{htb^2} \tag{5-7.22}$$

The value for α is given by Eq. (5-7.11), in which E and G are the tensile and shearing moduli of elasticity, respectively, I is the moment of inertia of the entire section with respect to a centroidal axis parallel to the web, and J is an equivalent polar moment of inertia of the section. Values of α calculated by this equation come close to values obtained from actual tests. For a section made up of slender, approximately rectangular areas, such as a rolled-steel channel, angle, or I-section, J is given by Eq. (5-5.9) if it is noted that b_i in this article replaces $2b_i$ in Eq. (5-5.9) and t_i replaces $2h_i$ in Eq. (5-5.9). All equations in this article have been derived for I-sections, but they apply as well to channels or z-sections.

Various Loads and Supports for Beams in Torsion/The solution of Eq. (5-7.12) given by Eq. (5-7.13) is for the particular beam shown in Fig. 5-7.3. However, solutions of the equation have been obtained for beams loaded and supported as shown in Figs. A, B, C, and D in Table 5-7.1 by arranging the particular solution of the differential equation to suit the conditions of loading and support for each beam. The values of the maximum lateral bending moment M_{max} given in Table 5-7.1 may be used in Eq. (5-7.22) to compute the maximum lateral bending stress in the beam. The formulas in Table 5-7.1 where I-sections are shown may also be used for channels or Z-bars.

Table 5-7.1

Beams Subjected to Torsion

Type of Loading and Support	*Maximum Lateral Bending Moment in Flange*	*Angle of Twist of Beam of Length L*
Fig. A	$M_{\max}=\frac{T\alpha}{h}\tanh\frac{L}{2\alpha}$ $=\frac{T\alpha}{h}$, if $\frac{L}{2\alpha}>2.5$	$\theta=\frac{T}{JG}\left(L-2\alpha\tanh\frac{L}{2\alpha}\right)$ $=\frac{T}{JG}(L-2\alpha)$ Error is small if $\frac{L}{2\alpha}>2.5$
Fig. B	$M_{\max}=\frac{T\alpha}{2h}\left(\coth\frac{L}{2\alpha}-\frac{2\alpha}{L}\right)$ $=\frac{T\alpha}{h}$, if $\frac{L}{2\alpha}$ is large	$\theta=\frac{T}{2JG}\left(\frac{L}{4}-\alpha\tanh\frac{L}{4\alpha}\right)$ $=\frac{T}{2JG}\left(\frac{L}{4}-\alpha\right)$ Error is small if $\frac{L}{4\alpha}>2.5$
Fig. C	$M_{\max}=\frac{T\alpha}{h}\left(\coth\frac{L}{\alpha}-\frac{\alpha}{L}\right)$ $=\frac{T\alpha}{h}$, if $\frac{L}{\alpha}$ is large	$\theta=\frac{T}{JG}\left(\frac{L}{2}-\alpha\tanh\frac{L}{2\alpha}\right)$ $=\frac{T}{JG}\left(\frac{L}{2}-\alpha\right)$ Error is small if $\frac{L}{2\alpha}>2.5$
Fig. D	$M_{\max}=\frac{T\alpha}{h}\,\frac{\sinh\frac{L_1}{\alpha}\sinh\frac{L_2}{\alpha}}{\sinh\frac{L}{\alpha}}$ $=\frac{T\alpha}{2h}$, if $\frac{L_1}{\alpha}$ and $\frac{L_2}{\alpha}>2$ Error is small	Approximate value $\theta=\frac{1}{2}\frac{T}{JG}\left(\frac{L}{2}-\alpha\tanh\frac{L}{2\alpha}\right)$ $=\frac{1}{2}\frac{T}{JG}\left(\frac{L}{2}-\alpha\right)$ Error is small if $\frac{L}{2\alpha}>2.5$

PROBLEM SET 5-7

1. A wide-flange steel ($E=200$ GPa and $G=77.5$ GPa) I-beam has a depth of 300 mm, a web thickness of 15 mm, a flange width of 270 mm, a flange thickness of 20 mm, and a length of 8.00 m. The I-beam is fixed at one end and is free at the other end. A twisting moment

$T = 7.00$ kN · m is applied at the free end. Determine the maximum normal stress and maximum shearing stress in the I-beam and the angle of rotation β of the free end of the I-beam.

2. The I-beam in Fig. P5-7.2 is an aluminum alloy ($E = 72.0$ GPa and $G = 27.1$ GPa) extruded section. It is fixed at the wall, and it is attached rigidly to the thick massive plate at the other end. Determine the magnitude of P for $\sigma_{max} = 160$ MPa.

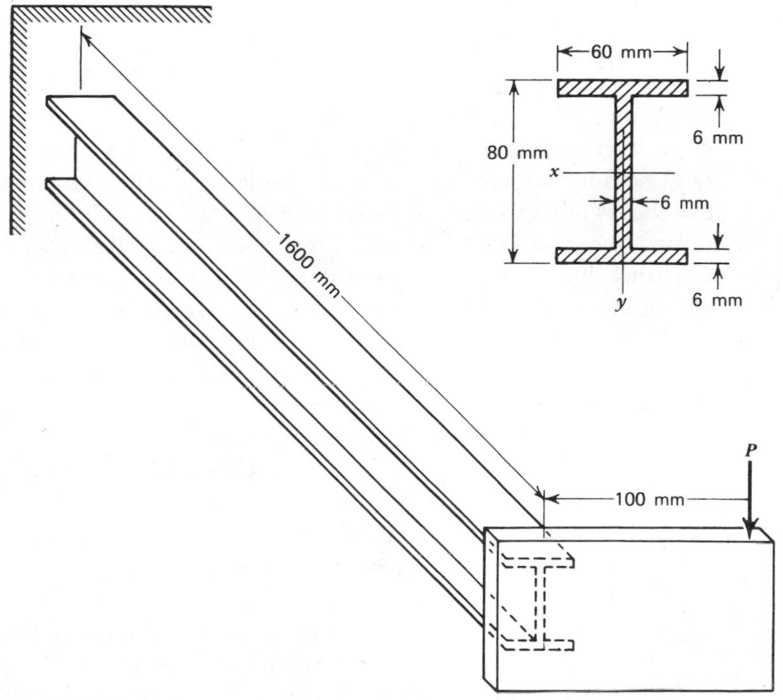

Fig. P5-7.2

Ans. $P = 1.069$ kN

3. Let the thick plate in Problem 2 be subjected to a torque $T = 150$ N · m directed along the axis of the I-beam. Determine the maximum shearing stress and the angle of twist β of the plate.

5-8 FULLY-PLASTIC TORSION

Consider a torsion member made of an elastic-perfectly plastic material, that is, one whose shearing stress-shearing strain diagram is flat topped at the shearing yield stress τ_Y. As the torque is gradually increased, yielding starts at one or more places on the boundary of the cross section and spreads inward with increasing torque. Finally, the entire cross section becomes plastic at the limiting, fully plastic, torque. The torsion analysis for the limiting torque or fully plastic torque is considered in this article.

Equations (5-2.8) and (5-2.14) are valid for both the elastic and plastic regions of each cross section of a torsion member. At the fully plastic torque, the resultant shearing stress is $\tau = \tau_Y$ at every point in the cross section. Thus, Eqs. (5-2.8) and (5-2.14) give

$$\sigma_{zx}^2 + \sigma_{zy}^2 = \left(\frac{\partial \phi}{\partial y}\right)^2 + \left(\frac{\partial \phi}{\partial x}\right)^2 = \tau_Y^2 \tag{5-8.1}$$

Equation (5-8.1) uniquely determines the stress function $\phi(x, y)$ for a given torsion member for fully plastic conditions. Since the unit angle of twist does not appear in Eq. (5-8.1), the deformation (twist) of the torsion member is not specified at the fully plastic torque.

Now we consider a procedure by which Eq. (5-8.1) may be used to construct the stress function surface for the cross section of a given torsion member at fully plastic torque. Equation (5-2.13) indicates that $\phi = 0$ on the boundary, and Eq. (5-8.1) indicates that the absolute value of the maximum slope of ϕ everywhere in the cross section is a constant equal to τ_Y; therefore, the magnitude of ϕ at a point is equal to τ_Y times its distance from the nearest boundary, measured along the perpendicular from the point to the nearest boundary. The contour curves of constant ϕ are perpendicular to the direction of maximum slope and, hence, are parallel to the nearest boundary.

Consider the problem of constructing the stress function ϕ for a square cross section with sides $2a$, as indicated in Fig. 5-8.1. At a given point P the resultant shearing stress is τ_Y and is directed along a contour curve of constant ϕ; the elevation of the stress function at point P is equal to τ_Y times its perpendicular distance to the nearest boundary. The stress

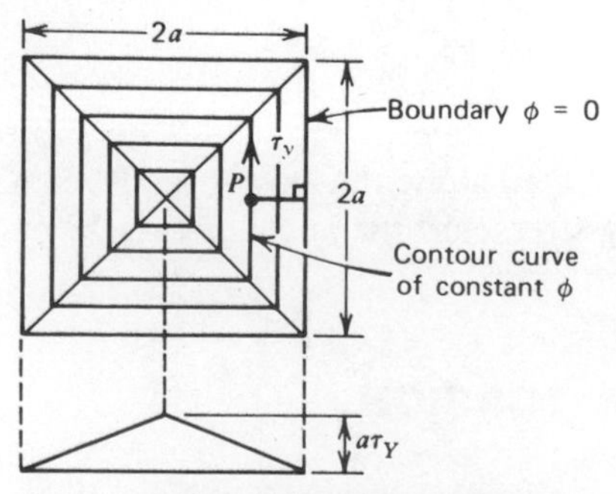

Fig. 5-8.1/Stress function surface for a fully plastic square cross section.

function ϕ for the square cross section is a pyramid of height $\tau_Y a$; this condition suggested to Nadai[5] the so-called *sand-heap analogy* because sand poured on a flat plate with the same dimensions as the cross section of the torsion member tends to form a pyramid similar to that indicated in Fig. 5-8.1.

The fully plastic torque T_P for the square cross section can be obtained by means of Eq. (5-2.21), which indicates that the torque is equal to twice the volume under the stress function. Since the volume of a pyramid is equal to one-third of the area of the base times the height, we have

$$T_P = 2\left[\tfrac{1}{3}(2a)^2 \tau_Y a\right] = \tfrac{8}{3}\tau_Y a^3 \tag{5-8.2}$$

The fully plastic torques for a few common cross sections are listed in Table 5-8.1 and are compared with the maximum elastic solutions for these cross sections. We see by Table 5-8.1 that the load carrying capacity for the cross sections considered are greatly increased when we make it possible for yielding to spread throughout the cross sections.

Expressions for the fully plastic torques for a number of common structural sections have been derived (Table 5-8.2). We remark that the expressions in Table 5-8.2 are not exact. When the cross section has a reentrant corner as indicated in Fig. 5-8.2, the correct stress surface is as shown in Fig. 5-8.2*a* and not as in Fig. 5-8.2*b*. Since the expressions in Table 5-8.2 are based on the assumption that Fig. 5-8.2*b* is correct, these expressions are not exact for sections with reentrant corners. Finally, we note that the expressions for T_P in the second column of Table 5-8.2 hold for each of the two cross sections in the first column. Since the cross sections in Table 5-8.2 are made up of long narrow rectangular sections, the error in the expressions in Table 5-8.2 are small; furthermore, the error makes the expressions conservative. The maximum elastic torques are not given in Table 5-8.2 because of the influence on initial yielding of the high stress concentration factors at reentrant corners. In summary, Table 5-8.2 has limited applicability to the design of such sections in an actual structure for two reasons. First, failure by buckling, at least for thin sections, is likely to be the basis for design. Second, torsion of such

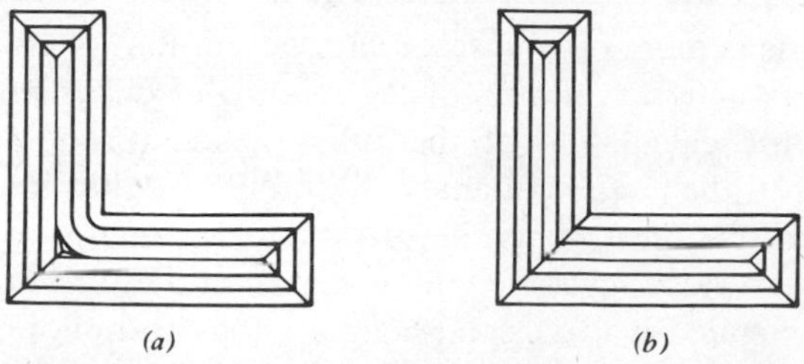

Fig. 5-8.2/Stress function surfaces for a fully plastic angle section. (*a*) Correct plastic-stress surface. (*b*) Incorrect plastic-stress surface.

Table 5-8.1

θ_Y, T_Y, T_P and T_P/T_Y for Five Common Cross Sections

Section	Maximum Elastic Torque T_Y and Unit Angle of Twist θ_Y	Fully Plastic Torque T_P	Ratio $\frac{T_P}{T_Y}$
Square 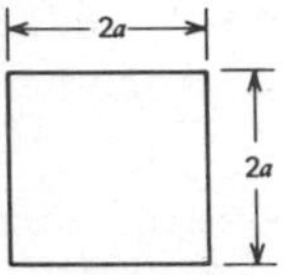	$T_Y = 1.664\tau_Y a^3$ $\theta_Y = \frac{1.475\tau_Y}{2Ga}$	$\frac{8}{3}\tau_Y a^3$	1.605
Rectangle 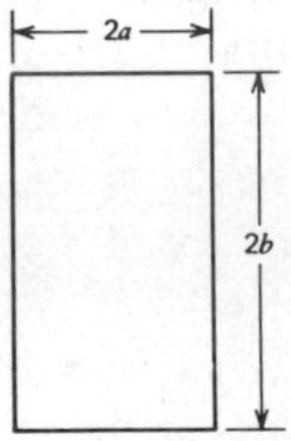			
$\frac{b}{a} = 2$	$T_Y = 3.936\tau_Y a^3$ $\theta_Y = \frac{1.074\tau_Y}{2Ga}$	$\frac{20}{3}\tau_Y a^3$	1.69
$\frac{b}{a} = \infty$	$T_Y = \frac{8}{3}\tau_Y ba^2$ $\theta_Y = \frac{\tau_Y}{2Ga}$	$4\tau_Y ba^2$	1.50
Equilateral triangle 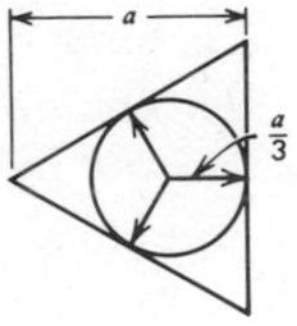	$T_Y = \frac{2}{15\sqrt{3}}\tau_Y a^3$ $\theta_Y = \frac{2\tau_Y}{Ga}$	$\frac{2\sqrt{3}}{27}\tau_Y a^3$	1.67
Circle 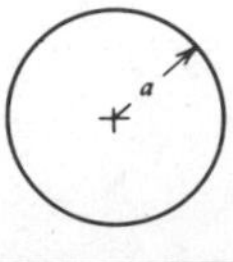	$T_Y = \frac{\pi}{2}\tau_Y a^3$ $\theta_Y = \frac{\tau_Y}{Ga}$	$\frac{2}{3}\pi\tau_Y a^3$	1.33

sections is more often a secondary action that is usually accompanied by primary action, which produces bending or direct stresses.

In the calculation of the fully plastic torque for a hollow torsion member, the method of analysis is similar to that for elastic torsion of the hollow torsion member, since the stress function $\phi(x, y)$ is flat-topped (has zero slope) over the hollow region of the torsion member. Sadowsky[6] has extended Nadai's sand-heap analogy to hollow torsion members. In order to simplify the analysis, only hollow torsion members of constant wall thickness are considered. For such torsion members, the fully plastic torque T_P is obtained by subtracting from the fully plastic torque T_{PS} of a solid torsion member having the boundary of the outer cross section,

Table 5-8.2

T_P for Common Structural Sections

Section	*Fully Plastic Torque* T_P
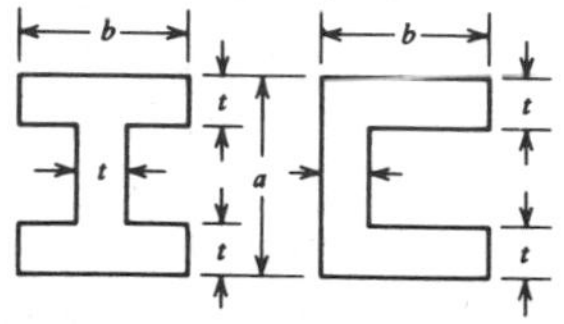 b b t t a t t t	$\tau_Y t^2\left[\frac{a}{2}+b-\frac{7}{6}t\right]$
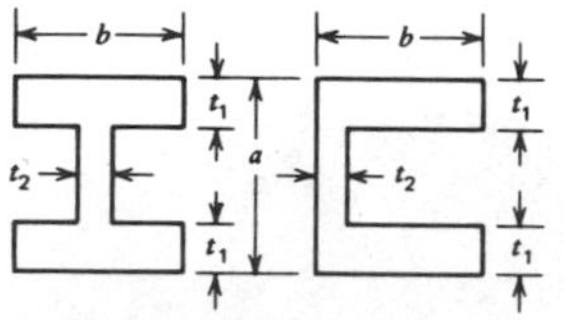 b b t_1 t_1 a t_2 t_2 t_1 t_1	$\tau_Y\left[t_1^2\left(b-\frac{t_1}{3}\right)+\frac{t_2^2}{2}\left(a+\frac{t_2}{3}\right)-t_1t_2^2\right]$
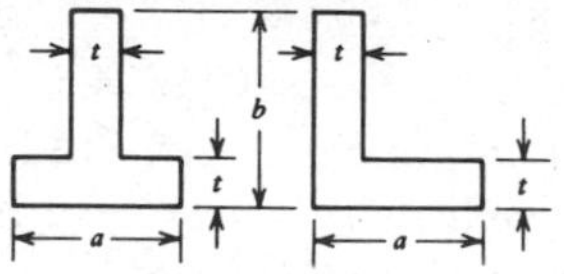 t t b t t a a	$\frac{\tau_Y t^2}{2}\left[a+b-\frac{4}{3}t\right]$
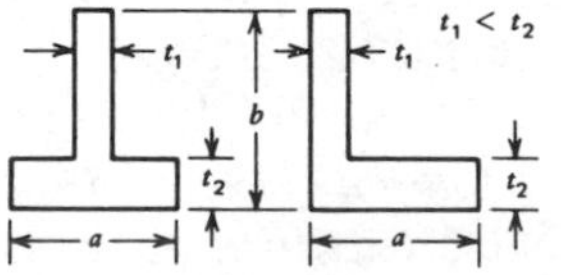 $t_1 < t_2$ t_1 t_1 b t_2 t_2 a a	$\frac{\tau_Y}{2}\left[at_2^2+bt_1^2-\frac{t_2^3}{3}-t_1^2t_2\right]$
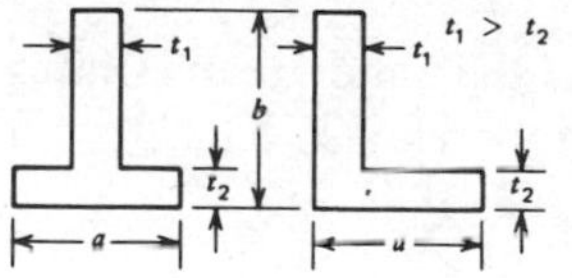 $t_1 > t_2$ t_1 t_1 b t_2 t_2 a u	$\frac{\tau_Y}{2}\left[at_2^2+bt_1^2-\frac{t_1^3}{3}-t_2^2t_1\right]$

the fully plastic torque T_{PH} of a solid torsion member having the cross section identical to the hollow region. That is, for such members

$$T_P = T_{PS} - T_{PH} \tag{5-8.3}$$

PROBLEM SET 5-8

1. Derive the relation for the fully plastic torque for a rectangular cross section having dimensions $2a$ by $2b$.

2. Derive the relation for the fully plastic torque for the cross section in the first row of Table 5-8.2.

3. A rectangular section torsion member has dimensions of 100 mm by 150 mm and is made of a steel for which the shearing yield point is $\tau_Y = 100.0$ MPa. Determine T_P for the cross section and the ratio of T_P to T_Y where T_Y is the maximum elastic torque.

4. A rectangular hollow torsion member has external dimensions of 200 mm by 400 mm. The cross section has a uniform thickness of 30 mm. For a material that has a shearing yield stress $\tau_Y = 120$ MPa, determine the fully plastic torque.

   Ans. $T_P = 455.0 \text{ kN} \cdot \text{m}$

REFERENCES

1. A. P. Boresi and P. P. Lynn, *Elasticity in Engineering Mechanics*, 2nd Ed., Prentice-Hall, Englewood Cliffs, New Jersey, 1974.
2. R. L. Thoms and F. D. Masch, "Membrane Analogy Studies Employing Visible Contour Lines," *Developments in Theoretical and Applied Mechanics*, Vol. 2, Pergamon Press, New York, 1965.
3. T. von Karmon and W. Z. Chien, "Torsion with Variable Twist," *Journal of Aeronautical Sciences*, Vol. 13, No. 10, 1946, pp. 503–510.
4. F. A. Smith, Jr., F. M. Thomas, and J. O. Smith, "Torsional Analysis of Heavy Box Beams in Structures," *Journal of the Structural Division*, ASCE, Vol. 96, No. ST3, Proc. Paper 7165, 1970, pp. 613–635.
5. A. Nadai, *Theory of Flow and Fracture of Solids*, Vol. 1, McGraw-Hill, New York, 1950.
6. M. A. Sadowsky, "An Extension of the Sand Heap Analogy in Plastic Torsion Applicable to Cross-Sections Having One or More Holes," *Journal of Applied Mechanics*, Vol. 62, 1941.

Additional References

1. C. F. Kollbrunner, *Torsion in Structures: An Engineering Approach*, Springer-Verlag, New York, 1969.
2. S. Timoshenko, "Theory of Bending, Torsion and Buckling of Thin-Walled Members of Open Cross Section," *Journal of the Franklin Institute*, Vol. 239, No. 3, March 1945, pp. 201–219.

CHAPTER 6

NONSYMMETRICAL BENDING OF STRAIGHT BEAMS

6-1 DEFINITION OF SHEAR CENTER IN BENDING. SYMMETRICAL AND NONSYMMETRICAL BENDING

The straight cantilever beam shown in Fig. 6-1.1 has a cross section of arbitrary shape. It is subjected to pure bending by the end couple $\mathbf{M}_0$. Let the origin 0 of the coordinate system (x, y, z) be chosen at the centroid of the beam cross section at the left end of the beam, with the z-axis directed along the centroidal axis of the beam, and the (x, y)-axes taken in the plane of the cross section. Generally, the orientation of the (x, y)-axis is arbitrary. However, we often choose the (x, y)-axes so that the moments of inertia of the cross section I_x, I_y, and I_{xy} are easily calculated, or we may take them to be principal axes (see the Appendix).

The bending moment, which acts at the left end of the beam (Fig. 6-1.1a), is represented by the vector $\mathbf{M}_0$ directed perpendicular to a plane that forms an angle ϕ $(0 \leq \phi < \pi)$ taken positive when measured counterclockwise from the x-z-plane as viewed from the positive z-axis. This plane is called *the plane of load* or *the plane of loads*. A more complete description of the plane of loads is given later on in this Article. Consider now a cross section of the beam at distance z from the left end. The free-body diagram of the part of the beam to the left of this section is shown in Fig. 6-1.1b. For equilibrium of this part of the beam, a moment $\mathbf{M}$, equal in magnitude but opposite in sense to $\mathbf{M}_0$ must act at section z. For the case shown $(\pi/2 < \phi < \pi)$, the (x, y)-components (M_x, M_y) of $\mathbf{M}$ are related to the signed magnitude M of $\mathbf{M}$ by the relations $M_x = M \sin\phi$, $M_y = -M\cos\phi$. Since $\pi/2 < \phi < \pi$, $\sin\phi$ is positive and $\cos\phi$ is negative. Since (M_x, M_y) are positive (Fig. 6-1.1b), the sign of M is positive. A more complete discussion of the sign convention for M is given in Art. 6-2, following Eq. (6-2.10).

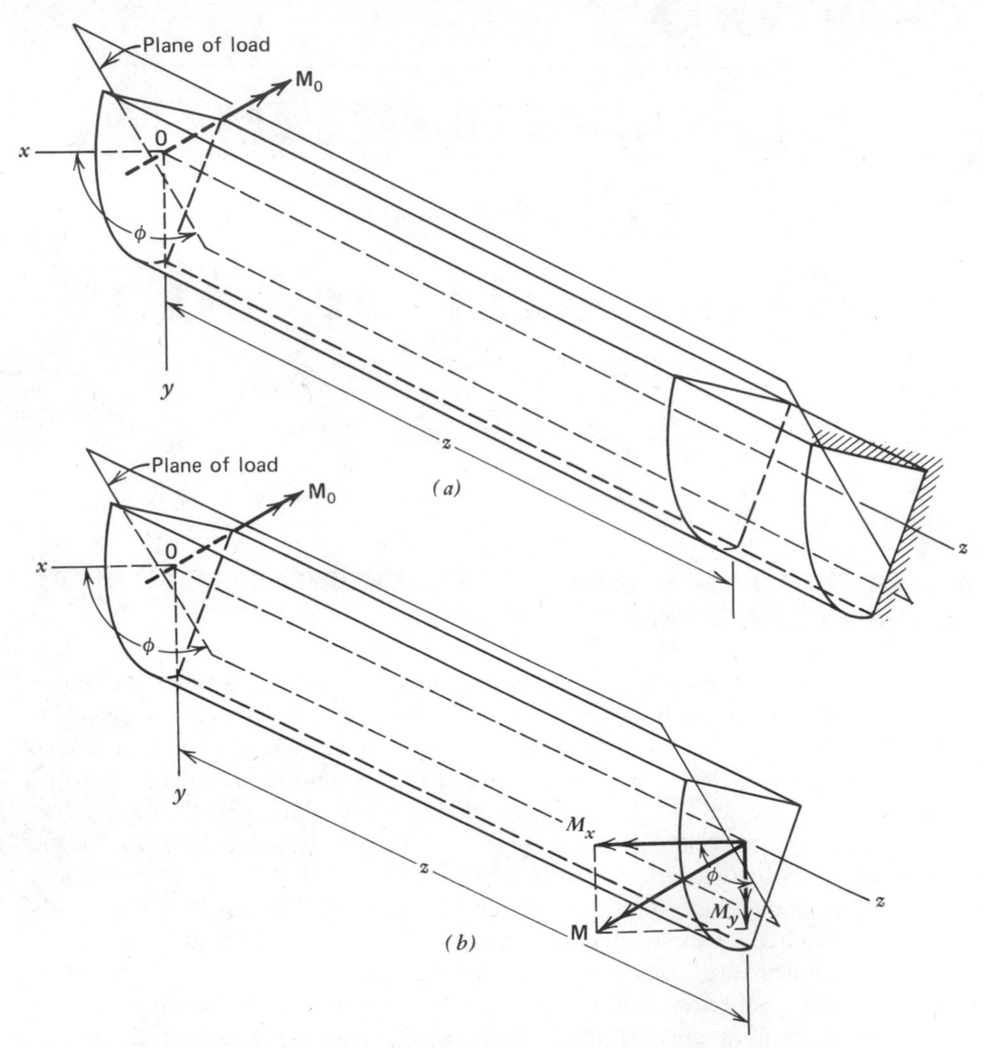

Fig. 6-1.1/Cantilever beam with an arbitrary cross section subjected to pure bending.

Shear Loading of a Beam. Shear Center Defined/Let the beam shown in Fig. 6-1.2*a* be subjected to a concentrated force **P** that lies in the end plane ($z = 0$) of the beam cross section. The vector representing **P** lies in a plane which forms angle ϕ ($0 \le \phi \le \pi$), taken positive when measured counterclockwise from the *z*-*x*-plane as viewed from the positive *z*-axis. This plane is called *the plane of the load*. Consider a cross section of the beam at distance *z* from the left end. The free body diagram of the part of the beam to the left of this section is shown in Fig. 6-1.2*b*. For equilibrium of this part of the beam, a moment **M**, with components M_x and M_y, shear components V_x and V_y, and in general, a twisting moment **T** (with vector directed along the positive *z*-axis) must act on the section at z. However, if the line of action of force **P** passes through a certain point C (the shear center) in the cross section, $\mathbf{T} = 0$. In this discussion we assume that the line of action of **P** passes through the shear center. Hence **T** is not shown in Fig. 6-1.2*b*. Note that in Fig. 6-1.2*b*, the force **P** requires V_x, V_y to be positive (directed along positive (x, y)-axes, respectively). The component M_x is also directed along the positive x-axis. However, since $\phi < \pi/2$, M_y is negative (directed along the negative y-axis).

There is a particular axial line in the beam called the *bending axis of the beam*, which is parallel to the centroidal axis of the beam (the line

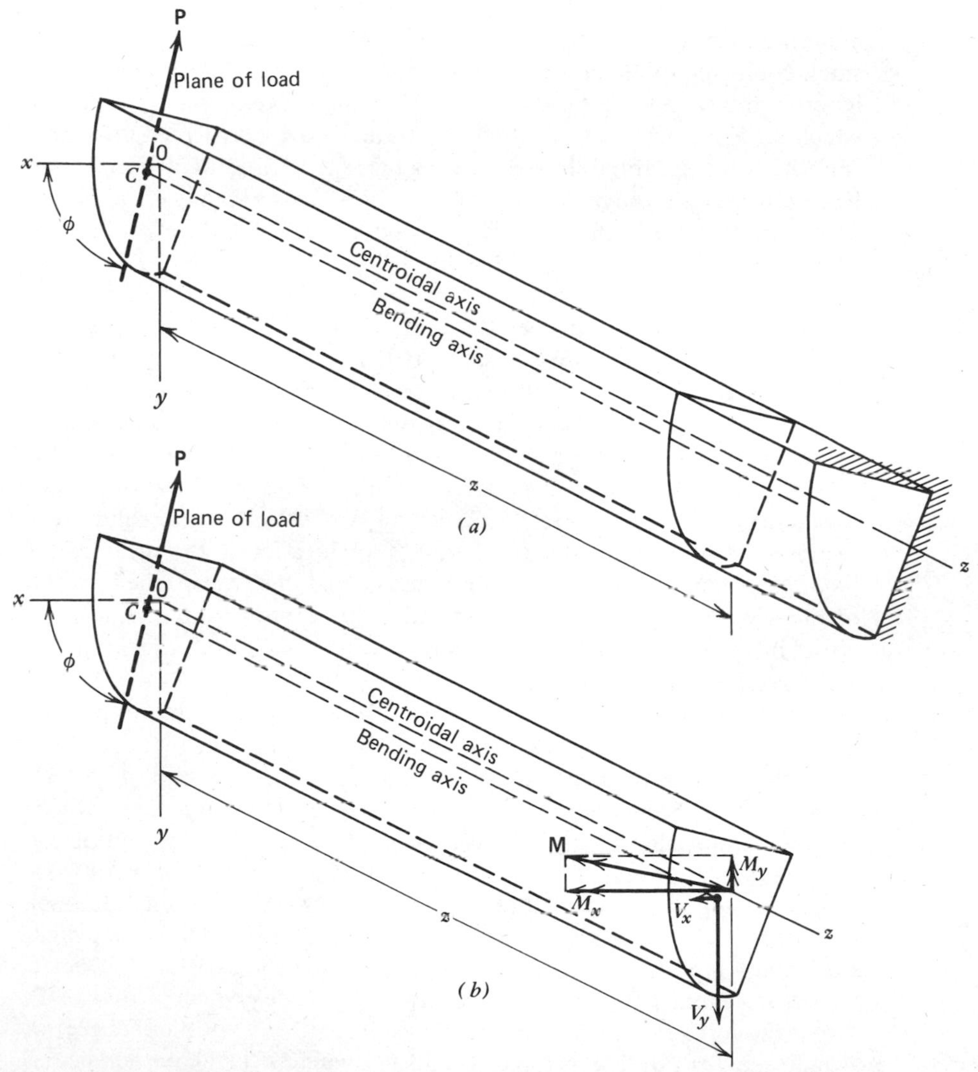

Fig. 6-1.2/Cantilever beam with an arbitrary cross section subjected to shear loading.

which passes through the centroids of all of the cross sections of the beam). Except for special cases, the bending axis does not coincide with the centroidal axis (Fig. 6-1.2).

The intersection of the bending axis with any cross section of the beam locates a point C in that cross section called *the shear center* of the cross section (see Art. 7-1). Thus, the bending axis passes through the shear centers of all the cross sections of the beam.

In Art. 6-2, formulas are derived for the normal stress component σ_{zz} that acts on the cross section at z in terms of the bending moment components (M_x, M_y). Also, one may derive formulas for the shearing stress components (τ_{zx}, τ_{zy}) due to the shearing forces (V_x, V_y). However, if the length L of the beam is large compared to the maximum cross section dimension D, such that $L/D > 5$, the maximum shearing stress is small compared to the maximum normal stress. Hence, in this chapter we ignore the shearing stresses due to (V_x, V_y); that is, we consider beams for which $L/D > 5$. Thus for bending of a beam by a concentrated force and for which the shearing stresses are negligible, the line of action of the force must pass through the shear center of a cross section of the beam; otherwise, the beam will be subjected to both bending and torsion (twist). Thus, the theory of pure bending of beams assures that the shearing stresses due to concentrated loads are negligible and that the lines of actions of concentrated forces that act on the beam pass through the shear center of a beam cross section. If the cross section of a beam has either an axis of symmetry or an axis of antisymmetry, it may be shown that the shear center C is located on that axis (Fig. 6-1.3). If the cross section has two or more axes of symmetry or antisymmetry, the shear center is located at the intersection of the axes (Figs. 6-1.3*a* and *d*). For a general cross section (Fig. 6-1.1) or for a relatively thick, solid cross section (Fig. 6-1.3*c*), the determination of the location of the shear center requires advanced computational methods.[1] For this reason, the location of the shear center is often determined in an approximate manner; the errors introduced by such approximations of the shear center location are discussed in the next paragraph.

Let the line of action of force **P**, Fig. 6-1.2, pass through an approximate location of the shear center of the beam, point B in Fig. 6-1.4. Let C be the location of the shear center. Since the line of action of force **P** does not pass through C, the force **P** is assumed to be replaceable by a couple (torque) that lies in the cross section and a force with an action line that passes through C. This representation or transformation of force **P** is assumed to be valid for the deformable beam cross section, although strictly speaking, it is applicable to rigid bodies only.[2] The transformation is accomplished by adding self-equilibrating forces **P**′ and **P**″ at C which are parallel to **P** and which have magnitudes equal to that of **P**. Thus, the force **P** is considered to be equivalent to a torque (couple) of magnitude $T = Pd$, due to forces **P** and **P**″, where d is the perpendicular distance between **P** and **P**″ and a force **P**′ acting at C.

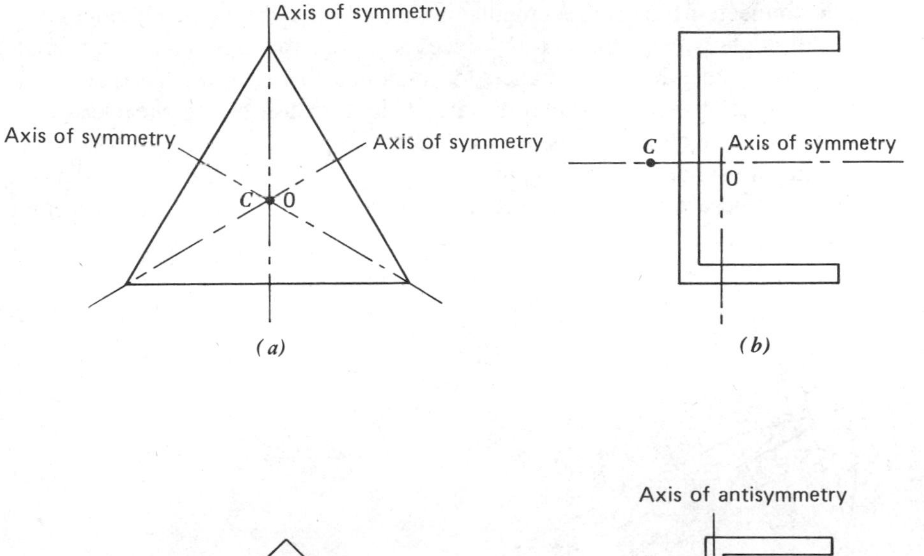

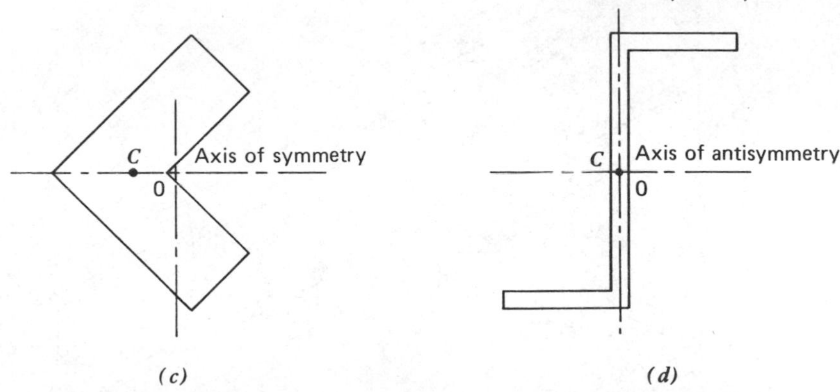

Fig. 6-1.3/(*a*) Equilateral triangle section. (*b*) Open channel section. (*c*) Angle section. (*d*) Z-section.

Now pass a cutting plane through the member at distance z from the left end. The free body diagram of the beam to the left of the cut is shown in Fig. 6-1.4*b*. For equilibrium, the forces at the cut include a bending moment **M** with components M_x and M_y, torque of magnitude $T = Pd$, and shears V_x and V_y. The normal stress distribution σ_{zz} due to M_x and M_y can be calculated by the formulas derived in Art. 6-2. The shearing stresses due to V_x and V_y are considered to be negligible ($L/D > 5$). The shearing stress due to torque T may be computed by the methods presented in Chapter 5. Cross sections with thick walls, Fig. 6-1.3*c*, require large torques if the maximum shearing stress due to the torque is to be significant. For such cross sections, an approximate location of the shear center will suffice, since shearing stresses due to T are small compared to the maximum value of σ_{zz}, provided Pz is large compared to Pd. However, caution must be used for cross sections made of connected narrow rectangular walls such as the open channel cross section shown in Fig. 6-1.3*b*, since, as noted in Chapter 5, such cross sections have little resistance to torsional loads. For these kinds of cross sections, an accurate estimate of the location of the shear center is necessary. Such problems are treated in Chapter 7.

In this chapter, the shear centers for many of the cross sections considered are not known exactly. Consequently, unless the shear center

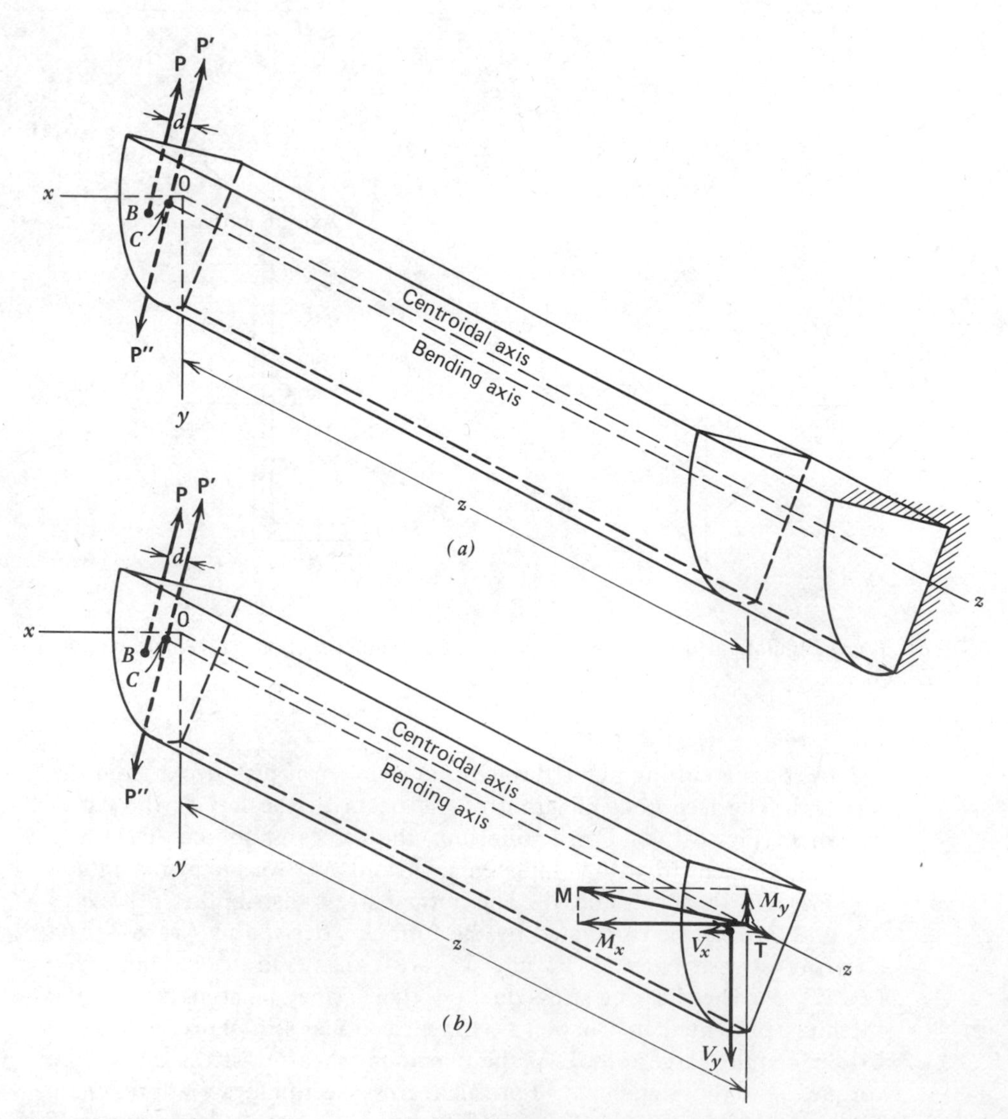

Fig. 6-1.4/Cantilever beam with arbitrary cross section subjected to shear loading not at shear center.

is located by intersecting axes of symmetry or antisymmetry, the location of the shear center is approximated. The reader should have a better understanding of such approximations after studying Chapter 7.

Symmetrical Bending. Nonsymmetrical Bending/In the Appendix, it is shown that every beam cross section has principal axes (X, Y). With respect to principal axes (X, Y), the product of inertia of the cross section is zero; that is, $I_{XY}=0$. The principal axes (X, Y) for the cross section of the cantilever beam of Fig. 6-1.1 are shown in Fig. 6-1.5. For convenience, axes (X, Y) are also shown at the section distance z from the left end of the beam. At the left end, let the beam be subjected to a couple $\mathbf{M}_0$ with sense in the negative X direction and a force $\mathbf{P}$ at the shear center C with sense in the negative Y direction (Fig. 6-1.5*a*). These loads produce a bending moment $\mathbf{M}=\mathbf{M}_X$ at the cut section with sense in the positive X direction. By Bernoulli beam theory,[1] the stress σ_{zz} normal to the cross section is given by the flexure formula

$$\sigma_{zz}=\frac{M_X Y}{I_X} \tag{6-1.1}$$

where Y is the distance from the principal axis X to the point in the cross section at which σ_{zz} acts, and I_X is the principal moment of inertia of the cross-sectional area relative to the X-axis. Equation (6-1.1) shows that σ_{zz} is zero for $Y=0$ (the X-axis). Consequently, the X-axis is called the *neutral axis of bending of the cross section*; that is, the axis for which $\sigma_{zz}=0$. We define the bending moment component M_X as positive when the sense of the vector representing $\mathbf{M}_X$ is in the positive X direction. Since M_X is related to σ_{zz} by Eq. (6-1.1), σ_{zz} is a tensile stress for positive values of Y and a compressive stress for negative values of Y. In addition to causing a bending moment component M_X, load P produces a positive shear V_Y at the cut section. It is assumed that the maximum shearing stress τ_{ZY} resulting from V_Y is small compared to the maximum value of σ_{zz}. Hence, since this chapter treats pure bending effects, we neglect shearing stresses in this chapter.

Likewise, if a load $\mathbf{Q}$ (applied at the shear center C), directed along the positive X-axis, and a moment $\mathbf{M}_0$ directed along the negative Y-axis, are applied to the left end of the beam (Fig. 6-1.5*b*), they produce a bending moment $\mathbf{M}=\mathbf{M}_Y$ directed along the positive Y-axis. The normal stress distribution σ_{zz}, due to the component M_Y, is also given by the flexure formula. Thus,

$$\sigma_{zz}=-\frac{M_Y X}{I_Y} \tag{6-1.2}$$

where X is the distance from the principal axis Y to the point in the cross section at which σ_{zz} acts, and I_Y is the principal moment of inertia of the

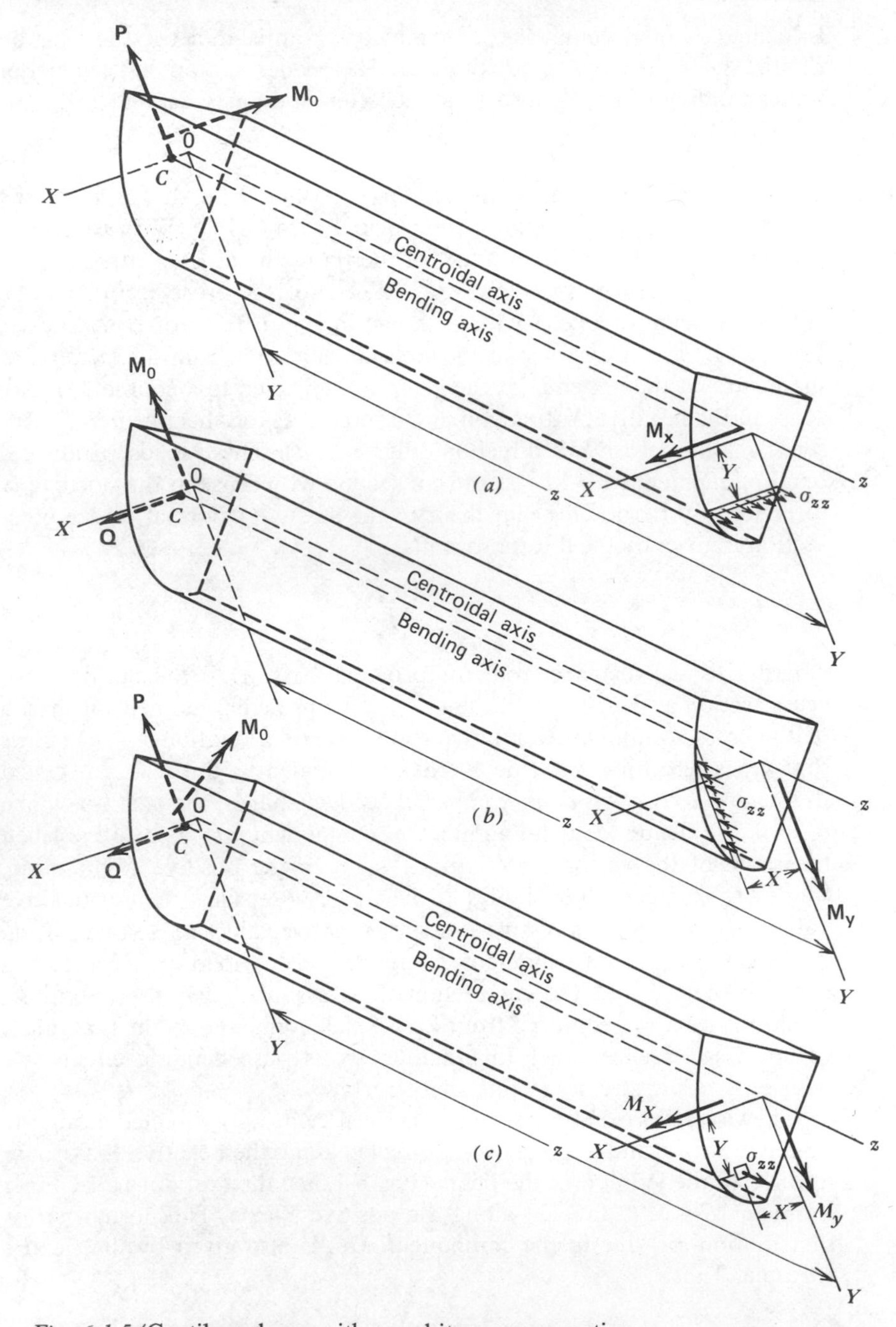

Fig. 6-1.5/Cantilever beam with an arbitrary cross section.

cross-sectional area relative to the Y axis. The negative sign arises from the fact that a positive M_Y produces compressive stresses on the positive side of the X axis. Now for $X=0$ (the Y-axis), $\sigma_{zz}=0$. Hence, in this case, the Y-axis is the *neutral axis of bending of the cross section*; that is, the axis for which $\sigma_{zz}=0$. In either case (Eq. (6-1.1) or (6-1.2)), the beam is subjected to *symmetrical bending*. (Bending occurs about a neutral axis in the cross section that coincides with the corresponding principal axis.)

In Fig. 6-1.5c, the beam is subjected to moment $\mathbf{M}_0$ with components in the negative directions of both axis (X, Y), as well as concentrated forces $\mathbf{P}$ and $\mathbf{Q}$ acting through the shear center C. These loads result in a bending moment $\mathbf{M}$ at the cut section with positive components (M_X, M_Y). For this loading, the stress σ_{zz} normal to the cross section may be obtained by superposition of Eqs. (6-1.1) and (6-1.2). Thus,

$$\sigma_{zz}=\frac{M_X Y}{I_X}-\frac{M_Y X}{I_Y} \tag{6-1.3}$$

In this case, the moment $\mathbf{M}=(M_X, M_Y)$ is not parallel to either of the principal axis (X, Y). Hence, the bending of the beam occurs about an axis that is not parallel to either the X- or Y-axis. When the axis of bending does not coincide with a principal axes direction, the bending of the beam is said to be nonsymmetrical. The determination of the *neutral axis of the cross section* for nonsymmetrical bending is discussed in Art. 6-2.

Plane of Loads. Symmetrical and Nonsymmetrical Loading/Often a beam is loaded by forces that lie in a plane which coincides with a plane of symmetry of the beam, Fig. 6-1.6. Since, then, the y-axis is an axis of symmetry for the cross section, it is a principal axis. Hence, if axes (x, y) are principal axes for the cross section, the beams in Figs. 6-1.6a and b undergo symmetrical bending; that is, bending about a principal axis of a cross section, since the moment vector in Fig. 6-1.6a and the force vectors in Fig. 6-1.6b are parallel to principal axes. (See the discussion above in the section entitled "Symmetrical Bending. Nonsymmetrical Bending.") We further observe that since the shear center lies on the y-axis, the plane of the load contains the axis of bending of the beam. More generally, it is shown later in this chapter that if the plane of loads does not coincide with a plane of symmetry of the beam, the beam may still deform symmetrically (bend about a principal axis), provided that the plane of loads contains the bending axis and is parallel to one of the principal planes [the (x, z)- and (y, z)-planes in Fig. 6-1.6].

Consider next two beams with cross sections shown in Fig. 6-1.7. Since a rectangular cross section (Fig. 6-1.7a) has two axes of symmetry that pass through its centroid 0, the shear center C of a rectangular cross section (which is located at the intersection of the two axes of symmetry)

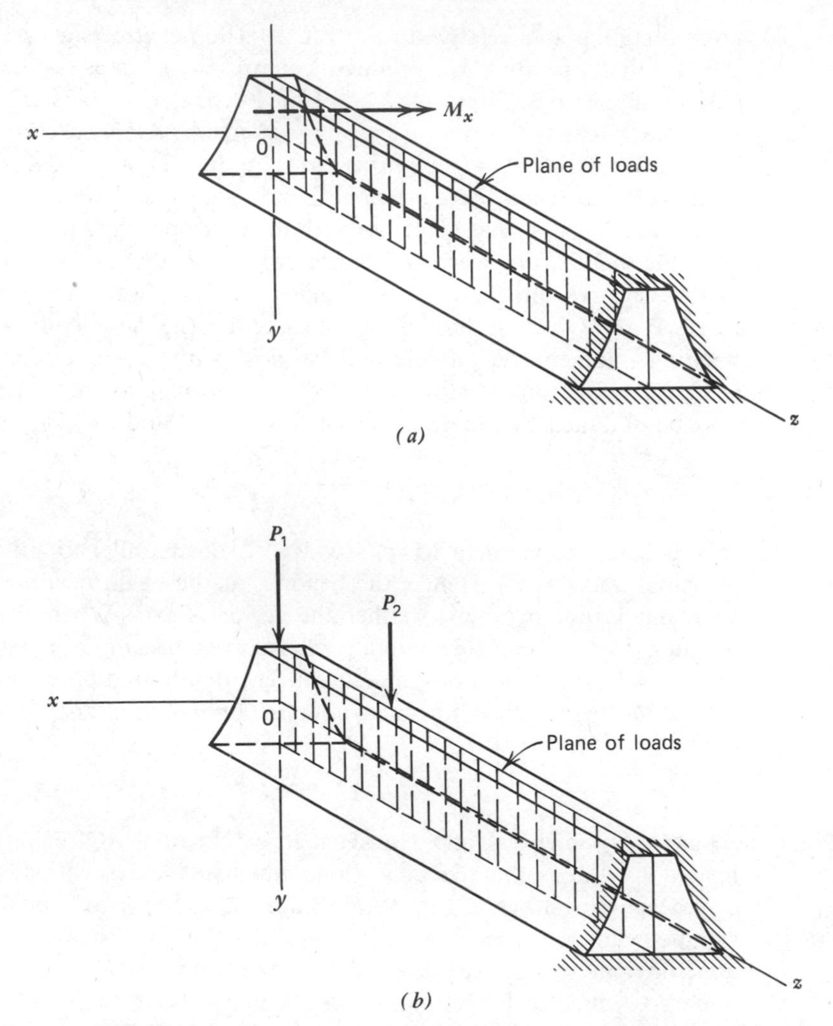

Fig. 6-1.6/Plane of loads coincident with the plane of symmetry of the beam. (a) Couple loads. (b) Lateral loads.

coincides with the centroid 0. Let the intersection of the plane of the loads and the plane of the cross section be denoted by line *L-L*, which forms angle ϕ $(0 \leq \phi < \pi)$ measured counterclockwise from the *x-z*-plane, and which passes through the shear center *C*. Since the plane of loads contains point *C*, the bending axis of the rectangular beam lies in the plane of the loads. If the angle ϕ equals 0 rad or $\pi/2$ rad, the rectangular beam will undergo symmetrical bending; that is, bending about a principal axis. For other values of ϕ, the beam undergoes nonsymmetrical bending; that is, bending for which the neutral axis of bending of the cross section does not coincide with either of the principal axes *x*, *y*.

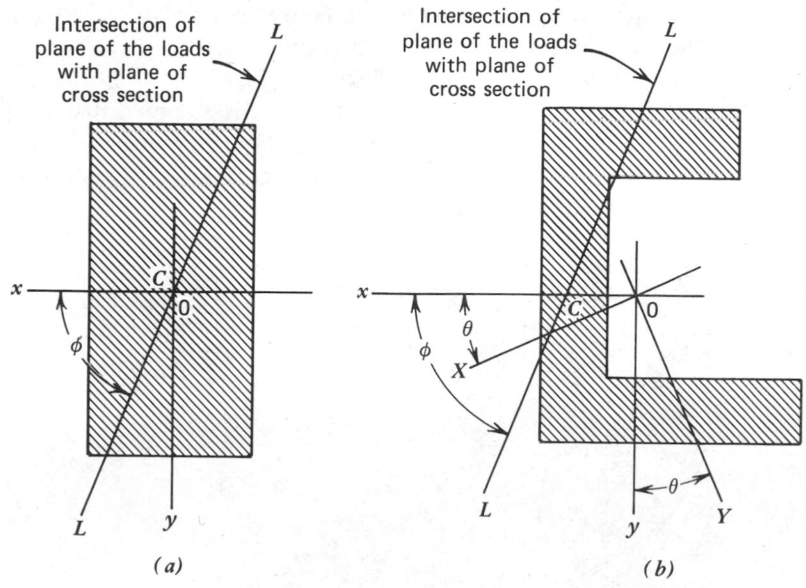

Fig. 6-1.7/Unsymmetrically loaded beams. (a) Rectangular cross sections. (b) Channel cross section.

In the case of a general channel section (Fig. 6-1.7*b*), the principal axes (X, Y) are located by a rotation through angle θ (positive θ is taken counterclockwise) from the (x, y)-axes as shown. The value of θ is determined by Eq. (A-3.4), in the Appendix. Although the plane of loads contains the shear center C (and, hence, the bending axis of the beam), it is not parallel to either of the principal planes (X, z), (Y, z). Hence, in general, the channel beam (Fig. 6-1.7*b*) undergoes nonsymmetrical bending; that is, bending about an axis that is not a principal axis. However, for the two special cases, $\phi = \theta$ or $\phi = \theta + \pi/2$, the channel beam undergoes symmetrical bending.

6-2 BENDING STRESSES IN BEAMS SUBJECTED TO NONSYMMETRICAL BENDING

Let a cutting plane be passed through a straight cantilever beam at section z. The free body diagram of the beam to the left of the cut is shown in Fig. 6-2.1*a*. The beam has constant cross section of arbitrary shape. The origin 0 of the coordinate axes is chosen at the centroid of the beam cross section at the left end of the beam with the z-axis taken parallel to the beam. The left end of the beam is subjected to a bending couple $\mathbf{M}_0$ which is equilibrated by bending moment $\mathbf{M}$ acting on the cross section at z, with positive components (M_x, M_y) as shown. The bending moment $\mathbf{M} = (M_x, M_y)$ is the resultant of the forces due to the normal stress σ_{zz} acting on the section (Fig. 6-2.1b). For convenience, we show (x, y)-axes at the cross section z. It is assumed that the (x, y)-axes are not principal axes for the cross section. In this article, we derive the load-stress formula which relates the normal stress σ_{zz} acting on the cross section to the components (M_x, M_y).

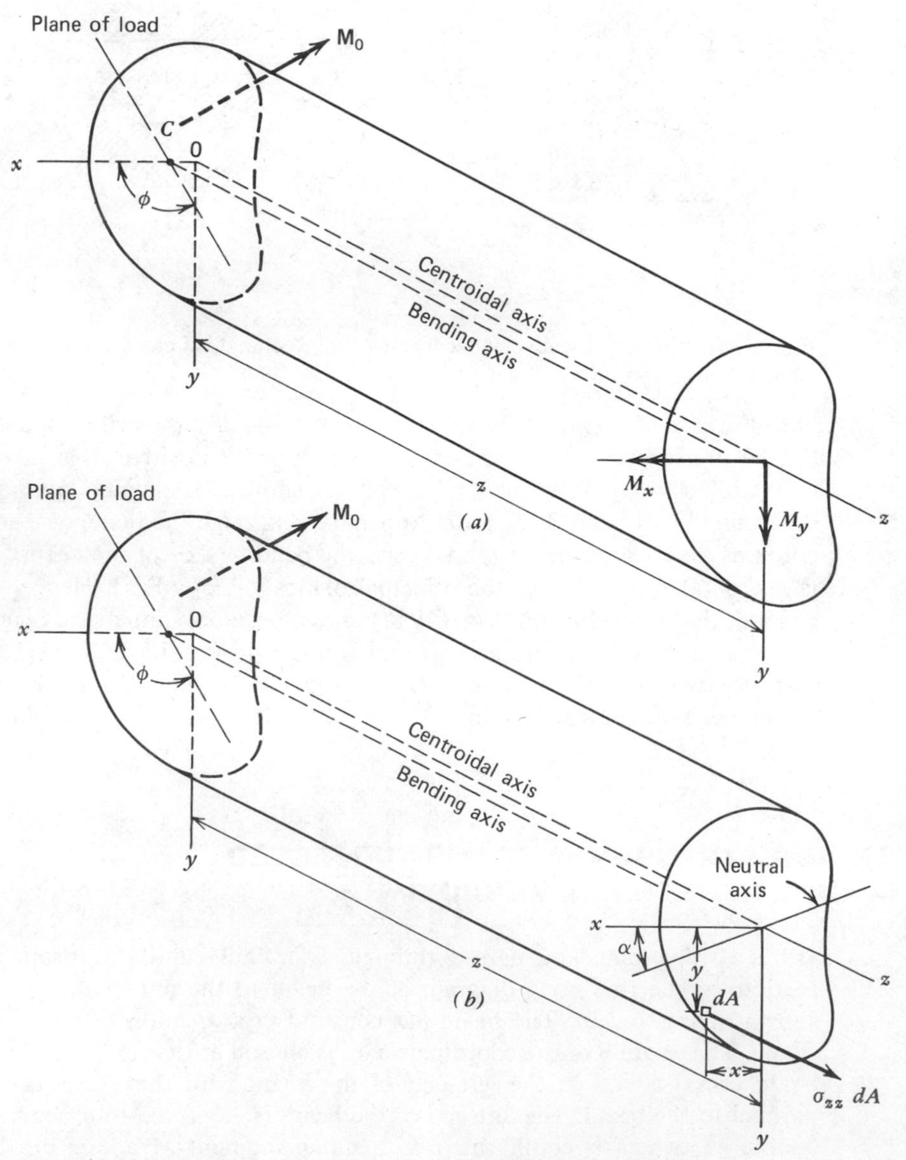

Fig. 6-2.1/Pure bending of a nonsymmetrically loaded cantilever beam.

As discussed in the introduction to Chapter 2, the derivation of load-stress and load-deformation relations for the beam requires that equations of equilibrium, compatibility conditions, and stress-strain relations be satisfied for the beam along with specified boundary conditions for the beam.

Equations of Equilibrium / Application of the equations of equilibrium to the free body diagram in Fig. 6-2.1*b* yields (since there is no net resultant force in the *z* direction)

$$
\begin{aligned}
0 &= \int \sigma_{zz}\, dA \\
M_x &= \int y\sigma_{zz}\, dA \\
M_y &= -\int x\sigma_{zz}\, dA
\end{aligned}
\tag{6-2.1}
$$

where dA denotes an element of area in the cross section and the integration is performed over the area A of the cross section. To evaluate the integrals in Eq. (6-2.1), it is necessary that the functional relation between σ_{zz} and (x, y) be known. The determination of σ_{zz} as a function of (x, y) is achieved by considering the geometry of deformation and the stress-strain relations.

Geometry of Deformation / We assume that plane sections of an unloaded beam remain plane after the beam is subjected to pure bending. Consider two plane cross sections perpendicular to the bending axis of an unloaded beam such that the centroids of the two sections are separated by a distance Δz. These two planes are parallel since the beam is straight. These planes rotate with respect to each other when moments M_x and M_y are applied. Hence, the extension e_{zz} of longitudinal fibers of the beam between the two planes can be represented as a linear function of (x, y); namely,

$$e_{zz} = a'' + b''x + c''y \tag{6-2.2}$$

where a'', b'', and c'' are constants. Since the beam is initially straight, all fibers have the same initial length Δz so that the strain ϵ_{zz} can be obtained by dividing Eq. (6-2.2) by Δz. Thus,

$$\epsilon_{zz} = a' + b'x + c'y \tag{6-2.3}$$

where $\epsilon_{zz} = e_{zz}/\Delta z$, $a' = a''/\Delta z$, $b' = b''/\Delta z$, and $c' = c''/\Delta z$.

Stress-Strain Relations / According to the theory of pure bending of straight beams, the only nonzero stress component in the beam is σ_{zz}. For linearly elastic conditions, Hooke's law states

$$\sigma_{zz} = E\epsilon_{zz} \tag{6-2.4}$$

Eliminating ϵ_{zz} between Eqs. (6-2.3) and (6-2.4), we obtain

$$\sigma_{zz} = a + bx + cy \tag{6-2.5}$$

where $a = Ea'$, $b = Eb'$, and $c = Ec'$.

Load-Stress Relation for Unsymmetrical Bending/Substitution of Eq. (6-2.5) into Eqs. (6-2.1) yields

$$\begin{aligned} 0 &= \int (a + bx + cy)\, dA = a\int dA + b\int x\, dA + c\int y\, dA \\ M_x &= \int (ay + bxy + cy^2)\, dA = a\int y\, dA + b\int xy\, dA + c\int y^2\, dA \qquad (6\text{-}2.6) \\ M_y &= -\int (ax + bx^2 + cxy)\, dA = -a\int x\, dA - b\int x^2\, dA - c\int xy\, dA \end{aligned}$$

Since the z-axis passes through the centroid of each cross section of the beam, $\int x\, dA = \int y\, dA = 0$. The other integrals in Eqs. (6-2.6) are defined in the Appendix. Equations (6-2.6) simplify to

$$\begin{aligned} 0 &= aA \\ M_x &= bI_{xy} + cI_x \qquad (6\text{-}2.7) \\ M_y &= -bI_y - cI_{xy} \end{aligned}$$

where I_x and I_y are the centroidal moments of inertia of the beam cross section with respect to the x- and y-axes, respectively, and I_{xy} is the centroidal product of inertia of the beam cross section. Solving Eqs. (6-2.7) for the constants a, b, and c, we obtain

$$\begin{aligned} a &= 0 \qquad (\text{because } A \neq 0) \\ b &= -\frac{M_y I_x + M_x I_{xy}}{I_x I_y - I_{xy}^2} \qquad (6\text{-}2.8) \\ c &= \frac{M_x I_y + M_y I_{xy}}{I_x I_y - I_{xy}^2} \end{aligned}$$

The substitution of Eqs. (6-2.8) into Eq. (6-2.5) gives the normal stress distribution σ_{zz} on a given cross section of a beam subjected to unsymmetrical bending in the form

$$\sigma_{zz} = -\left[\frac{M_y I_x + M_x I_{xy}}{I_x I_y - I_{xy}^2}\right] x + \left[\frac{M_x I_y + M_y I_{xy}}{I_x I_y - I_{xy}^2}\right] y \tag{6-2.9}$$

Equation (6-2.9) is not the most convenient form for the determination of the maximum value of the flexure stress σ_{zz}. Before the location of points of maximum tensile and compressive stresses in the cross section can be determined, it is necessary to locate the neutral axis. For this purpose, it is desirable that the neutral axis location include the angle ϕ between the plane of the loads and the x-axis; ϕ is measured positive counterclockwise (Fig. 6-1.7). The magnitude of ϕ is generally in the neighborhood of $\pi/2$ rad ($0 \le \phi < \pi$). The bending moments M_x and M_y

can be written in terms of ϕ as follows:

$$\begin{aligned} M_x &= M \sin \phi \\ M_y &= -M \cos \phi \end{aligned} \tag{6-2.10}$$

in which M is the signed magnitude of moment **M** at the cut section. The sign of M is positive if the x projection of the vector **M** is positive; it is negative if the x projection of **M** is negative. Since the (x, y)-axes are chosen for the convenience of the one making the calculations, they are chosen so that the magnitude of M_x is not zero. Therefore, by Eqs. (6-2.10),

$$\cot \phi = -\frac{M_y}{M_x} \tag{6-2.11}$$

Neutral Axis/The neutral axis of the cross section of a beam subjected to unsymmetrical bending is defined to be the axis in the cross section for which $\sigma_{zz} = 0$. Thus, by Eq. (6-2.9), the equation of the neutral axis of the cross section is

$$y = \frac{M_x I_{xy} + M_y I_x}{M_x I_y + M_y I_{xy}} x = x \tan \alpha \tag{6-2.12}$$

where α is the angle between the neutral axis of bending and the x-axis; α is measured positive counterclockwise (Fig. 6-2.1), and

$$\tan \alpha = \frac{M_x I_{xy} + M_y I_x}{M_x I_y + M_y I_{xy}} \tag{6-2.13}$$

Since $x = y = 0$ satisfies Eq. (6-2.12), the neutral axis passes through the centroid of the section. The right side of Eq. (6-2.13) can be expressed in terms of the angle ϕ by using Eq. (6-2.11). Thus,

$$\tan \alpha = \frac{I_{xy} - I_x \cot \phi}{I_y - I_{xy} \cot \phi} \tag{6-2.14}$$

More Convenient Form for the Flexure Stress σ_{zz}/Elimination of M_y between Eqs. (6-2.9) and (6-2.13) results in a more convenient form for the normal stress distribution σ_{zz} for beams subjected to unsymmetrical bending; namely

$$\sigma_{zz} = \frac{M_x(y - x \tan \alpha)}{I_x - I_{xy} \tan \alpha} \tag{6-2.15}$$

where $\tan \alpha$ is given by Eq. (6-2.14). Once the neutral axis is located on the cross sections at angle α as indicated in Fig. 6-2.1, points in the cross section where the tensile and compressive flexure stresses are maxima are easily determined. The coordinates of these points can be substituted into Eq. (6-2.15) to determine the magnitudes of these stresses. If M_x is zero, Eq. (6-2.9) may be used instead of Eq. (6-2.15) to determine magnitudes of these stresses, or axes (x, y) may be rotated by $\pi/2$ to obtain new reference axes (x', y').

Note: Equations (6-2.14) and (6-2.15) have been derived assuming that the beam is subjected to pure bending. These equations are exact for pure

6-2.1

FORTRAN Computer Program for Unsymmetrical Bending

```
      PROGRAM UNSYB(INPUT,OUTPUT,UNIN,UNOUT,TAPE5 = UNIN,TAPE6 = UNOUT)
    C READ NUMBER OF CASES
      READ (5,*) NN
      DO 23 J = 1,NN
      WRITE (6,24)
24    FORMAT ("  I(X)   I(Y)    I(XY)    I(MAX)  I(MIN)     M
     APHI   THETA   ALPHA")
    C INPUT N, MOMENTS OF INERTIA, MOMENT AND ANGLE OF PLANE OF LOADS
      READ (5,*) N,EIX,EIXY,EM,PHI
    C COMPUTE PRINCIPAL AXES ANGLE THETA
      B = ABS(EIX - EIY)
      IF (B.GT..001) GO TO 1
      B = .000001
1     THETA = .5*ATAN(-2.*EIXY/B)
    C COMPUTE NEUTRAL AXES ANGLE ALPHA
      EIMAX = .5*(EIX + EIY) + SQRT(.25*(EIX - EIY)**2 + EIXY**2)
      EIMIN = .5*(EIX + EIY) - SQRT(.25*(EIX - EIY)**2 + EIXY**2)
      EMX = EM*SIN(PHI)
      TANAL = (EIXY - EIX/TAN(PHI))/(EIY - EIXY/TAN(PHI))
      ALPHA = ATAN(TANAL)
      A = EMX/(EIX - EIXY*TANAL)
      JJ = J
      WRITE (6,30)JJ
      WRITE (6,26) EIX,EIY,EIXY,EIMAX,EIMIN,EM,PHI,THETA,ALPHA
30    FORMAT("CASE ",JJ)
26    FORMAT (6E9,3,3F7,3,/,/)
      WRITE(6,27)
    C COMPUTE SIGMA ZZ AT N COORDINATE PAIRS (X, Y)
```

```
27        FORMAT ("      X      Y      SIGZZ")
          DO 28 I = 1,N
          READ (5,*) X,Y
          SIGZZ = A*(Y - X*TANAL)
          WRITE (6,32) X,Y,SIGZZ
32        FORMAT (3F10.2)
28        CONTINUE
23        CONTINUE
          STOP
          END
```

INPUT

```
2
1,.5625E + 09,.3906E + 09,.0,.3142E + 08,1.7453
-125.,150.
2,.1244E + 09,.1244E + 09,.7076E + 08,-.4000E + 08,1.5708
-59.24,-215.76
-84.24,84.24
```

OUTPUT

```
     I(X)        I(Y)        I(XY)      I(MAX)      I(MIN)         M         PHI    THETA   ALPHA
 .563E + 09  .391E + 09     0.        .563E + 09  .391E + 09   .314E + 08   1.745   0.000    .249
N      X           Y         SIGZZ
1  -125.00      150.00       10.00
     I(X)        I(Y)        I(XY)      I(MAX)      I(MIN)         M         PHI    THETA   ALPHA
 .124E + 09  .124E + 09   .708E + 08  .195E + 09  .536E + 08  -.400E + 08   1.571   -.785    .517
N      X           Y         SIGZZ
1   -59.24     -215.76       86.54
2   -84.24       84.24      -62.82
```

bending. Although they are not exact for beams subjected to transverse shear loads, often the equations are assumed to be valid for such beams. The error in this assumption is usually small, particularly if the beam has a length of at least five times its maximum cross-sectional dimension.

In the derivation of Eqs. (6-2.14) and (6-2.15), the (x, y)-axes are any convenient set of orthogonal axes that have an origin at the centroid of the cross-sectional area. The equations are valid if (x, y) are principal axes; in this case $I_{xy} = 0$. If the axes are principal axes and $\phi = \pi/2$, Eq. (6-2.14) indicates that $\alpha = 0$ and Eq. (6-2.15) reduces to Eq. (6-1.1).

For convenience in deriving Eqs. (6-2.14) and (6-2.15), the origin for the x, y, z coordinate axes was chosen (see Fig. 6-2.1*b*) at the end of the free body diagram opposite from the cut section with the positive z-axis toward the cut section. The equations are equally valid if the origin is taken at the cut section with the positive z-axis toward the opposite end of the free body diagram. If ϕ_2 is the magnitude of ϕ for the second choice of axes and ϕ_1 is the magnitude of ϕ for the first choice of axes, then $\phi_2 = \pi - \phi_1$.

A FORTRAN digital computer program for the solution of the nonsymmetrical bending of beams (Eq. 6-2.15) is listed in Table 6-2.1. It can easily be run on a microcomputer.

EXAMPLE 6-2.1

Channel Section Beam

The cantilever beam in Fig. E6-2.1*a* has a channel section as shown in Fig. E6-2.1*b*. The concentrated load $P = 12.0$ kN lies in the plane (the plane of the loads) making an angle $\theta = \pi/3$ rad with the x-axis. Load P lies in the plane of the cross section of the free end of the beam and passes through shear center C; in Chapter 7 we find that the shear center lies on the y-axis as shown. Locate points of maximum tensile and compressive stresses in the beam and determine their magnitudes

SOLUTION

Several properties of the cross-sectional area are needed (see the Appendix).

$$A = 10{,}000 \text{ mm}^2 \qquad I_x = 39.69 \times 10^6 \text{ mm}^4$$

$$y_0 = 82.0 \text{ mm} \qquad I_y = 30.73 \times 10^6 \text{ mm}^4$$

$$I_{xy} = 0$$

The orientation of the neutral axis for the beam is given by Eq. (6-2.14). Thus,

$$\tan\alpha = -\frac{I_x}{I_y}\cot\phi = -\frac{39{,}690{,}000}{30{,}730{,}000}(0.5774) = -0.7457$$

$$\alpha = -0.6407 \text{ rad}$$

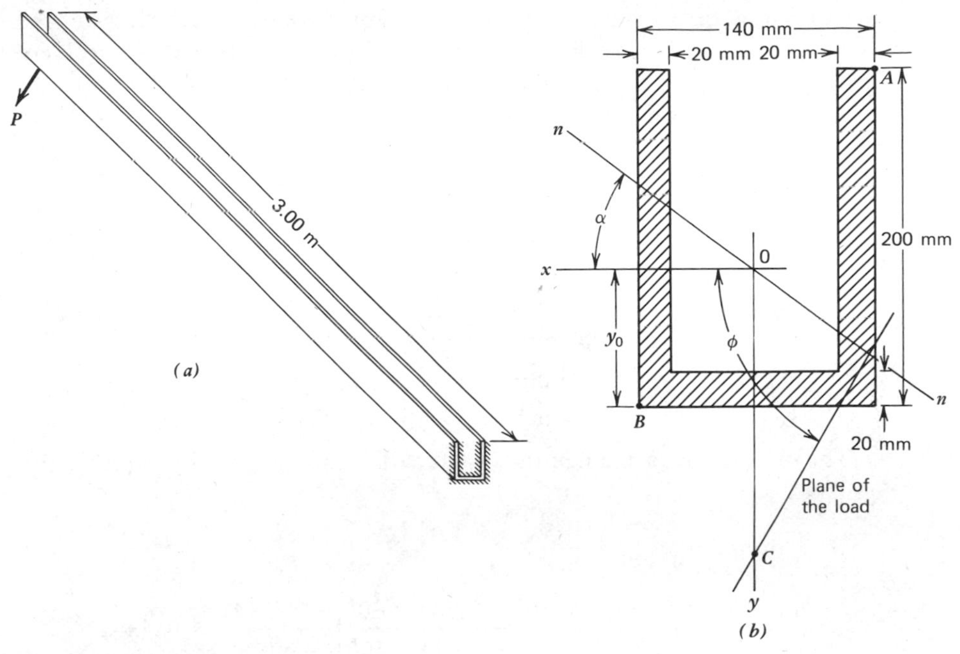

Fig. E6-2.1

The negative sign indicates that the neutral axis *n-n*, which passes through the centroid ($x = y = 0$), is located clockwise 0.6407 rad from the *x*-axis (Fig. E6-2.1*b*). The maximum tensile stress occurs at point *A* while the maximum compressive stress occurs at point *B*. These stresses are given by Eq. (6-2.15) after M_x has been determined. From Fig. E6-2.1*a*

$$M = -3.00P = -36.0 \text{ kN} \cdot \text{m}$$

$$M_x = M \sin\phi = -31.18 \text{ kN} \cdot \text{m}$$

$$\sigma_A = \frac{M_x(y_A - x_A \tan\alpha)}{I_x} = \frac{-31{,}180{,}000[-118-(-70)(-0.7457)]}{39{,}640{,}000}$$

$$= 133.7 \text{ MPa}$$

$$\sigma_B = \frac{-31{,}180{,}000[82 - 70(-0.7457)]}{39{,}640{,}000} = -105.6 \text{ MPa}$$

EXAMPLE 6-2.2
Angle Beam

Plates are welded together to form the 120 mm by 80 mm by 10 mm angle-section beam shown in Fig. E6-2.2*a*. The beam is subjected to a concentrated load $P = 4.00$ kN as shown. The load *P* lies in the plane (the plane of loads) making an angle $\phi = 2\pi/3$ rad with the *x*-axis. Load *P* passes through shear center *C*; in Chapter 7 we find that the shear center is located at the intersection of the two legs of the angle section.

Determine the maximum tensile and compressive bending stresses at the section of the beam where the load is applied. (a) Solve the problem using the load-stress relations derived for unsymmetrical bending. (b) Solve the problem using Eq. (6-1.3).

SOLUTION

(a) Several properties of the cross-sectional area are needed (see the Appendix).

$$A = 1900 \text{ mm}^2 \qquad I_x = 2.783 \times 10^6 \text{ mm}^4$$
$$x_0 = 19.74 \text{ mm} \qquad I_y = 1.003 \times 10^6 \text{ mm}^4$$
$$y_0 = 39.74 \text{ mm} \qquad I_{xy} = -0.973 \times 10^6 \text{ mm}^4$$

The orientation of the neutral axis for the beam is given by Eq. (6-2.14). Thus,

$$\tan\alpha = \frac{I_{xy} - I_x \cot\phi}{I_y - I_{xy}\cot\phi}$$
$$= \frac{-0.973 \times 10^6 - 2.783 \times 10^6(-0.5774)}{1.003 \times 10^6 - (-0.973 \times 10^6)(-0.5774)} = 1.4368$$
$$\alpha = 0.9628 \text{ rad}$$

The positive sign indicates that the neutral axis n-n, which passes through the centroid ($x = y = 0$), is located counterclockwise 0.9628 rad from the x-axis (Fig. E6-2.2b). The maximum tensile stress occurs at point A while the maximum compressive stress occurs at point B. These stresses are given by Eq. (6-2.15) after M_x has been determined. From Fig. E6-2.2a

$$M = 1.2P = 4.80 \text{ kN} \cdot \text{m}$$
$$M_x = M\sin\phi = 4.80 \times 10^3\ (0.8660) = 4.157 \text{ kN} \cdot \text{m}$$
$$\sigma_A = \frac{M_x[y_A - x_A \tan\alpha]}{I_x - I_{xy}\tan\alpha}$$
$$= \frac{4.157 \times 10^6[39.74 - (-60.26)(1.4368)]}{2.783 \times 10^6 - (-0.973 \times 10^6)(1.4368)} = 125.6 \text{ MPa}$$
$$\sigma_B = \frac{4.157 \times 10^6[-80.26 - 19.74(1.4368)]}{2.783 \times 10^6 - (-0.973 \times 10^6)(-0.5801)} = -108.0 \text{ MPa}$$

(b) To solve the problem using Eq. (6-1.3), it is necessary that the principal axes for the cross section be determined. The two values of the angle θ between the x-axis and the principal axes are given by Eq. (A-3.4). Thus, we obtain

$$\tan 2\theta = -\frac{2I_{xy}}{I_x - I_y} = -\frac{2(-0.973 \times 10^6)}{2.783 \times 10^6 - 1.003 \times 10^6} = 1.0933$$
$$\theta = 0.4150 \text{ rad}$$

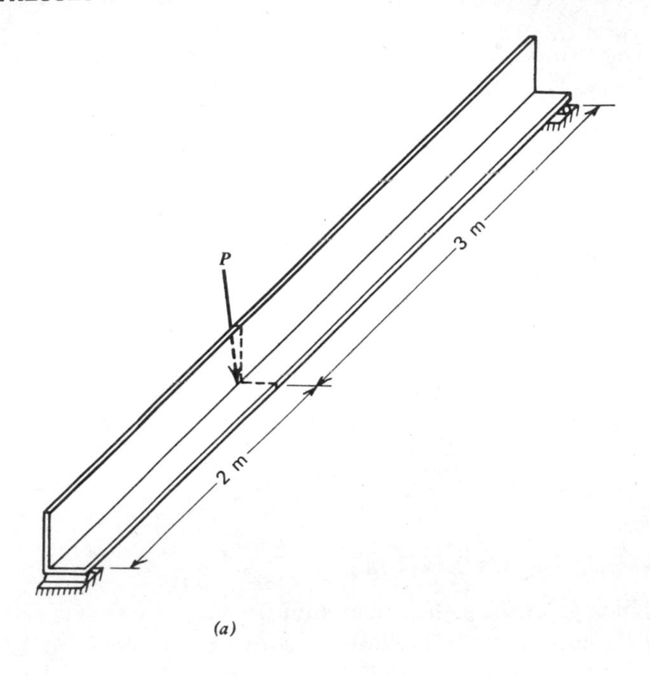
P
3 m
2 m

(a)

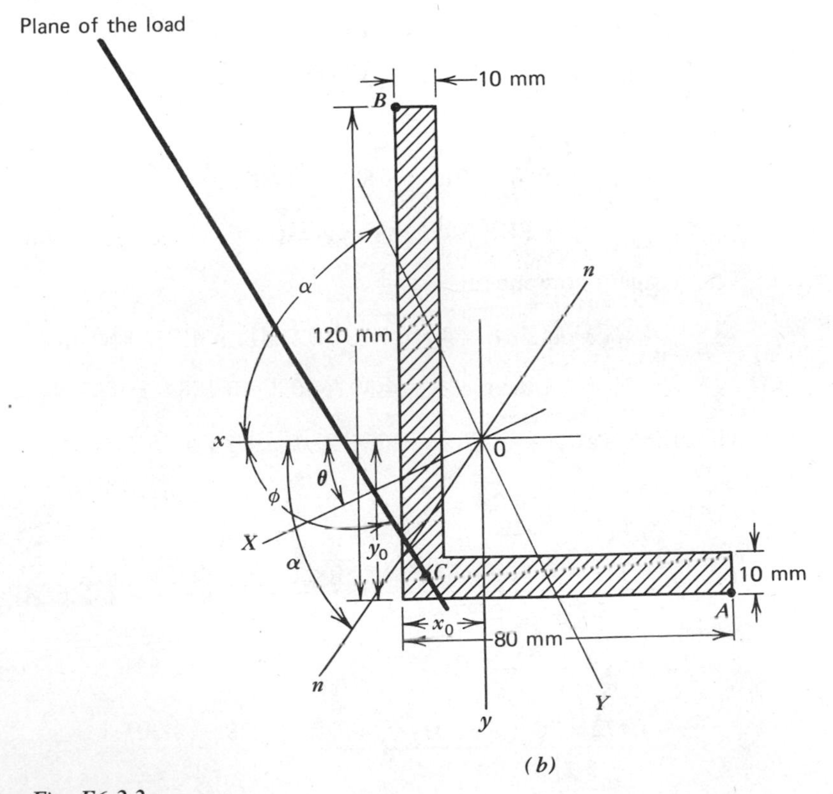
Plane of the load
10 mm
B
α
120 mm
n
x
0
θ
φ
X
y_0
α
C
10 mm
A
x_0
80 mm
n
Y
y

(b)

Fig. E6-2.2

The principal X- and Y-axes are shown in Fig. E6-2.2b. Thus (see Eq. A-3.2, the Appendix)

$$I_X = I_x \cos^2\theta + I_y \sin^2\theta - 2I_{xy}\sin\theta\cos\theta = 3.212 \times 10^6 \text{ mm}^4$$

$$I_Y = I_x + I_y - I_X = 0.574 \times 10^6 \text{ mm}^4$$

Note that now angle ϕ is measured from the X-axis and not from the x-axis as for part (a). Hence

$$\phi = 2\pi/3 - \theta = 1.6794 \text{ rad}$$

Angle α, determines the orientation of the neutral axis, is now measured from the X-axis and is given by Eq. (6-2.14). Hence, we find

$$\tan\alpha = -\frac{I_X \cot\phi}{I_Y} = -\frac{3.212 \times 10^6(-0.1090)}{0.574 \times 10^6} = 0.6099$$

$$\alpha = 0.5477 \text{ rad}$$

which gives the same orientation for the neutral axis as for part (a).

To use Eq. (6-1.3) relative to axes (X, Y), the X- and Y-coordinates of points A and B are needed. They are

$$X_A = x_A\cos\theta + y_A\sin\theta = -60.26(0.9151) + 39.74(0.4032) = -39.12 \text{ mm}$$

$$Y_A = y_A\cos\theta - x_A\sin\theta = 39.74(0.9151) - (-60.26)(0.4032) = 60.66 \text{ mm}$$

and

$$X_B = 19.74(0.9151) - 80.26(0.4032) = -14.30 \text{ mm}$$

$$Y_B = -80.26(0.9151) - 19.74(0.4032) = -81.41 \text{ mm}$$

The moment components are

$$M_X = M\sin\phi = 4.80 \times 10^3(0.9941) = 4.772 \text{ kN} \cdot \text{m}$$

$$M_Y = -M\cos\phi = -4.80 \times 10^3(-0.1084) = 520 \text{ N} \cdot \text{m}$$

The stresses at A and B are calculated using Eq. (6-1.3). Thus,

$$\sigma_A = \frac{M_X Y_A}{I_X} - \frac{M_Y X_A}{I_Y}$$

$$= \frac{4.772 \times 10^6(60.66)}{3.212 \times 10^6} - \frac{0.520 \times 10^6(-39.12)}{0.574 \times 10^6} = 125.6 \text{ MPa}$$

$$\sigma_B = \frac{M_X Y_B}{I_X} - \frac{M_Y X_B}{I_Y}$$

$$= \frac{4.772 \times 10^6(-81.41)}{3.212 \times 10^6} - \frac{0.520 \times 10^6(-14.30)}{0.574 \times 10^6} = -108.0 \text{ MPa}$$

These values for σ_A and σ_B agree with the values calculated in part (a). However, the computational work is greater in part (b) than in part (a).

PROBLEM SET 6-2

1. A timber beam 250 mm wide by 300 mm deep by 4.2 m long is used as a simple beam on a span of 4 m. It is subjected to a concentrated load P at the midsection of the span. The plane of the loads makes an angle $\phi = 5\pi/9$ rad with the horizontal x-axis. The beam is made of yellow pine with a yield stress $Y = 25.0$ MPa. If the beam has been designed with a factor of safety $SF = 2.50$ against initiation of yielding, determine the magnitude of P and the orientation of the neutral axis.

2. The plane of the loads for the rectangular section beam in Fig. P6-2.2 coincides with a diagonal of the rectangle. Show that the neutral axis for the beam cross section coincides with the other diagonal.

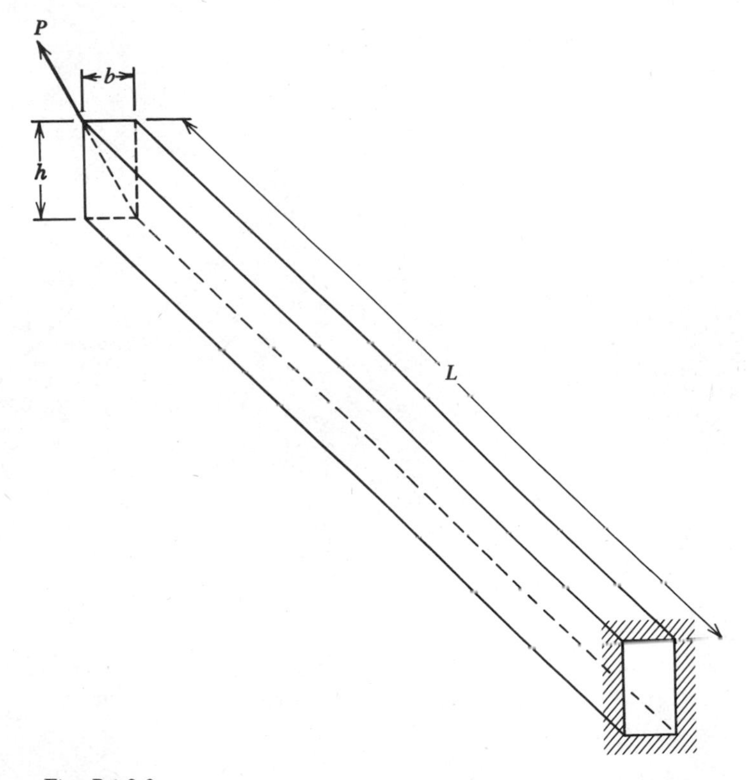

Fig. P6-2.2

3. In Fig. P6-2.3 let $b = 300$ mm, $h = 300$ mm, $t = 25.0$ mm, $L = 2.50$ m, and $P = 16.0$ kN. Calculate the maximum tensile and compressive stresses in the beam, and determine the orientation of the neutral axis.

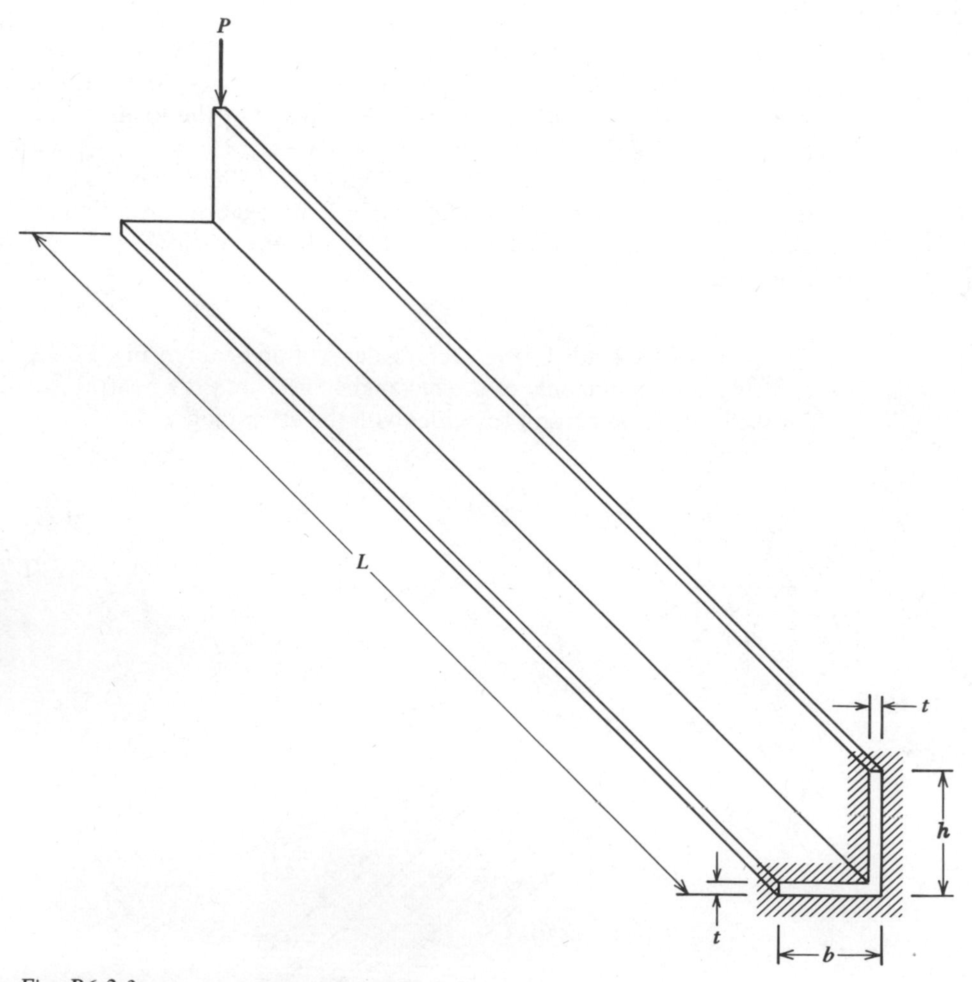

Fig. P6-2.3

4. In Fig. P6-2.3 let $b = 200$ mm, $h = 300$ mm, $t = 25.0$ mm, $L = 2.50$ m, and $P = 16.0$ kN. Calculate the maximum tensile and compressive stresses in the beam and determine the orientation of the neutral axis.

   Ans. $\sigma_{zz(\text{ten})} = 98.6$ MPa, $\sigma_{zz(\text{com})} = -81.9$ MPa

5. In Fig. P6-2.5 let $b = 150$ mm, $t = 50.0$ mm, $h = 150$ mm, and $L = 2.00$ m. The beam is made of a steel that has a yield point stress $Y = 240$ MPa. Using a factor of safety of $SF = 2.00$, determine the magnitude of P if $\phi = 2\pi/9$ rad from the horizontal x-axis.

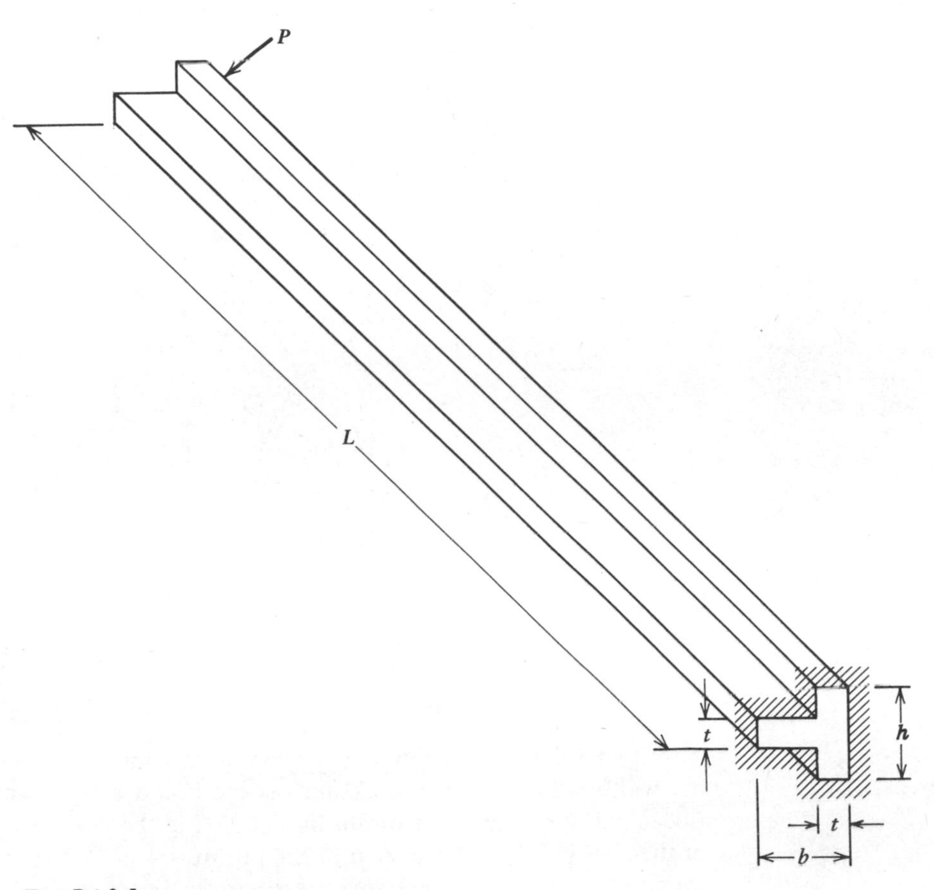

Fig. P6-2.5

6. A simple beam is subjected to a concentrated load $P = 4.00$ kN at the midlength of a span of 2.00 m. The beam cross section is formed by nailing two 50.0 mm by 150 mm boards together as indicated in Fig. P6-2.6. The plane of the loads passes through the centroid of the two boards as indicated. Determine the maximum flexure stress in the beam and the orientation of the neutral axis.

   Ans. $\sigma_{zz(\max)} = 4.17$ MPa, $\alpha = 1.3522$ rad

7. Solve Problem 6 if $\phi = 1.900$ rad.

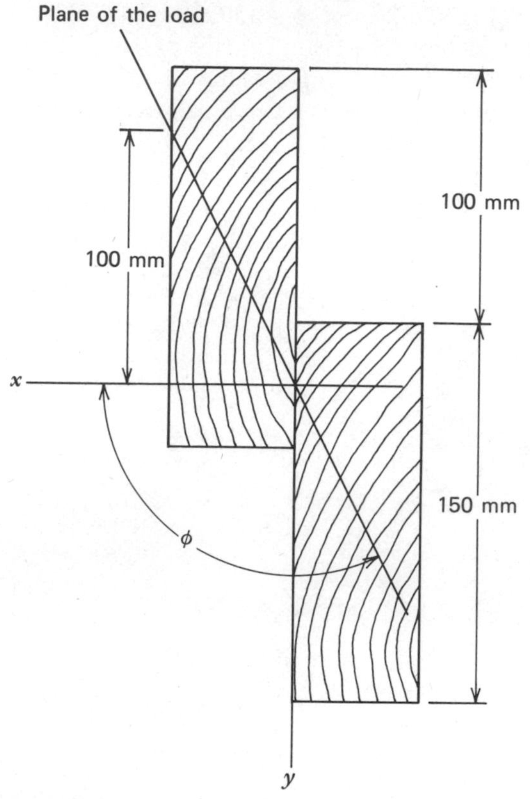

Fig. P6-2.6

8. A C-180 × 14.6 rolled steel channel ($I_x = 8.87 \times 10^6$ mm^4, depth = 178 mm, width = 53 mm, $x_B = 13.7$ mm) is used as a simply supported beam as, for example, a purlin in a roof (Fig. P6-2.8). If the slope of the roof is 1/2 and the span of the purlin is 4 m, determine the maximum tensile and compressive stresses in the beam caused by a uniformly distributed vertical load of 1.00 kN/m.

   Ans. $\sigma_{zz(\text{ten})} = 48.4$ MPa, $\sigma_{zz(\text{com})} = -105.2$ MPa

9. Two L-89 × 64 × 7.9 rolled steel angles ($I_{x_1} = 391 \times 10^3$ mm^4, $I_{y_1} = 912 \times 10^3$ mm^4, $I_{x_1y_1} = 349 \times 10^3$ mm^4, and $A = 1148$ mm^2) are welded to a 200 mm by 10 mm steel plate to form a composite z-bar (Fig. P6-2.9). The z-bar is a simply supported beam used as a purlin in a roof of slope $\frac{1}{2}$. The beam has a span of 4.00 m. The yield stress of the steel in the plate and angles is $Y = 300$ MPa. The beam has been designed using a factor of safety of $SF = 2.50$ against initiation of yielding. If the plane of the loads is vertical, determine the

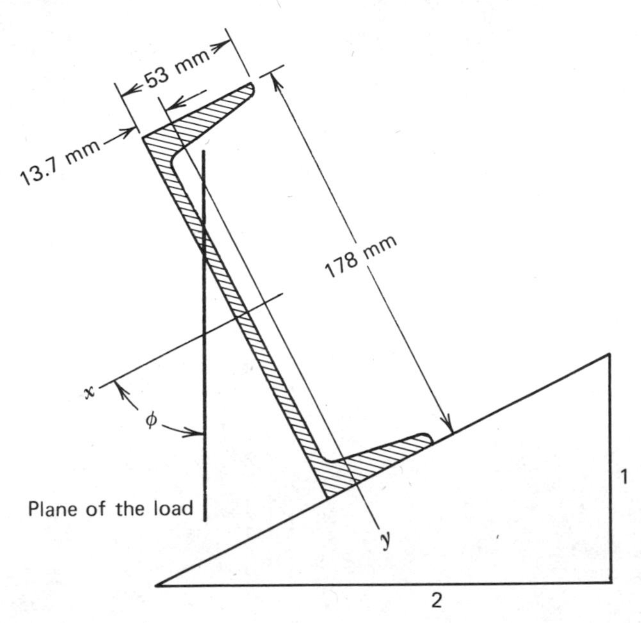

Fig. P6-2.8

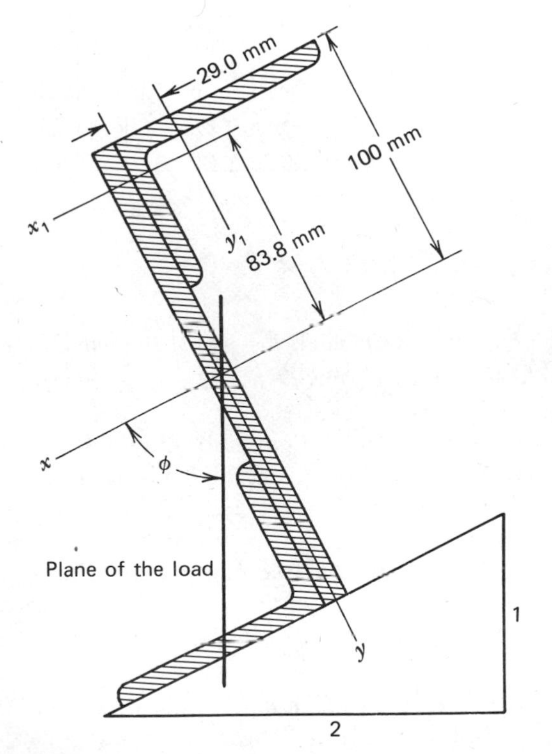

Fig. P6-2.9

magnitude of the maximum distributed load that can be applied to the beam.

10. A steel z-bar is used as a centilever beam having a length of 2.00 m. Looking from the free end toward the fixed end of the beam, the cross section has the orientation and dimensions shown in Fig. P6-2.10. A concentration load $P = 14.0$ kN acts at the free end of the beam at an angle $\phi = 1.25$ rad. Determine the maximum flexure stress in the beam.

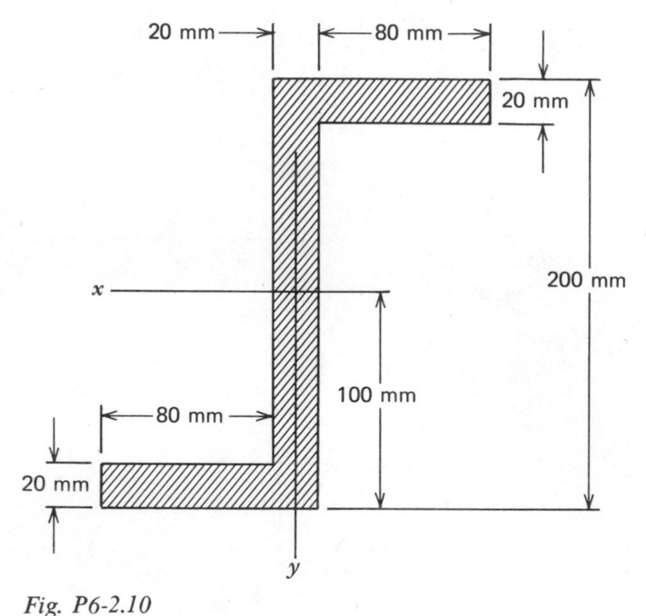

Fig. P6-2.10

Ans. $I_x = 39.36 \times 10^6$ mm^4, $I_y = 9.84 \times 10^6$ mm^4, $I_{xy} = 14.40 \times 10^6$ mm^4, $\alpha = 0.2557$ rad, $\sigma_{zz(\max)} = 76.6$ MPa.

11. An extruded bar of aluminum alloy has the cross section shown in Fig. P6-2.11. A 1.00 m length of this bar is used as a cantilever beam. A concentrated load $P = 1.25$ kN is applied at the free end and makes an angle of $\phi = 5\pi/9$ rad with the x-axis. The view in Fig. P6-2.11 is from the free end toward the fixed end of the beam. Determine the maximum tensile and compressive stresses in the beam.

12. An extruded bar of aluminum alloy has the cross section shown in Fig. P6-2.12. A 2.10 m length of this bar is used as a simple beam on a span of 2.00 m. A concentrated load $P = 5.00$ kN is applied at

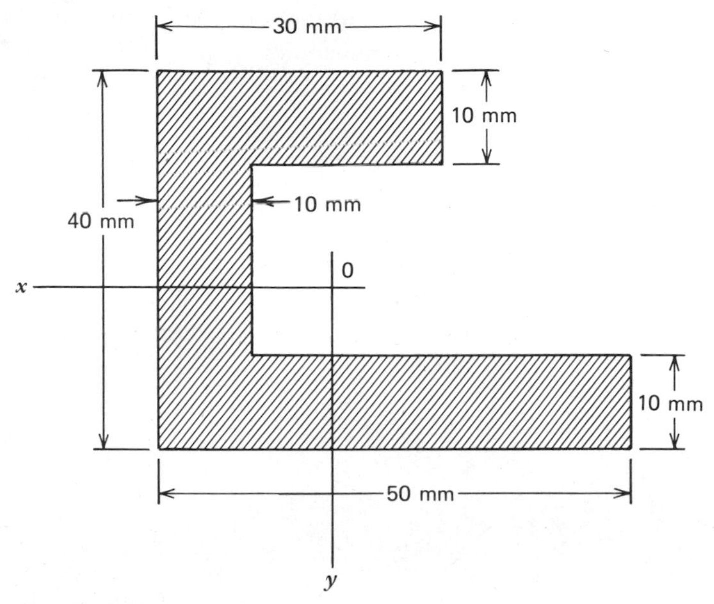

Fig. P6-2.11

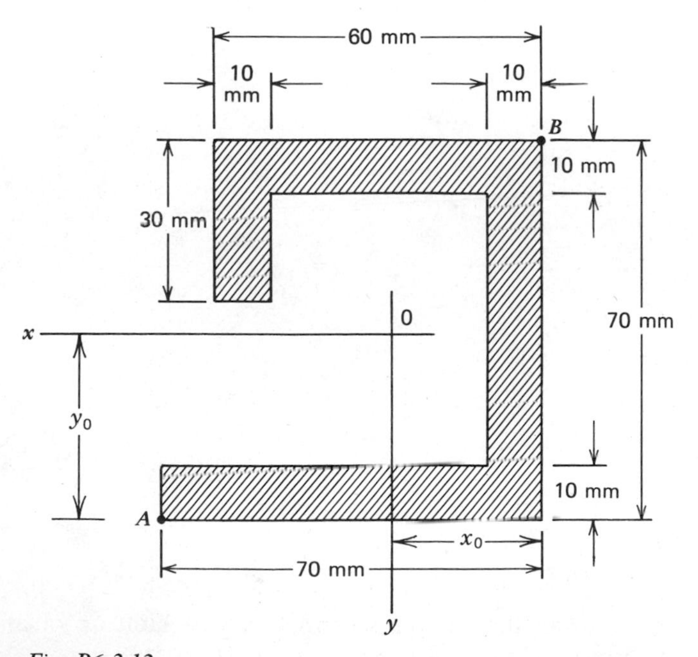

Fig. P6-2.12

midlength of the span and makes an angle of $\phi = 1.40$ rad with the x-axis. Determine the maximum tensile and compressive stresses in the beam.

Ans. $x_0 = 28.0$ mm, $y_0 = 35.0$ mm, $I_x = 1.330 \times 10^6$ mm^4, $I_y = 917 \times 10^3$ mm^4, $I_{xy} = 30.0 \times 10^3$ mm^4, $\alpha = -0.2153$ rad, $\sigma_A = 91.5$ MPa, $\sigma_B = -75.8$ MPa

13. A cantilever beam has a right triangular cross section and is loaded by a concentrated load P at the free end (Fig. P6-2.13). Solve for the stresses at points A and C at the fixed end if $P = 4.00$ kN, $h = 120$ mm, $b = 75.0$ mm, and $L = 1.25$ m.

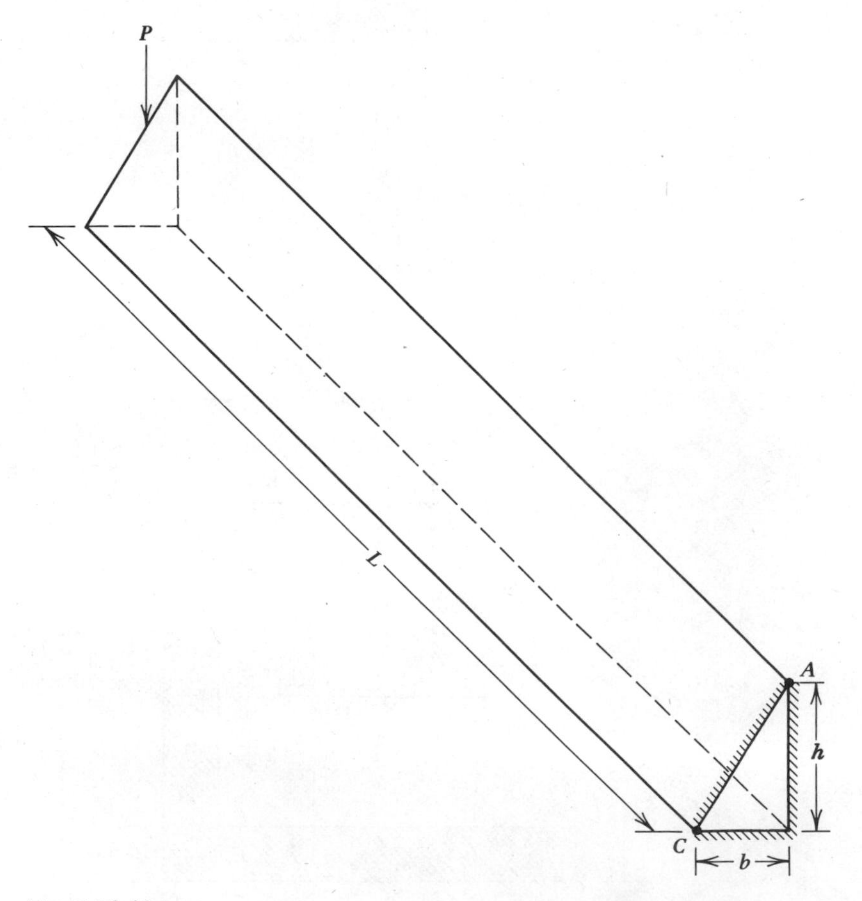

Fig. P6-2.13

14. A girder that supports a brick wall is built up of an S-310 × 47.3 I-beam ($A_1 = 6032$ mm^2, $I_{x_1} = 90.7 \times 10^6$ mm^4, $I_{y_1} = 3.90 \times 10^6$

mm^4), a C-310 × 30.8 channel ($A_2 = 3929$ mm^2, $I_{x_2} = 53.7 \times 10^6$ mm^4, $I_{y_2} = 1.61 \times 10^6$ mm^4), and a cover plate 300 mm by 10 mm riveted together (Fig. P6-2.14). The girder is 6.00 m long and is simply supported at its ends. The load is uniformly distributed such that $w = 20.0$ kN/m. Determine the orientation of the neutral axis and the maximum tensile and compressive stresses.

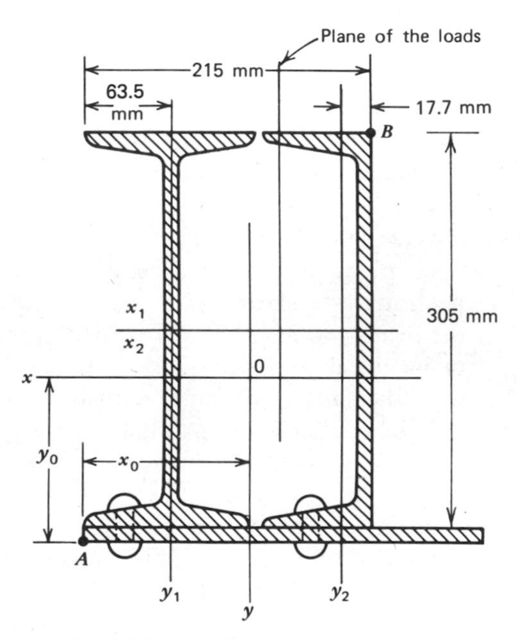

Fig. P6-2.14

Ans. $\alpha = -0.1653$ rad, $\sigma_A = 66.3$ MPa, $\sigma_B = -92.3$ MPa

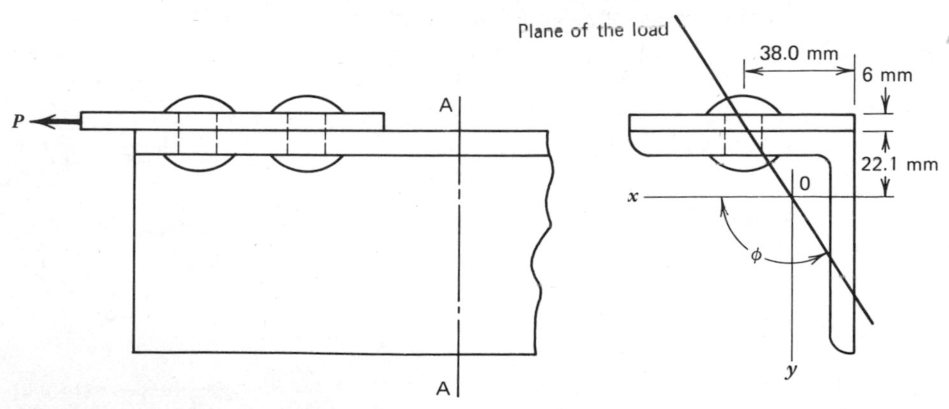

Fig. P6-2.15

15. A load $P = 50$ kN is applied to an L-76 × 76 × 7.9 rolled steel angle ($I_x = I_y = 570 \times 10^3$ mm^4, $I_{xy} = -332.5 \times 10^3$ mm^4, $A = 1148$ mm^2) by means of a 76 mm by 6 mm plate riveted to the angle (Fig. P6-2.15). The action line of load P coincides with the centroidal axis of the plate. Determine the maximum stress at a section, such as AA, of the angle. *Hint:* Resolve the load P into a load (equal to P) at the centroid of the angle and a bending couple.

6-3 DEFLECTIONS OF STRAIGHT BEAMS SUBJECTED TO UNSYMMETRICAL BENDING

If every transverse shear load applied to a straight beam lies in one plane and if every couple applied to the beam lies in a plane parallel to the plane of the transverse shear loads, the neutral axis for every cross section of the beam will have the same orientation as long as the beam material remains linearly elastic. The deflections of the beam will be in a direction perpendicular to the neutral axis. It is convenient to determine the component of the deflection parallel to an axis, say the y-axis. The total deflection is easily determined once one component has been determined.

Consider the intersection of the (y, z)-plane with the beam in Fig. 6-2.1. A side view of this section of the deformed beam is shown in Fig. 6-3.1. In the deformed beam, the two straight lines F^*G^* and H^*J^* represent the intersection of the (y, z)-plane with two planes perpendicular to the axis of the beam, a distance Δz apart at the neutral surface. Before deformation, the lines FG and HJ are parallel and distance Δz apart. Since plane sections remain plane, the extensions of F^*G^* and H^*J^* meet at the center of curvature $0'$. The distance from $0'$ to the neutral surface is the radius of curvature R_y of the beam in the (y, z)-plane. Since the center of curvature lies on the negative side of the y-axis, R_y is negative. We assume that the deflections are small so that $1/R_y \cong d^2v/dz^2$, where v is the y-component of displacement. Under deformation of the beam, a fiber at distance y below the neutral surface elongates an amount $e_{zz} = (\Delta z)\epsilon_{zz}$. Initially the length of the fiber is Δz. By geometry of similar triangles,

$$-\frac{\Delta z}{R_y} = \frac{(\Delta z)\epsilon_{zz}}{y}$$

Dividing by Δz, we obtain

$$-\frac{1}{R_y} = \frac{\epsilon_{zz}}{y} \qquad \frac{1}{R_y} \cong \frac{d^2v}{dz^2} \tag{6-3.1}$$

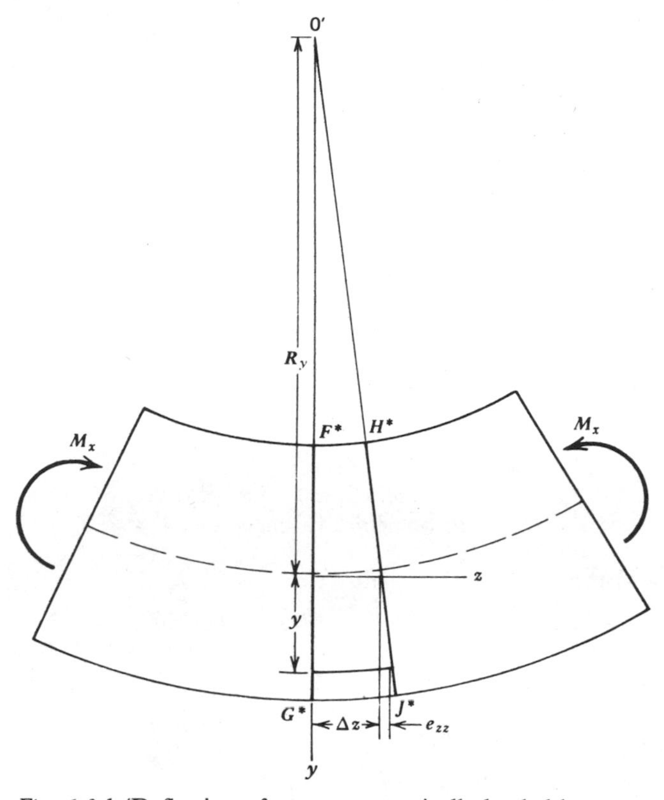

Fig. 6-3.1/Deflection of an unsymmetrically loaded beam.

For linearly elastic behavior, Eqs. (6-2.15) and (6-2.9), with $x = 0$, and Eq. (6-2.4) yield

$$\frac{\epsilon_{zz}}{y} = \frac{M_x}{E(I_x - I_{xy}\tan\alpha)} = \frac{M_x I_y + M_y I_{xy}}{E\left(I_x I_y - I_{xy}^2\right)}$$

which with Eq. (6-3.1) yields

$$\frac{d^2v}{dz^2} = -\frac{M_x}{E(I_x - I_{xy}\tan\alpha)} = -\frac{M_x I_y + M_y I_{xy}}{E\left(I_x I_y - I_{xy}^2\right)} \tag{6-3.2}$$

Note the similarity of Eq. (6-3.2) to the elastic curve equation for symmetrical bending. The only difference is that the term I has been replaced by $(I_x - I_{xy}\tan\alpha)$. The solution of the differential relation Eq. (6-3.2) gives the y-component of the deflection v at any section of the beam. As is indicated in Fig. 6-3.2, the total deflection of the neutral axis at any section of the beam is perpendicular to the neutral axis. Therefore,

$$u = -v\tan\alpha \tag{6-3.3}$$

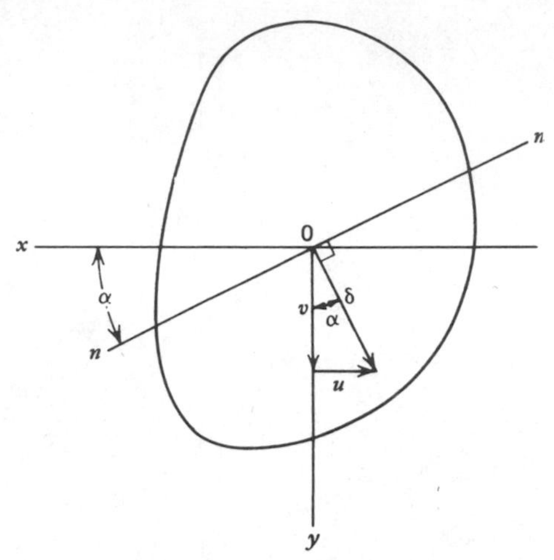

Fig. 6-3.2/Components of deflection of an unsymmetrically loaded beam.

and the total displacement is

$$\delta = \sqrt{u^2 + v^2} = \frac{v}{\cos \alpha} \tag{6-3.4}$$

EXAMPLE 6-3.1

Channel Section Simple Beam

Let the channel section beam in Fig. E6-2.1 be loaded as a simple beam with a concentrated load $P = 35.0$ kN acting at the center of the beam. Determine the maximum tensile and compressive stresses in the beam if $\phi = 5\pi/9$. If the beam is made of an aluminum alloy ($E = 72.0$ GPa), determine the maximum deflection of the beam.

SOLUTION

From Problem E6-2.1

$$\tan \alpha = -\frac{I_x}{I_y} \cot \phi = -\frac{39{,}690{,}000}{30{,}730{,}000} \cot \frac{5\pi}{9} = 0.2277$$

$$\alpha = 0.2239 \text{ rad}$$

$$M = \frac{PL}{4} = \frac{35.0(3.00)}{4} = 26.25 \text{ kN} \cdot \text{m}$$

$$M_x = M \sin \phi = 25.85 \text{ kN} \cdot \text{m}$$

$$\sigma_{\text{tension}} = \frac{25{,}850{,}000[82 - (-70)(0.2277)]}{39{,}690{,}000} = 63.8 \text{ MPa}$$

$$\sigma_{\text{compression}} = \frac{25{,}850{,}000[-118 - 70(0.2277)]}{39{,}690{,}000} = -87.2 \text{ MPa}$$

Since the deflection of the center of a simple beam subjected to a concentrated load in the center is given by the relation $PL^3/48EI$, the y-component of the deflection of the center of the beam is

$$v = \frac{PL^3 \sin\phi}{48EI_x} = \frac{35{,}000(3000)^3 \sin 5\pi/9}{48(72{,}000)(39{,}690{,}000)} = 6.78 \text{ mm}$$

$$u = -v \tan\alpha = -6.78(0.2277) = -1.54 \text{ mm}$$

$$\delta = \sqrt{u^2 + v^2} = 6.95 \text{ mm}$$

EXAMPLE 6-3.2

Cantilever I-Beam

A cantilever beam has a length of 3 m with cross section indicated in Fig. E6-3.2. The beam is constructed by welding two 40 mm by 40 mm steel ($E = 200$ GPa) bars longitudinally to the S-200 × 27.4 steel I-beam ($I_x = 24 \times 10^6$ mm^4 and $I_y = 1.55 \times 10^6$ mm^4). The bars and I-beam have the same yield stress, $Y = 300$ MPa. The beam is subjected to a concentrated load P at the free end at an angle $\phi = \pi/3$ rad with the x-axis. Determine the magnitude of P necessary to initiate yielding in the beam and the resulting deflection of the free end of the beam.

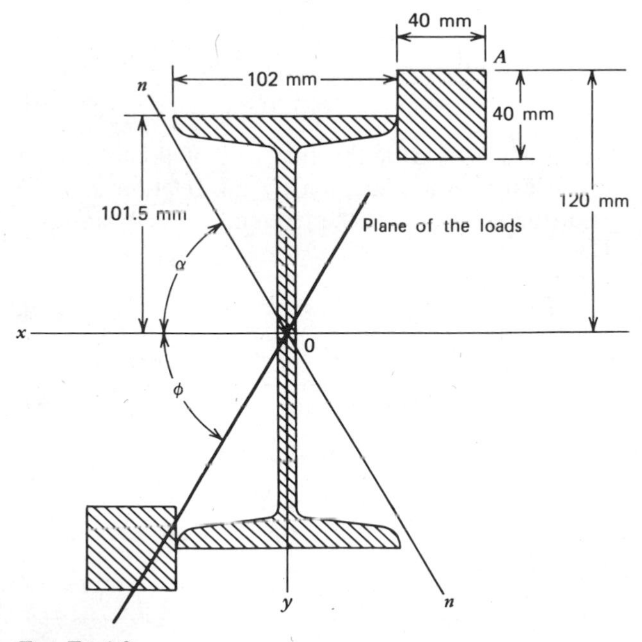

Fig. E6-3.2

SOLUTION

Values of I_x, I_y, and I_{xy} for the composite cross section can be obtained using the procedure outlined in the Appendix.

$$I_x = 56.43 \times 10^6 \text{ mm}^4 \qquad I_y = 18.11 \times 10^6 \text{ mm}^4$$

$$I_{xy} = 22.72 \times 10^6 \text{ mm}^4$$

The orientation of the neutral axis for the beam is given by Eq. (6-2.14). Hence, we find

$$\tan\alpha = \frac{I_{xy} - I_x \cot\phi}{I_y - I_{xy}\cot\phi} = \frac{22.72 \times 10^6 - 56.43 \times 10^6(0.5774)}{18.11 \times 10^6 - 22.72 \times 10^6(0.5774)} = -1.9759$$

$$\alpha = -1.023 \text{ rad}$$

The orientation of the neutral axis $n - n$ is indicated in Fig. E6-3.1. The maximum tensile stress occurs at point A; the magnitude of the stress is obtained using Eq. (6-2.15).

$$M = -3P \text{ N} \cdot \text{m}$$

$$M_x = M \sin\phi = -2.598P \text{ N} \cdot \text{m}$$

$$\sigma_A = Y = \frac{M_x(y_A - x_A \tan\alpha)}{I_x - I_{xy} \tan\alpha}$$

$$P = \frac{Y(I_x - I_{xy}\tan\alpha)}{(-2.598 \times 10^3)(y_A - x_A \tan\alpha)}$$

$$= \frac{300[56.43 \times 10^6 - 22.72 \times 10^6(-1.9759)]}{-2.598 \times 10^3[-120 - (-91)(-1.9759)]} = 39.03 \text{ kN}$$

Since the deflection of the free end of a cantilever beam subjected to symmetrical bending is given by the relation $P_y L^3/3EI$, the y-component of the deflection of the free end of the beam is

$$v = \frac{PL^3 \sin\phi}{3E(I_x - I_{xy}\tan\alpha)}$$

$$= \frac{39.03 \times 10^3(3 \times 10^3)^3(0.8660)}{3(200 \times 10^3)[56.43 \times 10^6 - 22.72 \times 10^6(-1.9759)]} = 17.33 \text{ mm}$$

Hence,

$$u = -v \tan\alpha = 34.25 \text{ mm}$$

and the total displacement of the free end of cantilever beam is

$$\phi = \sqrt{u^2 + v^2} = 38.39 \text{ mm}$$

PROBLEM SET 6-3

1. Determine the deflection of the beam in Problem 6-2.1 if $E = 12.0$ GPa for the yellow pine.

2. The beam in Problem 6-2.3 is made of 7075-T6 aluminum alloy for which $E = 71.7$ GPa. Determine the deflection of the free end of the beam.

   Ans. $v = 13.81$ mm, $u = -7.86$ mm, $\delta = 15.89$ mm

3. The beam in Problem 6-2.4 is made of 7075-T6 aluminum alloy for which $E = 71.7$ GPa. Determine the deflection of the free end of the beam.

4. The beam in Problem 6-2.6 is made of yellow pine for which $E = 12.0$ GPa. Determine the deflection at the center of the beam.

   Ans. $v = 0.33$ mm, $u = -1.49$ mm, $\delta = 1.53$ mm

5. Determine the deflection of the center of the beam in Problem 6-2.8. $E = 200$ GPa.

6. If the beam in Problem 6-2.9 is subjected to a distributed load of $w = 6.5$ kN/m, determine the deflection of the beam at the center of the beam. $E = 200$ GPa.

   Ans. $v = 1.58$ mm, $u = 8.25$ mm, $\delta = 8.40$ mm

7. Determine the deflection of the beam in Problem 6-2.10. $E = 200$ GPa.

8. Determine the deflection of the free end of the beam in Problem 6-2.11. $E = 72.0$ GPa.

   Ans. $v = 33.16$ mm, $u = 6.25$ mm, $\delta = 33.74$ mm

9. Determine the deflection of the midspan of the beam in Problem 6-2.12. $E = 72.0$ GPa.

10. Determine the deflection of the free end of the beam in Problem 6-2.13. $E = 200$ GPa.

    Ans. $v = 4.82$ mm, $u = -3.86$ mm, $\delta = 6.18$ mm

6-4
CHANGE IN DIRECTION OF NEUTRAL AXIS AND INCREASE IN STRESS AND DEFLECTION IN ROLLED SECTIONS DUE TO A VERY SMALL INCLINATION OF PLANE OF LOADS TO A PRINCIPAL PLANE

Some commonly rolled sections such as I-beams and channels are designed so that I_x is many times greater than I_y and $I_{xy}=0$. Equation (6-2.14) indicates that the angle α may be large even though ϕ is nearly equal to $\pi/2$ rad. Thus, the neutral axis of such I-beams and channels is steeply inclined to the horizontal axis (the x-axis) of symmetry when the plane of the loads deviates slightly from the vertical plane of symmetry. As a consequence, the maximum flexure stress and the maximum deflection may be quite large. These rolled sections should not be used as beams unless the lateral deflection is prevented. If lateral deflection of the beam is prevented, unsymmetrical bending cannot occur.

In general, however, I-beams and channels make very poor cantilever beams. The following example illustrates this fact.

EXAMPLE 6-4.1
An Unsuitable Cantilever Beam

An S-610 × 134 I-beam ($I_x = 937 \times 10^6$ mm^4 and $I_y = 18.7 \times 10^6$ mm^4) is subjected to a bending moment M in a plane with angle $\phi = 1.5533$ rad; the plane of the loads is 1^0 ($\pi/180$ rad) clockwise from the (y, z)-plane of symmetry. Determine the neutral axis and the ratio of the maximum tensile stress in the beam to the maximum tensile stress for symmetrical bending.

SOLUTION

The cross section of the I-beam with the plane of the loads is indicated in Fig. E6-4.1. The orientation of the neutral axis for the beam is given by Eq. (6-2.14).

$$\tan\alpha = \frac{-I_x \cot\phi}{I_y} = -\frac{937\times 10^6(0.01746)}{18.7\times 10^6} = -0.8749$$

$$\alpha = -0.7188 \text{ rad}$$

The orientation of the neutral axis is indicated in Fig. E6-4.1. If the beam is subjected to a positive bending moment, the maximum tensile stress is

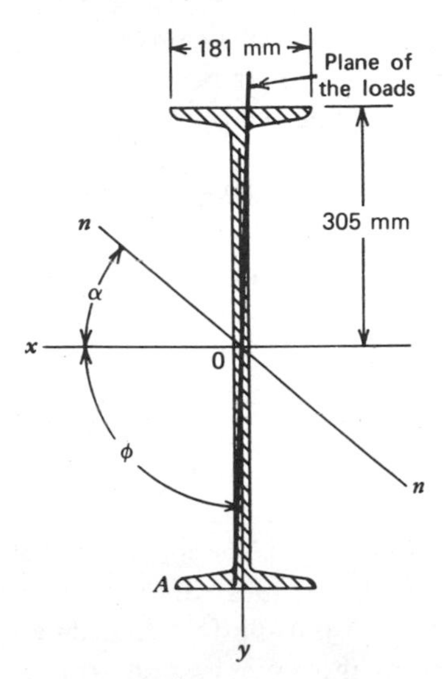

Fig. E6-4.1

located at point A. By Eq. (6-2.10),

$$M_x = M \sin\phi = 0.9998M$$

$$\sigma_A = \frac{0.9998M[305 - 90.5(-0.8749)]}{937 \times 10^6} = 4.099 \times 10^{-7} M \text{ MPa} \quad (1)$$

When the plane of the loads coincide with the y-axis (Fig. E6-4.1), the beam is subjected to symmetrical bending and the maximum bending stress is

$$\sigma_A = \frac{My}{I_x} = \frac{305M}{937 \times 10^6} = 3.255 \times 10^{-7} M \text{ MPa} \quad (2)$$

The ratio of the stress σ_A given by Eq. (1) to that given by Eq. (2) is 1.260. Hence the maximum stress in the I-beam is increased 25.2 percent when the plane of the loads is merely 1^0 from the symmetrical vertical plane.

6-5 FULLY PLASTIC LOAD FOR UNSYMMETRICAL BENDING

A beam of general cross section (Fig. 6-5.1) is subjected to pure bending. The material in the beam has a flat top stress-strain diagram with yield point Y in both tension and compression (Fig. 2-6.2a). At the fully plastic load, the deformations of the beam are unchecked and continue (until possibly the material begins to strain harden). The fully plastic load is the upper limit for failure loads (Art. 3-3) since the deformations of the beam at the outset of any strain hardening generally exceeds design limits for the deformations.

In contrast to the direct calculation of fully plastic load in symmetrical bending (Art. 3-3), an inverse method is required to determine the fully plastic load for a beam subjected to unsymmetrical bending. Although the plane of the loads is generally specified for a given beam, the orientation and location of the neutral axis, when the fully plastic moment is developed at a given section of the beam, must be determined by trial and error. The analysis is begun by assuming a value for the angle α (Fig. 6-5.1). The neutral axis is inclined to the x-axis by the angle α but does not necessarily pass through the centroid as in the case of linearly elastic conditions. The location of the neutral axis is determined by the condition that it must divide the cross-sectional area into equal parts. This follows from the fact that since the yield point stress is the same for tension and compression, the area A_T that has yielded in tension must be equal to the area A_C that has yielded in compression. In other words, the net resultant tension force on the section must be equal to the net resultant compression force.

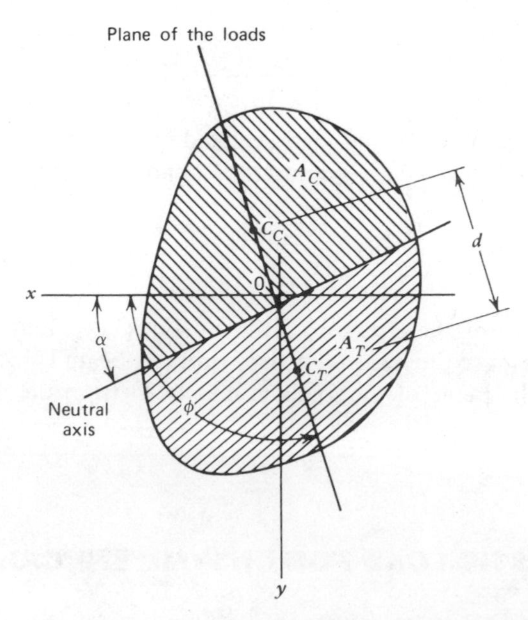

Fig. 6-5.1/Location of a neutral axis for fully plastic bending of an unsymmetrically loaded beam.

The yield point stress Y is uniform over the area A_T that has yielded in tension; the resultant tensile force $P_T = YA_T$ is located at the centroid C_T of A_T. Similarly the resultant compressive force $P_C = YA_C$ is located at the centroid C_C of A_C. The fully plastic moment M_P is given by

$$M_P = YA_T d = \frac{YAd}{2} \tag{6-5.1}$$

where d is the distance between the centroids C_T and C_C as indicated in Fig. 6-5.1. A plane through the centroids C_T and C_C is the plane of the loads for the beam. In case the calculated angle ϕ (Fig. 6-5.1) does not correspond to the plane of the applied loads, a new value is assumed for α and the calculations are repeated. Once the angle ϕ (Fig. 6-5.1) corresponds to the plane of the applied loads, the magnitude of the fully plastic load is calculated by setting the moment of the applied loads equal to M_P given by Eq. (6-5.1).

EXAMPLE 6-5.1

Fully Plastic Moment for Unsymmetrical Bending

A steel beam has the cross section shown in Fig. E6-5.1. The beam is made of a steel having a yield point stress $Y = 280$ MPa. Determine the fully plastic moment for the condition that the neutral axis passes through point B. Locate the resulting neutral axis and the plane of the loads.

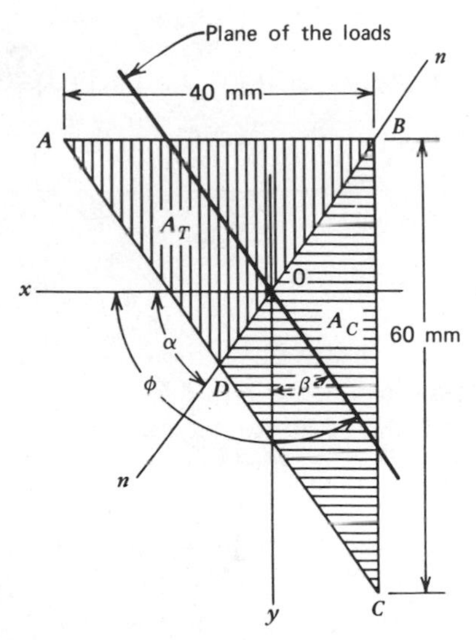

Fig. E6-5.1

SOLUTION

The neutral axis must divide the cross section into two equal areas since the area that has yielded in tension A_T must equal the area that has yielded in compression A_C. The neutral axis extends from point B to point D, the midpoint of AC:

$$\tan \alpha = \frac{30}{20} = 1.5$$

$$\alpha = 0.9828 \text{ rad}$$

The plane of the loads passes through the centroids of areas ABD and BCD. The centroids of these areas are located at $(\frac{20}{3}, -10)$ for ABD and $(-\frac{20}{3}, 10)$ for BCD.

$$\tan \beta = \frac{20/3 - (-20/3)}{10 - (-10)} = 0.6667$$

$$\beta = 0.5880 \text{ rad}$$

$$\phi = \pi/2 + \beta = 2.1588 \text{ rad}$$

The fully plastic moment M_P is equal to the product of the force on either of the two areas (A_T or A_C) and the distance d between the two centroids.

$$d = \sqrt{\left(20 - \frac{20}{3}\right)^2 + (30 - 10)^2} = 24.04 \text{ mm}$$

$$M_P = A_T Y d = \tfrac{1}{2}(40)(30)(280)(24.04) = 4.039 \times 10^6 \text{ N} \cdot \text{mm}$$

$$= 4.039 \text{ kN} \cdot \text{m}$$

PROBLEM SET 6-5

1. The cantilever beam in Problem 6-2.11 is made of a low carbon steel that has a yield stress $Y = 200$ MPa. (a) Determine the fully plastic load P_P for the beam for the condition that $\alpha = 0$. (b) Determine the fully plastic load P_P for the beam for the condition that $\alpha = \pi/6$ rad.

2. The cantilever beam in Problem 6-2.13 is made of a mild steel that has a yield point stress $Y = 240$ MPa. Determine the fully plastic load P_P for the condition that $\alpha = 0$.

   Ans. $P_P = 21.21$ kN at $\phi = 1.2679$ rad

REFERENCES

1. H. L. Langhaar and A. P. Boresi, *Engineering Mechanics*, McGraw-Hill, New York, 1959.
2. A. P. Boresi and P. P. Lynn, *Elasticity in Engineering Mechanics*, Prentice-Hall, Englewood Cliffs, New Jersey, 1974.

Additional References

1. J. N. Goodier, "A Theorem on the Shearing Stress in Beams with Applications to Multicellular Sections," *Journal of the Aeronautical Sciences*, Vol. 11, No. 3, July 1944, pp. 272–280.
2. K. Washizu, "Some Considerations on the Center of Shear," *Transactions of Japan Society for Aeronautical and Space Sciences*, Vol. 9, No. 15, 1966, pp. 77–83.
3. A. Weinstein, "The Center of Shear and the Center of Twist," *Quarterly of Applied Mathematics*, Vol. 5, No. 1, 1947, pp. 97–99.

CHAPTER 7

SHEAR CENTER FOR THIN-WALL BEAM CROSS SECTIONS

7-1 APPROXIMATIONS EMPLOYED FOR SHEAR IN THIN-WALL BEAM CROSS SECTIONS

The definition of the bending axis of a straight beam with constant cross section was given in Chapter 6; it is the axis through the shear centers of the cross sections of the beam. To bend a beam without twisting, the plane of the loads must contain the axis of bending, that is, the plane of the loads must pass through the shear center of every cross section of the beam.

For a beam with cross sections that possess two or more axes of symmetry or antisymmetry, the bending axis is the same as the longitudinal centroidal axis, because for each cross section the shear center and the centroid coincide. However, for cross sections with only one axis of symmetry, the shear center and the centroid do not coincide, as is shown later in this chapter. For example, consider the equal-leg angle section shown in Fig. 7-1.1. Let the beam cross section be oriented so that the principal axes of inertia (X, Y) are directed horizontally and vertically. The beam bends symmetrically without twist (Chapter 6), if it is loaded by a vertically directed force P that passes through the shear center C (Fig. 7-1.1*b*). As is shown later, the shear center C coincides approximately with the intersection of the center lines of the two legs of the angle section. When the load P is applied at the centroid 0 of the cross section, the beam bends and twists (Fig. 7-1.1*a*).

The determination of the exact location of the shear center for an arbitrary cross section is beyond the scope of this book (see reference 1). However, an approximate solution is presented in this chapter, which gives reasonably accurate results for "thin-wall" cross sections. The simplifying assumptions upon which the approximate solution is based may be clarified by reference to Fig. 7-1.2. In Fig. 7-1.2, the cross section shown is that of the beam in Fig. 7-1.1*b* and is obtained by passing a cutting plane perpendicular to the bending axis through the beam. The

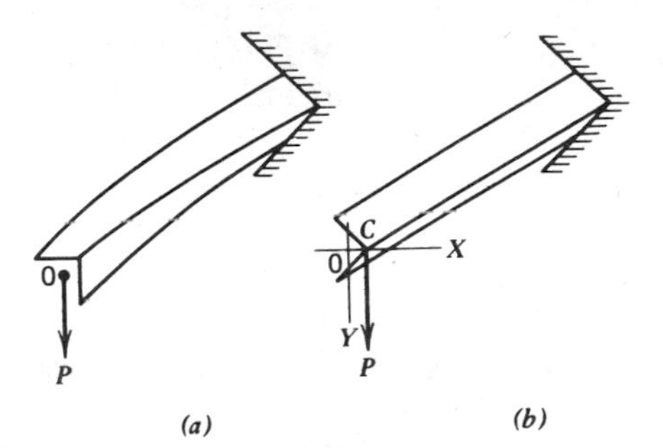

Fig. 7-1.1/Effect of applying load through shear center. (a) Load P applied at point 0 produces twist and bending. (b) Load P applied at point C produces bending only.

view shown is obtained by looking from the support toward the end of the beam at which P is applied.

For equilibrium of the beam element so obtained, the shearing stresses on the cut cross section must balance the load P. However, the shearing stresses in the cross section are difficult to compute exactly. Hence, simplifying approximations are employed. Accordingly, consider a portion of the legs of the cross section, shown enlarged in Fig. 7-1.2b. Let axes x, y, z be chosen so that the (x, y)-axes are tangent and normal, respectively to the upper leg, and let the z-axis be taken perpendicular to

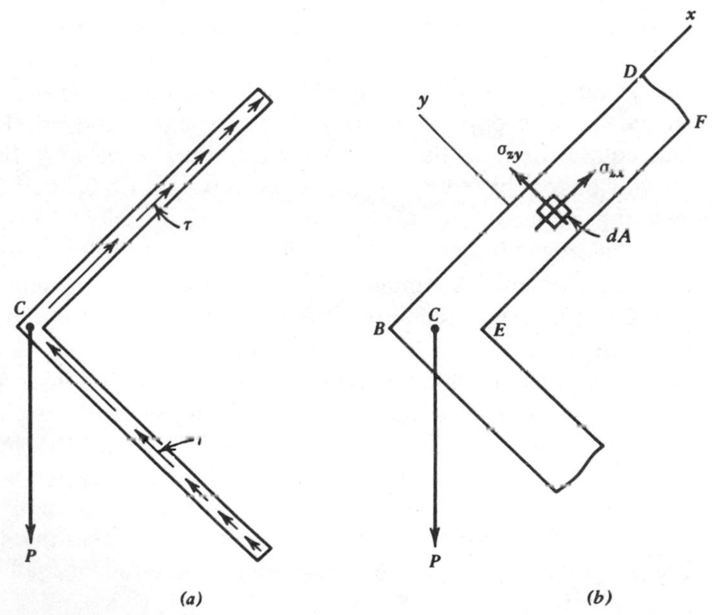

Fig. 7-1.2/Shearing stress distribution in an equal-leg angle section.

the cross section (the plane of Fig. 7-1.2*b*), and directed positively along the axis of the beam from the load P to the support. Then by the stress notation defined in Chapter 1, the shearing stress components in the cross section of the beam are σ_{zx} and σ_{zy} as shown. Since the shearing stresses on the lateral surfaces of the beam are zero, $\sigma_{yz} = 0$. Hence, σ_{zy} vanishes at BD and EF (since $\sigma_{yz} = \sigma_{zy}$; see Art. 1-3). Hence, since $\sigma_{zy} = 0$ at BD and EF, and since the wall thickness between BD and EF is small (thin-wall) with respect to the length of the legs of the cross section, we assume that σ_{zy} does not change greatly (remains approximately zero) through the wall. Hence, the effect of σ_{zy} (the shearing stress in the thickness direction) is ignored in the following discussion. In addition, it is assumed that the shearing stress component σ_{zx} (along the legs) is approximately constant through the wall thickness and is equal to the average tangential shearing stress τ in the wall (Fig. 7-1.2*a*). With these approximations for σ_{zy} and σ_{zx}, we find that a reasonably accurate simple estimate of the shear center may be obtained.

Since the above approximations are based on the concept of "thin-walls," the question as to what constitutes a thin wall arises. It so happens that the answer to this question is not of great importance, as the following example illustrates.

Consider the cross section, shown in Fig. 7-1.3, of a "thick-wall" cantilever beam. The view is taken looking from the support to the loaded (unrestrained) end. The approximate solution presented later in this chapter indicates that the shear center is located at point C', a distance $0.707t$ from the corner as shown. The applied load P is shown acting at point C'. By the theory of elasticity, the shear center is located[2] at point C, $0.237t$ to the right of point C'. To examine the effects of this discrepancy, let us apply at point C two forces P' and P'' equal in magnitude and opposite in sense. (The forces P' and P'' do not disturb the equilibrium of the beam.) Hence, the beam may be considered to be loaded by force P acting at point C' or, equivalently, by a bending moment due to P' (equal in magnitude to force P) and a twisting moment due to P and P''. For example at the support, the bending moment (maximum is equal to PL (L is the length of the beam) and the torque T is equal to $0.237tP$. In practice, the ratio of the bending moment to the twisting moment is equal to 100 or more. In addition, the torsion theory presented in Chapter 5 shows that a "thick-wall" section, such as the one shown in Fig. 7-1.3, has considerable torsional resistance. Therefore, in applications in which length L of the beam is large compared to t the shearing stresses due to the torque ($T = 0.237tP$) are so small compared to the maximum bending stress that they can be neglected. Thus, for these beams the shearing stresses due to the torque are no greater than the shearing stresses due to shear load P'; both are often neglected.

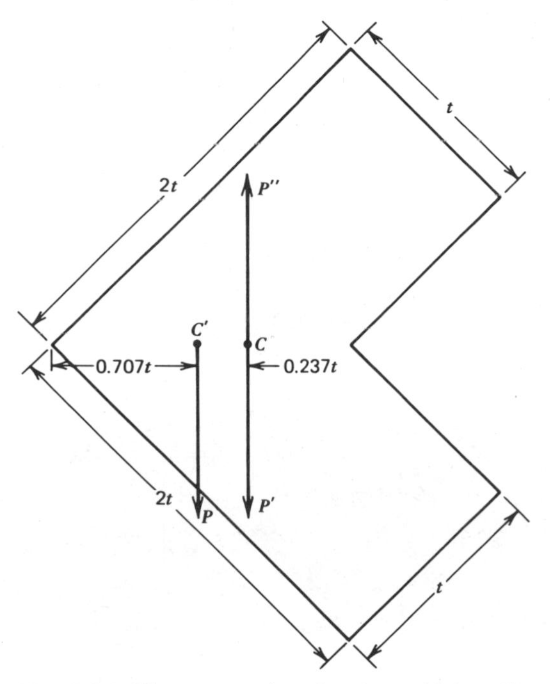

Fig. 7-1.3/Shear center location for a thick-wall angle section.

7-2
SHEAR FLOW IN THIN-WALL BEAM CROSS SECTIONS

The average shearing stress τ at each point in the walls of the beam cross section is assumed to have a direction tangent to the wall. The product of this shearing stress and wall thickness t defines the *shear flow* q (Chapter 5); thus,

$$q = \tau t \tag{7-2.1}$$

The equation that will be now derived for determining the shear flow q assumes that the beam material remains linearly elastic and that the flexure formula (Eq. 6-1.1) is valid. Hence, we assume that the plane of the loads contains the bending axis of the beam and is parallel to one of the principal axes of inertia. It is convenient to consider a beam cross section that has one axis of symmetry (Fig. 7-2.1). If the load P is parallel to the y-axis and passes through the shear center C'', the x-axis is the neutral axis for linearly elastic behavior and the flexure formula is valid. The derivation of the formula for q requires that both the bending moment M_x and total shear V_y be defined; load P is taken in the negative y-direction so that both M_x and V_y will be positive.

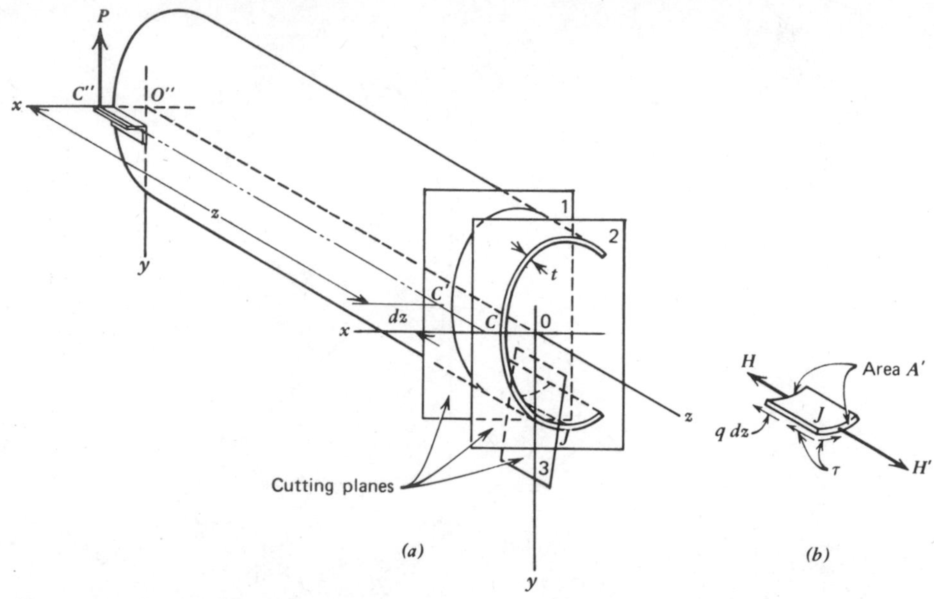

Fig. 7-2.1/Shear flow in a beam having a symmetrical cross section.

We wish to determine the shear flow q in the cross section of the beam in Fig. 7-2.1a at a distance $z + dz$ from load P at point J in the cross section. The free body diagram necessary to determine q is obtained by three cutting planes. Cutting planes 1 and 2 are perpendicular to the z-axis at distances z and $z + dz$ from the load P. Cutting plane 3 is parallel to the z-axis and perpendicular to the lateral surface of the beam at J. The free body diagram removed by the three cutting planes is indicated in Fig. 7-2.1b. The normal stress distributions σ_{zz} as given by the flexure formula act on the faces made by cutting planes 1 and 2. The resulting forces on these faces of area A' are parallel to the z-axis and are indicated in Fig. 7-2.1b as H and H', respectively. Since the forces H and H' are unequal in magnitude, equilibrium of forces in the z-direction is maintained by the force qdz on the face made by cutting plane 3. Therefore,

$$q\,dz = H' - H \tag{7-2.2}$$

Now, integrations of σ_{zz} over the faces with area A' at sections 1 and 2 yield (with the flexure formula)

$$H = \int_{A'} \sigma_{zz}\, dA = \int_{A'} \frac{M_x y}{I_x}\, dA$$

and

$$H' = \int_{A'} (\sigma_{zz} + d\sigma_{zz})\, dA = \int_{A'} \frac{(M_x + dM_x)\, y}{I_x}\, dA$$

Substitution of these two relations in Eq. (7-2.2) and solution for q yields

$$q = \frac{dM_x}{dz}\frac{1}{I_x}\int_{A'} y\,dA$$

According to beam theory, the total shear V_y in the cross section of a beam is given by $V_y = dM_x/dz$. Hence, since $\int_{A'} y\,dA = A'\bar{y}'$ where $\bar{y}'$ is the distance from the x-axis to the centroid of A', we may express q as

$$q = \frac{V_y A'\bar{y}'}{I_x}$$

Furthermore, since the value of the shearing stress τ in the longitudinal section cut by plane 3 (Fig. 7-2.1) is the same as the shearing stress in the cross section cut by plane 2, the shear flow in the cross section at point J is

$$q = \tau t = \frac{V_y A'\bar{y}'}{I_x} \tag{7-2.3}$$

where t is the wall thickness at point J.

Equation (7-2.3) is used to locate the shear center of thin-wall beam cross sections for both symmetrical and unsymmetrical bending. The method is demonstrated in Art. 7-3 for beam cross sections made up of moderately thin walls but with walls sufficiently thick so that local buckling is not a problem (see Art. 3-6). The relative dimensions necessary to prevent local buckling have been investigated both analytically[3,4] and experimentally.[5]

In many applications (girders, for example), the beam cross sections are built up by joining stiff longitudinal stringers by thin webs. The webs are generally stiffened at several locations along the length of the beam. The shear center location for beams of this type are considered in Art. 7-4.

7-3 SHEAR CENTER FOR A CHANNEL SECTION

A cantilever beam subjected to bending loads in a plane perpendicular to the axis x of symmetry of the beam is shown in Fig. 7-3.1. We wish to locate the plane of the loads so that the channel bends without twisting. In other words, we wish to locate the bending axis CC' of the beam, or the shear center C of any cross section AB.

The loads P_1 and P_2 in Fig. 7-3.1a are replaced by their resultant V; then V is transformed into a force and a couple at section AB by introducing, at the shear center C whose location is as yet unknown, two equal and opposite forces V' and V'', each equal in magnitude to V. The forces V and V'' constitute the external bending couple at section AB, which is held in equilibrium by the internal resisting moment at section AB in accordance with the flexure formula, Eq. (6-1.1); the distribution of the normal stress σ_{zz} on section AB is shown in Fig. 7-3.1*a*. The force V' is located at a distance e from the center of the web of the channel, as indicated in Figs. 7-3.1*a* and 7-3.1*b*. Force V' is resisted by shearing stress τ or shear flow q (Eq. 7-2.3), in cross section AB. Since the shear flow is directed along the straight sides of the channel, it produces forces F_1, F_2, and F_3, which lie in the cross section as indicated in Fig. 7-3.1*b*. Accordingly, by equilibrium

$$\sum F_x = F_2 - F_1 = 0 \tag{7-3.1}$$

$$\sum F_y = V' - F_3 = 0 \tag{7-3.2}$$

$$\sum M_z = V'e - F_1 h = 0 \tag{7-3.3}$$

The magnitude of V' is $P_1 + P_2$. Hence, it is known. Therefore, the determination of the distance e from the center line of the web to the shear center requires only that the force F_1 ($= F_2$) be determined.

To determine F_1, it is convenient to think of the beam cross section as made up of line segments (Fig. 7-3.1*a*) with specified thicknesses. Since the forces F_1, F_2, F_3 are assumed to lie along the center line of the walls, the cross section is idealized as three narrow rectangles of lengths b, h,

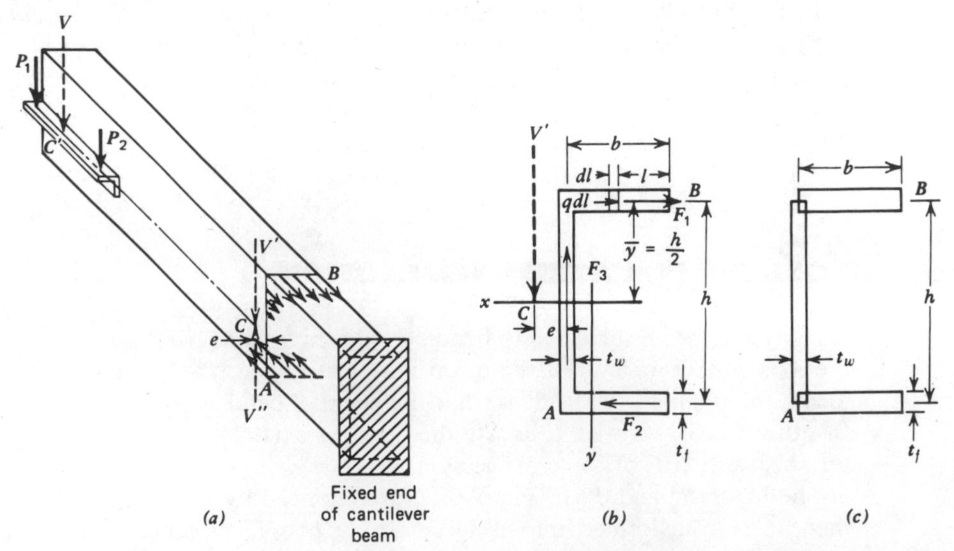

Fig. 7-3.1/Shear center for a channel section. (*a*) Channel section beam. (*b*) Location of *C*. (*c*) Idealized areas.

and b as indicated in Fig. 7-3.1c; note that the actual and idealized cross-sectional areas are equal since the three areas overlap. However, the moments of inertia of the actual and idealized cross sections differ from each other slightly. The moment of inertia of the idealized area is

$$I_x = \frac{1}{12} t_w h^3 + 2bt_f \left(\frac{h}{2}\right)^2 + 2\frac{1}{12} bt_f^3$$

This result may be simplified further by neglecting the third term, since

Table 7-3.1

Locations of Shear Centers for Sections Having One Axis of Symmetry

Section	Shear center location	
Fig. A	$\frac{e}{b} = \dfrac{1 + \frac{2b_1}{b}\left(1 - \frac{4b_1^2}{3h^2}\right)}{2 + \frac{h}{3b} + \frac{2b_1}{b}\left(1 + \frac{2b_1}{h} + \frac{4b_1^2}{3h^2}\right)}$	All three of these equations reduce to Eq. (7-3.6) when $b_1 = 0$ and $t_w = t_f = t$
Fig. B	$\frac{e}{b} = \dfrac{1 + \frac{2b_1}{b}\left(1 - \frac{4b_1^2}{3h^2}\right)}{2 + \frac{h}{3b} + \frac{2b_1}{b}\left(1 - \frac{2b_1}{h} + \frac{4b_1^2}{3h^2}\right)}$	
Fig. C	$\frac{e}{b} = \dfrac{1 - \frac{b_1^2}{b^2}}{2 + \frac{2b_1}{b} + \frac{t_w h}{3t_f b}}; \quad b_1 < b$	
Fig. D	$\frac{e}{b} = \dfrac{\frac{b_1^2}{\sqrt{2}\, b^2}\left(3 - \frac{2b_1}{b}\right)}{1 + \frac{3b_1}{b} - \frac{3b_1^2}{b^2} + \frac{b_1^3}{b^3}}$	
Fig. E	$\frac{e}{R} = \dfrac{2(\sin\theta - \theta\cos\theta)}{\theta - \sin\theta\cos\theta}$ For semicircle, $\theta = \frac{\pi}{2}$ and $\frac{e}{R} = 4/\pi$	

Table 7-3.1

Continued

Fig. F

$$\frac{e}{R}=\frac{12+6\pi\frac{b+b_1}{R}+6\left(\frac{b}{R}\right)^2+12\frac{b}{R}\cdot\frac{b_1}{R}+3\pi\left(\frac{b_1}{R}\right)^2-4\left(\frac{b_1}{R}\right)^3\frac{b}{R}}{3\pi+12\frac{b+b_1}{R}+4\left(\frac{b_1}{R}\right)^2\left(3+\frac{b_1}{R}\right)}$$

For $b_1=0$:

$$\frac{e}{R}=\frac{4+2\pi\frac{b}{R}+2\left(\frac{b}{R}\right)^2}{\pi+4\frac{b}{R}}$$

For $b=0$:

$$\frac{e}{R}=\frac{3\left[4+\frac{2b_1\pi}{R}+\pi\left(\frac{b_1}{R}\right)^2\right]}{3\pi+4\left(\frac{b_1}{R}\right)^3+12\frac{b_1}{R}+12\left(\frac{b_1}{R}\right)^2}$$

for the usual channel section t_f is small compared to b or h. Thus, we write

$$I_x=\tfrac{1}{12}t_wh^3+\tfrac{1}{2}t_fbh^2 \tag{7-3.4}$$

The force F_1 may be found from the shear flow equation

$$F_1=\int_0^b q\,dl=\frac{V_y}{I_x}\int_0^b A'\bar{y}'\,dl=\frac{V_yt_fh}{2I_x}\int_0^b l\,dl=\frac{V_yt_fb^2h}{4I_x} \tag{7-3.5}$$

where q is given by Eq. (7-2.3). The distance e to the shear center of the channel section is determined by substituting Eqs. (7-3.4) and (7-3.5) into Eq. (7-3.3) with the magnitude of V' set equal to that of V_y. Thus, we find

$$e=\frac{b}{2+\frac{1}{3}\frac{t_wh}{t_fb}} \tag{7-3.6}$$

Because of the assumptions employed and the approximations used, Eq. (7-3.6) gives an approximate location of the shear center for channel sections. The error is small for thin-wall sections. The approximate locations of the shear center for several other thin-wall sections with an axis of symmetry are given in Table 7-3.1.

EXAMPLE 7-3.1

Shear Center for Channel with Sloping Flanges

A 4 mm thickness plate of steel is formed into the cross section shown in Fig. E7-3.1*a*. Locate the shear center for the cross section.

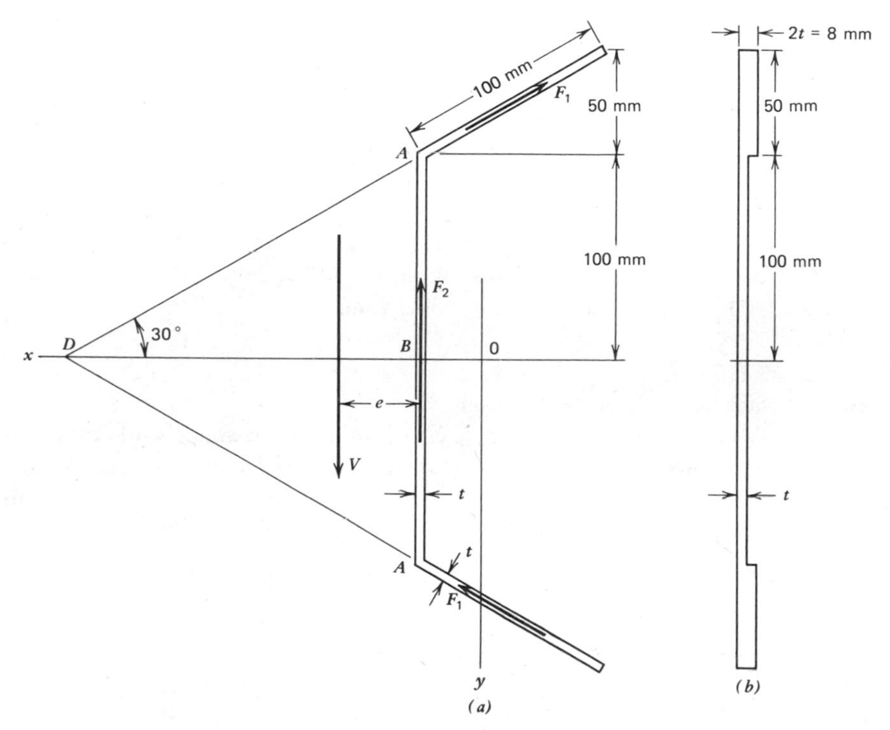

Fig. E7-3.1

SOLUTION

For simplicity, we approximate the actual cross section (Fig. E7-3.1*a*) by the cross section shown in Fig. E7-3.1*b*. The moment of inertia about the *x*-axis for the cross section in Fig. E7-3.1*b* closely approximates that for the actual cross section in Fig. E7-3.1*a*.

$$I_x = \frac{8(300)^3}{12} - \frac{4(200)^3}{12} = 15{,}330{,}000 \text{ mm}^4$$

Because of the shear flow, forces F_1 and F_2 are developed in the three legs of the cross section. The magnitude of force F_1 requires integration; therefore, it is convenient to use point D as the moment center so that the magnitude of F_1 is not required. Since the shear flow from A to B to A varies parabolically, the average shear flow is equal to the shear flow at

A plus 2/3 of the difference between the shear flow at B and the shear flow at A.

$$q_A = \frac{V}{I_x} A'\bar{y} = \frac{V}{I_x}(100)(4)(125) = 50{,}000\frac{V}{I_x}$$

$$q_B = q_A + \frac{V}{I_x}(100)(4)(50) = 70{,}000\frac{V}{I_x}$$

$$q_{ave} = q_A + \frac{2}{3}(q_B - q_A) = 63{,}330\frac{V}{I_x}$$

$$F_2 = 200 q_{ave} = 63{,}330\frac{V}{I_x}(200) = 12{,}670{,}000\frac{V}{I_x}$$

With point D as the moment center, the clockwise moment of V must equal the counterclockwise moment of F_2. Thus, we have $(173.2 - e)\ V = 173.2\ F_2$, and, hence, $e = 30.1$ mm.

EXAMPLE 7-3.2
Shear Center for Unequal-Leg Channel

A beam has an unsymmetrical section whose shape and dimensions are as shown in Fig. E7-3.2*a*. Locate the shear center in the plane of the cross section through which the plane of the loads of the beam must pass if the beam is to bend without twisting of its cross sections.

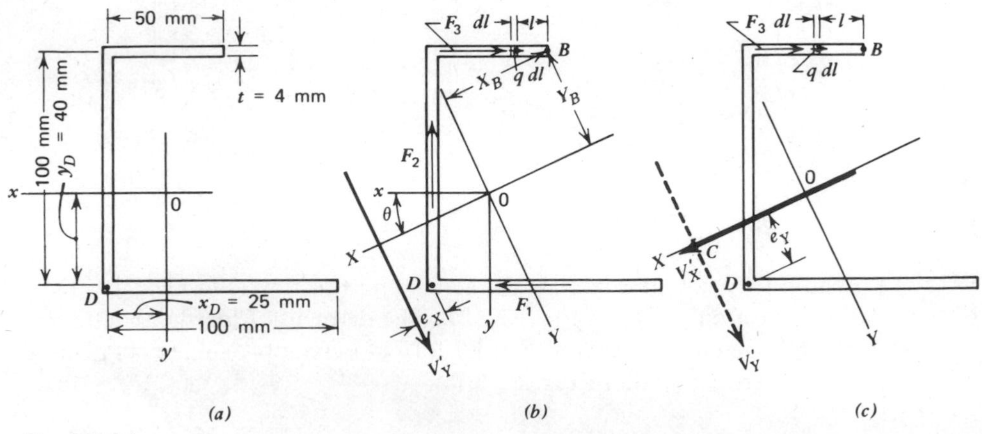

Fig. E7-3.2

SOLUTION

Centroidal x- and y-axes are chosen, which are parallel to the sides of the thin-wall legs of the cross section. The origin 0 of the coordinate axes is found to be located at $x_D = 25.0$ mm and $y_D = 40.0$ mm. To apply the theory to unsymmetrical sections, we use principal X- and Y-axes. As indicated in the Appendix, the principal axes may be described in terms of I_x, I_y, and I_{xy}. These values are $I_x = 1.734 \times 10^6$ mm^4, $I_y = 0.876 \times 10^6$ mm^4, and $I_{xy} = -0.500 \times 10^6$ mm^4. The angle θ between the x-axis and the X-axis is obtained by the relation (Eq. (A-3.4))

$$\tan 2\theta = -\frac{2I_{xy}}{I_x - I_y} = \frac{-2(-0.500 \times 10^6)}{1.734 \times 10^6 - 0.876 \times 10^6} = 1.166$$

from which $\theta = 0.4308$ rad. Since θ is positive, the X-axis is located counterclockwise from the x-axis. By using the equations in the Appendix, we find the principal moments of inertia to be $I_X = 1.964 \times 10^6$ mm^4 and $I_Y = 0.646 \times 10^6$ mm^4. The principal axes are shown in Figs. E7-3.2b and c.

The shear center C is located by considering two separate cases of loading (without twisting) in two orthogonal planes of the loads. The intersection of these two planes of loads determines the shear center C. Thus, assume that the resultant V'_Y of unbalanced loads on one side of the section in Fig. E7-3.2b is parallel to the Y-axis. Since V'_Y is assumed to pass through the shear center, the beam bends without twisting and the X-axis is the neutral axis; hence, the flexure formula and Eq. (7-2.3) apply. Because of the shear flow, forces F_1, F_2, and F_3 are developed in the three legs of the cross section (Fig. E7-3.2b). Only the magnitude of F_3 is required if point D is chosen as the moment center. In order to determine F_3, it is necessary that the shear flow q be determined as a function of l, the distance from point B. The coordinates of point B, the shear flow q, and force F_3 are determined as follows:

$$X_B = x_B \cos\theta + y_B \sin\theta = -25(0.9086) - 60(0.4176) = -47.77 \text{ mm}$$

$$Y_B = y_B \cos\theta - x_B \sin\theta = -60(0.9086) + 25(0.4176) = -44.08 \text{ mm}$$

$$q = \frac{V_Y}{I_X} A'\overline{Y}' = \frac{V_Y}{I_X} tl\left(|Y_B| + \frac{1}{2} l \sin\theta\right)$$

$$F_3 = \int_0^{50} q\, dl = \frac{V_Y t}{I_X} \int_0^{50} l\left(44.08 + \frac{0.4176}{2} l\right) dl = 0.1299 V_Y$$

Using the fact that $V'_Y = V_Y$ (the total shear at the section) we obtain the distance e_X from point D to force V'_Y, which passes through the shear center, from the equilibrium moment equation. Therefore,

$$V_Y e_X - 100 F_3$$

or

$$e_X = 12.99 \text{ mm}$$

Next assume that the resultant of the unbalanced loads on one side of the section in Fig. E7-3.2c is V'_X and it is parallel to the X-axis. Since V'_X is assumed to pass through the shear center, the beam bends without twisting and the Y-axis is the neutral axis. The shear flow q and force F_3

are given by

$$q = \frac{V_X}{I_Y} A'\bar{X}' = \frac{V_X}{I_Y} tl\left(|X_B| - \frac{1}{2} l \cos\theta\right)$$

$$F_3 = \int_0^{50} q\, dl = \frac{V_X t}{I_Y} \int_0^{50} l\left(47.77 - \frac{0.9086}{2} l\right) dl = 0.2525 V_X$$

Set $V_X' = V_X$ (the total shear at the section) and take moments about point D. Therefore,

$$V_X e_Y = 100 F_3$$

$$e_Y = 25.25 \text{ mm}$$

In terms of principal coordinates, the shear center C is located at

$$X_C = x_D \cos\theta + y_D \sin\theta + e_X = 52.41 \text{ mm}$$

$$Y_C = y_D \cos\theta - x_D \sin\theta - e_Y = 0.66 \text{ mm}$$

The x- and y-coordinates of the shear center C are

$$x_C = X_C \cos\theta - Y_C \sin\theta = 47.35 \text{ mm}$$

$$y_C = Y_C \cos\theta + X_C \sin\theta = 22.49 \text{ mm}$$

PROBLEM SET 7-3

1. Locate the shear center for the hat section beam shown in Fig. *A* of Table 7-3.1 by deriving the expression for e.

2. Verify the relation for e for the cross section shown in Fig. *B* of Table 7-3.1.

3. Locate the shear center for an unsymmetrical I-beam shown in Fig. *C* of Table 7-3.1 by deriving the expression of e.

4. Show that the shear center for the cross section in Fig. *D* of Table 7-3.1 is located at distance e from the tip as shown.

5. Derive the relation for e for the circular arc cross section shown in Fig. *E* of Table 7-3.1.

6. Derive the relation for e for the helmet cross section shown in Fig. *F* of Table 7-3.1.

7. An extruded bar of aluminum alloy has the cross section shown in Fig. P7-3.7. Locate the shear center for the cross section. *Note*: Small differences in the value of e may occur because of differences in the approximations of I_x.

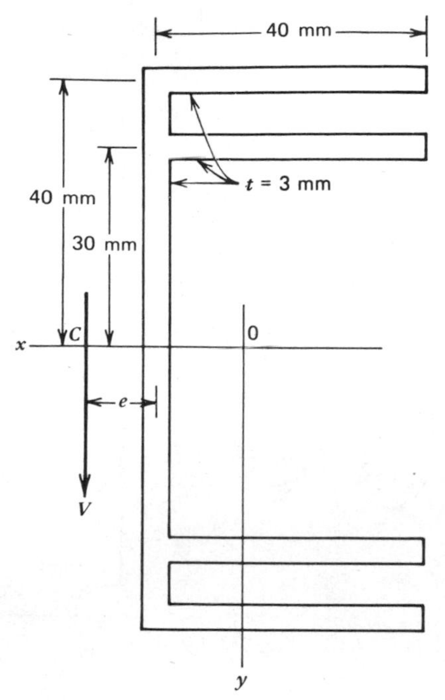

Fig. P7-3.7

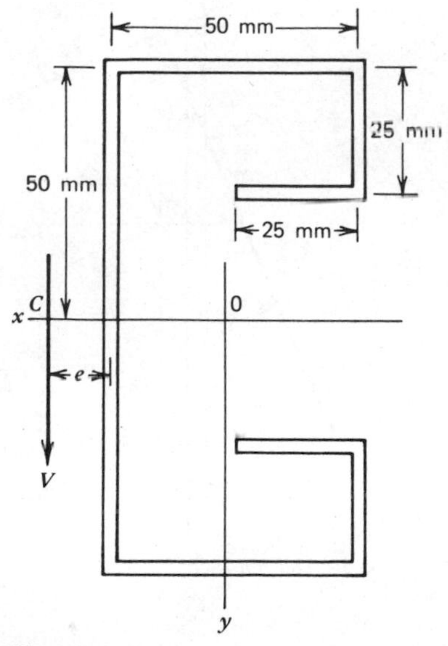

Fig. P7-3.8

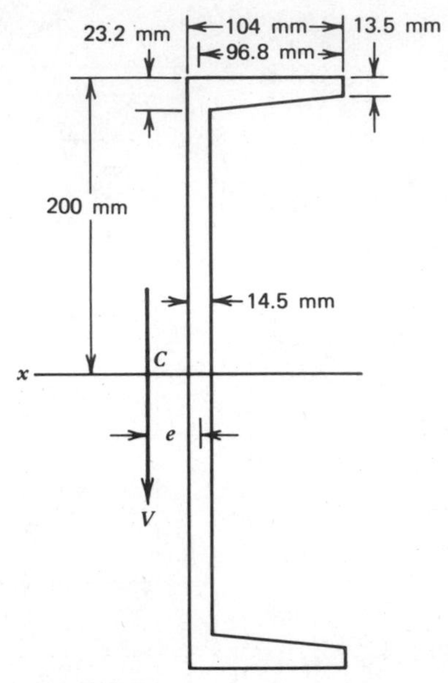

Fig. P7-3.9

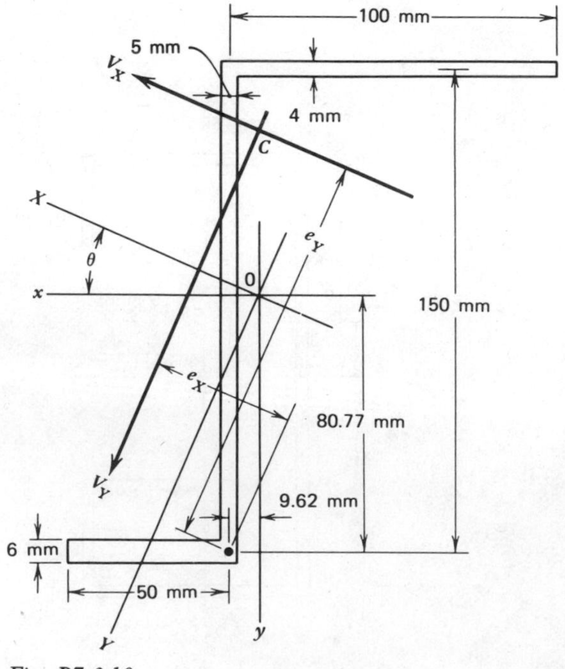

Fig. P7-3.10

8. A 2.50 mm thick plate of steel is formed into the cross section shown in Fig. P7-3.8. Locate the shear center for the cross section.

   Ans. $e = 37.14$ mm

9. A rolled steel channel (400 mm, 71.47 kg/m) has the dimensions shown in Fig. P7-3.9. Locate the shear center for the cross section.

10. A beam has the cross section shown in Fig. P7-3.10. Locate the shear center for the cross section.

    Ans. $\theta = -0.4215$ rad, $e_X = 45.50$ mm, $e_Y = 126.88$ mm

11. An extruded bar of aluminum alloy has the cross section shown in Fig. P7-3.11. Locate the shear center for the cross section.

12. A 4 mm thick plate of steel is formed into the cross section shown in Fig. P7-3.12. Locate the shear center for the cross section.

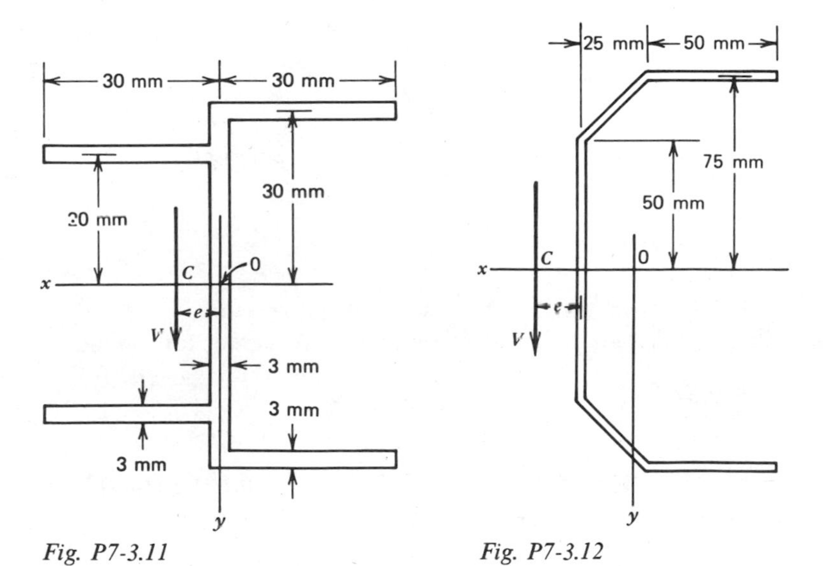

Fig. P7-3.11 *Fig. P7-3.12*

Ans. $e = 28.50$ mm

13. A 5 mm thick plate of steel is formed into the cross section shown in Fig. P7-3.13. Locate the shear center for the cross section.

14. A 5 mm thickness plate of steel is formed into the semicircular shape shown in Fig. P7-3.14. Locate the shear center for the cross section.

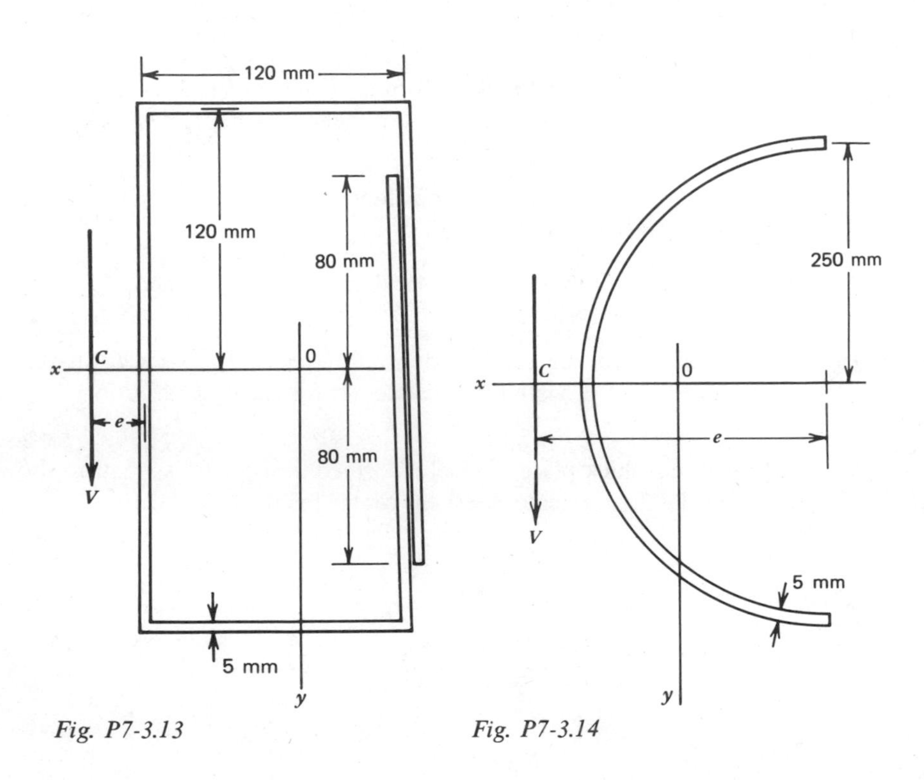

Fig. P7-3.13

Fig. P7-3.14

Ans. $e = 318.31$ mm

15. The two horizontal top and bottom arms of the extruded bar of Fig. P7-3.7 are removed. Locate the shear center for the modified section.

7-4 SHEAR CENTER OF COMPOSITE BEAMS FORMED FROM STRINGERS AND THIN WEBS

Often, particularly in the aircraft industry, beams are built up by welding or riveting longitudinal stiffeners, called stringers, to thin webs. Such beams are often designed to carry large bending loads and small shear loads. Two examples of cross sections of such beams are shown in Fig. 7-4.1. A beam whose cross section consists of two T-section stringers joined to a semicircular web is shown in Fig. 7-4.1*a*, and a beam whose cross section consists of a vertical web joined to two angle section

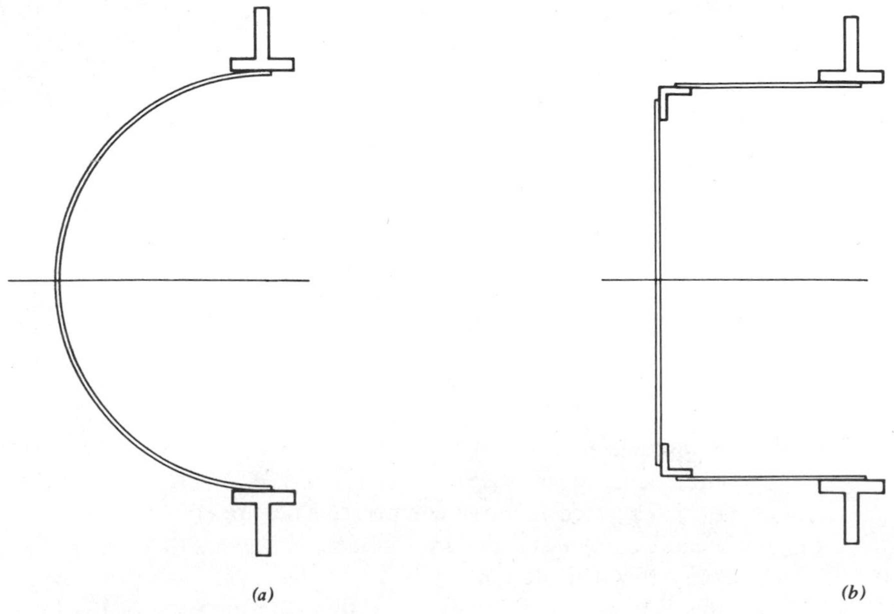

Fig. 7-4.1/Typical beam cross sections built up of stringers and thin webs.

stringers which, in turn, are joined to two horizontal webs that support two T-section stringers is shown in Fig. 7-4.1*b*.

In practice, beams with cross sections similar to those shown in Fig. 7-4.1*b* have webs so thin that they may buckle, particularly in aircraft applications. Consider, for example, a cantilever beam subject to end load (Fig. 7-4.2). Before buckling, the state of stress at the neutral surface is pure shear, as indicated on the volume element *A* in Fig. 7-4.2. After buckling, the state of stress is as indicated on volume element *B* in Fig. 7-4.2. A photograph[6] of a similar beam with a buckled web is indicated in Fig. 7-4.3. After buckling of the web, the shear in the beam is carried by diagonal tension (block *B*, Fig. 7-4.2).[3,4,5] To strengthen such beams, transverse stiffeners are placed at each end and along the beam, as indicated in Fig. 7-4.3. These stiffeners restrain relative motion between the longitudinal stiffeners so that the beam may develop resistance to the diagonal tension. In addition, transverse stiffeners are located at sections where loads are applied to the beam. In this chapter, we assume that the web thickness is sufficiently thick so that the shear flow does not cause web buckling.

The calculation of the shear center location for beam cross sections similar to those shown in Fig. 7-4.1 is based on two simplifying assumptions: (1) that the web does not support tensile or compressive stresses due to bending loads and (2) that the shear flow is constant in a web between pairs of transverse stringers. As noted above, the actual webs of

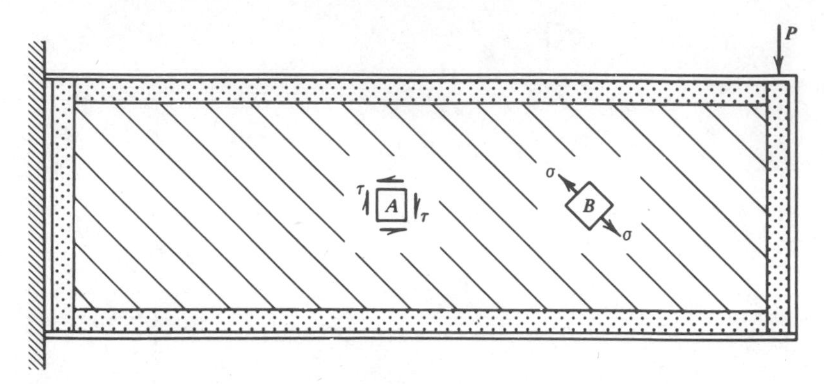

Fig. 7-4.2/Web buckling in a cantilever beam.

these composite beams are often so thin that they may buckle under small compressive stresses (Fig. 7-4.2 or 7-4.3). Therefore, the webs should not be expected to carry compressive flexure stresses. In general, the webs can carry tensile flexure stresses. However, this capability is usually ignored in their design.

Since the web walls are usually very thin, the moment of inertia for symmetrical cross sections of composite beams is approximated by the relation

$$I_x = 2\sum_{i=1}^{n} A_i \bar{y}_i^2 \tag{7-4.1}$$

where $2n$ is the number of stringers, A_i are the cross-sectional areas of the stringers on one side of the neutral axis (x-axis), and $\bar{y}_i$ are the distances from the neutral axis to the centroids of the areas A_i. Equation (7-4.1) discards the effect of the web. Hence, I_x is underestimated.

Fig. 7-4.3/Diagonal-tension beam.

Therefore, with this value of I_x the computed flexure stresses are overestimated (higher than the true stresses).

Note: Transverse shearing stresses are developed in the areas A_i of the stringers so that the stringers carry part of the total shear load V_y applied to the beam. However, the part of V_y carried by each stringer is usually ignored. This error is corrected in part by assuming that each web is extended to the centroid of the area of each stringer, thus increasing the contribution of the web. The procedure is demonstrated in the following example.

EXAMPLE 7-4.1

Shear Center for Composite Beam

A composite beam has a symmetrical cross section as shown in Fig. E7-4.1. A vertical web with a thickness of 2 mm is riveted to two square stringers. Two horizontal webs, with a thickness of 1 mm, are riveted to the square stringers and to the T-section stringers. Locate the shear center of the cross section.

SOLUTION

The centroid of each T-section is located 9.67 mm from its base. The distance from the x-axis to the centroid of each T-section is

$$\bar{y}_2 = 100 + 10 + 1 + 9.67 = 120.67 \text{ mm}$$

The approximate value of I_x (Eq. 7-4.1) is

$$I_x = 2A_1\bar{y}_1^2 + 2A_2\bar{y}_2^2 = 2(400)(100)^2 + 2(324)(120.67)^2$$
$$= 17.44 \times 10^6 \text{ mm}^4$$

In these kind of calculations, the shear flow q_1 is assumed to be constant from the centroid of the T-section to the centroid of the square stringers. The magnitude of q_1 is (Eq. 7-2.3)

$$q_1 = \frac{V_y}{I_x}A'\bar{y}' = \frac{V_y}{I_x}(324)(120.67) = 39.10 \times 10^3 \frac{V_y}{I_x} \text{ (N/mm)}$$

where V_y is the total shear at the section. The forces F_1, F_2, and F_3 are given by the relations

$$F_1 = (9.67 + 0.5)q_1 = 397.6 \times 10^3 \frac{V_y}{I_x} \text{ (N)}$$

$$F_2 = 60q_1 = 2.346 \times 10^6 \frac{V_y}{I_x} \text{ (N)}$$

$$F_3 = (10 + 0.5)q_1 = 410.5 \times 10^3 \frac{V_y}{I_x} \text{ (N)}$$

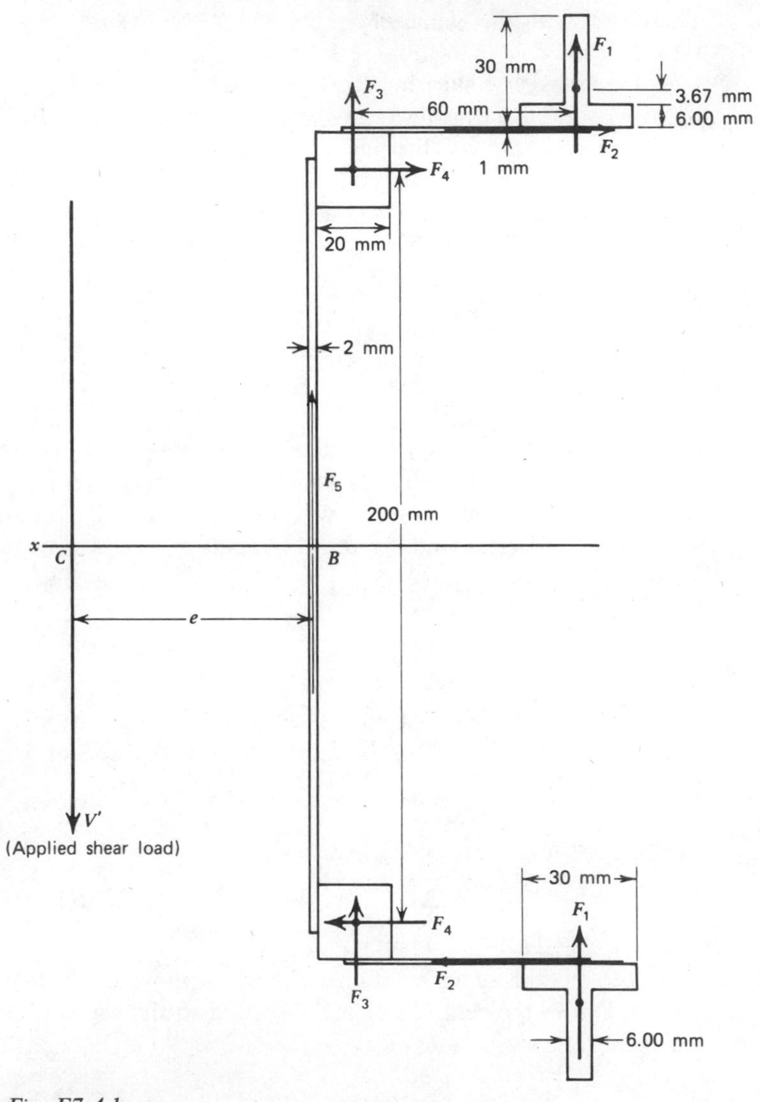

30 mm
F_1
3.67 mm
6.00 mm
F_3
60 mm
F_2
F_4
1 mm
20 mm
2 mm
F_5
200 mm
x
C
B
e
V'
(Applied shear load)
30 mm
F_1
F_4
F_2
F_3
6.00 mm

Fig. E7-4.1

The shear flow q_2 is also assumed to be constant between centroids of the square stringers. Hence,

$$q_2 = q_1 + \frac{V_y}{I_x}(400)(100) = 79.10 \times 10^3 \frac{V_y}{I_x} \text{ (N/mm)}$$

The forces F_4 and F_5 are given by the relations

$$F_4 = (10 + 1)q_2 = 870.1 \times 10^3 \frac{V_y}{I_x} \text{ (N)}$$

$$F_5 = 200 q_2 = 15.82 \times 10^6 \frac{V_y}{I_x} \text{ (N)}$$

These forces with V' (Fig. E7-4.1) must satisfy equilibrium in the y-direction, that is,

$$\sum F_y = V' - 2F_1 - 2F_3 - F_5 = 0$$

Hence,

$$V' = \frac{2(397.6 \times 10^3 V_y) + 2(410.5 \times 10^3 V_y) + 15.82 \times 10^6 V_y}{17.44 \times 10^6} = V_y$$

Thus, the applied shear load V' is equal to the total internal shear V_y in the section. The moment equilibrium equation for moments about point B determines the shear center location. Thus,

$$\sum M_B = V'e + 2F_1(71) + 2F_3(11) - F_2(221) - F_4(200) = 0$$

$$e = [2.346 \times 10^6(221) + 870.1 \times 10^3(200) - 2(397.6 \times 10^3)(71) - 2(410.5 \times 10^3)(11)]/17.44 \times 10^6$$

$$e = 35.95 \text{ mm}$$

This estimate of the location of the shear center C (Fig. E7-4.1) may be in error by several percent because of the simplifying assumptions. Hence, if the transverse bending loads are placed at C, they may introduce a small torque load in addition to bending loads. In most applications, the shearing stresses resulting from this small torque are relatively insignificant (Chapter 5).

PROBLEM SET 7-4

1. A beam is built up of a thin steel sheet of thickness $t = 0.60$ mm bent into a semicircle as shown in Fig. P7-4.1. Two 25 mm square stringers are welded to the thin web as shown. Locate the shear center for the cross section.

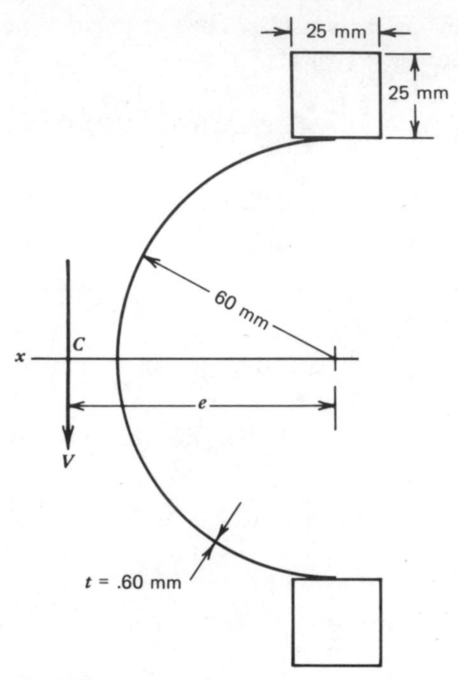

Fig. P7-4.1

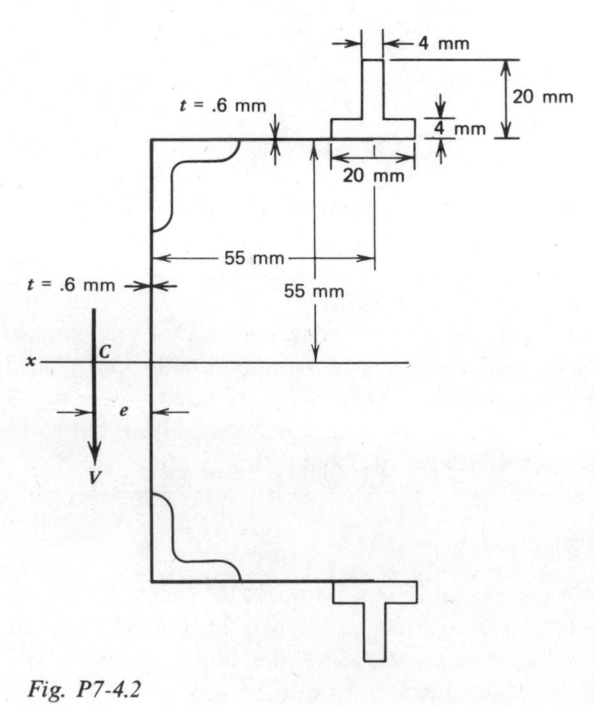

Fig. P7-4.2

2. A beam has a symmetrical cross section (Fig. P7-4.2). A vertical web with a thickness of 0.60 mm is welded to two 20 mm by 20 mm by 4 mm angle section ($A = 146$ mm^2 and centroid location 6.4 mm) stringers. The two horizontal webs have a thickness of 0.60 mm and are welded to the angle sections and 20 mm by 20 mm by 4 mm T-section stringers. Locate the shear center for the cross section.

   Ans. $e = 28.20$ mm

3. A composite beam has a symmetrical cross section as shown in Fig. P7-4.3. A vertical web with a thickness of 2 mm is welded to the center of the flange of two 50 mm by 60 mm by 10 mm T-section stringers. Two horizontal webs, with a thickness of 1 mm, are welded to these stringers and to two additional T-section stringers. Locate the shear center of the cross section.

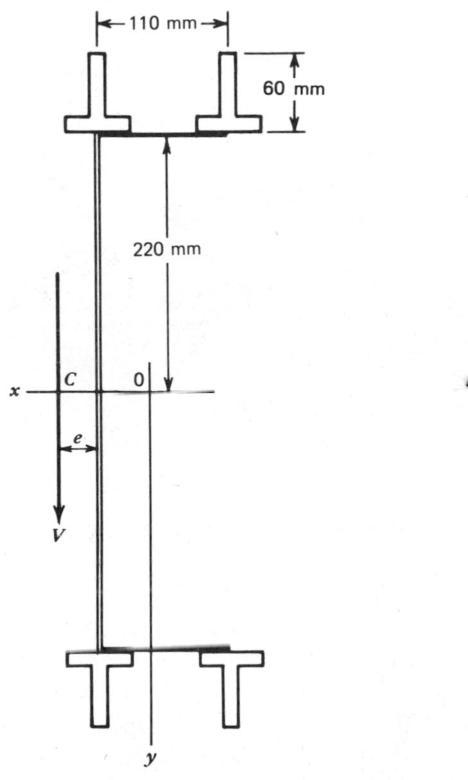

Fig. P7-4.3

4. A composite beam has a symmetrical cross section, as shown in Fig. P7-4.4. A vertical web with a thickness of 2 mm is riveted to four rolled 30 mm by 30 mm by 5 mm angle sections ($A = 278$ mm^2 and centroid location 7.7 mm). Two horizontal webs, with thickness of 1 mm, are riveted to the angles and to areas A_1 (25 mm by 25 mm) and A_2 (40 mm by 40 mm). Locate the shear center of the cross section.

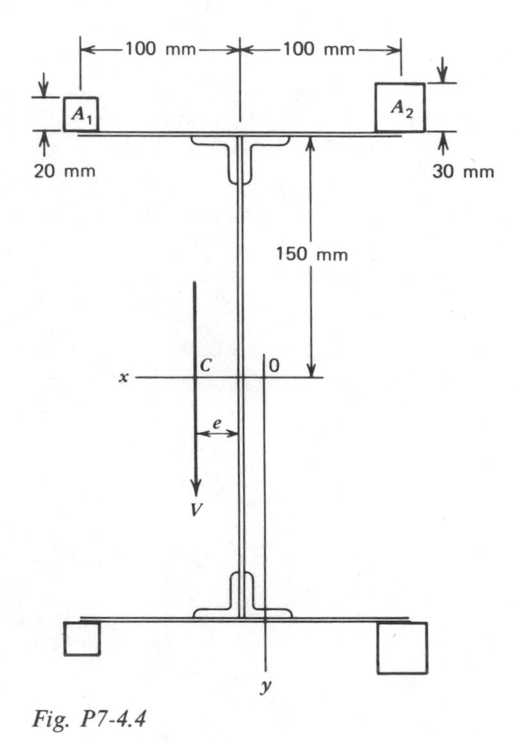

Fig. P7-4.4

Ans. $e = 28.44$ mm

7-5
SHEAR CENTER OF BOX BEAMS

Another class of practical beams is the box beam (with boxlike cross section) (Fig. 7-5.1). Box beams ordinarily have thin walls. However, they usually have walls sufficiently thick so that the walls will not buckle when subjected to elastic compressive stresses developed by bending (Fig. 7-5.1). Box beams may be composed of several legs of different thickness (Fig. 7-5.1) or they may be a composite of longitudinal stringers and very thin webs (Fig. 7-5.2). The beams in Figs. 7-5.1 and 7-5.2 are one

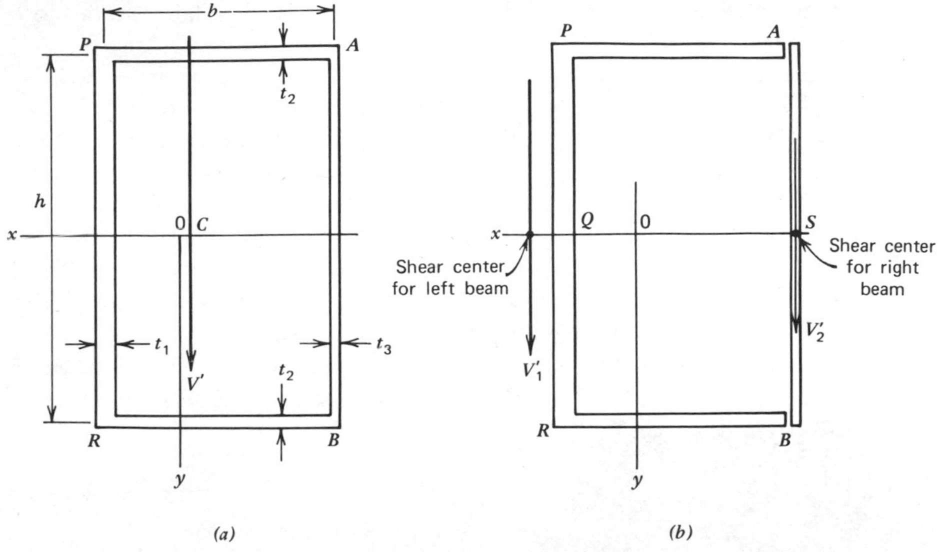

Fig. 7-5.1/Box beam.

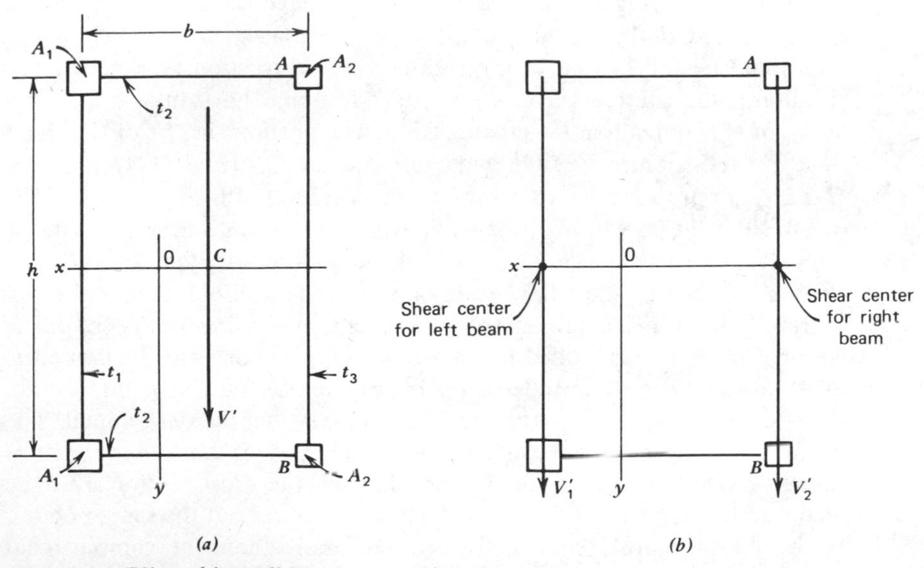

Fig. 7-5.2/Ultra thin-wall box beam with stringers.

compartment box beams. Generally box beams may contain two or more compartments.

For convenience, let the x-axis be an axis of symmetry in Figs. 7-5.1 and 7-5.2. Let the beams be subjected to symmetrical bending. Hence, let the plane of the loads be parallel to the y-axis and let it contain the shear center C. The determination of the location of the shear center requires that the shearing stress distribution in the cross section be known. However, the shearing stress distribution cannot be obtained using Eq. (7-2.3) alone, since area A' is not known. (A' is the area of the wall from a point of interest in the wall to a point in the wall where $q = 0$.) Consequently, an additional equation, Eq. (5-6.3), along with Eq. (7-2.3), is required to obtain the shearing stress distribution for a cross section of a box beam. Since there is no twisting, the unit angle of twist in the beam is zero and, hence, Eq. (5-6.3) yields

$$\int_0^l \frac{q}{t}\, dl = 0 \tag{7-5.1}$$

where dl is an infinitesimal length of the wall of the box beam cross section at a point where the thickness is t and the shear flow is q. The length l of the perimeter of the box beam cross section is measured counterclockwise from any convenient point in the wall.

The shear flow q_A at any point, say point A in Fig. 7-5.1 or Fig. 7-5.2, is an unknown. If this shear flow is subtracted from the actual shear flow at every point of the box beam wall, the resulting shear flow at A (and in this case at B because of symmetry) is zero. We refer to such a point (zero shear flow) as a *cut*. Then the resulting shear flow is the same as if the two beams (Fig. 7-5.1*b* and Fig. 7-5.2*b*) have no shear resistance at points A and B, but still have continuity of displacement at points A and B. Since the subtraction of q_A results in a subtraction of a zero force resultant, the subtraction produces no additional horizontal or vertical components of load on the cross section. The portions V_1', V_2' of the shear load V' acting on each of the two parts AB and BA (Fig. 7-5.1*b* and Fig. 7-5.2*b*) are proportional to the moments of inertia of the two parts of the beam since the curvature of the two parts must be continuous at points A and B. For convenience, let $V' = I$ (in magnitude) so that $V_1' = I_1$ and $V_2' = I_2$. Then, the shear flow at any point in the wall of either of the two parts of the beams (Fig. 7-5.1*b*) can be obtained using Eq. (7-2.3). The shear flow q_A is then added to the resulting shear flows for the two parts of the beam. The magnitude of q_A is obtained by satisfying Eq. (7-5.1). The force in each wall of the cross section can then be determined. The location of the shear center is obtained from the fact that the moment of these forces about any point in the plane of the cross section must be equal to the moment of the applied shear load V' about the same point.

For beams whose cross section contain more than one compartment (Fig. 7-5.3), the above procedure must be repeated for a point in the wall

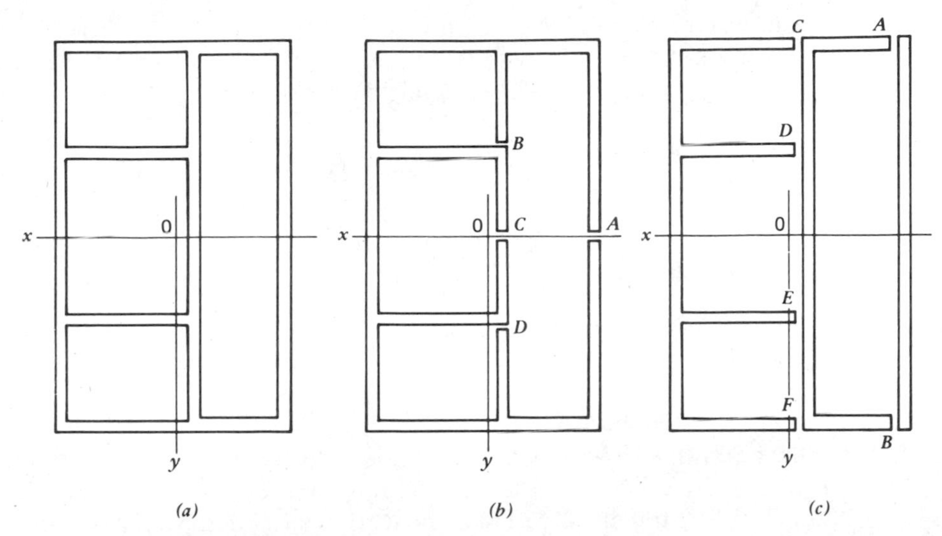

Fig. 7-5.3/Multicompartment box beam.

of each compartment, such as at A, B, C, and D in Fig. 7-5.3b or at A, B, C, D, E, and F in Fig. 7-5.3c. The magnitudes of the shear flows that must be subtracted for each compartment is obtained by satisfying Eq. (7-5.1) for each compartment.

Unsymmetrical box beam cross sections can also be treated by the above procedure. In this case, it is desirable to refer the calculation to principal axes, say X and Y. The method proceeds as follows: first, we locate the plane of the loads for bending about the X-axis; second, we locate the plane of the loads for bending about the Y-axis. The shear center of the cross section is given by the intersection of these two planes. The bending axis intersects each cross section of the box beam at the shear center.

EXAMPLE 7-5.1

Shear Center for Box Beam

For the box beam in Fig. 7-5.1, let $b = 300$ mm, $h = 500$ mm, $t_1 = 20$ mm and $t_2 = t_3 = 10$ mm. Determine the location of the shear center for the cross section.

SOLUTION

The moment of inertia for the x-axis is $I_x = 687.9 \times 10^6$ mm^4. Cuts are taken at points A and B to divide the beam into two parts (Fig. 7-5.1b). For convenience, let the magnitude of the shear load V' for the box beam be equal to the magnitude of I_x so that $V_1' = I_{x1}$ and $V_2' = I_{x2}$. The shear flow q is determined at points P, Q, and S for the two parts of the cut beam cross section (Fig. 7-5.1b) as follows: (with $V_1' = V_1$, $V_2' = V_2$)

$$q_P = \frac{V_1 A'\bar{y}'}{I_{x1}} = (bt_2)\frac{h}{2} = 300(10)(250) = 750.0 \text{ kN/mm}$$

$$q_Q = q_P + \left(\frac{h}{2}t_1\right)\frac{h}{4} = 1{,}375.0 \text{ kN/mm}$$

$$q_S = \left(\frac{h}{2}t_3\right)\frac{h}{4} = 312.5 \text{ kN/mm}$$

The senses of the shear flows oppose those of V_1' and V_2'. For the left side of the beam (Fig. E7-5.1b), the shear flow increases linearly from zero at B to q_P at R and decreases linearly from q_P at P to zero at A. The shear flow changes parabolically from q_P at R to q_Q at Q and back to q_P at P. For the right side of the beam, the shear flow changes parabolically from zero at B to q_S at S and back to zero at A. Now, we add q_A (assumed positive in a counterclockwise direction) to the value of q at every point in the cross section (Fig. E7-5.1b), and we require that Eq. (7-5.1) be satisfied. Starting at P, we find that

$$0 = \left[q_A - q_P - \frac{2}{3}(q_Q - q_P)\right]\frac{h}{t_1}$$

$$+ \left(q_A - \frac{q_P}{2}\right)\frac{b}{t_2} + \left(q_A + \frac{2}{3}q_S\right)\frac{h}{t_3} + \left(q_A - \frac{q_P}{2}\right)\frac{b}{t_2}$$

$$0 = 135.0q_A - (750.0 + \tfrac{2}{3}625.0)25 - (375.0)30$$

$$+ (\tfrac{2}{3}312.5)50 - (375.0)30$$

$$q_A = 305.6 \text{ kN/mm}$$

This value of q_A must be added to the values computed for the cross section with the cuts to give the shear flow (Fig. E7-5.1b). The equilibrium moment equation for moments about point B gives

$$0 = V'e - \left(444.4 + \frac{2}{3}625\right)(500)(300) - \frac{444.4}{2}(177.76)(500)$$

$$+ \frac{305.6}{2}(122.24)(500)$$

$$e = \frac{139.57 \times 10^9 \text{ N} \cdot \text{mm}}{687.9 \times 10^6 \text{ N}} = 202.9 \text{ mm}$$

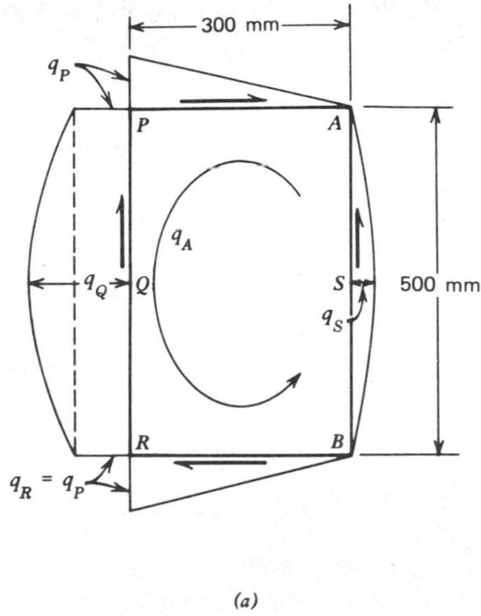

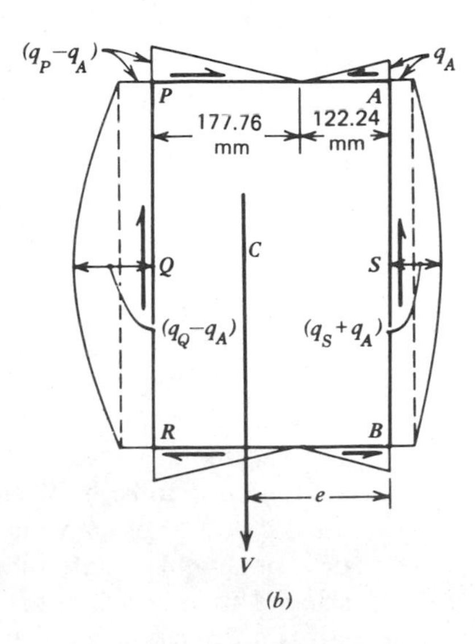

Fig. E7-5.1

The shear center C lies on the x-axis at a point 202.9 mm to the left of the center line of the right leg of the box section.

PROBLEM SET 7-5

1. For the box beam in Fig. 7-5.1, let $b = 100$ mm, $h = 200$ mm, $t_1 = 20$ mm, $t_2 = 10$ mm, and $t_3 = 5$ mm. Determine the location of the shear center for the cross section.

2. For the box beam in Fig. 7-5.2, let $b = 200$ mm, $h = 400$ mm, $t_1 = t_2 = t_3 = 1$ mm and $A_1 = 3A_2 = 900$ mm^2. Determine the location of the shear center for the cross section.

   Ans. $e = 83.33$ mm

3. Let $t_1 = 2$ mm with other dimensions from Problem 2 remaining unchanged. Determine the location of the shear center.

7-6 STRAIGHT SHAFTS SUBJECTED TO COMBINED LOADS

Straight shafts are often subjected to a combination of axial loads, bending moments, and twisting moments. In applications, most straight shafts have circular cross sections. Circular straight shafts subjected to combined loads are treated in Chapter 3 (ignoring the effects of stress concentrations) and in Chapter 13 (including the effects of stress concentrations).

In Chapter 5, the effects of torsion on shafts with noncircular cross sections are studied. It is noted that the cross section of such shafts subjected to torsion undergo warping, that is, points in the cross section of the shaft undergo unequal displacements along the axis of the shaft. Consequently, the cross section planes do not remain plane under the action of the torque. If the warping of the cross section is unrestrained, the state of stress in the shaft cross section is one of shear. However, in practice, members subjected to combined loads are often supported in a manner that prevents warping of the cross sections at the supports. When the warping of a cross section is constrained, normal stresses act on the cross section to keep the cross section plane (to prevent the points in the cross section from moving axially). Hence, if warping of a cross section of a noncircular shaft subjected to torsion is prevented, axial (normal) stresses are induced on the cross section, in addition to torsional shearing stresses. In particular, these normal stresses may be significant at and near the constrained section. Therefore, to obtain the total normal stress that acts on a cross section, the normal stress due to the prevention of warping must be added to the normal stresses due to axial loads and bending moments that may be acting on the member.

In Art. 5-7, the effects of preventing warping is examined for rolled cross sections, such as I-sections, Z-sections, and channel (⊏) sections. The results of that study may be incorporated in the study of members that are subjected to combined loads and that have certain sections constrained (e.g., supported sections, end sections, etc.). Consequently, let us consider two cases of rolled members subjected to combined loads and constrained (fixed) at a wall support (Fig. 7-6.1). Let the I-beam in Fig. 7-6.1*a* be subjected to a twisting moment T and an axial load P. This type of loading occurs occasionally in practice. In the analysis of such constrained members, it is common to assume that at the wall the torque is resisted mainly by shearing stresses that act over the areas of the flanges of the I-beam. This approximation is reasonably accurate for most I-beams, since the width of the flange is usually several times greater than the thickness of the web. Consequently, we assume that the torsional load is transmitted to the fixed end (support) as shear loads, equal to $V = T/h$, that act over the cross-sectional area of each flange (see Figs. 5-7.1 and 5-7.3). However, because of the constraining action of the wall, the torsional load also induces normal stresses on the flange cross section. In particular, the torsional load induces a tensile stress at point B (Fig. 7-6.1*a*), which must be added to the tensile stress ($\sigma = P/A$) due to the axial load P. There are two locations in the beam cross section at the wall at which failure may occur. For example, failure may occur at

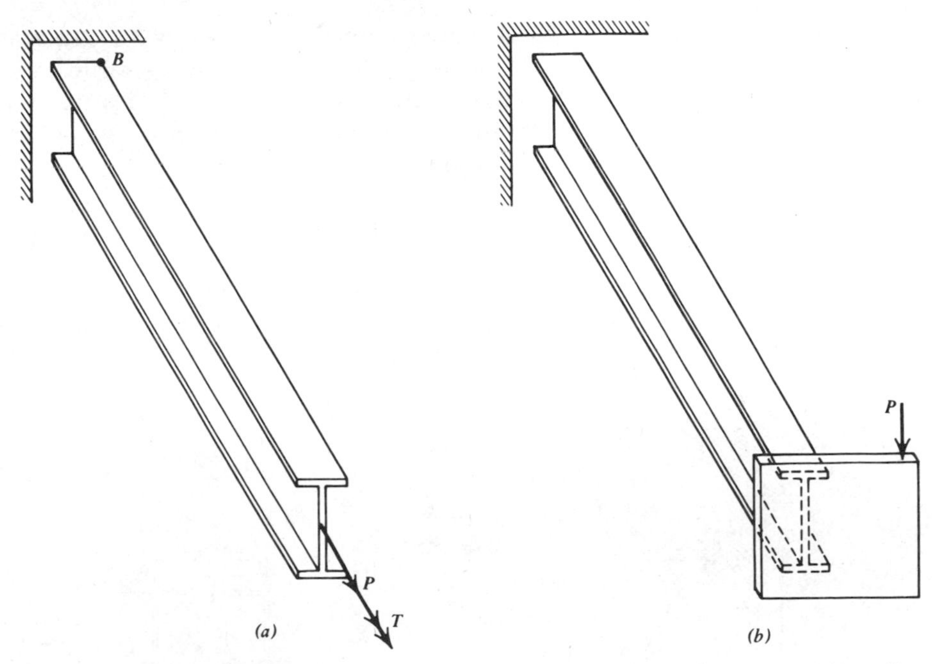

Fig. 7-6.1/Combined loading of an I-beam. (*a*) Axial tension and torsion. (*b*) Bending and torsion.

point B. It also may occur at the center of the flange where the stress state consists of a shearing stress (due to the total shear V) computed by means of Eq. (7-2.3) and a normal stress ($\sigma = P/A$) due to the axial load P. Failure may also possibly occur at some distance from the fixed end, where warping is not restricted. At distances far removed from the end, the state of stress is determined essentially by the torsional shearing stress (computed by means of Eq. 5-5.10) and the normal stress $\sigma = P/A$.

As noted above, the combination of loads indicated in Fig. 7-6.1*a* occurs occasionally in practice. A more frequently occurring combination of loads is shown in Fig. 7-6.1*b*. However, a rolled section (I-beam) is an extremely poor member to support such combinations, since as noted in Art. 5-5, rolled sections are not particularly suited to resist large torsional moments; in torsion, rolled section members often undergo excessively large rotations. For this reason box cross sections are usually employed to resist combinations of loads such as shown in Fig. 7-6.1*b*. (See also Fig. 5-7.2.)

Box Cross Sections/Structural members with box cross sections are often used to support combinations of axial loads, bending moments, and twisting moments. We consider members with box cross sections that do not warp under the action of torsional loads. As noted in Art. 5-7, two such classes of box cross sections exist; namely, (1) the hollow cross section with constant thickness walls and equilateral legs and (2) the hollow thin-wall cross section of rectangular shape (Fig. 7-6.2) with dimensions restricted by the condition (see Art. 5-7 and Fig. 7-6.2*b*)

$$bt_1 = ht_2 \tag{7-6.1}$$

The following illustrative problem indicates the method of solution when Eq. (7-6.1) is satisfied. For cases in which Eq. (7-6.1) is not satisfied, the reader is referred to reference 4 of Chapter 5.

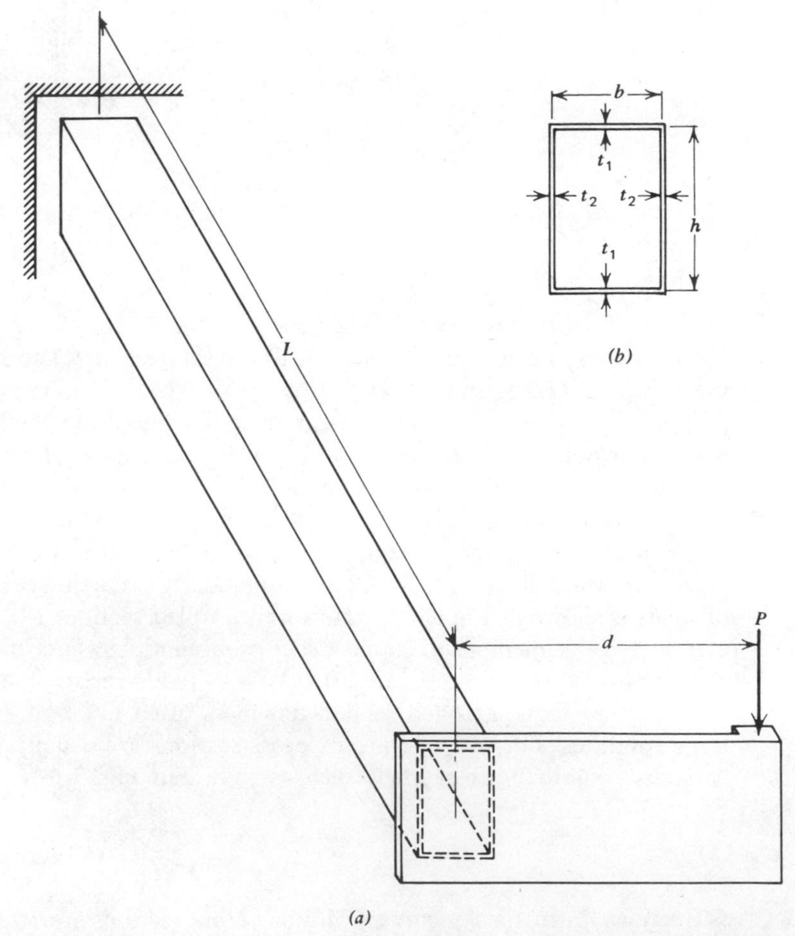

Fig. 7-6.2/Combined loading of a box section member. (*a*) Box section member. (*b*) Cross section of a box section.

EXAMPLE 7-6.1

Combined Loading of Box Cross Section Member

Let the box cross section member in Fig. 7-6.2 have the following dimensions: $b = 200$ mm, $t_1 = 6$ mm, $h = 300$ mm, $t_2 = 4$ mm, $L = 3$ m, and $d = 0.80$ m. The beam is made of a ductile steel with a yield stress of $Y = 520$ MPa. Determine the magnitude of P if the beam has been designed with a factor of safety $SF = 2.00$. Use the maximum shearing stress criterion of failure.

SOLUTION

Since the dimensions of the cross section satisfy Eq. (7-6.1), the cross section does not warp when subjected to a torsional load. Load P develops a torque $T = 800P$ (N · mm) in the box section and a bending moment $M = 3000P$ (N · mm) at the fixed end. The maximum bending stress occurs in the 200 mm by 6 mm flange and at the junction of the 300 mm by 4 mm sides with the flange. When the load P is increased by the factor of safety, failure by general yielding is assumed to occur. The maximum bending stress σ is calculated as follows:

$$\sigma = \frac{(SF)Mc}{I_x} = \frac{2.00(3000)P(150)}{72{,}030{,}00} = 0.01250P \text{ (MPa)} \tag{1}$$

where

$$I_x = \frac{204(306)^3}{12} - \frac{196(294)^3}{12} = 72.03 \times 10^6 \text{ mm}^4$$

The maximum shearing stress due to T occurs in the 300 mm by 4 mm sides and is given by Eq. (5-6.2)

$$\tau = \frac{(SF)T}{2At_{\min}} = \frac{2.00(800)P}{2(200)(300)(4)} = 0.00333P \text{ (MPa)} \tag{2}$$

By Eqs. (1) and (2) and Eq. (3-3.6), the maximum shearing stress is

$$\tau_{\max} = \frac{Y}{2} = \sqrt{(\sigma/2)^2 + \tau^2}$$

Hence

$$\frac{520}{2} = \sqrt{(0.01250P/2)^2 + (0.00333P)^2} = 0.00708P$$

and

$$P = 36.71 \text{ kN} \tag{3}$$

The load at which yielding first occurs in the member is, with Eq. (3), $P_Y = 2.00P = 2.00 \times 36.71 = 73.42$ kN. The region of the cross section in which yielding first occurs is localized in a small portion of the web near each flange. Hence, until the yielded region increases in size until it spreads to the flange, the gross load-deflection diagram for the member remains approximately linear. Consequently, the load $P_Y = 73.42$ kN should not be considered as the design load for failure by general yielding. Rather it is assumed that failure by general yielding occurs when the material in the flanges reaches yield. Since the shearing stress in each flange is given by Eq. (5-6.2), that is,

$$\tau = \frac{SF(T)}{2At_1} = \frac{2.00(800P)}{2(200)(300)6} = 0.00222P \text{ (MPa)} \tag{4}$$

the design load P is obtained by substituting the stress components given by Eqs. (1) and (4) into Eq. (3-3.6). Thus,

$$\frac{520}{2} = \sqrt{(0.01250P/2)^2 + (0.00222P)^2} = 0.00663P$$

or

$$P = 39.20 \text{ kN} \tag{5}$$

The load $P = 39.20$ kN is considered to be the design load for the box cross section member subjected to the combined loads shown in Fig. 7-6.2a if load P is static or load P is slowly applied a few times. If the load application is repeated a large number (N) of times, the possibility of fatigue failure should be considered. (For $N > 10^6$, see Art. 3-5.) For fatigue loading, the normal stress (Eq. 1) should also be multiplied by the stress concentration factor for tension (see Chapter 13). To complicate matters further, two different effects contribute to the shearing stress concentration, that is, act to increase the influence of the shearing stress given by Eq. (2). These are (1) a stress concentration resulting from an abrupt change in cross section, and (2) a stress concentration resulting from the reentrant corner at the junction of the web and the flange.

PROBLEM SET 7-6

1. A square box section member with mean dimensions of 200 mm and a wall thickness of 12 mm is made of an aluminum alloy with yield stress $Y = 305$ MPa. Choose centroidal axes (x, y) parallel to the edges of the section. If $M_x = 120$ kN · m, determine the magnitude of the torque T that will initiate yielding at points in the cross section not at stress concentrations. Use the maximum shearing stress criterion of failure.

2. Let the member in Problem 1 be subjected to the following loads: $M_x = 45$ kN · m, and $T = 50$ kN · m. What is the factor of safety against initiation of yielding at points in the cross section not at

stress concentrations. Use the maximum shearing stress criterion of failure.

Ans. $SF = 2.43$

REFERENCES

1. A. P. Boresi and P. P. Lynn, *Elasticity in Engineering Mechanics*, 2nd Ed., Prentice-Hall, Englewood Cliffs, New Jersey, 1974, Chapter 7.
2. C. C. Kelber, "Numerical Determination of the Shear Center Coordinates of Heavy L-Shaped Sections," M.S. Thesis, Department of Theoretical and Applied Mechanics, University of Illinois, Urbana, Illinois, 1948.
3. F. Bleich, *Buckling Strength of Metal Structures*, McGraw-Hill, New York, 1952, pp. 337–343.
4. S. P. Timoshenko and J. M. Gere, *Theory of Elastic Stability*, McGraw-Hill, New York, 1961, pp. 360–439.
5. P. Kuhn, *Stresses in Aircraft and Shell Structures*, McGraw-Hill, New York, 1956.
6. H. L. Langhaar, "Theoretical and Experimental Investigations of Thin-Webbed Plate-Girder Beams," Consolidated Aircraft Corporation, Report No. SG-895, San Diego, California, September 1, 1942.

Additional References

1. J. N. Goodier, "A Theorem on the Shearing Stress in Beams with Applications to Multicellular Sections," *Journal of the Aeronautical Sciences*, Vol. 11, No. 3, July 1944, pp. 272–280.
2. K. Washizu, "Some Considerations on the Center of Shear," *Transactions of Japan Society for Aeronautical and Space Sciences*, Vol. 9, No. 15, 1966, pp. 77–83.
3. A. Weinstein, "The Center of Shear and the Center of Twist," *Quarterly of Applied Mathematics*, Vol. 5, No. 1, 1947, pp. 97–99.

CHAPTER 8

CURVED BEAMS

8-1 INTRODUCTION

The flexure formula (Eq. 6-1.1) is accurate for symmetrically loaded straight beams subjected to pure bending. It is also generally used to obtain approximate results for the design of straight beams subjected to shear loads, when the plane of loads contains the shear center and is parallel to a principal axis of the beam; the resulting errors in the computed stresses are small enough to be negligible as long as the beam length is at least five times the maximum cross-sectional dimension. In addition, the flexure formula is used in the design of curved beams for which the radius of curvature is more than five times the beam depth for the same reason. However, for curved beams the error in the computed stress predicted by the flexure formula increases as the ratio of the radius of curvature of the beam to the depth of the beam decreases in magnitude. Hence, as this ratio decreases, one needs a more accurate solution for curved beams.

Timoshenko[1] has presented a solution based on the theory of elasticity for the linear elastic behavior of curved beams of rectangular cross sections for the loading shown in Fig. 8-1.1*a*. He used polar coordinates and obtained relations for the radial stress σ_{rr}, the circumferential stress $\sigma_{\theta\theta}$, and the shearing stress $\sigma_{r\theta}$ (Fig. 8-1.1*b*). However, most curved beams do not have rectangular cross sections. Therefore, in the following we present an approximate curved beam solution that is generally applicable to all symmetrical cross sections. This solution is based upon two simplifying assumptions: (1) plane sections before loading remain plane after loading and (2) the radial stress σ_{rr} and shearing stress $\sigma_{r\theta}$ are sufficiently small so that the state of stress is essentially one dimensional. The resulting formula for the circumferential stress $\sigma_{\theta\theta}$ is the curved beam formula.

8-2 CIRCUMFERENTIAL STRESS IN A CURVED BEAM

Consider the curved beam shown in Fig. 8-2.1*a*. The cross section of the beam has a plane of symmetry. We assume that the applied loads all lie

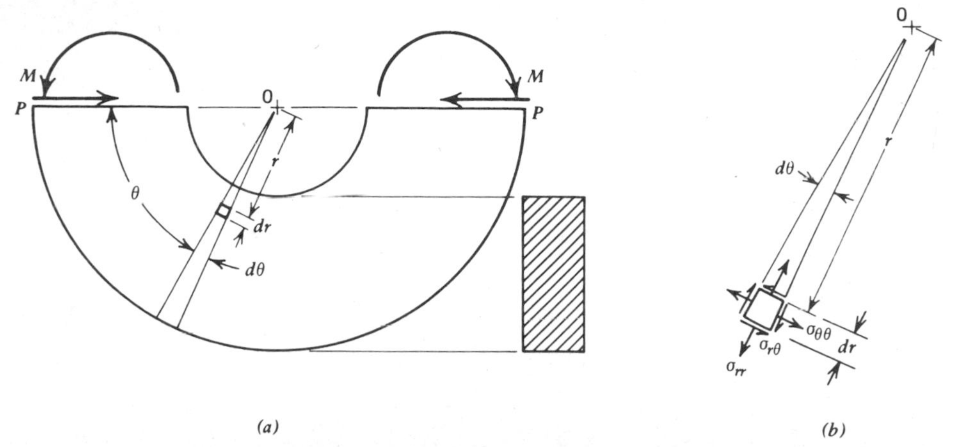

Fig. 8-1.1/Rectangular section curved beam. (*a*) Curved beam loading. (*b*) Stress components.

in one plane, which coincides with the plane of symmetry. The applied loads produce a positive moment, as shown in Fig. 8-2.1*b*, at each section of the curved beam. Thus, because of a positive moment, the radius of curvature, at each section of the beam, is increased in magnitude. We wish to determine an approximate formula for the circumferential stress distribution $\sigma_{\theta\theta}$ on section *BC*. A free body diagram of an element *FBCH* of the beam is shown in Fig. 8-2.1*b* (see Fig. 8-2.1*a*). The normal traction *N*, at the centroid of the cross section, the shear *V*, and moment M_x acting on face *FH* are shown in their positive directions. These forces must be balanced by the resultants due to the normal stress $\sigma_{\theta\theta}$ and the shearing stress $\sigma_{r\theta}$ that act on face *BC*. The effect of the shearing stress $\sigma_{r\theta}$ on the computation of $\sigma_{\theta\theta}$ is usually small, except for curved beams with very thin webs. However, since ordinarily, practical curved beams are not designed with thin webs because of the possibility of failure by excessive radial stresses (see Art. 8-3), in practice, neglecting the effect of $\sigma_{r\theta}$ on the computation of $\sigma_{\theta\theta}$ is reasonable.

Let the *z*-axis be normal to face *BC* (Fig. 8-2.1*b*). By equilibrium of forces in the *z*-direction and of moments about the centroidal *x*-axis, we find

$$\sum F_z = \int \sigma_{\theta\theta}\, dA - N = 0$$

$$\sum M_x = \int \sigma_{\theta\theta}(R - r)\, dA - M_x = 0$$

or

$$N = \int \sigma_{\theta\theta}\, dA \tag{8-2.1}$$

$$M_x = \int \sigma_{\theta\theta}(R - r)\, dA \tag{8-2.2}$$

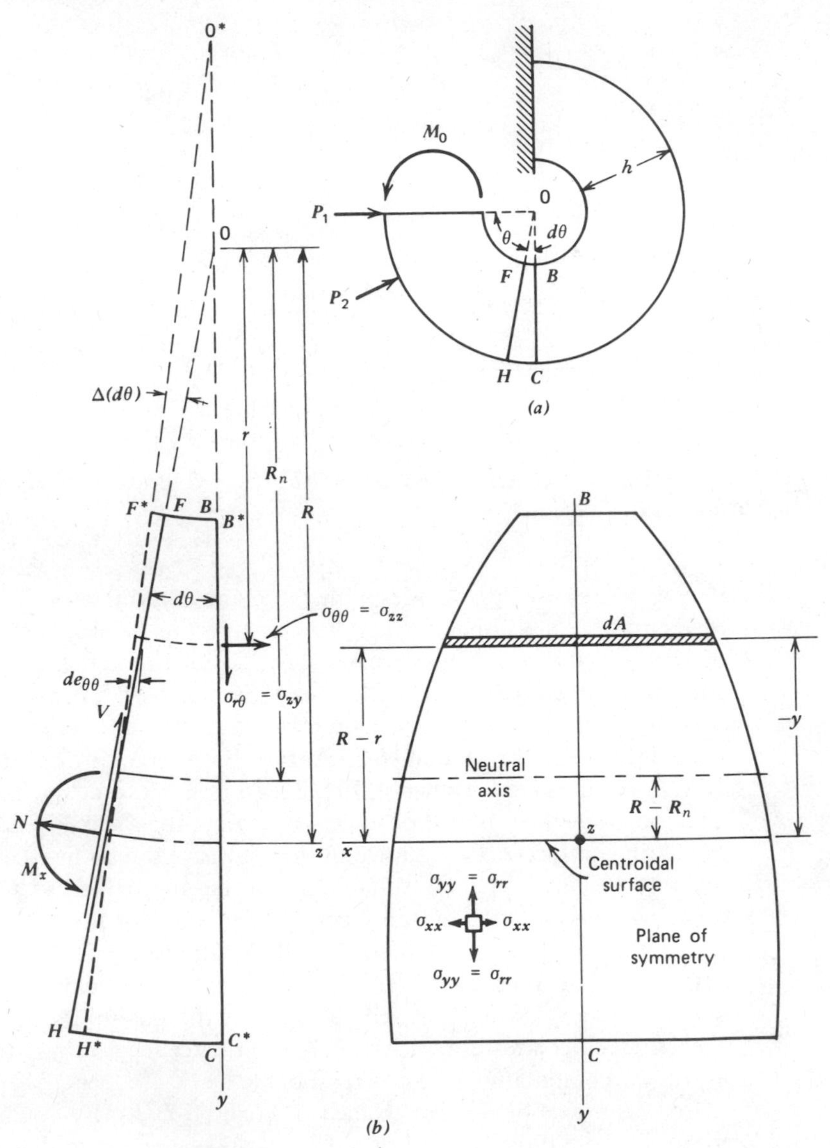

Fig. 8-2.1/Curved beam.

where R is the distance from the center of curvature of the curved beam to the centroid of the beam cross section and r locates the element of area dA from the center of curvature. The integrals of Eqs. (8-2.1) and (8-2.2) cannot be evaluated until $\sigma_{\theta\theta}$ is expressed in terms of r. The functional relationship between $\sigma_{\theta\theta}$ and r is obtained from the assumed geometry of deformation and the stress-strain relations for the material.

The curved beam element $FBCH$ in Fig. 8-2.1b represents the element in the undeformed state. The element $F^*B^*C^*H^*$ represents the element after it is deformed by the loads. For simplicity, we have positioned the deformed element so that face B^*C^* coincides with face BC. As in the case of straight beams, we assume that plane B^*C^* remains plane under the deformation. Face F^*H^* of the deformed curved beam element forms an angle $\Delta(d\theta)$ with respect to FH. Line F^*H^* intersects line FH at the neutral axis of the cross section (axis for which $\sigma_{\theta\theta} = 0$) at distance R_n from the center of curvature. The movement of the center of curvature from point 0 to point 0* is exaggerated in Fig. 8-2.1b in order to visualize the geometry changes. For infinitesimally small displacements, the movement of the center of curvature is infinitesimal. The elongation $de_{\theta\theta}$ of a typical element in the θ direction is equal to the distance between faces FH and F^*H^* and varies linearly with the distance $(R_n - r)$. The corresponding strain $\epsilon_{\theta\theta}$ however, is a nonlinear function of r, since the element length $r\,d\theta$ varies with r. This fact distinguishes a curved beam from a straight beam. Thus, by Fig. 8-2.1b, we obtain for the strain

$$\epsilon_{\theta\theta} = \frac{de_{\theta\theta}}{r\,d\theta} = \frac{(R_n - r)\,\Delta(d\theta)}{r\,d\theta} = \left(\frac{R_n}{r} - 1\right)\omega \tag{8-2.3}$$

where

$$\omega = \frac{\Delta(d\theta)}{d\theta} \tag{8-2.4}$$

It is assumed that σ_{xx} is sufficiently small so that it may be discarded. Hence, the curved beam is considered to be a problem in plane stress. Although radial stress σ_{rr} may, in certain cases, be of importance (see Art. 8-3), here we neglect its effect on $\epsilon_{\theta\theta}$. Then, by Hooke's law, we find

$$\sigma_{\theta\theta} = E\epsilon_{\theta\theta} = \frac{R_n - r}{r}E\omega = \frac{E\omega R_n}{r} - E\omega \tag{8-2.5}$$

Substituting Eq. (8-2.5) into Eqs. (8-2.1) and (8-2.2), we obtain

$$N = R_n E\omega \int \frac{dA}{r} - E\omega \int dA = R_n E\omega A_m - E\omega A \tag{8-2.6}$$

$$\begin{aligned} M_x &= R_n R E\omega \int \frac{dA}{r} - (R + R_n)E\omega \int dA + E\omega \int r\,dA \\ &= R_n R E\omega A_m - (R + R_n)E\omega A + E\omega R A = R_n E\omega (R A_m - A) \end{aligned} \tag{8-2.7}$$

where A is the cross sectional area of the curved beam and A_m has the dimensions of length and is defined by the relation

$$A_m = \int \frac{dA}{r} \tag{8-2.8}$$

Equation (8-2.7) can be rewritten in the form

$$R_n E\omega = \frac{M_x}{RA_m - A} \tag{8-2.9}$$

Then substitution into Eq. (8-2.6) gives

$$E\omega = \frac{A_m M_x}{A(RA_m - A)} - \frac{N}{A} \tag{8-2.10}$$

The circumferential stress distribution for the curved beam is obtained by substituting Eq. (8-2.9) and (8-2.10) into Eq. (8-2.5) to obtain the curved beam formula

$$\sigma_{\theta\theta} = \frac{N}{A} + \frac{M_x(A - rA_m)}{Ar(RA_m - A)} \tag{8-2.11}$$

The normal stress distribution given by Eq. (8-2.11) is hyperbolic in form; that is, it varies as $1/r$. For the case of a curved beam with rectangular cross section ($R/h = 0.75$) subjected to pure bending, the normal stress distribution is shown in Fig. 8-2.2.

Since Eq. (8-2.11) has been based on several simplifying assumptions, it is essential that its validity be verified. Results predicted by the curved beam formula can be compared with those obtained from the elasticity solution for curved beams with rectangular sections and with those obtained from the experiments on curved beams with other kinds of cross sections. The maximum value of the circumferential stress $\sigma_{\theta\theta(CB)}$ as given by the curved beam formula, may be computed from Eq. (8-2.11) for curved beams of rectangular cross sections subjected to pure bending and to shear (Fig. 8-2.3). The ratios of $\sigma_{\theta\theta(CB)}$ to the elasticity solution

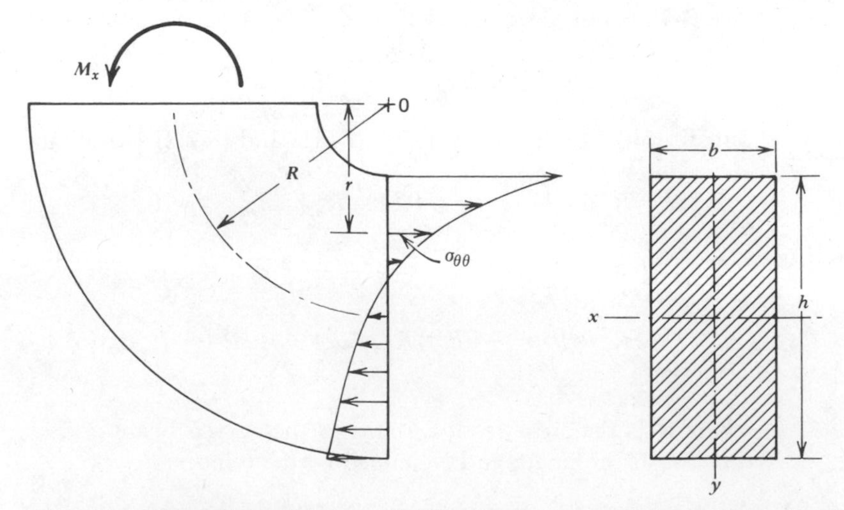

Fig. 8-2.2/Circumferential stress distribution in a rectangular section curved beam ($R/h = 0.75$).

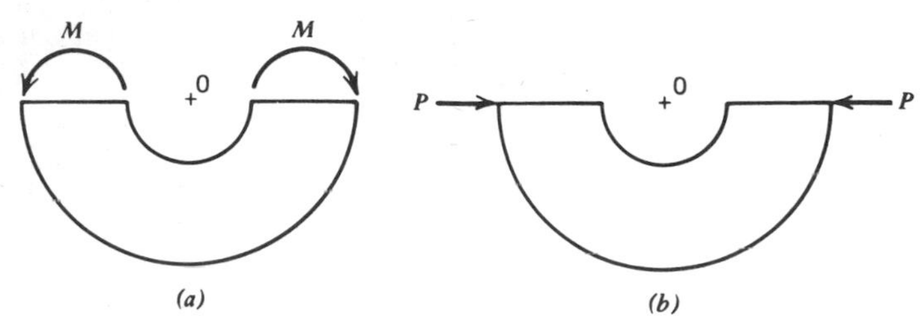

Fig. 8-2.3/Types of curved beam loadings. (*a*) Pure bending. (*b*) Shear loading.

$\sigma_{\theta\theta(\text{elast})}$ are listed in Table 8-2.1 for pure bending (Fig. 8-2.3*a*) and for shear loading (Fig. 8-2.3*b*), for several values of R/h, where h denotes the beam depth (Fig. 8-2.1*a*). The nearer these ratios are to one, the less error in Eq. (8-2.11). The curved beam formula is more accurate for pure bending than for shear loading. Most curved beams are subjected to a combination of bending and shear. The value of R/h is usually greater than 1.0 for curved beams, so that the error in the curved beam formula is not particularly significant. However, possible errors occur in the curved beam formula for I-section and T-section curved beams. These errors are discussed in Art. 8-4. Also listed in Table 8-2.1 are the ratios of the maximum circumferential stress $\sigma_{\theta\theta(st)}$ given by the straight beam flexure formula (Eq. 6-1.1) to the value $\sigma_{\theta\theta(\text{elast})}$. The straight beam solution is appreciably in error for small values of R/h and is in error by 7 percent for $R/h = 5.0$; the error is nonconservative. Generally, for curved beams with R/h greater than 5.0, the flexure formula is used.

As R becomes large compared to h, the right-hand term in Eq. (8-2.11) reduces to $-M_x y/I_x$. The negative sign results because the sign conven-

Table 8-2.1

Ratios of the Maximum Circumferential Stress in Rectangular Section Curved Beams as Computed by Elasticity Theory, by the Curved Beam Formula and by the Flexure Formula

	Pure Bending		Shear Loading	
$\frac{R}{h}$	$\frac{\sigma_{\theta\theta(CB)}}{\sigma_{\theta\theta(elast)}}$	$\frac{\sigma_{\theta\theta(st)}}{\sigma_{\theta\theta(elast)}}$	$\frac{\sigma_{\theta\theta(CB)}}{\sigma_{\theta\theta(elast)}}$	$\frac{\sigma_{\theta\theta(st)}}{\sigma_{\theta\theta(elast)}}$
0.65	1.046	0.439	0.855	0.407
0.75	1.012	0.526	0.898	0.511
1.0	0.997	0.654	0.946	0.653
1.5	0.996	0.774	0.977	0.776
2.0	0.997	0.831	0.987	0.834
3.0	0.999	0.888	0.994	0.890
5.0	0.999	0.933	0.998	0.934

tion for moments for curved beams is opposite to that for straight beams (see Eq. 6-1.1). To prove this reduction, note that $r = R + y$. Then the term RA_m in Eq. (8-2.11) may be written as

$$RA_m = \int \left(\frac{R}{R+y} + 1 - 1 \right) dA = A - \int \frac{y}{R+y}\, dA \tag{a}$$

Hence, the denominator of the right-hand term in Eq. (8-2.11) becomes, for $R/h \to \infty$,

$$\begin{aligned} Ar(RA_m - A) &= -A \int \left(\frac{Ry}{R+y} + y - y \right) dA - Ay \int \frac{y}{R+y}\, dA \\ &= A \int \frac{y^2}{R+y}\, dA - A \int y\, dA - Ay \int \frac{y}{R+y}\, dA \\ &= \frac{A}{R} \int \frac{y^2}{1 + \frac{y}{R}}\, dA - A \int y\, dA - \frac{Ay}{R} \int \frac{y}{1 + \frac{y}{R}}\, dA \\ &= \frac{AI_x}{R} \end{aligned} \tag{b}$$

since as $R/h \to \infty$, $y/R \to 0$, $1 + y/R \to 1$, $\int [y^2 dA/(1 + y/R)] \to I_x$, and $\int [y\, dA/(1 + y/R)] \to 0$. The right-hand term in Eq. (8-2.11) then

Table 8-2.2

Analytical Expressions for A, R, and $A_m = \int \frac{dA}{r}$

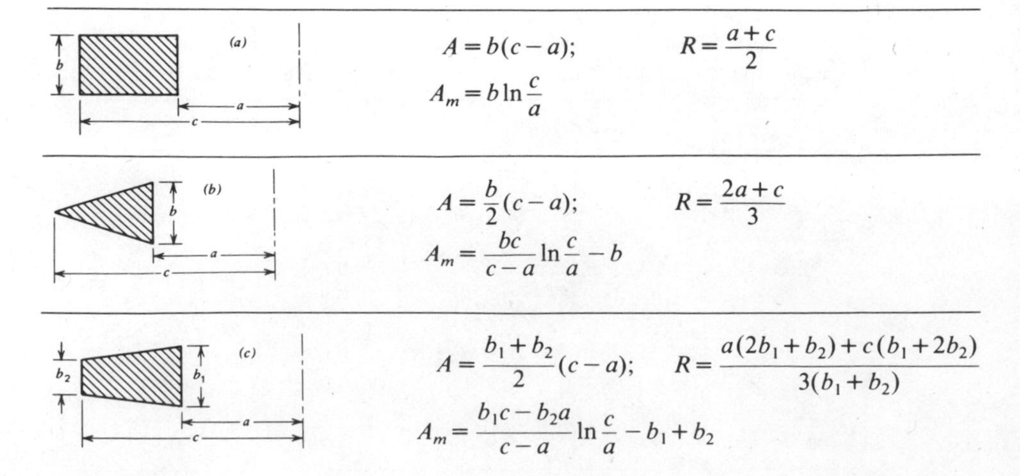

Section	A, A_m	R
(a)	$A = b(c-a)$; $A_m = b \ln \frac{c}{a}$	$R = \frac{a+c}{2}$
(b)	$A = \frac{b}{2}(c-a)$; $A_m = \frac{bc}{c-a} \ln \frac{c}{a} - b$	$R = \frac{2a+c}{3}$
(c)	$A = \frac{b_1+b_2}{2}(c-a)$; $A_m = \frac{b_1 c - b_2 a}{c-a} \ln \frac{c}{a} - b_1 + b_2$	$R = \frac{a(2b_1+b_2)+c(b_1+2b_2)}{3(b_1+b_2)}$

Table 8-2.2

Continued

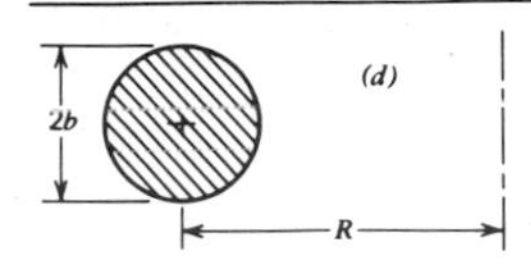

$$A = \pi b^2$$
$$A_m = 2\pi(R - \sqrt{R^2 - b^2})$$

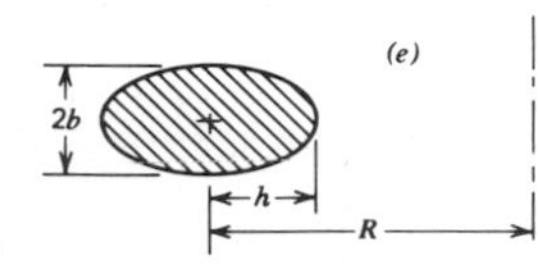

$$A = \pi bh$$
$$A_m = \frac{2\pi b}{h}(R - \sqrt{R^2 - h^2})$$

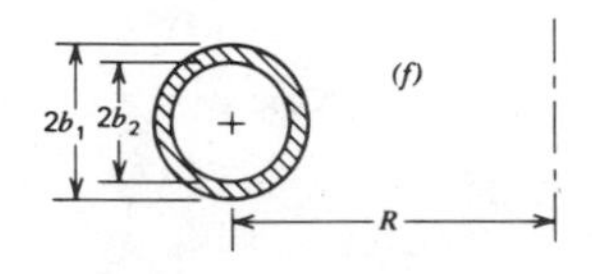

$$A = \pi(b_1^2 - b_2^2)$$
$$A_m = 2\pi\left(\sqrt{R^2 - b_2^2} - \sqrt{R^2 - b_1^2}\right)$$

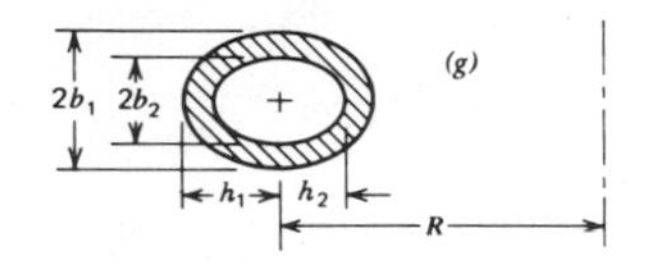

$$A = \pi(b_1h_1 - b_2h_2)$$
$$A_m = 2\pi\left(\frac{b_1R}{h_1} - \frac{b_2R}{h_2} - \frac{b_1}{h_1}\sqrt{R^2 - h_1^2} + \frac{b_2}{h_2}\sqrt{R^2 - h_2^2}\right)$$

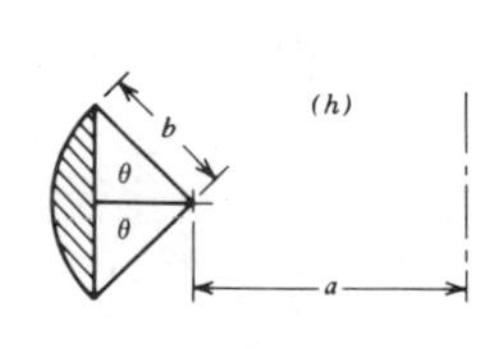

$$A = b^2\theta - \frac{b^2}{2}\sin 2\theta;\ R = a + \frac{4b\sin^3\theta}{3(2\theta - \sin 2\theta)}.$$

For $a > b$,

$$A_m = 2a\theta - 2b\sin\theta - \pi\sqrt{a^2 - b^2} + 2\sqrt{a^2 - b^2}\ \sin^{-1}\left[\frac{b + a\cos\theta}{a + b\cos\theta}\right]$$

For $b > a$,

$$A_m = 2a\theta - 2b\sin\theta + 2\sqrt{b^2 - a^2}\ \ln\left[\frac{b + a\cos\theta + \sqrt{b^2 - a^2}\ \sin\theta}{a + b\cos\theta}\right]$$

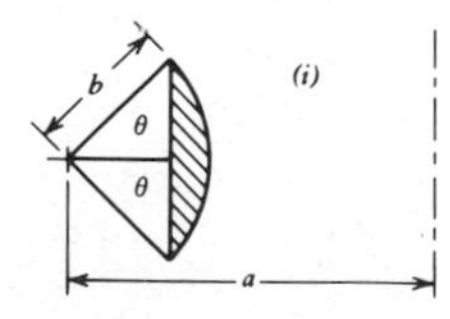

$$A = b^2\theta - \frac{b^2}{2}\sin 2\theta;\ R = a - \frac{4b\sin^3\theta}{3(2\theta - \sin 2\theta)}$$
$$A_m = 2a\theta + 2b\sin\theta - \pi\sqrt{a^2 - b^2} - 2\sqrt{a^2 - b^2}\ \sin^{-1}\left[\frac{b - a\cos\theta}{a - b\cos\theta}\right]$$

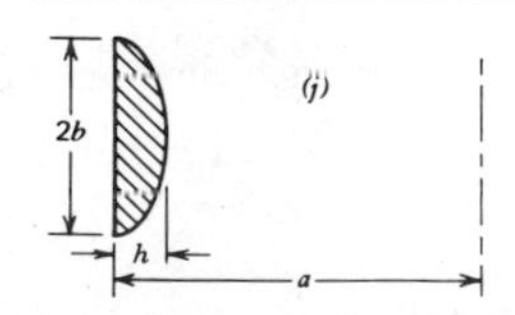

$$A = \frac{\pi bh}{2};\ R = a - \frac{4h}{3\pi}$$
$$A_m = 2b + \frac{\pi b}{h}(a - \sqrt{a^2 - h^2}) - \frac{2b}{h}\sqrt{a^2 - h^2}\ \sin^{-1}\left(\frac{h}{a}\right)$$

simplifies to

$$\frac{M_x R}{A I_x}(A - RA_m - yA_m) = \frac{M_x R}{A I_x}\left(\int \frac{y/R}{1+\frac{y}{R}}\, dA - \frac{y}{R}\int \frac{dA}{1+\frac{y}{R}}\right) = -\frac{M_x y}{I_x} \qquad \text{(c)}$$

The curved beam solution for curved beams (see Eq. 8-2.11) requires that A_m defined by Eq. (8-2.8) be calculated for cross sections of various shapes. The number of significant digits retained in calculating A_m must be greater than that required for $\sigma_{\theta\theta}$ since RA_m approaches the value of A as R/h becomes large (see Eq. a). Explicit formulas for A, A_m, and R for several curved beam cross sectional areas are listed in Table 8-2.2. Often the cross section of a curved beam is composed of two or more of the fundamental areas listed in Table 8-2.2. The values of A, A_m, and R for the composite area are given by summation. Thus, for composite cross sections,

$$A = \sum_{i=1}^{n} A_i \qquad \text{(8-2.12)}$$

$$A_m = \sum_{i=1}^{n} A_{m_i} \qquad \text{(8-2.13)}$$

$$R = \frac{\sum_{i=1}^{n} R_i A_i}{\sum_{i=1}^{n} A_i} \qquad \text{(8-2.14)}$$

where n is the number of fundamental areas that form the composite area.

Location of Neutral Axis of Cross Section/The neutral axis of bending of the cross section is defined by the conditions $\sigma_{\theta\theta} = 0$. The neutral axis is located at distance R_n from the center of curvature. The magnitude of R_n is obtained from Eq. (8-2.11) with the condition that $\sigma_{\theta\theta} = 0$ on the neutral surface $r = R_n$. Thus, Eq. (8-2.11) yields

$$R_n = \frac{AM_x}{A_m M_x + N(A - RA_m)} \qquad \text{(8-2.15)}$$

For pure bending, $N = 0$, and then Eq. (8-2.15) yields

$$R_n = \frac{A}{A_m} \qquad \text{(8-2.16)}$$

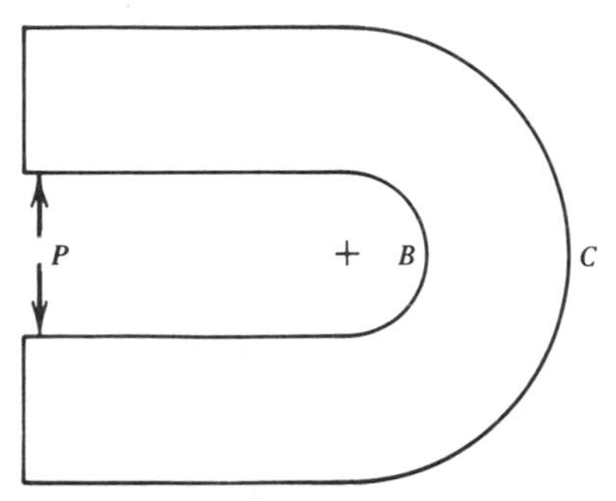

Fig. E8-2.1

EXAMPLE 8-2.1
Stress in Curved Beam Portion of a Frame

The frame shown in Fig. E8-2.1 has a square cross section with dimensions of 50.0 mm. The load P is located 100 mm from the center of curvature of the curved beam portion of the frame. The radius of curvature of the inner surface of the curved beam is $a = 30$ mm. If $P = 9.50$ kN, determine the values for the maximum tensile and compressive stresses in the frame.

SOLUTION

The circumferential stresses $\sigma_{\theta\theta}$ are calculated using Eq. (8-2.11). Required values for A, A_m, and R for the curved beam are calculated using the equations in Row a of Table 8-2.2. For the curved beam $a = 30$ mm and $c = 80$ mm.

$$A = b(c - a) = 50(80 - 30) = 2500 \text{ mm}^2$$

$$A_m = b \ln \frac{c}{a} = 50 \ln \frac{80}{30} = 49.04 \text{ mm}$$

$$R = \frac{a + c}{2} = \frac{80 + 30}{2} = 55 \text{ mm}$$

Hence, the maximum tensile stress is

$$\sigma_{\theta\theta B} = \frac{P}{A} + \frac{M_x(A - rA_m)}{Ar(RA_m - A)} = \frac{9500}{2500} + \frac{155(9500)[2500 - 30(49.04]}{2500(30)[55(49.04) - 2500]}$$

$$= 106.2 \text{ MPa}$$

The maximum compressive stress is

$$\sigma_{\theta\theta C} = \frac{9500}{2500} + \frac{155(9500)[2500 - 80(49.04)]}{2500(80)[55(49.04) - 2500]} = -49.3 \text{ MPa}$$

EXAMPLE 8-2.2

Stresses in a Crane Hook

Section BC is the critically stressed section of a crane hook (Fig. E8-2.2). For a large number of manufactured crane hooks, the critical section BC can be closely approximated by a trapezoidal area with half of an ellipse at the inner radius and an arc of a circle at the outer radius. Such a section is shown in Fig. E8-2.2b, including dimensions for the critical cross section. The crane hook is made of a ductile steel that has a yield stress of $Y = 500$ MPa. Assuming that the crane hook is to be designed with a factor of safety of $SF = 2.00$ against initiation of yielding, determine the maximum load P that can be carried by the crane hook.

SOLUTION

The circumferential stresses $\sigma_{\theta\theta}$ are calculated using Eq. (8-2.11). To calculate values of A, R, and A_m for the curved beam cross section, we divide the cross section into basic areas A_1, A_2, and A_3 (Fig. E8-2.2b).

For area A_1, $a = 84$ mm. Substituting this dimension along with other given dimensions into Table 8-2.2, row j, we find

$$A_1 = 1658.76 \text{ mm}^2 \qquad R_1 = 73.81 \text{ mm} \qquad A_{m_1} = 22.64 \text{ mm} \qquad (1)$$

For the trapezoidal area A_2, $a = 60 + 24 = 84$ mm and $c = a + 100 = 184$ mm. Substituting these dimensions along with other given dimensions into Table 8-2.2, row c, we find

$$A_2 = 6100.00 \text{ mm}^2 \qquad R_2 = 126.62 \text{ mm} \qquad A_{m_2} = 50.57 \text{ mm} \qquad (2)$$

For area A_3, $\theta = 0.5721$ rad, $b = 31.40$ mm, and $a = 157.60$ mm. When these values are substituted into Table 8-2.2, row h, we obtain

$$A_3 = 115.27 \text{ mm}^2 \qquad R_3 = 186.01 \text{ mm} \qquad A_{m_3} = 0.62 \text{ mm} \qquad (3)$$

Substituting values of A_i, R_i, and A_{mi} from Eqs. (1), (2), and (3) into Eqs. (8-2.12), (8-2.13), and (8-2.14), we calculate

$$A = 6100.00 + 115.27 + 1658.76 = 7874.03 \text{ mm}^2$$

$$A_m = 50.57 + 0.62 + 22.64 = 73.83 \text{ mm}$$

$$R = \frac{6100.00(126.62) + 115.27(186.01) + 1658.76(73.81)}{7874.03}$$

$$= 116.37 \text{ mm}$$

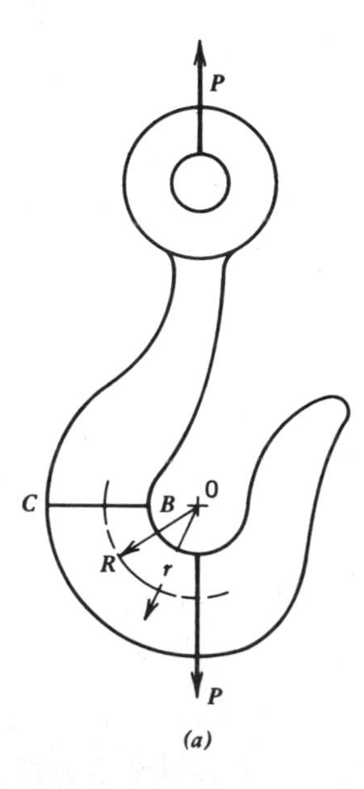

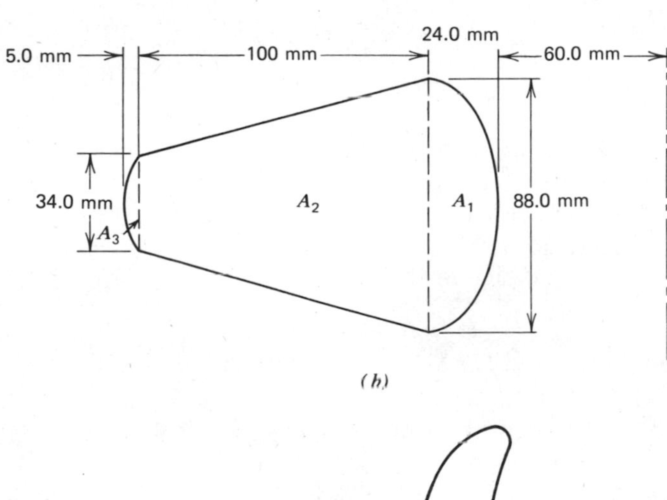

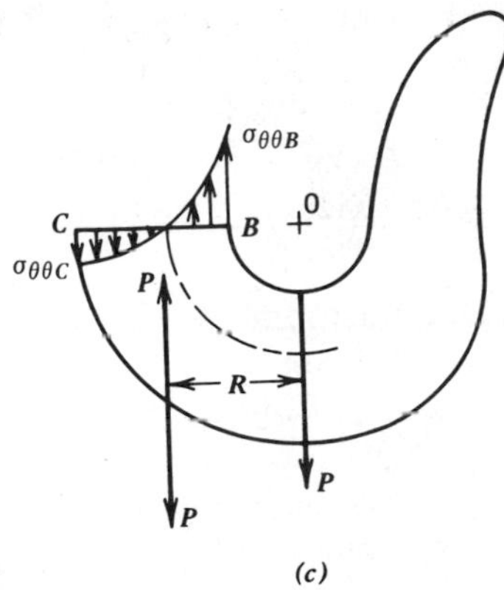

Fig. E8-2.2/Crane hook.

As indicated in Fig. E8-2.2c, the circumferential stress distribution $\sigma_{\theta\theta}$ is due to the normal load $N = P$ and the moment $M_x = PR$. The maximum tension and compression values of $\sigma_{\theta\theta}$ occur at points B and C, respectively. For points B and C, Fig. E8-2.2b yields

$$r_B = 60 \text{ mm}$$

$$r_C = 60 + 24 + 100 + 5 = 189 \text{ mm}$$

Substituting the required values into Eq. (8-2.11), we find

$$\sigma_{\theta\theta B} = \frac{P}{7874.03} + \frac{116.37P[7874.03 - 60(73.83)]}{7874.03(60)[116.37(73.83) - 7874.03]}$$

$$= 0.000127P + 0.001182P$$

$$= 0.001309P \text{ (tension)}$$

$$\sigma_{\theta\theta C} = \frac{P}{7874.03} + \frac{116.37P[7874.03 - 189(73.83)]}{7874.03(189)[116.37(73.83) - 7874.03]}$$

$$= 0.000127P - 0.000662P$$

$$= -0.000535P \text{ (compression)}$$

Since the absolute magnitude of $\sigma_{\theta\theta B}$ is greater than $\sigma_{\theta\theta C}$ initiation of yield occurs when $\sigma_{\theta\theta B}$ equals the yield stress Y. The corresponding value of the failure load (P_f) is the load at which yield occurs. Dividing the failure load $P_f = Y/(0.001309)$ by the factor of safety $SF = 2.00$, we obtain the design load P; namely,

$$P = \frac{500}{2.00(0.001309)} = 190{,}900 \text{ N}$$

Thus, based upon a factor of safety of 2.00 at sea level on earth, the crane hook can support a weight of mass $m = P/g$, where g denotes the acceleration due to gravity. In other words, a mass as large as $m = 190{,}900/9.81 = 19{,}460$ kg, can be hoisted by the crane with a factor of safety of 2.00 against yield initiation.

Computer Program for Crane Hooks/To expedite the solution for crane hooks, a FORTRAN computer program CRAHK is presented in Table E8-2.2, along with a set of input data and the corresponding output. As noted from the program and data, several cases can be calculated with one computer run. Consequently, parametric studies of the crane hook dimensions can be rapidly performed.

Table E8-2.2

FORTRAN Program for Crane Hooks

```
1           PROGRAM CRAHK(INPUT,OUTPUT,CRIN,CROUT,TAPE5 =
2         ACRIN,TAPE6 = CROUT)
3           READ (5,*) NN
4           DO 2 J = 1,NN
5           WRITE (6,22)
6  22       FORMAT ("        H1        H2        H3        B1        B2")
7           READ (5,*) N,EH1,EH2,EH3,B1,B2
8           WRITE (6,23) EH1,EH2,EH3,B1,B2
9  23       FORMAT (5F10.2)
10          WRITE (6,24)
11 24       FORMAT ("        RB        R        AREA        AM"
12        C"     SIG(MAX)/P   SIG(MIN)/P")
13          DO 3 I = 1,N
14          READ (5,*) RB
15          A1 = .7854*B1*EH1
16          AA = RB + EH1
17          C = AA + EH2
18          R1 = AA - .42441*EH1
19          D = SQRT(AA**2 - EH1**2)
20          AM1 = B1 + 1.5708*B1*(AA - D)/EH1 - B1*D*ASIN(EH1/AA)
21        A/EH1
22          A2 = .5*(B1 + B2)*EH2
23          R2 = (AA*(2.*B1 + B2) + C*(B1 + 2.*B2))/(3.*(B1 + B2))
24          AM2 = (B1*C - B2*AA)*ALOG(C/AA)/EH2 - B1 + B2
25          B3 = (B2**2 + EH3**2*4.)/(8.*EH3)
26          THETA = ASIN(.5*B2/B3)
27          A3 = B3**2*THETA - .5*B3**2*SIN(2.*THETA)
28          A = C + EH3 - B3
29          R3 = A + 1.3333*B3*SIN(THETA)**3/(2.*THETA - SIN(2.*
30        ATHETA))
31          DD = SQRT(A**2 - B3**2)
32          Q = (B3 + A*COS(THETA))/(A + B3*COS(THETA))
33          AM3 = 2.*A*THETA - 2.*B3*SIN(THETA) - 3.14159*DD +
34        A2.*DD*ASIN(Q)
35          AREA = A1 + A2 + A3
36          AM = AM1 + AM2 + AM3
37          R = (R1*A1 + R2*A2 + R3*A3)/AREA
38          SIG1 = 1./AREA + R*(AREA - RB*AM)/(AREA*RB*(R*AM
39        A - AREA))
40          SIG2 = 1./AREA + R*(AREA - (C + EH3)*AM)/(AREA*(C +
41        AEH3)*(R*AM - AREA))
42          WRITE (6,25) RB,R,AREA,AM,SIG1,SIG2
43 25       FORMAT (4F12.2,2F12.7)
44 3        CONTINUE
```

Table 8-2.2

Continued

```
45 2      CONTINUE
46        STOP
47        END
```

INPUT FOR PROGRAM CRAHK

```
3
1,63.,303.,28.33,184.,74.
165.
5,.93,3.97,.18,3.46,1.35
2.25
2.375
2.5
2.675
2.75
5,.83,4.07,.18,3.50,1.35
2.25
2.375
2.50
2.675
2.75
```

OUTPUT FOR PROGRAM CRAHK

H1	H2	H3	B1	B2	
63.00	303.00	28.33	184.00	74.00	
RB	R	AREA	AM	SIG(MAX)/P	SIG(MIN)/P
165.00	335.05	49741.25	163.95	0.0001986	−0.0000773
H1	H2	H3	B1	B2	
0.93	3.97	0.18	3.46	1.35	
RB	R	AREA	AM	SIG(MAX)/P	SIG(MIN)/P
2.25	4.47	12.24	3.00	0.8336250	−0.3283893
2.38	4.60	12.24	2.90	0.8407182	−0.3437211
2.50	4.72	12.24	2.81	0.8483529	−0.3590766
2.68	4.90	12.24	2.70	0.8598112	−0.3806141
2.75	4.97	12.24	2.65	0.8649603	−0.3898584
H1	H2	H3	B1	B2	
0.83	4.07	0.18	3.50	1.35	
RB	R	AREA	AM		SIG(MIN)/P
2.25	4.46	12.32	3.03	0.8118828	−0.3236749
2.38	4.58	12.32	2.93	0.8189578	−0.3388556
2.50	4.71	12.32	2.84	0.8265510	−0.3540606
2.68	4.88	12.32	2.72	0.8379177	−0.3753882
2.75	4.96	12.32	2.68	0.8430174	−0.3845428

PROBLEM SET 8-2

1. The frame shown in Fig. E8-2.1 has a rectangular cross section with a width of 10 mm and a depth of 40 mm. The load P is located 120 mm from the centroid of section BC. The frame is made of a steel having a yield stress of $Y = 430$ MPa. The frame has been designed using a factor of safety of $SF = 1.75$ against initiation of yielding. Determine the maximum allowable magnitude of P, if the radius of curvature at section BC is $R = 40$ mm.

2. Solve Problem 1 for the condition that $R = 35$ mm.

   Ans. $P = 3.174$ kN

3. The curved beam in Fig. P8-2.3 has a circular cross section 50 mm in diameter. The inside diameter of the curved beam is 40 mm. Determine the stress at B when $P = 20$ kN.

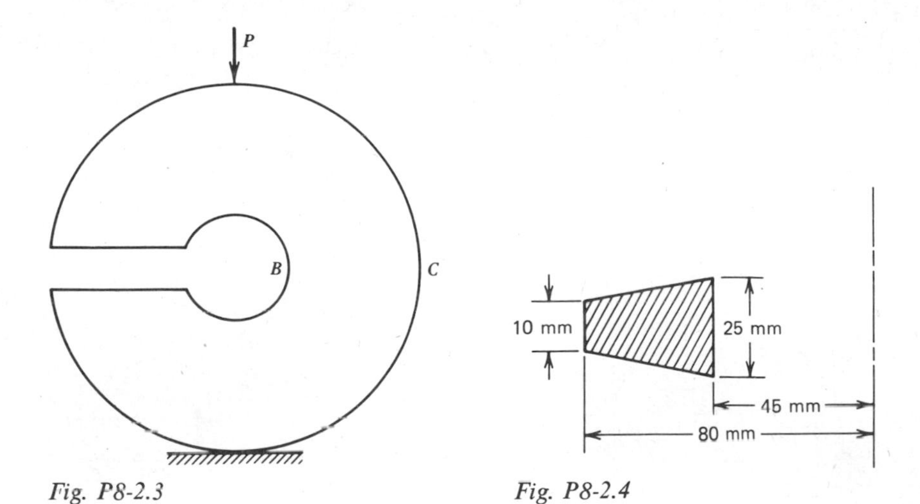

Fig. P8-2.3　　*Fig. P8-2.4*

4. Let the crane hook in Fig. E8-2.2 have a trapezoidal cross section as shown in Row c of Table 8-2.2 with (see Fig. P8-2.4) $a = 45$ mm, $c = 80$ mm, $b_1 = 25$ mm, and $b_2 = 10$ mm. Determine the maximum load to be carried by the hook if the working stress is 150 MPa.

   Ans. $P = 7.34$ kN

5. A curved beam is built up by welding rectangular and elliptical cross section curved beams together; the cross section is shown in Fig. P8-2.5. The center of curvature is located 20 mm from B. The curved

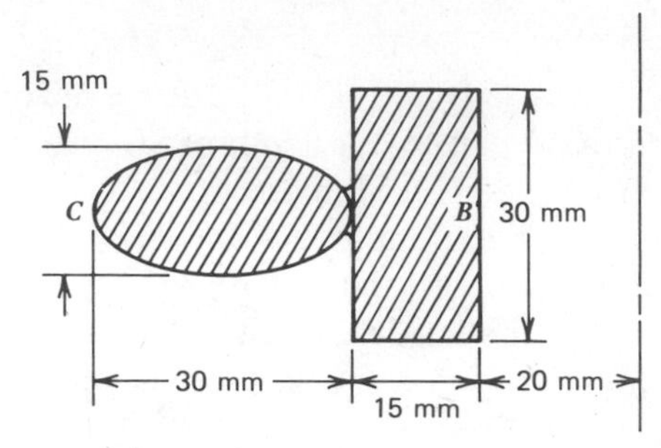

Fig. P8-2.5

beam is subjected to a positive bending moment M_x(N · m). Determine the stresses at points B and C in terms of M_x.

6. A commercial crane hook has the cross-sectional dimensions shown in Fig. P8-2.6 at the critical section that is subjected to an axial load $P = 100$ kN. Determine the circumferential stresses at the inner and outer radii for this load. Assume that area A_1 is half of an ellipse (see Row j in Table 8-2.2).

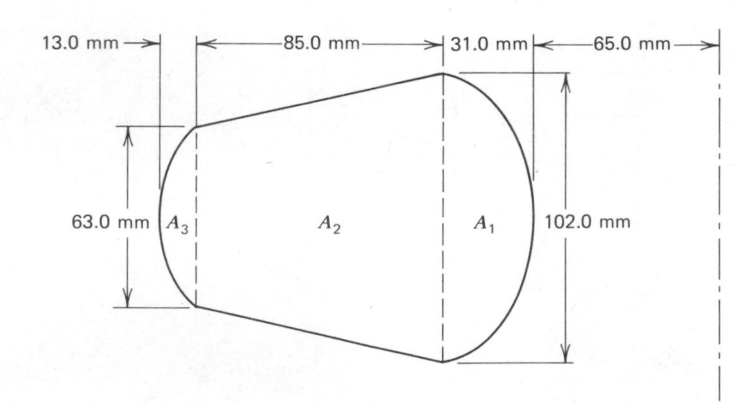

Fig. P8-2.6

Ans. $\sigma_{\theta\theta B} = 113.5$ MPa, $\sigma_{\theta\theta C} = -43.6$ MPa

7. A crane hook has the cross-sectional dimensions shown in Fig. P8-2.7 at the critical section that is subjected to an axial load $P = 90.0$ kN. Determine the circumferential stresses at the inner and outer radii for this load. Note that A_1 and A_3 are enclosed by circular arcs.

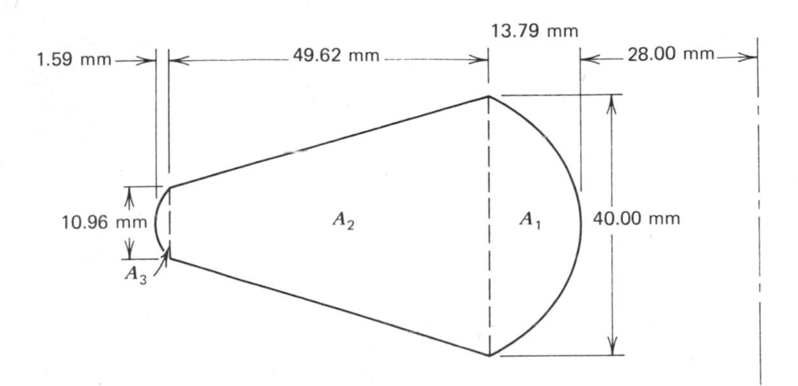

Fig. P8-2.7

8. The curved beam in Fig. P8-2.8 has a triangular cross section with the dimensions shown. If $P = 40$ kN, determine the circumferential stresses at B and C.

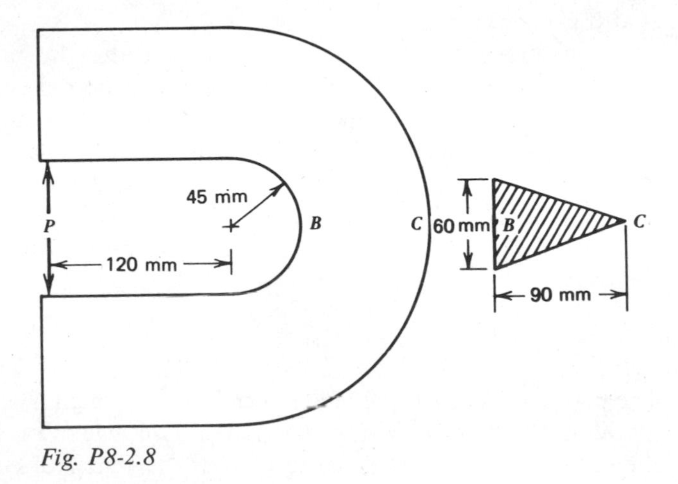

Fig. P8-2.8

Ans. $\sigma_{\theta\theta B} = 297.8$ MPa, $\sigma_{\theta\theta C} = -238.1$ MPa

8-3 RADIAL STRESSES IN CURVED BEAMS

The curved beam formula for circumferential stress $\sigma_{\theta\theta}$ (Eq. 8-2.11) is based on the assumption that the effect of radial stress is negligible. This assumption is quite accurate for curved beams with circular, rectangular, or trapezoidal cross sections; that is, cross sections that do not possess thin webs. However, in curved beams with cross sections in the form of

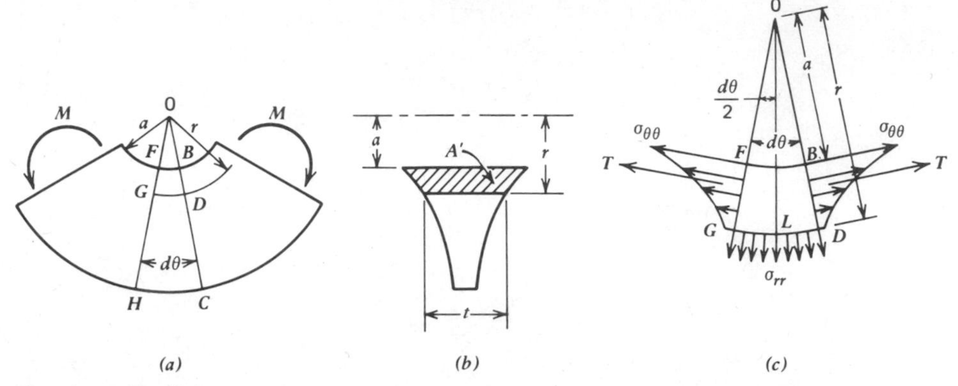

Fig. 8-3.1/Radial stress in a curved beam. (*a*) Side view. (*b*) Cross section shape. (*c*) Element *BDGF*.

an *H*, *T*, or *I*, the webs may be so thin that deformation of the cross section may produce a maximum radial stress in the web that may exceed the maximum circumferential stress. The beam should be designed so that this condition does not occur.

To illustrate the above remarks, we consider the radial stress that occurs in a curved beam at radius r from the center of curvature 0 of the beam (Fig. 8-3.1*a*). Consider equilibrium of the element *BDGF* of the beam shown enlarged in the free body diagram in Fig. 8-3.1*c*. The faces *BD* and *GF*, which subtend the infinitesimal angle $d\theta$, have the area A' shown shaded in Fig. 8-3.1*b*. The distribution of $\sigma_{\theta\theta}$ on each of these areas produces a resultant circumferential force T (Fig. 8-3.1*c*) given by the expression

$$T = \int_a^r \sigma_{\theta\theta}\, dA \tag{8-3.1}$$

The components of the circumferential forces along line $0L$ are balanced by the radial stress σ_{rr} acting on the area $tr\,d\theta$, where t is the thickness of the curved beam at the distance r from the center of curvature 0 (Fig. 8-3.1*b*). Thus for equilibrium in the radial direction along $0L$, $\Sigma F_r = 0 = \sigma_{rr} tr\, d\theta - 2T \sin(d\theta/2) = (\sigma_{rr} tr - T)\, d\theta$, since for infinitesimal angle $d\theta/2, \sin(d\theta/2) = d\theta/2$. Thus,

$$\sigma_{rr} = \frac{T}{tr} \tag{8-3.2}$$

The force T is obtained by substitution of Eq. (8-2.11) into Eq. (8-3.1). Thus,

$$T = \frac{N}{A}\int_a^r dA + \frac{M_x}{RA_m - A}\int_a^r \frac{dA}{r} - \frac{M_x A_m}{A(RA_m - A)}\int_a^r dA$$

$$T = \frac{A'}{A}N + \frac{AA'_m - A'A_m}{A(RA_m - A)}M_x \tag{8-3.3}$$

where

$$A'_m = \int_a^r \frac{dA}{r} \qquad \text{and} \qquad A' = \int_a^r dA \tag{8-3.4}$$

Substitution of Eq. (8-3.3) into Eq. (8-3.2) yields the relation for the radial stress. For rectangular cross section curved beams subjected to shear loading (Fig. 8-2.3*b*), a comparison of the resulting approximate solution with the elasticity solution indicates that the approximate solution is conservative. Furthermore, for such beams it remains conservative to within 6 percent for values of $R/h > 1.0$ even if the term involving N in Eq. (8-3.3) is discarded. Consequently, if we retain only the moment term in Eq. (8-3.3), the expression for the radial stress may be approximated by the formula,

$$\sigma_{rr} = \frac{AA'_m - A'A_m}{trA(RA_m - A)} M_x \tag{8-3.5}$$

to within 6 percent of the elasticity solution for rectangular cross section curved beams subjected to shear loading (Fig. 8-2.3*b*).

EXAMPLE 8-3.1

Radial Stress in T-Section

The curved beam in Fig. E8-3.1 is subjected to a load $P = 120$ kN. The dimensions of section BC are also shown. Determine the circumferential stress at B and the radial stress at the junction of the flange and web at section BC.

SOLUTION

The magnitudes of A, A_m, and R are given by Eqs. (8-2.12), (8-2.13), and (8-2.14), respectively. They are

$$A = 48(120) + 120(24) = 8640 \text{ mm}^2$$

$$R = \frac{48(120)(96) + 120(24)(180)}{8640} = 124.0 \text{ mm}$$

$$A_m = 120 \ln \frac{120}{72} + 24 \ln \frac{240}{120} = 77.93 \text{ mm}$$

The circumferential stress is given by Eq. (8-2.11). It is

$$\sigma_{\theta\theta B} = \frac{120{,}000}{8640} + \frac{364.0(120{,}000)[8640 - 72(77.93)]}{8640(72)[124.0(77.93) - 8640]}$$

$$= 13.9 + 207.8 = 221.7 \text{ MPa}$$

The radial stress at the junction of the flange and web is given by Eq.

(8-3.5), with $r = 120$ mm and $t = 24$ mm. Magnitudes of A' and A'_m are

$$A' = 48(120) = 5760 \text{ m}^2$$

$$A'_m = 120 \ln \frac{120}{72} = 61.30 \text{ mm}$$

Substitution of these values in Eq. (8-3.5) gives

$$\sigma_{rr} = \frac{364.0(120{,}000)[8640(61.30) - 5760(77.93)]}{24(120)(8640)[124.0(77.93) - 8640]} = 138.5 \text{ MPa}$$

Hence, the magnitude of this radial stress is appreciably less than the maximum circumferential stress ($|\sigma_{\theta\theta B}| > |\sigma_{\theta\theta C}|$) and may not be of

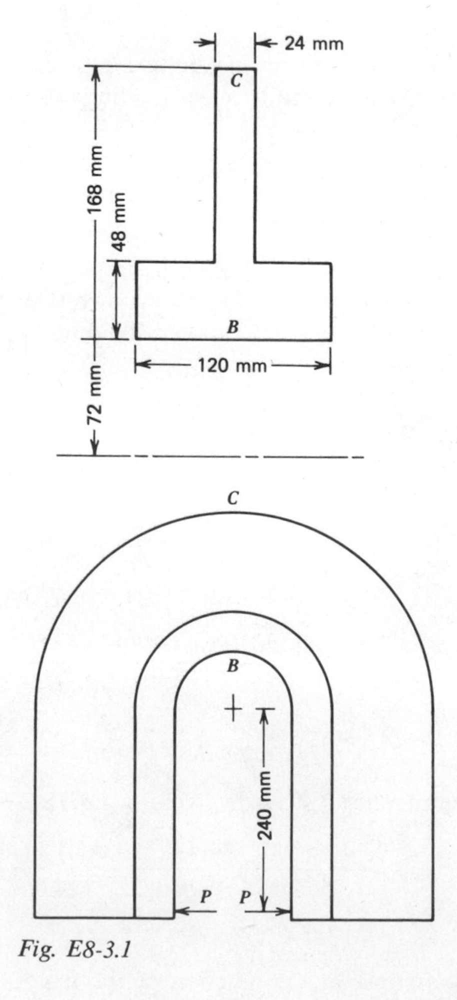

Fig. E8-3.1

concern for the design engineer. However, in the solution of this problem, the effect of the stress concentration at the fillet joining the flange to the web has not been considered. This stress concentration increases the magnitude of the radial stress at the junction of the flange and web. However, the increase in stress is localized. Hence, it is not significant for curved beams made of ductile metal and subjected to static loads. However, for curved beams made of brittle materials or for curved beams of ductile material subjected to repeated loads, the localized stresses are significant. The effect of stress concentrations at fillets are considered in Chapter 12.

PROBLEM SET 8-3

1. For the curved beam in Problem 8-2.5, determine the radial stress in terms of the moment M_x if the thickness of the web is 10 mm.

2. In Fig. P8-3.2 is shown a cast iron frame with a U-shaped cross section. The ultimate tensile strength of the case iron is $\sigma_u = 320$ MPa. (a) Determine the maximum value of P based on a factor of safety $SF = 4.00$ which is based on the ultimate strength. (b) Neglecting the effect of stress concentrations for the fillet at the junction of the web and flange, determine the maximum radial stress when this load is applied. (c) Is the maximum radial stress less than the maximum circumferential stress?

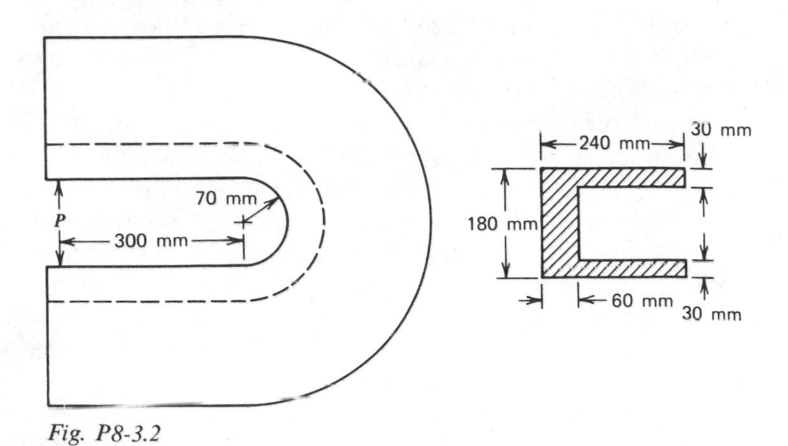

Fig. P8-3.2

Ans. (a) $P = 110.8$ kN (b) $\sigma_{rr} = 42.4$ MPa (c) Yes

8-4
CORRECTION OF CIRCUMFERENTIAL STRESSES IN CURVED BEAMS HAVING I, T, OR SIMILAR CROSS SECTIONS

If the curved beam formula is used to calculate circumferential stresses in curved beams having thin flanges, the computed stresses are considerably in error and the error is nonconservative. The error arises because the radial forces developed in the curved beam causes the outer portion of the flanges to deflect radially, thereby distorting the cross section of the curved beam. The resulting effect is to decrease the stiffness of the curved beam, to decrease the circumferential stresses in the outer portion of the flanges, and to increase the circumferential stresses in the inner portion of the flanges.

Consider a short length of a thin flange I-section curved beam included between faces BC and FH which form an infinitesimal angle $d\theta$ as indicated in Fig. 8-4.1*a*. If the curved beam is subjected to a positive moment M_x the circumferential stress distribution results in a tensile force T acting on the inner flange and a compressive force C acting on the outer flange, as shown. The components of these forces in the radial direction are $T\,d\theta$ and $C\,d\theta$. If the cross section of the curved beam did not distort, these forces would be uniformly distributed along each flange, as indicated in Fig. 8-4.1*b*. However, the two portions of the tension and compression flanges act as cantilever beams fixed at the web. The resulting bending due to cantilever beam action causes the flanges to distort, as indicated in Fig. 8-4.1*c*.

The effect of the distortion of the cross section on the circumferential stresses in the curved beam can be determined by examining the portion of the curved beam $ABCD$ in Fig. 8-4.1*d*. Sections AC and BD are separated by angle θ in the unloaded beam. When the curved beam is subjected to a positive moment, the center of curvature moves from 0 to 0*, section AC moves to A^*C^*, section BD moves to B^*D^*, and the included angle becomes θ^*. If the cross section does not distort, the inner tension flange AB elongates to length A^*B^*. Since the outer portion of the inner flange moves radially inward relative to the undistorted position (Fig. 8-4.1*c*), the circumferential elongation of the outer portion of the inner flange is less than that indicated in Fig. 8-4.1*d*. *Therefore,* $\sigma_{\theta\theta}$ *in the outer portion of the inner flange is less than that calculated using the curved beam formula.* In order to satisfy equilibrium, it is necessary that $\sigma_{\theta\theta}$ for the portion of the flange near the web be greater than that calculated using the curved beam formula. Now consider the outer compression flange. As indicated in Fig. 8-4.1*d*, the outer flange shortens from CD to C^*D^* if the cross section does not distort. Because of the distortion (Fig. 8-4.1*c*), the outer portion of the compressive flange moves radially outward requiring less compressive contraction. *Therefore, the magnitude*

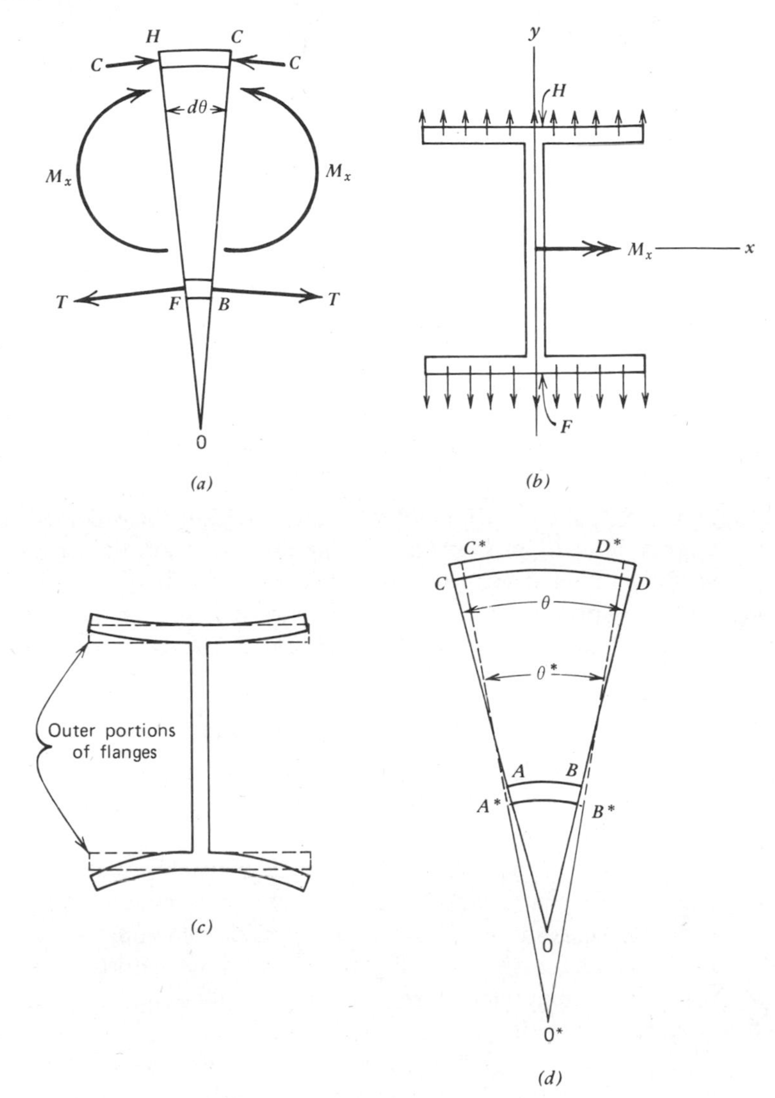

Fig. 8-4.1/Distortion of cross section of an I-section curved beam.

of $\sigma_{\theta\theta}$ in the outer portion of the compression outer flange is less than that calculated by the curved beam formula, and the magnitude of $\sigma_{\theta\theta}$ in the portion of the compression flange near the web is larger than that calculated by the curved beam formula.

The resulting circumferential stress distribution is indicated in Fig. 8-4.2. Since the curved beam formula assumes that the circumferential stress is independent of x (Fig. 8-2.1), corrections are required if the formula is to be used in the design of curved beams having I or T cross sections and similar cross sections. There are two approaches (approxi-

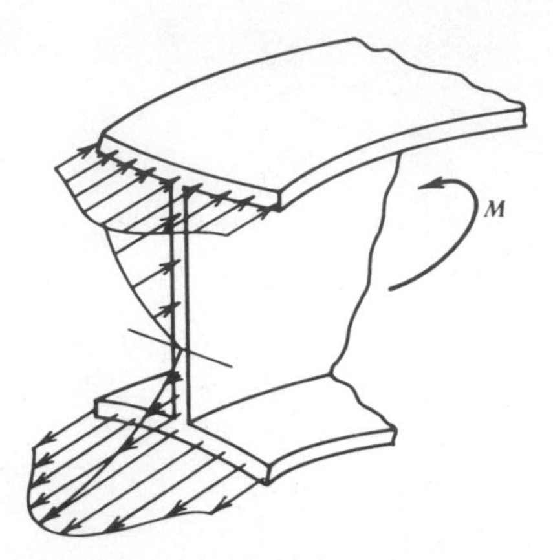

Fig. 8-4.2/Stresses in I-section of curved beam.

mations) that can be employed in the design of these curved beams. One approach is to prevent the radial distortion of the cross section by welding radial stiffeners to the curved beams. If distortion of the cross section is prevented, the use of the curved beam formula is appropriate. A second approach, suggested by H. Bleich,[2] is discussed below.

Bleich's Correction Factors/Bleich reasoned that the actual maximum circumferential stresses in the tension and compression flanges for the I-section curved beam (Fig. 8-4.3*a*) may be calculated by the curved beam formula applied to an I-section curved beam with *reduced flange widths*, as indicated in Fig. 8-4.3*b*. By Bleich's method, if the same bending moment is applied to the two cross sections in Fig. 8-4.3, the computed maximum circumferential tension and compression stresses for the cross section shown in Fig. 8-4.3*b*, with no distortion, is equal to the actual maximum circumferential tension and compression stresses for the cross section in Fig. 8-4.3*a*, with distortion.

The approximate solution proposed by Bleich gives the results presented in tabular form in Table 8-4.1. In order to use the table, the ratio $b_p^2/\bar{r}t_f$ must be calculated where

b_p = projecting width of flange (see Fig. 8-4.3*a*)

$\bar{r}$ = radius of curvature to the center of flange

t_f = thickness of flange

The reduced width b_p' of the projecting part of each flange (Fig. 8-4.3*b*) is given by the relation

$$b_p' = \alpha b_p \tag{8-4.1}$$

where α is obtained from Table 8-4.1 for the computed value of the ratio

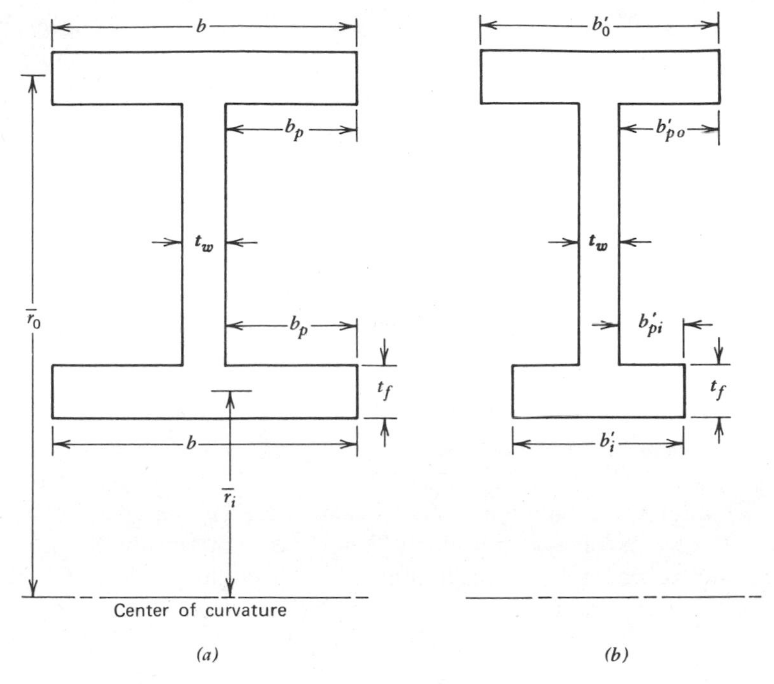

Fig. 8-4.3/Original and modified I-section for a curved beam.

$b_p^2/\bar{r}t_f$. The reduced width of each flange (Fig. 8-4.3*b*) is given by

$$b' = 2b_p' + t_w \tag{8-4.2}$$

where t_w is the thickness of the web. The curved beam formula (Eq. 8-2.11) when applied to an undistorted cross section corrected by Eq. (8-4.2) predicts the maximum circumferential stress in the actual (distorted) cross section. This maximum stress occurs at the center of the

Table 8-4.1

Table for Calculating the Effective Width and the Lateral Bending Stress of Curved I- or T-Beams

$b_p^2/\bar{r}t$	0.2	0.3	0.4	0.5	0.6	0.7	0.8	0.9	1.0
α	0.977	0.950	0.917	0.878	0.838	0.800	0.762	0.726	0.693
β	0.580	0.836	1.056	1.238	1.382	1.495	1.577	1.636	1.677
$b_p^2/\bar{r}t$	1.1	1.2	1.3	1.4	1.5	2.0	3.0	4.0	5.0
α	0.663	0.636	0.611	0.589	0.569	0.495	0.414	0.367	0.334
β	1.703	1.721	1.728	1.732	1.732	1.707	1.671	1.680	1.700

inner flange. It should be noted that the state of stress at this point in the curved beam is not uniaxial. Because of the bending of the flanges (Fig. 8-4.1*c*), an *x*-component of stress σ_{xx} (Fig. 8-2.1) is developed; the sign of σ_{xx} is opposite to that of $\sigma_{\theta\theta(\max)}$. Bleich obtained an approximate solution for σ_{xx} for the inner flange. It is given by the relation

$$\sigma_{xx} = -\beta\bar{\sigma}_{\theta\theta} \tag{8-4.3}$$

where β is obtained from Table 8-4.1 for the computed value of the ratio $b_p^2/\bar{r}t_f$, and where $\bar{\sigma}_{\theta\theta}$ is the magnitude of the circumferential stress at mid-thickness of the inner flange; the value of $\bar{\sigma}_{\theta\theta}$ is calculated based on the corrected cross section.

Although Bleich's analysis was developed for curved beams with relatively thin flanges, the results obtained agree closely with a similar solution obtained by C. G. Anderson[3] for I-beams and box beams in which the analysis was not restricted to thin-flanged sections. Similar analyses of tubular curved beams with circular and rectangular cross sections have been made by T. von Kármán[4] and by S. Timoshenko.[5] An experimental investigation by D. C. Broughton, M. E. Clark, and H. T. Corten[6] showed that another type of correction is needed if the curved beam has extremely thick flanges and thin webs. For such beams each flange tends to rotate about a neutral axis of its own in addition to the rotation about the neutral axis of the curved beam cross section as a whole. Curved beams for which the circumferential stresses are apprecia-bly increased by this action probably fail by excessive radial stresses.

Note: The radial stress can be calculated using either the original or the modified cross section.

EXAMPLE 8-4.1

Bleich Correction Factors for T-Section

A T-section curved beam has the dimensions indicated in Fig. E8-4.1*a* and is subjected to pure bending. The curved beam is made of a steel having a yield stress $Y = 280$ MPa. (a) Determine the magnitude of the moment which indicates yielding in the curved beam if Bleich's correc-tion factors are not used. (b) Use Bleich's correction factors to obtain a modified cross section. Determine the magnitude of the moment that initiates yielding for the modified cross section and compare with the result of Part (a).

SOLUTION

(a) The magnitudes of A, A_m, and R for the original cross section are given by Eq. (8-2.12), (8-2.13) and (8-2.14), respectively, as follows: $A = 4000\ \text{mm}^2$, $A_m = 44.99$ mm, and $R = 100.0$ mm. By comparison of

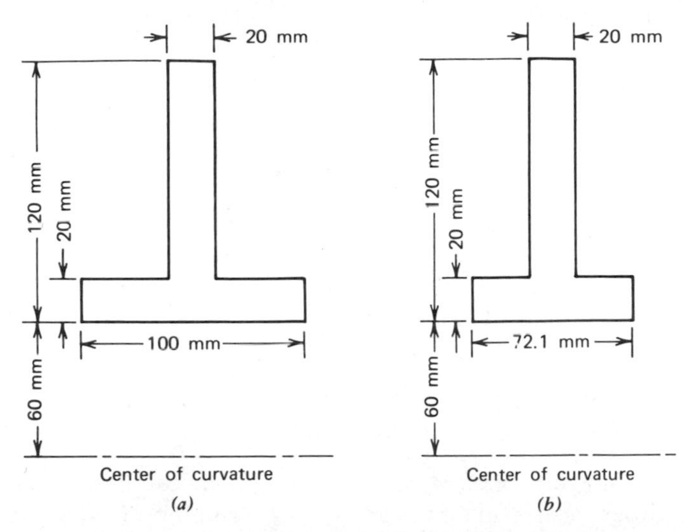

Fig. E8-4.1/(a) Original section. (*b*) Modified section.

the stresses at the locations $r = 180$ mm and at $r = 60$ mm, we find that the maximum magnitude of $\sigma_{\theta\theta}$ occurs at the outer radius ($r = 180$ mm). See Eq. (8-2.11). Thus,

$$\sigma_{\theta\theta(\max)} = \left| \frac{M_x[4000 - 180(44.99)]}{4000(180)[100.0(44.99) - 4000]} \right|$$

$$= |-1.141 \times 10^{-5} M_x| \text{ (MPa)}$$

where M_x has the dimensions of N · mm. Since the state of stress is assumed to be uniaxial, the magnitude of M_x to initiate yielding is obtained by setting $\sigma_{\theta\theta} = -Y$. Thus,

$$M_x = \frac{280}{1.141 \times 10^{-5}} = 24{,}540{,}000 \text{ N} \cdot \text{mm} = 24.54 \text{ kN} \cdot \text{m}$$

(b) The dimensions of the modified cross section are computed by Bleich's method; hence, $b_p^2/\bar{r}t_f$ must be calculated. It is

$$\frac{b_p^2}{\bar{r}t_f} = \frac{40(40)}{70(20)} = 1.143$$

A linear interpolation in Table 8-4.1 yields $\alpha = 0.651$ and $\beta = 1.711$. Hence, by Eq. (8-4.2), the modified flange width given is $b_p' = \alpha b_p = 0.651(40) = 26.04$ mm and $b' = 2b_p' + t_w = 2(26.04) + 20 = 72.1$ mm (Fig. E8-4.1*b*). For this cross section, by means of Eqs. (8-2.12), (8-2.13), and

(8-2.14), we find

$$A = 72.1(20) + 20(100) = 3442 \text{ mm}^2$$

$$R = \frac{72.1(20)(70) + 20(100)(130)}{3442} = 104.9 \text{ mm}$$

$$A_m = 72.1 \ln \frac{80}{60} + 20 \ln \frac{180}{80} = 36.96 \text{ mm}$$

Now by means of Eq. (8-2.11), we find that the maximum magnitude of $\sigma_{\theta\theta}$ occurs at the inner radius of the modified cross section. Thus, with $r = 60$ mm, Eq. (8-2.11) yields

$$\sigma_{\theta\theta(\max)} = \frac{M_x[3442 - 60(36.96)]}{3442(60)[104.9(36.96) - 3442]} = 1.363 \times 10^{-5} M_x \text{ (MPa)}$$

The magnitude of M_x that causes yielding can be calculated by means of either the maximum shearing stress criterion of failure or the maximum octahedral shearing stress criterion of failure. If the maximum shearing stress criterion is used, the minimum principal stress must be computed. The minimum principal stress is σ_{xx}. Hence, by Eq. (8-4.3), we find

$$\bar{\sigma}_{\theta\theta} = \frac{M_x[3442 - 70(36.96)]}{3442(70)[104.9(36.96) - 3442]} = 8.15 \times 10^{-6} M_x \text{ (MPa)}$$

$$\sigma_{xx} = -\beta\bar{\sigma}_{\theta\theta} = -1.711(8.15 \times 10^{-6} M_x) = -1.394 \times 10^{-5} M_x \text{ (MPa)}$$

and

$$\tau_{\max} = \frac{\sigma_{\max} - \sigma_{\min}}{2} = \frac{Y}{2} = \frac{\sigma_{\theta\theta(\max)} - \sigma_{xx}}{2}$$

$$M_x = 10{,}140{,}000 \text{ N} \cdot \text{mm} = 10.14 \text{ kN} \cdot \text{m}$$

A comparison of the moment M_x as determined in Parts (a) and (b) above indicates that the computed M_x required to initiate yielding is reduced by 58.8 percent because of the distortion of the cross section. Since the yielding is highly localized, its effect is not of concern unless the curved beam is subjected to fatigue loading (see Art. 3-5). If the second principal stress σ_{xx} is neglected, the moment M_x is reduced by 16.5 percent because of the distortion of the cross section. The distortion is reduced if the flange thickness is increased.

PROBLEM SET 8-4

1. A T-section curved beam has the cross section shown in Fig. P8-4.1. The center of curvature lies 40 mm from the flange. If the curved beam is subjected to a positive bending moment $M_x = 2.50$ kN · m, determine the stresses at the inner and outer radii. Use Bleich's

correction factors. What is the maximum shearing stress in the curved beam?

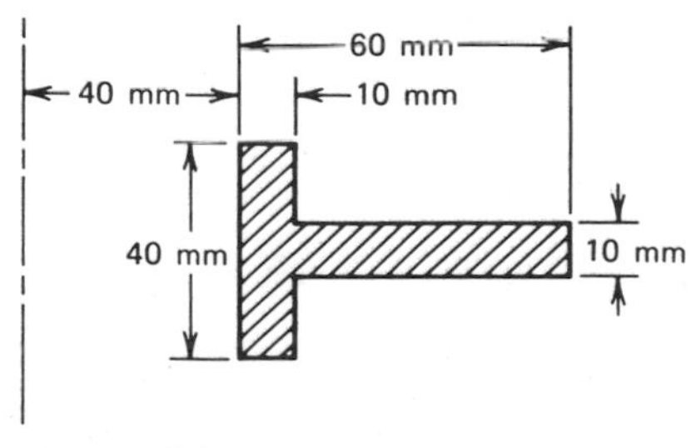

Fig. P8-4.1

2. Determine the radial stress at the junction of the web and the flange for the curved beam in Problem 1. Neglect stress concentrations.

Ans. $\sigma_{rr} = 118.1$ MPa

3. A load $P = 12.0$ kN is applied to the clamp shown in Fig. P8-4.3. Determine the circumferential stresses at points B and C assuming that the curved beam formula is valid at that section.

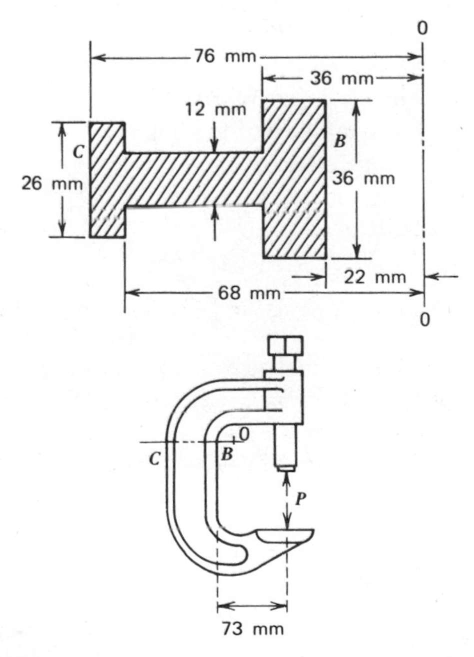

Fig. P8-4.3

4. Determine the radial stress at the junction of the web and inner flange of the curved beam portion of the clamp in Problem 3. Neglect stress concentrations.

   Ans. $\sigma_{rr} = 69.7$ MPa

8-5 DEFLECTIONS OF CURVED BEAMS

A convenient method of determining the deflections of a linearly elastic curved beam is by the use of Castigliano's theorem (Chapter 4). For example, the deflections of the free end of the curved beam in Fig. 8-2.1*a* are given by the relations

$$\delta_{P_1} = \frac{\partial U}{\partial P_1} \tag{8-5.1}$$

$$\theta = \frac{\partial U}{\partial M_0} \tag{8-5.2}$$

where δ_{P_1} is the component of the deflection of the free end of the curved beam in the direction of load P_1, θ is the angle of rotation of the free end of the curved beam in the direction of M_0, and U is the total elastic strain energy in the curved beam. The total strain energy U (see Eq. 4-3.3) is equal to the integral of the strain energy density U_0 over the volume of the curved beam (see Eqs. 2-4.10 and 4-3.4).

Consider the strain energy density U_0 for a curved beam (Fig. 8-2.1). Because of the symmetry of loading relative to the (y, z)-plane, $\sigma_{xy} = \sigma_{xz} = 0$, and since the effect of the transverse normal stress σ_{xx} (Fig. 8-2.1*b*) is ordinarily neglected, the formula for the strain energy density U_0 reduces to the form

$$U_0 = \frac{1}{2E}\sigma_{\theta\theta}^2 + \frac{1}{2E}\sigma_{rr}^2 - \frac{\nu}{E}\sigma_{rr}\sigma_{\theta\theta} + \frac{1}{2G}\sigma_{r\theta}^2$$

where the radial normal stress σ_{rr}, the circumferential normal stress $\sigma_{\theta\theta}$, and the shearing stress $\sigma_{r\theta}$ are, relative to the (x, y, z)-axes of Fig. 8-2.1*b*, $\sigma_{rr} = \sigma_{yy}$, $\sigma_{\theta\theta} = \sigma_{zz}$, and $\sigma_{r\theta} = \sigma_{yz}$. In addition, the effect of σ_{rr} is often small for curved beams of practical dimensions. Hence, the effect of σ_{rr} is often discarded from the expression for U_0. Then,

$$U_0 = \frac{1}{2E}\sigma_{\theta\theta}^2 + \frac{1}{2G}\sigma_{r\theta}^2$$

The stress components $\sigma_{\theta\theta}$ and $\sigma_{r\theta}$, respectively, contribute to the strain energies U_N and U_S because of the normal traction N and the shear V (Fig. 8-2.1*b*). In addition, $\sigma_{\theta\theta}$ contributes to the bending strain energy U_M, as well as to a strain energy U_{MN} because of a coupling effect between the moment M and the traction N, as we shall see in the derivation below.

If curved beams have a small length to depth ratio, the curved beam formula should be used to calculate the circumferential stress $\sigma_{\theta\theta}$. Ordinarily it is sufficiently accurate to approximate the strain energies U_S and U_N that are due to shear V and traction N, respectively, by the formulas for straight beams (see Art. 4-3). However, the strain energy U_M due to bending must be modified. To compute the strain energy due to bending, consider the curved beam shown in Fig. 8-2.1*b*. Since the strain energy increment dU for a linearly elastic material undergoing small displacement is independent of the order in which loads are applied, let the shear load V and normal load N be applied first. Next, let the moment be increased from zero to M_x. The strain energy increment due to bending is

$$dU_M = \tfrac{1}{2} M_x \Delta(d\theta) = \tfrac{1}{2} M_x \omega \, d\theta \tag{8-5.3}$$

where $\Delta(d\theta)$, the change in $d\theta$, and $\omega = \Delta(d\theta)/d\theta$ are due to M_x alone. Hence, ω is determined from Eq. (8-2.10) with $N = 0$. Consequently, Eqs. (8-5.3) and Eq. (8-2.10) yield (with $N = 0$)

$$dU_M = \frac{A_m M_x^2}{2A(RA_m - A)E} \, d\theta \tag{8-5.4}$$

During the application of M_x, additional work is done by N because the centroidal (middle) surface, (Fig. 8-2.1*b*) is stretched an amount $d\bar{e}_{\theta\theta}$. Let the corresponding strain energy increment due to the stretching of the middle surface be denoted by dU_{MN}. This strain energy increment dU_{MN} is equal to the work done by N as it moves through the distance $d\bar{e}_{\theta\theta}$. Thus,

$$dU_{MN} = N \, d\bar{e}_{\theta\theta} = N \bar{\epsilon}_{\theta\theta} R \, d\theta \tag{8-5.5}$$

where $d\bar{e}_{\theta\theta}$ and $\bar{\epsilon}_{\theta\theta}$ refer to the elongation and strain of the centroidal axis, respectively. The strain $\bar{\epsilon}_{\theta\theta}$ is given by Eq. (8-2.3) with $r = R$. Thus, Eq. (8-2.3) (with $r = R$) and Eqs. (8-5.5), (8-2.9), and (8-2.10) (with $N = 0$) yield the strain energy increment dU_{MN} due to coupling of the moment M_x and traction N.

$$dU_{MN} = \frac{N}{E}\left[\frac{M_x}{RA_m - A} - R\frac{A_m M_x}{A(RA_m - A)}\right] d\theta = -\frac{M_x N}{EA} \, d\theta \tag{8-5.6}$$

By Eqs. (4-3.6), (4-3.11), (8-5.4), and (8-5.6), the total strain energy U for the curved beam is obtained in the form

$$U=\int\frac{kV^2R}{2AG}\,d\theta+\int\frac{N^2R}{2AE}\,d\theta+\int\frac{A_mM_x^2}{2A(RA_m-A)E}\,d\theta-\int\frac{M_xN}{EA}\,d\theta \tag{8-5.7}$$

Equation (8-5.7) is an approximation, since it is based on the assumptions that plane sections remain plane and that the effect of the radial stress σ_{rr} on U is negligible. It might be expected that the radial stress increases the strain energy. Hence, Eq. (8-5.7) yields a low estimate of the actual strain energy. However, if M_x and N have the same sign, the coupling energy U_{MN}, the last term in Eq. (8-5.7), is negative. Ordinarily, U_{MN} is small and, in many cases, it is negative. Hence, we recommend that U_{MN}, the coupling strain energy be discarded from Eq. (8-5.7) when it is negative. The discarding of U_{MN} from Eq. (8-5.7) raises the estimate of the actual strain energy when U_{MN} is negative, and compensates to some degree for the lower estimate due to discarding σ_{rr}.

The deflection δ_{elast} of rectangular cross section curved beams has been given by Timoshenko[1] for the two types of loading shown in Fig. 8-2.3. The ratio of the deflection δ_U given by Castigliano's theorem and the deflection δ_{elast} is presented in Table 8-5.1 for several values of R/h. The shear coefficient k (see Eqs. 4-3.11 and Table 4-3.1) was taken to be 1.5 for the rectangular section, and Poisson's ratio ν was assumed to be 0.30.

Note: The deflection of curved beams is much less influenced by the curvature of the curved beam than is the circumferential stress $\sigma_{\theta\theta}$. If

Table 8-5.1

Ratios of Deflections in Rectangular Section Curved Beams as Computed by Elasticity Theory and by Approximate Strain Energy Solution

	Neglecting U_{MN}		*Including U_{MN}*	
	Pure Bending	*Shear Loading*	*Pure Bending*	*Shear Loading*
$\left(\frac{R}{h}\right)$	$\left(\frac{\delta_U}{\delta_{elast}}\right)$	$\left(\frac{\delta_U}{\delta_{elast}}\right)$	$\left(\frac{\delta_U}{\delta_{elast}}\right)$	$\left(\frac{\delta_U}{\delta_{elast}}\right)$
0.65	0.923	1.563	0.697	1.215
0.75	0.974	1.381	0.807	1.123
1.0	1.004	1.197	0.914	1.048
1.5	1.006	1.085	0.968	1.016
2.0	1.004	1.048	0.983	1.008
3.0	1.002	1.021	0.993	1.003
5.0	1.000	1.007	0.997	1.001

R/h is greater than 2.0, the strain energy due to bending can be approximated by that for a straight beam. Thus, for $R/h > 2.0$, for computing deflections the third and fourth terms on the right-hand side of Eq. (8-5.7) may be replaced by

$$U_M = \int \frac{M_x^2}{2EI_x} R\,d\theta \tag{8-5.8}$$

In particular, we note that the deflection of a rectangular cross section curved beam with $R/h = 2.0$ is 7.7 percent greater when the curved beam is assumed to be straight than when it is assumed to be curved.

Deflections of Curved Beam. Cross Sections in the Form of an I, T, etc./ As discussed in Art. 8-4, the cross sections of curved beams in the form of an I, T, etc., undergo distortion when loaded. One effect of the distortion is to decrease the stiffness of the curved beam. As a result, deflections calculated on the basis of the undistorted cross section are less than the actual deflections. Therefore, the deflection calculations should be based upon modified cross sections determined by Bleich's correction factors (Table 8-4.1). The strain energy terms U_N and U_M for the curved beams should also be calculated using the modified cross section. We recommend that the strain energy U_S be calculated with $k = 1.0$, and with the cross-sectional area A replaced by the area of the web $A_w = th$, where t is the thickness of the web and h is the curved beam depth. Also, as a working rule, we recommend that the coupling energy U_{MN} be neglected if it is negative, and that it be doubled if it is positive.

EXAMPLE 8-5.1

Deformations in a Curved Beam Subjected to Pure Bending

The curved beam in Fig. E8-5.1 is made of an aluminum alloy ($E = 72.0$ GPa), has a rectangular cross section with a depth of 60 mm, and is

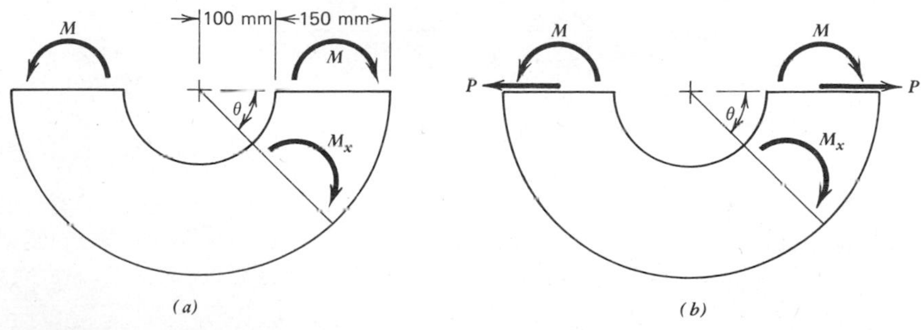

Fig. E8-5.1

subjected to a pure bending moment $M = 24.0$ kN · m. (a) Determine the angle change between the two horizontal faces where M is applied. (b) Determine the relative displacement of the centroids of the horizontal faces of the curved beam.

SOLUTION

Required values for A, A_m, and R for the curved beam are calculated using equations in Row a of Table 8-2.2.

$$A = 60(150) = 9000 \text{ mm}^2$$

$$A_m = 60 \ln \frac{250}{100} = 54.98 \text{ mm}$$

$$R = 100 + 75 = 175 \text{ mm}$$

(a) The angle change between the two faces where M is applied is given by Eq. (8-5.2). As indicated in Fig. E8-5.1a, the magnitude of M_x at any angle θ is $M_x = M$. Thus, by Eq. (8-5.2), we obtain

$$\theta = \frac{\partial U}{\partial M} = \int_0^{\pi} \frac{A_m M_x}{A(RA_m - A)E}(1)\, d\theta$$

$$= \frac{54.98(24{,}000{,}000)\pi}{9000[175(54.98) - 9000](72{,}000)}$$

$$= 0.01029 \text{ rad}$$

(b) In order to determine the deflection of the curved beam, a load P must be applied as indicated in Fig. E8-5.1b. In this case $M_x = M + PR\sin\theta$ and $\partial U/\partial P = R\sin\theta$. Then the deflection is given by Eq. (8-5.1), in which the integral is evaluated with $P = 0$. Thus, the relative displacement is given by the relation

$$\delta_P = \frac{\partial U}{\partial P} = \int_0^{\pi} \left. \frac{A_m M_x}{A(RA_m - A)E} \right|_{P=0} (R\sin\theta)\, d\theta$$

or

$$\delta_P = \frac{54.98(24{,}000{,}000)(175)(2)}{9000[175(54.98) - 9000](72{,}000)} = 1.147 \text{ mm}$$

EXAMPLE 8-5.2
Deflections in a Press

The press (Fig. E8-5.2a) has the cross section shown in Fig. E8-5.2b. It is subjected to a load $P = 11.2$ kN. The press is made of a steel with $E = 200$ GPa, and $\nu = 0.30$. Determine the separation of the jaws of the press due to the load.

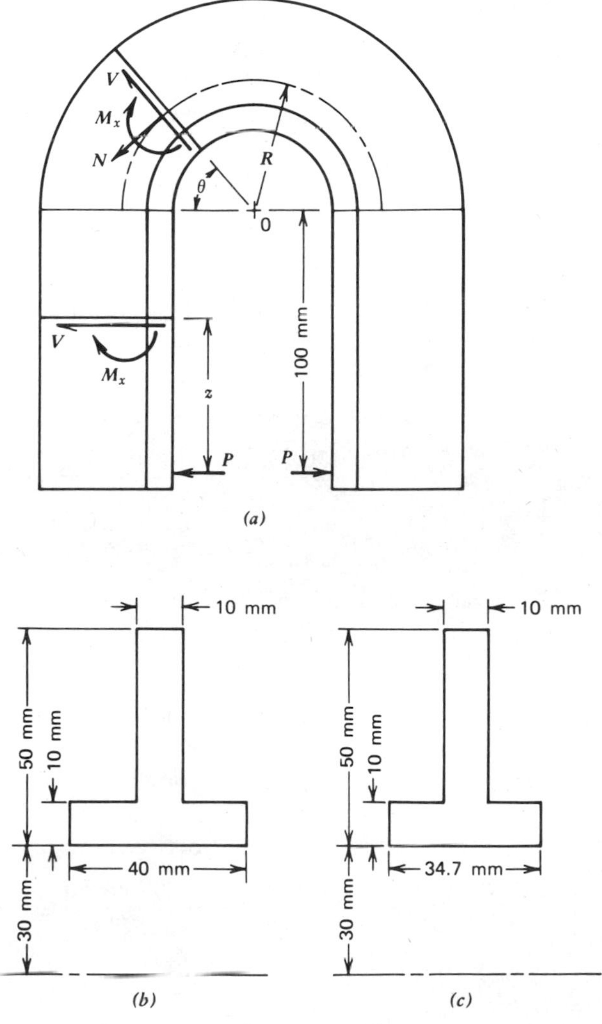

Fig. E8-5.2/(*b*) Original section. (*c*) Modified section.

SOLUTION

The press is made up of two straight members and a curved member. We compute the strain energies due to bending and shear in the straight beams, without modification of the cross sections. The moment of inertia of the cross section is $I_x = 181.7 \times 10^3$ mm^4. We choose the origin of the coordinate axes at load P, with z measured from P toward the curved beam. Then the applied loads shear V and moment M_x at a section in the straight beam is

$$V = P$$
$$M_x = Pz$$

In the curved beam portion of the press, we employ Bleich's correction factor to obtain a modified cross section. With the dimensions in Fig. E8-5.2*b*, we find

$$\frac{b_p^2}{\bar{r}t_f} = \frac{15^2}{35(10)} = 0.643$$

A linear interpolation in Table 8-4.1 yields the result $\alpha = 0.822$. The modified cross section is shown in Fig. E8-5.2*c*. Equations (8-2.12), (8-2.13), and (8-2.14) give

$$A = 34.7(10) + 10(40) = 747 \text{ mm}^2$$
$$R = \frac{34.7(10)(35) + 10(40)(60)}{747} = 48.4 \text{ mm}$$
$$A_m = 10 \ln\frac{80}{40} + 34.7 \ln\frac{40}{30} = 16.9 \text{ mm}$$

With θ defined as indicated in Fig. E8-5.2*a*, the applied loads shear V, normal load N, and moment M_x for the curved beam are

$$V = P\cos\theta$$
$$N = P\sin\theta$$
$$M_x = P(100 + R\sin\theta)$$

Summing the strain energy terms for the two straight beams and the curved beam and taking the derivative with respect to P (Eq. 8-5.1), we compute the increase in distance δ_P between the load points as

$$\delta_P = 2\int_0^{100} \frac{P}{A_w G}\,dz + 2\int_0^{100} \frac{Pz^2}{EI_x}\,dz + \int_0^{\pi} \frac{P\cos^2\theta}{A_w G} R\,d\theta + \int_0^{\pi} \frac{P\sin^2\theta}{AE} R\,d\theta$$
$$+ \int_0^{\pi} \frac{P(100 + R\sin\theta)^2 A_m}{A(RA_m - A)E}\,d\theta$$

The shearing modulus is $G = E/[2(1+\nu)] = 76{,}900$ MPa. Hence,

$$\delta_P = \frac{2(11{,}200)(100)}{76{,}900(500)} + \frac{2(11{,}200)(100)^3}{3(200{,}000)(181{,}700)}$$
$$+ \frac{11{,}200(48.4)\pi}{500(76{,}900)(2)} + \frac{11{,}200(48.4)\pi}{747(200{,}000)(2)}$$
$$+ \frac{16.9(11{,}200)}{747[48.4(16.9) - 747](200{,}000)}\left[(100)^2\pi + (48.4)^2\frac{\pi}{2} + 2(100)(48.4)(2)\right]$$

or

$$\delta_P = 0.058 + 0.205 + 0.022 + 0.006 + 0.972 = 1.263 \text{ mm}$$

PROBLEM SET 8-5

1. The curved beam in Fig. P8-5.1 is made of a steel ($E = 200$ GPa) that has a yield point stress $Y = 420$ MPa. Determine the magnitude of the bending moment $M_x = M_Y$ required to initiate yielding in the curved beam, the angle change of the free end, and the horizontal and vertical components of the deflection of the free end.

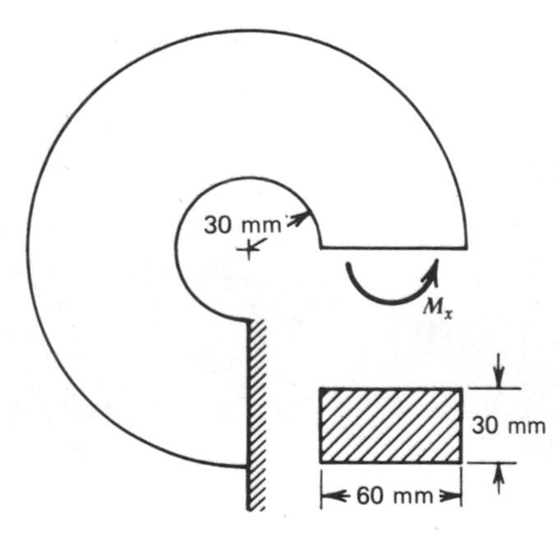

Fig. P8-5.1

2. Determine the deflection of the curved beam in Problem 8-2.3 at the point of load application. The curved beam is made of an aluminum alloy for which $E = 72.0$ GPa and $G = 27.1$ GPa. Let $k = 1.3$.

   Ans. $\delta_P = 0.1629$ mm

3. The triangular cross section curved beam in Problem 8-2.8 is made of steel ($E = 200$ GPa, $G = 77.5$ GPa). Determine the separation of the points of application of the load. Let $k = 1.5$.

4. Determine the deflection across the center of curvature of the cast iron curved beam in Problem 8-3.2 when $P = 126$ kN. $E = 102.0$ GPa and $G = 42.5$ GPa. Let $k = 1.0$ with the area in shear equal to the product of the web thickness and the depth.

   Ans. $\delta_Q = 0.3449$ mm

8.6 STATICALLY INDETERMINATE CURVED BEAMS. CLOSED RING SUBJECTED TO A CONCENTRATED LOAD

Many loaded curved members, such as closed rings and chain links, are statically indeterminate (see Art. 4-5). For such members, equations of equilibrium are not sufficient to determine all of the loads (V, N, M_x) at a section of the member. The additional relations needed to solve for the loads are obtained using Castigliano's theorem with known boundary conditions on the deformations. Since closed rings are commonly used in engineering, we present the computational procedure for a closed ring.

Consider a closed ring subjected to a central load P (Fig. 8-6.1*a*). From the condition of symmetry, the deformation of each quadrant of the ring is identical. Hence, we need consider only one quadrant. The quadrant (Fig. 8-6.1*b*) may be considered fixed at section FH with a load $P/2$ and moment M_0 at section BC. Because of the symmetry of the ring, as the ring deforms, section BC remains perpendicular to section FH. Therefore, by Castigliano's theorem, we have for the rotation of face BC

$$\phi_{BC} = \frac{\partial U}{\partial M_0} = 0 \tag{8-6.1}$$

The applied loads V, N, and M_x at a section forming angle θ with the face BC are

$$\begin{aligned} V &= \frac{P}{2}\sin\theta \\ N &= \frac{P}{2}\cos\theta \\ M_x &= M_0 - \frac{PR}{2}(1-\cos\theta) \end{aligned} \tag{8-6.2}$$

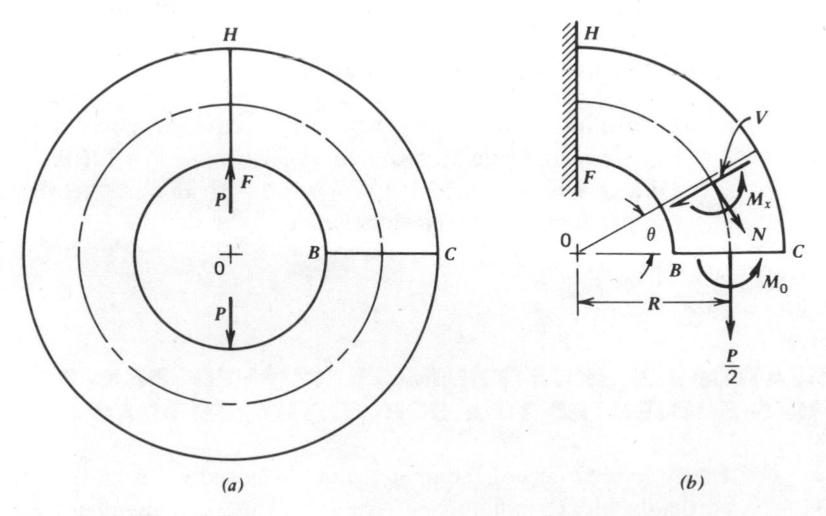

Fig. 8-6.1/Closed ring.

Substituting Eq. (8-5.7) and Eq. (8-6.2) into Eq. (8-6.1), we find

$$0 = \int_0^{\pi/2} \frac{\left[M_0 - \frac{PR}{2}(1 - \cos\theta)\right] A_m}{A(RA_m - A)E} d\theta - \int_0^{\pi/2} \frac{\frac{P}{2}\cos\theta}{AE} d\theta \qquad (8\text{-}6.3)$$

where U_{MN} has been included. The solution of Eq. (8-6.3) is

$$M_0 = \frac{PR}{2}\left(1 - \frac{2A}{RA_m\pi}\right) \qquad (8\text{-}6.4)$$

If R/h is greater than 2.0, we take the bending energy U_M as given by Eq. (8-5.8) and ignore the coupling energy U_{MN}. Then, M_0 is given by the relation

$$M_0 = \frac{PR}{2}\left(1 - \frac{2}{\pi}\right) \qquad (8\text{-}6.5)$$

With M_0 known, the loads at every section of the closed ring (Eqs. 8-6.2) are known. The stresses and deformations of the closed ring may be calculated by the methods of Arts. 8-2 to 8-5.

PROBLEM SET 8-6

1. The ring in Fig. P8-6.1 has an inside diameter of 100 mm, an outside diameter of 180 mm, and a circular cross section. The ring is made of a steel having a yield stress $Y = 520$ MPa. Determine the maximum allowable magnitude of P if the ring has been designed with a factor of safety $SF = 1.75$ against initiation of yielding.

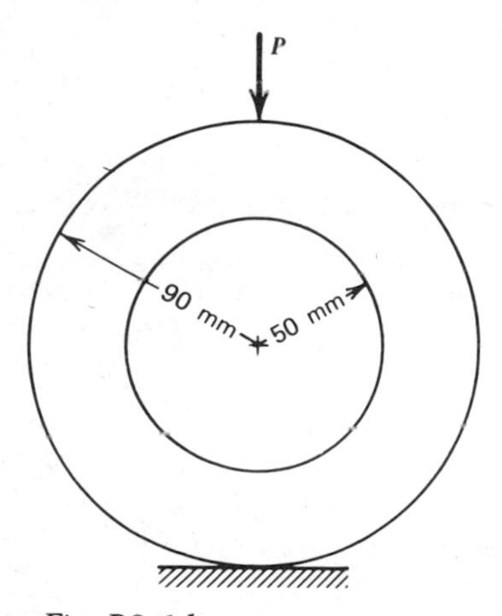

Fig. P8-6.1

2. If $E = 200$ GPa and $G = 77.5$ GPa for the steel in Problem 1, determine the deflection of the ring for a load $P = 60$ kN. Let $k = 1.3$.

   Ans. $\delta_P = 2.088$ mm

3. An aluminum alloy ring has a mean diameter of 600 mm and has a rectangular cross section with 200 mm width and a depth of 300 mm (radial direction). The ring is loaded by diametrically opposite radial loads $P = 4.00$ MN. Determine the maximum tensile and compressive circumferential stresses in the ring.

4. If $E = 72.0$ GPa and $G = 27.1$ GPa for the aluminum alloy ring in Problem 3, determine the separation of the points of application of the loads. Let $k = 1.5$.

   Ans. $\delta_P = 8.742$ mm

5. The link in Fig. P8-6.5 has a circular cross section and is made of a steel having a yield point stress of $Y = 250$ MPa. Determine the magnitude of P that will initiate yielding in the link.

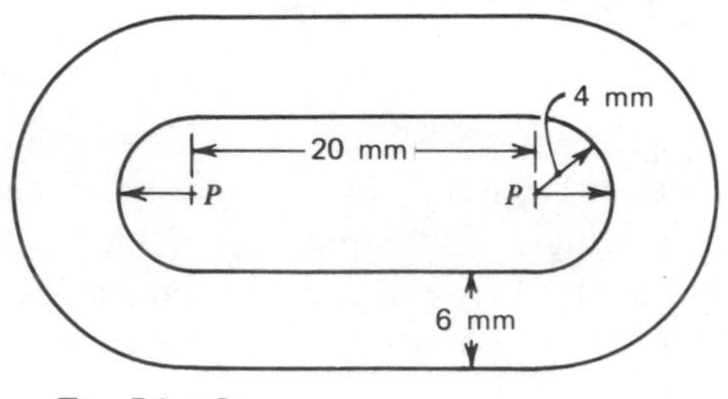

Fig. P8-6.5

8-7 FULLY PLASTIC LOADS FOR CURVED BEAMS

In this article we consider curved beams made of elastic-perfectly plastic materials with yield point stress Y (Fig. 2-6.2a). For a curved beam made of elastic-perfectly plastic material, the fully plastic moment M_P under pure bending is the same as that for a straight beam with identical cross section and material. However, because of the nonlinear distribution of the circumferential stress $\sigma_{\theta\theta}$ in a curved beam, the ratio of the fully plastic moment M_P under pure bending to maximum elastic moment M_Y is much greater for a curved beam than for a straight beam with the same cross section.

Most curved beams are subjected to complex loading other than pure bending. The stress distribution for a curved beam at the fully plastic load P_P for a typical loading condition is indicated in Fig. 8-7.1. Since the tension stresses must balance the compression stresses and the load P_P, the part A_T of the cross-sectional area A that has yielded in tension is greater than the part A_C of area A that has yielded in compression. In addition to the unknowns A_T and A_C, a third unknown is P_P, the load at the fully plastic condition. This follows from the fact that R can be calculated, and D is generally specified rather than P_P. The three equations necessary to determine the three unknowns A_T, A_C, and P_P are obtained from the equations of equilibrium and the fact that the sum of A_T and A_C must equal the cross sectional area A, that is,

$$A = A_T + A_C \tag{8-7.1}$$

The equilibrium equations are (Fig. 8-7.1)

$$\sum F_z = 0 = A_T Y - A_C Y - P_P \tag{8-7.2}$$

$$\sum M_x = 0 = P_P D - A_T Y \bar{y}_T - A_C Y \bar{y}_C \tag{8-7.3}$$

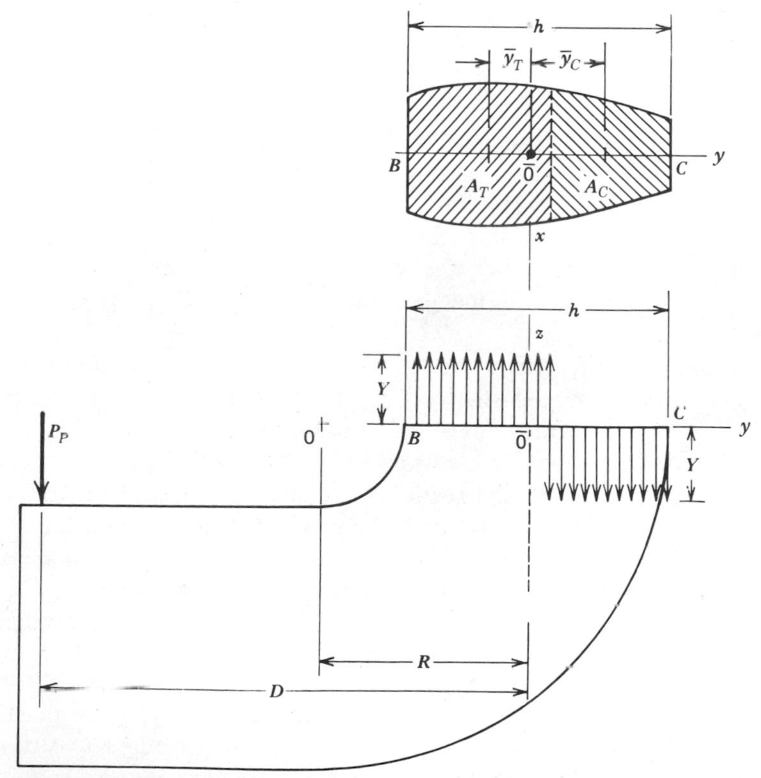

Fig. 8-7.1/Stress distribution for a fully plastic load on a curved beam.

In Eq. (8-7.3), $\bar{y}_T$ and $\bar{y}_C$ locate the centroids of A_T and A_C, respectively, as measured from the centroid of the cross-sectional area of the curved beam (Fig. 8-7.1). Let M be the moment, about the centroid axis x, resulting from the stress distribution on section BC (Fig. 8-7.1). Then,

$$M = P_P D = A_T Y \bar{y}_T + A_C Y \bar{y}_C \tag{8-7.4}$$

The most convenient method of solving Eqs. (8-7.1), (8-7.2), and (8-7.4) for the magnitudes of A_T, A_C, and P_P is often a trial and error procedure, since $\bar{y}_T$ and $\bar{y}_C$ are not known until A_T and A_C are known.[7]

The moment M (Eq. 8-7.4) is generally less than the fully plastic moment M_P for pure bending. It is desirable to know the conditions under which M due to load P_P can be assumed equal to M_P, since for pure bending A_T is equal to A_C, and the calculations are greatly simplified. When the distance D is greater than the depth h of the curved beam, M is approximately equal to M_P for some common sections. For example, for $D = h$, we note that $M = 0.94M_P$ for curved beams with rectangular sections and $M = 0.96M_P$ for curved beams with circular sections. However, for curved beams with T-sections, M may be greater than M_P. Other exceptions are curved beams with I-sections and box sections, for which D should be greater than $2h$ in order for M to be approximately equal to M_P.

Fully Plastic versus Maximum Elastic Loads for Curved Beams / A linearly elastic analysis of a load carrying member is required in order to predict the load-deflection relation of linearly elastic behavior of the member up to the load P_Y that initiates yielding in the member. The fully plastic load is also of interest since it is often considered to be the limiting load that can be applied to the member before the deformations become excessively large.

The fully plastic load P_P for a curved beam is often more than twice the maximum elastic load P_Y. Fracture loads for curved beams that are made of ductile metals and that are subjected to static loading may be 4 to 6 times P_Y. Dimensionless load-deflection experimental data for a uniform rectangular section hook made of a structural steel are shown in Fig. 8-7.2. The deflection is defined as the change in distance ST between points S and T on the hook. The hook did not fracture even for loads such that $P/P_Y > 5$. A computer program written by J. C. McWhorter, H. R. Wetenkamp, and O. M. Sidebottom[7] gave the predicted curve in Fig. 8-7.2. The experimental data agree well with predicted results.

As noted in Fig. 8-7.2, the ratio of P_P to P_Y is 2.44. Furthermore, the load-deflection curve does not level off at the fully plastic load, but continues to rise. This behavior may be attributed to strain-hardening. Because of the steep stress gradient in the hook, the strains in the most strained fibers become so large that the material in the most strained fibers begins to strain harden before yielding can penetrate to sufficient depth at section BC in the hook to develop the fully plastic load.

The usual practice in predicting the deflection of a structure at the fully plastic load is to calculate the deflection of the structure at the fully plastic load, assuming that the structure behaves in a linearly elastic manner up to the fully plastic load (point Q in Fig. 8-7.2) and multiplying this deflection by the ratio P_P/P_Y (in this case 2.44). In this case, with this procedure (Fig. 8-7.2) the resulting calculated deflection (approximately calculated as $2.44(2.4) = 5.9$) is greater than the measured deflection.

Usually, curved members such as crane hooks and chains are not subjected to a sufficient number of repetitions of peak loads during their life for fatigue failure to occur. Therefore, the working loads for these members are often obtained by application of a factor of safety to the fully plastic loads. It is not uncommon to have the working load as great or greater than the maximum elastic load P_Y.

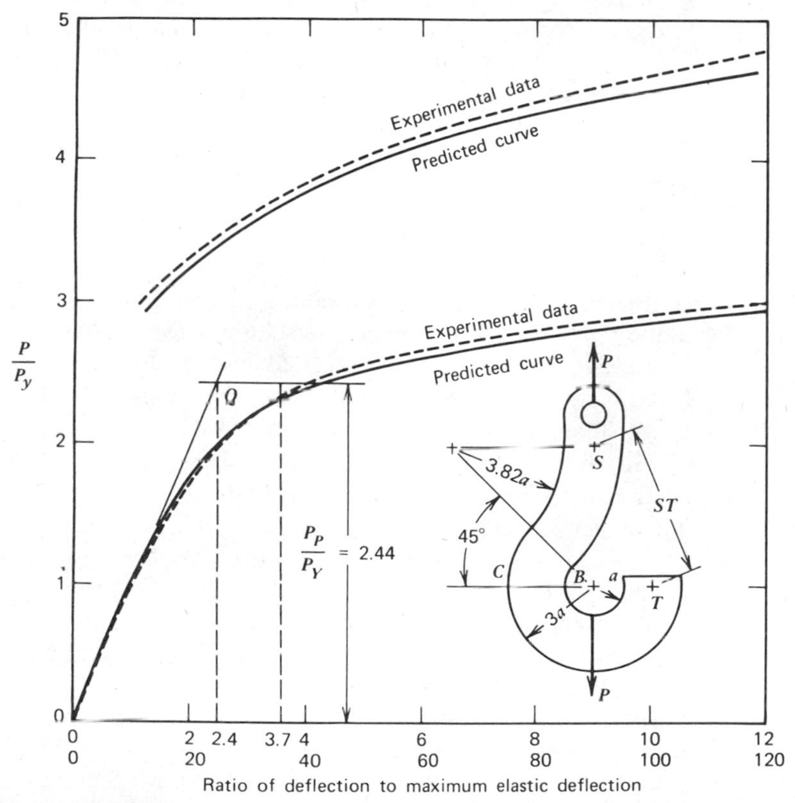

Fig. 8-7.2/Dimensionless load-deflection curves for a uniform rectangular section hook made of structural steel.

PROBLEM SET 8-7

1. Let the curved beam in Fig. 8-7.1 have a rectangular cross section with depth h and width b. Show that the ratio of the bending moment for fully plastic load P_P to the fully plastic moment for pure bending $M_P = Ybh^2/4$ is given by the relation,

$$\frac{M}{M_P} = \frac{4D}{h}\sqrt{1 + \frac{4D^2}{h^2}} - \frac{8D^2}{h^2}$$

2. Let the curved beam in Problem 8-2.1 be made of a steel that has a flat top stress-strain diagram at the yield point stress $Y = 430$ MPa. From the answer to Problem 8-2.1 the load that initiates yielding is equal to $P_Y = SF(P) = 6.05$ kN. Since $D = 3h$, assume $M = M_P$ and calculate P_P. Determine the ratio P_P/P_Y.

   Ans. $P_P = 14.33$ kN, $P_P/P_Y = 2.37$

3. Let the steel in the curved beam in Example 8-4.1 have a flat top at the yield point stress $Y = 280$ MPa. Determine the fully plastic moment for the curved beam. Note that the original cross section must be used. The distortion of the cross section increases the fully plastic moment for a positive moment.

REFERENCES

1. S. Timoshenko and J. Goodier, *Theory of Elasticity*, 3rd Ed., McGraw-Hill, 1970.
2. H. Bleich, "Die Spannungsverteilung in den Gurtungen gekrümmter Stabe mit T- und I-formigem Querschnitt," *Der Stahlbau, Beilage zur Zeitschrift, Die Bautechnik*, Vol. 6, No. 1, Jan. 6, 1933, pp 3–6. English translation, Navy Department, The David W. Taylor Model Basin, Translation 228, January 1950.
3. C. G. Anderson, "Flexural Stresses in Curved Beams of I- and Box Sections," presented to the Institution of Mechanical Engineers, November 3, 1950.
4. T. von Kármán, *Zeitschrift des Vereines deutscher Ingenieure*, Vol. 55, 1911, p. 1889.
5. S. Timoshenko, "Bending Stresses in Curved Tubes of Rectangular Cross-Section," *Trans. ASME*, Vol. 45, 1923, pp. 135–140.
6. D. C. Broughton, M. E. Clark, and H. T. Corten, "Tests and Theory of Elastic Stresses in Curved Beams Having I- and T-Sections," *Experimental Mechanics*, Vol. 8, No. 1, 1950, pp. 143–155.
7. J. C. McWhorter, H. R. Wetenkamp, and O. M. Sidebottom, "Finite Deflections of Curved Beams," *Journal of the Engineering Mechanics Division, Proceedings of ASCE*, April 1971, pp. 345–358.

Additional References

1. Y. Enda, "Analysis of Thin-Walled Curved Beams by the Transfer Matrix Method," *Advances in Computational Methods in Structural Mechanics and Design*, University of Alabama press, Huntsville, 1972, pp. 757–774.
2. V. Z. Vlasov, *Theory of Thin-Walled Elastic Beams*, Gihodo Press, Tokyo, 1967.
3. F. Wansleben, "Die Berechnung drehfester gekrümmten dünnwandigen Trägern," *Der Stahlbau*, Vol. 33, 1964, pp. 364–372.

CHAPTER 9

BEAMS ON ELASTIC FOUNDATIONS

In certain applications, a beam of relatively small bending stiffness is placed on an elastic foundation and loads are applied to the beam. The loads are transferred through the beam to the foundation. The beam and foundation must be designed to resist the loads without failing. Often, failure occurs in the beam before it occurs in the foundation. Accordingly, in this chapter we assume that the foundation has sufficient strength to prevent failure. Furthermore, we assume that the foundation resists the loads transmitted by the beam, in a linearly elastic manner; that is, the pressure developed at any point between the beam and the foundation is proportional to the deflection of the beam at that point. This assumption is fairly accurate for small deflections. However, if the deflections are large, the resistance of the foundation generally does not remain linearly proportional to the beam deflection. For large deflections, the resistance of the foundation is larger than that for small deflections and is related in a nonlinear way to the beam deflection. The increased resistance due to the nonlinear response of the foundation tends to reduce the deflections and stresses in the beam compared to those due to a linear foundation response. Since we consider small displacements, the solution presented in this chapter for the beam on an elastic foundation is generally conservative for the range of deflections treated. Furthermore, since we consider only a linear response of the foundation, we drop the term linear in our discussion.

The solution presented in this chapter for beams on elastic foundations can be used to obtain a simple approximate solution for beams supported by identical elastic springs that are uniformly spaced along the beam. This approximate solution is used widely in practice.

9-1 GENERAL THEORY

The response to loads of a beam resting on an elastic foundation is described by a single differential equation subject to different boundary conditions for the beam, depending on how the beam is supported at its ends. For instance, consider a beam of infinite length attached along its length to an elastic foundation (Fig. 9-1.1). Let the origin of coordinate axes (y, z) be located at the centroid of the beam cross section, and let a concentrated lateral load P be applied to the beam at the origin of the (x, y, z)-axes. The z-axis coincides with the axis of the beam, and the y-axis is normal to the elastic foundation. The load P causes the beam to deflect, which in turn displaces the elastic foundation. As a result, a distributed force is developed between the beam and the foundation. Thus, relative to the beam, the resistance of the foundation produces a laterally distributed force q (force per unit length) on the beam (Fig. 9-1.1). In the solution of the deflection problem, we shall see that in certain regions the deflection of the beam may be negative. Hence, since the beam is assumed to be attached to the foundation, the foundation may in certain regions exert a tensile force on the beam.

A free body diagram of an element Δz of the beam is shown in Fig. 9-1.1*b*, with positive sign conventions for the total shear V_y and moment M_x indicated. For the indicated sign convention and for the condition of

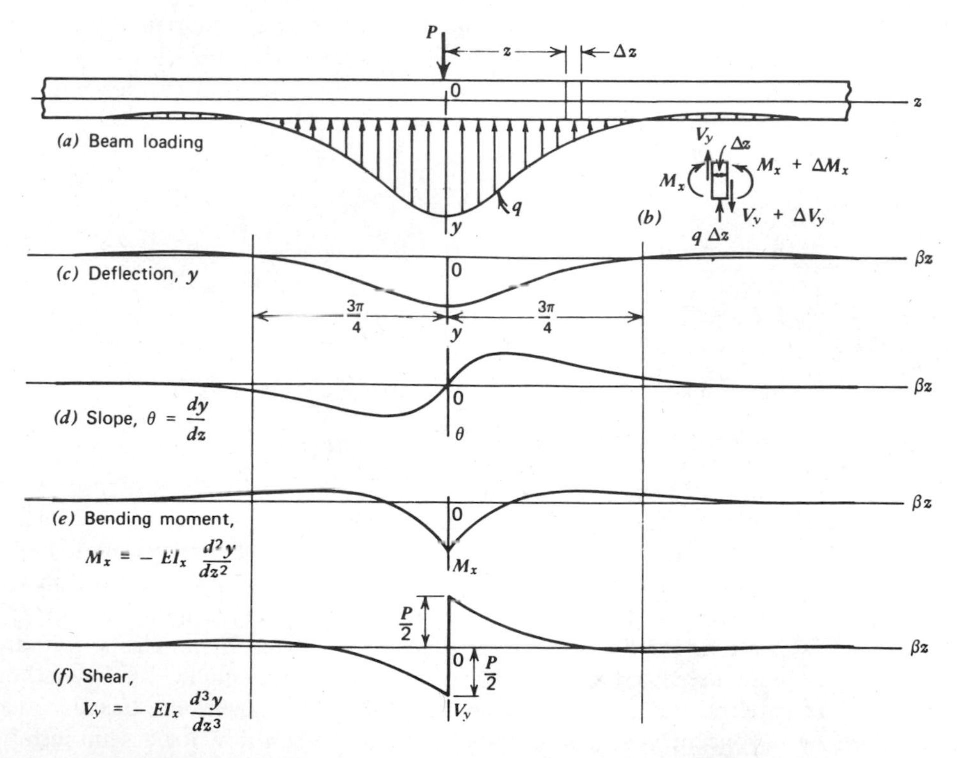

Fig. 9-1.1/Infinite beam on an elastic foundation and loaded at origin.

small displacements, we obtain the differential relations

$$\frac{dy}{dz} = \theta$$

$$EI_x \frac{d^2y}{dz^2} = -M_x$$

$$EI_x \frac{d^3y}{dz^3} = -V_y \tag{9-1.1}$$

$$EI_x \frac{d^4y}{dz^4} = -q$$

where q is taken to be positive if it pushes up on the beam, that is, q is positive if it acts in the negative y-direction.

For the linearly elastic foundation, the distributed load q is linearly proportional to the deflection y of the beam; thus,

$$q = ky \tag{9-1.2}$$

where the spring coefficient k may be written in the form

$$k = bk_0 \tag{9-1.3}$$

in which b is the beam width and k_0 is the elastic spring constant for the foundation. The dimensions of k_0 are Newtons per square millimeter per millimeter of deflection (N/mm^3). Substitution of Eq. (9-1.2) into the fourth of Eqs. (9-1.1) yields the differential equation of the bending axis of the beam on an elastic foundation.

$$EI_x \frac{d^4y}{dz^4} = -ky \tag{9-1.4}$$

With the notation,

$$\beta = \sqrt[4]{\frac{k}{4EI_x}} \tag{9-1.5}$$

the general solution y of Eq. (9-1.4) may be expressed as

$$y = e^{\beta z}(C_1 \sin \beta z + C_2 \cos \beta z) + e^{-\beta z}(C_3 \sin \beta z + C_4 \cos \beta z) \tag{9-1.6}$$

Equation (9-1.6) represents the general solution to the response of an infinite beam on an elastic foundation subjected to a concentrated lateral load. The fact that Eq. (9-1.6) is the solution of Eq. (9-1.4) may be verified by substitution; the magnitudes of the constants of integration C_1, C_2, C_3, and C_4 are determined by the boundary conditions.

Solutions for the response of a beam supported by an elastic foundation and subjected to specific lateral loads can be obtained by the method of superposition, by employing the solution for an infinite beam loaded by a concentrated load (Fig. 9-1.1) and the solution for a semi-infinite

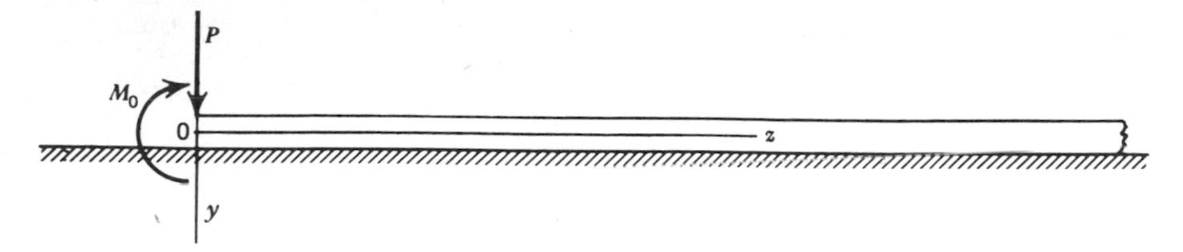

Fig. 9-1.2/Semi-infinite beam on an elastic foundation and loaded at the end.

beam loaded at the end by a concentrated load P and moment M_0 as indicated in Fig. 9-1.2. In either of the cases shown in Figs. 9-1.1 and 9-1.2, the deflection of the beam goes to zero for large positive values of z. Consequently, the constants C_1 and C_2 in Eq. (9-1.6) must be set equal to zero, and the equation for the displacement y of the bending axis of the beam reduces to

$$y = e^{-\beta z}(C_3 \sin \beta z + C_4 \cos \beta z); \qquad z \geq 0 \tag{9-1.7}$$

Because of symmetry, the displacement of the beam in Fig. 9-1.1 for negative values of z can be obtained from the solution for positive values of z, that is, $y(-z) = y(z)$. For the case of the semi-infinite beam (Fig. 9-1.2), $z \geq 0$, so that Eq. (9-1.7) applies directly.

9-2 INFINITE BEAM SUBJECTED TO CONCENTRATED LOAD: BOUNDARY CONDITIONS

Consider a beam of infinite length, resting on an infinite foundation, and loaded at the origin 0 of coordinate axes (y, z) with concentrated load P (Fig 9-1.1). To determine the two constants of integration, C_3 and C_4 in Eq. (9-1.7), we employ the conditions (a) that the slope of the beam remains zero under the load because of symmetry and (b) that half of the load P must be supported by the elastic foundation under the half of the beam specified by positive values of z. The other half of P is supported by elastic foundation where $z < 0$. Thus, we obtain the relations

$$\frac{dy}{dz} = 0 \quad \text{for} \quad z = 0; \quad \text{and} \quad 2\int_0^{\infty} ky\,dz = P \tag{9-2.1}$$

The condition of vanishing slope at $z = 0$ yields, with Eq. (9-1.7),

$$C_3 = C_4 = C$$

Hence, Eq. (9-1.7) becomes

$$y = Ce^{-\beta z}(\sin \beta z + \cos \beta z) \tag{9-2.2}$$

Substituting Eq. (9-2.2) into the second of Eqs. (9-2.1), we obtain

$$C = \frac{P\beta}{2k} \tag{9-2.3}$$

Consequently, the equation of the deflected axis of the beam is

$$y = \frac{P\beta}{2k} e^{-\beta z}(\sin \beta z + \cos \beta z) \qquad z \geq 0 \tag{9-2.4}$$

Equation (9-2.4) holds for positive values of z. The deflections for negative values of z are obtained by the condition that $y(-z) = y(z)$, that is, by symmetry. Values for the slope, moment, and shear are obtained by substitution of Eq. (9-2.4) into Eqs. (9-1.1). Thus, we find

$$y = \frac{P\beta}{2k} A_{\beta z} \qquad z \geq 0 \tag{9-2.5}$$

$$\theta = -\frac{P\beta^2}{k} B_{\beta z} \qquad z \geq 0 \tag{9-2.6}$$

$$M_x = \frac{P}{4\beta} C_{\beta z} \qquad z \geq 0 \tag{9-2.7}$$

$$V_y = -\frac{P}{2} D_{\beta z} \qquad z \geq 0 \tag{9-2.8}$$

where

$$A_{\beta z} = e^{-\beta z}(\sin \beta z + \cos \beta z) \qquad B_{\beta z} = e^{-\beta z} \sin \beta z$$

$$C_{\beta z} = e^{-\beta z}(\cos \beta z - \sin \beta z) \qquad D_{\beta z} = e^{-\beta z} \cos \beta z \tag{9-2.9}$$

For convenience, values of $A_{\beta z}$, $B_{\beta z}$, $C_{\beta z}$, and $D_{\beta z}$ are listed in Table 9-2.1 for $0 \leq \beta z \leq 5\pi/2$.

Values of deflection, slope, bending moment, and shear at any point along the beam are given by Eqs. (9-2.5), (9-2.6), (9-2.7), and (9-2.8), respectively. By using the summetry conditions, $y(-z) = y(z)$, $\theta(-z) = -\theta(z)$, $M_x(-z) = M_x(z)$, and $V_y(-z) = -V_y(z)$, these quantities are plotted versus βz in Figs. 9-1.1*c*, *d*, *e*, and *f*. Since all of these quantities approach zero as βz becomes large, the above solutions may be used as approximations for beams of finite length. In particular, in Table 9-2.1, we note that $A_{\beta z} = 0$ for $\beta z = 3\pi/4$; therefore, the beam has zero deflection at a distance $3\pi/(4\beta)$ from the load. A beam with a length $L = 3\pi/(2\beta)$ loaded at the center has a maximum deflection 5.5 percent greater[1] and a maximum bending moment 1.9 percent greater than for a

Table 9-2.1

βz	$A_{\beta z}$	$B_{\beta z}$	$C_{\beta z}$	$D_{\beta z}$
0	1	0	1	1
0.001	1.0000	0.0010	0.9980	0.9990
0.002	1.0000	0.0020	0.9960	0.9980
0.003	1.0000	0.0030	0.9940	0.9970
0.004	1.0000	0.0040	0.9920	0.9960
0.005	1.0000	0.0050	0.9900	0.9950
0.006	1.0000	0.0060	0.9880	0.9940
0.007	0.9999	0.0070	0.9861	0.9930
0.008	0.9999	0.0080	0.9841	0.9920
0.009	0.9999	0.0087	0.9821	0.9910
0.010	0.9999	0.0099	0.9801	0.9900
0.011	0.9999	0.0109	0.9781	0.9890
0.012	0.9999	0.0119	0.9761	0.9880
0.013	0.9998	0.0129	0.9742	0.9870
0.014	0.9998	0.0138	0.9722	0.9860
0.015	0.9998	0.0148	0.9702	0.9850
0.016	0.9997	0.0158	0.9683	0.9840
0.017	0.9997	0.0167	0.9663	0.9830
0.018	0.9997	0.0177	0.9643	0.9820
0.019	0.9996	0.0187	0.9624	0.9810
0.02	0.9996	0.0196	0.9604	0.9800
0.03	0.9991	0.0291	0.9409	0.9700
0.04	0.9984	0.0384	0.9216	0.9600
0.05	0.9976	0.0476	0.9025	0.9501
0.10	0.9906	0.0903	0.8100	0.9003
0.15	0.9796	0.1283	0.7224	0.8510
0.20	0.9651	0.1627	0.6398	0.8024
0.25	0.9472	0.1927	0.5619	0.7546
0.30	0.9267	0.2189	0.4888	0.7078
0.35	0.9036	0.2416	0.4204	0.6620
0.40	0.8784	0.2610	0.3564	0.6174
0.45	0.8515	0.2774	0.2968	0.5742
0.50	0.8231	0.2908	0.2414	0.5323
0.55	0.7934	0.3016	0.1902	0.4918
0.60	0.7628	0.3099	0.1430	0.4529
0.65	0.7315	0.3160	0.0996	0.4156
0.70	0.6997	0.3199	0.0599	0.3798
0.75	0.6676	0.3220	0.0237	0.3456
$\frac{1}{4}\pi$	0.6448	0.3224	0	0.3224
0.80	0.6353	0.3223	−0.0093	0.3131
0.85	0.6032	0.3212	−0.0391	0.2821
0.90	0.5712	0.3185	−0.0658	0.2527
0.95	0.5396	0.3146	−0.0896	0.2250
1.00	0.5083	0.3096	−0.1109	0.1987

Table 9-2.1

Continued

1.05	0.4778	0.3036	−0.1294	0.1742
1.10	0.4476	0.2967	−0.1458	0.1509
1.15	0.4183	0.2890	−0.1597	0.1293
1.20	0.3898	0.2807	−0.1716	0.1091
1.25	0.3623	0.2719	−0.1815	0.0904
1.30	0.3355	0.2626	−0.1897	0.0729
1.35	0.3098	0.2530	−0.1962	0.0568
1.40	0.2849	0.2430	−0.2011	0.0419
1.45	0.2611	0.2329	−0.2045	0.0283
1.50	0.2384	0.2226	−0.2068	0.0158
1.55	0.2166	0.2122	−0.2078	0.0044
$\frac{1}{2}\pi$	0.2079	0.2079	−0.2079	0
1.60	0.1960	0.2018	−0.2077	−0.0059
1.65	0.1763	0.1915	−0.2067	−0.0152
1.70	0.1576	0.1812	−0.2046	−0.0236
1.75	0.1400	0.1720	−0.2020	−0.0310
1.80	0.1234	0.1610	−0.1985	−0.0376
1.85	0.1078	0.1512	−0.1945	−0.0434
1.90	0.0932	0.1415	−0.1899	−0.0484
1.95	0.0795	0.1322	−0.1849	−0.0527
2.00	0.0667	0.1230	−0.1793	−0.0563
2.05	0.0549	0.1143	−0.1737	−0.0594
2.10	0.0438	0.1057	−0.1676	−0.0619
2.15	0.0337	0.0975	−0.1613	−0.0638
2.20	0.0244	0.0895	−0.1547	−0.0652
2.25	0.0157	0.0820	−0.1482	−0.0663
2.30	0.0080	0.0748	−0.1416	−0.0668
2.35	0.0008	0.0679	−0.1349	−0.0671
$\frac{3}{4}\pi$	0	0.0671	−0.1342	−0.0671
2.40	−0.0056	0.0613	−0.1282	−0.0669
2.45	−0.0114	0.0550	−0.1215	−0.0665
2.50	−0.0166	0.0492	−0.1149	−0.0658
2.55	−0.0213	0.0435	−0.1083	−0.0648
2.60	−0.0254	0.0383	−0.1020	−0.0637
2.65	−0.0289	0.0334	−0.0956	−0.0623
2.70	−0.0320	0.0287	−0.0895	−0.0608
2.75	−0.0347	0.0244	−0.0835	−0.0591
2.80	−0.0369	0.0204	−0.0777	−0.0573
2.85	−0.0388	0.0167	−0.0721	−0.0554
2.90	−0.0403	0.0132	−0.0666	−0.0534
2.95	−0.0415	0.0100	−0.0614	−0.0514
3.00	−0.0422	0.0071	−0.0563	−0.0493
3.05	−0.0427	0.0043	−0.0515	−0.0472
3.10	−0.0431	0.0019	−0.0469	−0.0450
π	−0.0432	0	−0.0432	−0.0432

Table 9-2.1

Continued

3.15	−0.0432	−0.0004	−0.0424	−0.0428
3.20	−0.0431	−0.0024	−0.0383	−0.0407
3.25	−0.0427	−0.0042	−0.0343	−0.0385
3.30	−0.0422	−0.0058	−0.0306	−0.0365
3.35	−0.0417	−0.0073	−0.0271	−0.0344
3.40	−0.0408	−0.0085	−0.0238	−0.0323
3.45	−0.0399	−0.0097	−0.0206	−0.0303
3.50	−0.0388	−0.0106	−0.0177	−0.0283
3.55	−0.0378	−0.0114	−0.0149	−0.0264
3.60	−0.0366	−0.0121	−0.0124	−0.0245
3.65	−0.0354	−0.0126	−0.0101	−0.0227
3.70	−0.0341	−0.0131	−0.0079	−0.0210
3.75	−0.0327	−0.0134	−0.0059	−0.0193
3.80	−0.0314	−0.0137	−0.0040	−0.0177
3.85	−0.0300	−0.0139	−0.0023	−0.0162
3.90	−0.0286	−0.0140	−0.0008	−0.0147
$\frac{5}{4}\pi$	−0.0278	−0.0140	0	−0.0139
3.95	−0.0272	−0.0139	0.0005	−0.0133
4.00	−0.0258	−0.0139	0.0019	−0.0120
4.50	−0.0132	−0.0108	0.0085	−0.0023
$\frac{3}{2}\pi$	−0.0090	−0.0090	0.0090	0
5.00	−0.0046	−0.0065	0.0084	0.0019
$\frac{7}{4}\pi$	0	−0.0029	0.0058	0.0029
5.50	0.0000	−0.0029	0.0058	0.0029
6.00	0.0017	−0.0007	0.0031	0.0024
2π	0.0019	0	0.0019	0.0019
6.50	0.0018	0.0003	0.0012	0.0018
7.00	0.0013	0.0006	0.0001	0.0007
$\frac{9}{4}\pi$	0.0012	0.0006	0	0.0006
7.50	0.0007	0.0005	−0.0003	0.0002
$\frac{5}{2}\pi$	0.0004	0.0004	−0.0004	0

beam with infinite length. Although the error in using the solution for a beam of length $L = 3\pi/(2\beta)$ is nonconservative, the error is not large; therefore, the infinite beam solution yields reasonable results for beams as short as $L = 3\pi/(2\beta)$ when loaded at the center. The infinite beam solution also yields reasonable results for much longer beams for any location of the concentrated load as long as the distance from the load to either end of the beam is equal to or greater than $3\pi/(4\beta)$. Since the deflection y remains positive over the length $L = 3\pi/(2\beta)$, the distributed load q (Fig. 9-1.1) does not change sign over this length; therefore, it is not necessary when $L \geq 3\pi/(2\beta)$, in applications under statical loading, for the beam to be attached to the elastic foundation.

EXAMPLE 9-2.1

Diesel Locomotive Wheels on Rail

A railroad uses steel rails ($E = 200$ GPa) with a depth of 184 mm. The distance from the top of the rail to its centroid is 99.1 mm, and the moment of inertia of the rail is 36.9×10^6 mm^4. The rail is supported by ties, ballast, and a road bed that together are assumed to act as an elastic foundation with spring constant $k = 14.0$ N/mm^2. (a) Determine the maximum deflection, maximum bending moment, and maximum flexure stress in the rail for a single wheel load of 170 kN. (b) A particular Diesel locomotive has three wheels per truck equally spaced at 1.70 m. Determine the maximum deflection, maximum bending moment, and maximum flexure stress in the rail if the load on each wheel is 170 kN.

SOLUTION

The equations for the bending moment and the deflection require the value of β. From Eq. (9-1.5), we find that

$$\beta = \sqrt[4]{\frac{k}{4EI_x}} = \sqrt[4]{\frac{14}{4(200 \times 10^3)(36.9 \times 10^6)}} = 0.000830 \text{ mm}^{-1}$$

(a) The maximum deflection and maximum bending moment occur under the load where $A_{\beta z} = C_{\beta z} = 1.00$. Equations (9-2.5) and (9-2.7) give

$$y_{\max} = \frac{P\beta}{2k} = \frac{170 \times 10^3(0.000830)}{2(14)} = 5.039 \text{ mm}$$

$$M_{\max} = \frac{P}{4\beta} = \frac{170 \times 10^3}{4(0.000830)} = 51.21 \text{ kN} \cdot \text{m}$$

$$\sigma_{\max} = \frac{M_{\max} c}{I_x} = \frac{51.21 \times 10^6(99.1)}{36.9 \times 10^6} = 137.5 \text{ MPa}$$

(b) The deflection and bending moment at any section of the beam are obtained by superposition of the effects of each of the three wheel loads. With superposition, an examination of Figs. 9-1.1*c* and *e* indicates that the maximum deflection and the maximum bending moment occur either under the center wheel or under one of the end wheels. Let the origin be located under one of the end wheels. The distance from the origin to the next wheel is $z_1 = 1.7 \times 10^3$ mm. Hence, $\beta z_1 = 0.000830(1.7 \times 10^3) = 1.411$. The distance from the origin to the second wheel is $z_2 = 2(1.7 \times 10^3)$ mm. Hence, $\beta z_2 = 0.000830(2)(1.7 \times 10^3) = 2.882$. From Table 9-2.1, we find

$$A_{\beta z1} = 0.2797 \qquad C_{\beta z1} = -0.2018$$
$$A_{\beta z2} = -0.0377 \qquad C_{\beta z2} = -0.0752$$

The deflection and bending moment at the origin (under one of the end wheels) are

$$y_{end} = \frac{P\beta}{2k}(A_{\beta z0} + A_{\beta z1} + A_{\beta z2}) = 5.039(1 + 0.2797 - 0.0377)$$
$$= 6.258 \text{ mm}$$
$$M_{end} = \frac{P}{4\beta}(C_{\beta z0} + C_{\beta z1} + C_{\beta z2}) = 51.20 \times 10^6(1 - 0.2018 - 0.0752)$$
$$= 37.02 \text{ kN} \cdot \text{m}$$

Now, let the origin be located under the center wheel. The distance between the center wheel and either of the end wheels is $z_1 = 1.7 \times 10^3$ mm. Therefore,

$$y_{center} = \frac{P\beta}{2k}(A_{\beta z0} + 2A_{\beta z1}) = 5.039[1 + 2(0.2797)] = 7.858 \text{ mm}$$
$$M_{center} = \frac{P}{4\beta}(C_{\beta z0} + 2C_{\beta z1}) = 51.20 \times 10^6[1 - 2(0.2018)]$$
$$= 30.54 \text{ kN} \cdot \text{m}$$

Thus, we find

$$y_{center} = y_{max} = 7.858 \text{ mm}$$
$$M_{end} = M_{max} = 37.02 \text{ kN} \cdot \text{m}$$

and

$$\sigma_{max} = \frac{M_{max}c}{I_x} = \frac{37.02 \times 10^6(99.1)}{36.9 \times 10^6} = 99.4 \text{ MPa}$$

Beam Supported on Equally Spaced Separate Elastic Supports / Long beams are sometimes supported by elastic springs equally spaced along the beam (Fig. 9-2.1*a*). Although coil springs are shown in Fig. 9-2.1, each spring support may be due to the resistance of a linearly elastic member or a structure such as a tension member, a straight beam, or a curved beam. It is possible to obtain an exact solution for the spring-supported beam of Fig. 9-2.1*a* by energy methods (see Art. 4-5); however, the computational work becomes prohibitive as the number of springs becomes large.

Alternatively, we may proceed as follows: Let each spring in Fig. 9-2.1*a* have the same constant K. The force R that each spring exerts on the beam is directly proportional to the deflection y of the beam at the section where the spring is attached. Thus, we write

$$R = Ky \tag{9-2.10}$$

We assume that the load R is distributed uniformly over a spacing l, a distance $l/2$ to the right and to the left of each spring. Thus, we obtain

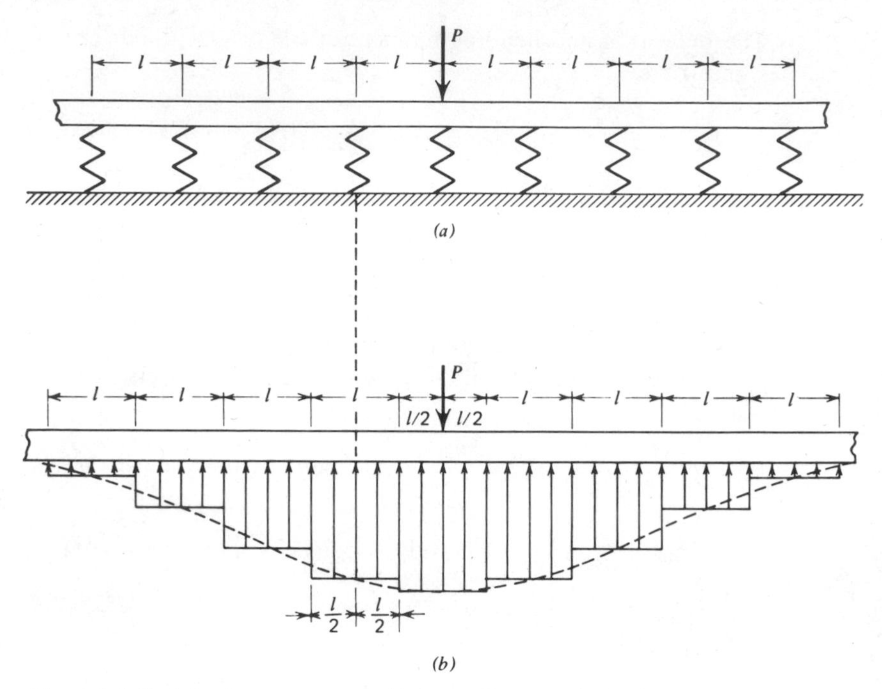

Fig. 9-2.1/Infinite beam supported by equally spaced elastic springs.

the stepped distributed loading shown in Fig. 9-2.1*b*. If the stepped distributed loading is approximated by a smooth average curve (dashed curve in Fig. 9-2.1*b*), the approximate distributed load is similar to the distributed load q of Fig. 9-1.1*a*. Since the dashed curve in Fig. 9-2.1*b* intersects each of the steps near its center, we assume that the dashed curve does indeed intersect each of the steps directly beneath the spring. Thus, we assume that an equivalent spring constant k exists, such that

$$k = \frac{K}{l} \tag{9-2.11}$$

Hence, substitution of Eq. (9-2.11) into Eq. (9-1.5) yields an equivalent β for the springs. Next, we assume that Eqs. (9-2.5) through (9-2.8) are valid for an infinite beam supported by equally spaced elastic supports and loaded in the center. The resulting approximate solution becomes more accurate as the spacing l between springs becomes small. However, we note that this approximate solution becomes greatly in error when the spacing l between springs becomes large. It has been found that the error in the solution is not excessive if we require that the spacing l between springs satisfies the condition

$$l \leq \frac{\pi}{4\beta} \tag{9-2.12}$$

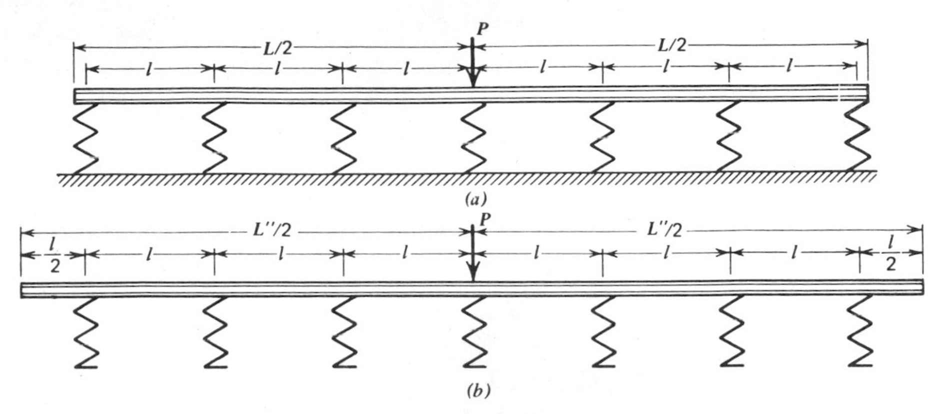

Fig. 9-2.2

The magnitude of the error for an l spacing that satisfies Eq. (9-2.12) is discussed in the example problem that follows this article.

The approximate solution for a beam of infinite length, with equally spaced elastic supports, may be used to obtain a reasonable approximate solution for a sufficiently long finite length beam. We note that the load exerted by each elastic spring has been assumed to be distributed over a distance l, the distribution being uniform over a distance $l/2$ to the left and right of the spring (Figs. 9-2.1*a* and *b*). Hence, consider a beam of length L supported by discrete elastic springs (Fig. 9-2.2*a*). In general, the end springs do not coincide with the ends of the beam but lie at some distance less than $l/2$ from the beam ends. Since the distributed effect of the end springs are assumed to act over length l, $l/2$ to the left and to the right of the end springs, we extend the beam of length L to a beam of length L'', where (Fig. 9-2.2*b*)

$$L'' = ml \tag{9-2.13}$$

and the integer m denotes the number of spring supports. If $L'' \geq 3\pi/(2\beta)$, the approximate solution for a spring supported infinite beam yields a reasonably good approximation for a spring supported finite beam of length L.

EXAMPLE 9-2.2

Finite Length Beam Supported by Seven Springs

An aluminum alloy I-beam (depth = 100 mm, $I_x = 2.45 \times 10^6$ mm^4, $E = 72.0$ GPa) has a length $L = 6.8$ m and is supported by 7 springs ($K = 110$ N/mm) spaced at distance $l = 1.10$ m center to center along the beam. A load $P = 12.0$ kN is applied at the center of the beam over one of the springs. Using the approximate solution method described in

Art. 9-2, determine the load carried by each spring, the deflection of the beam under the load, the maximum bending moment, and the maximum bending stress in the beam. The exact solution of this problem has been presented in Example 4-5.3.

SOLUTION

The magnitude of the factor β is estimated by means of Eqs. (9-2.11) and (9-1.5). Thus, we find

$$k = \frac{K}{l} = \frac{110}{1.1 \times 10^3} = 0.100 \text{ N/mm}^2$$

$$\beta = \sqrt[4]{\frac{0.100}{4(72 \times 10^3)(2.45 \times 10^6)}} = 0.000614 \text{ mm}^{-1}$$

By Eqs. (9-2.12) and (9-2.13), we see that

$$l = 1.10 \times 10^3 < \frac{\pi}{4\beta} = \frac{\pi}{4(0.000614)} = 1279 \text{ mm}$$

$$L'' = 7(1.10 \times 10^3) = 7700 \text{ mm} > \frac{3\pi}{2\beta} = \frac{3\pi}{2(0.000614)} = 7675 \text{ mm}$$

Hence, the limiting conditions on l and L'' are satisfied. The maximum deflection and the maximum bending moment occur under the load where $A_{\beta z} = C_{\beta z} = 1.00$. Equations (9-2.5) and (9-2.7) give

$$y_{\max} = \frac{P\beta}{2k} = \frac{12 \times 10^3(0.000614)}{2(0.10)} = 36.84 \text{ mm}$$

$$M_{\max} = \frac{P}{4\beta} = \frac{12 \times 10^3}{4(0.000614)} = 4.886 \times 10^6 \text{ N} \cdot \text{mm}$$

$$\sigma_{\max} = \frac{M_{\max} c}{I_x} = 99.7 \text{ MPa}$$

The deflection $y_{\max}$ ($y_D = y_{\max}$ in Fig. E4-5.3*b*) occurs at the origin at the center of the beam under the load. The magnitude of βz for the first, second, and third springs to the right and left of the load are $\beta l = 0.6754$, $2\beta l = 1.3508$, and $3\beta l = 2.0262$, respectively. From Table 9-2.1, $A_{\beta l} = 0.7153$, $A_{2\beta l} = 0.3094$, and $A_{3\beta l} = 0.0605$. The deflections of the springs C, B, and A (see Fig. E4-5.3*b*) are given by Eq. (9-2.5).

$$y_C = \frac{P\beta}{2k} A_{\beta l} = 36.84(0.7153) = 26.35 \text{ mm}$$

$$y_B = \frac{P\beta}{2k} A_{2\beta l} = 36.84(0.3094) = 11.40 \text{ mm}$$

$$y_A = \frac{P\beta}{2k} A_{3\beta l} = 36.84(0.0605) = 2.23 \text{ mm}$$

Table E9-2.2

Quantity	*Exact Solution Example 4-5.3*	*Approximate Solution*
Reaction *A*	−454 N	245 N
Reaction *B*	1216 N	1254 N
Reaction *C*	3094 N	2899 N
Reaction *D*	4288 N	4052 N
M_{max}	4.580 kN · m	4.886 kN · m
y_{max}	38.98 mm	36.84 mm

The reaction for each spring may be computed by means of Eq. (9-2.10). A comparison of the approximate solution presented here with the exact solution of Example 4-5.3 is given in Table E9-2.2. Although the reaction at *A* is considerably in error, the results in Table E9-2.2 indicate that the approximate maximum deflection is 5.50 percent less than the exact deflection, while the approximate maximum bending moment is 6.68 percent greater than the exact bending moment. These errors in the maximum deflection and maximum moment are not large when one considers the simplicity of the present solution compared to that of Example 4-5.3.

PROBLEM SET 9-2

1. The ballast and roadbed under railroad rails may vary appreciably from location to location. If the magnitude of k is 50 percent less than the value in Example 9-2.1, determine the percentage increase in the maximum deflection and the maximum bending moment for the rail for the same wheel load.

2. An S-130 × 15 steel I-beam ($E = 200$ GPa) has a depth of 127 mm, a width of 76 mm, a moment of inertia of $I_x = 5.12 \times 10^6$ mm^4, and a length of 4 m. It rests on a hard rubber foundation. The value of the spring constant for the hard rubber is $k_0 = 0.270$ N/mm^3. If the beam is subjected to a concentrated load, $P = 60.0$ kN, at the center of the beam, determine the maximum deflection and the maximum flexure stress at the center of the beam.

   Ans. $y_{max} = 2.187$ mm; $\sigma_{max} = 124.4$ MPa

3. Solve Problem 2 if the steel beam is replaced by an aluminum alloy beam for which $E = 72.0$ GPa.

4. A heavy machine has a mass of 60,000 kg. Its mass center is equidistance from each of four ground supports located at the four corners of a square 2 m on a side. Before it is moved to its permanent location, temporary support must be designed to hold the machine on a level horizontal surface on the ground. The surface layer of the ground is silt above a thick layer of inorganic clay. By the theory of soil mechanics, it is estimated that the spring constant of the soil is $k_0 = 0.029$ N/mm^3. The machine is placed centrally on two long timber beams ($E = 12.4$ GPa), 200 mm wide and 300 mm deep. The beams are parallel to one another, with centers 1.50 m apart. Determine the maximum deflection of the beams, the maximum flexure stress in the beams, and the minimum required length L for the beams.

   Ans. $y_{max} = 13.27$ mm; $\sigma_{max} = 14.84$ MPa; $L > 8.10$ m

5. A 60 kN capacity hoist may be moved along a steel S-150 × 25.7 I-beam ($E = 200$ GPa). The I-beam has a depth of 152 mm and a moment of inertia, $I_x = 11.0 \times 10^6$ mm^4. The beam is supported by a series of vertical steel rods ($E = 200$ GPa) of length 2.50 m, of diameter 18.0 mm, and spaced 500 mm center to center. (a) For capacity load at the center of the beam, located under one of the rods, determine the maximum stress in the beam and in the rods. (b) Does l satisfy Eq. (9-2.12)?

6. After installation of the I-beam of Problem 5, it becomes necessary to lower the I-beam 800 mm. This was done by adding 18.0 mm diameter aluminum alloy bars ($E = 72$ GPa) of this length to the steel bars. For capacity load at the center of the beam located under one of the composite bars, determine the maximum stress in the beam and in the rods.

   Ans. $\sigma_{max\,(beam)} = 82.8$ MPa; $\sigma_{max\,(rod)} = 73.7$ MPa

7. A long wooden beam, ($E = 12.4$ GPa) of depth 200 mm, and width 60 mm, is supported by 100 mm rubber cubes placed equidistance along the beam at $l = 600$ mm. The cube edges are parallel and perpendicular to the axis of the beam. The rubber has a spring constant of $k_0 = 0.330$ N/mm^3. A load P is applied to the center of the beam located over one of the rubber cubes. (a) If the wood has a yield stress of $Y = 40.0$ MPa, determine the magnitude of P based on a factor of safety $SF = 2.50$. What is the maximum pressure developed between the rubber and beam? (b) Does l satisfy Eq. (9-2.12)?

8. A long 50 mm diameter steel bar ($E = 200$ GPa and $Y = 300$ MPa) is supported by a number of pairs of 2 mm diameter high-strength steel

wires ($E = 200$ GPa and $Y = 1200$ MPa). An end view of the beam and wires is shown in Fig. P9-2.8. The pairs of wires are equally spaced at $l = 900$ mm. A load P is applied to the center of the long beam at the same location as one pair of wires. (a) Determine the magnitude of P if both the beam and the wires are designed with factor of safety, $SF = 2.00$. (B) Does l satisfy Eq. (9-2.12)?

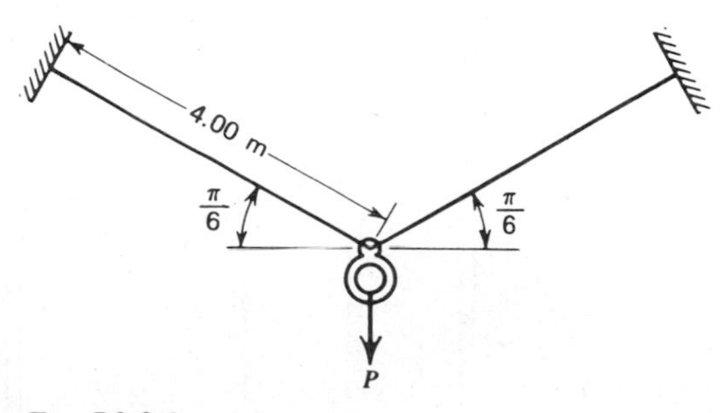

Fig. P9-2.8

Ans. (a) $P = 5.428$ kN; (b) Yes

9. A long 40 mm diameter steel beam ($E = 200$ GPa) is supported by a number of semicircular curved beams. (See end view in Fig. P9-2.9.) The curved beams are spaced along the beam with spacing $l = 550$ mm. Each curved beam is made of steel, has a circular cross section of diameter 30 mm, and a radius of curvature $R = 300$ mm. A load

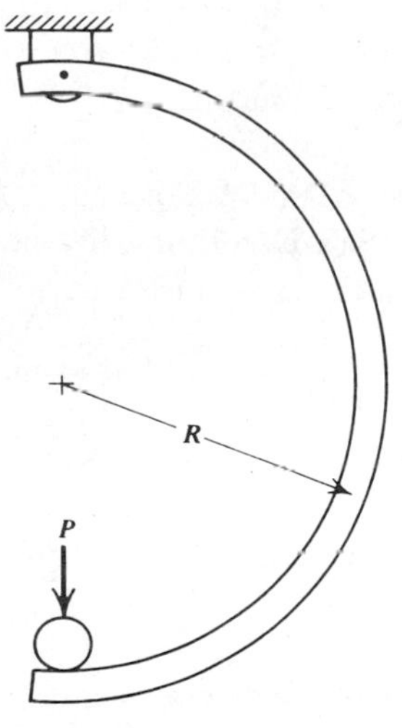

Fig. P9-2.9

$P = 3.00$ kN is applied to the center of the long beam located at one of the curved beams. Determine the maximum stress in the long beam and in the curved beams.

10. The beams in Fig. P9-2.10 are S-200 × 27.4 steel I-beams (203 mm deep, $I_x = 24.0 \times 10^6$ mm^4, $E = 200$ GPa). If a load $P = 90.0$ kN is applied to the center of the long beam located over one of the cross beams, determine the maximum flexure stress in the long beam and in the cross beams.

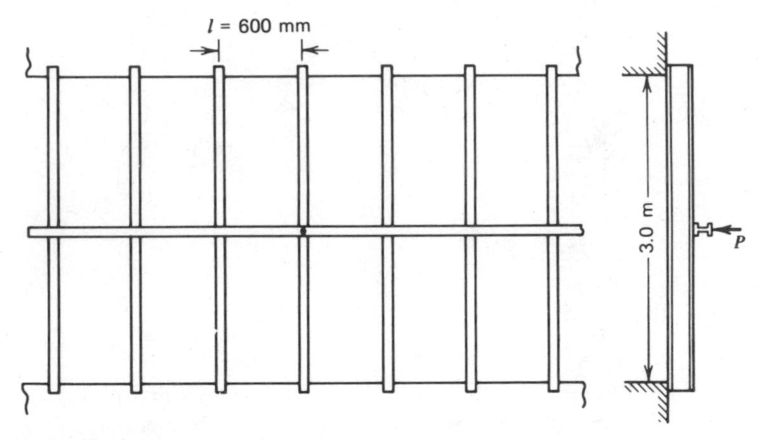

Fig. P9-2.10

Ans. $\sigma_{\max(\text{long})} = 102.6$ MPa; $\sigma_{\max(\text{cross})} = 79.4$ MPa

11. Let the curved beams in Problem 9 be made of an aluminum alloy ($E = 72.0$ GPa). Determine the maximum stress in the long beam and in the curved beams.

12. Let the long beam in Problem 10 be made of an aluminum alloy ($E = 72.0$ GPa). Determine the maximum flexure stress in the long beam and in the cross beams.

Ans. $\sigma_{\max(\text{long})} = 79.4$ MPa, $\sigma_{\max(\text{cross})} = 102.6$ MPa

9-3 INFINITE BEAM SUBJECTED TO A DISTRIBUTED LOAD SEGMENT

The solution for the problem of a concentrated load at the center of an infinite beam on an infinite elastic foundation can be used to obtain

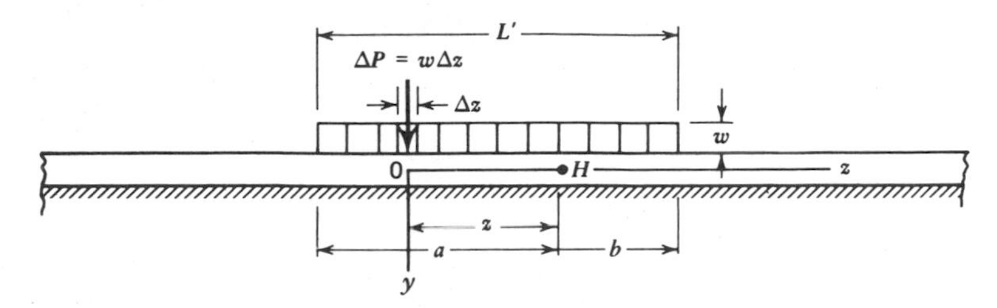

Fig. 9-3.1/Uniformly distributed load segment on an infinite beam resting on an elastic foundation.

solutions for distributed loads. Only segments of uniformly distributed loads are considered in this article. Consider an infinite beam resting on an infinite elastic foundation and subjected to a uniformly distributed load w over a segment of length L' (Fig. 9-3.1). The deflection, slope, bending moment, and shear of the beam can be determined with the solution presented in Art. 9-2. Since the maximum values of these quantities generally occur within the segment of length L', we obtain the solution only in this segment.

Hence, consider an infinitesimal length Δz of the beam within the segment of length L'. In this segment, the beam is subjected to a uniformly distributed load w (Fig. 9-3.1). Hence, a load $\Delta P = w\Delta z$ acts on the element Δz. We treat the load $\Delta P = w\Delta z$ as a concentrated load. We choose the origin of the coordinate axes under load ΔP. Next, consider any point H at distance z from the load $\Delta P = w\Delta z$; note that H is located at distances a and b from the left and right ends of segment L', respectively. The deflection Δy_H at H due to the concentrated load $\Delta P = w\Delta z$ is given by Eq. (9-2.4) with $P = \Delta P = w\Delta z$. Thus, we have

$$\Delta y_H = \frac{w\Delta z\beta}{2k} e^{-\beta z}(\cos\beta z + \sin\beta z) \tag{9-3.1}$$

The total deflection y_H due to the distributed load over the entire length L' is obtained by superposition. It is the algebraic sum of increments given by Eq. (9-3.1). Hence, by the integration process, we obtain

$$y_H = \sum_{\lim \Delta z \to 0} \Delta y_H = \int_0^a \frac{w\beta}{2k} e^{-\beta z}(\cos\beta z + \sin\beta z)\, dz$$

$$+ \int_0^b \frac{w\beta}{2k} e^{-\beta z}(\cos\beta z + \sin\beta z)\, dz$$

$$= \frac{w}{2k}(2 - e^{-\beta a}\cos\beta a - e^{-\beta b}\cos\beta b) \tag{9-3.2}$$

Values of slope, bending moment, and shear at point H may also be obtained by superposition. These expressions may be simplified by means of Eqs. (9-2.9). Thus, we obtain the results

$$y_H = \frac{w}{2k}(2 - D_{\beta a} - D_{\beta b}) \tag{9-3.3}$$

$$\theta_h = \frac{w\beta}{2k}(A_{\beta a} - A_{\beta b}) \tag{9-3.4}$$

$$M_H = \frac{w}{4\beta^2}(B_{\beta a} + B_{\beta b}) \tag{9-3.5}$$

$$V_H = \frac{w}{4\beta}(C_{\beta a} - C_{\beta b}) \tag{9-3.6}$$

Generally, the maximum values of deflection and of bending moment are of greatest interest. The maximum deflection occurs at the center of segment L'. The maximum bending moment may or may not occur at the center of segment L'. In general, the location of the maximum bending moment depends on the magnitude of $\beta L'$.

$\beta L' \leq \pi$/For $\beta L'$ less than or equal to π, the data for $B_{\beta z}$ in Table 9-2.1 indicate that the maximum bending moment occurs at the center of segment L'.

$\beta L' \to \infty$/As $\beta L'$ becomes large,

$$\theta \to 0 \qquad M_x \to 0 \qquad V_y \to 0 \quad \text{and} \quad y \to \frac{w}{k} \tag{9-3.7}$$

everywhere, except near the ends of segment L'. The data in Table 9-2.1 indicate that the maximum bending moment occurs when either βa or βb is equal to $\pi/4$.

Intermediate Values of $\beta L'$/For $\beta L'$ greater than π the location of the maximum bending moment may lie outside of segment L'. (See Problem 9-3.3 and the example problem that follows this article.) However, the maximum moment value outside of segment L' for the example problem is only 3.0 percent greater than the maximum bending moment within segment L'. The location of the maximum bending moment can be obtained by trial and error; however, because of the small difference, sufficient accuracy can be obtained by taking the location of the maximum moment to be $\pi/(4\beta)$ from either end of the uniformly distributed load within length L'.

EXAMPLE 9-3.1

Uniformly Distributed Load on a Segment of Wood Beam

A long wood beam ($E = 10.0$ GPa) has a rectangular cross section with a depth of 200 mm and a width of 100 mm. It rests on an earth foundation. The spring constant for the foundation is $k_0 = 0.040$ N/mm^3. A uni-

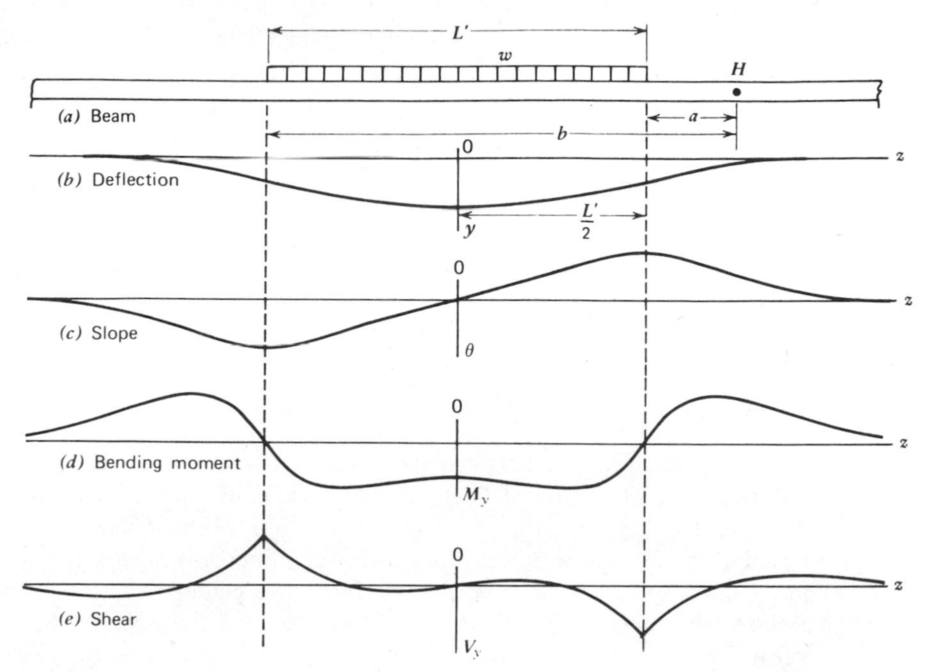

Fig. E9-3.1

formly distributed load $w = 35.0$ N/mm extends over a length $L' = 3.61$ m of the beam (Fig. E9-3.1*a*). Determine the maximum deflection, the maximum flexure stress, and the maximum pressure between the beam and foundation. Take the origin of coordinates at the center of segment L'.

SOLUTION

The magnitude of β is obtained by means of Eqs. (9-1.3) and (9-1.5). Thus, we find

$$k = bk_0 = 100(0.040) = 4.00 \text{ N/mm}^2;$$

$$I_x = \frac{bh^3}{12} = \frac{100(200)^3}{12} = 66.67 \times 10^6 \text{ mm}^4$$

$$\beta = \sqrt[4]{\frac{k}{4EI_x}} = \sqrt[4]{\frac{4}{4(10 \times 10^3)(66.67 \times 10^6)}} = 0.001107 \text{ mm}^{-1}$$

The magnitude of $\beta L'$, needed to determine where the maximum bending moment occurs, is $\beta L' = 0.001107(3.61 \times 10^3) = 4.00$. Since $\beta L'$ is greater than π, the maximum bending moment does not occur at the center of segment L'. With values of βa and $\beta b = 4.00 - \beta a$, values of the quantities y, θ, M_x, and V_y may be calculated by means of Eqs. (9-3.3),

(9-3.4), (9-3.5), and (9-3.6), and the data from Table 9-2.1. These quantities are plotted in Figs. E9-3.1*b*, *c*, *d*, and *e*, respectively. Values are also shown for points in the beam outside the distributed load segment where $\beta b = \beta(a + L')$ and a is the distance from H to the nearest edge of the distributed load (Fig. E9-3.1*a*). Equations (9-3.4) and (9-3.6), for θ_H and V_H respectively, are valid for points away from the distributed load; however, different equations are needed for y_H and M_H as indicated in Problem 9-3.3.

The maximum deflection occurs at the center of segment L' where $\beta a = \beta b = 2.00$. Equation (9-3.3), with $D_{\beta a} = -0.0563$ from Table 9-2.1, gives

$$y_{\max} = \frac{w}{k}(1 - D_{\beta a}) = \frac{35}{4}(1 + 0.0563) = 9.243 \text{ mm}$$

The maximum pressure between the beam and the foundation occurs at the point of maximum deflection; thus, we find that the maximum pressure is $q_{\max} = y_{\max} k_0 = 9.243(0.040) = 0.370$ MPa. There are four possible locations at which the largest bending moment may occur. They are located symmetrically with respect to the center of segment L'. Relative maximum bending moments occur at locations where $V_H = 0$. From Table 9-2.1, it is found that $V_H = 0$ ($C_{\beta a} = C_{\beta b}$) when $\beta a = 0.858$ and $\beta b = 3.142$ and also when $\beta a = 0.777$ and $\beta b = 4.777$. These conditions locate the position of relative maximum bending moments inside segment L' and outside of segment L', respectively. However, the value of the largest bending moment is located outside of segment L' and is given by the equation indicated in Problem 9-3.3. Thus, we find (outside of segment L')

$$M_{\max} = \left| \frac{-w}{4\beta^2}(B_{\beta a} - B_{\beta b}) \right|$$

$$= \frac{35}{4(0.001107)^2}[0.3223 - (-0.0086)] = 2.363 \text{ kN} \cdot \text{m}$$

This value is 3 percent greater than the bending moment calculated by means of Eq. (9-3.5) with $\beta a = 0.858$ and $\beta b = 3.142$. In practice, this difference is not especially significant.

The corresponding flexure stress is

$$\sigma_{\max} = \frac{M_{\max} c}{I_x} = \frac{2.363 \times 10^6 (100)}{66.67 \times 10^6} = 3.544 \text{ MPa}$$

If the maximum bending moment is assumed to occur at $\pi/(4\beta)$ (see end of Art. 9-3), $\beta a = \pi/4$ and $\beta b = 4 - \pi/4$ (inside of segment L'). Substituting these values for βa and βb in Eq. (9-3.5), we find

$$M_H = \frac{w}{4\beta^2}(B_{\beta a} + B_{\beta b}) = \frac{35}{4(0.001107)^2}[0.3224 + (-0.0029)]$$

$$= 2.362 \text{ kN} \cdot \text{m}$$

which is 3.5 percent less than the largest moment M_{max} computed above. Generally, the value M_H for $\beta a = \pi/4$ and $\beta b = \beta L' - \pi/4$ gives a good approximation of M_{max}.

PROBLEM SET 9-3

1. Let the load of 60.0 kN in Problem 9-2.2 be distributed over a length of 1.00 m. Determine the maximum deflection and maximum flexure stress in the beam.

2. The long wood beam in Problem 9-2.7 is subjected to a distributed load w over length $L' = 3.00$ m. Determine the magnitude of w based on a factor of safety $SF = 2.00$.

   Ans. $w = 123.2$ kN/m

3. Show that, for point H located outside segment L' (Fig. E9-3.1), the following equations are valid; $y_H = w(D_{\beta a} - D_{\beta b})/2k$ and $M_H = -w(B_{\beta a} - B_{\beta b})/(4\beta^2)$.

9-4 SEMI-INFINITE BEAM SUBJECTED TO LOADS AT ITS END

A semi-infinite beam resting on an infinite linearly elastic foundation is loaded at its end by a concentrated load P and a positive bending moment M_0 (Fig. 9-1.2). The boundary conditions that determine the two constants of integration C_3 and C_4 in Eq. (9-1.7) are

$$EI_x \left.\frac{d^2y}{dz^2}\right|_{z=0} = -M_0$$

$$EI_x \left.\frac{d^3y}{dz^3}\right|_{z=0} = -V_y = P \tag{9-4.1}$$

Substitution of Eq. (9-1.7) into these boundary conditions yields two linear equations in C_3 and C_4. Solving these equations for C_3 and C_4, we obtain

$$C_3 = \frac{2\beta^2 M_0}{k} \qquad C_4 = \frac{2\beta P}{k} - C_3 \tag{9-4.2}$$

Substituting these into Eq. (9-1.7), we find

$$y = \frac{2\beta e^{-\beta z}}{k}[P\cos\beta z - \beta M_0(\cos\beta z - \sin\beta z)] \tag{9-4.3}$$

Values of slope, bending moment, and shear are obtained by substitution of Eq. (9-4.3) into Eqs. (9-1.1). These equations are simplified with the definitions given by Eqs. (9-2.9). Thus, we have

$$y = \frac{2P\beta}{k} D_{\beta z} - \frac{2\beta^2 M_0}{k} C_{\beta z} \tag{9-4.4}$$

$$\theta = -\frac{2P\beta^2}{k} A_{\beta z} + \frac{4\beta^3 M_0}{k} D_{\beta z} \tag{9-4.5}$$

$$M_x = -\frac{P}{\beta} B_{\beta z} + M_0 A_{\beta z} \tag{9-4.6}$$

$$V_y = -PC_{\beta z} - 2M_0 \beta B_{\beta z} \tag{9-4.7}$$

These results are valid provided that the beam is attached to the foundation everywhere along its length.

EXAMPLE 9-4.1
I-Beam Loaded at Its End

An S-100 × 11.5 steel I-beam ($E = 200$ GPa) has a depth of 102 mm, a width of 68 mm, a moment of inertia of $I_x = 2.53 \times 10^6$ mm^4, and a length of 4 m. It is attached to a rubber foundation for which $k_0 = 0.350$ N/mm^3. A concentrated load $P = 30.0$ kN is applied at one end of the beam. Determine the maximum deflection and maximum flexure stress in the beam and determine the location of each.

SOLUTION

The spring coefficient k is equal to the product of the beam width and the elastic spring constant k_0 for the foundation; that is, $k = 68(0.350) = 23.8$ N/mm^2. From Eq. (9-1.5), we find that

$$\beta = \sqrt[4]{\frac{k}{4EI_x}} = \sqrt[4]{\frac{23.8}{4(200{,}000)(2{,}530{,}000)}} = 0.001852 \text{ mm}^{-1}$$

Since

$$L = 4000 \text{ mm} > \frac{3\pi}{2\beta} = \frac{3\pi}{2(0.001852)} = 2540 \text{ mm},$$

the beam can be considered to be a long beam. Values for deflection y and moment M_x are given by Eqs. (9-4.4) and (9-4.6). The maximum deflection occurs at the end where load P is applied, since $D_{\beta z}$ is maximum where $\beta z = 0$. The maximum moment occurs at $z = \pi/4\beta$, where $B_{\beta z}$ is a maximum. Thus, the maximum deflection is

$$y_{\max} = \frac{2P\beta}{k} = \frac{2(30{,}000)(0.001852)}{23.8} = 4.67 \text{ mm}$$

The location of y_{max} is at $z = 0$. The maximum moment is

$$M_{max} = -\frac{0.3224P}{\beta} = -\frac{0.3224(30{,}000)}{0.001852} = -5.22 \text{ kN} \cdot \text{m}$$

and, therefore, the maximum stress is

$$\sigma_{max} = \frac{M_{max}c}{I_x} = \frac{5{,}220{,}000(51)}{2{,}530{,}000} = 105.3 \text{ MPa}$$

The location of σ_{max} is at $z = \pi/4\beta = 424$ mm.

PROBLEM SET 9-4

1. Let load P be moved to one end of the beam in Problem 9-2.2. Determine the maximum deflection and maximum flexure stress in the beam and give the location of the maximum flexure stress.

2. Let the hoist in Problem 9-2.5 be moved to one end of the beam. Each rod supporting the I-beam is a spring exerting an influence over length l. If the end of the beam is $l/2 = 250$ mm from the nearest tension rod, determine the maximum stress in the rods and in the beam.

   Ans. $\sigma_{max(rod)} = 149.3$ MPa; $\sigma_{max(beam)} = 60.7$ MPa

3. A long rectangular section brass beam ($E = 82.7$ GPa) has a depth of 20 mm, a width of 15 mm, and rests on a hard rubber foundation (Fig. P9-4.3). The value of the spring constant for the hard rubber foundation is 0.200 N/mm³. If the beam is subjected to a concentrated load $P = 700$ N at the location shown, determine the maximum deflection of the beam and maximum flexure stress in the beam.

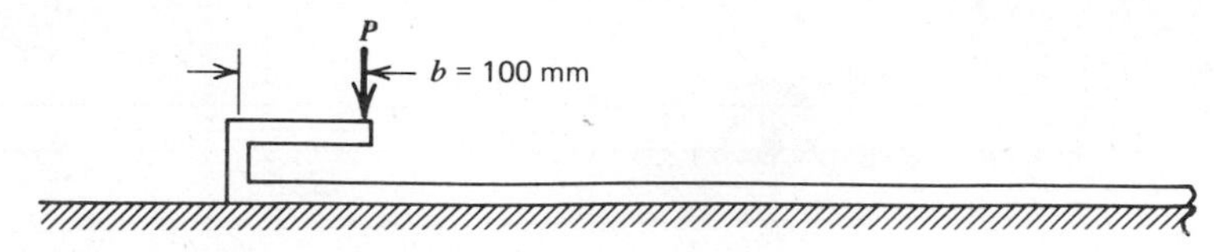

Fig. P9-4.3

4. Solve Problem 3 if $b = 200$ mm.

   Ans. $\sigma_{max} = 140$ MPa, $y_{max} = 0.833$ mm at $z = 159$ mm

5. An S-100 × 11.5 steel I-beam (depth = 102 mm, $I_x = 2.53 \times 10^6$ mm⁴, $E = 200$ GPa) is long and is supported by many springs ($K = 100$

N/mm) spaced at distance $l = 500$ mm center to center along the beam. A load $P = 3.50$ kN is applied to the left end of the beam at a distance of 2.00 m from the first spring. Determine the maximum flexure stress in the beam and the maximum tension load and the maximum compression load in the springs. *Hint*: $M_0 = -P(2000 - l/2)$.

6. Solve Problem 5 for the case where the steel beam is replaced by an aluminum alloy beam for which $E = 72.0$ GPa.

   Ans. $\sigma_{max} = 141.1$ MPa, Compression = 4.23 kN (1st spring), Tension = 720 N (6th spring)

9-5 SEMI-INFINITE BEAM WITH CONCENTRATED LOAD NEAR ITS END

The solution for a semi-infinite beam resting on an infinite linearly elastic foundation with a concentrated load P near its end may be obtained from the solutions presented in Arts. 9-2 and 9-4. Consider a beam subjected to load P at distance a from its end (Fig. 9-5.1a). Let the beam be extended to infinity to the left as indicated by the dashed line. For the beam so extended, Eqs. (9-2.7) and (9-2.8) give magnitudes for $M_{x(z=-a)} = PC_{\beta a}/(4\beta)$ and $V_{y(z=-a)} = PD_{\beta a}/2$ at distance a to the left of the origin (Fig. 9-5.1a). Now let the beam (Fig. 9-5.1a) be loaded at the left end (Fig. 9-5.1b), by loads Q and M with magnitudes

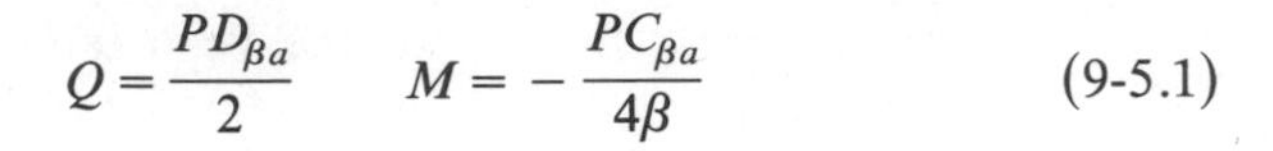

$$Q = \frac{PD_{\beta a}}{2} \qquad M = -\frac{PC_{\beta a}}{4\beta} \qquad (9\text{-}5.1)$$

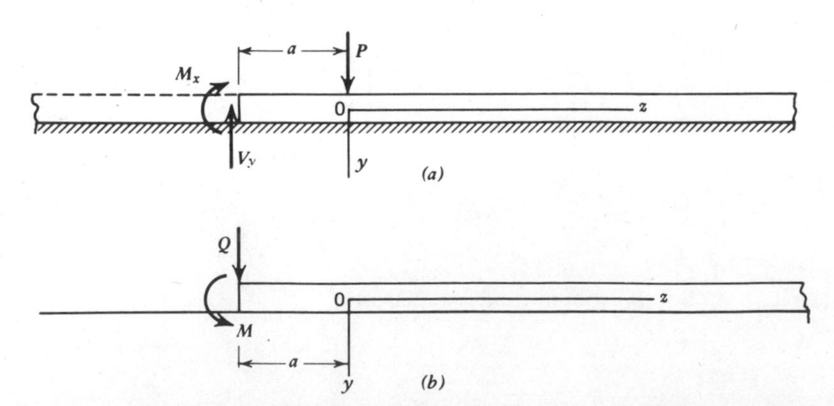

Fig. 9-5.1/Semi-infinite beam on an elastic foundation loaded near its end.

Table 9-5.1

Computer Program: Beams on Elastic Foundations

```
      PROGRAM BELFO(INPUT,OUTPUT,BEIN,BEOUT,TAPE5 = BEIN,TAPE6 = BEOUT)
      READ (5,*) NN
      DO 22 J = 1,NN
      WRITE (6,23)
23    FORMAT ("   M(0)        SPRK        E        I        P        L
     CL(SP)        A        K")
      READ (5,*) N,E,EI,EMO,P,SPRK,EL,ELL,A,EK
      IF (SPRK.LT.0.01) GO TO 2
      EK = SPRK/ELL
2     BETA = (EK/(4.*E*EI))**.25
      IF (EL.GT.0.01) GO TO 42
      EL = 6.0/BETA
      IF(N.NE.1) GO TO 42
      A = .5*EL
42    WRITE(6,24) EMO,SPRK,E,EI,P,EL,ELL,A,EK
24    FORMAT (5E9.3,4F8.2)
      IF (ELL.LT..785/BETA) GO TO 53
      WRITE (6,34)
34    FORMAT ("SPRING SPACING TOO FAR APART")
      GO TO 22
53    IF (EL.GT.4.7/BETA) GO TO 43
      WRITE (6,35)
35    FORMAT ("SHORT BEAM")
      GO TO 22
43    IF (N.NE.3) GO TO 3
      BETAA = BETA*A
      EXA = EXP(-BETAA)
      CA = EXA*(COS(BETAA) - SIN(BETAA))
      DA = EXA*COS(BETAA)
      DELZ = A/5.
      IF (BETAA.GT..6) GO TO 44
3     K = IFIX(.017/BETA)
      DELZ = 10.*FLOAT(K)
44    Z = -DELZ
      WRITE(6,55)
55    FORMAT ("   Z(END)        Z(P)        Y        M(X)")
      DO 25 I = 1,10
      Z = Z + DELZ
      BETAZ = Z*BETA
      EXZ = EXP(-BETAZ)
      AZ = EXZ*(SIN(BETAZ) + COS(BETAZ))
      BZ = EXZ*SIN(BETAZ)
      CZ = EXZ*(COS(BETAZ) - SIN(BETAZ))
      DZ = EXZ*COS(BETAZ)
      IF (N.EQ.1) GO TO 4
```

Table 9-5.1

(*Continued*)

```
      IF (N.EQ.2) GO TO 5
      X = ABS(A - Z)
      BETAX = X*BETA
      EXX = EXP(-BETAX)
      AX = EXX*(SIN(BETAX) + COS(BETAX))
      CX = EXX*(COS(BETAX) - SIN(BETAX))
C     EQUATIONS (9-5.2) and (9-5.3).
      Y = .5*P*BETA/EK*(AX + 2.*DA*DZ + CA*CZ)
      EMX = .25*P/BETA*(CX - 2.*DA*BZ - CA*AZ)
      ZE = Z
      ZP = -A + Z
      GO TO 6
C     EQUATIONS (9-4.4) and (9-4.6).
5     Y = 2.*P*BETA*DZ/EK - 2.*BETA**2*EMO*CZ/EK
      EMX = -P*BZ/BETA + EMO*AZ
      ZP = Z
      ZE = Z
      GO TO 6
C     EQUATIONS (9-2.5) and (9-2.7).
 4    Y = .5*P*BETA*AZ/EK
      EMX = .25*P*CZ/BETA
      ZE = A + Z
      ZP = Z
6     WRITE (6,33) ZE,ZP,Y,EMX
33    FORMAT (3F10.2,E15.4)
25    CONTINUE
22    CONTINUE
      STOP
      END
```

INPUT

```
1 5
2 3,7.200E + 04,2.450E + 06,0.,1.200E + 04,0.,7700.,0.,1000.,2.
3 3,7.200E + 04,2.450E + 06,0.,1.200E + 04,0.,7700.,0.,800.,2.
4 3,7.200E + 04,2.450E + 06,0.,1.200E + 04,0.,7700.,0.,550.,2.
5 3,7.200E + 04,2.450E + 06,0.,1.200E + 04,0.,7700.,0.,300.,2.
6 2,8.270E + 04,1.000E + 04,8.400E + 04,7.000E + 02,0.,0.,0.,0.,3.75
```

OUTPUT

M(0)	SPRK	E	I	P	L	L(SP)	A	K
0.	0.	.720E + 05	.245E + 07	.120E + 05	7700.00	0.00	1000.00	2.00

Z(END)	Z(P)	Y	M(X)
0.00	−1000.00	1.15	0.
200.00	−800.00	1.91	.5612E + 05
400.00	−600.00	2.66	.2651E + 06

Table 9-5.1

(*Continued*)

600.00	−400.00	3.34	.6865E + 06
800.00	−200.00	3.87	.1374E + 07
1000.00	0.00	4.07	.2370E + 07
1200.00	200.00	3.83	.1288E + 07
1400.00	400.00	3.28	.5102E + 06
1600.00	600.00	2.62	−.5624E + 04
1800.00	800.00	1.95	−.3120E + 06

M(0)	SPRK	E	I	P	L	L(SP)	A	K
0.	0.	.720E + 05	.245E + 07	.120E + 05	7700.00	0.00	800.00	2.00

Z(END)	Z(P)	Y	M(X)
0.00	−800.00	2.80	0.
160.00	−640.00	3.20	.7515E + 05
320.00	−480.00	3.60	.3143E + 06
480.00	−320.00	3.94	.7374E + 06
640.00	−160.00	4.17	.1362E + 07
800.00	0.00	4.21	.2199E + 07
960.00	160.00	3.96	.1330E + 07
1120.00	320.00	3.53	.6630E + 06
1280.00	480.00	2.99	.1764E + 06
1440.00	640.00	2.43	−.1573E + 06

M(0)	SPRK	E	I	P	L	L(SP)	A	K
0.	0.	.720E + 05	.245E + 07	.120E + 05	7700.00	0.00	550.00	2.00

Z(END)	Z(P)	Y	M(X)
0.00	−550.00	5.77	0.
110.00	−440.00	5.66	.6932E + 05
220.00	−330.00	5.54	.2755E + 06
330.00	−220.00	5.40	.6157E + 06
440.00	−110.00	5.22	.1087E + 07
550.00	0.00	4.97	.1684E + 07
660.00	110.00	4.61	.1081E + 07
770.00	220.00	4.18	.5895E + 06
880.00	330.00	3.71	.1994E + 06
990.00	440.00	3.23	−.1009E + 06

M(0)	SPRK	E	I	P	L	L(SP)	A	K
0.	0.	.720E + 05	.245E + 07	.120E + 05	7700.00	0.00	300.00	2.00

Z(END)	Z(P)	Y	M(X)
0.00	−300.00	9.76	0.
130.00	−170.00	8.79	.1595E + 06
260.00	−40.00	7.80	.6160E + 06
390.00	90.00	6.76	.2559E + 06
520.00	220.00	5.69	−.3560E + 06
650.00	350.00	4.65	−.7755E + 06
780.00	480.00	3.69	−.1038E + 07
910.00	610.00	2.82	−.1175E + 07

Table 9-5.1

(Continued)

1040.00	740.00	2.07	−.1216E+07					
1170.00	870.00	1.43	−.1187E+07					
M(0)	SPRK	E	I	P	L	L(SP)	A	K
.840E+05	0.	.827E+05	.100E+05	.700E+03	1034.03	0.00	0.00	3.75
Z(END)	Z(P)	Y	M(X)					
0.00	0.00	.66	.8400E+05					
20.00	20.00	.74	.7052E+05					
40.00	40.00	.78	.5813E+05					
60.00	60.00	.80	.4692E+05					
80.00	80.00	.79	.3691E+05					
100.00	100.00	.77	.2808E+05					
120.00	120.00	.73	.2041E+05					
140.00	140.00	.69	.1384E+05					
160.00	160.00	.63	.8295E+04					
180.00	180.00	.58	.3702E+04					

Since the origin of the coordinate axes is distance a to the right of the loaded end, the deflection and bending moment for this loading are given by Eqs. (9-4.4) and (9-4.6), respectively, if the coordinate z is replaced by $(a+z)$. Superposing the two loadings for the two beams in Fig. 9-5.1 cancels the moment and shear at the left end. Thus, superposition of the two results yields the solution for a semi-infinite beam loaded by a concentrated load P at distance a from the left end. Using Eqs. (9-2.5), (9-4.4), and (9-5.1), we obtain the deflection y for $z \geq -a$. Thus, we find for y the formula

$$y = \frac{P\beta}{2k}\left(A_{\beta z} + 2D_{\beta a}D_{\beta(a+z)} + C_{\beta a}C_{\beta(a+z)}\right) \tag{9-5.2}$$

Similarly Eqs. (9-2.7), (9-4.6), and (9-5.1) give the bending moment M_x for $z \geq -a$ as follows:

$$M_x = \frac{P}{4\beta}\left(C_{\beta z} - 2D_{\beta a}B_{\beta(a+z)} - C_{\beta a}A_{\beta(a+z)}\right) \tag{9-5.3}$$

Since the quantities $A_{\beta z}$ and $C_{\beta z}$ in Eqs. (9-5.2) and (9-5.3) are symmetrical in z, for negative values of z, we use the conditions $A_{\beta z}(-z) = A_{\beta z}(z)$ and $C_{\beta z}(-z) = C_{\beta z}(z)$.

A FORTRAN computer program for the solutions of beams on elastic foundations given by Eqs. (9-2.5), (9-2.7), (9-4.4), (9-4.6), (9-5.2), and (9-5.3) is listed in Table 9-5.1, along with sample input and output data. Incorporated in the program are the approximation concepts discussed in Art. 9-2, following Eq. (9-2.9).

EXAMPLE 9-5.1
I-Beam Loaded near One End

Let the load in Example 9-4.1 be moved to a location 500 mm from one end of the beam. Determine the maximum deflection and maximum flexure stress in the beam and determine the location of each.

SOLUTION

From Example 9-4.1, we find that $k = 23.8\ \text{N/mm}^2$ and $\beta = 0.001852\ \text{mm}^{-1}$. The deflection y and bending moment M_x are given by Eqs. (9-5.2) and (9-5.3). Since $\beta a = 0.001852(500) = 0.9260$, Table 9-2.1 gives $C_{\beta a} = -0.0782$ and $D_{\beta a} = 0.2383$. Hence,

$$y = \frac{P\beta}{2k}\left[A_{\beta z} + 2D_{\beta a}D_{\beta(a+z)} + C_{\beta a}C_{\beta(a+z)}\right]$$
$$= 1.1672\left[A_{\beta z} + 0.4766D_{\beta(a+z)} - 0.0782C_{\beta(a+z)}\right]$$
$$M_x = \frac{P}{4\beta}\left[C_{\beta z} - 2D_{\beta a}B_{\beta(a+z)} - C_{\beta a}A_{\beta(a+z)}\right]$$
$$= 4{,}050{,}000\left[C_{\beta z} - 0.4766B_{\beta(a+z)} + 0.0782A_{\beta(a+z)}\right]$$

By trial and error, it is found that the maximum deflection $y_{\max}$ occurs at 424 mm from the end of the beam, where $z = -76$ mm ($\beta z = 0.1808$ and $\beta(a+z) = \pi/4 = 0.7854$). From Table 9-2.1, $A_{\beta z} = 0.9816$, $D_{\beta(a+z)} = 0.3224$, and $C_{\beta(a+z)} = 0$. Thus,

$$y_{\max} = 1.1672[0.9816 + 0.4766(0.3224) - 0.0782(0)]$$
$$= 1.3251\ \text{mm}$$

By trial and error, it is found that the maximum bending moment $M_{\max}$ occurs at 500 mm from the end of the beam ($\beta z = 0$ and $\beta(a+z) = 0.9260$). From Table 9-2.1, $C_{\beta z} = 1.0000$, $A_{\beta(a+z)} = 0.5548$, and $B_{\beta(a+z)} = 0.3165$. Hence,

$$M_{\max} = 4{,}050{,}000[1.0000 - 0.4766(0.3165) + 0.0782(0.5548)]$$
$$= 3{,}615{,}000\ \text{N}\cdot\text{mm}$$

and, therefore,

$$\sigma_{\max} = \frac{M_{\max}c}{I_x} = \frac{3{,}615{,}000(51)}{2{,}530{,}000} = 72.9\ \text{MPa}$$

PROBLEM SET 9-5

1. Let the load $P = 60.0$ kN in Problem 9-2.2 be moved to one of the quarter points in the beam. Determine the maximum deflection and

maximum flexure stress in the beam and determine locations for each.

2. Let the load $P = 60.0$ kN in Problem 9-2.2 be moved to a location 500 mm from one end of the beam. Determine the maximum deflection and maximum flexure stress in the beam and determine locations for each.

   Ans. $y_{max} = 3.036$ mm at free end; $\sigma_{max} = 94.5$ MPa under load

3. Let the hoist in Problem 9-2.5 with a capacity load of 60 kN be located under the second rod from one end. Since each spring is assumed to exert an influence over a length $l = 500$ mm, the load acts at distance $a = 750$ mm from the end of the beam. Determine the maximum deflection of the beam, the maximum flexure stress in the beam, the maximum stress in the rods, and the locations for each.

4. Let the hoist in Problem 9-2.5 with a capacity load of 60 kN be located under the first rod from one end. Since each spring is assumed to exert an influence over a length $l = 500$ mm, the load acts at distance $a = 250$ mm from the end of the beam. Determine the maximum deflection of the beam, the maximum flexure stress in the beam, the maximum stress in the rods, and the locations for each.

   Ans. $y_{max} = 2.80$ mm at free end; $\sigma_{max\,(beam)} = 39.2$ MPa at 880 mm from free end; $\sigma_{max\,(rod)} = 172.7$ MPa under load

5. A four-wheel car runs on steel rails ($E = 200$ GPa). The rails have a depth of 120 mm. The distance from the top of a rail to its centroid is 69 mm, and its moment of inertia is 17.07×10^6 mm^4. The rail rests on an elastic foundation with spring constant $k = 12.0$ N/mm^2. The two wheels on each side of the car are spaced 2.50 m center to center. If each wheel load is 80.0 kN, determine the maximum deflection and maximum flexure stress when a car wheel is located at one end of the rail and the other car wheel on the same rail is 2.50 m from the end.

9-6 SHORT BEAMS

The solutions that have been presented in the foregoing articles are good approximations for a beam that is supported by an elastic foundation and has a length greater than $3\pi/(2\beta)$. However, for a beam whose length is less than $3\pi/(2\beta)$, so-called *short beams*, special solutions are required. The reader is referred to the book by M. Hetényi[1] for a solution applicable for short beams. For the special case of a concentrated load

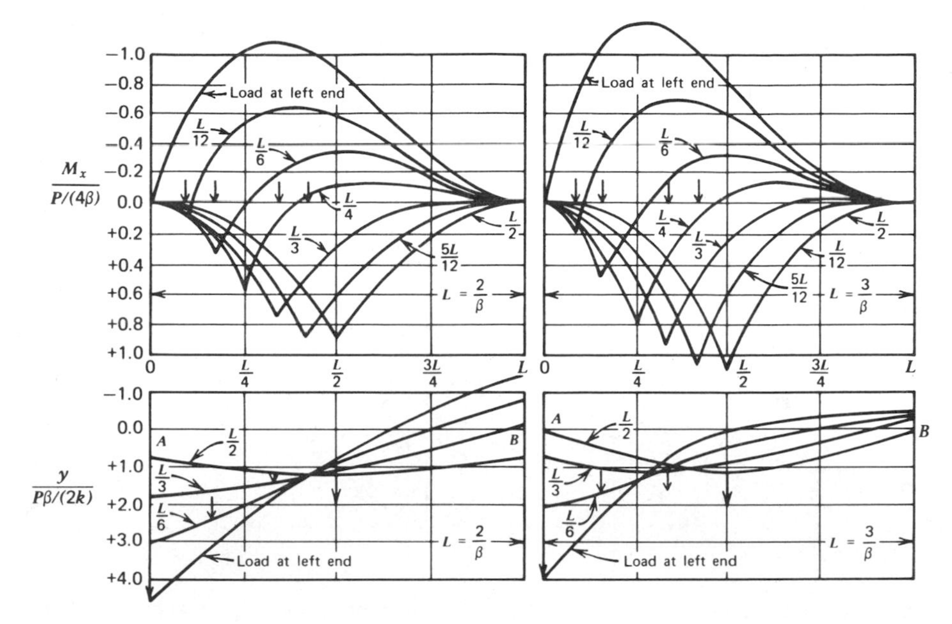

Fig. 9-6.1a/Bending moment diagrams and deflection curves for a short beam on elastic supports subjected to concentrated load located as shown on each curve. The ends of the beams are unrestrained (free).

located at the center of a short beam, the maximum deflection y_{max} and the maximum bending moment M_{max} occur under the load; their magnitudes are given by the following equations:

$$y_{max} = \frac{P\beta}{2k} \frac{\cosh \beta L + \cos \beta L + 2}{\sinh \beta L + \sin \beta L} \tag{9-6.1}$$

$$M_{max} = \frac{P}{4\beta} \frac{\cosh \beta L - \cos \beta L}{\sinh \beta L + \sin \beta L} \tag{9-6.2}$$

in which L is the length of the beam. Magnitudes of the deflection y and the bending moment M_x for other locations of the concentrated load are beyond the scope of the book. However, solutions have been calculated for several load locations for three short beams and one long beam. The results are presented in Fig. 9-6.1.

Design tables for finite beams with free ends on a Winkler foundation have been given by Iyengar and Ramu.[2] The cases of simply supported ends and clamped ends may be treated by appropriate superposition techniques. A solution for finite beams with elastic end restraints on a Winkler foundation has been given by Ting.[3] This solution can be used to simulate a beam on elastic foundations with various boundary conditions, including initial settlement of an end of the beam. The effect of other structural members connected to a beam on a Winkler foundation

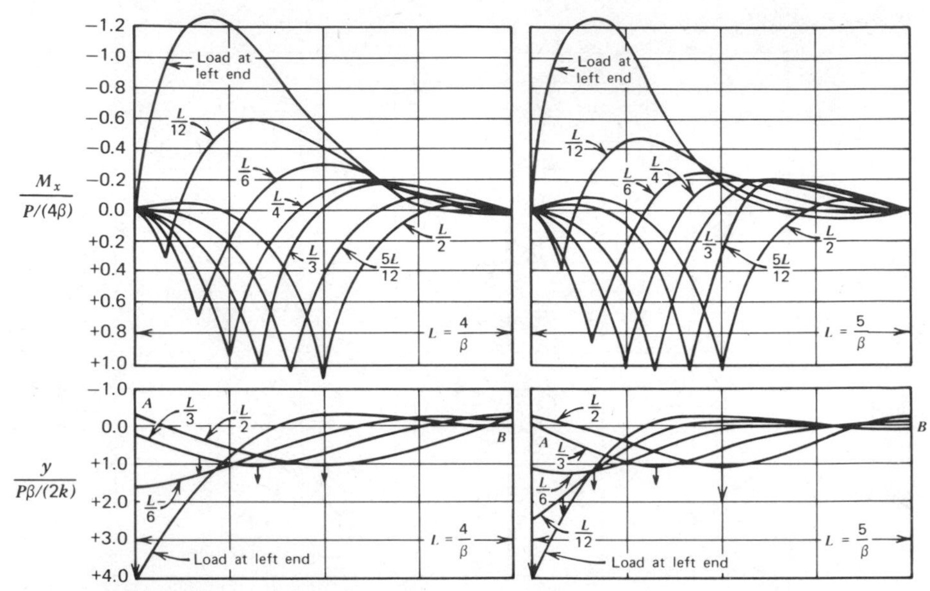

Fig. 9-6.1b/Bending moment diagrams and deflection curves for a short beam on elastic supports subjected to concentrated load located as shown on each curve. The ends of the beams are unrestrained (free).

can also be assessed by using proper values of the elastic end restraints. The solution is in a form that can be coded easily into computer language.

PROBLEM SET 9-6

1. An S-310 × 52 steel I-beam ($E = 200$ GPa) has a length of $L = 3.00$ m, a depth of 305 mm, a flange width of 129 mm, and a moment of inertia $I_x = 95.3 \times 10^6$ mm^4. The beam rests on a hard rubber elastic foundation whose spring constant is $k_0 = 0.300$ N/mm^3. If the beam is subjected to a concentrated lateral load $P = 270$ kN at its center, determine the maximum deflection and the maximum flexure stress in the beam.

2. The magnitude of βL for the beam in Problem 1 is 2.532. Determine the maximum deflection and the maximum flexure stress in the beam if the load is moved to one end of the beam. Use linear interpolation with the curves in Fig. 9-6.1.

   Ans. $y_{max} = 12.37$ mm; $\sigma_{max} = 147.2$ MPa

9-7
THIN-WALL CIRCULAR CYLINDERS

The concept of a beam on an elastic foundation may be used to approximate the response of thin-wall circular cylinders subjected to loads that are rotationally symmetrical (Fig. 9-7.1). We use cylindrical coordinates r, θ, z for radial, circumferential, and axial directions. The dimensions of a long thin-wall cylinder may be represented by the mean radius a and the wall thickness h. Let a long thin-wall cylinder be subjected to a ring load w having dimensions N/mm where the length dimension is measured in the circumferential direction. We show that the response of the cylinder is similar to that of a corresponding beam, on an elastic foundation, subjected to a concentrated load at its center (Art. 9-2).

In developing an analogy between a thin-wall circular cylinder and a beam on an elastic foundation, we specify the analogous beam and elastic foundation as follows: Cut a strip from the cylinder of width $a\,\Delta\theta$ (Fig. 9-7.2b). For convenience let the width $a\,\Delta\theta$ be unity (say, 1 mm). We consider this strip of length L and width $a\Delta\theta = 1$ as a beam. We consider the remainder of the cylinder to act as the elastic foundation. The spring constant k for the elastic foundation is obtained by imagining the open-ended cylinder to be subjected to an external pressure p_2. This pressure p_2 produces a uniaxial state of stress for which the only nonzero stress component is $\sigma_{\theta\theta} = ap_2/h$. Hence, by Hooke's law, the circumferential strain is $\epsilon_{\theta\theta} = \sigma_{\theta\theta}/E$. In turn, by strain-displacement relations, we can express $\epsilon_{\theta\theta}$ in terms of the radial displacement u as follows (see Eqs. 1-9.6):

$$u = a\epsilon_{\theta\theta} = \frac{a\sigma_{\theta\theta}}{E} = \frac{a^2 p_2}{Eh} \tag{9-7.1}$$

Since u is constant along the length of the cylinder, the magnitude of k is

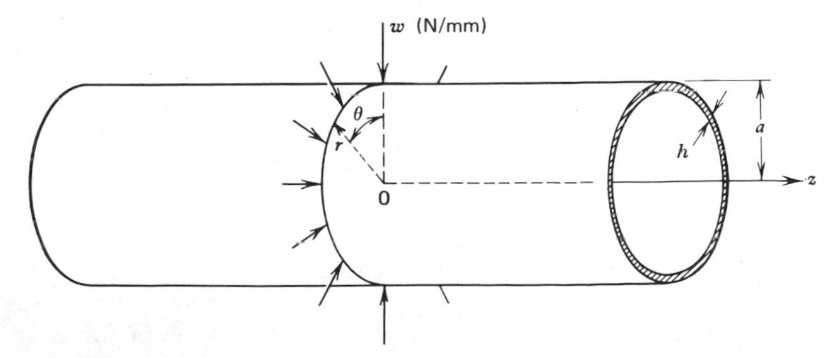

Fig. 9-7.1/Ring load on a thin-wall cylinder.

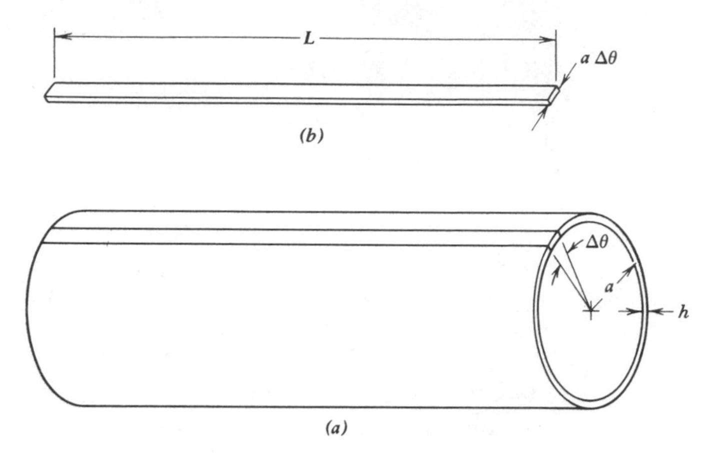

Fig. 9-7.2/Thin-wall cylinder.

given by Eq. (9-3.7) where u replaces y and $w = p_2(a\,\Delta\theta) = p_2$, since $a\,\Delta\theta = 1$. Hence, we have

$$k = \frac{w}{u} = \frac{Eh}{a^2} \tag{9-7.2}$$

Note that the narrow strip (Fig. 9-7.2*b*), which represents the beam on an elastic foundation, has a different state of stress (and strain) than other beams considered in this chapter. The beam in Fig. 9-1.1 was assumed to be free to deform in the x-direction, thus developing anticlastic curvature.[4] Each of the two sides of the beam in Fig. 9-7.2*b* lie in a radial plane of the cylinder; these sides are constrained to lie in the same planes after deformation. Therefore, since the beam cannot deform anticlastically, EI_x in Eq. (9-1.5) must be replaced by $D = Eh^3/[12(1-\nu^2)]$ (see Eq. 6 of Example 2-4.1). Replacing EI_x by D and using Eq. (9-7.2), we express β in the form

$$\beta = \sqrt[4]{\frac{3(1-\nu^2)}{a^2h^2}} \tag{9-7.3}$$

With the value of β given by Eq. (9-7.3), the solution for any of the loadings considered in Arts. 9-2 through 9-5 are applicable for thin-wall circular cylinders subjected to circumferential line loads. They may also be used to obtain estimates of the response of a thin-wall cylinder subjected to rotationally symmetric loads that vary along the axis of the cylinder.

Note: The analogous elastic foundation for the strip taken from a thin-wall circular cylinder is very stiff compared to the usual elastic foundation. Hence, the analogy is applicable even for a cylinder with length less than the radius a. If we assume that $\nu = 0.30$, the minimum length L for which the analogy is applicable is

$$L = \frac{3\pi}{2\beta} = 3.67a\sqrt{\frac{h}{a}} \qquad \left(L = 0.82a \text{ for } \frac{h}{a} = \frac{1}{20}\right) \tag{9-7.4}$$

Generally for thin-wall cylinders, h/a is less than $1/20$, and the length L of the cylinder influcnced by the concentrated ring load is less than $0.82a$. Often the beam analogy can be employed to obtain estimates of the response of noncylindrical circular shell segments (for instance, conical shells) if the change in radius for a given length L is small compared to the average radius a in the length L.

EXAMPLE 9-7.1

Stresses in Storage Tank

A closed end thin-wall cylinder is used as an oil storage tank which rests on one of its ends (see Fig. E9-7.1). The tank has a diameter of 30 m, a

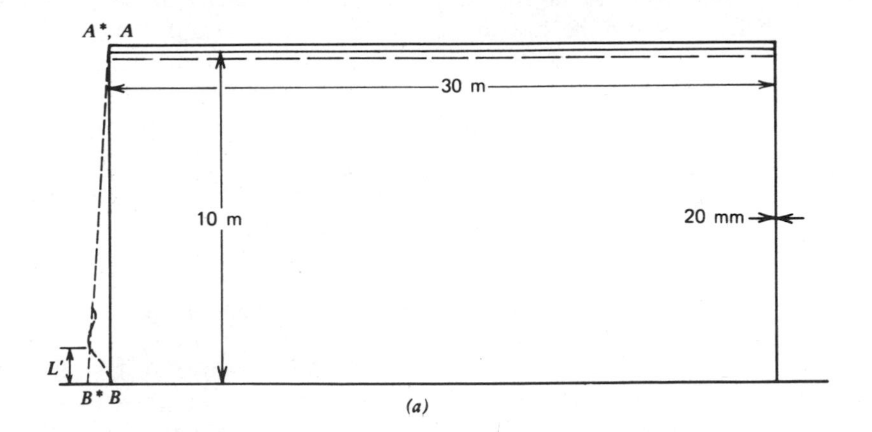

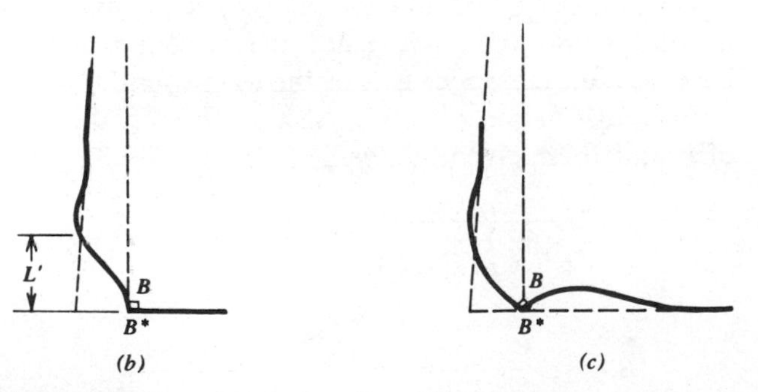

Fig. E9-7.1/(a) Thin-wall cylinder oil storage tank.

depth of 10 m, and a wall thickness of 20 mm. The tank is made of steel for which $E = 200$ GPa and $\nu = 0.29$. Determine the maximum shearing stress in the tank if it is filled with oil having a mass density of 900 kg/m^3 under the following different conditions: (a) Assume that the bottom of the tank does not influence the circumferential stress in the cylindrical walls. (b) Assume that the radial displacement of the junction between the cylinder and bottom remains zero during loading and that the bottom has infinite stiffness. (c) Assume that the radial displacement of the junction between the cylinder and bottom remains zero and that the bottom plate is sufficiently flexible that the moment at the junction can be considered to be zero.

SOLUTION

(a) Choose cylindrical coordinates r, θ, and z. The pressure in the cylinder increases linearly with depth. If the bottom does not exert moments or radial forces on the cylinder walls, wall AB in Fig. E9-7.1a deforms into the straight line A^*B^*. The stresses in the cylinder walls at B^* are σ_{rr}, $\sigma_{\theta\theta}$, and σ_{zz}. The radial stress σ_{rr} and axial stress σ_{zz} at the bottom are small and are neglected compared to the circumferential stress $\sigma_{\theta\theta}$. By the solution for thin-wall cylinders, we find

$$\sigma_{\theta\theta} = \frac{pa}{h} = \frac{(10 \times 10^3)(9.807)(900 \times 10^{-9})(15 \times 10^3)}{20} = 66.20 \text{ MPa}$$

The maximum shearing stress is given by Eq. (1-4.27)

$$\tau_{\max} = \frac{\sigma_{\max} - \sigma_{\min}}{2} = \frac{\sigma_{\theta\theta}}{2} = 33.10 \text{ MPa}$$

(b) In part b the bottom of the tank is assumed to have infinite stiffness. As indicated in Fig. E9-7.1b, the bottom prevents both a radial displacement and a change in slope of the cylinder wall at B. Although the cylinder is not uniformly loaded, we consider it to be a uniformly loaded long cylinder with a ring load w applied at its center. The cylinder center is taken as the junction between the cylinder and the bottom, and it is cut at this line load. Hence, the bottom of the tank produces a ring load $w/2$ on the upper half of the long cylinder and a bending moment to prevent rotation of the cut section. Hence, the associated magnitudes of k and β are given by Eqs. (9-7.2) and (9-7.3). Thus, we find

$$k = \frac{Eh}{a^2} = \frac{(200 \times 10^3)(20)}{(15 \times 10^3)^2} = 0.0178 \text{ N/mm}^2$$

$$\beta = \sqrt[4]{\frac{3(1 - \nu^2)}{h^2 a^2}} = \sqrt[4]{\frac{3\left[1 - (0.29)^2\right]}{(20)^2 (15 \times 10^3)^2}} = 0.00235 \text{ mm}^{-1}$$

Since the cylinder is subject to internal pressure due to the oil, it is not uniformly loaded as assumed in the proposed solution. For the analogy to be valid the minimum uniformly loaded length (Eq. 9-7.4) needs to be

$$L' = \frac{L}{2} = \frac{3\pi}{4\beta} = \frac{3\pi}{4(0.00235)} = 1003 \text{ mm}$$

which corresponds to the distance L' in Figs. E9-7.1*a* and *b*. Thus, only 10 percent of the cylinder height needs to be uniformly loaded; the variation of pressure over this height is considered small enough to be neglected. The radial displacement u of the walls of the cylinder away from end effects is given by Eq. (9-7.1)

$$u = \frac{\sigma_{\theta\theta} a}{E} = \frac{66.20(15 \times 10^3)}{200 \times 10^3} = 4.965 \text{ mm}$$

Since the radial displacement of the bottom plate of the tank is assumed to be zero, the ring load w causes a radial displacement inward of 4.965 mm. The magnitude of w is obtained by substituting the known value of u (equal y) into Eq. (9-2.5) for $\beta z = 0$. Hence, by

$$u = \frac{w\beta}{2k} = \frac{w(0.00235)}{2(0.0178)} = 4.965 \text{ mm}$$

we find

$$w = 75.21 \text{ N/mm}$$

The maximum bending moment is given by Eq. (9-2.7) for $\beta z = 0$. Thus, we obtain

$$M_{\max} = \frac{w}{4\beta} = \frac{75.21}{4(0.00235)} = 8001 \text{ N} \cdot \text{mm}$$

and hence,

$$\sigma_{zz(\max)} = \frac{M_{\max} c}{I} = \frac{M_{\max}(h/2)}{h^3/12} = \frac{8001(6)}{(20)^2} = 120.0 \text{ MPa}$$

The radial stress σ_{rr} is small and is neglected. Since the radial displacement of the cylindrical wall at the bottom is the same as for the unloaded cylinder, the average value of $\sigma_{\theta\theta}$ through the wall thickness is zero. However, due to bending, the ratio of $\sigma_{\theta\theta}$ to σ_{zz} is proportional to Poisson's ratio (see Eq. 1 of Example 2-4.1). Therefore,

$$\sigma_{\theta\theta(\max)} = \nu\sigma_{zz(\max)} = 0.29(120.0) = 34.8 \text{ MPa}$$

The maximum shearing stress at the junction between the bottom of the tank and the cylindrical walls of the tank is

$$\tau_{\max} = \frac{\sigma_{\max} - \sigma_{\min}}{2} = \frac{120.0}{2} = 60.0 \text{ MPa}$$

which is 81 percent greater than for part (a).

The radial displacement u for the junction between the bottom of the tank and the cylindrical walls of the tank have been neglected. However, its magnitude may be computed by the following relation:

$$u_{\text{bottom}} = \frac{w(1-\nu)a}{2Eh} = \frac{75.21(0.71)(15 \times 10^3)}{2(200 \times 10^3)(20)} = 0.100 \text{ mm}$$

This value is only 2 percent of the displacement of the unrestrained cylinder wall.

(c) If the bending moment at the junction of the cylindrical walls and the tank bottom is zero, the thin-walled cylinder can be treated as a beam on an elastic foundation loaded at one end. The bottom of the tank is assumed to prevent a radial displacement as indicated in Fig. E9-7.1*c*. Let w be the ring load produced by the bottom of the tank. The radial displacement u is given by Eq. (9-4.4) for $\beta z = 0$.

$$u = \frac{2w\beta}{k} = \frac{2w(0.00235)}{0.0178} = 4.965 \text{ mm}$$

$$w = 18.80 \text{ N/mm}$$

The maximum moment occurs at a distance $\pi/(4\beta) = 334$ mm from the bottom and has a magnitude given by Eq. (9-4.6)

$$M_{\max} = -\frac{w}{\beta} B_{\beta z} = -\frac{18.80(0.3224)}{0.00235} = -2579 \text{ N} \cdot \text{mm}$$

$$\sigma_{zz(\max)} = \left| \frac{M_{\max} c}{I} \right| = \left| -\frac{2579(6)}{20^2} \right| = |-38.69| \text{ MPa}$$

This bending stress causes a circumferential stress $\sigma_{\theta\theta 1}$, which is part of the resultant circumferential stress.

$$\sigma_{\theta\theta 1} = \nu \sigma_{zz(\max)} = 0.29(-38.69) = -11.22 \text{ MPa}$$

Another part of the circumferential stress $\sigma_{\theta\theta 2}$ comes from the fact that the maximum bending stress occurs at a location ($\beta z = \pi/4$) where the displacement is not maximum. The radial displacement given by Eq. (9-4.4) is

$$u = \frac{2w\beta}{k} D_{\beta z} = 0.3224 u_{\max}$$

Since $\sigma_{\theta\theta(\max)} = 66.20$ MPa is the uniform circumferential stress in the thin-walled cylinder when $u = 0$, the average circumferential stress for $u = 0.3224 u_{\max}$ is

$$\sigma_{\theta\theta 2} = (1 - 0.3224)\sigma_{\theta\theta(\max)} = 0.6776(66.10) = 44.86 \text{ MPa}$$

The circumferential stress at the point where $\sigma_{zz(\min)}$ occurs is

$$\sigma_{\theta\theta} = \sigma_{\theta\theta 1} + \sigma_{\theta\theta 2} = -11.22 + 44.86 = 33.64 \text{ MPa}$$

and

$$\tau_{max} = \frac{\sigma_{max} - \sigma_{min}}{2} = \frac{33.64 - (-38.69)}{2} = 36.17 \text{ MPa}$$

which is 9 percent greater than for part (a).

If the maximum shearing stress theory of failure is used, the maximum shearing stress indicates the severity of the loading conditions. If the bottom of the tank is rigid (one limiting condition), the maximum shearing stress is 81 percent greater than that for unrestrained cylindrical walls. If the bottom does not offer any resistance to bending (a second limiting condition), the shearing stress is 9 percent greater than that for unrestrained cylindrical walls. The actual condition of loading for most flat bottom tanks would be between the two limiting conditions but nearer to the condition of a rigid bottom.

PROBLEM SET 9-7

1. A steel ($E = 200$ GPa and $\nu = 0.29$) thin-wall cylinder has an inside diameter of 40 mm and a wall thickness of 1 mm. The cylinder may be considered fixed where it enters the stiffened end of a pressure vessel. The residual stress of installation may be considered negligible. Determine the bending stresses resulting from an internal pressure of 3 MPa.

2. A thin-wall cylinder is made of an aluminum alloy ($E = 72.0$ MPa and $\nu = 0.33$), has an outside diameter of 1 m, and a wall thickness of 5 mm. A split ring with square cross section 20 mm on a side is tightened on the cylinder until the stress in the split ring is 100 MPa. Assume that the split ring applies two line loads separated by the 20 mm dimension of the ring. Determine the principal stresses at the inner radius of the cylinder below the center line of the split ring.

   Ans. $\sigma_{zz} = 103.0$ MPa; $\sigma_{\theta\theta} = -62.1$ MPa

3. Let the split ring in Problem 2 be rounded on the inside surface so as to apply a line load at the center of the ring. Determine the maximum principal stresses at the inner radius of the cylinder.

4. A closed end steel cylinder ($E = 200$ GPa and $\nu = 0.29$) has an inside radius $a = 2.00$ m, a wall thickness $h = 10$ mm, and hemispherical ends. Since the state of stress is different for cylinder and hemisphere, their radial displacements will be different. Show that the length $L/2$ is small compared to a so that the short length of the hemisphere can be considered as another cylinder. Determine the shear force w in terms of internal pressure p_1 at the junction of

the cylinder and hemisphere (assumed to be another cylinder). Note that the bending moment at the junction is zero because of symmetry. Determine the maximum bending stress $\sigma_{zz(\text{bending})}$ in the cylinder, the axial stress σ_{zz}, and the circumferential stress $\sigma_{\theta\theta}$ at the outside of the cylinder at that location, and the ratio of the maximum shearing stress at that location to the maximum shearing stress in the cylinder.

Ans. $w = 13.73p_1$; $\sigma_{zz(\text{bending})} = 29.17p_1$; $\sigma_{zz} = 129.2p_1$; $\sigma_{\theta\theta} = 174.6p_1$; ratio $= 0.874$

REFERENCES

1. M. Hetényi, *Beams on Elastic Foundation*, University of Michigan Press, Ann Arbor, Michigan, 1946.
2. K. T. Sundara Raja Iyengar and S. Anantha Ramu, *Design Tables for Beams on Elastic Foundations and Related Problems*, Applied Science Publishers, London, 1979.
3. Bing-Yuan Ting, "Finite Beams on Elastic Foundation with Restraints," *Journal of the Structural Division*, Proc. of the American Society of Civil Engineers, Vol. 108, No. ST 3, March 1982, pp. 611–621.
4. A. P. Boresi and P. P. Lynn, *Elasticity in Engineering Mechanics*, Prentice-Hall, Englewood Cliffs, New Jersey, 1974, Art. 4-13.

Additional References

1. K. T. Sundara Raja Iyengar and S. Anantha Ramu, "Analysis of Finite Beams on Elastic Foundation," *Journal of Institution of Engineers, India*, Vol. 44, 11, PtC16, July 1964.
2. D. Young and R. P. Felgar, *Tables of Characteristic Functions Representing the Normal Modes of Transverse Vibrations of a Beam*, University of Texas Publication, No. 4913, July 1949.
3. K. T. Sundara Raja Iyengar and S. Anantha Ramu, "Influence Lines for Beams on Elastic Foundations," *Journal of the Structural Division, Proc. of the American Society of Civil Engineers*, Vol. 91, ST3, June 1965.
4. S. Timoshenko and S. Woinowsky-Krieger, *Theory of Plates and Shells*, 2nd Ed., McGraw-Hill, New York, 1960, p. 259.

CHAPTER 10

FLAT PLATES

In this chapter, we present the theory of elastic plates and a number of examples. We develop the basic equations (Arts. 10-1 through 10-7), and we outline methods of solution for rectangular plates (Art. 10-8) and circular plates (Art. 10-9) subjected to simple loading.

10-1 INTRODUCTION

A flat plate is a structural element or member whose middle surface lies in a flat plane. The dimension of a flat plate in a direction normal to the plane of its middle surface is called the *thickness* of the plate. A plate is characterized by the fact that its lateral dimension, the thickness, is relatively small compared to the plate dimensions in the plane of the middle surface. As a consequence, the bending behavior of a plate depends strongly on the plate thickness, as compared to the in-plane dimensions of the plate.

Plates may be classified according to the magnitude of the thickness compared to the magnitude of its other dimensions and according to the magnitude of the lateral deflection compared to the thickness. Thus, we may speak of (1) relatively thick plates with small deflections, (2) relatively thin plates with small deflections, (3) very thin plates with large deflections, (4) extremely thin plates (membranes) that may undergo either large or small deflections, and so on. There are no sharp lines of distinction between these classifications; rather there are gradual transition regions between two categories, in which the response of the plate exhibits some of the characteristics of both categories.

Additional descriptions are applied to describe plates. For example, if the distance (thickness) between the two surfaces (faces) that bound the thickness is constant, the plate is said to be of constant thickness; if not, it is said to be of variable thickness. The faces of the plate are taken to be equidistant from the middle surface of the plate. Further descriptions of a plate pertain, as we shall see, to the manner in which the plate edges are constrained and to the manner in which the plate material responds to load.

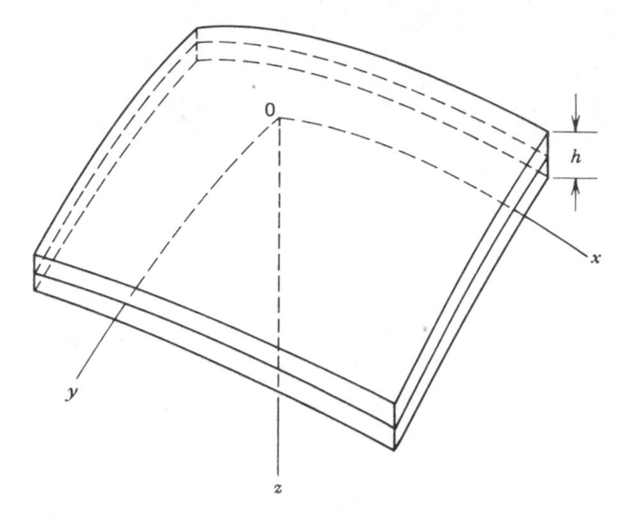

Fig. 10-1.1

Because thermal stresses are of major importance in aircraft and in nuclear power systems, the effects of temperature are included in the stress-strain relations. Some attention is given to nonisotropic material behavior. However, the treatment presented here is largely a study of the small-deflection theory of thin constant thickness isotropic elastic plates with temperature effects included. In the general development of the theory, (x, y, z)-coordinates are taken to be *orthogonal curvilinear plate coordinates*, where (x, y) are orthogonal curvilinear coordinates that lie in the middle surface of the plate and the z-coordinate is perpendicular to the middle surface of the plate (Fig. 10-1.1).

In the development of the theory, kinematic relations and associated strain-displacement relations are presented. Stress resultants for a plate are defined and the equations of equilibrium are derived by employing the principle of virtual work. Temperature effects are included in the elastic stress-strain relations. In turn, the stress resultant-displacement relations are derived including temperature effects. The boundary conditions for the plate follow directly from the principle of virtual work (Art. 10-7).

10-2 STRESS RESULTANTS IN A FLAT PLATE

The concepts of stress and stress notation have been introduced in Chapter 1. Although the major results were developed for rectangular coordinates, results were presented for orthogonal space coordinates

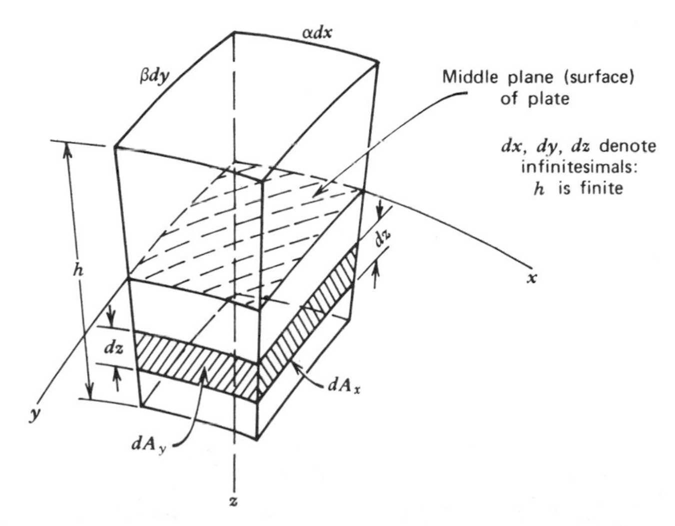

Fig. 10-2.1

(x, y, z) (see Eqs. 1-5.2 and 1-5.3). In particular, we recall that σ_{xx} denotes the tensile stress on a plane element that is normal to an x-coordinate line and $(\sigma_{xy}, \sigma_{xz})$ denote (y, z)-components, respectively, of the shearing stress that acts on a plane element normal to the x-coordinate line. Similar interpretations apply for σ_{yy}, σ_{zz}, and σ_{yz}. As for rectangular coordinates, $\sigma_{yz} = \sigma_{zy}$, $\sigma_{xz} = \sigma_{zx}$, and $\sigma_{xy} = \sigma_{yx}$ for orthogonal curvilinear coordinates (x, y, z). For nonorthogonal curvilinear coordinates the symmetry of shears does not hold.[1]

It is convenient to introduce special notations for in-plane forces (tractions), bending moments, twisting moments, and shears in a plate. Thus, with respect to orthogonal curvilinear coordinates (x, y, z), consider a differential element of the plate cut out by the surfaces $x =$ constant and $y =$ constant (Fig. 10-2.1), where (x, y) are orthogonal curvilinear coordinates in the middle plane of the plate and coordinate z is the straight line coordinate perpendicular to the middle plane. The elements of area of these cross sections of a flat plate are

$$dA_y = \alpha\, dx\, dz \qquad dA_x = \beta\, dy\, dz \tag{10-2.1}$$

where

$$\alpha = \alpha(x, y) \qquad \beta = \beta(x, y) \qquad \gamma = 1$$

(See Eq. 1-5.3.)

Let N_{xx} denote the tensile force on a cross sectional face of the element ($x =$ constant), *per unit length of the y-coordinate line* on the middle surface (Fig. 10-2.2). Then, the total tensile force on the *differential* element in the x-direction is $N_{xx}\beta\, dy$. Hence, since $dA_x = \beta\, dy\, dz$, we

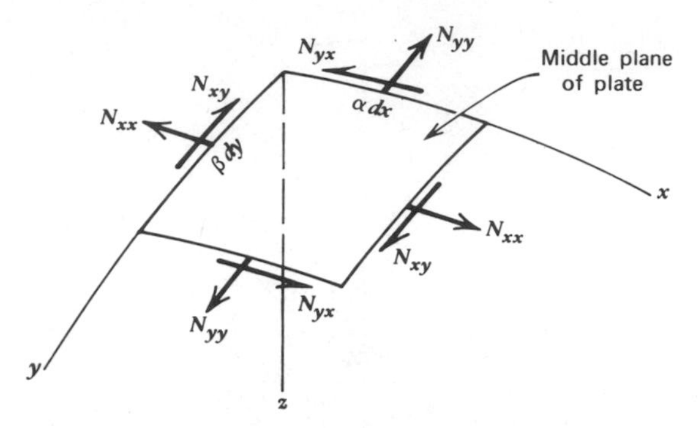

Fig. 10-2.2/Resultant tractions on a reference surface.

have

$$N_{xx}\beta\,dy = dy \int_{-h/2}^{h/2} \beta\sigma_{xx}\,dz$$

where the middle surface has been taken as the reference surface. More generally, the reference surface ($z = 0$) may be taken as any plane (for example, the upper face of the plate). Then the integral in the above equation is determined by the thickness h and the location of the reference surface. Hence, we have, since $\beta = \beta(x, y)$

$$N_{xx} = \int_{-h/2}^{h/2} \sigma_{xx}\,dz \tag{a}$$

In a similar manner, the tractive force N_{yy} *per unit length of the x-coordinate line* on the middle surface (Fig. 10-2.2) is

$$N_{yy}\alpha\,dx = dx \int_{-h/2}^{h/2} \alpha\sigma_{yy}\,dz$$

or

$$N_{yy} = \int_{-h/2}^{h/2} \sigma_{yy}\,dz \tag{b}$$

Likewise, the shearing force N_{xy} *per unit length of the y-coordinate line* is given by

$$N_{xy}\beta\,dy = dy \int_{-h/2}^{h/2} \beta\sigma_{xy}\,dz$$

or

$$N_{xy} = \int_{-h/2}^{h/2} \sigma_{xy}\,dz \tag{c}$$

and for N_{yx}, the shearing force *per unit length of a x-coordinate line*,

$$N_{yx} = \int_{-h/2}^{h/2} \sigma_{xy}\,dz = N_{xy} \tag{d}$$

We let (Q_x, Q_y) be the transverse shears per unit length of a y-coordinate line and x-coordinate line, respectively. Hence, for the transverse shearing forces (Q_x, Q_y) *per unit length of a coordinate line*, we find (Fig. 10-2.3) that

$$Q_x = \int_{-h/2}^{h/2} \sigma_{xz}\, dz$$
$$Q_y = \int_{-h/2}^{h/2} \sigma_{yz}\, dz \qquad \text{(e)}$$

We let M_{xx} be the bending moment *per unit length of a y-coordinate line*. Then by Fig. 10-2.3, we obtain, with positive directions indicated by the right-hand rule for moments (arrows),

$$M_{xx} = \int_{-h/2}^{h/2} z\sigma_{xx}\, dz \qquad \text{(f)}$$

For the twisting moment M_{xy} *per unit length of a y-coordinate line*, we find

$$M_{xy} = \int_{-h/2}^{h/2} z\sigma_{xy}\, dz \qquad \text{(g)}$$

Similarly for bending moment and twisting moment *per unit length of a x-coordinate line*,

$$M_{yy} = \int_{-h/2}^{h/2} z\sigma_{yy}\, dz$$
$$M_{yx} = \int_{-h/2}^{h/2} z\sigma_{xy}\, dz = M_{xy} \qquad \text{(h)}$$

In summary, we have the tractions $(N_{xx}, N_{yy}, N_{xy} = N_{yx})$, the transverse shears (Q_x, Q_y), the bending moments (M_{xx}, M_{yy}), and the twisting

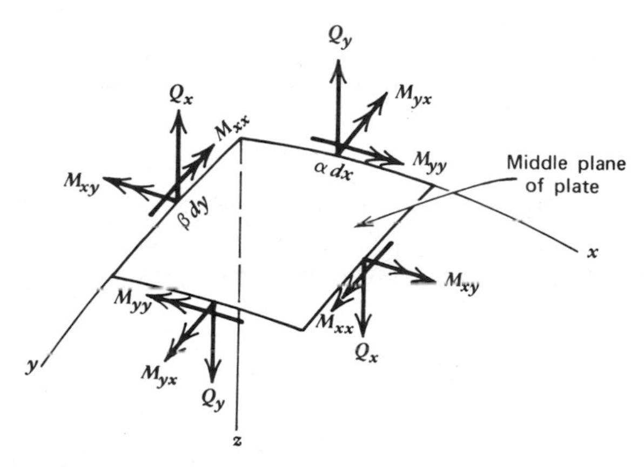

*10-2.3/*Resultant moments and shears on a reference surface.

moments ($M_{xy}=M_{yx}$) in the form (for the reference surface, $z=0$, coincident with the plate middle surface)

$$
\begin{aligned}
N_{xx} &= \int_{-h/2}^{h/2} \sigma_{xx}\, dz \\
N_{yy} &= \int_{-h/2}^{h/2} \sigma_{yy}\, dz \\
N_{xy} = N_{yx} &= \int_{-h/2}^{h/2} \sigma_{xy}\, dz \\
Q_{x} &= \int_{-h/2}^{h/2} \sigma_{xz}\, dz \\
Q_{y} &= \int_{-h/2}^{h/2} \sigma_{yz}\, dz \\
M_{xx} &= \int_{-h/2}^{h/2} z\sigma_{xx}\, dz \\
M_{yy} &= \int_{-h/2}^{h/2} z\sigma_{yy}\, dz \\
M_{xy} = M_{yx} &= \int_{-h/2}^{h/2} z\sigma_{xy}\, dz
\end{aligned}
\tag{10-2.2}
$$

The positive senses of forces and moments are shown in Figs. (10-2.2) and (10-2.3). However, there is no universal agreement between authors on the sign conventions for the shears (Q_x, Q_y) and twisting moments ($M_{xy}=M_{yx}$).*

10-3 KINEMATICS: STRAIN-DISPLACEMENT RELATIONS FOR PLATES

In this article, we let (U, V, W) be the components of the displacement vector, of a point P in the plate, on tangents to the local coordinate lines at P (Fig. 10-3.1). In the following development, the notation (u, v, w) is reserved for the displacement components of the corresponding point P' on the middle surface of the plate (Fig. 10-3.1). Then, by Eqs. (1-9.5), for orthogonal coordinates [for a flat plate $\alpha=\alpha(x, y)$, $\beta=\beta(x, y)$, $\gamma=1$;

*Here, we follow the convention employed by H. L. Langhaar, reference 2. See also reference 3.

see Eq. 1-5.3], we have the small-displacement strain-displacement relations for point P

$$
\begin{aligned}
\epsilon_{xx} &= \frac{1}{\alpha}\left(U_x + \frac{\alpha_y V}{\beta}\right) \\
\epsilon_{yy} &= \frac{1}{\beta}\left(V_y + \frac{\beta_x U}{\alpha}\right) \\
\epsilon_{zz} &= W_z \\
2\epsilon_{xy} &= \frac{U_y}{\beta} + \frac{V_x}{\alpha} \\
2\epsilon_{xz} &= U_z + \frac{W_x}{\alpha} \\
2\epsilon_{yz} &= V_z + \frac{W_y}{\beta}
\end{aligned}
\tag{10-3.1}
$$

where the (x, y, z) subscripts on U, V, W, α, and β denote partial differentiation. Equations (10-3.1) are exact linear strain-displacement relations for the three-dimensional kinematical problem of the plate. However, the purpose of plate theory is to reduce the three-dimensional problem to a more tractable two-dimensional problem. The approximation that is usually used to achieve this reduction is due to Kirchhoff; namely, it is assumed that the straight-line normals to the undeformed middle plane (reference plane) of the plate remain approximately straight and inextensional under the deformation of the plate. In addition, it is

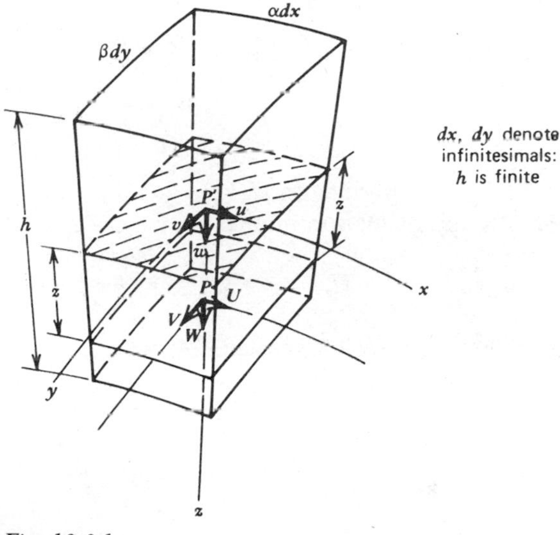

Fig. 10-3.1

assumed that the normals in the deformed plate remain normal to the deformed middle plane. Since under this assumption line elements normal to the middle surface do not extend and no angular distortion occurs between normals and the reference surface, it follows that the Kirchhoff approximation is equivalent to assuming that $\epsilon_{zz} = \epsilon_{xz} = \epsilon_{yz} = 0$. The Kirchhoff assumption is not limited to problems of small displacements, since it is purely kinematical in form. It does not depend upon material properties. Hence, it may be employed in plasticity studies of plates, etc. However, since ϵ_{xz} and ϵ_{yz} are discarded, for an isotropic material, it implies that σ_{xz}, σ_{yz} are zero and, hence, the shears Q_x and Q_y are implied to be zero (Eq. 10-2.2). If values of (Q_x, Q_y) are needed, they are reintroduced into the theory through the equations of equilibrium (Art. 10-4). However, some inconsistencies are inevitable when the Kirchhoff approximation is employed in plates. It is nevertheless more accurate than the membrane theory (approximation) of plates, which requires not only (Q_x, Q_y) to be zero, but also that $M_{xx}, M_{yy}, M_{xy} = M_{yx}$ vanish.

The Kirchhoff approximation implies that U, V, W are linear functions of z, irrespective of the magnitude of the displacement. General expressions for (U, V, W) in terms of the displacement components (u, v, w) of the middle surface are very complicated for large displacements. The resulting nonlinear relations for the strains $(\epsilon_{xx}, \epsilon_{yy}, \ldots \epsilon_{yz})$ in terms of (u, v, w) are even more complicated. However, it is feasible to derive general strain equations in terms of (u, v, w) for any plate, if we employ the small-displacement strain relations (Eqs. 10-3.1).

Since, by the Kirchhoff approximation, z is not changed by the deformation (Fig. 10-3.1), the difference W-w is a small quantity of second order. Therefore, we make the approximation $W = w$. Then by the last two of Eqs. (10-3.1), we obtain

$$2\epsilon_{xz} = \frac{w_x}{\alpha} + U_z$$
$$2\epsilon_{yz} = \frac{w_y}{\beta} + V_z \tag{10-3.2}$$

where (x, y, z) subscripts on U, V, w denote partial differentiation. Since Kirchhoff's approximation implies that $\epsilon_{xz} = \epsilon_{yz} = 0$, Eqs. (10-3.2) yield

$$U_z + \frac{w_x}{\alpha} = 0 \qquad V_z + \frac{w_y}{\beta} = 0 \tag{10-3.3}$$

Integrations of Eqs. (10-3.3) yields

$$U = -\frac{w_x}{\alpha} z + f(x, y)$$
$$V = -\frac{w_y}{\beta} z + g(x, y) \tag{10-3.4}$$

The additive functions $f(x, y)$ and $g(x, y)$ are determined by the conditions $U=u$, $V=v$ for $z=0$. Hence, by Eqs. (10-3.4),

$$\begin{aligned} U &= u - z\frac{w_x}{\alpha} \\ V &= v - z\frac{w_y}{\beta} \\ W &= w \end{aligned} \tag{10-3.5}$$

where (u, v, w) are functions of (x, y) only. Equation (10-3.5) determines how U, V, W vary through thickness of the plate in accord with Kirchhoff's approximation and small-displacement theory. Substitution of Eqs. (10-3.5) into Eqs. (10-3.1) yields

$$\begin{aligned} \epsilon_{xx} &= \frac{1}{\alpha}\frac{\partial}{\partial x}\left(\frac{\alpha u - zw_x}{\alpha}\right) + \frac{\alpha_y}{\alpha\beta}\left(\frac{\beta v - zw_y}{\beta}\right) \\ \epsilon_{yy} &= \frac{1}{\beta}\frac{\partial}{\partial y}\left(\frac{\beta v - zw_y}{\beta}\right) + \frac{\beta_x}{\alpha\beta}\left(\frac{\alpha u - zw_x}{\alpha}\right) \\ \epsilon_{zz} &= 0 \\ \gamma_{xy} &= 2\epsilon_{xy} = \frac{\beta}{\alpha}\frac{\partial}{\partial x}\left(\frac{\beta v - zw_y}{\beta^2}\right) + \frac{\alpha}{\beta}\frac{\partial}{\partial y}\left(\frac{\alpha u - zw_x}{\alpha^2}\right) \\ \gamma_{xz} &= \gamma_{yz} = 2\epsilon_{xz} = 2\epsilon_{yz} = 0 \end{aligned} \tag{10-3.6}$$

Alternatively, Eqs. (10-3.6) may be written by separating the z term as follows:

$$\begin{aligned} \epsilon_{xx} &= \frac{u_x}{\alpha} + \frac{\alpha_y v}{\alpha\beta} - \frac{z}{\alpha}\left[\frac{\partial}{\partial x}(w_x/\alpha) + \frac{\alpha_y w_y}{\beta^2}\right] \\ \epsilon_{yy} &= \frac{v_y}{\beta} + \frac{\beta_x u}{\alpha\beta} - \frac{z}{\beta}\left[\frac{\partial}{\partial y}(w_y/\beta) + \frac{\beta_x w_x}{\alpha^2}\right] \\ \gamma_{xy} &= 2\epsilon_{xy} = \frac{\beta}{\alpha}\frac{\partial}{\partial x}\left(\frac{v}{\beta}\right) + \frac{\alpha}{\beta}\frac{\partial}{\partial y}\left(\frac{u}{\alpha}\right) \\ &\quad - \frac{z}{\alpha}\left[\frac{\partial}{\partial x}\left(\frac{w_y}{\beta}\right) - \frac{\beta_x w_y}{\beta^2}\right] \\ &\quad - \frac{z}{\beta}\left[\frac{\partial}{\partial y}\left(\frac{w_x}{\alpha}\right) - \frac{\alpha_y w_x}{\alpha^2}\right] \\ \epsilon_{zz} &= \gamma_{xz} = \gamma_{yz} = 0 \end{aligned} \tag{10-3.7}$$

Equations (10-3.6) or (10-3.7) are approximations of Eqs. (10-3.1) that result from application of the Kirchhoff approximation. These approximations form the basis of classical small-displacement plate theory.

Rotation of a Plate Surface Element/To obtain boundary conditions at the junctions of two plates, it is sometimes necessary to compute the rotations of the plate (middle) surface at the junction. As noted in the theory of deformation,[1] the small-displacement rotation $\boldsymbol{\omega}$ of a volume element is a vector quantity given by the relation

$$\boldsymbol{\omega} = \tfrac{1}{2}\,\text{curl}\,\mathbf{q} \tag{10-3.8}$$

where $\mathbf{q} = (U, V, W)$ is the displacement vector, and where the operation curl $\mathbf{q}(= \nabla \times \mathbf{q})$ must be expressed in terms of the appropriate coordinate system (recall that here we employ orthogonal curvilinear coordinates). The expression for the curl in curvilinear orthogonal coordinates is[4]

$$\text{curl}\,\mathbf{q} = \nabla \times \mathbf{q} = \frac{1}{\alpha\beta\gamma}\begin{vmatrix} \alpha\mathbf{i} & \beta\mathbf{j} & \gamma\mathbf{k} \\ \dfrac{\partial}{\partial x} & \dfrac{\partial}{\partial y} & \dfrac{\partial}{\partial z} \\ \alpha U & \beta V & \gamma W \end{vmatrix} \tag{10-3.9}$$

where the displacement vector $\mathbf{q}$ is

$$\mathbf{q} = U\mathbf{i} + V\mathbf{j} + W\mathbf{k} \tag{10-3.10}$$

and $(\mathbf{i}, \mathbf{j}, \mathbf{k})$ denote unit vectors tangent to (x, y, z)-coordinate lines, respectively. With Eqs. (10-3.5) and (10-3.9), Eqs. (10-3.8) yield with $\gamma = 1$ and with $z = 0$ (after differentiation)

$$\omega_x = \frac{w_y}{\beta}, \qquad \omega_y = -\frac{w_x}{\alpha}, \qquad \omega_z = \frac{1}{2\alpha\beta}\left[\frac{\partial}{\partial x}(\beta v) - \frac{\partial}{\partial y}(\alpha u)\right] \tag{10-3.11}$$

where $(\omega_x, \omega_y, \omega_z)$ are the components of $\boldsymbol{\omega}$ along tangents to the (x, y, z)-coordinate lines, respectively.

In terms of (ω_x, ω_y), we may rewrite Eqs. (10-3.6) and (10-3.7) in the forms

$$\begin{aligned} \epsilon_{xx} &= \frac{1}{\alpha}\frac{\partial}{\partial x}(u + z\omega_y) + \frac{\alpha_y}{\alpha\beta}(v - z\omega_x) \\ \epsilon_{yy} &= \frac{1}{\beta}\frac{\partial}{\partial y}(v - z\omega_x) + \frac{\beta_x}{\alpha\beta}(u + z\omega_y) \\ 2\epsilon_{xy} = \gamma_{xy} &= \frac{\beta}{\alpha}\frac{\partial}{\partial x}\left(\frac{v - z\omega_x}{\beta}\right) + \frac{\alpha}{\beta}\frac{\partial}{\partial y}\left(\frac{u + z\omega_y}{\alpha}\right) \end{aligned} \tag{10-3.12}$$

and

$$\begin{aligned} \epsilon_{xx} &= \frac{u_x}{\alpha} + \frac{\alpha_y v}{\alpha\beta} - \frac{z}{\alpha}\left[-\frac{\partial\omega_y}{\partial x} + \frac{\alpha_y}{\beta}\omega_x\right] \\ \epsilon_{yy} &= \frac{v_y}{\beta} + \frac{\beta_x u}{\alpha\beta} - \frac{z}{\beta}\left[\frac{\partial\omega_x}{\partial y} - \frac{\beta_x\omega_y}{\alpha}\right] \\ \gamma_{xy} = 2\epsilon_{xy} &= \frac{\beta}{\alpha}\frac{\partial}{\partial x}\left(\frac{v}{\beta}\right) + \frac{\alpha}{\beta}\frac{\partial}{\partial y}\left(\frac{u}{\alpha}\right) - \frac{z}{\alpha}\left[\frac{\partial\omega_x}{\partial x} - \frac{\beta_x\omega_x}{\beta}\right] \\ &\quad - \frac{z}{\beta}\left[-\frac{\partial\omega_y}{\partial y} + \frac{\alpha_y}{\alpha}\omega_y\right] \end{aligned} \tag{10-3.13}$$

For rectangular coordinates, $\alpha = \beta = 1$ and $\alpha_x = \alpha_y = \beta_x = \beta_y = 0$. Then Eqs. (10-3.12) and (10-3.13) reduce to

$$\epsilon_{xx} = \frac{\partial}{\partial x}(u + z\omega_y) \qquad \epsilon_{yy} = \frac{\partial}{\partial y}(v - z\omega_x)$$
$$2\epsilon_{xy} = \gamma_{xy} = \frac{\partial}{\partial x}(v - z\omega_x) + \frac{\partial}{\partial y}(u + z\omega_y) \tag{10-3.14}$$

and

$$\epsilon_{xx} = u_x + z\frac{\partial \omega_y}{\partial x} \qquad \epsilon_{yy} = v_y - z\frac{\partial \omega_x}{\partial y}$$
$$2\epsilon_{xy} = \gamma_{xy} = \frac{\partial v}{\partial x} + \frac{\partial u}{\partial y} - z\left(\frac{\partial \omega_x}{\partial x} - \frac{\partial \omega_y}{\partial y}\right) \tag{10-3.15}$$

where

$$\omega_x = w_y \qquad \omega_y = -w_x \tag{10-3.16}$$

Alternatively in terms of (u, v, w), we may write

$$\epsilon_{xx} = u_x - zw_{xx} \qquad \epsilon_{yy} = v_y - zw_{yy}$$
$$2\epsilon_{xy} = \gamma_{xy} = v_x + u_y - 2zw_{xy} \tag{10-3.17}$$

where we recall that (x, y)-subscripts on (u, v, w) denote partial differentiation.

The strain-displacement relations derived above are employed in the classical small-displacement theory of plates. For an alternative derivation of these relations see Marguerre and Woernle.[3]

10-4 EQUILIBRIUM EQUATIONS FOR SMALL-DISPLACEMENT THEORY OF FLAT PLATES

The equations of equilibrium for a plate may be derived by several methods. For example, they may be derived (1) by considering the equilibrium requirements for an infinitesimal plate element (dx, dy, dz) (Fig. 10-2.1), (2) by integrating the pointwise equilibrium equations (Eqs. 1-5.1 or 1-5.2) through the plate thickness and employing the definitions of Eqs. (10-2.2), or (3) by a direct application of the principle of virtual work. In the following derivation, we employ Method 2. Similar results have been obtained by Marguerre and Woernle by Methods 1 and 3 for rectangular coordinates.[3]

We consider an element of the plate generated by all normals erected on an element $dx\,dy$ of the middle surface. This element may be subjected to external forces caused by gravity and by external shears and pressures applied to the faces of the plate. Since the area of the element $dx\,dy$ of the middle surface is $\alpha\beta\,dx\,dy$, the resultant external force on the element of the plate is denoted by $\mathbf{P}\ \alpha\beta\,dx\,dy$. The vector $\mathbf{P}$ is the resultant force per unit area of the middle surface. It is a function of the coordinates (x, y) of the middle surface. The vector $\mathbf{P}$ is considered to act at the middle surface of the plate, and it is resolved into components P_x, P_y, P_z along (x, y, z)-coordinate lines, respectively. Often the component P_z is denoted by p or q, since usually it results from normal pressures on the faces of the plate. Since vector $\mathbf{P}$ is moved to the midsurface of the plate, in addition to the external force $\mathbf{P}\ \alpha\beta\,dx\,dy$, an external couple $\mathbf{R}\ \alpha\beta\,dx\,dy$ acts on the midsurface element of the plate. In addition, externally applied couples may act. However, we consider a couple that results only from shearing stresses on the external faces of the plate. Hence, relative to the midsurface $R_z = 0$, and

$$\begin{aligned} \alpha\beta R_x &= -\alpha\beta z\sigma_{yz}\big|_{-h/2}^{h/2} \\ \alpha\beta R_y &= \alpha\beta z\sigma_{xz}\big|_{-h/2}^{h/2} \end{aligned} \tag{10-4.1}$$

To employ Method 2, we require the pointwise equilibrium equations. Thus, for $\alpha = \alpha(x, y)$, $\beta = \beta(x, y)$, $\gamma = 1$, we obtain by Eqs. (1-5.2)

$$\begin{aligned} &\frac{\partial}{\partial x}(\beta\sigma_{xx}) + \frac{\partial}{\partial y}(\alpha\sigma_{xy}) + \frac{\partial}{\partial z}(\alpha\beta\sigma_{xz}) + \alpha_y\sigma_{xy} - \beta_x\sigma_{yy} + \alpha\beta B_x = 0 \\ &\frac{\partial}{\partial x}(\beta\sigma_{xy}) + \frac{\partial}{\partial y}(\alpha\sigma_{yy}) + \frac{\partial}{\partial z}(\alpha\beta\sigma_{yz}) + \beta_x\sigma_{xy} - \alpha_y\sigma_{xx} + \alpha\beta B_y = 0 \\ &\frac{\partial}{\partial x}(\beta\sigma_{xz}) + \frac{\partial}{\partial y}(\alpha\sigma_{yz}) + \frac{\partial}{\partial z}(\alpha\beta\sigma_{zz}) + \alpha\beta B_z = 0 \end{aligned} \tag{10-4.2}$$

The force equilibrium equations for N_{xx}, N_{yy}, N_{xy}, Q_x, and Q_y are obtained by integrating the differential equations of equilibrium (Eqs. 10-4.2) through the thickness h of the plate. For example, the first term in the first of Eqs. (10-4.2) is $\partial(\beta\sigma_{xx})/\partial x$. Integrating this term with respect to z between the limits $-h/2$ and $h/2$, and utilizing Eqs. (10-2.2), we obtain

$$\int_{-h/2}^{h/2} \frac{\partial}{\partial x}(\beta\sigma_{xx})\,dz = \frac{\partial}{\partial x}\int_{-h/2}^{h/2} \beta\sigma_{xx}\,dz = \frac{\partial}{\partial x}(\beta N_{xx})$$

The second term in Eq. (10-4.2) is integrated similarly. For the integral of the third term, we obtain

$$\int_{-h/2}^{h/2} \frac{\partial}{\partial z}(\alpha\beta\sigma_{xz})\,dz = \alpha\beta\sigma_{xz}\big|_{-h/2}^{h/2} = \alpha\beta P_x$$

The fourth integral obtained from Eq. (10-4.2) is

$$\int_{-h/2}^{h/2} \alpha_y\sigma_{xy}\,dz = \alpha_y N_{xy}$$

Similarly, the other terms can be integrated.

To obtain the moment equilibrium equations, we multiply Eqs. (10-4.2) by z and then integrate through the thickness and employ the definitions of Eq. (10-2.2).

The complete set of equilibrium equations obtained is thus

$$\frac{\partial}{\partial x}(\beta N_{xx}) + \frac{\partial}{\partial y}(\alpha N_{xy}) + \alpha_y N_{xy} - \beta_x N_{yy} + \alpha\beta P_x + \alpha\beta h B_x = 0$$

$$\frac{\partial}{\partial x}(\beta N_{xy}) + \frac{\partial}{\partial y}(\alpha N_{yy}) + \beta_x N_{xy} - \alpha_y N_{xx} + \alpha\beta P_y + \alpha\beta h B_y = 0$$

$$\frac{\partial}{\partial x}(\beta Q_x) + \frac{\partial}{\partial y}(\alpha Q_y) + \alpha\beta P_z + \alpha\beta h B_z = 0$$

$$\frac{\partial}{\partial x}(\beta M_{xx}) + \frac{\partial}{\partial y}(\alpha M_{xy}) + \alpha_y M_{xy} - \beta_x M_{yy} - \alpha\beta Q_x + \alpha\beta R_y = 0$$

$$\frac{\partial}{\partial x}(\beta M_{xy}) + \frac{\partial}{\partial y}(\alpha M_{yy}) + \beta_x M_{xy} - \alpha_y M_{xx} - \alpha\beta Q_y - \alpha\beta R_x = 0$$

$$N_{xy} = N_{yx} \tag{10-4.3}$$

For rectangular coordinates $\alpha = \beta = 1$. Then Eqs. (10-4.3) yield

$$\frac{\partial N_{xx}}{\partial x} + \frac{\partial N_{xy}}{\partial y} + P_x + hB_x = 0$$

$$\frac{\partial N_{xy}}{\partial x} + \frac{\partial N_{yy}}{\partial y} + P_y + hB_y = 0$$

$$\frac{\partial Q_x}{\partial x} + \frac{\partial Q_y}{\partial y} + P_z + hB_z = 0 \tag{10-4.4}$$

$$\frac{\partial M_{xx}}{\partial x} + \frac{\partial M_{xy}}{\partial y} - Q_x + R_y = 0$$

$$\frac{\partial M_{xy}}{\partial x} + \frac{\partial M_{yy}}{\partial y} - Q_y - R_x = 0$$

$$N_{xy} = N_{yx}$$

Equations (10-4.3) are exact relations, provided that (x, y, z) are orthogonal curvilinear plate coordinates for the deformed plate. They are approximations for the small-displacement theory of plates if (x, y, z) are orthogonal curvilinear plate coordinates in the undeformed plate, since Eqs. (10-4.2) are approximations for such axes.* Therefore, we shall use them as the equilibrium relations for the small-displacement theory of plates relative to orthogonal curvilinear plate axes in the undeformed plate.

*See Appendix 3C of reference 1.

The last of Eqs. (10-4.3) is an identity that follows from Eqs. (10-2.2). Often R_x and R_y are zero; in any case, they may usually be discarded from Eqs. (10-4.3). However, if they are retained, we obtain from the third, fourth, and fifth of Eqs. (10-4.3), by elimination of Q_x and Q_y,

$$\frac{\partial}{\partial x}\left\{\frac{1}{\alpha}\left[\frac{\partial}{\partial x}(\beta M_{xx}) + \frac{\partial}{\partial y}(\alpha M_{xy}) + \alpha_y M_{xy} - \beta_x M_{yy} + \alpha\beta R_y\right]\right\}$$
$$+ \frac{\partial}{\partial y}\left\{\frac{1}{\beta}\left[\frac{\partial}{\partial x}(\beta M_{xy}) + \frac{\partial}{\partial y}(\alpha M_{yy}) - \alpha_y M_{xx} + \beta_x M_{xy} - \alpha\beta R_x\right]\right\}$$
$$+ h\alpha\beta B_z + \alpha\beta P_z = 0 \quad (10\text{-}4.5)$$

Equation (10-4.5) is called the moment equilibrium equation of plates. For rectangular axes, $\alpha = \beta = 1$, and Eq. (10-4.5) reduces to (discarding R_x and R_y)

$$\frac{\partial^2 M_{xx}}{\partial x^2} + 2\frac{\partial^2 M_{xy}}{\partial x \partial y} + \frac{\partial^2 M_{yy}}{\partial y^2} + hB_z + P_z = 0 \quad (10\text{-}4.6)$$

10-5 STRESS-STRAIN-TEMPERATURE RELATIONS FOR ISOTROPIC ELASTIC PLATES

The preceding equations, derived in Arts. 10-2, 10-3, and 10-4, are independent of material properties. They are equally applicable to problems of elasticity, plasticity, and creep, irrespective of the effects of temperature.

In conventional plate theory, it is assumed that the plate is in a state of plane stress; that is, $\sigma_{xz} = \sigma_{yz} = \sigma_{zz} = 0$. For isotropic elastic planes, the relations $\sigma_{xz} = \sigma_{yz} = 0$ are consistent with the Kirchhoff approximation, which signifies that $\epsilon_{xz} = \epsilon_{yz} = 0$. However, the Kirchhoff approximation has been criticized since it includes the approximation $\epsilon_{zz} = 0$. The condition $\epsilon_{zz} = 0$ conflicts with the assumption that $\sigma_{zz} = 0$. The condition $\epsilon_{zz} = 0$ is incorrect; however, the strain ϵ_{zz} has little effect on the strains $\epsilon_{xx}, \epsilon_{yy}, \epsilon_{xy}$. Thus, the approximation $\epsilon_{zz} = 0$ is merely expedient. In the stress-strain relations, the condition of plane stress $\sigma_{zz} = 0$ is commonly used instead of $\epsilon_{zz} = 0$, and this circumstance is often regarded as an inconsistency. However, in approximations, the significant question is not the consistency of the assumptions, but rather the magnitude of the error that results, since nearly all approximations lead to inconsistencies. In plate theory, the values of ϵ_{zz} and σ_{zz} are not of particular importance. Viewed in this light, the Kirchhoff approximation merely implies that ϵ_{zz} has small effects upon σ_{xx} and σ_{yy}, and that σ_{xz} and σ_{yz} are not very significant. We observe further that the Kirchhoff approximation need not be restricted to linearly elastic plates; it is also applicable to studies of plasticity and creep of plates, and it is not restricted to small displacements.

For linearly elastic isotropic materials and plane stress relative to the (x, y) plane ($\sigma_{zz} = \sigma_{xz} = \sigma_{yz} = 0$), stress-strain-temperature relations are[1]

$$
\begin{aligned}
\sigma_{xx} &= \frac{E}{1-\nu^2}(\epsilon_{xx} + \nu\epsilon_{yy}) - \frac{EkT}{1-\nu} \\
\sigma_{yy} &= \frac{E}{1-\nu^2}(\nu\epsilon_{xx} + \epsilon_{yy}) - \frac{EkT}{1-\nu} \\
\sigma_{xy} &= 2G\epsilon_{xy} = G\gamma_{xy}
\end{aligned}
\tag{10-5.1}
$$

where E is Young's modulus, ν is Poisson's ratio, k is the coefficient of linear thermal expansion, G is the shear modulus, and T is the temperature measured above an arbitrary zero. It may be assumed without complication that k is a function of temperature T.

By Eqs. (10-3.7) and (10-5.1), $\sigma_{xx}, \sigma_{yy}, \sigma_{xy}$ may be expressed in terms of u, v, w, and T. Then, by Eqs. (10-2.2), the quantities N_{xx}, N_{yy}, N_{xy}, M_{xx}, M_{yy}, M_{xy} may be expressed in terms of u, v, w, and T. Then the first two of Eqs. (10-4.3) and Eq. (10-4.5) become differential equations in u, v, w. Thus, the equilibrium equations are expressed in terms of the displacement vector of the reference surface of the plate. For homogeneous plates, it is convenient to take the reference surface midway between the plate faces. However, for layered or reinforced plates, some other reference surface may be more appropriate. Then the integral limits $(-h/2, h/2)$ in Eqs. (10-2.2) would be modified accordingly. In the following, we take the reference surface as the middle surface of the plate. Hence, the faces of the plate are located at $z = \pm h/2$.

Although the Kirchhoff approximation implies that Q_x, Q_y vanish for isotropic linearly elastic plates, estimates of Q_x, Q_y may be obtained from the fourth and fifth of Eqs. (10-4.3).

Substitution of Eqs. (10-3.7) into Eqs. (10-5.1) and substitution of the results into Eq. (10-2.2) yields

$$
\begin{aligned}
N_{xx} &= \frac{Eh}{\alpha(1-\nu^2)}\left[\frac{\nu\beta_x}{\beta}u + u_x + \frac{\alpha_y}{\beta}v + \frac{\nu\alpha}{\beta}v_y\right] - \frac{E}{1-\nu}\int_{-h/2}^{h/2} kT\,dz \\
N_{yy} &= \frac{Eh}{\alpha(1-\nu^2)}\left[\frac{\beta_x}{\alpha}u + \frac{\nu\beta}{\alpha}u_x + \frac{\nu\alpha_y}{\alpha}v + v_y\right] - \frac{E}{1-\nu}\int_{-h/2}^{h/2} kT\,dz \\
N_{xy} &= Gh\left[\frac{\alpha}{\beta}\frac{\partial}{\partial y}\left(\frac{u}{\alpha}\right) + \frac{\beta}{\alpha}\frac{\partial}{\partial x}\left(\frac{v}{\beta}\right)\right] \\
M_{xx} &= -\frac{D}{\alpha}\left[\frac{\partial}{\partial x}\left(\frac{w_x}{\alpha}\right) + \frac{\alpha_y w_y}{\beta^2} + \frac{\nu\alpha}{\beta}\frac{\partial}{\partial y}\left(\frac{w_y}{\beta}\right) + \frac{\nu\beta_x}{\alpha\beta}w_x\right] \\
&\quad - \frac{E}{1-\nu}\int_{-h/2}^{h/2} zkT\,dz
\end{aligned}
\tag{10-5.2}
$$

$$M_{yy} = -\frac{D}{\beta}\left[\frac{\partial}{\partial y}\left(\frac{w_y}{\beta}\right) + \frac{\beta_x w_x}{\alpha^2} + \frac{\nu\beta}{\alpha}\frac{\partial}{\partial x}\left(\frac{w_x}{\alpha}\right) + \frac{\nu\alpha_y}{\alpha\beta}w_y\right]$$
$$-\frac{E}{1-\nu}\int_{-h/2}^{h/2} zkT\,dz$$

$$M_{xy} = -\frac{Gh^3}{6\alpha\beta}\left(w_{xy} - \frac{\alpha_y}{\alpha}w_x - \frac{\beta_x}{\beta}w_y\right)$$

where

$$G = \frac{E}{2(1+\nu)} \qquad D = \frac{Eh^3}{12(1-\nu^2)} \tag{10-5.2a}$$

The quantity D is called the *flexural rigidity* of the plate.

Alternatively, with Eqs. (10-3.7) and (10-5.2), we may write

$$N_{xx} = \frac{Eh}{1-\nu^2}\left(\epsilon_{xx}^0 + \nu\epsilon_{yy}^0 - T^0\right)$$

$$N_{yy} = \frac{Eh}{1-\nu^2}\left(\nu\epsilon_{xx}^0 + \epsilon_{yy}^0 - T^0\right)$$

$$N_{xy} = 2Gh\epsilon_{xy}^0 = Gh\gamma_{xy}^0 \qquad \gamma_{xy}^0 = 2\epsilon_{xy}^0$$

$$M_{xx} = -D\left(\frac{\kappa_{xx}}{\alpha^2} + \frac{\nu\kappa_{yy}}{\beta^2} + T^1\right) \tag{10-5.3}$$

$$M_{yy} = -D\left(\nu\frac{\kappa_{xx}}{\alpha^2} + \frac{\kappa_{yy}}{\beta^2} + T^1\right)$$

$$M_{xy} = -\frac{(1-\nu)D}{\alpha\beta}\kappa_{xy}$$

where $\epsilon_{xx}^0, \epsilon_{yy}^0, \epsilon_{xy}^0$ are the strain components in the plate middle surface $(z=0)$,

$$T^0 = \frac{1+\nu}{h}\int_{-h/2}^{h/2} kT\,dz$$
$$T^1 = \frac{12(1+\nu)}{h^3}\int_{-h/2}^{h/2} zkT\,dz \tag{10-5.4}$$

are the zeroth and first moments of T with respect to z, and

$$\kappa_{xx} = -\frac{\alpha_x w_x}{\alpha} + \frac{\alpha_y\alpha}{\beta^2}w_y + w_{xx}$$

$$\kappa_{yy} = \frac{\beta_x\beta}{\alpha^2}w_x - \frac{\beta_y}{\beta}w_y + w_{yy} \tag{10-5.5}$$

$$\kappa_{xy} = -\frac{\alpha_y w_x}{\alpha} - \frac{\beta_x}{\beta}w_y + w_{xy}$$

are the curvatures of the middle surface relative to (x, y)-axes. Hence,

$$\epsilon_{xx}^0 = \frac{1}{Eh}\left(N_{xx} - \nu N_{yy}\right) + \frac{T^0}{1+\nu}$$

$$\epsilon_{yy}^0 = \frac{1}{Eh}\left(N_{yy} - \nu N_{xx}\right) + \frac{T^0}{1+\nu} \tag{10-5.6}$$

$$\gamma_{xy}^0 = 2\epsilon_{xy}^0 = \frac{1}{Gh} N_{xy}$$

and

$$\kappa_{xx} = -\frac{12\alpha^2}{Eh^3}\left(M_{xx} - \nu M_{yy}\right) - \frac{\alpha^2 T^1}{1+\nu}$$

$$\kappa_{yy} = -\frac{12\beta^2}{Eh^3}\left(M_{yy} - \nu M_{xx}\right) - \frac{\beta^2 T^1}{1+\nu} \tag{10-5.7}$$

$$\kappa_{xy} = -\frac{12(1+\nu)\alpha\beta}{Eh^3} M_{xy}$$

For rectangular coordinates $\alpha = \beta = 1$. Then, the moment curvature relations (the last three of Eqs. 10-5.3) reduce to

$$M_{xx} = -D\left(\kappa_{xx} + \nu\kappa_{yy} + T^1\right)$$

$$M_{yy} = -D\left(\nu\kappa_{xx} + \kappa_{yy} + T^1\right) \tag{10-5.8}$$

$$M_{xy} = -(1-\nu) D\kappa_{xy}$$

where (by Eqs. 10-5.5),

$$\kappa_{xx} = w_{xx} \qquad \kappa_{yy} = w_{yy} \qquad \kappa_{xy} = w_{xy} \tag{10-5.9}$$

Stress Components in Terms of Tractions and Moments / Equations (10-3.7) and (10-5.1) lead to the conclusion that $\sigma_{xx}, \sigma_{yy}, \sigma_{xy}$ vary linearly through the thickness of the plate; that is, $\sigma_{xx} = a + bz, \ldots, \ldots$. Hence, by Eqs. (10-2.2), $a = N_{xx}/h$, $b = 12M_{xx}/h^3$. Similarly, the coefficients in the linear expressions for σ_{yy} and σ_{xy} are determined. Thus, we find

$$\sigma_{xx} = \frac{N_{xx}}{h} + \frac{12zM_{xx}}{h^3}$$

$$\sigma_{yy} = \frac{N_{yy}}{h} + \frac{12zM_{yy}}{h^3} \tag{10-5.10}$$

$$\sigma_{xy} = \frac{N_{xy}}{h} + \frac{12zM_{xy}}{h^3}$$

Pure Bending of Plates/If a plate is subjected to bending moments (M_{xx}, M_{yy}) only, we refer to the plate problem as one of *pure bending of plates*. In particular, for pure bending of plates, $N_{xx} = N_{yy} = N_{xy} = Q_x = Q_y = M_{xy} = 0$ in the preceding equations.

10-6 STRAIN ENERGY OF A PLATE

For plane stress theory, the strain energy density of a homogeneous isotropic elastic plate, referred to orthogonal plate coordinates is (see Art. 2-5).

$$U_0 = \frac{G}{1-\nu}\left[\epsilon_{xx}^2 + \epsilon_{yy}^2 + 2\nu\epsilon_{xx}\epsilon_{yy} + 2(1-\nu)\epsilon_{xy}^2 - 2(1+\nu)(\epsilon_{xx}+\epsilon_{yy})kT\right] \tag{10-6.1}$$

where U_0 has the dimensions of energy per unit volume. Since the volume element of a plate is $\alpha\beta\, dx\, dy\, dz$, the total strain energy U of the plate is

$$U = \iiint U_0 \alpha\beta\, dx\, dy\, dz \tag{10-6.2}$$

The integrations with respect to x and y extend over the middle surface of the plate, while the integration with respect to z extends between the limits $-h/2$ and $h/2$. By Eqs. (10-3.7), (10-6.1), and (10-6.2), we find after integration with respect to z, the total strain energy

$$U = U_m + U_b + U_t \tag{10-6.3}$$

where U_m, the *membrane energy* of the plate, is linear in the thickness h, and U_b the *bending energy* of the plate, is cubic in h. The term U_t represents the strain energy that results from the temperature T (heating). Hence, if G and ν are taken independent of z, integration with respect to z yields

$$\begin{aligned} U_m &= \iint \frac{Gh}{1-\nu}\left[(\epsilon_{xx}^0)^2 + (\epsilon_{yy}^0)^2 + 2\nu\epsilon_{xx}^0\epsilon_{yy}^0 + 2(1-\nu)(\epsilon_{xy}^0)^2\right]\alpha\beta\, dx\, dy \\ U_b &= \iint \frac{Gh^3}{12(1-\nu)}\left[\left(\frac{\kappa_{xx}}{\alpha^2}\right)^2 + \left(\frac{\kappa_{yy}}{\beta^2}\right)^2 + 2\nu\left(\frac{\kappa_{xx}}{\alpha^2}\right)\left(\frac{\kappa_{yy}}{\beta^2}\right)\right. \\ &\qquad \left. + 2(1-\nu)\left(\frac{\kappa_{xy}}{\alpha\beta}\right)^2\right]\alpha\beta\, dx\, dy \\ U_t &= -\iint \frac{Eh}{1-\nu^2}\left[(\epsilon_{xx}^0 + \epsilon_{yy}^0)T^0 - \left(\frac{\kappa_{xx}}{\alpha^2} + \frac{\kappa_{yy}}{\beta^2}\right)\frac{h^2T^1}{12}\right]\alpha\beta\, dx\, dy \end{aligned} \tag{10-6.4}$$

By means of Eqs. (10-3.7), with $z = 0$, and Eqs. (10-5.5), the strain energy (Eqs. 10-6.3 and 10-6.4) may be expressed as a function of the middle surface displacement components (u, v, w). The strain energy is employed in conjunction with the Rayleigh-Ritz procedure to obtain approximate solutions of plate problems.[5] The strain energy also serves, by means of variational principles, to determine plate boundary conditions.[6] In addition, the differential equations of equilibrium, in terms of (u, v, w), are obtained from the total *potential energy* expression by means of Euler's equation of the calculus of variations. In the next article, we employ the principle of stationary potential energy to determine boundary conditions for a plate.

10-7 BOUNDARY CONDITIONS FOR PLATES

In this article, we employ the principle of stationary potential energy (Art. 4-1) to obtain boundary conditions for the classical theory of plates. For simplicity, we consider rectangular coordinates ($\alpha = \beta = 1$) and a rectangular plate that lies in the region $0 \leq x \leq a$, $0 \leq y \leq b$ (Fig. 10-7.1). Also, for purposes of demonstration, we discard temperature effects and consider the effects of tractions N_{xx}, N_{yy}, N_{xy} to be negligible compared to the moments M_{xx}, M_{yy}, M_{xy}. Furthermore, we recall that in classical plate theory, the effects of Q_x, Q_y are also discarded.

The principle of stationary potential energy states

$$\delta W_e = \delta U \tag{10-7.1}$$

where the first variation δU of the strain energy is

$$\delta U = \int_0^b \int_0^a \delta \overline{U} \, dx \, dy \tag{10-7.2}$$

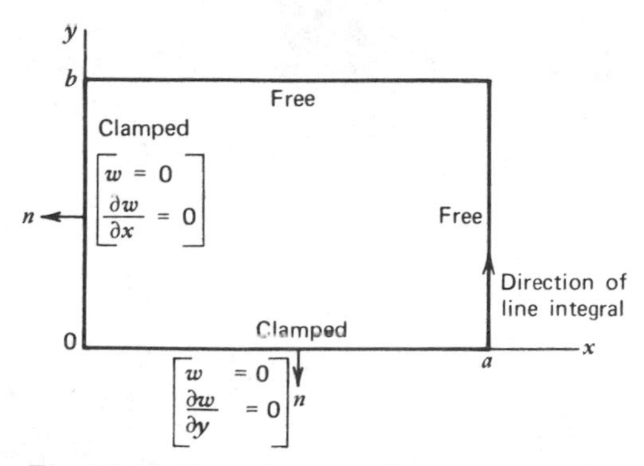

Fig. 10-7.1/Boundary conditions at a reference surface edge.

with

$$\delta \bar{U} = \int_{-h/2}^{h/2} \left(\sigma_{xx} \delta \epsilon_{xx} + \sigma_{yy} \delta \epsilon_{yy} + \sigma_{zz} \delta \epsilon_{zz} + 2\sigma_{xy} \delta \epsilon_{xy} + 2\sigma_{xz} \delta \epsilon_{xz} + 2\sigma_{yz} \delta \epsilon_{yz} \right) dz \tag{10-7.3}$$

and for $(P_x = P_y = 0,\ P_z = p)$, $(R_x = R_y = 0)$ and $(B_x = B_y = B_z = 0)$ (see Art. 10-4)

$$\delta W_e = \int_0^b \int_0^a p\, \delta w\, dx\, dy \tag{10-7.4}$$

Thus, Eqs. (10-7.1), (10-7.2), and (10-7.4) yield, with the Kirchhoff approximations $\epsilon_{zz} = \epsilon_{xz} = \epsilon_{yz} = 0$ (and, hence, $\delta\epsilon_{zz} = \delta\epsilon_{xz} = \delta\epsilon_{yz} = 0$),

$$\int_0^b \int_0^a (\delta \bar{U} - p\, \delta w)\, dx\, dy = 0 \tag{10-7.5}$$

Since (N_{xx}, N_{yy}, N_{xy}) and temperature effects have been discarded, Eqs. (10-5.6) and (10-3.17) yield $\epsilon_{xx}^0 = u_x = 0$, $\epsilon_{yy}^0 = v_y = 0$ and $2\epsilon_{xy}^0 = v_x + u_y = 0$. Hence, by Eqs. (10-3.17),

$$\epsilon_{xx} = -zw_{xx} \qquad \epsilon_{yy} = -zw_{yy} \qquad \epsilon_{xy} = -zw_{xy} \tag{10-7.6}$$

Substitution of Eqs. (10-7.6) into Eq. (10-7.3) yields, with Eqs. (10-2.2) and (10-7.5),

$$\int_0^b \int_0^a \left(M_{xx} \delta w_{xx} + M_{yy} \delta w_{yy} + 2M_{xy} \delta w_{xy} + p\, \delta w \right) dx\, dy = 0 \tag{10-7.7}$$

Now successive integration by parts of Eq. (10-7.7) yields[1]

$$\int_0^b \int_0^a \left(\frac{\partial M_{xx}}{\partial x} \delta w_x + \frac{\partial M_{xy}}{\partial y} \delta w_y \right) dx\, dy - \oint (M_{xx} \delta w_x + M_{xy} \delta w_y)\, dy$$
$$+ \int_0^b \int_0^a \left(\frac{\partial M_{xy}}{\partial y} \delta w_x + \frac{\partial M_{yy}}{\partial y} \delta w_y \right) dx\, dy + \oint (M_{xy} \delta w_x + M_{yy} \delta w_y)\, dx$$
$$- \int_0^b \int_0^a q\, \delta w\, dx\, dy = 0$$

and

$$\int_0^b \int_0^a \left[\frac{\partial^2 M_{xx}}{\partial x^2} + 2 \frac{\partial^2 M_{xy}}{\partial x \partial y} + \frac{\partial^2 M_{yy}}{\partial y^2} + p \right] \delta w\, dx\, dy$$
$$+ \oint \left[M_{xx} \delta w_x + M_{xy} \delta w_y - \left(\frac{\partial M_{xx}}{\partial x} + \frac{\partial M_{xy}}{\partial y} \right) \delta w \right] dy$$
$$+ \oint \left[- M_{xy} \delta w_x - M_{yy} \delta w_y + \left(\frac{\partial M_{xy}}{\partial x} + \frac{\partial M_{yy}}{\partial y} \right) \delta w \right] dx = 0 \tag{10-7.8}$$

where the line integrals are taken along the boundary in a counterclockwise direction (Fig. 10-7.1).

We note that the integral over the area of the plate leads to the moment equilibrium equation (Eq. 10-4.6, with $B_z = 0$ and $P_z = p$).

To be specific, consider the rectangular plate to be *clamped* along the edges $y = 0$ and $x = 0$. Let the edges $x = a$ and $y = b$ be *free* of forces and moments. Then, we have the *forced boundary conditions*

$$w = \frac{\partial w}{\partial n} = 0 \quad \text{for } x = 0 \text{ and } y = 0 \tag{10-7.9}$$

where n denotes the normal direction to the edge. Since the variations must satisfy the forced boundary conditions, we also have

$$\delta w = \delta \frac{\partial w}{\partial n} = 0 \quad \text{for } x = 0 \text{ and } y = 0 \tag{10-7.10}$$

Consequently, Eq. (10-7.8) reduces to

$$\int_0^b \left[M_{xx}\,\delta w_x + M_{xy}\,\delta w_y - \left(\frac{\partial M_{xx}}{\partial x} + \frac{\partial M_{xy}}{\partial y} \right) \delta w \right] dy \Bigg|^{x=a}$$
$$+ \int_0^a \left[M_{xy}\,\delta w_x + M_{yy}\,\delta w_y - \left(\frac{\partial M_{xy}}{\partial x} + \frac{\partial M_{yy}}{\partial y} \right) \delta w \right] dx \Bigg|^{y=b} = 0 \tag{10-7.11}$$

The line integrals of Eq. (10-7.11) lead to additional boundary conditions (natural conditions) for the free edge after further integration by parts. In this connection, we note that for $x = a$, the functions $\delta w_x(y)$ and $\delta w(y)$ are independent, where we recall that $w_x = \partial w / \partial x$. However, the functions $\delta w_y = \delta(\partial w / \partial y)$ and δw are not independent for $x = a$. Hence, the second term of the integral in dy must again be integrated by parts. Thus, integrating by parts and noting that $\partial M_{xx}/\partial x + \partial M_{xy}/\partial y = Q_x$ (see the fourth of Eqs. 10-4.4, with $R_y = 0$), we obtain

$$\int_0^b \left[M_{xx}\,\delta w_x - \left(Q_x + \frac{\partial M_{xy}}{\partial y} \right) \delta w \right] dy \Bigg|^{x=a} + M_{xy}\,\delta w \,|^{x=a,\, y=b}$$

and similarly for $y = b$,

$$\int_0^a \left[M_{yy}\,\delta w_x - \left(Q_y + \frac{\partial M_{xy}}{\partial x} \right) \delta w \right] dx \Bigg|^{y=b} + M_{xy}\,\delta w \,|^{x=a,\, y=b}$$

Hence, for the free edges $x = a$ and $y = b$, we must have the natural boundary conditions

$$M_{xx}=0 \qquad V_x=Q_x+\frac{\partial M_{xy}}{\partial y}=0 \qquad \text{for } x=a$$
$$M_{yy}=0 \qquad V_y=Q_y+\frac{\partial M_{xy}}{\partial x}=0 \qquad \text{for } y=b \tag{10-7.12}$$

In addition, at the corner of two free edges, we have the additional natural boundary condition

$$M_{xy}=0 \qquad \text{for } x=a,\ y=b \tag{10-7.13}$$

Consequently, at a free edge of a classical plate, say $x=a$, the shear Q_x and twisting moment M_{xy} do not vanish separately, but rather the combination $Q_x+\partial M_{xy}/\partial y=V_x$, the so-called Kirchhoff shear, vanishes. Alternatively, we may express V_x and V_y in the form (with the fourth and fifth of Eqs. 10-4.4, with $R_x=R_y=0$)

$$V_x=\frac{\partial M_{xx}}{\partial x}+2\frac{\partial M_{xy}}{\partial y}$$
$$V_y=\frac{\partial M_{yy}}{\partial y}+2\frac{\partial M_{xy}}{\partial x} \tag{10-7.14}$$

Hence, in summary in terms of the displacement w and its derivatives, we may write (see Eqs. 10-5.8)

$$M_{xx}=-D(w_{xx}+\nu w_{yy}) \qquad M_{yy}=-D(w_{yy}+\nu w_{xx})$$
$$M_{xy}=-(1-\nu)Dw_{xy}$$
$$V_x=-D\left[w_{xxx}+(2-\nu)w_{xyy}\right]$$
$$V_y=-D\left[w_{yyy}+(2-\nu)w_{xxy}\right] \tag{10-7.15}$$

Consequently, substitution for M_{xx}, M_{xy}, M_{yy}, in terms of w, in Eq. (10-4.6) yields, with $B_z=0$ and $P_z=p$,

$$\nabla^2\nabla^2 w=\frac{p}{D} \tag{10-7.16}$$

where $\nabla^2\nabla^2 w=w_{xxxx}+2w_{xxyy}+w_{yyyy}$. Equation (10-7.16) is the plate equation; it is one of the main results of classical plate theory. It is a fourth order partial differential equation. Hence, the plate problem is to find solutions of Eq. (10-7.16) that satisfy the boundary conditions (clamped, free, simply supported, etc.) at the edges of the plate. Fortunately, the most important plate shapes are rectangular and circular, which may be treated most readily.

10-8 SOLUTION OF RECTANGULAR PLATE PROBLEM

A large collection of solved rectangular plate problems has been presented by Timoshenko and Woinowsky-Krieger.*

In this article, initially, we treat the small displacement theory of simply supported rectangular plates for certain simple loadings. Thus, initially, we consider bending effects only, since in the case of small displacements, these effects dominate. Fourier series methods of solutions are employed. We also present results of an approximate solution due to Westergaard.[8]

Solution of $\nabla^2\nabla^2 w = p/D$. Rectangular Plate / In Art. 10-7, for bending effects dominant, we obtained the plate equation

$$\nabla^2\nabla^2 w = p/D \tag{10-8.1}$$

where p denotes lateral pressure and D is the flexural rigidity. The plate theory based upon Eq. (10-8.1) is often referred to as the flexural (or bending) theory of plates. For this case, the solution of the plate problem requires that the lateral displacement w satisfies Eq. (10-8.1) and appropriate boundary conditions. We note that since $\nabla^2\nabla^2$ is an invariant vector operator, Eq. (10-8.1) holds for all coordinate systems, provided that proper expressions for $\nabla^2\nabla^2$ are employed.

For simplicity, we consider here a simply supported rectangular plate of thickness h and in-plane dimensions a and b (Fig. 10-8.1). Then, we observe that any function[9]

$$w(x, y) = X_n(x)\sin\frac{n\pi y}{b} \tag{10-8.2}$$

satisfies the simple support boundary conditions at $y = 0$ and $y = b$†

$$\left.\begin{aligned} &w = 0 \\ &M_{yy} = -D(w_{yy} + \nu w_{xx}) = 0 \end{aligned}\right\} y = 0, b \tag{10-8.3}$$

*See reference 5. See also K. Marguerre and Hans-Theo Woernle, reference 3, in which isotropic and orthotropic plate solutions are presented for rectangular and circular plates for a wide variety of boundary conditions. Marguerre and Woernle have presented a systematic treatment that clarifies the effects of shear deformation and hence clarifies the boundary conditions for the classical plate, in which shear deformation is discarded. In addition, the treatment by Marguerre and Woernle emphasizes the orthotropic plate, which is more interesting and more important practically, than the isotropic plate. M. Naruoka, reference 7, has presented an extensive bibliography on the theory of plates indexed by author and subject matter.

†One advantage of this single series method (the Levy method) is that the subsequent series solution (see Eq. 10-8.8) converges quite rapidly compared to a double series representation for w (the Navier method), i.e., a solution form of the type

$$w = \sum_{m=1}^{\infty}\sum_{n=1}^{\infty} A_{mn}\sin\frac{m\pi x}{a}\sin\frac{n\pi y}{b}.$$

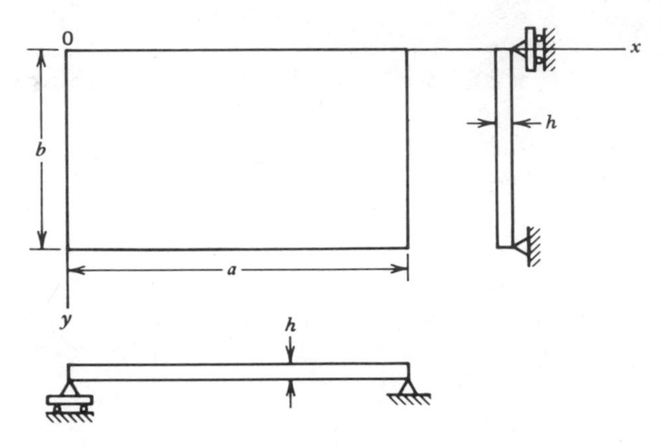

Fig. 10-8.1

Similarly, we may also write $w(x, y)$ in the form

$$w(x, y) = Y_n(y) \sin \frac{n\pi x}{a} \tag{10-8.2a}$$

which in turn satisfies the simple support boundary conditions at $x = 0$ and $x = a$; that is,

$$\left.\begin{aligned} w &= 0 \\ M_{xx} &= -D(w_{xx} + \nu w_{yy}) = 0 \end{aligned}\right\} x = 0, a \tag{10-8.3a}$$

For our purposes here, we employ Eq. (10-8.2). Thus, substitution into Eq. (10-8.1) yields an ordinary differential equation for $X_n(x)$. Its solution contains four constants of integration, which may be selected to satisfy the four boundary conditions at the edges $x = 0$ and $x = a$ (two at $x = 0$ and two at $x = a$). However, before this procedure may be carried out, the lateral pressure p must be expressed in appropriate form. Corresponding to the solution form (Eq. 10-8.2), we express p in the form

$$p(x, y) = p_0 \sum_{n=1}^{\infty} f_n(x) \sin \frac{n\pi y}{b} \tag{10-8.4}$$

In many practical cases, p may be written in the product form

$$p(x, y) = p_0 f(x) g(y) \tag{10-8.5}$$

Then, Eqs. (10-8.4) and (10-8.5) yield

$$p(x, y) = f(x) \sum_{n=1}^{\infty} p_n \sin \frac{n\pi y}{b} \tag{10-8.6}$$

where

$$p_n = \frac{2p_0}{b} \int_0^b g(y) \sin \frac{n\pi y}{b} \, dy \tag{10-8.7}$$

Consequently, to satisfy Eq. (10-8.1), we must generalize $w(x, y)$ to

$$w(x, y) = \sum_{n=1}^{\infty} X_n(x) \sin \frac{n\pi y}{b} \tag{10-8.8}$$

Then substitution of Eqs. (10-8.6) and (10-8.8) into Eq. (10-8.1) yields the set of ordinary differential equations

$$D\left[X_n'''' - 2\left(\frac{n\pi}{b}\right)^2 X_n'' + \left(\frac{n\pi}{b}\right)^4 X_n\right] = p_n f(x) \qquad n = 1, 2, \ldots \tag{10-8.9}$$

for the functions $X_n(x)$. The solution of Eq. (10-8.9) for the X_n, and substitution into Eq. (10-8.8) yields the solution of the simply supported rectangular plate subjected to pressure p (Eq. 10-8.6). The resulting series solution gives good results (converges well) for $a > b$, and often for $a = b$. If $a < b$, it is better to use the form of solution of Eq. (10-8.2a) or simply interchange the labels a, b, so that again $a > b$.

In the treatment of Eq. (10-8.9), for simplicity, we take $f(x) = 1$. Then, Eq. (10-8.9) yields

$$X_n''''(x) - 2\left(\frac{n\pi}{b}\right)^2 X_n''(x) + \left(\frac{n\pi}{b}\right)^4 X_n(x) = \frac{p_n}{D} \tag{10-8.10}$$

By the theory of ordinary differential equations, the general solution of Eq. (10-8.10) is

$$X_n(x) = \frac{p_n}{D}\left(\frac{b}{n\pi}\right)^4 \left[1 + (A_{1n} + xA_{2n}) \cosh \frac{n\pi x}{b} + (B_{1n} + xB_{2n}) \sinh \frac{n\pi x}{b}\right] \qquad n = 1, 2, \ldots \tag{10-8.11}$$

The constants A_{1n}, A_{2n}, B_{1n}, B_{2n} are selected to satisfy the four boundary conditions

$$\left.\begin{aligned} w &= 0 \\ M_{xx} &= -D(w_{xx} + \nu w_{yy}) = 0 \end{aligned}\right\} x = 0, a \tag{10-8.12}$$

Substitution of Eqs. (10-8.11) into Eq. (10-8.8) and then substitution of the results into Eq. (10-8.12) yields, after considerable algebra,[3]

$$X_n(x) = \frac{p_n}{D}\left(\frac{b}{n\pi}\right)^4 \left\{1 - \cosh \frac{n\pi x}{b} + \frac{n\pi x}{b} \sinh \frac{n\pi x}{b} + \frac{1}{1 + \cosh \frac{n\pi a}{b}} \left[\left(\sinh \frac{n\pi a}{b} - \frac{n\pi a}{b}\right) \sinh \frac{n\pi x}{b} - \frac{n\pi a}{b} \sinh \frac{n\pi a}{b} \cosh \frac{n\pi x}{b}\right]\right\} \tag{10-8.13}$$

With $X_n(x)$ and hence $w(x, y)$ known, Eqs. (10-7.15) may be used to compute $M_{xx}, M_{yy}, M_{xy}, V_x, V_y$.

EXAMPLE 10-8.1

Square Plate Subject to Sinusoidally Distributed Pressure

A square plate is simply supported on all edges (Fig. 10-8.1) and is loaded by gravel such that

$$p(x, y) = p_0 \sin\frac{\pi x}{a} \sin\frac{\pi y}{b}; \qquad a = b \tag{1}$$

(a) Determine the maximum deflection and its location. (b) Determine the maximum values of the moments M_{xx}, M_{yy}. (c) Determine the maximum values of the Kirchhoff shear forces V_x, V_y.

SOLUTION

The boundary conditions for simply supported edges are

$$\begin{aligned} &w = 0, \qquad M_{xx} = 0 \qquad \text{for } x = 0, a \\ &w = 0, \qquad M_{yy} = 0 \qquad \text{for } y = 0, b(=a) \end{aligned} \tag{2}$$

Since $w = 0$ around the plate boundary, $\partial^2 w/\partial x^2 = 0$ for edges parallel to the x-axis and likewise $\partial^2 w/\partial y^2 = 0$ for edges parallel to the y-axis. Hence, noting the expressions for M_{xx}, M_{yy} in Eq. (10-7.15), we may rewrite the boundary conditions, Eqs. (2) in the form

$$\begin{aligned} &w = 0, \qquad \frac{\partial^2 w}{\partial x^2} = 0 \qquad \text{for } x = 0, a \\ &w = 0, \qquad \frac{\partial^2 w}{\partial y^2} = 0 \qquad \text{for } y = 0, a \end{aligned} \tag{3}$$

(a) Equations (3) may be satisfied by taking w in the form

$$w = w_0 \sin\frac{\pi x}{a} \sin\frac{\pi y}{a} \tag{4}$$

where w_0 is a constant that must be chosen to satisfy the plate equation [Eq. (10-7.16)], namely, with Eq. (1),

$$\frac{\partial^4 w}{\partial x^4} + 2\frac{\partial^4 w}{\partial x^2 \partial y^2} + \frac{\partial^4 w}{\partial y^4} = \frac{p_0}{D} \sin\frac{\pi x}{a} \sin\frac{\pi y}{a} \tag{5}$$

Substitution of Eq. (4) into Eq. (5) yields

$$w_0 = \frac{p_0 a^4}{4\pi^4 D} \tag{6}$$

By Eq. (4), we see that the maximum deflection of the plate occurs at

$x = y = a/2$. Thus, the maximum deflection of the plate is

$$w_{max} = w_0 = \frac{p_0 a^4}{4\pi^4 D} \qquad \text{at } x = y = \frac{a}{2} \tag{7}$$

(b) To determine the maximum values of moments M_{xx}, M_{yy}, we find from Eqs. (10-7.15), with Eqs. (4) and (6)

$$M_{xx} = M_{yy} = \frac{p_0 a^2(1+\nu)}{4\pi^2} \sin\frac{\pi x}{a} \sin\frac{\pi y}{a} \tag{8}$$

It is seen that the maximum values of M_{xx} and M_{yy} occur at $x = y = a/2$. Thus,

$$(M_{xx})_{max} = (M_{yy})_{max} = \frac{p_0 a^2(1+\nu)}{4\pi^2} \qquad \text{at } x = y = \frac{a}{2} \tag{9}$$

(c) To calculate the Kirchhoff shear forces, we have by Eqs. (10-7.15), with Eqs. (4) and (6)

$$\begin{aligned} V_x &= \frac{p_0 a}{4\pi}(3-\nu)\cos\frac{\pi x}{a}\sin\frac{\pi y}{a} \\ V_y &= \frac{p_0 a}{4\pi}(3-\nu)\sin\frac{\pi x}{a}\cos\frac{\pi y}{a} \end{aligned} \tag{10}$$

We see that the maximum values of V_x, V_y occur along the edges of the plate. Thus, by Eqs. (10),

$$\begin{aligned} (V_x)_{max} &= \frac{p_0 a}{4\pi}(3-\nu) \qquad \text{at } y = \frac{a}{2},\ x = 0, a \\ (V_y)_{max} &= \frac{p_0 a}{4\pi}(3-\nu) \qquad \text{at } x = \frac{a}{2},\ y = 0, a \end{aligned} \tag{11}$$

Westergaard Approximate Solution for Rectangular Plates. Uniform Load/The solution of the simply supported rectangular plate subjected to pressure was indicated above. By the results of the bending (flexural) theory of plates for uniform pressure it may be shown that at the center of the plate the stress is always greater in the direction of the shorter span than in the direction of the larger span (Fig. 10-8.2). This fact may be made plausible by physical considerations. For example, consider the two strips *EF* and *GH* (Fig. 10-8.2). The deflections of the two strips at the center of the plate are, of course, equal. However, the shorter strip (*GH*), being the stiffer, carries the greater load, and hence, a greater stress is developed in it.

Rectangular Plate with Simply Supported Edges/In Fig. 10-8.3, the bending moment *per unit width across the diagonal* at the corner (denoted by M_{diag}), the bending moment per unit width at the center of the strip

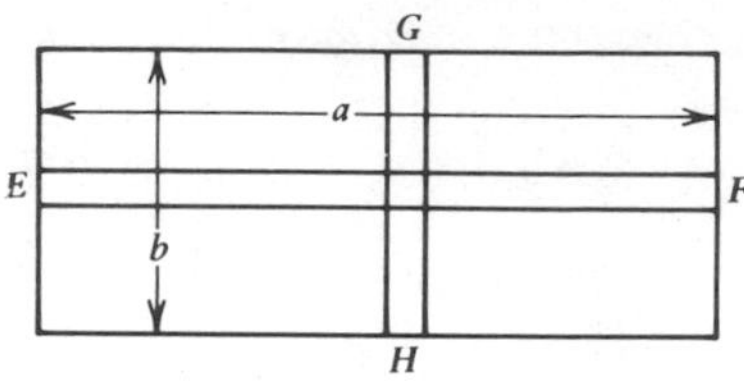

Fig. 10-8.2

GH (Fig. 10-8.2) in the short span b (denoted by M_{bc}) and the bending moment per unit width at the center of the strip EF (Fig. 10-8.2) in the long span a (denoted by M_{ac}) are plotted.

The curves and equations in Fig. 10-8.3 were obtained by Westergaard[8] with slight modifications in the results obtained from the theory of flexure of plates. The modifications were made in order to obtain relatively simple expressions and, in doing so, allowance was made for some redistribution of stress accompanying slight yielding at the portions of high (and more or less localized) stresses. Note that the moment coefficient for a square slab ($b/a = 1$) is $1/24 = 0.0417$, and that for a long narrow slab ($b/a = 0$) the moment coefficient for the short span is $1/8 = 0.125$. For intermediate values of b/a, the moment coefficient is always greater in the short span than elsewhere, and its value is intermediate between the limiting values of 1/24 and 1/8.

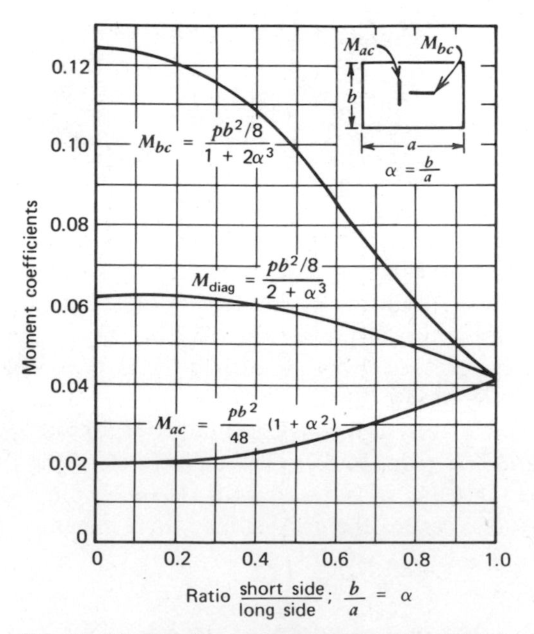

Fig. 10-8.3/Bending moment per unit width in rectangular plates with simply supported edges. Poisson's ratio ν is assumed to be zero.

Rectangular Plate with Fixed Edges / If the plate is rigidly held (fixed) at the edges and is subjected to a uniformly distributed load, the maximum moment per unit width occurs at the centers of the long edges, that is, at the fixed ends of the central strip of the short span.

Two limiting cases of a fixed-edged rectangular slab will be considered first. If the plate is very long and narrow ($b/a = 0$), the forces at the short ends of the plate will have negligible effect on the moment in the central part of the plate and, hence, the plate may be considered to be a fixed-end beam with a span equal to the short dimension of the plate; therefore, the negative moment per unit width M_{be} at the fixed edges of the short span is $pb^2/12$, and the positive moment M_{bc} at the center of the short span is $pb^2/24$. The other limiting case is that of the square slab ($b/a = 1$) for which the moment coefficient at the center of the edges is approximately 0.05 and the moment coefficient at the center is 0.018.

For plates having other values of b/a, the maximum negative moment M_{be} and the maximum positive moment M_{bc} are given in Fig. 10-8.4. These values were obtained by Westergaard[8] by simplifying the results obtained from the theory of flexure of flat plates. Owing to the advantageous redistribution of stresses accompanying slight yielding of the plate at points of maximum stress, the plate is somewhat stronger than is indicated by the results obtained from the theory.

For plates made of ductile metal, the maximum moment used in design should probably be about the average of the values of M_{be} and M_{bc} given in Fig. 10-8.4. Bach,[10] from the results of experiments, recommends the moment coefficients given by the dotted line in Fig. 10-8.4. Experi-

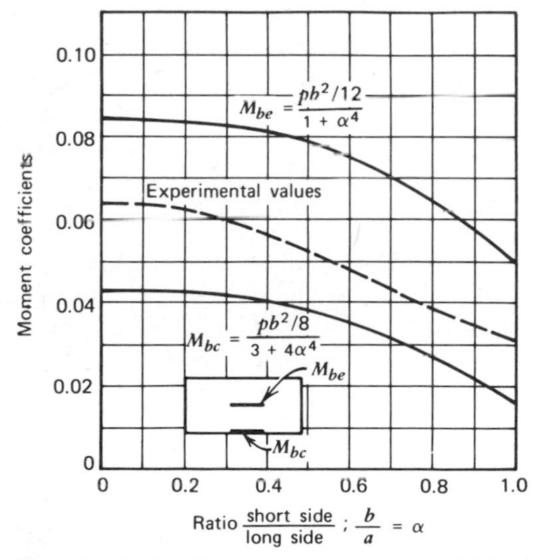

Fig. 10-8.4 / Bending moment per unit width in rectangular plates with fixed edges. Poisson's ratio ν is assumed to be zero.

mental results for steel plates 0.61 m by 1.22 m ($b/a = 0.5$) with the thicknesses varying from 3 mm to 19 mm indicate the maximum moment per unit width to be approximately 0.042 pb^2. Results indicate that there is not much difference in the value of the stress at the center and at the end of the short span.

Other Types of Edge Conditions / Formulas obtained by Westergaard,[8] giving approximate values of the moments per unit width in rectangular plates, including some of the formulas discussed in the preceding articles, are shown in Table 10-8.1. These formulas give results fairly close to those found from the theory of flexure of slabs, in which for convenience the value of Poisson's ratio $\nu = 0$ has been assumed. The effect of Poisson's ratio is to increase the bending moment per unit width in the plate. Let $M_{ac\nu}$ and $M_{bc\nu}$ represent the values of the bending moments at the center of a rectangular plate when the material has a Poisson's ratio ν not assumed to be zero. Approximate values of these bending moments are given by the expressions

$$\begin{aligned} M_{ac\nu} &= M_{ac} + \nu M_{bc} \\ M_{bc\nu} &= M_{bc} + \nu M_{ac} \end{aligned} \tag{10-8.14}$$

in which M_{ac} and M_{bc} are values of the bending moments as given in Table 10-8.1, or subsequent tables, in which ν has been assumed to be zero. In using these formulas for plates made of ductile material, it should be borne in mind that they give results that probably err somewhat on the side of safety.

Deflection of Rectangular Plate; Load Uniformly Distributed / The differential equation for plates has been solved only for relatively simple shapes of plates and for certain simple types of loading. From the solution of this equation for rectangular plates subjected to uniformly distributed loads the maximum deflection w_{max} at the center of the plate is given by the equation

$$w_{max} = C(1 - \nu^2)(pb^4/Eh^3) \tag{10-8.15}$$

where p is the uniformly distributed load per unit of area, b is the short span length, E is the modulus of elasticity of the material in the plate, h is the plate thickness, ν is Poisson's ratio, and C is a dimensionless constant whose value depends upon the ratio b/a of the sides of the plate and upon the type of support at the edge of the plate.

Several investigators have computed values of the constant C in Eq. (10-8.15); some of the values are as follows: For a uniformly loaded square ($b/a = 1$) plate simply supported at its edges, $C = 0.047$; if the plate is very long and narrow ($b/a = 0$, approximately), $C = 0.16$. Thus the deflection of a long narrow plate is more than three times that of a square plate having the same thickness as the narrow plate; in fact, the supports at the short ends of a narrow plate ($b/a < \frac{1}{3}$) have very little effect in preventing deflection at the center of the plate. If all the edges of

Table 10-8.1

Formulas Obtained by the Theory of Flexure of Slabs, Giving Approximate Values of Bending Moments per Unit Width and Maximum Deflections in Rectangular and Elliptical Slabs under Uniform Load (Given by Westergaard; Reference 8)†

		Moments in Span b		*Moments in Span a*		*Maximum Deflection* $w_{max} = C(1-\nu^2) \times (pb^4/Eh^3)$
		At Center of Edge $-M_{be}$	*At Center of Slab* M_{bc}	*At Center of Edge* $-M_{ae}$	*Along Center Line of Slab* M_{ac}	*Values of C*
Rectangular Slabs	Four edges simply supported	0	$\dfrac{\frac{1}{8}pb^2}{1+2\alpha^3}$	0	$\dfrac{pb^2}{48}(1+\alpha^2)$	$\dfrac{0.16}{1+2.4\alpha^3}$
	Span b fixed; span a simply supported.	$\dfrac{\frac{1}{12}wb^2}{1+0.2\alpha^4}$	$\dfrac{\frac{1}{24}pb^2}{1+0.4\alpha^4}$	0	$\dfrac{pb^2}{80}(1+0.3\alpha^2)$	$\dfrac{0.032}{1+0.4\alpha^3}$
	Span a fixed; span b simply supported	0	$\dfrac{\frac{1}{8}pb^2}{1+0.8\alpha^2+6\alpha^4}$	$\dfrac{\frac{1}{8}pb^2}{1.08\alpha^4}$	$0.015pb^2\left(\dfrac{1+3\alpha^2}{1+\alpha^4}\right)$	$\dfrac{0.16}{1+\alpha^2+5\alpha^4}$
	All edges fixed	$\dfrac{\frac{1}{12}wb^2}{1+\alpha^4}$	$\dfrac{\frac{1}{8}pb^2}{3+4\alpha^4}$	$\dfrac{1}{24}wb^2$	$0.009pb^2(1+2\alpha^2-\alpha^4)$	$\dfrac{0.032}{1+\alpha^4}$
	Elliptical slab with fixed edges; diameters a and b $b/a=\alpha$	$\dfrac{\frac{1}{12}wb^2}{1+\frac{2}{3}\alpha^2+\alpha^4}$	$\dfrac{\frac{1}{24}pb^2}{1+\frac{2}{3}\alpha^2+\alpha^4}$	$\dfrac{\frac{1}{12}pb^2\alpha^2}{1+\frac{2}{3}\alpha^2+\alpha^4}$	$\dfrac{\frac{1}{24}pb^2\alpha^2}{1+\frac{2}{3}\alpha^2+\alpha^4}$	

†Poisson's Ratio $\nu = 0$ (see Eq. 10-8.14). b = shorter side; a = longer side; $b/a = \alpha$.

a uniformly loaded square plate are fixed, the constant in Eq. (10-8.15) is $C = 0.016$. A comparison of this value of C with the value 0.047 for simply supported edges shows that if the edges of a square plate are held rigidly (fixed) the deflection at the center of the plate is about one-third of the deflection for simply supported edges. However, the edges of a plate are seldom if ever clamped rigidly and, therefore, the deflection at the center of a plate having partial restraint at its edges would be given by a value of C between 0.016 and 0.047.

Values of the constant C in Eq. (10-8.15) for various ratios of b/a and for various conditions at the supports are given in Table 10-8.1. From experiments on plates 0.61 m by 1.22 m with the edges carefully clamped, the measured deflections on relatively thin plates ($h/b \leq 0.02$) agree very closely up to values of deflections not greater than about one-half the plate thickness with those given by the formulas for deflections in Table 10-8.1. The formulas for deflections in this table give values that are too large when the direct tensile stresses in the plate are appreciable; this condition begins when the maximum deflection of the plate reaches a value of about one-half the thickness of the plate.

EXAMPLE 10-8.2
Water Tank

A water tank 3.60 m deep and 2.70 m square is to be made of structural steel plate. The sides of the tank are divided into nine panels by two vertical supports or stiffeners and two horizontal supports; that is, each

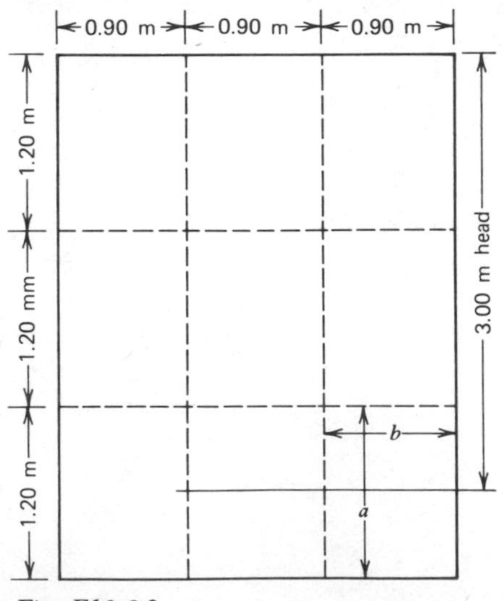

Fig. E10-8.2

panel is 0.90 m wide and 1.20 m high, and the average head of water on a lower panel is 3.00 m (Fig. E10-8.2). Determine the thickness of the plate for the lower panels, using a working stress of $\sigma_w = 124.0$ MPa. Calculate the maximum deflection of the panel.

SOLUTION

The mean pressure on a bottom panel is $p = (3.00 \text{ m})(9.80 \text{ kPa/m}) = 29.4$ kPa. We assume this pressure to be uniformly distributed over the panel. We also assume that the edges of the panel are fixed.

(a) For fixed edges, by Fig. 10-8.4 with $b/a = 0.75$, we have approximately,

$$M = 0.042\,pb^2 = (0.042)\left(29.4 \times 10^3 \frac{\text{N}}{\text{m}^2}\right)(0.90^2 \text{m}^2) = 1000 \frac{\text{N} \cdot \text{m}}{\text{m}}$$

and hence,

$$\sigma = M\frac{c}{I} = \frac{6M}{h^2}$$

Thus,

$$h = \sqrt{\frac{6M}{\sigma_w}} = \sqrt{\frac{6(1000)}{124}} = 6.96 \text{ mm}$$

Probably a slightly higher value of h should be used for adequate strength, considering the error in the assumption of uniformly distributed pressure. In addition, available commercial sizes and other factors, such as rusting, might dictate a larger thickness.

(b) For the displacement, we have from Table 10-8.1, for fixed edges, $C = 0.032/[1 + (0.75)^4] = 0.0243$. With $\nu = 0.29$ and $E = 200$ GPa, we find

$$w_{\text{max}} = 0.0243(1 - 0.29^2)\frac{(29.4 \times 10^3 \text{ Pa})(900 \text{ mm})^4}{200 \times 10^9 \text{ Pa}(6.96 \text{ mm})^3}$$

or

$$w_{\text{max}} = 6.37 \text{ mm}$$

This deflection is more than one-half the thickness of the plate. Hence, probably direct tensile stress would reduce the value of w_{max}. See Art. 10-9.

PROBLEM SET 10-8

1. Repeat Example 10-8.1 for the case of a rectangular plate $a \neq b$.

2. Repeat Example 10-8.1 for the case $p = p_0 \sin \frac{m\pi x}{a} \sin \frac{n\pi y}{a}$, where m and n are integers.

3. Repeat Problem 2 for the case of a rectangular plate $a \neq b$,

4. Determine the twisting moment M_{xy} and the stress σ_{xy} for the plate of Example 10-8.1.

5. Compute the stresses $\sigma_{xx}, \sigma_{yy}, \sigma_{xy}$ for the plate of Example 10-8.1.

6. Let the simply supported plate be subjected to load $p(x, y)$ given in the form of a double trigonometric series

$$p(x, y) = \sum_{m=1}^{\infty} \sum_{n=1}^{\infty} A_{mn} \sin \frac{m\pi x}{a} \sin \frac{n\pi y}{b} \tag{a}$$

Let the displacement $w(x, y)$ be represented in terms of a double trigonometric series

$$w(x, y) = \sum_{m=1}^{\infty} \sum_{n=1}^{\infty} W_{mn} \sin \frac{m\pi x}{a} \sin \frac{n\pi y}{b} \tag{b}$$

This double series method was used by Navier (the Navier method) in a lecture presented to the French Academy in 1820.

(a) Show that

$$W_{mn} = \frac{1}{\pi^4 D} \frac{A_{mn}}{\left(\frac{m^2}{a^2} + \frac{n^2}{b^2}\right)^2} \tag{c}$$

(b) For the case $p(x, y) = p_0$, show by the method of Fourier series that

$$\begin{aligned} A_{mn} &= \frac{4p_0}{ab} \int_0^a \int_0^b \sin \frac{m\pi x}{a} \sin \frac{n\pi y}{b}\, dx\, dy \\ &= \frac{16 p_0}{\pi^2 mn} \end{aligned} \tag{d}$$

(c) and hence, that

$$w = \frac{16 p_0}{\pi^6 D} \sum_{m=1}^{\infty} \sum_{n=1}^{\infty} \frac{\sin \frac{m\pi x}{a} \sin \frac{n\pi y}{b}}{mn\left(\frac{m^2}{a^2} + \frac{n^2}{b^2}\right)^2} \tag{e}$$

(d) and that the maximum deflection is given by

$$w_{max} = \frac{16p_0}{\pi^6 D} \sum_{m=1}^{\infty} \sum_{n=1}^{\infty} \frac{(-1)^{(m+n-2)/2}}{mn\left(\dfrac{m^2}{a^2} + \dfrac{n^2}{b^2}\right)^2}$$

This series converges extremely rapidly. Using only the first term for a square plate ($a = b$), we obtain

$$w_{max} \cong \frac{4p_0 a^4}{\pi^6 D}$$

$$\cong 0.0042 \frac{p_0 a^4}{D}$$

7. A rectangular steel plate ($E = 200$ GPa, $\nu = 0.29$, $Y = 280$ MPa) has a length of 2 m, a width of 1 m, and fixed edges. The plate is subjected to a uniform pressure $p = 270$ kPa. Assume that the design pressure for the plate is limited by the maximum stress in the plate; this would be the case for fatigue loading, for instance. For a working stress $\sigma_w = Y/2$, determine the plate thickness and the maximum deflection.

8. If the pressure for the plate in Problem 7 is increased, yielding will be initiated by moment M_{be} at the fixed edge of the plate; however, the pressure-deflection curve for the plate will remain nearly linear until after the pressure has been increased to initiate yielding due to bending at the center of the plate. Determine the plate thickness and maximum deflection for the plate in Problem 7 if the plate has a factor of safety $SF = 2.00$ against initiation of yielding at the center of the plate.

   Ans. $h = 22.1$ mm, $w_{max} = 3.45$ mm

9. A square structural steel trap door ($E = 200$ GPa, $\nu = 0.29$, $Y = 240$ MPa) has a side length of 1.50 m and a thickness of 15 mm. The plate is simply supported and subjected to a uniform pressure. Determine the yield pressure p_Y and the maximum deflection when this pressure is applied.

   Ans. $p_Y = 74.4$ kPa, $w_{max} = 24.1$ mm

10-9
SOLUTION OF CIRCULAR PLATE PROBLEM

In this article, we consider solutions for circular plates undergoing small elastic displacements. We also present some results for large elastic deflections of circular plates; that is, for maximum deflections which are large compared to the plate thickness h. In the case of large deflections, direct tensile forces (tractions) which, though small for deflections less than one-half the plate thickness, become relatively much larger for deflections greater than the thickness.

Solution of $\nabla^2\nabla^2 w = p/D$. Circular Plate / For the circular plate with radius a and thickness h, we employ polar coordinates with origin at the center of the plate (Fig. 10-9.1). Then, Eq. (10-8.1) may be written in the form[1]

$$\nabla^2\nabla^2 w = \left(\frac{\partial^2}{\partial r^2} + \frac{1}{r}\frac{\partial}{\partial r} + \frac{1}{r^2}\frac{\partial^2}{\partial \theta^2}\right)\left(\frac{\partial^2 w}{\partial r^2} + \frac{1}{r}\frac{\partial w}{\partial r} + \frac{1}{r^2}\frac{\partial^2 w}{\partial \theta^2}\right) = \frac{p}{D} \tag{10-9.1}$$

The general solution of Eq. (10-9.1) is presented by Marguerre and Woernle.[3] Here, we consider only the axial symmetric case, in which the plate is loaded and supported symmetrically. Then, Eq. (10-9.1) reduces to (since dependency upon θ vanishes),

$$\left(\frac{d^2}{dr^2} + \frac{1}{r}\frac{d}{dr}\right)\left(\frac{d^2 w}{dr^2} + \frac{1}{r}\frac{dw}{dr}\right) = \frac{p}{D} \tag{10-9.2}$$

The solution of Eq. (10-9.2), with $p = p_0 =$ constant, is

$$w = \frac{p_0 r^4}{64D} + A_1 + A_2 \ln r + B_1 r^2 + B_2 r^2 \ln r \tag{10-9.3}$$

where A_1, A_2, B_1, B_2 are constants of integration. The constants A_1, A_2, B_1, B_2 are determined by the boundary conditions at $r = a$ and

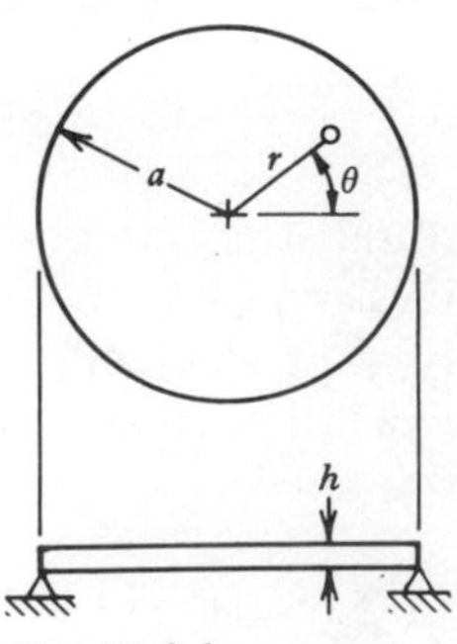

Fig. 10-9.1

the regularity conditions that w, ω_r, M_{rr}, and V_r must be finite at the center of the plate [origin $r = 0$ of the (r, θ)-coordinate system].

Analogous to the expressions for the rectangular plate, we have (Eqs. 10-3.11, 10-5.3, 10-5.5 with $\alpha = 1$, $\beta = r$; see also Eq. 10-7.12)

$$
\begin{aligned}
M_{rr} &= -D\left[w_{,rr} + \nu\left(\frac{w_{,r}}{r} + \frac{w_{,\theta\theta}}{r^2}\right)\right]\\
M_{\theta\theta} &= -D\left[\frac{w_{,r}}{r} + \frac{w_{,\theta\theta}}{r^2} + \nu w_{,rr}\right]\\
M_{rr} + M_{\theta\theta} &= -D(1+\nu)\nabla^2 w\\
M_{r\theta} &= -D(1-\nu)\frac{\partial}{\partial r}\left(\frac{w_{,\theta}}{r}\right)\\
V_r &= -D\left[\frac{\partial}{\partial r}(\nabla^2 w) + (1-\nu)\frac{1}{r}\frac{\partial}{\partial r}\left(\frac{w_{,\theta\theta}}{r}\right)\right]\\
V_\theta &= -D\left[\frac{1}{r}\frac{\partial}{\partial \theta}(\nabla^2 w) + (1-\nu)\frac{\partial^2}{\partial r^2}\left(\frac{w_{,\theta}}{r}\right)\right]\\
\omega_r &= \frac{1}{r}w_{,\theta} \qquad \omega_\theta = -w_{,r}
\end{aligned}
\tag{10-9.4}
$$

Accordingly, for the solid plate, by Eqs. (10-9.3) and (10-9.4), we conclude that $A_2 = B_2 = 0$.

Circular Plates with Simply Supported Edges / For a solid circular plate simply supported at the edges $r = a$, the boundary conditions are, with Eqs. (10-9.3) and (10-9.4) with $A_2 = B_2 = 0$,

$$w(a) = A_1 + B_1 a^2 + \frac{p_0 a^4}{64D} = 0$$

$$-\frac{1}{D}M_{rr}(a) = 2(1+\nu)B_1 + (3+\nu)\frac{p_0 a^2}{16D} = 0$$

Hence, solving these equations for A_1 and B_1, we obtain with Eqs. (10-9.3) and (10-9.4) the following results for the simply supported solid circular plate with uniform lateral pressure $p = p_0$:

$$
\begin{aligned}
w &= \frac{p_0 a^4}{64D}\left[1 - \left(\frac{r}{a}\right)^2\right]\left[\frac{5+\nu}{1+\nu} - \left(\frac{r}{a}\right)^2\right]\\
M_{rr} &= \frac{p_0 a^2}{16}(3+\nu)\left[1 - \left(\frac{r}{a}\right)^2\right]\\
M_{\theta\theta} &= \frac{p_0 a^2}{16}\left[3 + \nu - (1+3\nu)\left(\frac{r}{a}\right)^2\right]
\end{aligned}
\tag{10-9.5}
$$

Circular Plates with Fixed Edges / For a solid circular plate with fixed edge at $r = a$, the boundary conditions with $A_2 = B_2 = 0$ and with Eqs. (10-9.3) and (10-9.4), are

$$w(a) = A_1 + B_1 a^2 + \frac{p_0 a^4}{64D} = 0$$

$$w_\theta(a) = -w_r(a) = -2B_1 a - \frac{p_0 a^3}{15D} = 0$$

Solving these equations for A_1 and B_1, we obtain by Eqs. (10-9.3) and (10-9.4) the following results for the solid circular plate with fixed edges at $r = a$, subject to uniform lateral pressure $p = p_0$:

$$w = \frac{p_0 a^4}{64D}\left[1 - \left(\frac{r}{a}\right)^2\right]^2$$

$$M_{rr} = \frac{p_0 a^2}{16}\left[1 + \nu - (3 + \nu)\left(\frac{r}{a}\right)^2\right] \tag{10-9.6}$$

$$M_{\theta\theta} = \frac{p_0 a^2}{16}\left[1 + \nu - (1 + 3\nu)\left(\frac{r}{a}\right)^2\right]$$

Equations (10-9.3), (10-9.4), (10-9.5), and (10-9.6) summarize the bending theory of simply supported and clamped circular plates subject to uniform lateral pressure. Numerous solutions for other types of plates, loadings and boundary conditions have been presented by Marguerre and Woernle.[3] In particular, Marguerre and Woernle have presented extensive results for orthotropic plates.

Circular Plate with Circular Hole at the Center / For a simply supported circular plate of radius a with circular hole of radius b at the center and subjected to uniform lateral pressure $p = p_0$ (see Case 7 of Table 10-9.2 and Fig. 10-9.2), the boundary conditions are (see Eqs. 10-9.3 and 10-9.4)

$$V_r(b) = -D\left(\frac{4B_2}{b} + \frac{p_0 b}{2D}\right) = 0$$

$$M_{rr}(b) = -D\left\{-(1-\nu)\frac{A_2}{b^2} + 2B_1(1+\nu)\right. \tag{10-9.7}$$

$$\left. + B_2[3 + \nu + 2(1+\nu)\ln b] + \frac{(3+\nu)p_0 b^2}{16D}\right\} = 0$$

and

$$w(a) = A_1 + A_2 \ln a + B_1 a^2 + B_2 a^2 \ln a + \frac{p_0 a^4}{64D} = 0$$

$$M_{rr}(a) = -D\left\{ -(1-\nu)\frac{A_2}{a^2} + 2B_1(1+\nu) \right.$$

$$\left. + B_2[3+\nu+2(1+\nu)\ln a] + \frac{(3+\nu)p_0 a^2}{16D} \right\} = 0 \quad (10\text{-}9.8)$$

Solving these equations for A_1, A_2, B_1, and B_2, we obtain

$$A_1 = -\frac{p_0 a^4}{4D}\left\{ \frac{(1+\nu)\ln\frac{a}{b}\ln a}{(1-\nu)\left(\frac{a}{b}\right)^2\left[\left(\frac{a}{b}\right)^2 - 1\right]} - \frac{(5-\nu)\ln a}{4(1-\nu)\left(\frac{a}{b}\right)^2} \right.$$

$$\left. + \frac{\left(\frac{a}{b}\right)^2 \ln a - \ln b}{2\left(\frac{a}{b}\right)^2\left[\left(\frac{a}{b}\right)^2 - 1\right]} - \frac{(3+\nu)\left[\left(\frac{a}{b}\right)^2 - 1\right]}{8(1+\nu)\left(\frac{a}{b}\right)^2} + \frac{1}{16} \right\}$$

$$A_2 = \frac{p_0 a^4}{4D}\left\{ \frac{(1+\nu)\ln\frac{a}{b}}{(1-\nu)\left(\frac{a}{b}\right)^2\left[\left(\frac{a}{b}\right)^2 - 1\right]} - \frac{(3+\nu)}{4(1-\nu)\left(\frac{a}{b}\right)^2} \right\}$$

$$B_1 = \frac{p_0 a^2}{8D}\left\{ \frac{\left(\frac{a}{b}\right)^2 \ln a - \ln b}{\left(\frac{a}{b}\right)^2\left[\left(\frac{a}{b}\right)^2 - 1\right]} - \frac{(3+\nu)\left[\left(\frac{a}{b}\right)^2 - 1\right]}{4(1+\nu)\left(\frac{a}{b}\right)^2} \right\}$$

$$B_2 = -\frac{p_0 b^2}{8D} \quad (10\text{-}9.9)$$

With these coefficients and Eqs. (10-9.3) and (10-9.4), the displacement and the stress resultants may be computed.

For example, for $a/b = 2$, and $\nu = 0.30$, the maximum displacement is

$$w(b) = w_{max} - 0.682\frac{p_0 a^4}{Eh^3} \quad (10\text{-}9.10)$$

This result is 2.7 percent greater than that for Case 7 of Table 10-9.2.

Except for simple types of loading and shapes of plates, such as a circular shape, the method of finding the bending moment by solving the plate equation (Eq. 10-8.1) is somewhat complicated. However, the results obtained can be reduced to tables or curves of coefficients for the maximum bending moments per unit width of a plate and for the maximum deflections of the plate, and some of these results are presented below.

The bending theory of elastic plates, however, does not make allowance for adjustments that take place when slight local yielding at portions of high stress causes a redistribution of stress. This redistribution of stress, in turn, may result in additional strength of the plate, which may often be incorporated into the design of plates, particularly plates of ductile material. We also observe that the bending theory of plates based upon Eq. (10-8.1) does not take into account the added resistance of the plate resulting from direct tensile stresses that accompany relatively large deflections.

Summary for Circular Plates with Simply Supported Edges /Consider a circular plate with simply supported edges, so that no displacement occurs at the edge. The lateral displacement w and bending moments M_{rr}, $M_{\theta\theta}$ for uniform lateral pressure p are given by Eqs. (10-9.5). The maximum displacement occurs at the center of the plate ($r = 0$). The corresponding maximum stress σ_{max} also occurs at the center of the plate. The value of σ_{max} is tabulated in Table 10-9.1. Results are given in Table 10-9.1 also for the case of a spot load ($P = \pi r_0^2 p$) at the center of the plate, where the solution is reasonably accurate, provided r_0 is a sufficiently small (nonzero) value.

Summary for Circular Plates with Fixed Edges /Consider a circular plate rigidly held (fixed) so that no rotation or displacement occurs at the edge. We observe that under service conditions the edges of plates are seldom completely "fixed," although usually they are subject to some restraint; furthermore, a slight amount of yielding at the fixed edge may destroy much of the effect of the restraint and thereby transfer the moment to the central part of the plate. For these reasons, the restraint at the edges of a plate is considered of less importance, particularly if the plate is made of relatively ductile material, than would be indicated by the results of the theory of flexure of plates with fixed edges. In general, a medium-thick plate with a restrained (so-called fixed) edge will be intermediate in strength between the plate with a simply supported edge and the plate with an ideally fixed edge.

Formulas are given in Table 10-9.1 for the maximum deflection of clamped circular plates of ideal, elastic material.[11] Experiments have verified the formulas for uniformly distributed loads and a simply supported edge. These experiments with fixed-edged plates under uniformly distributed loads show that the formula for the deflection is correct for thin and medium-thick plates $[(h/a) < 0.1]$ for deflections not

Table 10-9.1

Formulas for Values of the Maximum Principal Stresses and Maximum Deflections in Circular Plates as Obtained by Theory of Flexure of Plates†

Support and Loading	*Principal Stress,* σ_{max}	*Point of Maximum Stress*	*Maximum Deflection,* w_{max}
Edge simply supported; load uniform ($r_0 = a$)	$\frac{3}{8}(3+\nu)p\frac{a^2}{h^2}$	Center	$\frac{3}{16}(1-\nu)(5+\nu)\frac{pa^4}{Eh^3}$
Edge fixed; load uniform ($r_0 = a$)	$\frac{3}{4}p\frac{a^2}{h^2}$	Edge	$\frac{3}{16}(1-\nu^2)\frac{pa^4}{Eh^3}$ ‡
Edge simply supported; load at center. $P = \pi r_0^2 p$, $r_0 \to 0$, but $r_0 > 0$	$\frac{3(1+\nu)}{2\pi h^2}P\left(\frac{1}{\nu+1} + \ln\frac{a}{r_0} - \frac{1-\nu}{1+\nu}\frac{r_0^2}{4a^2}\right)$	Center	$\frac{3(1-\nu)(3+\nu)Pa^2}{4\pi Eh^3}$
Fixed edge; load at center. $P = \pi r_0^2 p$, $r_0 \to 0$, but $r_0 > 0$	$\frac{3(1+\nu)P}{2\pi h^2}\left(\ln\frac{a}{r_0} + \frac{r_0^2}{4a^2}\right)$; a must be $> 1.7r_0$	Center	$\frac{3(1-\nu^2)Pa^2}{4\pi Eh^3}$

†a = radius of plate; r_0 = radius of central loaded area; h = thickness of plate; p = uniform load per unit area; ν = Poisson's ratio.
‡For thicker plates ($h/r > 0$) the deflection is $w_{max} = C\frac{3}{16}(1-\nu^2)(pa^4/Eh^3)$, where the constant C depends upon the ratio h/a as follows: $C = 1 + 5.72(h/a)^2$.

larger than about one-half the plate thickness. For thicker plates the measured values of deflection are much larger than those computed by the formula. Two reasons for this discrepancy exist: (a) lack of ideal fixity at the edge and (b) additional deflection in the thicker plates due to the shearing stresses. These experiments suggested that for thicker [$(h/a) > 0.1$] circular plates with fixed edges subjected to uniform loads the values of w_{max} given in Table 10-9.1 be multiplied by a factor that depends on the ratio of the thickness h to the radius r. This factor is $C = 1 + 5.72(h/a)^2$. Experiments on plates with edges securely clamped gave deflections that agreed closely with values computed by the use of the bending theory formula and the constant C.

Formulas for deflections by the bending theory give values that are too large for thin to medium-thick plates when loaded so that the deflections are larger than about one-half the plate thickness.

Summary for Stresses and Deflections in Flat Circular Plates with Central Holes / Circular plates of radius a with circular holes of radius b at their center are commonly used in engineering systems. For example, they occur in thrust-bearing plates, telephone and loudspeaker diaphragms, steam turbines, diffusers, piston heads, etc. Several cases of practical importance have been studied by Wahl and Lobo.[12] In all these cases, the maximum stress is given by simple formulas of the type

$$\sigma_{\max} = k_1 \frac{pa^2}{h^2} \quad \text{or} \quad \sigma_{\max} = \frac{k_1 P}{h^2} \tag{10-9.10}$$

depending on whether the applied load is uniformly distributed over the plate or concentrated along the edge of the central hole. Likewise, the maximum deflections are given by simple formulas of the type

$$w_{\max} = k_2 \frac{pa^4}{Eh^3} \quad \text{or} \quad w_{\max} = k_2 \frac{Pa^2}{Eh^2} \tag{10-9.11}$$

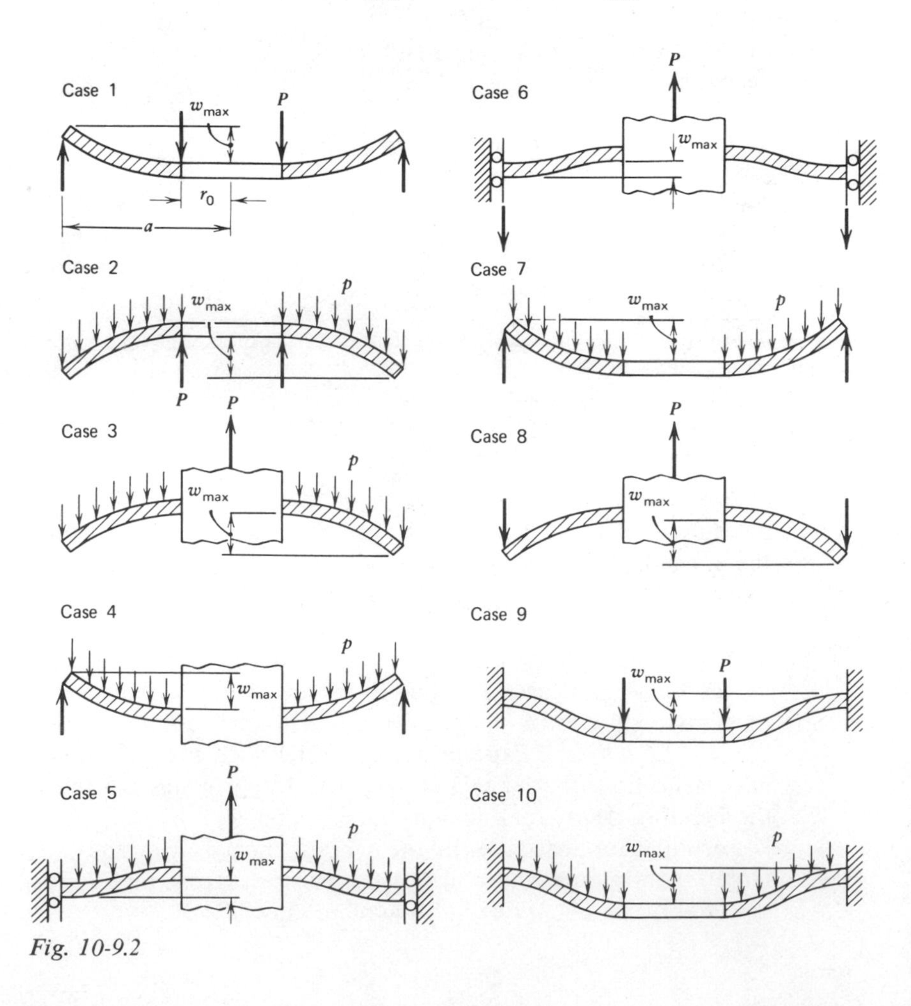

Fig. 10-9.2

Table 10-9.2

Coefficients k_1 and k_2 in Eqs. (10-9.10) and (10-9.11) for the Ten Cases Shown in Fig. 10-9.2: Poisson's Ratio $\nu = 0.30$

$\frac{a}{r_0} =$	1.25		1.5		2		3		4		5	
Case	k_1	k_2	k_1	k_2	k_1	k_2	k_1	k_2	k_1	k_2	k_1	k_2
1	1.10	0.341	1.26	0.519	1.48	0.672	1.88	0.734	2.17	0.724	2.34	0.704
2	0.66	0.202	1.19	0.491	2.04	0.902	3.34	1.220	4.30	1.300	5.10	1.310
3	0.135	0.00231	0.410	0.0183	1.04	0.0938	2.15	0.293	2.99	0.448	3.69	0.564
4	0.122	0.00343	0.336	0.0313	0.74	0.1250	1.21	0.291	1.45	0.417	1.59	0.492
5	0.090	0.00077	0.273	0.0062	0.71	0.0329	1.54	0.110	2.23	0.179	2.80	0.234
6	0.115	0.00129	0.220	0.0064	0.405	0.0237	0.703	0.062	0.933	0.092	1.13	0.114
7	0.592	0.184	0.976	0.414	1.440	0.664	1.880	0.824	2.08	0.830	2.19	0.813
8	0.227	0.00510	0.428	0.0249	0.753	0.0877	1.205	0.209	1.514	0.293	1.745	0.350
9	0.194	0.00504	0.320	0.0242	0.454	0.0810	0.673	0.172	1.021	0.217	1.305	0.238
10	0.105	0.00199	0.259	0.0139	0.480	0.0575	0.657	0.130	0.710	0.162	0.730	0.175

Wahl and Lobo have calculated numerical values for k_1 and k_2 for several values of the ratio a/b and for a Poisson's ratio of $\nu = 0.30$. The cases that they studied are shown in Fig. 10-9.2 and the corresponding values of k_1 and k_2 are tabulated in Table 10-9.2. For other solutions for symmetrical bending of circular plates, the interested reader is referred to Timoshenko and Woinowsky-Krieger.[5]

EXAMPLE 10-9.1

Circular Plate Fixed at Edges

A mild steel plate ($E = 200$ GPa, $\nu = 0.29$, $Y = 315$ MPa) has a thickness $h = 10$ mm and covers a circular opening having a diameter of 200 mm. The plate is fixed at the edges and is subjected to a uniform pressure p. (a) Determine the magnitude of the yield pressure p_Y and the deflection w_{max} at the center of the plate when this pressure is applied. (b) Determine a working pressure based on a factor of safety of $SF = 2.00$ relative to p_Y.

SOLUTION

(a) The maximum stress in the plate is a radial flexure stress at the outer edge of the plate given either by Eq. (10-9.5) and the flexure formula or by the appropriate equation in Table 10-9.1.

$$\sigma_{max} = \frac{3}{4} p_Y \frac{a^2}{b^2} = \frac{3p_Y(100)^2}{4(10)^2} = 75p_Y$$

The magnitude of p_Y by the maximum shearing stress theory of failure is obtained by setting σ_{max} equal to Y

$$p_Y = \frac{Y}{75} = \frac{315}{75} = 4.20 \text{ MPa}$$

The maximum deflection of the plate when this pressure is applied is given by the appropriate equation in Table 10-9.1. Thus,

$$w_{max} = \frac{3}{16}(1 - \nu^2)\frac{p_Y a^4}{Eh} = \frac{3(1 - 0.29^2)(4.20)(100)^4}{16(200 \times 10^3)(10)^3}$$

$$= 0.361 \text{ mm}$$

(b) Let p_w be the working pressure; its value is, based on p_Y,

$$p_w = \frac{p_Y}{SF} = \frac{4.20}{2.00} = 2.10 \text{ MPa}$$

Summary for Large Elastic Deflections of Circular Plates. Clamped Edge and Uniformly Distributed Load/Consider a circular plate of radius a and thickness h (Fig. 10-9.3a). Let the plate be loaded by lateral pressure p that causes a maximum deflection w_{max} that is large compared to the thickness h (Fig. 10-9.3c). Let the edge of the plate be clamped so that rotation and radial displacement are prevented (Fig. 10-9.3b). In Fig. 10-9.3d a diametral strip of one unit width is cut from the plate to show the bending moments per unit of width and the direct tensile forces that act in this strip at the edge and at the center of the plate. The direct tensile forces arise from two sources: first, the fixed support at the edge prevents the edge at opposite ends of a diametral strip from moving radially, thereby causing the strip to stretch as it deflects. Second, if the plate is not clamped at its edge but is simply supported as shown in Figs. 10-9.3e, f, and g, radial stresses arise out of the tendency for outer concentric rings of the plate, such as shown in Fig. 10-9.3h, to retain their original diameter as the plate defects. In Fig. 10-9.3h the concentric ring at the outer edge is shown cut from the plate. This ring tends to

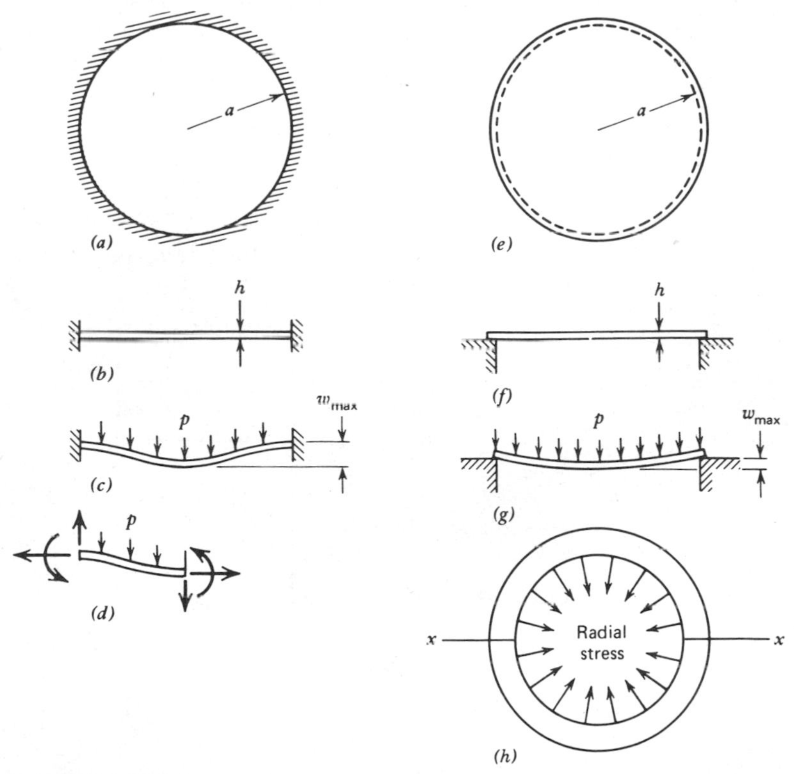

Fig. 10-9.3/Thin plates having large deflections in which direct tension is significant.

retain the original outside diameter of the unloaded plate; the radial tensile stresses acting on the inside of the ring, as shown in Fig. 10-9.3*h*, cause the ring diameter to decrease, and in doing so they introduce compressive stresses on every diametral section such as *xx*. These compressive stresses in the circumferential direction sometimes cause the plate to wrinkle or buckle near the edge, particularly if the plate is simply supported. The radial stresses are usually larger in the central portion of the plate than near the edge.

Thus when the plate is deflected more than about one-half the thickness, there are direct tensile stresses in addition to bending stresses; as will be indicated later, the significant values of these stresses occur either at the edge or at the center of the plate. Let the bending stresses in a radial plane at the edge and center of the plate be designated by σ_{be} and σ_{bc}, respectively, and let the corresponding direct tensile stresses be σ_{te} and σ_{tc}, respectively. Values of these stresses for a plate with clamped edges having a radius a and thickness h and made of a material having a modulus of elasticity E are given in Fig. 10-9.4. In Fig. 10-9.4 the ordinates are values of the stress multiplied by the quantity a^2/Eh^2 (to make dimensionless ordinates), and the abscissas are values of the maximum deflection w_{max} divided by the thickness h.[13] Note that the dimensionless ordinates and abscissas make it possible to use the curves for plates of any dimensions, provided that other conditions are the same. Also note that the bending stress σ_{be} at the fixed edge is the largest of these four stresses. The direct tensile stresses, though small for small deflections (deflections less than about one-half the plate thickness), become relatively much larger as the deflection increases. For example, if the deflection is equal to twice the plate thickness, the direct tensile stress

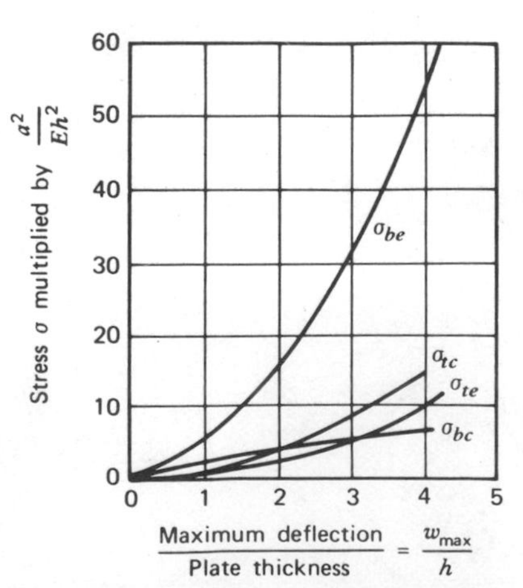

Fig. 10-9.4/Stresses in thin plates having large deflections; circular plate with clamped edges.

σ_{tc} at the center of the plate is equal to the bending stress σ_{bc} at the center; if the deflection is four times the thickness, the stress σ_{tc} is twice σ_{bc}.

Significant Stress; Edges Clamped / The maximum stress in the plate is at the edge and is the sum of the values of the bending stress σ_{be} and the direct tensile stress σ_{te} associated with the curves in Fig. 10-9.4. Values of this maximum stress σ_{max} multiplied by the quantity a^2/Eh^2 are shown as ordinates to the upper curve in Fig. 10-9.5*a*. The values of σ_{max} at points in the plate a short distance radially from the edge are very much smaller than at the edge; a minimum value occurs near the edge, and the stresses gradually approach another maximum value that occurs at the center of the plate. The maximum stress at the center of the plate is indicated by the lower curve in Fig. 10-9.5*a*, which represents the sum of the stresses σ_{bc} and σ_{tc} as given by the curves in Fig. 10-9.4. If the failure of the plate is by general yielding, the maximum stress at the center is the significant stress, since the effect of the maximum stress at the edge is localized. However, if the failure of the plate is by progressive fracture resulting from repeated applications of loads, or if the plate is made of brittle material and hence fails by sudden fracture under static loads, the stress at the edge would be the significant stress.

Load on Plate; Edges Clamped / In Fig. 10-9.5*b* the values of the load p on the plate with fixed edges multiplied by the quantity a^4/Eh^4 are represented

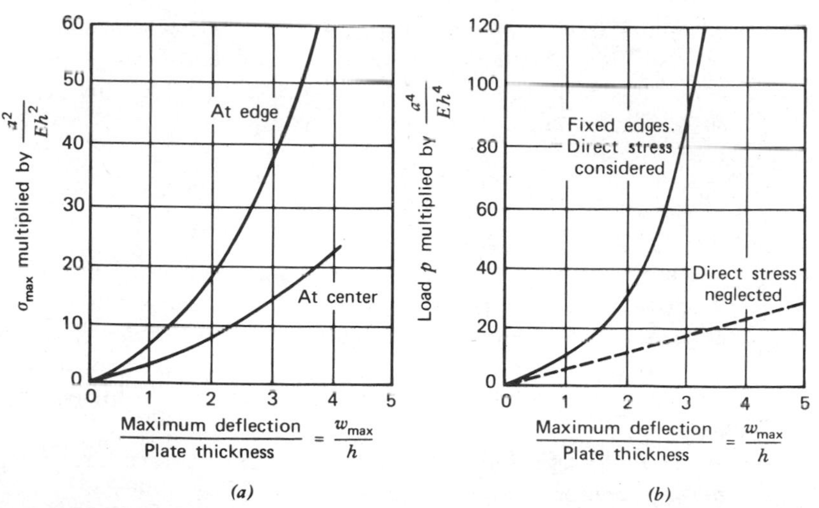

Fig. 10-9.5 / Maximum stresses and deflections in thin plates having large deflections; circular plate with clamped edges.

as ordinates, and maximum deflections divided by the plate thickness are abscissas, thus giving a dimensionless curve. The dashed line represents values of load and maximum deflection as computed by neglecting the effect of direct tensile stresses. A significant increase in the load p is indicated by the upward trend of the curve above the straight line for deflection larger than about one-half the plate thickness, which shows that the plate is much stronger than is indicated by the analysis in which the strength contributed by the direct tensile stress is neglected.

The relation between the load p and the stresses in the plate is obtained by using Figs. 10-9.5a and b jointly. For example, if the dimensions and the modulus of elasticity of the plate and the load p are given, the quantity pa^4/Eh^4 can be computed. In Fig. 10-9.5b the abscissa w_{max}/h corresponding to this value of pa^4/Eh^4 is found from the curve. The value of w_{max}/h thus found is now used as the abscissa in Fig. 10-9.5a, and the stress at the center or the edge of the plate is found by reading the ordinate corresponding to this abscissa to the appropriate curve in Fig. 10-9.5a and dividing it by a^2/Eh^2. This procedure is used in the following example.

EXAMPLE 10-9.2

Large Deflection of a Uniformly Loaded Circular Plate with Clamped Edge

A circular plate of aluminum alloy is 500 mm in diameter and 5 mm thick. The plate is subjected to a uniformly distributed pressure p, and is fixed at its edge. If the maximum pressure that the plate can support is assumed to be that pressure that causes a significant tensile stress equal to the tensile yield stress of the material (say, 288 MPa), (a) determine the allowable magnitude of the pressure p that develops not more than one-half the maximum pressure that the plate can support, and (b) compute the maximum deflection corresponding to this allowable pressure. The modulus of elasticity of the aluminum alloy is $E = 72.0$ GPa.

SOLUTION

We note by Fig. 10-9.5, that neither pressure p nor stress σ are linearly proportional to the deflections. In addition, the stress σ, either at the edge or at the center of the plate, is not linearly proportional to the pressure p. We must therefore apply the reduction factor (factor of safety $SF = 2$) to the load rather than to the stress (see Introduction, Chapter 3).

The factor of safety is applied to the failure pressure for the plate. The plate is assumed to fail by general yielding. As indicated in Fig. 10-9.5a, yielding initiates at the edge of the plate when σ_{max} at the edge is equal to Y. We assume that general yielding failure occurs shortly after the maximum stress at the center of the plate reaches the yield stress of the material; the pressure-deflection curve in Fig. 10-9.5*b* is assumed not to be influenced by the localized yielding at the edge. Hence, we seek the value of the pressure p that will cause a stress of 288 MPa at the center of the plate; this value of p is then to be reduced by the factor $SF = 2$.

Accordingly, we compute the factor

$$\frac{\sigma a^2}{Eh^2} = \frac{288(250^2)}{72 \times 10^3(5^2)} = 10$$

With the value 10 as the ordinate in Fig. 10-9.5*a* for the curve σ_{max} at the center of the plate, we read the corresponding abscissa $w_{max}/h = 2.4$. By Fig. 10-9.5*b*, with the abscissa equal to 2.4, we find

$$\frac{pa^4}{Eh^4} = 50$$

The value of p determined from this ratio ($p = 576$ kPa) represents the maximum pressure that the plate can support without yielding over a large portion of its volume. Therefore, one-half p or 288 kPa is considered the allowable magnitude of pressure that the plate may support.

(b) To determine the maximum deflection, we compute first the quantity

$$\frac{pa^4}{Eh^4} = \frac{0.288(250^4)}{72 \times 10^3(5^4)} = 25$$

By Fig. 10-9.5*b*, we find the corresponding abscissa $w_{max}/h = 1.8$. Hence, the deflection of the center of the plate is $w_{max} = 1.8h = 9.00$ mm.

Summary for Large Elastic Deflections of Circular Plates. Simply Supported Edge and Load Uniformly Distributed / It was found that, when the edge of a circular plate as shown in Fig. 10-9.3 is fixed and the plate is subjected to a uniformly distributed load, there exist direct radial tensile stresses in addition to the bending stresses. If a circular plate has its edge simply supported instead of fixed, the direct tensile stresses have somewhat smaller magnitudes, but they are still very effective in increasing the load resistance of the plate, particularly when the deflections are large relative to the thickness of the plate.

In Fig. 10-9.6*a* the ordinates to the curve marked σ_{tc} represent the direct tensile stresses at the center of the simply supported plate where these stresses are a maximum, and the ordinates to the curve marked σ_{bc}

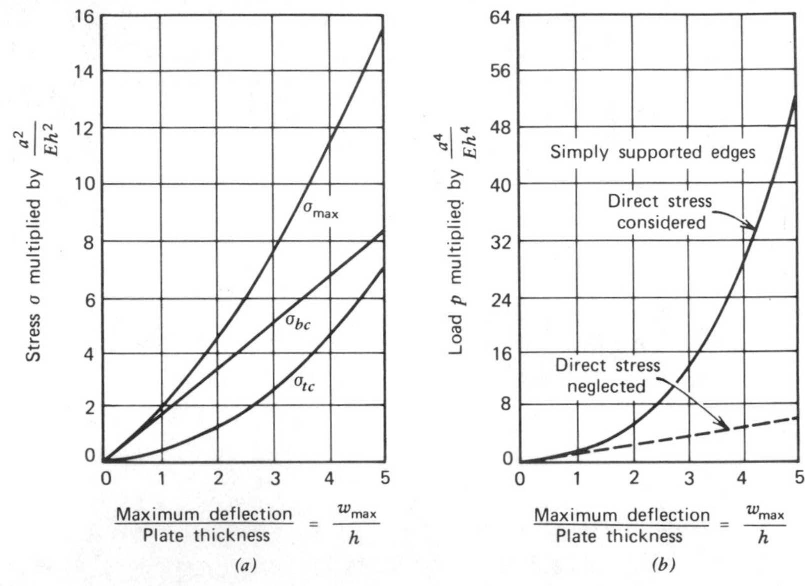

Fig. 10-9.6/Stresses in thin circular plates having large deflections; edges simply supported.

represent the bending stresses at the center of the plate that also have a maximum value at the center. The coordinates to the curves in Figs. 10-9.6*a* and *b* have the same meaning as those for Figs. 10-9.5*a* and *b* for a plate whose edge is fixed. In Fig. 10-9.6*a* the ordinates to the curve marked σ_{max} represent the sum of the stresses σ_{tc} and σ_{bc} that occur on the tensile side of the plate at the center. In Fig. 10-9.6*b* the curve represents the relation between the load and the maximum deflection, and the dashed line represents this relationship if the direct tensile stresses are neglected in the analysis. The solid curve in Fig. 10-9.6*b*, which rises above the dashed line when the maximum deflection becomes greater than one-half to one times the thickness of the plate, shows the influence of the direct tensile stress in increasing the elastic load resistance, especially of relatively thin plates for which the deflections are likely to be large in comparison with the thickness. Figures 10-9.6*a* and *b* are used in solving problems in a manner similar to the use of Figs. 10-9.5*a* and *b* as described in Example 10-9.2.

Rectangular or Other Shaped Plates with Large Deflections/The general behavior described for circular plates when the deflections are large also apply to rectangular, elliptical, or other shapes of plates. Curves giving data for rectangular plates similar to those given in Figs. 10-9.5 and 10-9.6 for circular plates are given by Ramberg et al.[14]

PROBLEM SET 10-9

1. Verify Eq. (10-9.3).

2. With Eqs. (10-9.3) and (10-9.4) and the boundary conditions for a solid circular plate simply supported at the outer edge, $r = a$, derive the results of Eqs. (10-9.5).

3. Repeat Problem 2 for the case of the solid circular plate with fixed edge at $r = a$; that is, derive Eqs. (10-9.6).

4. Derive Eqs. (10-9.9) and, hence, verify Eq. (10-9.10).

5. The cylinder of a steam engine is 400 mm in diameter, and the maximum steam pressure is 690 kPa. Find the thickness of the cylinder head which is a steel flat plate, assuming that the working stress is $\sigma_w = 82.0$ MPa. Determine the maximum deflection of the cylinder head. The plate has fixed edges. For the steel, $E = 200$ GPa and $\nu = 0.29$.

6. A cast-iron disk valve is a flat circular plate 300 mm in diameter and is simply supported. The plate is subjected to uniform pressure supplied by a head of 60 m of water. Find the thickness of the disk using a working stress of $\sigma_w = 14$ MPa. Determine the maximum deflection of the plate. For case iron, $E = 100$ GPa and $\nu = 0.20$.

   Ans. $h = 33.7$ mm, $w_{max} = 0.061$ mm

7. A circular plate is made of steel ($E = 200$ GPa, $\nu = 0.29$, and $Y = 276$ MPa), has a radius $a = 250$ mm, and has a thickness $h = 25$ mm. The plate is simply supported and is subjected to a uniform pressure $p = 1.38$ MPa. (a) Determine the maximum bending stress in the plate and the maximum deflection. (b) Determine the pressure p_Y that is required to initiate yielding in the plate and the factor of safety against initiation of yielding in the plate.

8. A circular steel plate with central hole is fixed at the central hole and uniformly loaded as indicated in Case 7 of Fig. 10-9.2. For the plate $a = 300$ mm, $r_0 = 100$ mm, $h = 10$ mm, $p = 100$ kPa, $E = 200$ GPa, and $Y = 290$ MPa. (a) Determine the maximum bending stress and the maximum deflection. (b) What is the factor of safety against initiation of yielding?

   Ans. (a) $\sigma_{max} = 194$ MPa, $w_{max} = 1.19$ mm
   (b) $SF = 1.50$

9. A circular opening in the flat end of a nuclear reactor pressure vessel is 254 mm in diameter. A circular steel plate 2.54 mm thick, with tensile yield stress $Y = 241$ MPa is used as a cover for the opening. When the cover plate is inserted in the opening, its edges are clamped securely. Determine the maximum internal pressure to which the vessel may be subjected if it is limited by the condition that it must not exceed one-third the pressure that will cause general yielding of the cover plate. $E = 200$ GPa for steel.

10. A circular plate made of aluminum alloy ($E = 72.0$ GPa and $Y = 276$ MPa) is to have a 254 mm diameter. The edge of the plate is to be clamped, and a pressure of $p = 73.8$ kPa is to be applied. Determine the required thickness h of the plate, so that this pressure (73.8 kPa) is two-thirds of the pressure that will cause the plate to just reach yield.

    Hint: Here, the stress at the edge of the plate is the significant stress, since no yield of the plate is permitted. Use Figs. 10-9.5*a* and *b* to solve for h by trial and error, with a value $p = \frac{3}{2} \times 73.8$ kPa $= 110.7$ kPa, and $\sigma = 276$ MPa.

    Ans. $h = 2.0$ mm

11. A circular steel plate whose diameter is 2.54 m and whose thickness is 12.7 mm is simply supported at its edge and is subjected to a uniformly distributed pressure p. The tensile yield point stress of the steel is 207 MPa. Determine the pressure p_Y that produces a maximum stress in the plate equal to the tensile yield point stress. Determine the maximum deflection for this pressure.

12. In Problem 11, determine the pressure p that produces a maximum stress at the center of the plate equal to one-half the yield point stress. Compare this pressure to that determined in Problem 11. Explain the result.

    Ans. $p = 14.0$ kPa

13. Rework Problem 9 for the case of a simply supported edge.

14. Let the aluminum plate in Problem 10 have a thickness of 2.0 mm and have simply supported edges. Determine the magnitude of the internal pressure p that can be applied to the plate if the pressure is two-thirds the pressure that will cause the plate to just reach yield.

    Ans. $p = 154$ kPa

REFERENCES

1. A. P. Boresi and P. P. Lynn, *Elasticity in Engineering Mechanics*, 2nd Ed., Prentice-Hall, Englewood Cliffs, New Jersey, 1974.
2. H. L. Langhaar, *Foundations of Practical Shell Analysis*, revised ed., Department of Theoretical and Applied Mechanics, University of Illinois, Urbana, 1964.
3. K. Marguerre and H.-T. Woernle, *Elastic Plates*, Blaisdell, Waltham, Massachusetts, 1969.
4. H. E. Newell, Jr., *Vector Analysis*, McGraw-Hill, New York, 1955.
5. S. Timoshenko and S. Woinowsky-Krieger, *Theory of Plates and Shells*, 2nd Ed., McGraw-Hill, New York, 1959.
6. H. L. Langhaar, *Energy Methods in Applied Mechanics*, Wiley, New York, 1962.
7. M. Naruoka, *Bibliography on Theory of Plates*, 1st Ed., Gihodo, Tokyo, 1981.
8. H. M. Westergaard, "Moments and Stresses in Slabs," *Proceedings of the American Concrete Institute*, Vol. 17, 1921.
9. M. Levy, *Comptes Rendes*, Vol. 129, 1899, pp. 535–539.
10. C. Bach, *Elastizität und Festigkeit*, 8th Ed., Julius Springer, Berlin, 1920, p. 598.
11. A. Morley, *Strength of Materials*, 8th Ed., Longmans, Green, London, 1935.
12. A. M. Wahl and G. Lobo, "Stresses and Deflections in Flat Circular Plates with Central Holes," *Transactions of the American Society of Mechanical Engineers*, Vol. 52, 1930, pp. 29–43.
13. J. Prescott, *Applied Elasticity*, Dover, New York, 1946, pp. 455–469.
14. W. A. Ramberg, A. E. McPherson, and S. Levey, "Normal Pressure Tests of Rectangular Plates," Report 748, National Advisory Committee for Aeronautics, 1942.

Additional References

1. D. O. Brush and B. O. Almroth, *Buckling of Bars, Plates, and Shells*, McGraw-Hill, New York, 1975.
2. E. H. Mansfield, *The Bending and Stretching of Plates*, Macmillan, New York, 1964.
3. R. Szilard, *Theory and Analysis of Plates*, Prentice-Hall, Englewood Cliffs, New Jersey, 1974.
4. A. C. Ugural, *Stresses in Plates and Shells*, McGraw-Hill, New York, 1981.
5. J. R. Vinson, *Structural Mechanics: The Behavior of Plates and Shells*, Wiley, New York, 1974.

CHAPTER 11

THE THICK-WALL CYLINDER

11-1 BASIC RELATIONS

In this article, we derive basic relations for the axisymmetric deformation of a thick-wall cylinder. Thick-wall cylinders are used widely in industry as pressure vessels, pipes, gun tubes, etc. In many applications the cylinder wall thickness is constant, and the cylinder is subjected to a uniform internal pressure p_1, a uniform external pressure p_2, an axial load P, and a temperature change T (measured from an initial uniform reference cylinder temperature, see Art. 2-5), (Fig. 11-1.1). Often the temperature change T is a function of the radial coordinate r only (Fig. 11-1.1). Under such conditions, the deformations of the cylinder are symmetrical with respect to the axis of the cylinder (axisymmetric). Furthermore, the deformations at a cross section sufficiently far removed from the junction of the cylinder and its end caps (Fig. 11-1.1) are practically independent of the axial coordinate z. In particular, if the cylinder is open (no end caps) and unconstrained, it undergoes axisymmetric deformations due to pressures p_1 and p_2 and temperature change $T = T(r)$, which are independent of z. If the cylinder's deformation is constrained by supports or by end caps, then in the vicinity of the supports or the junction between the cylinder and the end caps, the deformation and the stresses will depend on the axial coordinate z. For example, consider a pressure tank formed by welding together hemispherical caps and a cylinder (Fig. 11-1.2). Under the action of an internal pressure p_1, the tank deforms as indicated by the dotted inside boundary and the long dashed outside boundary (the deformations are exaggerated in Fig. 11-1.2). If the cylinder were not constrained by the end caps, it would be able to undergo a larger radial displacement. However, at the junctions between the hemispherical caps and the cylinder, the cylinder displacement is constrained by the stiff hemispherical caps. Consequently, the radial displacement (hence the strains and stresses) at cylinder cross sections near the end cap junctions differ from those at sections far removed from the end cap junctions. In this article, we consider the displacement, strains, and stresses at sections far removed from the end caps. The difficult problem of the determination of deformations, strains, and stresses near the junction of the thick-wall end caps and the thick-wall cylinder lies outside of the scope of our treatment. This problem often is treated by experimental methods, since its analytical solution depends upon a general three-dimensional study in the theory of elasticity (or plasticity if yielding occurs near the end cap junctions). For thin-wall cylinders, the stress near the end cap junctions may be estimated by the procedure outline in Art. 9-7 (see Problem 9-7.4).

Consequently, the solution presented in this chapter for thick-wall cylinders is applicable to sections sufficiently far from the end cap

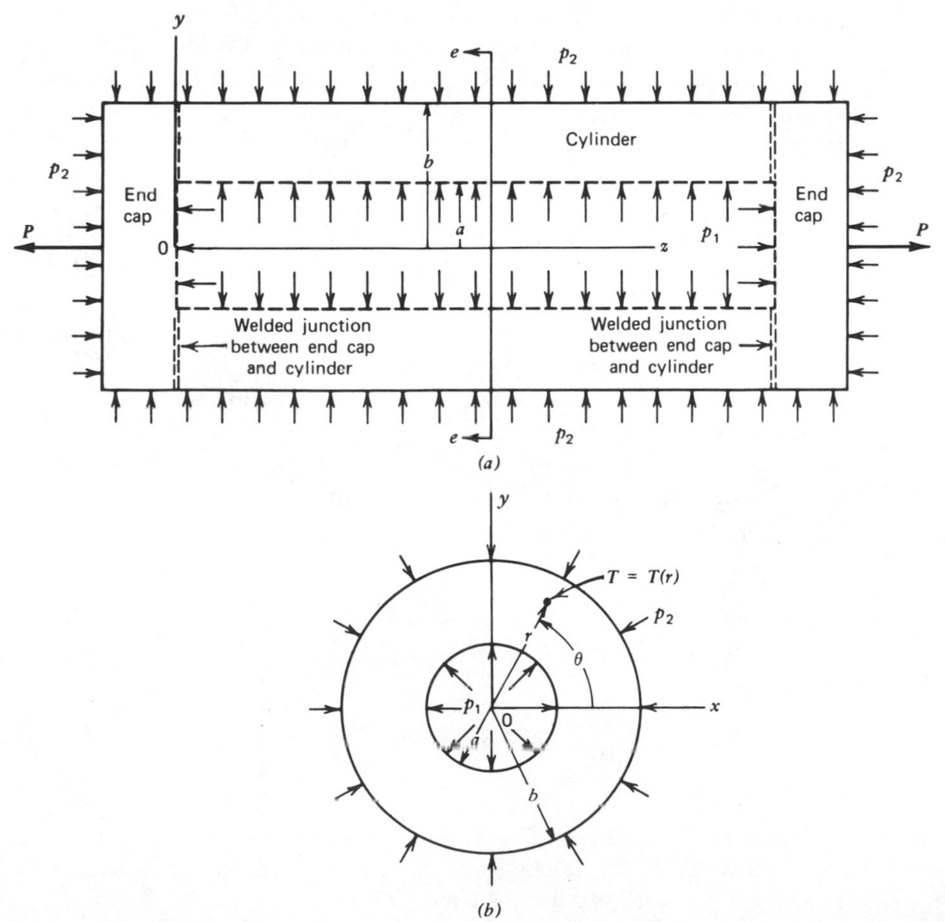

Fig. 11-1.1/Closed cylinder with internal pressure, external pressure, and axial loads. (*a*) Closed cylinder. (*b*) Section $e-e$.

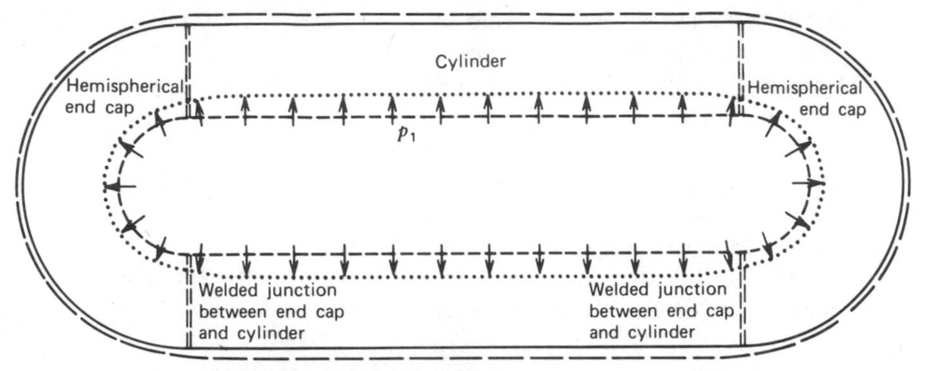

Fig. 11-1.2/Closed cylinder with hemispherical ends.

junctions so that the effects of the constraints imposed by the end caps are negligible. The solution is applicable also to thick-wall cylinders that do not have end caps, so-called open cylinders. Since only axially symmetrical loads and constraints are admitted, the solution is axisymmetrical, that is, a function of radial coordinate r, only.

We use cylindrical coordinates r, θ, z for radial, circumferential, and axial directions (Fig. 11-1.1). Let the cylinder be loaded as shown in Fig. 11-1.1. For analysis purposes, we remove a thin annulus of thickness dz from the cylinder (far removed from the end junctions) by passing two planes perpendicular to the z-axis, a distance dz apart (Fig. 11-1.3a). The cylindrical volume element $dr(r\,d\theta)\,dz$ shown in Fig. 11-1.3b is removed from the annulus. Because of radial symmetry, no shearing stresses act on the volume element and normal stresses are functions of r only. The nonzero stress components are principal stresses σ_{rr}, $\sigma_{\theta\theta}$, and σ_{zz}. The distributions of these stresses through the wall thickness are determined by the equations of equilibrium, compatibility relations, stress-strain relations, and by material response data.

Equation of Equilibrium/We neglect body force components. Hence, the equations of equilibrium for cylindrical coordinates (Eqs. 1-5.6) reduce to the single equation

$$r\frac{d\sigma_{rr}}{dr} = \sigma_{\theta\theta} - \sigma_{rr}, \quad \text{or} \quad \frac{d}{dr}(r\sigma_{rr}) = \sigma_{\theta\theta} \tag{11-1.1}$$

Strain-Displacement Relations and Compatibility Condition/The strain-displacement relations for the thick-walled cylinder (Eqs. 1-9.6) yields the three relations for extensional strains (since $v = 0$)

$$\epsilon_{rr} = \frac{\partial u}{\partial r} \qquad \epsilon_{\theta\theta} = \frac{u}{r} \qquad \epsilon_{zz} = \frac{\partial w}{\partial z} \tag{11-1.2}$$

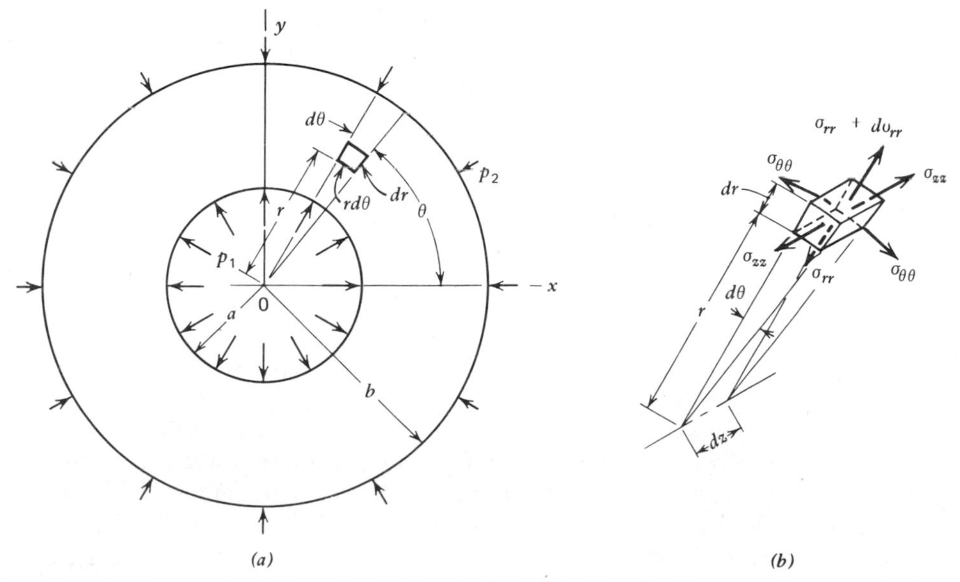

Fig. 11-1.3/Stresses in thick-wall cylinder. (*a*) Thin annulus of thickness *dz* perpendicular to the plane of the figure. (The *z*-axis is perpendicular to the plane of the figure.) (*b*) Cylindrical volume element of thickness *dz*.

where $u = u(r, z)$, $w = w(r, z)$ denote displacement components in the r- and z-directions, respectively. At sections far removed from the ends, the dependency upon z in u and w is considered to be small. Hence, at sections far from the ends, the shearing strain components are zero because of radial symmetry; furthermore, we assume ϵ_{zz} is constant. Eliminating the displacement $u = u(r)$ from the first two of Eqs. (11-1.2), we obtain

$$r\frac{d\epsilon_{\theta\theta}}{dr} = \epsilon_{rr} - \epsilon_{\theta\theta}, \quad \text{or} \quad \frac{d}{dr}(r\epsilon_{\theta\theta}) = \epsilon_{rr} \tag{11-1.3}$$

Equation (11-1.3) is the strain compatibility condition for the thick-wall cylinder.

Stress-Strain Relations/The material of the cylinder is taken to be isotropic and linearly elastic. Hence, the stress-strain-temperature relations are (see Eqs. 2-5.5), since $\epsilon_{r\theta} = \epsilon_{rz} = \epsilon_{\theta z} = 0$

$$\begin{aligned}
\epsilon_{rr} &= \frac{1}{E}[\sigma_{rr} - \nu(\sigma_{\theta\theta} + \sigma_{zz})] + kT \\
\epsilon_{\theta\theta} &= \frac{1}{E}[\sigma_{\theta\theta} - \nu(\sigma_{rr} + \sigma_{zz})] + kT \\
\epsilon_{zz} &= \frac{1}{E}[\sigma_{zz} - \nu(\sigma_{rr} + \sigma_{\theta\theta})] + kT = \text{constant}
\end{aligned} \tag{11-1.4}$$

where E, ν, and k denote the modulus of elasticity, Poisson's ratio, and the coefficient of linear thermal expansion, respectively. The term T in Eq. (11-1.4) represents the change in temperature measured from a uniform reference temperature (constant throughout the cylinder initially); see Art. 2-5.5, and reference 3 of Chapter 2.

Material Response Data / For a cylinder made of isotropic linearly elastic material, the material response data are represented by the results of tests required to determine the elastic constants (modulus of elasticity E and Poisson's ratio ν) and the coefficient of linear thermal expansion k. In order to determine the maximum elastic loads for the cylinder, the material data must include either the yield stress Y, obtained from a tension test, or the shearing yield stress τ_Y obtained from a torsion test of a hollow thin-wall tube. If the material response indicates that the material has a yield point (Fig. 2-1.1*b*), the value of either the yield point stress Y or shearing yield point stress τ_Y is needed to calculate the fully plastic pressure for the cylinder.

11-2 STRESS COMPONENTS FOR A CYLINDER WITH CLOSED ENDS

In this article, we obtain expressions for the stress components $\sigma_{rr}, \sigma_{\theta\theta}, \sigma_{zz}$, for a cylinder with closed ends; the cylinder is subjected to internal pressure p_1, external pressure p_2, axial load P, and temperature change T (Fig. 11-1.1).

We may express Eq. (11-1.3) in terms of $\sigma_{rr}, \sigma_{\theta\theta}, \sigma_{zz}$ and their derivatives with respect to r, by substitution of the first two of Eqs. (11-1.4) into Eq. (11-1.3). Since $\epsilon_{zz} = \text{constant}$, the last of Eqs. (11-1.4) may be used to express the derivative $d\sigma_{zz}/dr$ in terms of the derivatives of σ_{rr}, $\sigma_{\theta\theta}$, and T with respect to r. By means of this expression, we may eliminate $d\sigma_{zz}/dr$ from Eq. (11-1.3) to rewrite Eq. (11-1.3) in terms of σ_{rr}, $\sigma_{\theta\theta}$, and derivatives of σ_{rr}, $\sigma_{\theta\theta}$, and T. Since the undifferentiated terms in σ_{rr} and $\sigma_{\theta\theta}$ occur in the form $\sigma_{rr} - \sigma_{\theta\theta}$, Eq. (11-1.1) may be used to eliminate $\sigma_{rr} - \sigma_{\theta\theta}$. Hence, we obtain the differential expression

$$\frac{d}{dr}\left[\sigma_{rr} + \sigma_{\theta\theta} + \frac{kET}{1-\nu}\right] = 0 \tag{11-2.1}$$

Incorporated in Eq. (11-2.1) is the equation of equilibrium, Eq. (11-1.1), the strain compatibility equation, Eq. (11-1.3), and the stress-strain-temperature relations, Eqs. (11-1.4).

Integration of Eq. (11-2.1) yields the result

$$\sigma_{rr} + \sigma_{\theta\theta} + \frac{kET}{1-\nu} = 2C_1 \tag{11-2.2}$$

where $2C_1$ is a constant of integration (the factor 2 is included for simplicity of form in subsequent expressions). Elimination of the stress component $\sigma_{\theta\theta}$ between Eqs. (11-1.1) and (11-2.2) yields the following differential expression for σ_{rr}

$$\frac{d}{dr}\left(r^2\sigma_{rr}\right) = -\frac{kETr}{1-\nu} + 2C_1 r \tag{11-2.3}$$

Integration of Eq. (11-2.3) yields the result

$$\sigma_{rr} = -\frac{kE}{r^2(1-\nu)}\int_a^r Tr\,dr + \left(1 - \frac{a^2}{r^2}\right)C_1 + \frac{C_2}{r^2} \tag{11-2.4}$$

where the integration is carried out from the inner radius a of the cylinder (Fig. 11-1.1) to the radius r, and C_2 is a second constant of integration. Substitution of Eq. (11-2.4) into Eq. (11-2.2) yields the result

$$\sigma_{\theta\theta} = \frac{kE}{r^2(1-\nu)}\int_a^r Tr\,dr - \frac{kET}{1-\nu} + \left(1 + \frac{a^2}{r^2}\right)C_1 - \frac{C_2}{r^2} \tag{11-2.5}$$

By Eqs. (11-2.4) and (11-2.5), we obtain

$$\sigma_{rr} + \sigma_{\theta\theta} = 2C_1 - \frac{kET}{1-\nu} \tag{11-2.6}$$

Equation (11-2.6) serves as a check upon the computations (see Eq. 11-2.2). The constants of integration C_1 and C_2 are obtained from the boundary conditions $\sigma_{rr} = -p_1$ at $r = a$ and $\sigma_{rr} = -p_2$ at $r = b$ (Fig. 11-1.1). Substituting these boundary conditions into Eq. (11-2.4), we find

$$C_1 = \frac{1}{b^2 - a^2}\left[p_1 a^2 - p_2 b^2 + \frac{kE}{1-\nu}\int_a^b Tr\,dr\right], \qquad C_2 = -p_1 a^2 \tag{11-2.7}$$

Hence, Eq. (11-2.6) may be written as

$$\frac{\sigma_{rr} + \sigma_{\theta\theta}}{2} = \frac{p_1 a^2 - p_2 b^2}{b^2 - a^2} - \frac{kET}{2(1-\nu)} + \frac{kE}{(1-\nu)(b^2 - a^2)}\int_a^b Tr\,dr \tag{11-2.8}$$

To obtain σ_{zz}, we integrate each term of the last of Eqs. (11-1.4) over the cross sectional area of the cylinder. Thus, we have

$$\int_a^b \epsilon_{zz} 2\pi r\,dr = \frac{1}{E}\int_a^b \sigma_{zz} 2\pi r\,dr - \frac{2\nu}{E}\int_a^b \frac{\sigma_{\theta\theta} + \sigma_{rr}}{2} 2\pi r\,dr + k\int_a^b T 2\pi r\,dr \tag{11-2.9}$$

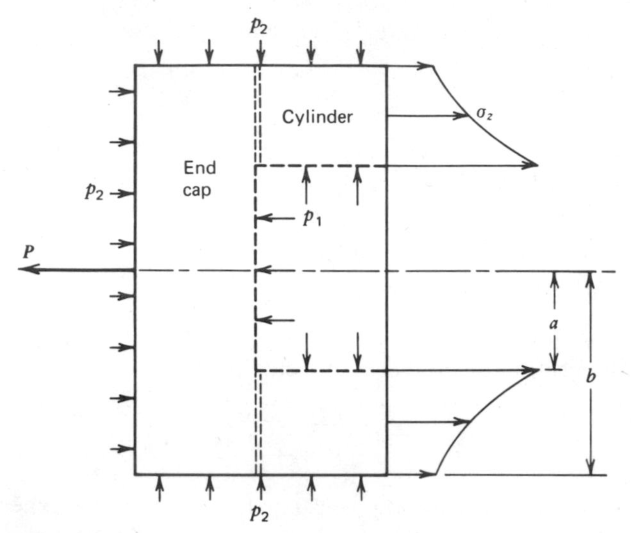

Fig. 11-2.1

For sections far removed from the end section ϵ_{zz} is a constant, and the integral of σ_{zz} over the cross sectional area is equal to the applied loads. Hence, because of pressures p_1, p_2, and axial load P applied to an end plate (Fig. 11-2.1), overall equilibrium in the axial direction requires

$$\int_a^b \sigma_{zz} 2\pi r\, dr = P + \pi\left(p_1 a^2 - p_2 b^2\right) \tag{11-2.10}$$

If there is no axial load P applied to the closed ends, $P = 0$.

Since the temperature change T does not appear in Eq. (11-2.10), the effects of temperature on σ_{zz} are self-equilibrating. With Eqs. (11-2.8), (11-2.9), and (11-2.10), the expression for ϵ_{zz} at a section far removed from the ends can be written in the form

$$\epsilon_{zz} = \frac{1-2\nu}{E(b^2-a^2)}\left(p_1 a^2 - p_2 b^2\right) + \frac{P}{\pi(b^2-a^2)E} + \frac{2k}{b^2-a^2}\int_a^b Tr\, dr \tag{11-2.11}$$

Substitution of Eq. (11-2.11) into the last of Eqs. (11-1.4), with Eq. (11-2.8), yields the following expression for σ_{zz} for a section far removed from the closed ends of the cylinder

$$\sigma_{zz} = \frac{p_1 a^2 - p_2 b^2}{b^2 - a^2} + \frac{P}{\pi(b^2-a^2)} - \frac{kET}{1-\nu} + \frac{2kE}{(1-\nu)(b^2-a^2)}\int_a^b Tr\, dr \tag{11-2.12}$$

Open Cylinder / If a cylinder has open ends and if there is no axial load applied on its ends, overall equilibrium of an axial portion of the cylinder (Fig.

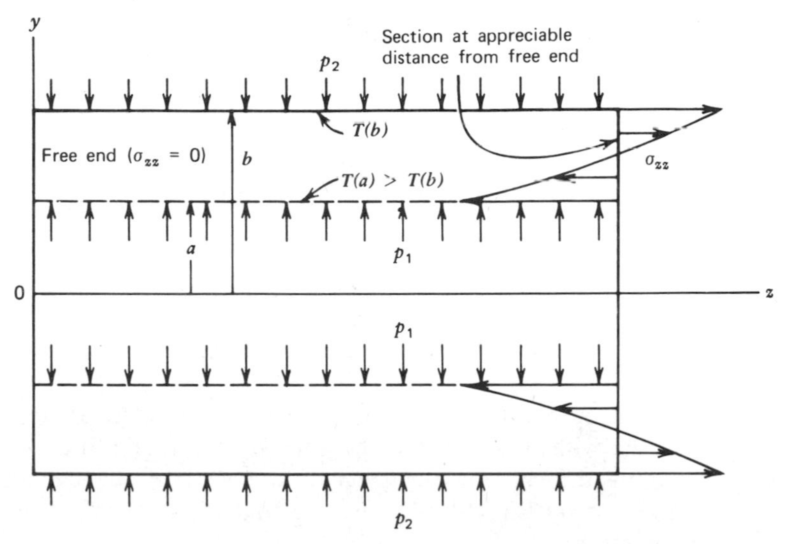

Fig. 11-2.2/Self-equilibrating axial stress distribution in an open cylinder.

11-2.2) requires that

$$\int_a^b 2\pi r\sigma_{zz}\,dz = 0 \tag{11-2.13}$$

Then by Eqs. (11-2.8), (11-2.9), and (11-2.13), the expression for ϵ_{zz} may be written in the form (also by Eqs. 11-2.10, 11-2.11, and 11-2.13)

$$\epsilon_{zz} = \frac{2\nu\left(p_2b^2 - p_1a^2\right)}{(b^2 - a^2)E} + \frac{2k}{b^2 - a^2}\int_a^b Tr\,dr \tag{11-2.14}$$

and for σ_{zz}, we obtain by Eqs. (11-1.4), (11-2.8), and (11-2.14),

$$\sigma_{zz} = \frac{kE}{1-\nu}\left(\frac{2}{b^2 - a^2}\int_a^b Tr\,dr - T\right) \tag{11-2.15}$$

We note, by Eq. (11-2.15), that if the temperature change $T = 0$, $\sigma_{zz} = 0$. However, $\epsilon_{zz} \neq 0$ (see Eq. 11-2.14), since the Poisson ratio $\nu \neq 0$.

11-3 STRESS COMPONENTS AND RADIAL DISPLACEMENT FOR CONSTANT TEMPERATURE

Stress Components/In the absence of temperature change we set $T = 0$. Then Eqs. (11-2.4), (11-2.5), (11-2.6), (11-2.7), and (11-2.12) may be used to obtain the following expressions for the stress components in a closed cylinder (cylinder with end caps)

$$\sigma_{rr} = \frac{p_1 a^2 - p_2 b^2}{b^2 - a^2} - \frac{a^2 b^2}{r^2(b^2 - a^2)}(p_1 - p_2) \qquad (11.3.1)$$

$$\sigma_{\theta\theta} = \frac{p_1 a^2 - p_2 b^2}{b^2 - a^2} + \frac{a^2 b^2}{r^2(b^2 - a^2)}(p_1 - p_2) \qquad (11\text{-}3.2)$$

$$\sigma_{zz} = \frac{p_1 a^2 - p_2 b^2}{b^2 - a^2} + \frac{P}{\pi(b^2 - a^2)} = \text{constant} \qquad (11\text{-}3.3)$$

$$\sigma_{rr} + \sigma_{\theta\theta} = \frac{2(p_1 a^2 - p_2 b^2)}{b^2 - a^2} = \text{constant} \qquad (11\text{-}3.4)$$

For an open cylinder in the absence of axial force P, $\sigma_{zz} = 0$ by Eq. (11-2.15) with $T = 0$. Since the sum $\sigma_{rr} + \sigma_{\theta\theta}$ and the stress σ_{zz} are constants through the thickness of the wall of the cylinder, by Eq. (11-2.9), or by Eq. (11-2.11), we see that ϵ_{zz} is constant (extension or compression).

Radial Displacement for Closed Cylinder/For no temperature change, $T = 0$. Then the radial displacement u for a point in a thick-wall closed cylinder (cylinder with end caps) may be obtained by the second of Eqs. (11-1.2), the second of Eqs. (11-1.4) and Eqs. (11-3.1), (11-3.2), and (11-3.3). The resulting expression for u is

$$u_{(\text{closed end})} = \frac{r}{E(b^2 - a^2)}\left[(1 - 2\nu)(p_1 a^2 - p_2 b^2) + \frac{(1+\nu)a^2 b^2}{r^2}(p_1 - p_2) - \nu\frac{P}{\pi}\right] \qquad (11\text{-}3.5)$$

Radial Displacement for Open Cylinder/Of special interest are open cylinders (cylinder without end caps), since an open inner cylinder is often shrunk to fit inside an open outer cylinder to increase the strength of the resulting composite cylinder. For an open cylinder, in the absence of temperature changes ($T = 0$), Eq. (11-2.15) yields $\sigma_{zz} = 0$. Hence, proceeding as for the closed cylinder, we obtain

$$u_{(\text{open end})} = \frac{r}{E(b^2 - a^2)}\left[(1 - \nu)(p_1 a^2 - p_2 b^2) + \frac{(1+\nu)a^2 b^2}{r^2}(p_1 - p_2)\right] \qquad (11\text{-}3.6)$$

EXAMPLE 11-3.1

Stresses in Hollow Cylinder

A thick-wall cylinder is made of steel ($E = 200$ GPa and $\nu = 0.29$), has an inside diameter of 20 mm, and an outside diameter of 100 mm. The cylinder is subjected to an internal pressure of 300 MPa. Determine the stress components σ_{rr} and $\sigma_{\theta\theta}$ at $r = a = 10$ mm, $r = 25$ mm, and $r = b = 50$ mm.

SOLUTION

The external pressure $p_2 = 0$. Equations (11-3.1) and (11-3.2) simplify to

$$\sigma_{rr} = p_1 \frac{a^2(r^2 - b^2)}{r^2(b^2 - a^2)}$$

$$\sigma_{\theta\theta} = p_1 \frac{a^2(r^2 + b^2)}{r^2(b^2 - a^2)}$$

Substitution of values for r equal to 10 mm, 25 mm, and 50 mm, respectively, into these equations yields the following results:

Stress	*r = 10 mm*	*r = 25 mm*	*r = 50 mm*
σ_{rr}	−300.0 MPa	−37.5 MPa	0.0
$\sigma_{\theta\theta}$	325.0 MPa	62.5 MPa	25.0 MPa

EXAMPLE 11-3.2
Stresses and Deformations in Hollow Cylinder

A thick-wall closed end cylinder is made of an aluminum alloy ($E = 72$ GPa and $\nu = 0.33$), has an inside diameter of 200 mm, and has an outside diameter of 800 mm. The cylinder is subjected to an internal pressure of 150 MPa. Determine the principal stresses and the maximum shearing stress at the inner radius ($r = a = 100$ mm) and the increase in the inside diameter due to the internal pressure.

SOLUTION

The principal stresses are given by Eqs. (11-3.1), (11-3.2), and (11-3.3). For the conditions that $p_2 = 0$ and $r = a$, these equations give

$$\sigma_{rr} = p_1 \frac{a^2 - b^2}{b^2 - a^2} = -p_1 = -150 \text{ MPa}$$

$$\sigma_{\theta\theta} = p_1 \frac{a^2 + b^2}{b^2 - a^2} = 150 \frac{100^2 + 400^2}{400^2 - 100^2} = 170 \text{ MPa}$$

$$\sigma_{zz} = p_1 \frac{a^2}{b^2 - a^2} = 150 \frac{100^2}{400^2 - 100^2} = 10 \text{ MPa}$$

The maximum shearing stress is given by Eq. (1-4.27).

$$\tau_{max} = \frac{\sigma_{max} - \sigma_{min}}{2} = \frac{170 - (-150)}{2} = 160 \text{ MPa}$$

The increase in the inside diameter due to the internal pressure is equal to twice the radial displacement given by Eq. (11-3.5) for the conditions $p_2 = P = 0$ and $r = a$.

$$u_{(r=a)} = \frac{p_1 a}{E(b^2 - a^2)}[(1 - 2\nu)a^2 + (1 + \nu)b^2]$$

$$= \frac{150(100)}{72{,}000(400^2 - 100^2)}[(1 - 0.66)100^2 + (1 + 0.33)400^2]$$

$$= 0.3003 \text{ mm}$$

The increase in the inside diameter due to the internal pressure is 0.6006 mm.

EXAMPLE 11-3.3
Stresses in a Composite Cylinder

Let the cylinder in Example 11-3.1 be a composite cylinder made by shrinking an outer cylinder on an inner cylinder. The inner cylinder has dimensions of $a = 10$ mm and $c_i = 25.072$ mm. The outer cylinder has dimensions of $c_0 = 25.000$ mm and $b = 50$ mm. Determine the stress components σ_{rr} and $\sigma_{\theta\theta}$ at $r = a = 10$ mm, $r = 25$ mm, and $r = b = 50$ mm for the composite cylinder. The inner cylinder is cooled to a uniform temperature T_1 and the outer cylinder is heated to a uniform temperature T_2 in order for the outer cylinder to slide freely over the inner cylinder. It is assumed that the two cylinders will slide freely if we allow an additional 0.025 mm to the required minimum difference in radii of 0.072 mm. Determine how much the temperature (degrees Celsius) must be raised in the outer cylinder above the temperature in the inner cylinder in order to freely assemble the two cylinders. $k = 0.0000117$ per degree Celsius.

SOLUTION

After the composite cylinder has been assembled, the change in stresses due to the internal pressure $p_1 = 300$ MPa are the same as for the cylinder in Example 11-3.1. These stresses are added to the residual stresses in the composite cylinder caused by shrinking the outer cylinder on the inner cylinder.

The initial difference between the outer radius of the inner cylinder and the inner radius of the outer cylinder is 0.072 mm. After the two cylinders have been assembled and allowed to cool to their initial uniform temperature, a pressure p_s is developed between the two cylinders. The pressure p_s is an external pressure for the inner cylinder and an internal pressure for the outer cylinder. The magnitude of p_s is obtained from the fact that the sum of the radial displacement of the inner surface of the outer cylinder and the absolute magnitude of the radial displacement of the outer surface of the inner cylinder must equal 0.072 mm. Hence, by Eq. (11-3.6),

$$\frac{c_0}{E(b^2 - c_0^2)}\left[(1-\nu)p_s c_0^2 + (1+\nu)p_s b^2\right]$$

$$-\frac{c_i}{E(c_i^2 - a^2)}\left[-(1+\nu)p_s c_i^2 - (1+\nu)p_s a^2\right] = 0.072$$

Solving for p_s, we obtain

$$p_s = 189.1 \text{ MPa}$$

The pressure p_s produces stresses (so-called residual stresses) in the nonpressurized composite cylinder. For the inner and outer cylinders, the residual stresses σ_{rr}^R and $\sigma_{\theta\theta}^R$ at the inner and outer radii are given by Eqs. (11-3.1) and (11-3.2). For the inner cylinder, $p_1 = 0$, $p_2 = p_s$, $a = 10$ mm, and $b = 25$ mm. For the outer cylinder $p_1 = p_s$, $p_2 = 0$, $a = 25$ mm, and $b = 50$ mm. The residual stresses are found to be

	Inner Cylinder		*Outer Cylinder*	
Residual Stress	*r = 10 mm*	*r = 25 mm*	*r = 25 mm*	*r = 50 mm*
σ_{rr}^R	0	−189.1 MPa	−189.1 MPa	0
$\sigma_{\theta\theta}^R$	−450.2 MPa	−261.1 MPa	351.1 MPa	126.0 MPa

The stresses in the composite cylinder after an internal pressure of 300 MPa has been applied are obtained by adding these residual stresses to the stresses calculated in Example 11-3.1. Thus, we find

	Inner Cylinder		*Outer Cylinder*	
Stress	*r = 10 mm*	*r = 25 mm*	*r = 25 mm*	*r = 50 mm*
σ_{rr}	−300.0 MPa	−226.6 MPa	−226.6 MPa	0
$\sigma_{\theta\theta}$	−125.2 MPa	−198.6 MPa	413.6 MPa	151.0 MPa

A comparison of these stresses for the composite cylinder with those for the solid cylinder in Example 11-3.1 indicates that the stresses have been greatly changed. The determination of possible improvements in the design of the open end cylinder necessitates consideration of particular criteria of failure (see Art. 11-4).

In order to have the inner cylinder slide easily into the outer cylinder during manufacture of the composite cylinder, the difference in temperature between the two cylinders is given by the relation

$$\Delta T = T_2 - T_1 = \frac{u}{rk} = \frac{0.72 + 0.025}{rk} = \frac{0.097}{25(0.0000117)} = 331.6°\text{C}$$

since for uniform temperatures T_1, T_2, we have $\sigma_{rr} = \sigma_{\theta\theta} = \sigma_{zz} = 0$ in each cylinder, and since then Eqs. (11-1.2) and (11-1.4) yield $\epsilon_{\theta\theta} = u/r = k\,\Delta T$, where $r = c_0 = c_i$.

PROBLEM SET 11-3

1. A long closed cylinder has an internal radius $a = 100$ mm and an external radius $b = 250$ mm. It is subjected to an internal pressure $p_1 = 80.0$ MPa ($p_2 = 0$). Determine the maximum radial, circumferential, and axial stresses in the cylinder.

2. Determine the radial and circumferential stress distributions for the cylinder in Problem 1.

3. Consider a one meter length of the unloaded cylinder in Problem 1 at a location in the cylinder some distance from the ends. What are the dimensions of this portion of the cylinder after $p_1 = 80.0$ MPa is applied? The cylinder is made of a steel for which $E = 200$ GPa and $\nu = 0.29$.

4. A closed cylinder has an inside diameter of 20 mm and an outside diameter of 40 mm. It is subjected to an external pressure $p_2 = 40$ MPa and an internal pressure of $p_1 = 100$ MPa. Determine the axial stress and the circumferential stress at the inner radius.

   Ans. $\sigma_{zz} = -20.0$ MPa; $\sigma_{\theta\theta} = 60.0$ MPa

5. A composite aluminum alloy ($E = 72.0$ GPa and $\nu = 0.33$) cylinder is made up of an inner cylinder with inner and outer diameters of 80 mm and 120 + mm, respectively, and an outer cylinder with inner and outer diameters of 120 mm and 240 mm, respectively. The composite cylinder is subjected to an internal pressure of 160 MPa. What must be the outside diameter of the inner cylinder if the circumferential stress at the inside of the composite cylinder is equal to 130 MPa?

6. What must be the outside diameter of the inner cylinder for the composite cylinder in Problem 5 if the maximum shearing stress at

the inner radius of the inner cylinder is equal to the maximum shearing stress at the inner radius of the outer cylinder? What are the values for the circumferential stress at the inside of the composite cylinder and the maximum shearing stress?

Ans. Diameter = 120.2271 mm; $\sigma_{\theta\theta(a)} = 85.1$ MPa; $\tau_{\max} = 122.6$ MPa

7. A gray cast iron ($E = 103$ Gpa and $\nu = 0.20$) cylinder has an outside diameter of 160 mm and an inside diameter of 40 mm. Determine the circumferential stress at the inner radius of the cylinder when the internal pressure is 60.0 MPa.

8. Let the cast iron cylinder in Problem 7 be a composite cylinder made up of an inner cylinder with inner and outer diameters of 40 mm and 80 + mm, respectively, and an outer cylinder with inner and outer diameters of 80 mm and 160 mm, respectively. What must be the outside diameter of the inner cylinder if the circumferential stress at the inside of the inner cylinder is equal to the circumferential stress at the inside of the outer cylinder? What is the magnitude of the circumferential stress at the inside of the composite cylinder?

Ans. Diameter = 80.0287 mm; $\sigma_{\theta\theta(a)} = 38.5$ MPa

11-4 CRITERIA OF FAILURE

The criterion of failure used in the design of a thick-wall cylinder depends on the type of material in the cylinder. As discussed in Art. 3-3, the maximum principal stress criterion should be used in the design of members made of brittle isotropic materials if the principal stress of largest magnitude is a tensile stress. Either the maximum shearing stress or maximum octahedral shearing stress criteria of failure should be used in the design of members made of ductile isotropic materials (see Arts. 3-3 and 2-6).

(a) Maximum Principal Stress Criterion of Failure/If a thick-wall cylinder is made of a brittle material, the material property associated with fracture is the tensile ultimate strength σ_u. At the failure loads, the maximum principal stress in the cylinder is equal to σ_u. If the maximum principal stress occurs at the constrained ends of the cylinder, it cannot be computed using the relations derived in Art. 11-3. At sections far removed from the ends, the maximum principal stress is either the circumferential stress $\sigma_{\theta\theta(r=a)}$ or the axial stress σ_{zz}. If the cylinder is loaded so that the magnitude of the maximum compressive principal stress is appreciably larger than the magnitude of the maximum tensile principal stress, the appropriate criterion of failure to be used in design is questionable. Such conditions are not considered in this book.

(b) Maximum Shearing Stress and Maximum Octahedral Shearing Stress Criteria of Failure/If excessive elastic deformation is not a design factor, failure of members made of ductile materials may be initiated as the result of inelastic deformation or fatigue (only high cycle fatigue is considered in this book, see Art. 3-5). Failure of these members are predicted either by the maximum shearing stress criterion of failure or the maximum octahedral shearing stress criterion of failure. The failure of the member may be either a general yielding failure or a fatigue failure at a large number of stress cycles.

General Yielding Failure/Thick-wall cylinders, which are subjected to static loads or subjected to peak loads only a few times during the life of the cylinder, usually fail by general yielding. General yielding may be defined to occur when yielding is initiated in the member at some point other than at a stress concentration. This definition is used in the illustrative problem at the end of this article (see also Art. 3-3). However, yielding may be initiated in the region of stress concentrations at the ends of the cylinder or at an opening for pipe connections. Yielding in such regions in usually highly localized and general yielding is unlikely. However, the possibility of failure by fatigue still may exist (see Art. 3-5). General yielding sometimes is considered to occur only after the member has yielded over an extensive region, such as occurs with fully plastic loads. Fully plastic loads for thick-wall cylinders are discussed in Art. 11-5.

Fatigue Failure/In practice, a thick-wall cylinder may be subjected to repeated pressurizations (loading and unloading) that may lead to fatigue failure (fracture). Since fatigue cracks often occur in the neighborhood of stress concentrations, every region of stress concentration must be considered in the design. In particular, the maximum shearing stress must be determined in the region of stress concentrations, since fatigue cracking usually originates at a point where either the maximum shearing stress or the maximum octahedral shearing stress occurs. The equations derived in Arts. 11-2 and 11-3 cannot be used to compute the design stresses, unless the maximum stresses occur at sections of the cylinder far removed from end constraints or other stress concentration regions.

Material Response Data for Design/If a member fails by general yielding, the material property associated with failure is the yield stress, which places a limit either on the value of the maximum shearing stress, if the maximum shearing stress criterion of failure is used, or on the value of the maximum octahedral shearing stress, if the maximum octahedral shearing stress criterion of failure is used. If the member fails by fatigue, the material property associated with the failure is the fatigue strength. For high cycle fatigue, both the maximum shearing stress criterion of failure and the maximum octahedral shearing stress criterion of failure are widely used in conjunction with the fatigue strength (see Art. 3-5, Example 3-5.1). The yield stress and the fatigue strength may be obtained by tests of either a tension specimen or a hollow thin-wall tube. It has been found that the values of these properties, as determined from tests of a hollow thin-wall tube in torsion, lead to a more accurate prediction of the material response for thick-wall cylinders than do the values obtained from a tension specimen. This result is because the critical state of stress in the cylinder is usually at the inner wall of the cylinder, and for the usual pressure loading it is essentially one of pure shear (as occurs in the torsion test) plus a hydrostatic state of stress. Since in many materials a hydrostatic stress does not greatly affect the yield, the material responds (yields) as if it were subjected to a state of pure shear. Consequently, if the material properties are determined by means of a torsion test of a hollow thin-wall tube, the maximum shearing stress criterion and the maximum octahedral shearing stress criterion predict failure loads that differ by less than one percent for either closed or open cylinders. The difference in these predictions may be as much as 15.5 percent if the material properties are obtained from tension specimen tests (Art. 2-6). These conclusions pertain in general to most metals. However the yield of most plastic materials is influenced by the hydrostatic state of stress. Hence, for most plastic materials, these conclusions may not generally hold.

The deviatoric state of stress (see Art. 1-4) in a closed cylinder is identical to that for pure shear. Hence, the maximum shearing stress criterion of failure and the octahedral shearing stress criterion of failure predict nearly identical factors of safety for the design of a closed cylinder if the yield stress for the material is obtained from torsion tests of hollow thin-wall tubes. Let the shearing yield stress obtained from a torsion test of a thin-wall hollow tube specimen be designated at τ_Y. If the maximum shearing stress for the inner radius of a closed cylinder is set equal to τ_Y, the pressure p_Y required to initiate yielding is obtained. (The reader is asked to derive the formula for p_Y in Problem 11-4.2.) For the special case of a closed cylinder with internal pressure only and with dimensions $b = 2a$, the yield pressure is found to be $p_Y = 0.75\ \tau_Y$; the corresponding dimensionless stress distribution is shown in Fig. 11-4.1

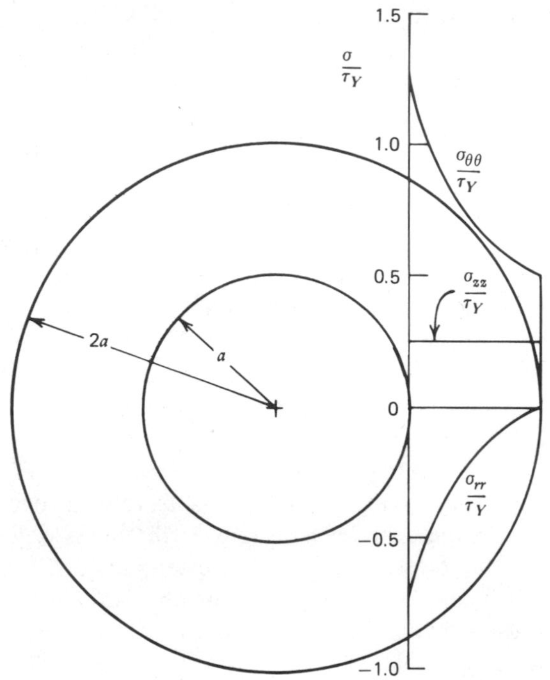

Fig. 11-4.1/Stress distributions in a closed cylinder at initiation of yielding ($b = 2a$).

Ideal Residual Stress Distributions for Composite Open Cylinders/It is possible to increase the strength of a thick-wall cylinder by introducing beneficial residual stress distributions. The introduction of beneficial residual stresses can be accomplished in several ways. In particular, there are two common ways of producing residual stresses in cylinders. One method consists of forming a composite cylinder from two or more open cylinders. For example, in the case of two cylinders, one cylinder (the inner cylinder) has an outer radius that is slightly larger than the inner radius of the other cylinder (the outer cylinder). The inner cylinder is slipped inside the outer cylinder after first heating the outer cylinder and/or cooling the inner cylinder. When the cylinders are allowed to return to their initially equal uniform temperatures (say room temperature), a pressure (the so-called shrink pressure) is created between the cylinder surfaces in contact. This pressure introduces stresses (residual stresses) in the cylinders. As a result, the strength of the composite cylinder under additional internal and external pressure loading is increased (Example 11-4.1). For more than two cylinders this process is repeated for each cylinder that is added to form the composite cylinder.

A second method consists of pressurizing a single cylinder until it deforms inelastically to some distance into the wall from the inner

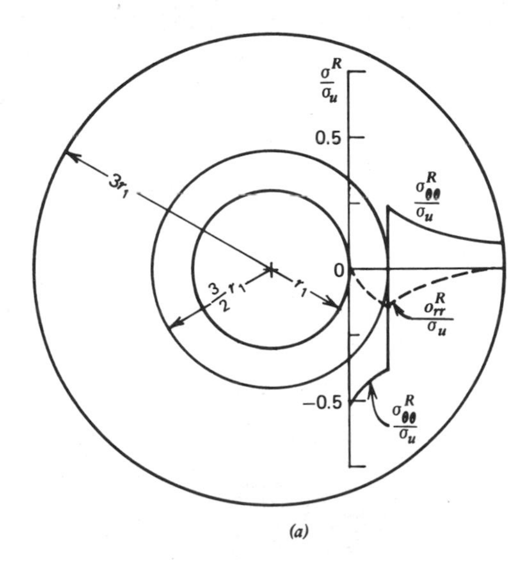

(a)

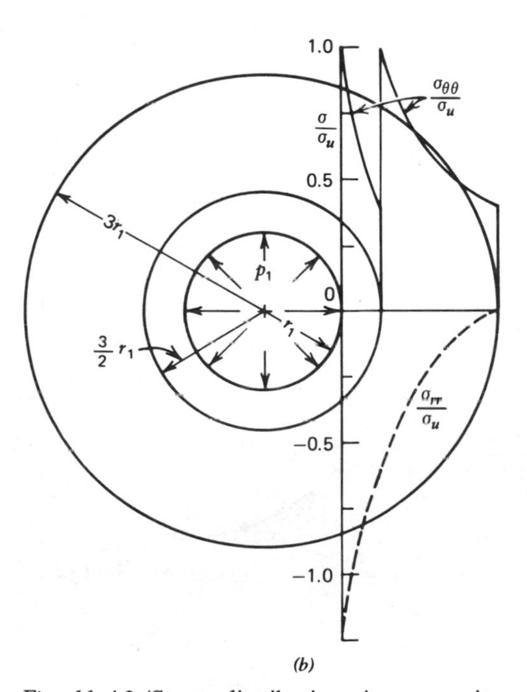

(b)

Fig 11-4.2/Stress distributions in composite cylinder made of brittle material that fails at inner radius of both cylinders simultaneously. (*a*) Residual stress distributions. (*b*) Stress distributions.

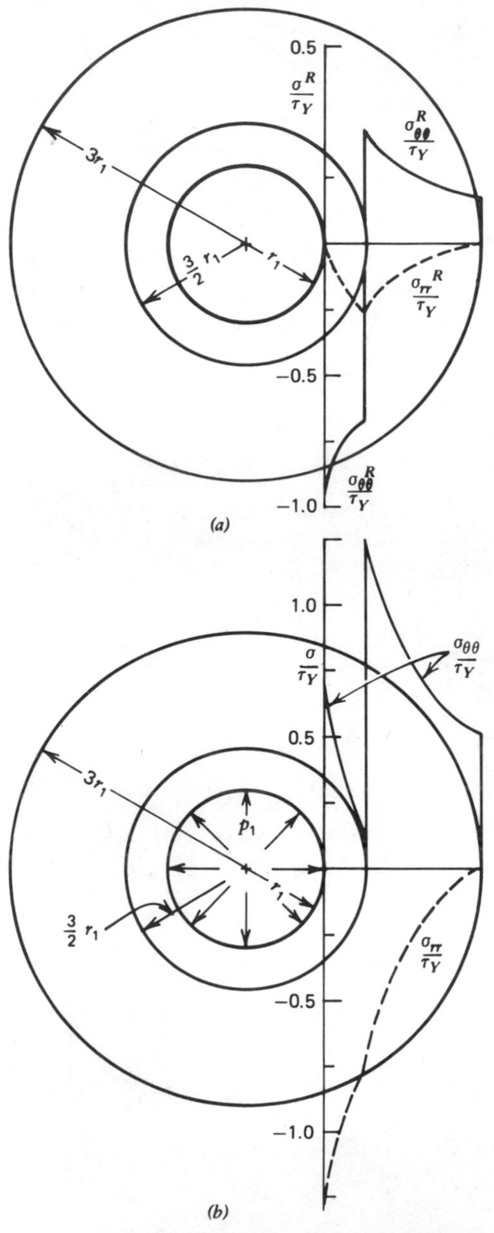

Fig. 11-4.3/Stress distributions in composite cylinder made of ductile material that fails at inner radius of both cylinders simultaneously. (*a*) Residual stress distributions. (*b*) Stress distributions.

surface (the process is called autofrettage). When the pressure is removed, a beneficial residual stress distribution remains in the cylinder (see Art. 11-5).

For a composite cylinder formed by two cylinders under a shrink fit and subject to internal pressure p_1, it may be shown that the most beneficial residual stress distribution is that which results in the composite cylinder failing (yielding or fracturing) simultaneously at the inner radii of the inner and the outer cylinders. Consider, for example, a composite cylinder formed by inner and outer cylinders made of a brittle material whose stress-strain diagram remains linear up to its ultimate strength σ_u. The inner cylinder has inner radius r_1 and outer radius $1.5r_1+$ (that is, the outer radius is slightly larger than $1.5r_1$). The outer cylinder has an inner radius of $1.5r_1$ and an outer radius of $3r_1$. See Fig. 11-4.2. Failure (fracture) of the brittle material occurs when the maximum principal stress reaches the ultimate strength σ_u. Since the maximum principal stress in the composite cylinder is the circumferential stress component $\sigma_{\theta\theta}$, for the most beneficial residual (dimensionless) stress distribution (Fig. 11-4.2*a*), failure of the composite cylinder occurs when $\sigma_{\theta\theta} = \sigma_u$, simultaneously at the inner radii of the inner and the outer cylinders (Fig. 11-4.2*b*). The most beneficial (ideal) residual stress distribution requires a specific difference between the inner radius of the outer cylinder and the outer radius of the inner cylinder, which produces a shrink pressure p_s (see Problem 11-4.9). This shrink pressure produces a residual stress distribution (Fig. 11-4.2*a*) such that the application of an internal pressure p_1 produces the (dimensionless) stress distribution of Fig. 11-4.2*b* at failure.

If the composite cylinder is made of a ductile metal, either the maximum shearing stress criterion of failure or the maximum octahedral shearing stress criterion of failure can be used. For example, let the composite cylinder of Fig. 11-4.3 be made of a ductile metal. Based upon the maximum shearing stress criterion of failure, the ideal residual stress distribution due to the shrink pressure p_s is shown in Fig. 11-4.3*a*. (In this case, the interference fit is different from the cylinder of Fig. 11-4.2; see Prob. 11-4.8.) For an internal pressure p_1 at failure of the cylinder, yield occurs simultaneously at the inner radii of the inner and outer cylinders, and the associated dimensionless stress distribution is shown in Fig. 11-4.3*b*.

EXAMPLE 11-4.1

Yield Failure of Thick-Wall Cylinder

The thick-wall cylinder in Example 11-3.1 is made of a ductile steel whose general yielding failure is accurately predicted by the octahedral shearing stress yield criteria. Determine the minimum yield stress for the steel for a factor of safety of $SF = 1.75$.

SOLUTION

The stress components calculated in Example 11-3.1 are for a cylinder that has been designed with a factor of safety of $SF = 1.75$. Yielding impends in the cylinder when the internal pressure is increased to $(SF)p_1 = 525$ MPa. The yield stress Y for the steel is obtained by setting the maximum octahedral shearing stress in the cylinder [when the pressure in the cylinder is $(SF)p_1$] equal to the octahedral shearing stress that occurs in a tension specimen made of the steel when the tension specimen axial stress is Y. The octahedral shearing stress in the tension specimen is given by the relation (see Eq. 2-6.6a). Hence,

$$\tau_{oct} = \frac{1}{3}\sqrt{(Y-0)^2 + (0-0)^2 + (0-Y)^2} = \frac{\sqrt{2}\,Y}{3} \qquad (1)$$

The octahedral shearing stress at any point in the thick-wall cylinder is given by the relation (see Eq. 1-4.8)

$$\tau_{oct} = \frac{1}{3}\sqrt{(\sigma_{\theta\theta} - \sigma_{rr})^2 + (\sigma_{rr} - \sigma_{zz})^2 + (\sigma_{zz} - \sigma_{\theta\theta})^2} \qquad (2)$$

For the open cylinder, the axial stress σ_{zz} is zero and the radial and circumferential stresses are

$$\sigma_{rr} = -1.75(300) = -525 \text{ MPa}$$

$$\sigma_{\theta\theta} = 1.75(325) = 568.8 \text{ MPa}$$

Substituting these stress components into Eq. (2) and setting Eq. (1) equal to Eq. (2), we obtain

$$Y = \frac{1}{\sqrt{2}}\sqrt{(568.8 + 525)^2 + (525)^2 + (568.8)^2} = 947.5 \text{ Mpa}$$

EXAMPLE 11-4.2

Yield Failure of Composite Thick-Wall Cylinder

The inner and outer cylinders of the composite thick-wall cylinder in Example 11-3.3 is made of the same ductile steel as the cylinder in Example 11-4.1. Determine the minimum yield stress for the steel in the composite cylinder for a factor of safety of $SF = 1.75$.

SOLUTION

Note Equations (1) and (2) in Example 11-4.1 are valid for this problem also.

For the composite open cylinder, it is necessary to consider initiation of yielding for the inside of the inner cylinder, as well as for the inside of the outer cylinder. The axial stress σ_{zz} is zero for both cylinders. At the inside of the inner cylinder, the radial and circumferential stresses for a pressure $(SF)p_1$ are

$$\sigma_{rr} = (1.75)(300) = -525 \text{ MPa}$$

$$\sigma_{\theta\theta} = (1.75)(325) - 450.2 = 118.6 \text{ MPa}$$

Substituting these stress components into Eq. (2) and setting Eq. (1) equal to Eq. (2), we obtain

$$Y = \frac{1}{\sqrt{2}}\sqrt{(118.6 + 525)^2 + (525)^2 + (118.6)^2} = 593.3 \text{ MPa}$$

At the inside of the outer cylinder, the radial and circumferential stresses for a pressure $(SF)p_1$ are

$$\sigma_{rr} = -(1.75)(37.5) - 189.1 = -254.7 \text{ MPa}$$

$$\sigma_{\theta\theta} = (1.75)(62.5) + 315.1 = 424.5 \text{ MPa}$$

Substituting these stress components into Eq. (2) and setting Eq. (1) equal to Eq. (2), we find

$$Y = \frac{1}{\sqrt{2}}\sqrt{(424.5 + 254.7)^2 + (254.7)^2 + (424.5)^2}$$

$$= 594.3 \text{ MPa} > 593.3 \text{ MPa}$$

For the composite cylinder, the yield stress should be $Y = 594.3$ MPa. An ideal design for a composite cylinder should cause yield to be the same for the inner and outer cylinders. (Note that the design above is nearly ideal.)

A comparison of the required yield stress for the single cylinder in Example 11-4.1 and the required yield stress for the composite cylinders indicates the advantage of the composite cylinder. The yield stress of the single cylinder material must be 59.4 percent greater than that of the composite cylinder, if both cylinders are subjected to the same initial pressure and are designed for the same factor of safety against initiation of yielding.

PROBLEM SET 11-4

1. (a) Derive the expression for the maximum shearing stress in a thick-wall cylinder subjected to internal pressure p_1, external pressure p_2, and axial load P, assuming that σ_{zz} is the intermediate principal stress, that is, $\sigma_{rr} < \sigma_{zz} < \sigma_{\theta\theta}$. (b) Derive an expression for the limiting value of the axial load P for which the expression in part (a) is valid.

2. Let σ_{zz} be the intermediate principal stress in a thick-wall cylinder ($\sigma_{rr} < \sigma_{zz} < \sigma_{\theta\theta}$). Using the maximum shearing stress criterion of failure, derive an expression for the internal pressure p_Y necessary to initiate yielding in the cylinder. The shearing yield stress for the material is τ_Y.

   Ans. $p_Y = \tau_Y(b^2 - a^2)/b^2 + p_2$

3. For a closed cylinder subjected to internal pressure p_1 only, show that the octahedral shearing stress τ_{oct} at the inner radius is given by the relation $\tau_{\text{oct}} = \sqrt{2}\, p_1 b^2/[\sqrt{3}\,(b^2 - a^2)]$.

4. A closed cylinder is made of a ductile steel that has a yield stress $Y = 600$ MPa. The inside diameter of the cylinder is 80 mm. Determine the outside diameter of the cylinder if the cylinder is subjected to an internal pressure only, of $p_1 = 140$ MPa, and the cylinder is designed using a factor of safety of $SF = 1.75$ based on the maximum shearing stress criterion of failure.

5. Solve Problem 4 using the octahedral shearing stress criterion of failure.

6. A closed cylinder with inner and outer radii of 60 mm and 80 mm, respectively, is subjected to an internal pressure $p_1 = 30.0$ MPa and an axial load $P = 650$ kN. The cylinder is made of a steel that has a yield stress stress of $Y = 280$ MPa. Determine the factor of safety SF used in design of the cylinder based on (a) the maximum shearing stress criterion of failure and (b) the maximum octahedral shearing stress criterion of failure.

   Ans. (a) $SF = 1.96$; (b) $SF = 2.00$

7. A closed cylinder with inner and outer diameters of 30 mm and 60 mm, respectively, is subjected to an internal pressure only. The cylinder is made of a brittle material having an ultimate strength of $\sigma_u = 160$ MPa. The outer diameter has been gradually reduced as we move away from each end so that stress concentrations at the ends can be neglected. Determine the magnitude of p_1 based on a factor of safety of $SF = 3.00$.

8. Two cylinders are slip-fit together to form a composite open cylinder. Both cylinders are made of a steel having a yield stress $Y = 700$ MPa. The inner cylinder has inner and outer diameters of 100 mm and 150 + mm, respectively. The outer cylinder has inner and outer diameters of 150 mm and 300 mm, respectively. (a) Determine the shrinking pressure p_s and the maximum internal pressure p_1 that can

be applied to the cylinder if it has been designed with a factor of safety of $SF = 1.85$ for simultaneous initiation of yielding at the inner radii of the inner and outer cylinders. Use the maximum shearing stress criterion of failure. (b) Determine the outer diameter of the inner cylinder required for the design. For the steel $E = 200$ GPa and $\nu = 0.29$.

Ans. (a) $p_s = 91.2$ MPa, $p_1 = 247.0$ MPa; (b) diameter $= 150.292$ mm

9. Two cylinders are slip-fit together to form a composite open cylinder. Both cylinders are made of a brittle material whose stress-strain diagram is linear up to the ultimate strength $\sigma_u = 480$ MPa. The inner cylinder has inner and outer radii of 50 mm and 75 + mm, respectively. The outer cylinder has inner and outer radii of 75 mm and 150 mm, respectively. Determine the shrinking pressure p_s and the maximum internal pressure p_1 that results in initiation of fracture simultaneously at the inner radii of both cylinders. Use the maximum principal stress criterion of failure.

11.5 FULLY PLASTIC PRESSURE. AUTOFRETTAGE

Thick-wall cylinders can be strengthened by introducing beneficial residual stress distributions. In Arts. 11-3 and 11-4, it was found that beneficial residual stress distributions may be produced in a composite cylinder formed by shrinking one cylinder on another. Beneficial residual stress distributions may also be introduced into a single cylinder by initially subjecting the cylinder to high internal pressure so that inelastic deformations occur in the cylinder. As a result an increase in load carrying capacity of the cylinder occurs because of the beneficial residual stress distributions that remain in the cylinder after the high pressure is removed. The residual stress distribution in the unloaded cylinder depends upon the depth of yielding produced by the high pressure, the shape of the inelastic portion of the stress-strain diagram for loading of a tensile specimen of the material, and the shape of the stress-strain diagram for unloading of the tensile specimen followed by compression loading of the specimen. If the material in the cylinder is a strain hardening material, a part (usually a small part) of the increase in load carrying capacity is due to the strengthening of the material resulting from strain hardening of the material. If the material exhibits a flat top stress-strain diagram at the lower yield point (i.e., elastic-perfectly plastic), all of the increase in load carrying capacity is due to the beneficial residual stress distributions.

The process of increasing the strength of open and closed cylinders by increasing the internal pressure until the cylinder is deformed inelastically is called *autofrettage*. The beneficial effect of the autofrettage process increases rapidly with the spread of inelastic deformation through-the-wall thickness of the cylinder. Once yielding has spread through the wall thickness, any further increases in load carrying capacity resulting from increased inelastic deformation is due to strain hardening of the material and increases slowly with the increased inelastic deformations. The minimum internal pressure p_1 required to produce yielding through the wall of the cylinder is an important pressure to be determined, since most of the increase in load carrying capacity is produced below this pressure, and the deformation of the cylinder remains small up to this pressure. For the special case where the stress-strain diagram of the material is flat topped at the lower yield point Y, the internal pressure p_1 is called the fully plastic pressure p_P.

We derive the fully plastic pressure by assuming that the maximum shearing stress criterion of failure is valid. Assuming that σ_{zz} is the intermediate principal stress ($\sigma_{rr} < \sigma_{zz} < \sigma_{\theta\theta}$) for the cylinder, $\sigma_{\theta\theta} - \sigma_{rr} = 2\tau_Y$ where τ_Y is the shearing yield stress. This result may be substituted into the equation of equilibrium, Eq. (11-1.1), to obtain

$$d\sigma_{rr} = \frac{2\tau_Y}{r}\,dr \tag{11-5.1}$$

Integration yields

$$\sigma_{rr} = 2\tau_Y \ln r + C \tag{11-5.2}$$

The constant of integration C is obtained from the boundary condition that $\sigma_{rr} = -p_2$ when $r = b$. Thus, we obtain

$$\sigma_{rr} = -2\tau_Y \ln \frac{b}{r} - p_2 \tag{11-5.3}$$

the radial stress distribution at the fully plastic pressure p_P. The magnitude of p_P is given by Eq. (11-5.3) since the internal pressure is then $p_1 = p_P = -\sigma_{rr}$, $r = a$. Thus, we obtain

$$p_P = 2\tau_Y \ln \frac{b}{a} + p_2 \tag{11-5.4}$$

In practice, p_2 is ordinarily taken equal to zero, since for $p_2 = 0$ the required internal pressure p_1 is smaller than for nonzero p_2. The circumferential stress distribution for the cylinder at the fully plastic pressure is obtained by substituting Eq. (11-5.3) into the relation $\sigma_{\theta\theta} - \sigma_{rr} = 2\tau_Y$ to obtain

$$\sigma_{\theta\theta} = 2\tau_Y\left(1 - \ln \frac{b}{r}\right) - p_2 \tag{11-5.5}$$

If the material in the cylinder is a *Tresca material*, that is, a material satisfying the maximum shearing stress criterion of failure, $\tau_Y = Y/2$, and the fully plastic pressure given by Eq. (11-5.4) is valid for cylinders subjected to axial loads in addition to internal and external pressures as long as σ_{zz} is the intermediate principal stress, that is, $\sigma_{rr} < \sigma_{zz} < \sigma_{\theta\theta}$. If the material in the cylinder is a von Mises material, that is a material satisfying the maximum octahedral shearing stress criterion of failure $\tau_Y = Y/\sqrt{3}$ (see Art. 2-6, Eq. 2-6.21), the fully plastic pressure given by Eq. (11-5.4) is valid for closed cylinders subjected to internal and external pressures only. For this loading, the maximum octahedral shearing stress criterion of failure requires that the axial stress be given by the relation

$$\sigma_{zz} = \frac{\sigma_{\theta\theta} + \sigma_{rr}}{2} \tag{11-5.6}$$

The proof of Eq. (11-5.6) is left to the reader.

In many applications, the external pressure p_2 is zero. In this case the ratio of the fully plastic pressure p_P (Eq. 11-5.4) to the pressure p_Y that initiates yielding in the cylinder at the inner wall (see Problem 11-4.2) is given by the relation

$$\frac{p_P}{p_Y} = \frac{2b^2}{b^2 - a^2} \ln \frac{b}{a} \tag{11-5.7}$$

In particular, this ratio becomes large as the ratio b/a becomes large. For $b = 2a$, Eq. (11-5.7) gives $p_P = 1.85 p_Y$; dimensionless radial, circumferential, and axial stress distributions for this cylinder are shown in Fig. 11-5.1. A comparison of these stress distributions with those at initiation of yielding (see Fig. 11-4.1) indicates that yielding throughout the wall thickness of the cylinder greatly alters the stress distributions. If the cylinder in Fig. 11-5.1 unloads elastically, the residual stress distributions can be obtained by multiplying the stresses in Fig. 11-4.1 by the factor 1.85, and subtracting them from the stresses in Fig. 11-5.1. For instance, the residual circumferential stress $\sigma_{\theta\theta}^R$ at the inner radius is calculated to be $\sigma_{\theta\theta}^R = -1.72\tau_Y$. This maximum circumferential residual stress can be expressed in terms of the tensile yield stress Y as follows: for a Tresca material $\sigma_{\theta\theta}^R = -0.86Y$ and for a von Mises material $\sigma_{\theta\theta}^R = -0.99Y$. However, one cannot always rely on the presence of this large compressive residual stress in the unloaded cylinder. In particular, all metals behave inelastically (the Bauschinger effect) when the cylinder is unloaded, resulting in a decrease in the beneficial effects of the residual stresses. For example, an investigation[1] indicated that the beneficial effect of the residual stresses at the inside of the cylinder (when $b = 2a$) is decreased to about 50 percent of that calculated based on the assumption that the cylinder unloads elastically. Consequently, the cylinder will

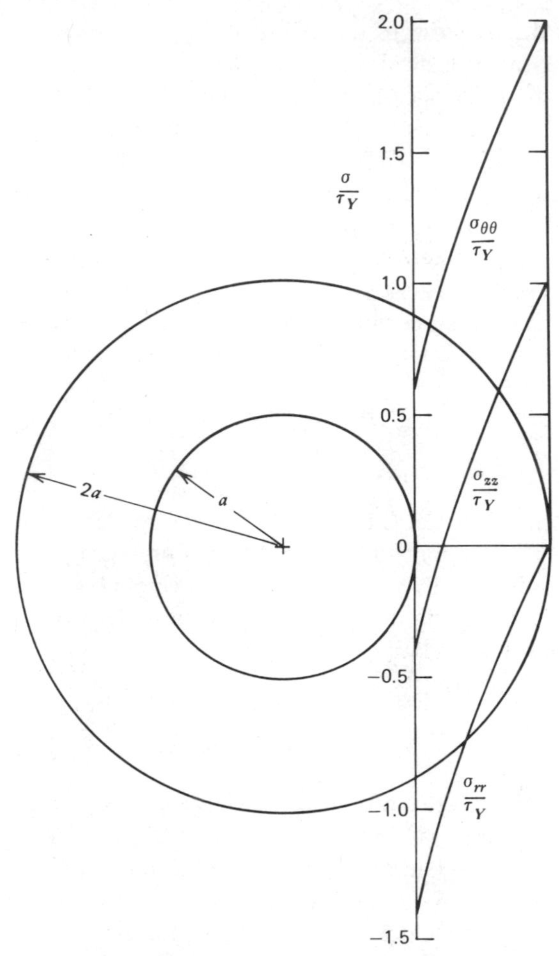

Fig. 11-5.1/Stress distributions in a closed cylinder at fully plastic pressure ($b = 2a$). Cylinder made of von Mises material.

respond inelastically rather than elastically the next time it is loaded to the fully plastic pressure.

EXAMPLE 11-5.1

Fully Plastic Pressure for Cylinder

A closed cylinder has an inner radius of 20 mm and an outer radius of 40 mm. It is made of a von Mises steel that has a yield point stress of $Y = 450$ MPa. (a) Determine the fully plastic internal pressure p_P for the cylinder. (b) Determine the maximum circumferential and axial residual stresses when the cylinder is unloaded from p_P, assuming that the values based on linear elastic unloading are decreased by 50 percent because of the inelastic deformation upon unloading. (c) Assuming that the elastic range of the octahedral shearing stress has not been altered by the inelastic deformation, determine the internal pressure p_1 that can be applied to the cylinder based on a factor of safety $SF = 1.80$. For $SF = 1.80$, compare this result with the pressure p_1 for a cylinder without residual stresses.

SOLUTION

(a) The shearing yield stress τ_Y for the von Mises steel is obtained using the octahedral shearing stress yield condition

$$\tau_Y = \frac{Y}{\sqrt{3}} = 259.8 \text{ MPa}$$

The magnitude of p_P is given by Eq. (11-5.4). Thus, we find

$$p_P = 2\tau_Y \ln\frac{b}{a} = 2(259.8)\ln\frac{40}{20} = 360.21 \text{ MPa}$$

The circumferential and axial stresses at the inner radius for fully plastic conditions are given by Eq. (11-5.5) and (11-5.6). They are

$$\sigma_{\theta\theta} = 2\tau_Y\left(1 - \ln\frac{b}{a}\right) = 2(259.8)\left(1 - \ln\frac{40}{20}\right) = 159.4 \text{ MPa}$$

$$\sigma_{zz} = \frac{\sigma_{\theta\theta} + \sigma_{rr}}{2} = \frac{159.4 - 360.2}{2} = -100.4 \text{ MPa}$$

(b) Assuming linearly elastic unloading, we compute the circumferential and axial residual stresses at $r = a$ as

$$\sigma_{\theta\theta}^R = 159.4 - \frac{p_P(b^2 + a^2)}{b^2 - a^2} = 159.4 - \frac{360.2(40^2 + 20^2)}{40^2 - 20^2}$$

$$= 440.9 \text{ MPa}$$

$$\sigma_{zz}^R = -100.4 - \frac{p_P a^2}{b^2 - a^2} = -100.4 - \frac{360.2(20^2)}{40^2 - 20^2} = 220.5 \text{ MPa}$$

The actual residual stresses may be as much as 50 percent less than these computed values. Thus,

$$\sigma_{\theta\theta}^R = 0.50(-440.9) = -220 \text{ MPa}$$

$$\sigma_{zz}^R = 0.50(-220.5) = -110.2 \text{ MPa} \tag{1}$$

(c) Yielding is initiated in the cylinder at a pressure $(SF)p_1 = 1.80p_1$. If the residual stresses are neglected, the stresses at the inner radius due to pressure $(SF)p_1$ are

$$\sigma_{rr} = -(SF)(p_1) = -1.80p_1$$

$$\sigma_{\theta\theta} = (SF)(p_1)\frac{b^2 + a^2}{b^2 - a^2} = (1.80)(p_1)\frac{40^2 + 20^2}{40^2 - 20^2}$$

$$= 3.000p_1 \tag{2}$$

$$\sigma_{zz} = (SF)(p_1)\frac{a^2}{b^2 - a^2} = (1.80)(p_1)\frac{20^2}{40^2 - 20^2}$$

$$= 0.6000p_1$$

The actual stresses at the inner radius are obtained by adding the residual stresses given by Eqs. (1) to those given by Eqs. (2). Thus,

$$\sigma_{rr} = -1.80p_1$$

$$\sigma_{\theta\theta} = 3.0000p_1 - 220.4 \tag{3}$$

$$\sigma_{zz} = 0.6000p_1 - 110.2$$

The octahedral shearing stress yield condition requires that the octahedral shearing stress in the tension specimen at yield be equal to the octahedral shearing stress at the inner radius of the cylinder. That is,

$$\frac{\sqrt{2}\,Y}{3} = \frac{1}{3}\sqrt{(\sigma_{\theta\theta} - \sigma_{rr})^2 + (\sigma_{rr} - \sigma_{zz})^2 + (\sigma_{zz} - \sigma_{\theta\theta})^2} \tag{4}$$

Substituting the values for the stress components given by Eq. (3) into Eq. (4), we find that

$$p_1 = 154.2 \text{ MPa}$$

is the working internal pressure for the cylinder that was preloaded to the fully plastic pressure. Substituting the values for the stress components given by Eq. (2) into Eq. (4), we obtain the working internal pressure for the cylinder without residual stresses

$$p_1 = 108.3 \text{ MPa}$$

Hence, the working pressure for the cylinder that is preloaded to the fully plastic pressure is 42.4 percent greater than the working pressure for the elastic cylinder without residual stresses.

PROBLEM SET 11-5

1. A thick-wall cylinder has an inside diameter of 180 mm and an outside diameter of 420 mm. It is made of a Tresca steel having a yield point stress of $Y = 460$ MPa. Determine the fully plastic pressure for the cylinder if $p_2 = 0$.

2. (a) Determine the working pressure p_1 for the thick-wall cylinder in Problem 1 if it is designed with a factor of safety of $SF = 3.00$ based on the fully plastic pressure. (b) What is the factor of safety based on the maximum elastic pressure p_Y.

   Ans. (a) $p_1 = 129.9$ MPa; (b) $SF = 1.45$

3. A composite open cylinder has an inner cylinder with inner and outer radii of 20 mm and 30 mm and is made of a steel with yield point stress $Y_1 = 400$ MPa. The outer cylinder has inner and outer radii of 30 mm and 60 mm and is made of a steel with yield point stress $Y_2 = 600$ MPa. Determine the fully plastic pressure for the composite cylinder if both steels are von Mises steels.

4. The closed cylinder in Example 11-5.1 is made of a Tresca material instead of a von Mises material. Obtain the cylinder solution for the Tresca material.

   Ans. $p_P = 311.9$ MPa; $p_1 = 133.5$ MPa (including residual stresses)
   $p_1 = 93.8$ MPa (without residual stresses)

11.6 CYLINDER SOLUTION FOR TEMPERATURE CHANGE ONLY

The stress distribution in a thick-wall cylinder subjected to uniform internal and external pressures p_1 and p_2, axial load P, and temperature change T that depends upon the radial coordinate r only, that is, $T = T(r)$, may be obtained from Eqs. (11-2.4), (11-2.5), (11-2.6), (11-2.7), and (11-2.12). The special case of constant uniform temperature was considered in Art. 11-3. In this article, the case of a cylinder subjected to a temperature change $T = T(r)$, in the absence of pressures and axial load, is treated. If internal and external pressures and temperature changes occur simultaneously, the resulting stresses may be obtained by superposition of the results of this article and Art. 11-3. As in Art 11-3, the results of this article are restricted to the static, steady-state, problem. Accordingly, the steady-state temperature change $T = T(r)$ is required input to the problem.

Steady-Steady Temperature Change (Distribution)/The temperature distribution in a homogeneous body in the absence of heat sources is given by Fourier's heat equation

$$\beta \nabla^2 T = \frac{\partial T}{\partial t} \tag{11-6.1}$$

in which β is the thermal diffusivity for the material in the body, where we consider T to be the temperature change measured from the uniform reference temperature of the unstressed state, and t is the time. For steady-state conditions, $\partial T/\partial t = 0$, and Eq. (11-6.1) reduces to

$$\nabla^2 T = 0 \tag{11-6.2}$$

In cylindrical coordinates (r, θ, z), Eq. (11-6.2) takes the form

$$\frac{\partial^2 T}{\partial r^2} + \frac{1}{r}\frac{\partial T}{\partial r} + \frac{1}{r^2}\frac{\partial^2 T}{\partial \theta^2} + \frac{\partial^2 T}{\partial z^2} = 0 \tag{11-6.3}$$

Since T is assumed to be a function of r only, for the cylinder Eq. (11-6.3) simplifies to

$$\frac{d^2 T}{dr^2} + \frac{1}{r}\frac{dT}{dr} = 0 \tag{11-6.4}$$

The solution of Eq. (11-6.4) is

$$T = C_1 \ln r + C_2 \tag{11-6.5}$$

where C_1 and C_2 are constants of integration. With Eq. (11-6.5), the boundary conditions $T = T_b$ for $r = b$ and $T = T_a$ for $r = a$ determine C_1 and C_2. The solution of Eq. (11-6.5) then takes the form

$$T = \frac{T_0}{\ln \dfrac{b}{a}} \ln \frac{b}{r} \tag{11-6.6}$$

where

$$T_0 = T_a - T_b$$

Stress Components/If $p_1 = p_2 = P = 0$, Eq. (11-6.6) can be used with Eqs. (11-2.4), (11-2.5), (11-2.6), (11-2.7), and (11-2.12) to obtain stress components for steady-state temperature distributions in a thick-wall cylinder. The results are

$$\sigma_{rr} = \frac{kET_0}{2(1-\nu)\ln \dfrac{b}{a}}\left[-\ln \frac{b}{r} + \frac{a^2(b^2 - r^2)}{r^2(b^2 - a^2)} \ln \frac{b}{a}\right] \tag{11-6.7}$$

$$\sigma_{\theta\theta} = \frac{kET_0}{2(1-\nu)\ln \dfrac{b}{a}}\left[1 - \ln \frac{b}{r} - \frac{a^2(b^2 + r^2)}{r^2(b^2 - a^2)} \ln \frac{b}{a}\right] \tag{11-6.8}$$

$$\sigma_{zz} = \sigma_{rr} + \sigma_{\theta\theta} = \frac{kET_0}{2(1-\nu)\ln \dfrac{b}{a}}\left[1 - 2\ln \frac{b}{r} - \frac{2a^2}{b^2 - a^2} \ln \frac{b}{a}\right] \tag{11-6.9}$$

Thus, the stress distributions for linearly elastic behavior of a thick-wall cylinder subjected to a steady-state temperature distribution are given by Eqs. (11-6.7), (11-6.8), and (11-6.9). When $T_0 = T_a - T_b$ is positive, the

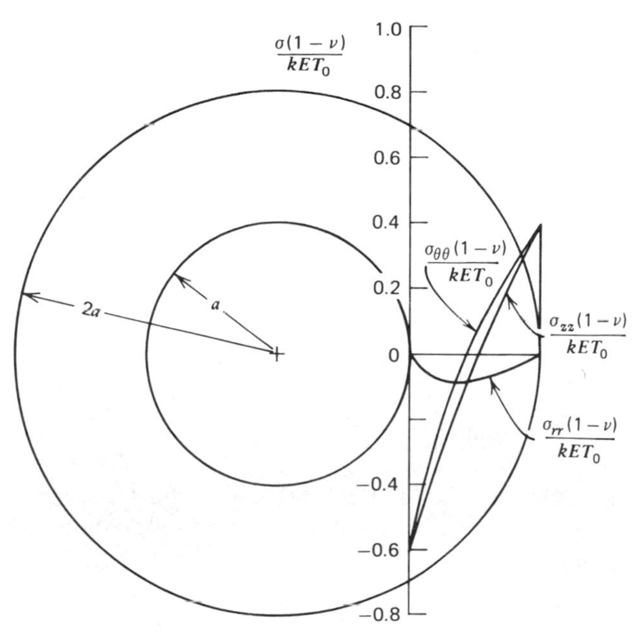

Fig. 11-6.1/Stress distributions in a cylinder subjected to a temperature gradient ($b = 2a$).

temperature at the inner radius T_a is greater than the temperature at the outer radius T_b. For the case of positive T_0, dimensionless stress distributions for a cylinder with $b = 2a$ are shown in Fig. 11-6.1. Since for this case, the stress components $\sigma_{\theta\theta}$ and σ_{zz} are compressive, a positive temperature difference T_0 is beneficial for a cylinder that is subjected to a combination of internal pressure p_1 and temperature since the compressive stresses due to T_0 balance tensile stresses due to p_1. The stresses in cylinders subjected to internal pressure p_1, external pressure p_2, axial load P, and steady-state temperature may be obtained as follows: the radial stress is given by adding Eq. (11-3.1) to Eq. (11-6.7), the circumferential stress is given by adding Eq. (11-3.2) to Eq. (11-6.8), and the axial stress is given by adding Eq. (11-3.3) to Eq. (11-6.9). One can show that yielding in the cylinder always initiates at the inner radius for all combinations of loads and steady-state temperature distributions (changes).

PROBLEM SET 11-6

1. An unloaded closed cylinder has an inner radius of 100 mm and an outer radius of 250 mm. The cylinder is made of a steel for which $k = 0.0000117$ per degree Celsius, $E = 200$ GPa, and $\nu = 0.29$. Determine the stress components at the inner radius for steady-state

temperature change with the temperature at the inner radius 100°C greater than the temperature at the outer radius.

2. Let the steel in the cylinder in Problem 1 have a yield stress of $Y = 500$ MPa. Determine the magnitude of T_0 necessary to initiate yielding in the cylinder based on (a) the maximum shearing stress criterion of failure and (b) the maximum octahedral shearing stress criterion of failure.

   Ans. (a) $T_0 = 235.3$°C; (b) $T_0 = 235.3$°C

3. The cylinder in Problem 1 is subjected to a temperature difference of $T_0 = 50$°C and to an internal pressure $p_1 = 100$ MPa. Determine the stress components at the inner radius.

4. A closed brass ($Y = 240$ MPa, $E = 96.5$ GPa, $\nu = 0.35$, $k = 0.0000166$ per degree Celsius) cylinder has an inside diameter of 70 mm and an outside diameter of 150 mm. It is subjected to a temperature difference $T_0 = T_a - T_b = 70$°C. For this value of T_0, (a) determine the magnitude p_1 of internal pressure required to initiate yield in the cylinder, and (b) determine the magnitude p_2 of external pressure required to initiate yield. (c) Repeat parts (a) and (b) for the case $T_0 = 0$. Use the maximum shearing stress criterion of failure.

   Ans. (a) $p_1 = 135.9$ MPa, (b) $p_2 = 51.9$ MPa, (c) $p_1 = p_2 = 93.9$ MPa

REFERENCES

1. O. M. Sidebottom, S. C. Chu, and H. S. Lamba, "Unloading of Thick-Walled Cylinders that have been Plastically Deformed," *Experimental Mechanics*, Vol. 16, No. 12, 1976, pp. 454–460.

Additional References

1. W. T. Koiter, "On Partially Plastic Thick-Walled Tubes," *Biezeno Anniversary Volume on Applied Mechanics*, Hoarlem, Holland, 1953, pp. 233–251.
2. S. P. Timoshenko and J. N. Goodier, *Theory of Elasticity*, 3rd Ed., McGraw-Hill, New York, 1970.

CHAPTER 12

STRESS CONCENTRATIONS. BASIC CONCEPTS

As noted in previous chapters, the formulas for determining stresses in simple structural elements and machine members are based on the assumption that the distribution of stress on any section of a member can be expressed by a mathematical law or equation of relatively simple form. For example, in a tension member subjected to an axial load the stress is assumed to be distributed uniformly over each cross section; in an elastic beam the stress on each cross section is assumed to increase directly with the distance from the neutral axis; etc.

The assumption that the distribution of stress on a section of a simple member may be expressed by relatively simple laws may be in error in many cases. The conditions that may cause the stress at a point in a simple member, such as a bar or beam, to be radically different from the value calculated from ordinary formulas include effects such as

(a) abrupt changes in section such as occur at the roots of the threads of a bolt, at the bottom of a tooth on a gear, at a section of a plate or beam containing a hole, at the corner of a keyway in a shaft;

(b) pressure at the points of application of the external forces, as, for example, at bearing blocks near the ends of a beam, at the points of contact of the wheels of a locomotive and the rail, at points of contact of gear teeth or of ball bearings on the races;

(c) discontinuities in the material itself, such as nonmetallic inclusions in steel, air holes in concrete, pitch pockets and knots in timber, or variations in the strength and stiffness of the component elements of which the member is made, such as crystalline grains in steel, fibers in wood, ingredients in concrete;

(d) initial stresses in a member that result, for example, from overstraining and cold working of metals during erection or fabrication, to heat treatment of metals, to shrinkage in castings and in concrete, or to residual stress resulting from welding operations; and

(e) cracks that exist in the member, which may be the result of fabrication, such as welding, cold working, grinding, or of other causes.

The conditions that cause the stresses to be greater than those given by the ordinary stress equations of mechanics of materials are frequently called *discontinuities* and *stress raisers* since they destroy the assumed regularity of stress distribution by sudden increases in the stress, frequently called *stress peaks*, at points near the stress raiser. The term *stress gradient* is used frequently to indicate the rate of increase of stress as a stress raiser is approached. The stress gradient may have an influence on the damaging effect of the peak value of the stress.

Often, large stresses due to stress concentrations are developed in only a small portion of a member. Hence, these stresses are called *localized stresses* or simply, *stress concentrations*. In many cases, particularly in which the stress is highly localized, a mathematical analysis is difficult or impracticable. Then, experimental or mechanical methods of stress analysis are used.

Whether the *significant stress* (stress associated with structural damage) in a metal member under a given type of loading is the localized stress *at a point*, or a somewhat smaller value representing the average stress over a small area including the point, depends on the internal state of the metal such as grain type and size, state of stress, stress gradient, temperature, and rate of straining; all of these factors may influence the ability of the material to make local adjustments in reducing somewhat the damaging effect of the stress concentration at the point.

The solution for the values of stresses by the theory of elasticity, as applied to members with known discontinuities or stress raisers, leads in general to differential equations that are difficult to solve. However, the elasticity method has been used with success to obtain stress concentrations in members containing changes of section such as that caused by a circular hole in a wide plate (see Art. 12-2). In addition, the use of numerical methods, such as finite elements,[1] has lead to *approximate* solutions to a wide range of stress concentration problems. Experimental methods of determining stress concentrations may also prove of value in cases for which the elasticity method becomes excessively difficult to apply.

Some experimental methods are primarily mechanical methods of solving the equation for stress obtained from the elasticity analysis; see, for example, the first three of the list of methods given in the next paragraph. These three methods tend to give values comparable with the elasticity method. Likewise the elastic strain (strain gage) method, when a very short gage length is used over which the strain is measured with high precision, gives values of stress concentration closely approximating the elasticity value. In the other methods mentioned, the properties of the materials used in the models usually influence the stress concentration obtained, causing values somewhat less than the elasticity values.

Each experimental method, however, has limitations, but at least one method usually yields useful results in a given situation. The names of some commonly used experimental methods[2,3,4,5] are (1) photoelastic (polarized light), (2) elastic membrane (soap film), (3) electrical analogy, (4) elastic strain (strain gage), (5) brittle coating, (6) brittle material (plaster model), (7) ductile material (Lüders' line), (8) rubber model, and (9) repeated stress.

In this chapter, we consider large stress gradients that arise in the vicinity of holes, notches, and cracks in a structural member or solid. In many practical engineering situations the failure of a structural member or system is due to the propagation of a crack or cracks that occur in the presence of large stress gradients. Unfortunately, the state of stress in the neighborhood of such geometrical irregularities, is usually three dimensional in form, thus increasing the difficulty of obtaining complete analytical solutions. Generally, powerful mathematical methods are required to describe the stress concentrations. We present some general concepts and basic techniques of stress concentration calculations and fracture mechanics. For more explicit and more advanced solutions the reader is referred to specialized works.

The results for computation of stress gradients play a fundamental role in the analysis of fracture and the establishment of fracture criteria (Chapter 3). In particular, stress concentrations coupled with repeated loading (fatigue loading) cause a large number of the failures in structures subjected to repeated loading. The reason for this fact is fairly clear, since stress concentrations lead to applied local stresses that exceed the nominal or average stress by large amounts, and repeated loading often causes fracture at loads far below those determined by a single test to fracture.

The concept of a stress-concentration factor is often employed by designers to account for the increase in stress at a concentration point, the nominal stress being multiplied by a stress concentration factor to obtain an estimate of the local stress at the point. A few examples of the use of stress concentration factors are given in the following articles. Several applications are presented in Chapter 13.

12-1 NATURE OF A STRESS CONCENTRATION PROBLEM. STRESS CONCENTRATION FACTOR

In the tension test of an isotropic homogeneous bar of constant cross sectional area A, the stress σ is assumed to be uniformly distributed over the cross section provided the section is sufficiently far removed from the ends of the bar, where the load may be applied in a nonuniform manner

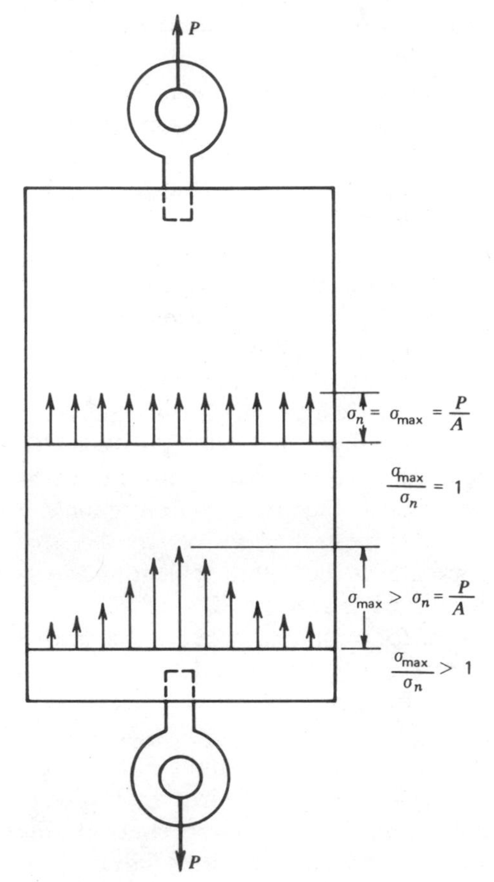

Fig. 12-1.1a

(Fig. 12-1.1*a*). At the end sections, ordinarily the stress distribution is not uniform. Nonuniformity of stress may also occur because of geometric changes (holes or notches) in the cross section of a specimen (Figs. 12-1.1*b* and *c*). This nonuniformity in stress distribution may result in a maximum stress σ_{max} at a section (local stress) that is considerably larger than the average (nominal) stress ($\sigma_n = P/A$, where P is the total tension load).† The ratio S_c defined as

$$S_c = \sigma_{max}/\sigma_n \tag{12-1.1}$$

† When the dimension of the hole or notch is small compared to the width of the bar, the area A is considered to be the cross sectional area of the bar away from the load application region or from the hole or notch in the member. For bars with finite widths, we take A to be the area of the bar at the hole or notch section. When the width of the bar is large compared to the diameter of the hole, the difference in the two definitions of A is small.

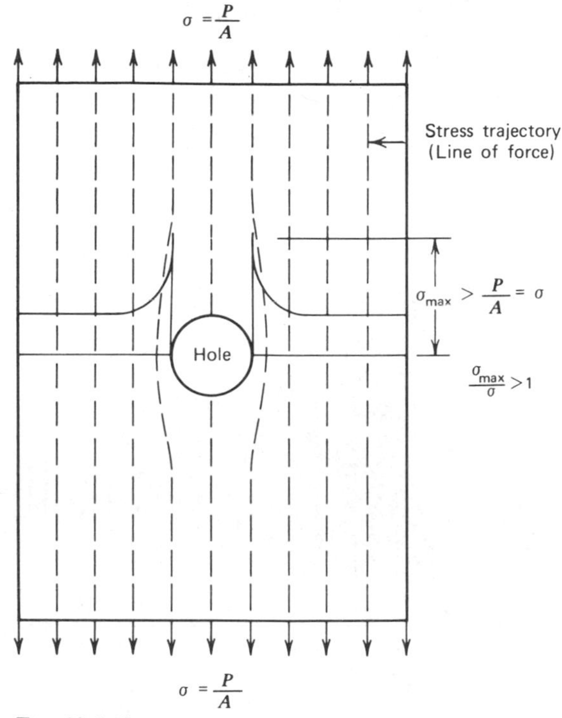

Fig. 12-1.1b

is called the *stress concentration factor* for the section (point); the more abrupt the cross sectional area transition in the tension specimen, the larger the stress-concentration factor (Fig. 12-1.1*d*).

If $\sigma_{\max}$ is the calculated value σ_c of the localized stress as found from the theory of elasticity, or the photoelasticity method, etc., S_c is given an additional subscript c and is written S_{cc}. Then, S_{cc} is called the *calculated stress concentration factor*; it is also sometimes referred to as a *form factor*. If, on the other hand, $\sigma_{\max}$ is the effective value σ_e found from tests of the actual material under the conditions of use, as for example under repeated stress by determining first the effective stress σ_e (fatigue strength) from specimens that contain the abrupt change in section or notch, and then obtaining the fatigue strength from specimens free from the notch, S_c is given the additional subscript e, and S_{ce} is called the *effective* or *significant stress concentration factor*; the term *strength reduction factor* is also used, especially in connection with repeated loads (fatigue). Thus we may write

$$\sigma_c = S_{cc}\sigma_n \quad \text{and} \quad \sigma_e = S_{ce}\sigma_n \tag{12-1.2}$$

The significance of values of S_{ce} are discussed in Chapter 13.

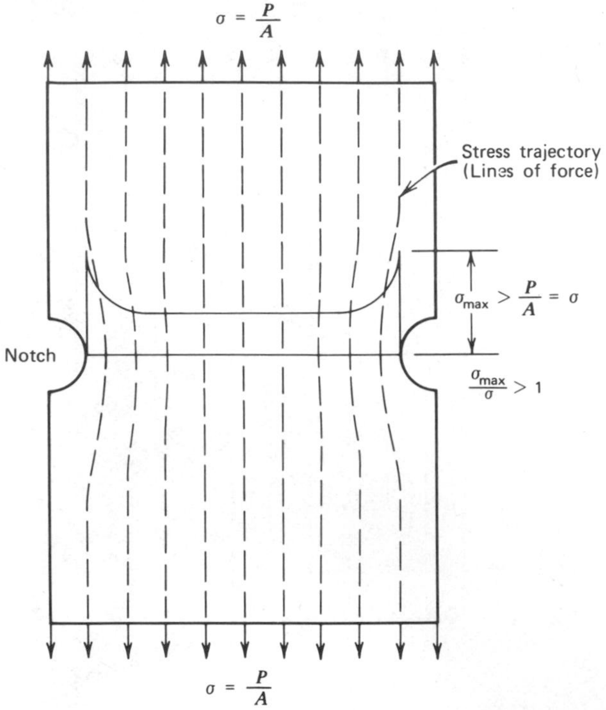

Fig. 12-1.1c

The values of calculated stress concentrations given in this chapter are not meant to be exhaustive but rather to be illustrative of the effects of different discontinuities as computed by the various methods of determining calculated stress concentrations or localized stresses.

A pictorial representation of stress trajectories (Fig. 12-1.1*b*, *c*, *d*) is often employed as an approximate model in the physics of solids to explain the nature of the strain (stress) in the neighborhood of a geometrical discontinuity (crack, dislocation, etc.) in a solid. This representation is based upon the analogy between magnetic lines of forces and stress trajectories.

For example, analogous to magnetic lines of forces, the *stress trajectories* (lines of force), whose paths must lie in the material, cluster together in passing around a geometric hole or discontinuity. In doing so the average spacing between the lines of force is reduced and, therefore, there results a stress concentration (stress gradient) or an increase in local stress (more lines of force are squeezed into the same area). To expand this idea further, consider a geometrical discontinuity (crack) and sketch the hypothetical local arrangement of atoms around one end of the crack (Fig. 12-1.2). The lines of force may be considered to be transmitted from one row of atoms to another. Therefore, the transmission of force around

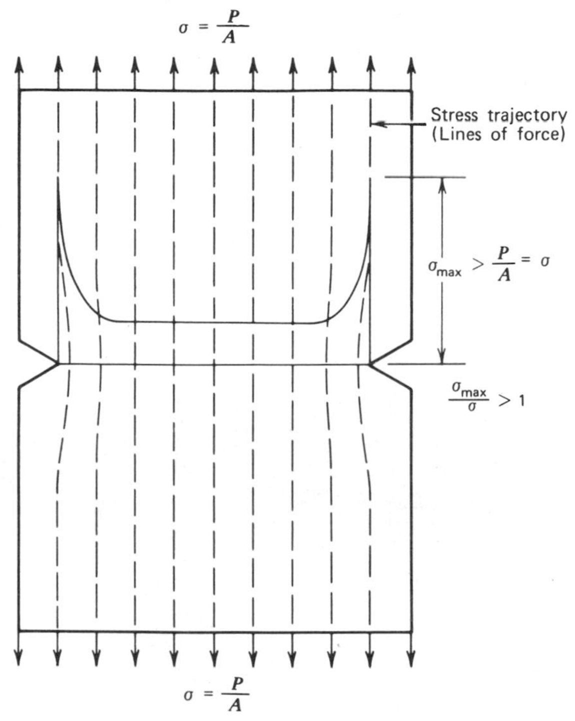

Fig. 12-1.1d

the end of the crack (say in a small crack in an infinite plate) entails heavy loading and straining of the bonds (AB, CD, AC, etc.). Smaller loads and strains are carried by bonds away from the crack (the strain of bond MN is much less than that of AB). For bonds sufficiently far removed from AB, for example, bond MN, the associated stress is essentially $\sigma = P/A$. The conceptual model of Fig. 12-1.2 leads to the conclusion that for bond AB to be extended, bonds AC and BD also must be extended. Hence, the uniaxial loading of the plate causes the region around the crack tip to have not only a high tensile strain in the y-direction but also a high tensile strain in the x-direction. The concept of lines of force also suggests a redistribution of strain energy from regions above the crack (regions R and Q in Fig. 12-1.2) to the highly strained region at the crack tip (see also Figs. 12-1.1*b*, *c*, *d*). Also because of the distortion of rectangular elements (Fig. 12-1.2), high shearing stresses exist in the neighborhood of a stress concentration.

In practical problems of stress concentrations, the state of stress in the neighborhood of the crack is three dimensional in nature. For such complex situations few complete analytical solutions exist. Indeed the majority of mathematical solutions to stress concentration problems are at best approximate two-dimensional solutions of plane stress cases, the

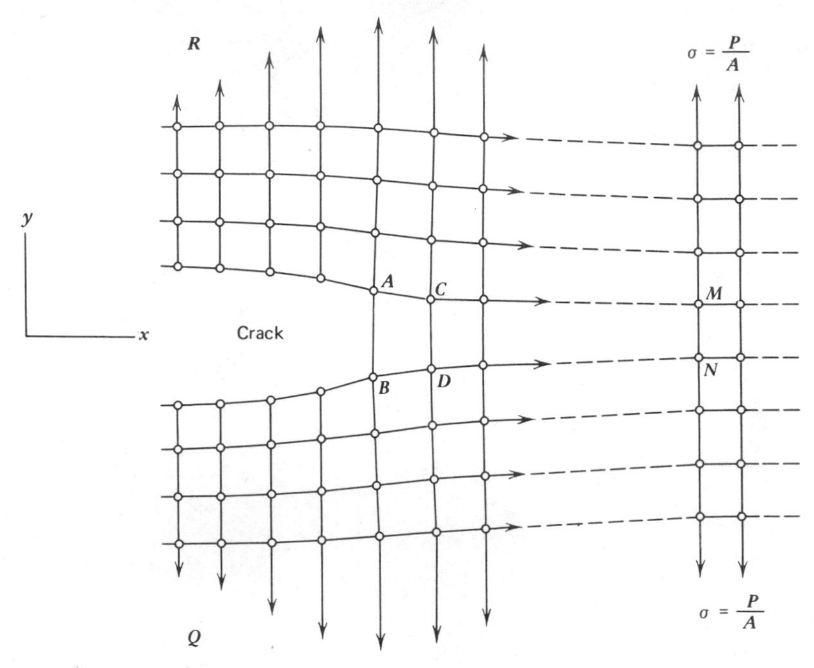

Fig. 12-1.2/Atomic model of crack in a solid.

case of plane strain being derived from the plane stress case.[5] Consequently, experimental methods of determining stress concentration factors are often employed to supplement or verify analytical predictions. Unfortunately, experimental methods are also limited in accuracy and particularly in generality. For this reason stress concentration factors are usually determined by several methods.

Stress concentrations may also arise because of concentrated loads such as point loads, line loads, spot loads, etc. (see Art 12-3, Chapter 13 and Chapter 14).

12-2 STRESS CONCENTRATION FACTORS. THEORY OF ELASTICITY

(a) Circular Hole in an Infinite Plane under Uniaxial Tension/Consider first the case of an infinite plate or sheet with a small circular hole of radius a under uniaxial tension σ (Fig. 12-2.1).

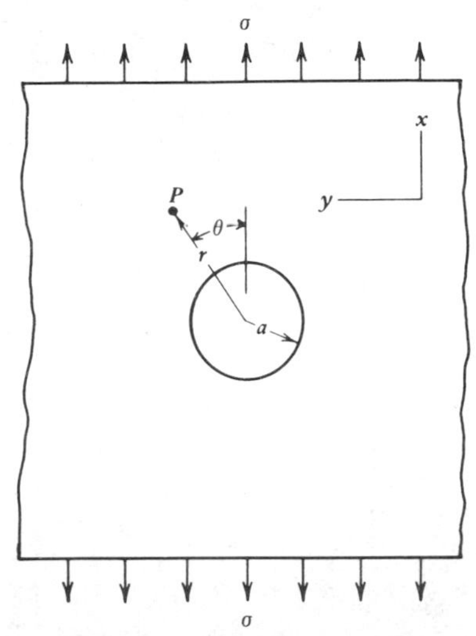

Fig. 12-2.1/Infinite plate with small circular hole.

With respect to polar coordinates (r, θ), the plane stress components at any point P are given by the formulas[6]

$$
\begin{aligned}
\sigma_{rr} &= \frac{\sigma}{2}\left(1-\frac{a^2}{r^2}\right)+\frac{\sigma}{2}\left(1-\frac{a^2}{r^2}\right)\left(1-\frac{3a^2}{r^2}\right)\cos 2\theta \\
\sigma_{\theta\theta} &= \frac{\sigma}{2}\left(1+\frac{a^2}{r^2}\right)-\frac{\sigma}{2}\left(1+\frac{3a^4}{r^4}\right)\cos 2\theta \\
\sigma_{r\theta} &= -\frac{\sigma}{2}\left(1-\frac{a^2}{r^2}\right)\left(1+\frac{3a^2}{r^2}\right)\sin 2\theta
\end{aligned}
\tag{12-2.1}
$$

We note that the stress state given by Eqs. (12-2.1) satisfies the boundary conditions at $r = a(\sigma_{rr} = \sigma_{r\theta} = 0$ for all $\theta)$ and at $r = \infty$ $(\sigma_{xx} = \sigma, \sigma_{xy} = 0$ for $\theta = 0, \pi$ and $\sigma_{yy} = 0$, $\sigma_{xy} = 0$ for $\theta = \pi/2, 3\pi/2)$. For $r = a$,

$$\sigma_{\theta\theta} = \sigma(1 - 2\cos 2\theta) \tag{12-2.2}$$

Hence, for $\theta = \pi/2, 3\pi/2$, $\sigma_{\theta\theta}$ attains its maximum value of $\sigma_{\theta\theta(\max)} = 3\sigma$. For $\theta = 0, \pi$, $\sigma_{\theta\theta}$ attains a compressive value $-\sigma$. Thus, $\sigma_{\theta\theta}$ attains a maximum tensile value of three times the uniformly distributed stress σ, at the hole $r = a$ for $\theta = \pi/2, 3\pi/2$ (Fig. 12-2.2). This value (3σ) is the largest normal stress that occurs in the plate. Hence, the stress concentration factor at the hole (Eq. 12-1.1) is $S_{cc} = 3$. It is the ratio of the maximum normal stress at the hole divided by the calculated normal stress that would exist at the same point in the absence of the hole, the

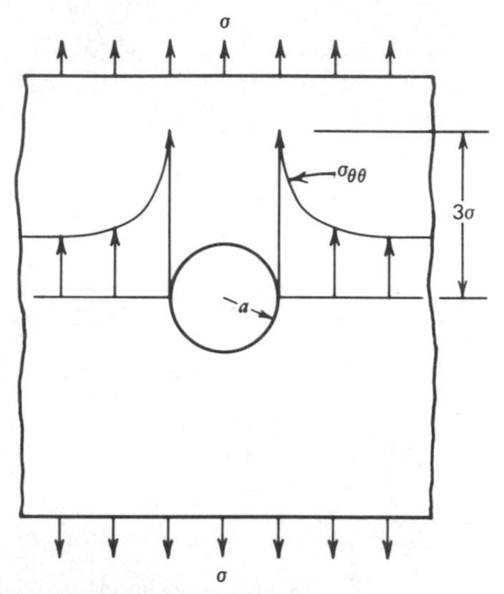

Fig. 12-2.2/$\sigma_{\theta\theta}$ distribution for $\theta = \pi/2$, $3\pi/2(-\pi/2)$.

stress σ being maintained the same. Figure 12-2.2 shows the fact that as r increases ($> a$), the maximum value of $\sigma_{\theta\theta}$ decreases rapidly (see Eqs. 12-2.1). Thus, the high stress gradient or stress concentration is quite localized in effect. For this reason Eqs. (12-2.1) are often used to estimate the stress concentration effect of a hole in a plate of finite width in the direction normal to the direction of tension σ. However, when the diameter of the hole is comparable to the width of the plate, Eqs. (12-2.1) are considerably in error. Several authors have studied the problem of a plate strip with circular hole by theoretical and experimental (photo elastic and strain-gage) methods. The results are summarized by the formula

$$S_{cc} = \frac{\sigma_{\max}}{\sigma_n} = \frac{3\kappa - 1}{\kappa + 0.3} \tag{12-2.3}$$

where κ is the ratio, width of strip/diameter of hole, and σ_n is the average stress over the weakened cross sectional area (the cross sectional area of the plate remaining at the section containing the hole).

(b) Elliptic Hole in an Infinite Plane Stressed in Direction Perpendicular to Major Axis of the Hole/Consider an infinite plate or sheet with elliptic hole of major axis $2a$ and minor axis $2b$ (Fig. 12-2.3). A uniform tensile stress σ is applied at a large distance from the hole and is directed perpendicular to the major axis of the elliptical hole; that is, $\sigma_{yy} = \sigma$ at infinity. For this problem, it is desirable to express the stress components relative to orthogonal curvilinear coordinates[6] (elliptic coordinates) (Fig. 12-2.4). In

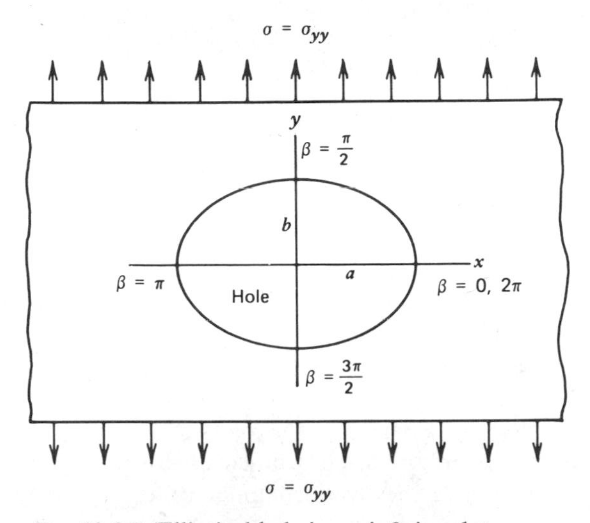

Fig. 12-2.3/Elliptical hole in an infinite plate.

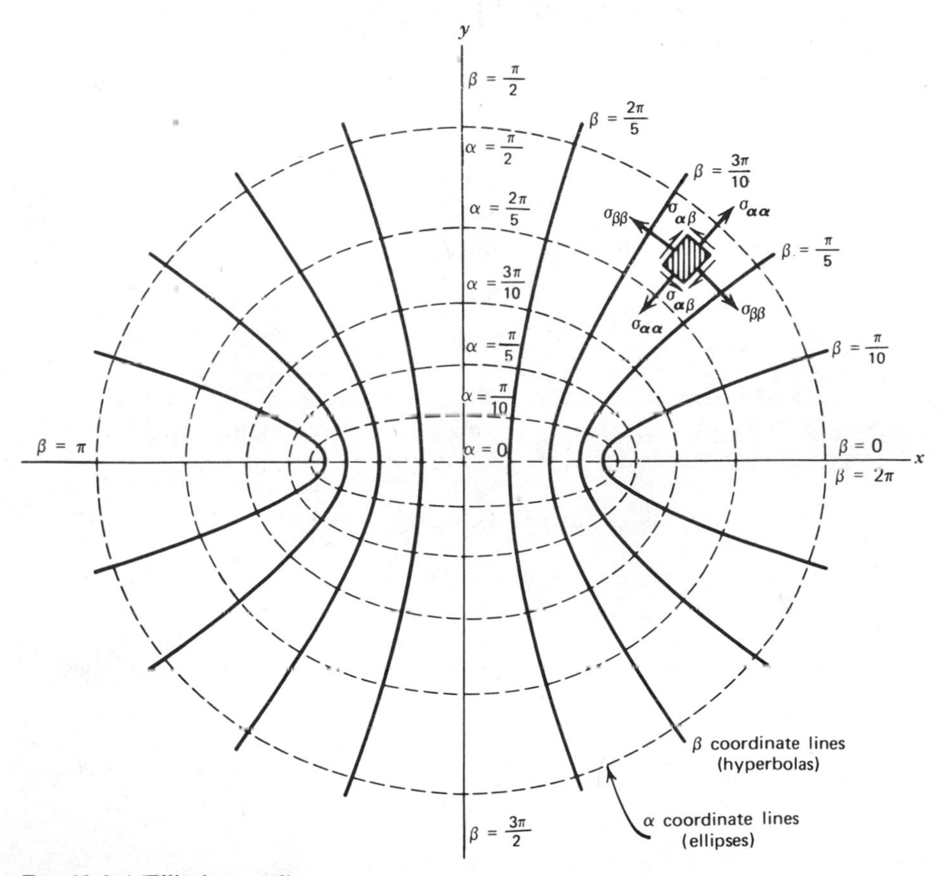

Fig. 12-2.4/Elliptic coordinates.

terms of elliptic coordinates (α, β), the equation of an ellipse is

$$\frac{x^2}{\cosh^2\alpha} + \frac{y^2}{\sinh^2\alpha} = c^2 \tag{12-2.4}$$

where for the ellipse with semi-axes (a, b), we have (Fig. 12-2.3)

$$a = c\cosh\alpha_0 \qquad b = c\sinh\alpha_0 \tag{12-2.5}$$

Thus, in the limit as $\alpha_0 \to 0$, the elliptical hole becomes a crack (an ellipse of zero height and of length $2a = 2c$). Because of this condition, the solution for the stresses in a plate with elliptical hole is employed to study the stresses in a plate with narrow crack of length $2a$.

The elastic stress distribution in a plate with elliptical hole has been determined by Inglis[7] by the method of complex potentials[5,8,9]. For uniaxial tension stress, perpendicular to the major axis of the elliptical hole, the sum of the stress components $\sigma_{\alpha\alpha}, \sigma_{\beta\beta}$ is given by the formula

$$\sigma_{\alpha\alpha} + \sigma_{\beta\beta} = \sigma e^{2\alpha_0}\left[\frac{(1 + e^{-2\alpha_0})\sinh 2\alpha}{\cosh 2\alpha - \cos 2\beta} - 1\right] \tag{12-2.6}$$

Since the stress $\sigma_{\alpha\alpha} = 0$ at the hole $(\alpha = \alpha_0)$, Eq. (12-2.6) yields the stress $\sigma_{\beta\beta}$ at the hole as

$$\sigma_{\beta\beta}|_{\alpha=\alpha_0} = \sigma e^{2\alpha_0}\left[\frac{(1 + e^{-2\alpha_0})\sinh 2\alpha_0}{\cosh 2\alpha_0 - \cos 2\beta} - 1\right] \tag{12-2.7}$$

where (α, β) are elliptic coordinates $(\alpha = \alpha_0$ at the hole) and by Eqs. (12-2.5)

$$\tanh\alpha_0 = \frac{b}{a} \tag{12-2.8}$$

where a is the semimajor axis of the ellipse and b is the semiminor axis. Therefore, by Eq. (12-2.7), the maximum value of $\sigma_{\beta\beta}$ is (for $\beta = 0, \pi$; $\cos 2\beta = 1$; this occurs at the ends of the major axis)

$$\sigma_{\beta\beta(\max)} = \sigma(1 + 2\coth\alpha_0) = \sigma\left(1 + \frac{2a}{b}\right) \tag{12-2.9}$$

Thus, the maximum value of $\sigma_{\beta\beta}$ increases without bound as $b/a \to 0$, that is, as the semiminor axis b becomes smaller and smaller relative to a. It is noteworthy that for $a = b$ (a circular hole) the maximum value of $\sigma_{\beta\beta}$ is 3σ, which agrees with the results given by Eq. (12-2.2). The distribution of $\sigma_{\beta\beta}$ around a circular hole $(a/b = 1)$ is shown in Fig. 12-2.5. The distribution of $\sigma_{\beta\beta}$ at the hole for $a/b = 5$ is shown in Fig. 12-2.6. By geometry, the radius of curvature of an ellipse at the end of the major axis is (Eq. 12-2.4)

$$\rho = \frac{b^2}{a} \tag{12-2.10}$$

where (a, b) are major and minor semiaxes lengths, respectively. Hence,

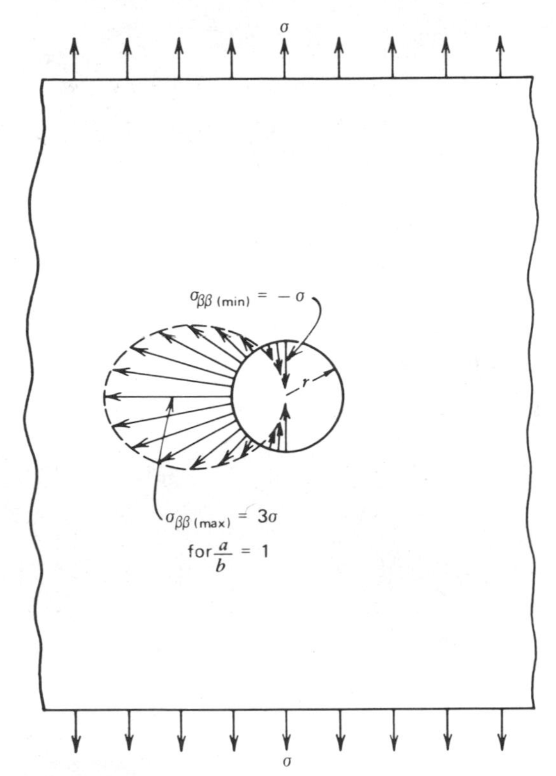

Fig. 12-2.5/Circumferential stress distribution around an edge of a circular hole in an infinite plate.

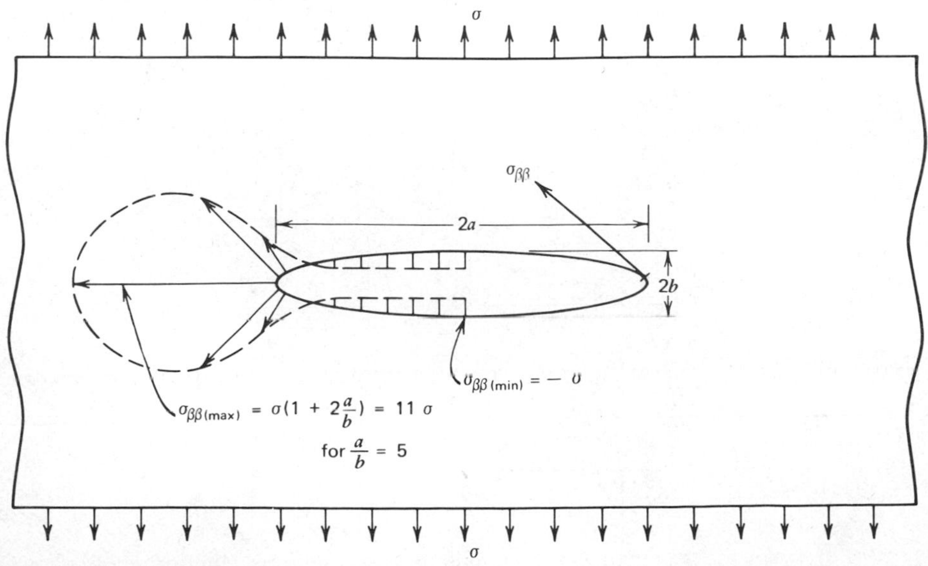

Fig. 12-2.6/Distribution of $\sigma_{\beta\beta}$ around an elliptical hole in an infinite plate loaded perpendicular to a major axis.

Eqs. (12-2.9) and (12-2.10) yield

$$\sigma_{\beta\beta(\max)} = \sigma\left(1 + 2\sqrt{a/\rho}\,\right) \tag{12-2.11}$$

Also by Eq. (12-2.7), the minimum value of $\sigma_{\beta\beta}$ is $\sigma_{\beta\beta(\min)} = -\sigma$ (at the ends of the minor axis, where $\beta = \pi/2, -\pi/2$).

(c) Elliptical Hole in Infinite Plate Stressed in Direction Perpendicular to the Minor Axis of the Hole/Let the plate be subjected to stress σ as indicated in Fig. 12-2.7, where, as in case (b), the dimensions (a, b) are very small compared to the length and width dimensions of the plate. By a transformation of Eq. (12-2.7), the value of $\sigma_{\beta\beta}$ at any point on the perimeter of the hole is

$$\sigma_{\beta\beta}\big|_{\alpha=\alpha_0} = \sigma\left[\frac{1 + \sinh 2\alpha_0 - e^{2\alpha_0}\cos 2\beta}{\cosh 2\alpha_0 - \cos 2\beta}\right] \tag{12-2.12}$$

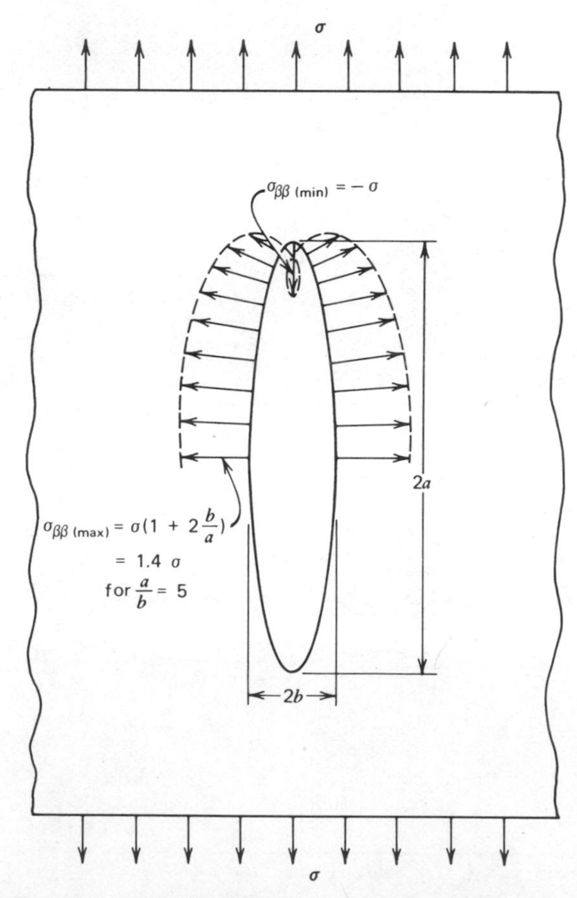

Fig. 12-2.7/Distribution of $\sigma_{\beta\beta}$ around an elliptical hole in an infinite plate loaded perpendicular to a minor axis.

For $\beta = \pi/2, -\pi/2$, $\sigma_{\beta\beta}$ attains the maximum value

$$\sigma_{\beta\beta(\max)} = \sigma(1 + 2\tanh\alpha_0) = \sigma\left(1 + \frac{2b}{a}\right) \tag{12-2.13}$$

at the ends of the minor axis. Again as in case (b), for $\beta = 0, \pi$, $\sigma_{\beta\beta}$ attains the minimum value $\sigma_{\beta\beta(\min)} = -\sigma$ (which now occurs at the ends of the major axis). The distribution of $\sigma_{\beta\beta}$ is given in Fig. 12-2.7, for $a/b = 5$.

(d) Crack in Plate / As $b \to 0$, the elliptical hole in an infinite plate becomes very flat and approaches the shape of a line crack (see Arts. 12-4 and 3-4). The maximum value of $\sigma_{\beta\beta}$ may become quite large compared to the applied stress for nonzero values of b as $b \to 0$, depending upon the nature of the load. For example, for the case of Fig. 12-2.3, Eq. (12-2.9) yields with $a/b = 100$, $(\sigma_{\beta\beta})_{\max} = 201\sigma$, which corresponds to a stress concentration factor of $S_{cc} = 201$. For the loading case of Fig. 12-2.7, with $a/b = 100$, Eq. (12-2.13) yields $(\sigma_{\beta\beta})_{\max} = 1.02\sigma$ or $S_{cc} = 1.02$. The case $b = 0$ leads to a special study of stress singularities (see Arts. 12-5 and 12-6). The practical significance of very large stress concentrations is discussed in Arts. 12-5 and 12-6 and in Chapter 13.

(e) Ellipsoidal Cavity / In a member subjected to axial tension the theoretical stress at the edge of an internal cavity having the shape of an ellipsoid has been obtained by Sadowsky and Sternberg.[10] The stress concentration factors for two special cases of such an internal discontinuity will be considered; namely for ellipsoids of revolution of the prolate spheroid type (football shape) and the oblate spheroid type (door-knob shape). The data for a prolate spheroid are given in Table 12-2.1. For this case, the semimajor axis a of the ellipsoid, which is the axis of revolution, is oriented so that it is perpendicular to the direction of the axial pull in the member, and the semiminor axis b always lies in a plane parallel to the axial pull. Axes a and b are considered to be very small compared to the cross sectional dimensions of the axial member. If the nominal (average) stress in the member is σ_n, the maximum stress occurs at the end of the semimajor axis a and has values for various ratios of b/a as given in Table 12-2.1.

The ellipsoid of revolution having the shape of the oblate spheroid has its semiminor axis b, which is the axis of revolution, oriented in the direction of the uniaxial pull in the member, and the semimajor axis a always lies in a plane perpendicular to the load. If the nominal (average) stress in the member is σ_n, the maximum stress occurs at the end of a semimajor axis a and has values for various ratios b/a as given in Table 12-2.2. These values of the calculated maximum elastic stress show that an internal flaw or cavity of spherical shape such as a gas bubble (an

Table 12-2.1

Stress at End of Semimajor Axis a of Internal Ellipsoidal Cavity of Prolate Spheroid Shape

Ratio b/a	1.0	0.8	0.6	0.4	0.2	0.1
Calculated stress	$2.05\sigma_n$	$2.17\sigma_n$	$2.33\sigma_n$	$2.52\sigma_n$	$2.70\sigma_n$	$2.83\sigma_n$

ellipsoid for which $b/a = 1$) raises the stress from σ_n to $2.05\sigma_n$; a long, narrow, stringlike internal flaw or cavity ($b/a = 0$) oriented in a direction perpendicular to the load raises the stress from σ_n to $2.83\sigma_n$; and a very flat, round cavity oriented so that the flat plane is perpendicular to the load raises the stress from σ_n to values as high or higher than $13.5\sigma_n$ if the material remains elastic; this value is comparable to the value for a narrow elliptical hole as given by Eq. (12-2.9).

(f) Grooves and Holes / The values of the calculated stress concentration factors for grooves as shown in Figs. *A* to *D* of Table 12-2.3 may be obtained from the diagram given by Neuber[11] (Fig. 12-2.8).

For example, let it be assumed that a member contains the groove shown in Fig. *A* of Table 12-2.3 and is subjected to an axial load P.

Let the calculated stress concentration factor be S_{cs} when the groove is very shallow. Then from Neuber,[11]

$$S_{cs} = 1 + 2\sqrt{\frac{t}{\rho}} \tag{12-2.14}$$

Let the calculated stress concentration factor be S_{cd} when the groove is very deep. Then from Neuber,[11]

$$S_{cd} = \frac{2\left(\dfrac{b}{\rho} + 1\right)\sqrt{\dfrac{b}{\rho}}}{\left(\dfrac{b}{\rho} + 1\right)\arctan\sqrt{\dfrac{b}{\rho}} + \sqrt{\dfrac{b}{\rho}}} \tag{12-2.15}$$

Table 12-2.2

Stress at End of Semimajor Axis a of Internal Ellipsoidal Cavity of Oblate Spheroid Shape

Ratio b/a	1.0	0.8	0.6	0.4	0.2	0.1
Calculated stress	$2.05\sigma_n$	$2.50\sigma_n$	$3.3\sigma_n$	$4.0\sigma_n$	$7.2\sigma_n$	$13.5\sigma_n$

Table 12-2.3

Directions for Use of Fig. 12-2.8 (Neuber) in Finding Calculated Stress Concentration Factor S_{cc} in Bars

Type of Notch	*Type of Load*	*Formula for Nominal Stress*	*Scale for* $\sqrt{\frac{t}{\rho}}$	*Curve for Finding* S_{cc}
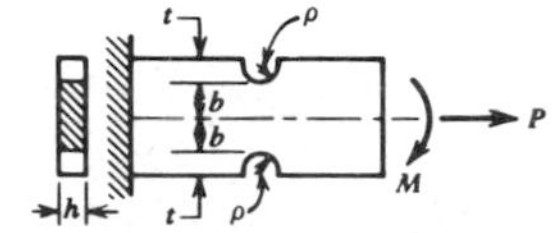 Fig. *A*	Tension	$\frac{P}{2bh}$	f	1
	Bending	$\frac{3M}{2b^2h}$	f	2
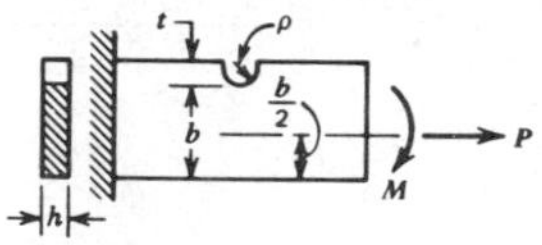 Fig. *B*	Tension	$\frac{P}{bh}$	f	3
	Bending	$\frac{6M}{b^2h}$	f	4
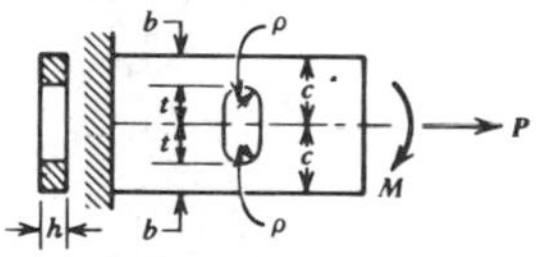 Fig. *C*	Tension	$\frac{P}{2bh}$	f	5
	Bending	$\frac{3Mt}{2h(c^3-t^3)}$	e	5
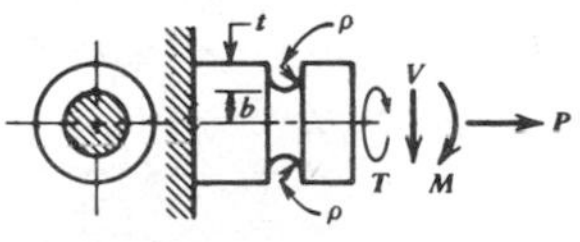 Fig. *D*	Tension	$\frac{P}{\pi b^2}$	f	6
	Bending	$\frac{4M}{\pi b^3}$	f	7
	Direct shear	$\frac{1.23V}{\pi b^2}$	e	8
	Torsional shear	$\frac{2T}{\pi b^3}$	e	9

Let S_{cc} represent the calculated stress concentration for any depth of groove. Then, according to Neuber, an approximate, and usually quite accurate, value of S_{cc} is given by the following equation:

$$S_{cc} = 1 + \frac{(S_{cs}-1)(S_{cd}-1)}{\sqrt{(S_{cs}-1)^2+(S_{cd}-1)^2}} \tag{12-2.16}$$

When the groove is very shallow, Eq. (12-2.16) reduces to $S_{cc} = S_{cs}$, and when the groove is very deep Eq. (12-2.16) reduces to $S_{cc} = S_{cd}$. Curve number 1 in Fig. 12-2.8 has been plotted by making use of Eqs. (12-2.14),

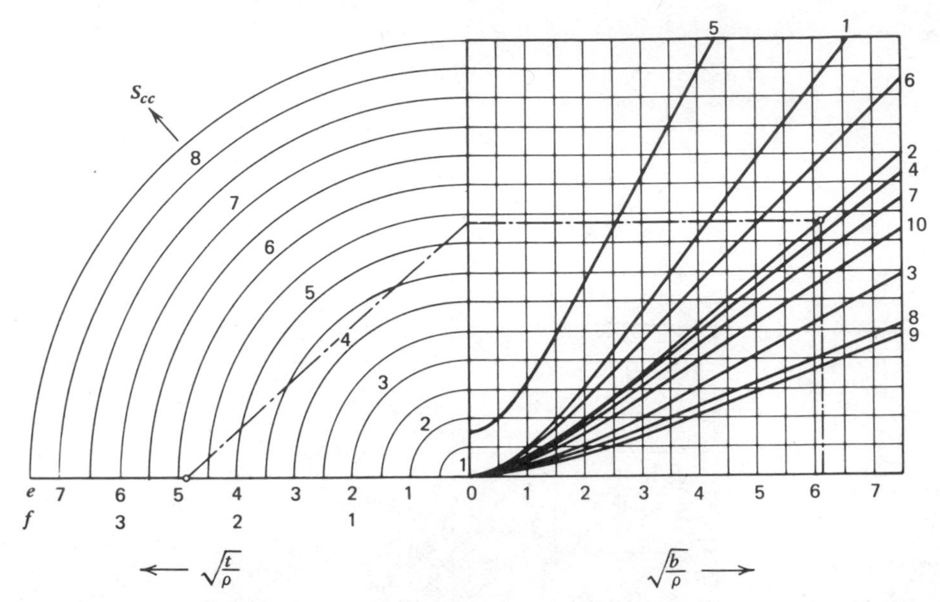

Fig. 12-2.8/Neuber's diagram (nomograph) for a calculated stress concentration factor at the root of a notch.

(12-2.15), and (12-2.16). The other curves were obtained in a similar manner.

Let it be assumed that $\rho = 6.35$ mm, $t = 38.0$ mm, and $b = 241.0$ mm in Fig. *A* of Table 12-2.3 and that the bar is subjected to a bending moment M. From these values, $\sqrt{t/\rho} = 2.45$ and $\sqrt{b/\rho}) = 6.16$. As indicated in Table 12-2.3, scale f applies for $\sqrt{t/\rho}$ and curve 2 for $\sqrt{b/\rho}$. Thus, to find the value of the calculated stress concentration factor, we enter Fig. 12-2.8 with $\sqrt{b/\rho} = 6.16$, proceed vertically upward to cut curve 2, then horizontally to the left to the axis of ordinates. We join this point to point $\sqrt{t/\rho} = 2.45$ on the left-hand axis of abscissas (on which scale f is applicable) by a straight line. This line is tangent to the circle corresponding to the appropriate calculated stress concentration factor; thus $S_{cc} = 4.25$.

Some values of calculated stress concentration factor for bending obtained from Neuber's diagram (Fig. 12-2.8) as found by Moore and Jordan are given in Fig. 12-2.9.

PROBLEM SET 12-2

1. For the flat bar in Fig. *C* of Table 12-2.3, let $b = 16t$, $c = 17t$, $\rho = t$ (circular hole). By means of Neuber's nomograph (Fig. 12-2.8) show that for the bar loaded in tension, S_{cc} is approximately 3. Note that

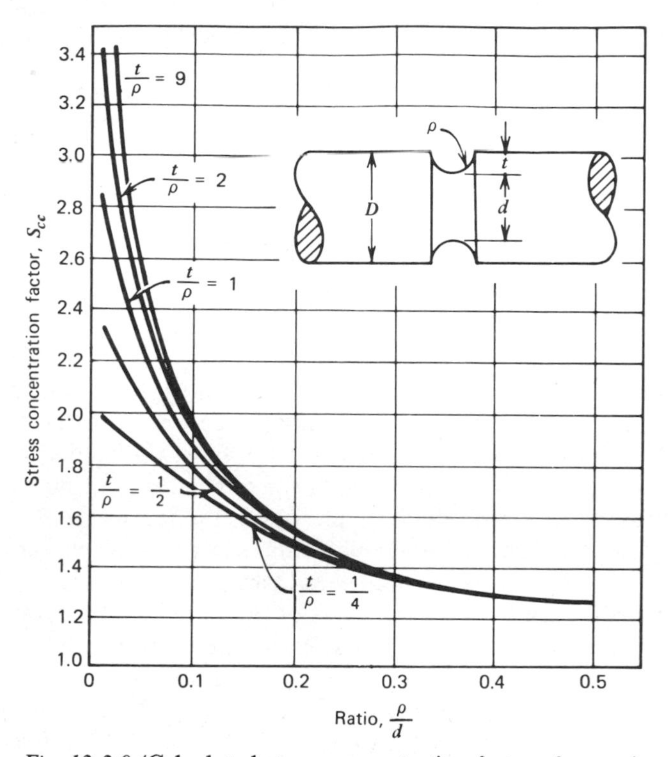

Fig. 12-2.9/Calculated stress concentration factors for semicircular grooves in a cylindrical member subjected to bending only as obtained from Neuber's diagram. (From Moore and Jordan.)

for this case, the half-width c of the bar is large compared to the radius of the hole.

2. With $\rho = t$, solve Problem 1 for (a) $b = 4t$ and $c = 5t$ and (b) $b = t$ and $c = 2t$.

   Ans. (a) $S_{cc} = 2.7$, (b) $S_{cc} = 2.3$

3. For the flat bar in Fig. A of Table 12-2.3, let $t = 4\rho$ and $b = 16\rho$. By Neuber's nomograph (Fig. 12-2.8) determine the value of S_{cc} for the cases where the bar is subjected to (a) axial tensile load and (b) bending.

4. A cylindrical shaft has a circular groove, the depth of groove is $t = 6.00$ mm, and the radius at the root of the groove is $\rho = 2.20$ mm. The radius of the cross section at the root of the groove is $b = 60$ mm. By Neuber's nomograph (Fig. 12-2.8) determine the value of S_{cc}

for the cases where the shaft is subjected to (a) axial tensile load, (b) bending, and (c) torsion.

Ans. (a) $S_{cc(P)} = 3.7$, (b) $S_{cc(M)} = 3.3$, (c) $S_{cc(T)} = 2.1$

12-3 STRESS CONCENTRATION FACTORS EXPERIMENTAL TECHNIQUES

(a) Photoelastic Method/The values of calculated stress concentration factors found by the photoelastic method agree well with results obtained from the theory of elasticity. Thus the photoelastic method may be used as a check, and it may be applied to some members in which the stress cannot be obtained mathematically; however, the technique of obtaining reliable results with the photoelastic method is acquired only after considerable experience. In particular, special care must be exercised to obtain trustworthy results when the radius of the notch is very small.

Values of the calculated stress concentration factors obtained by the photoelastic method for three forms of abrupt changes in section in flat specimens are shown as reported by Frocht in Fig. 12-3.1. In each specimen the stress distribution is uniform at distant sections on either side of the abrupt change; when the stress distribution is variable on either side of the abrupt change in section, as in bending, the calculated stress concentration factor is found to be somewhat smaller. These curves show that the value of S_{cc} varies with the ratio ρ/d. However, S_{cc} depends also on the ratio D/d. For the particular groove, hole, and fillet shown in Fig. 12-3.1, the value of ρ/d and D/d are related by the equation $D/d = 1 + 2\rho/d$.

The values of S_{cc} for the hole and groove in Fig. 12-3.1 can be found also by Neuber's solution, as obtained from Fig. 12-2.8 for various values of ρ/d. These values obtained from Neuber's nomograph agree satisfactorily with those found by the photoelastic method. The elasticity solution for the calculated stress concentration factor for the fillet is achieved by a numerical method that is an approximation. Hence, the photoelastic method is of special value for this type of discontinuity. For the fillet in Fig. 12-3.1 for which $t = \rho$, the curve marked $t = \rho$ gives values of S_{cc}. For members in which t is not equal to ρ the values of S_{cc} will be different as shown, for example, by the curve marked $t = 3\rho$. The influence of t/ρ on the values of S_{cc} for a fillet subjected to axial tension and to bending have been studied by Frocht.[12]

The distribution of stress shown in Fig. 12-3.2 was obtained by Coker[13] by the photoelastic method. The maximum stress at the edge of the groove in Fig. 12-3.2*a* is 1.37 times the average stress on the reduced

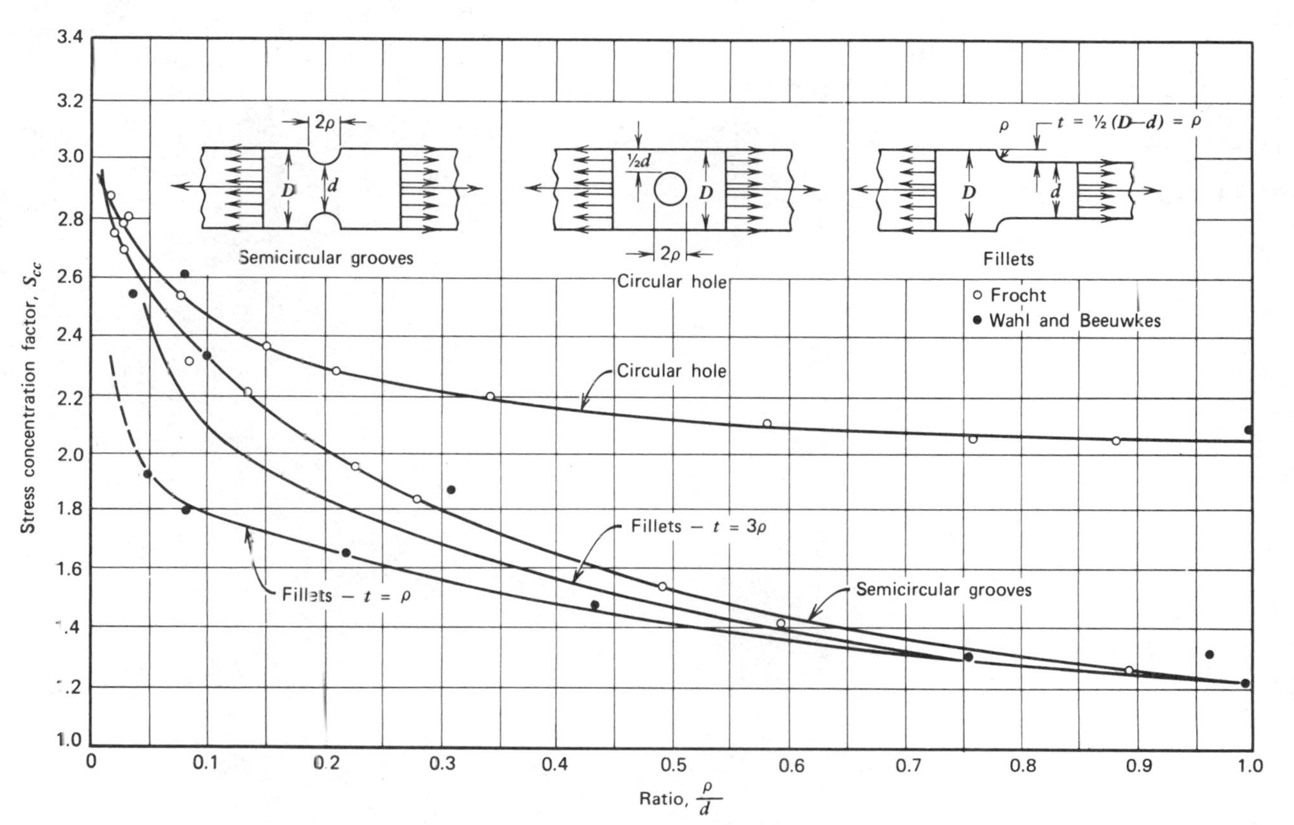

Fig. 12-3.1/Stress concentration factors obtained by use of the photoelastic method.

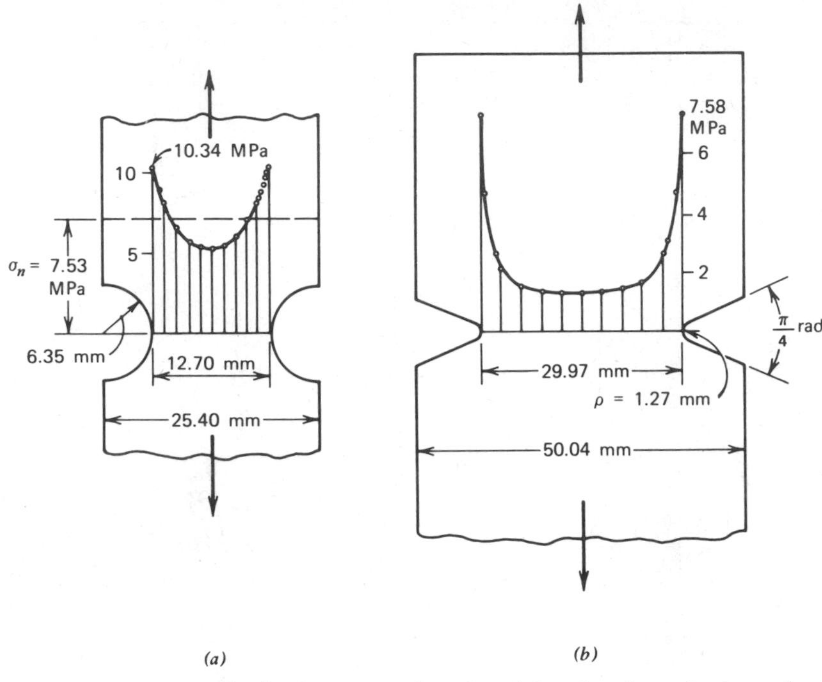

Fig. 12-3.2/Stress distribution at notches found by the photoelastic method (from Coker).

section, that is, $S_{cc} = 1.37$, by the photoelastic method. The value as found by using Neuber's nomograph is $S_{cc} = 1.45$. In Fig. 12-3.2*b* the groove has a much smaller radius and the plate is much wider. The photoelastic method gives a maximum stress of 7.58 MPa, whereas the nominal or average stress was 1.59 MPa, that is, $S_{cc} = 4.77$. The value as found by Neuber's nomograph is $S_{cc} = 5.50$. The rather sharp notch gives a high concentration of stress. However, the stress concentration depends on the relative depth of the notch. For example, if in Fig. 12-3.2*b* the notch geometry and dimensions are kept as shown and the outer width of the plate is reduced to 29.97 mm (the width of the root section is then 9.90 mm.) the value of $S_{cc} = 2.6$, from Neuber's solution.

(b) Strain Gage Method/Two examples are presented to indicate the use of the strain gage method to determine calculated stress concentration factors for a hole in a shaft and to determine the effect of a concentrated load or the strain (stress) distribution in a beam at the section where the load is applied.

Transverse Hole in Shaft/By using a specially designed, highly sensitive, and accurate mechanical strain gage that measured elastic strains in a

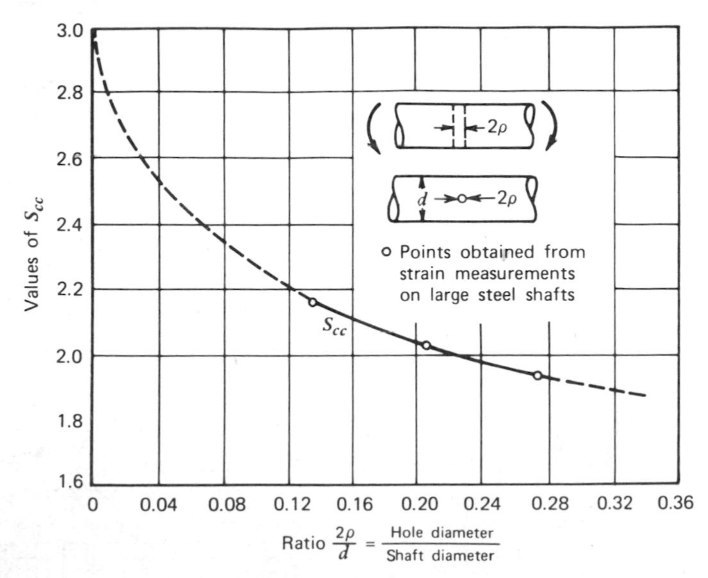

Fig. 12-3.3/Calculated stress concentration factors for a shaft in bending with a transverse hole as found by elastic strain method. (From Peterson and Wahl.)

2.54 mm gage length, Peterson and Wahl[14] obtained elastic stress concentration factors for a shaft containing a transverse hole and subjected to bending loads. Their results are shown in Fig. 12-3.3. With the same instrument they obtained the stress at a fillet in a large steel shaft tested as a beam. These values checked closely with the values found by Frocht by the photoelastic method for fillets of the same proportions (Fig. 12-3.2).

Effect of Local Pressure on Strain (Stress) Distributions in a Beam/The effect on the longitudinal bending strains (stresses) in a beam caused by the bearing pressure of a concentrated load applied at the midspan section of a steel rail beam is shown in the upper part of Fig. 12-3.4. The load was applied approximately along a line across the top of the rail section. It will be observed that the effect of the bearing pressure on the longitudinal stress extends well below the middle of the depth of the rail. The point of zero longitudinal stress is about 25 mm above the calculated position of the neutral axis for the section beneath the load, and the strain (stress) on the cross section does not vary directly as the distance from the neutral axis, as is usually assumed for such a beam. The results for the section underneath the load, however, are approximate because relatively long gage lengths were used and the two-dimensional aspect of the state of stress was neglected.

If, however, the same beam is loaded as shown in the lower part of Fig. 12-3.4, the strains (stresses) in the central portion, which is subjected to

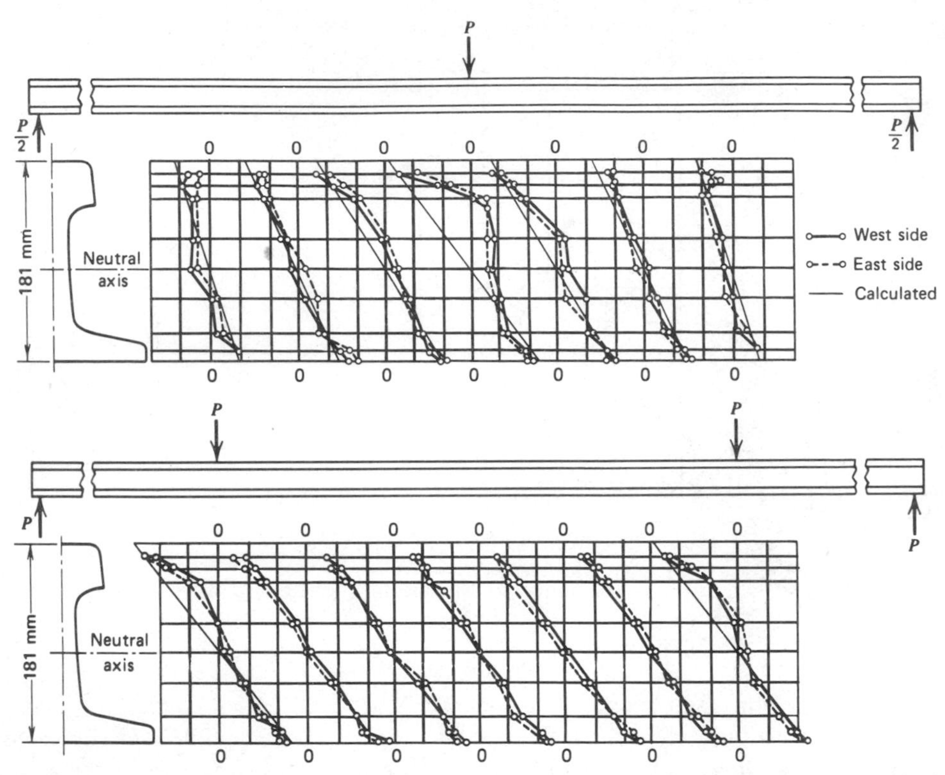

Fig. 12-3.4/The effect of bearing pressure of load at the center of beam on longitudinal strains in beam. (From A. N. Talbot, *Bull. A.R.E.A.*, Vol. 31.)

constant bending moment free from the influence of the bearing pressure of the loads, are in agreement with the usual assumptions for simple bending.

(c) Torsional Elastic Stress Concentration at Fillet in Shaft. Electrical Analogy Method/If all cross sections of a shaft are circular but the shaft contains a rather abrupt change in diameter, a localized stress occurs at the abrupt change of section. Jacobsen has investigated the concentration of torsional shearing stress at a fillet where the diameter of a shaft changes more or less abruptly, depending on the radius of the fillet. The electrical analogy method was used.

The results of the investigation are given in Fig. 12-3.5. For example, if the radius of a circular shaft changes from 52 mm to 39 mm by means of a fillet having a radius of 3.25 mm, $R/r = 1.33$, and $\rho/r = 1/12 = 0.083$; the maximum elastic shearing stress at the fillet as given by Fig. 12-3.5 is approximately 1.7 times the maximum shearing stress in the small shaft

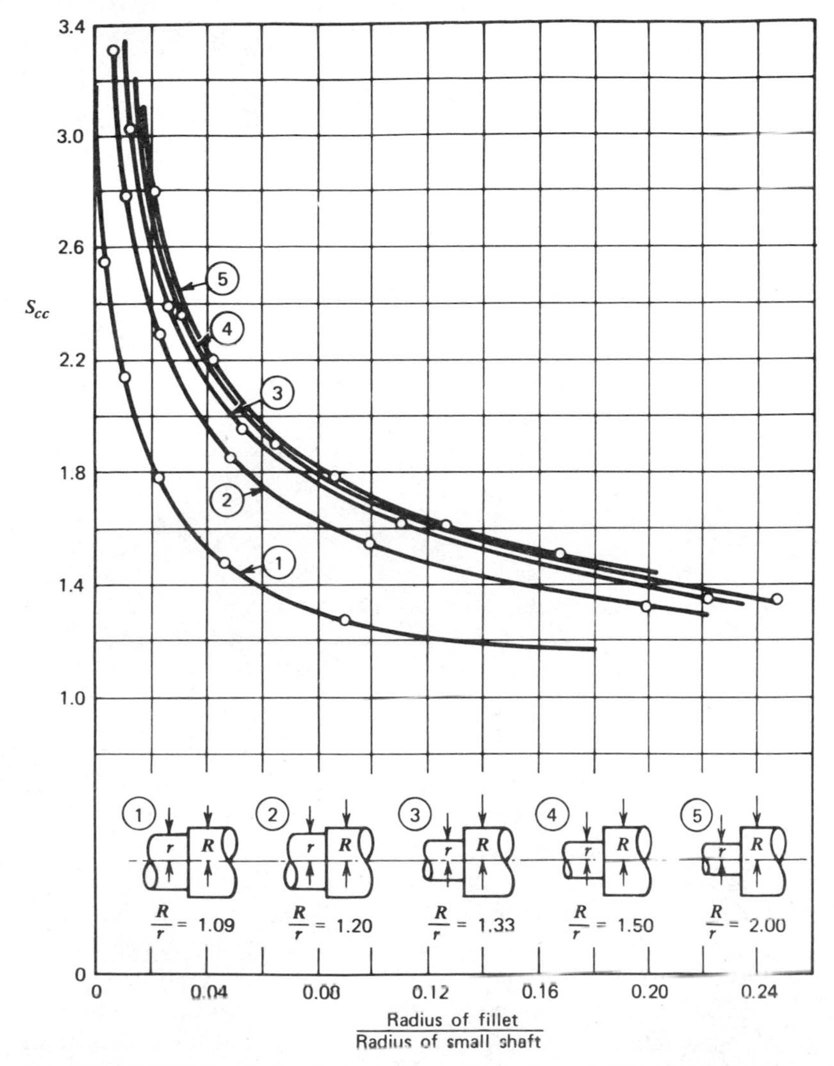

Fig. 12-3.5/Torsional shearing stress concentration at fillet in shaft of two diameters.

as found by the equation $\tau = Tr/J$, where T is the twisting moment, and J is the polar moment of inertia of the cross section of the smaller shaft ($J = \pi r^4/2$).

(d) Elastic Membrane Method. Torsional Stress Concentration/Griffith and Taylor,[16] by using a soap film as the elastic membrane (see Art. 5-4), found the torsional shearing stress in a hollow shaft at the filleted corner

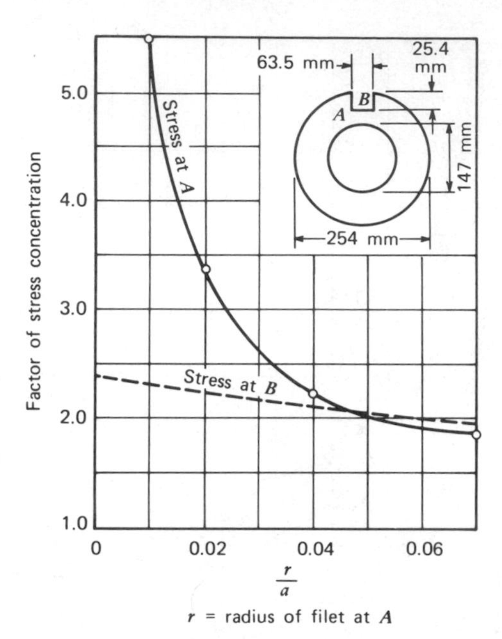

Fig. 12-3.6/Factors of torsional shearing stress concentration at keyway in hollow shaft.

of a keyway and also at the center of the flat bottom of the keyway. The external and internal diameters were $2a = 254$ mm and 147 mm, respectively, and the keyway was 25.4 mm deep and 63.5 mm wide.

Figure 12-3.6 shows the value of the ratios of the maximum torsional shearing stress at the fillet for various radii r of fillet to the maximum shearing stress that would be developed in the shaft if the shaft had no keyway. In other words, the ordinates to the curve give the elastic calculated stress concentration factors S_{cc} due to the keyway.

Ordinates to the dotted line in Fig. 12-3.6 are the elastic stress concentration factors for the shearing stress at the center of the bottom of the keyway; the stress at this point is approximately twice as great as would be the maximum shearing stress in the shaft if it had no keyway.

Torsional Stress at Fillet in Angle Section/The torsional shearing stress at a sharp internal corner of a bar subject to torque is infinitely large if the material does not yield when the stress becomes sufficiently high. If the corner is rounded off by means of a fillet, the stress is reduced; the amount of reduction corresponding to fillets of different radii in an angle section was found by Griffith and Taylor by use of the soap-film method. They used a section 25.4 mm wide (Fig. 12-3.7) and the straight portions or arms of the section were long.

The ratios of the maximum shearing stress at the fillet to the shearing stress in the straight portion or arm of the angle section for various radii

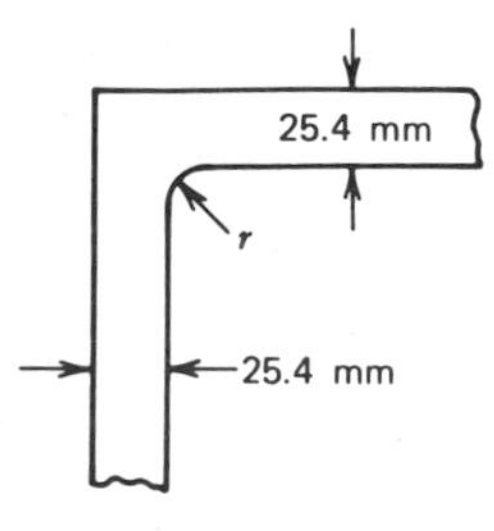

Fig. 12-3.7

of fillets are given in Table 12-3.1. These values show that a small fillet has a large influence in reducing the stress at the corner, and that practically no advantage is gained by making the radius of the fillet larger than about 6 mm.

Note: The stress concentration factors given in the foregoing discussions are for particular forms of discontinuities. Values of stress concentration factors for many other forms of discontinuities are available in the technical literature.

Table 12-3.1

Radius r of Fillet, mm (see Fig. 12-3.7)	*Ratio: Maximum Stress / Stress in Arm*
2.54	1.89
5.08	1.54
7.62	1.48
10.16	1.44
12.70	1.43
15.24	1.42
17.78	1.41

12-4 STRESS GRADIENTS DUE TO CONCENTRATED LOAD

In this article, we consider high stress gradients that occur in the neighborhood of a concentrated load. In a certain sense these high stress gradients may be thought of as a concentration of stress (lines of forces). The particular application employed, the point load on the free surface of the half plane (the so-called Boussinesq problem, Fig. 12-4.1), plays a role in the study of stress distributions close to a crack tip (determination of stress intensity factor defined in Art. 12-6) for a crack in tension.[17]

Consider a concentrated force P directed perpendicularly to the free edge of an infinitely large half-plane (sheet). The thickness of the sheet

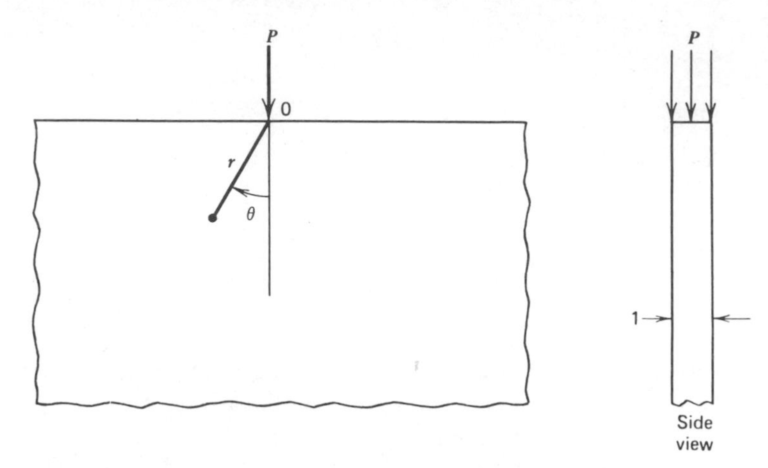

Fig. 12-4.1

may be taken as unity, so that P denotes the force per unit sheet thickness. In terms of polar coordinates (r, θ), the stress components are[6]

$$\sigma_{rr} = -\frac{2P\cos\theta}{\pi r} \qquad \sigma_{\theta\theta} = \sigma_{r\theta} = 0 \tag{12-4.1}$$

This distribution of stress is called a *simple radial distribution*, every polar coordinate element at a distance r from the point of application of P, being in simple compression in the radial direction. As seen by Eq. (12-4.1), σ_{rr} becomes very large as r becomes small and is undefined for $r = 0$ (the stress is said to be singular). For all real materials, yielding will occur in the neighborhood of the load, resulting in a plastic zone. If it is assumed that the plastic zone is sufficiently small so that the presence of the yielded material does not significantly alter the elastic solution, Eq. (12-4.1) yields

$$\frac{r}{\cos\theta} = \frac{2P}{\pi Y} = d_Y = \text{constant} \tag{12-4.2}$$

Thus, under the load, there is a circular plastic zone of diameter d_Y (Fig. 12-4.2). In general, there exists a family of circles of radii $d = r/\cos\theta$ that are tangent to the free boundary and for which σ_{rr} is a constant, since by Eq. (12-4.1),

$$\sigma_{rr} = -\frac{2P}{\pi d}$$

and for constant P and d, σ_{rr} is constant (Fig. 12-4.3). This conclusion has also been verified by simple photoelastic tests since the circles for constant σ_{rr} are isoclinics.[18]

Solution for the stress distributions due to concentrated forces, moments, etc. acting on the tip of a wedge ($-\pi/2 < \theta < \pi/2$ in Fig. 12-4.1) or at a wedge crack ($-\pi < \theta < \pi$ in Fig. 12-4.1) are also available.[6]

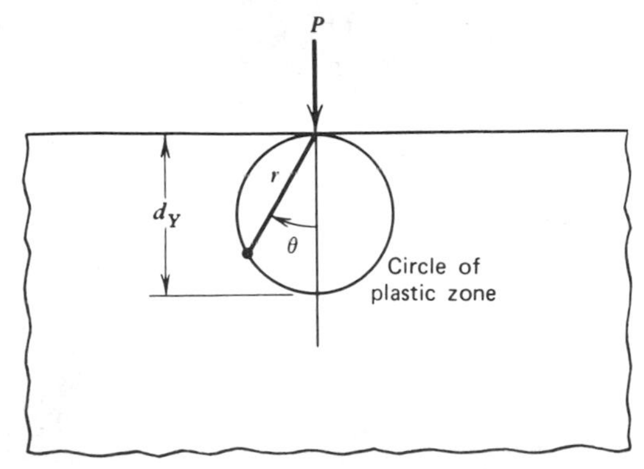

Fig. 12-4.2

12-5 THE STATIONARY CRACK

As noted in Art. 12-2, the tangential stress component $\sigma_{\beta\beta}$ around an elliptical hole in an infinite plate (sheet) subjected to uniform tensile stress σ in a direction perpendicular to the major axis of the hole depends upon the ratio a/b (Eq. 12-2.9). Hence, as $b/a \to 0$ (the elliptical hole becomes very flat), the maximum value of $\sigma_{\beta\beta}$ becomes very large. For example, for $a/b = 100$, $\sigma_{\beta\beta(\max)} = 201\sigma$; for $a/b = 1000$, $\sigma_{\beta\beta(\max)} = 2001\sigma$, etc. For sufficiently large ratios of b/a, the radius ρ of curvature at the edge of the major axis of the elliptical hole remains finite (nonzero). If we take the radius ρ very small (but nonzero), the elliptical hole solution may be used to obtain an approximate estimate of the stress distribution in the neighborhood of the crack edge (the end of the major

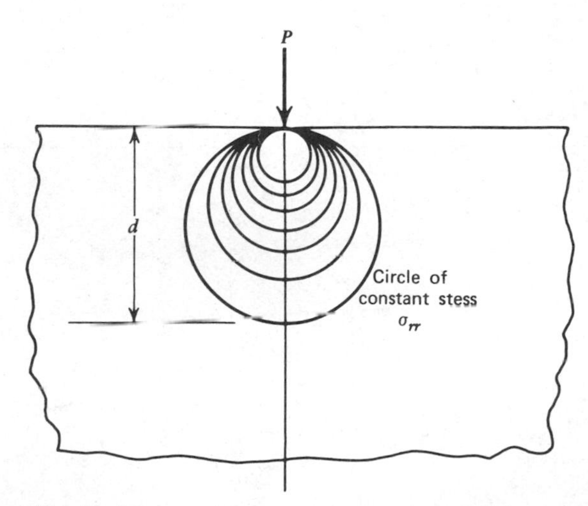

Fig. 12-4.3

axis of the hole). As the ratio $b/a \to 0$, we consider that $\rho \to 0$, and we are led to the case of a sharp crack of length $2a$ in an infinite plate with uniform stress σ applied at infinity in a direction perpendicular to the length $2a$. The stress distribution in the neighborhood of a *sharp crack edge* may be obtained directly from the elliptical hole problem by considering the case $b \to 0$.

In terms of ρ, the stress concentration factor S_{cc} is (Eq. 12-2.11),

$$S_{cc} = \frac{\sigma_{\beta\beta(\max)}}{\sigma} = 1 + 2\sqrt{\frac{a}{\rho}} \tag{12-5.1}$$

Thus, since many geometrical holes, notches, flaws, cracks, etc., may be approximated by an elliptical hole, it is to be expected that as $\rho \to 0$, $S_{cc} \to \infty$. All tabulated solutions of the crack problem exhibit this behavior. Most of the studies of fracture mechanics (see Chapter 3) are directed toward the behavior of the stress solution in the neighborhood of a crack edge, as $\rho \to 0$.

To examine the normal stresses in the neighborhood of a crack edge, it is convenient to represent the stress components in terms of (x, y)-axes (Fig. 12-5.1). Thus, for the elliptical hole, for stresses along the major axes of the elliptical hole, $y = 0$, $x > a$, we obtain by transforming the

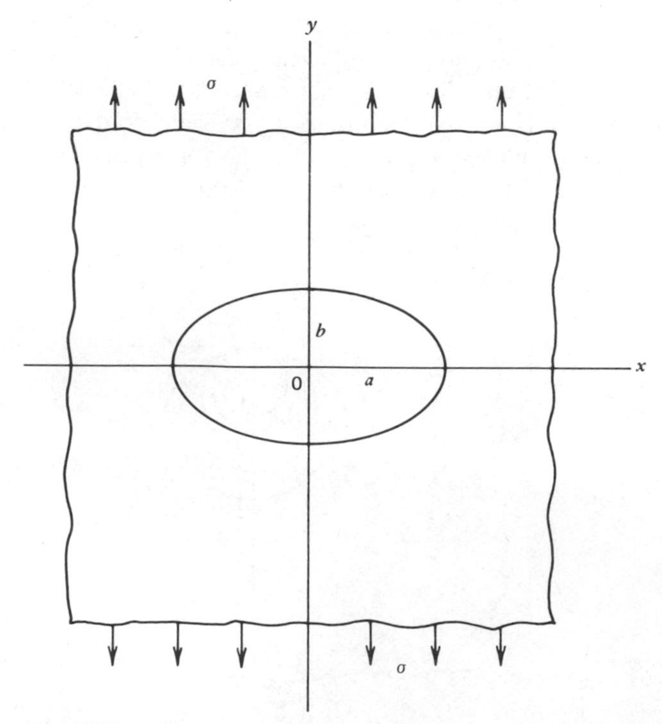

Fig. 12-5.1

stress components relative to (α, β) axes (Fig. 12-2.4) into stress components relative to (x, y)-axes[7]

$$\begin{aligned} \sigma_{xx} &= F_1(s) - F_2(s) \\ \sigma_{yy} &= F_1(s) + F_2(s) \end{aligned} \tag{12-5.2}$$

where s is a distance parameter

$$s = \frac{x}{2B} + \sqrt{\left(\frac{x}{2B}\right)^2 - m} \tag{12-5.3}$$

and

$$\begin{aligned} F_1(s) &= \frac{\sigma}{2}\left[1 + \frac{2(1+m)}{s^2 - m}\right] \\ F_2(s) &= \frac{\sigma}{2}\left\{1 + \frac{m^2 - 1}{s^2 - m}\left[1 + \left(\frac{m-1}{s^2 - m}\right)\left(\frac{3s^2 - m}{s^2 - m}\right)\right]\right\} \end{aligned} \tag{12-5.4}$$

with

$$B = \frac{1}{2}(a+b), \qquad m = \frac{a-b}{a+b} \tag{12-5.5}$$

By means of Eqs. (12-5.2), the stresses along the major axes in the neighborhood of the end of the major axes of an elliptical hole may be examined. Assuming that the radius ρ of curvature of a crack may be approximated by the radius of curvature of an equivalent elliptical hole (Eq. 12-2.10), approximations of the stress in the neighborhood of the crack may be obtained, provided $\rho \neq 0$, that is, for a blunt crack.

Blunt Crack / For the elliptical hole, let the radius of curvature be $\rho \ll a$ at the end of the major axis. Let $r = x - a$ be a parameter that measures distance from the end of the major axis (in the major axis direction, Fig. 12-5.2). In the neighborhood of $x = a$, $r \ll a$. Hence, in terms of r and ρ, we may write, with Eqs. (12-2.10), (12-5.2), (12-5.3), (12-5.4), and (12-5.5),

$$2B = a\left(1 + \sqrt{\frac{\rho}{a}}\right) \qquad m = \frac{1 - \sqrt{\rho/a}}{1 + \sqrt{\rho/a}}$$

$$s = 1 - \sqrt{\rho/a} + \sqrt{(2r+\rho)/a}$$

$$F_1(s) = F_1(r) \approx \frac{\sigma}{\sqrt{(2r+\rho)/a}}$$

$$F_2(s) = F_2(r) \approx \frac{\sigma(\rho/a)}{\left(\dfrac{2r+\rho}{a}\right)^{3/2}} \tag{12-5.6}$$

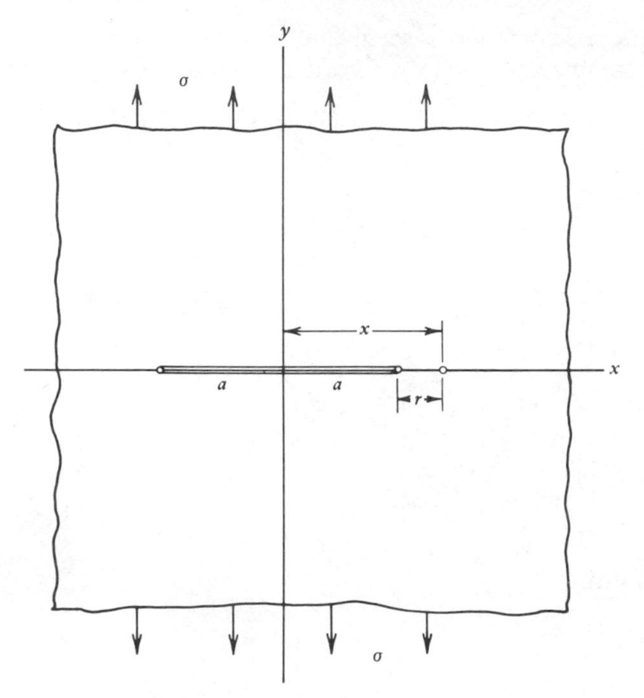

Fig. 12-5.2

Hence, in the neighborhood of the end of the major axis,

$$\sigma_{yy} = \frac{\sigma\sqrt{a}}{\sqrt{2r}} \frac{1 + \frac{\rho}{r}}{\left(1 + \frac{\rho}{2r}\right)^{3/2}} \qquad (12\text{-}5.7)$$

At the tip of the crack, $r = 0$, and then Eq. (12-5.7) reduces to

$$\sigma_{yy} = 2\sqrt{\frac{a}{\rho}}\,\sigma \qquad (12\text{-}5.8)$$

which agrees with Eq. (12-5.1) for $a \gg \rho$.

Sharp Crack/For the sharp crack, we may estimate the stress distribution from that of the elliptical hole by letting $b \to 0$. Then, we have

$$\begin{aligned} &2B = a, \qquad m = 1, \qquad s = \frac{x}{a} + \sqrt{\left(\frac{x}{a}\right)^2 - 1} \\ &x = r + a, \qquad r \ll a, \qquad b = 0 \end{aligned} \qquad (12\text{-}5.9)$$

and by Eqs. (12-5.4)

$$F_1(s) = F_1(r) \approx \frac{\sigma}{\sqrt{\dfrac{2r}{a}}} \qquad (12\text{-}5.10)$$

$$F_2(s) = F_2(r) \approx \frac{\sigma}{2}$$

Hence,

$$\sigma_{yy} = \frac{\sigma\sqrt{a}}{\sqrt{2r}} \qquad (12\text{-}5.11)$$

Clearly at the crack tip ($r = 0$) the stress is singular ($\sigma_{yy} \to \infty$ as $r \to 0$).

Alternatively in terms of x, it may be shown that

$$\sigma_{yy} = \frac{\sigma x}{\sqrt{x^2 - a^2}}$$

$$\sigma_{xx} = \sigma\left(\frac{x}{\sqrt{x^2 - a^2}} - 1\right) \qquad (12\text{-}5.12)$$

Again at the crack tip ($x = a$), σ_{yy} (and σ_{xx}) become infinite. For large values of x, $\sigma_{yy} \to \sigma$, and $\sigma_{xx} \to 0$ as expected (Fig. 12-5.2).

As we shall see in Art. 12-6, in describing crack propagation, it is conventional to introduce the combination $\sigma_{yy}\sqrt{2r}$ since this factor remains finite as $r \to 0$. More specifically, a factor π is introduced so that

$$K_{\mathrm{I}} = \sigma_{yy}\sqrt{2\pi r} = \sigma\sqrt{\pi a} \qquad (12\text{-}5.13)$$

The factor K_{I} is called the *stress intensity factor*. In certain fracture theories, it is assumed that the material fractures (the crack propagates) if K_{I} exceeds a critical value

$$K_{\mathrm{IC}} = \sigma_c\sqrt{\pi a} \qquad (12\text{-}5.14)$$

where σ_c is the corresponding critical tension stress. The term stress intensity factor should not be confused with the term stress concentration factor (Eq. 12-1.1), which represents the ratio between the maximum stress in a region of stress concentration and the average stress.

The results of this article are of importance in fracture mechanics, failure theories, and crack propagation studies. See Chapter 3 and Art. 12-6.

12-6 CRACK PROPAGATION. STRESS INTENSITY FACTOR

Elastic Stress at Tip of Sharp Crack/In Art. 12-2, we noted that the maximum stress at the ends of an elliptical hole in an infinite plate may be quite large. For example, when the plate is subjected to an edge tensile stress σ in the direction perpendicular to the major axis of the elliptic hole, the stress is given by Eq. (12-2.9). Hence, if the ratio $a/b = 100$, the value of $\sigma_{\beta\beta(\max)}$ is 201σ; if $a/b = 1000$, $\sigma_{\beta\beta(\max)} = 2001$; etc. The elliptical hole becomes very narrow and approaches the shape of an internal line crack as $a/b \to \infty$. In this case $\sigma_{\beta\beta(\max)} \to \infty$, and we can no longer utilize the concept of stress concentration factor in describing the behavior around the crack tip.

Physically, one might expect that when loads are applied to a member that contains a line crack, the extremely large stress at the edge of the crack will cause the crack to extend or propagate. Experiments bear out this expectation in that it has been observed that the crack may[†] propagate when the load attains a critical nominal value. In general, under lower values of the applied stress σ, the crack may propagate slowly a short distance and stop; whereas, under higher values of σ, the crack may propagate rapidly and continuously until a catastrophic separation of parts of the member occurs.

For a given member made of a given material, cracks may propagate under conditions such that the material is in the ductile state or under conditions such that the material is in the brittle state (Art. 3-4). If the dimensions of the member are such that the state of stress over most of the length of the crack tip is plane strain, the crack will propagate with minimum plastic deformation occurring at the crack tip. The material in such members is considered to be loaded in the brittle state. The state of stress outside the small plastic zone is assumed to be characterized by the elasticity solution presented in Art. 12-5. The discussion that follows assume that the materials are loaded in the brittle state.

Investigators have attempted to explain the mechanism of crack propagation in terms of the distribution of stress in the neighborhood of the crack tip. However, in addition, to help explain the crack propagation, another concept is required. Earlier investigators, particularly Griffith, introduced the concept of strain energy release rate, $\mathscr{G}$. The quantity $\mathscr{G}$ represents the amount of strain energy lost by the member per unit area of the newly formed crack area as the crack propagates. This strain energy is used up in forming the new surface area of the crack. In other words, the energy required to form the surface area of the extended crack is obtained from the strain energy of the body. Thus, the units of $\mathscr{G}$ are $\text{N} \cdot \text{m}/\text{m}^2$. Since the dimensions of $\mathscr{G}$ may be written $[F/L]$ (or N/m), $\mathscr{G}$ is referred to as the crack-extension force.

Stress Intensity Factor. Definition and Derivation/To examine the stress distribution near the edge of a crack in a flat plate, consider a crack of length $2a$, which is very small compared to the width and length of the plate (Fig.

[†] The capacity of the material to absorb relatively large amounts of energy per unit volume by plastic flow before fracture determines the level of nominal stress at which the crack propagates. In mild steel, a crack may not propagate until catastrophic fracture is imminent.

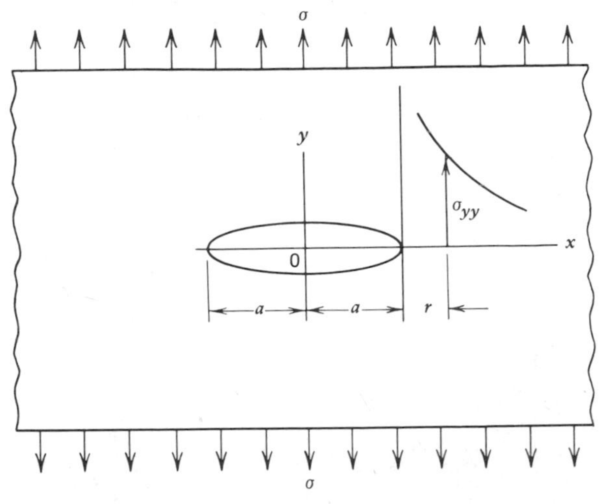

Fig. 12-6.1

12-6.1). Let the plate be subjected to uniformly distributed stress in a direction perpendicular to the crack length $2a$. As noted in Art. 12-5, Eq. (12-5.8), the elastic stress at the tip of the crack becomes infinitely large as the radius of curvature ρ at the tip goes to zero (Eq. 12-5.8). As shown in Art. 12-5, see Eq. (12-5.13) and Figs. 12-5.2 and 12-6.1, the stress σ_{yy} along the extension of the major axis (the expected path of crack propagation) is given by

$$\sigma_{yy} = \frac{K_{\mathrm{I}}}{\sqrt{2\pi r}} \tag{12-6.1}$$

where r is the distance from the crack tip measured along the x-axis and K_{I} is the stress intensity factor for a mode I crack. (See Art. 3-4 and Fig. 3-4.1.)

Following Irwin,[19] we define the stress intensity factor by means of the following limit:

$$K_{\mathrm{I}} = \lim_{\rho \to 0} \frac{\sqrt{\pi \rho}}{2} \sigma_{\max} \tag{12-6.2}$$

where ρ is the radius of curvature at the crack tip (Fig. 12-6.1), and $\sigma_{\max}$, the maximum stress at the crack tip, is a function of ρ; see Eqs. (12-5.7) and (12-2.11) for the case of an elliptical hole. Consequently, if we consider the line crack to be the limiting case of an elliptical hole, as $b \to 0$, we obtain by Eqs. (12-2.11) and (12-6.2)

$$K_{\mathrm{I}} = \lim_{\rho \to 0} \frac{\sqrt{\pi \rho}}{2} \left[\sigma \left(1 + 2\sqrt{\frac{a}{\rho}} \right) \right] = \sigma \sqrt{\pi a} \tag{12-6.3}$$

for mode I propagation of an internal crack in a flat plate; see also Eq. (12-5.13). Values of K_{I} for some other types of cracked members are listed in Table 3-4.2 and are repeated in Table 12-6.1.

Table 12-6.1

Stress Intensity Factors K_I

Case 1. Infinite Sheet with Uniform Tension at Infinity and Elliptic Hole b→0. Griffith's Crack	*Case 2. Periodic Array of Elliptic Holes (b→0) in Infinite Sheet with Uniform Tension at Infinity.*	*Case 3. Central Crack in Finite Width Strip Subject to Uniform Tension at Infinity b→0.*

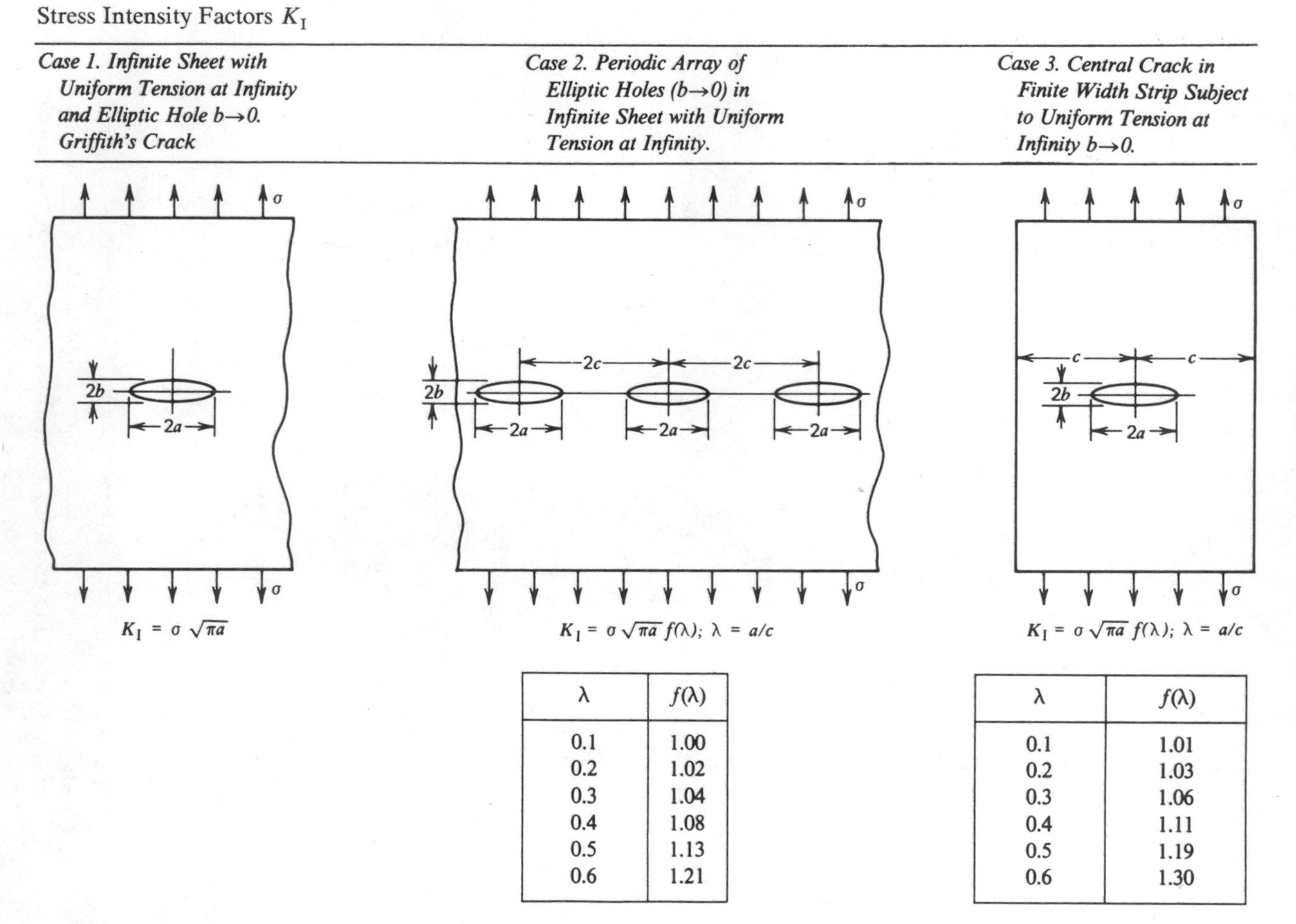

$K_I = \sigma\sqrt{\pi a}$

$K_I = \sigma\sqrt{\pi a}\, f(\lambda);\ \lambda = a/c$

λ	$f(\lambda)$
0.1	1.00
0.2	1.02
0.3	1.04
0.4	1.08
0.5	1.13
0.6	1.21

$K_I = \sigma\sqrt{\pi a}\, f(\lambda);\ \lambda = a/c$

λ	$f(\lambda)$
0.1	1.01
0.2	1.03
0.3	1.06
0.4	1.11
0.5	1.19
0.6	1.30

Table 12-6.1

Continued

Case 4. Single Edge Crack in Finite-Width Sheet.

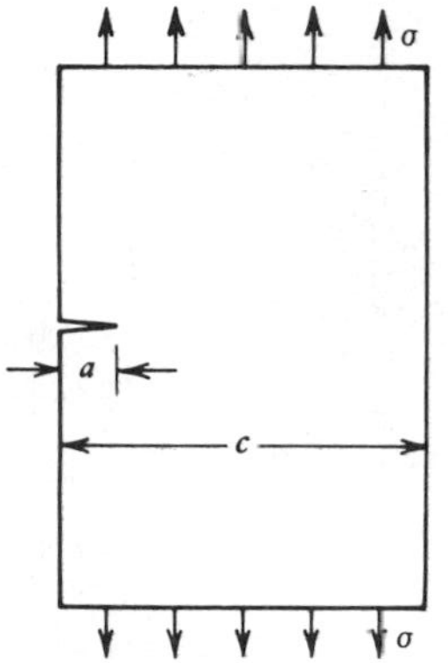

$K_I = \sigma \sqrt{\pi a} f(\lambda)$; $\lambda = a/c$

λ	$f(\lambda)$
0 ($c \to \infty$)	1.12
0.2	1.37
0.4	2.11
0.5	2.83

Case 5. Double Edge Crack in Finite-Width Sheet.

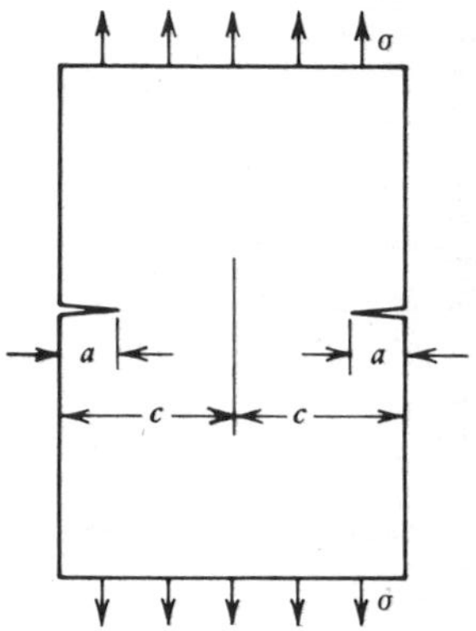

$K_I = \sigma \sqrt{\pi a} f(\lambda)$; $\lambda = a/c$

λ	$f(\lambda)$
0 ($c \to \infty$)	1.12
0.2	1.12
0.4	1.14
0.5	1.15
0.6	1.22

Case 6. Edge Crack in Beam in Bending.

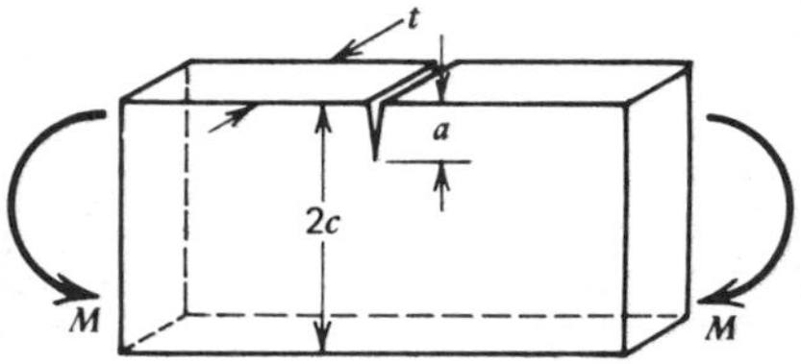

$K_I = \sigma \sqrt{\pi a} f(\lambda)$

$\lambda = a/2c$

$\sigma = \frac{3M}{2tc^2}$

λ	$f(\lambda)$
0.1	1.02
0.2	1.06
0.3	1.16
0.4	1.32
0.5	1.62
0.6	2.10

Crack Extension Force $\mathscr{G}$. Derivation / Following concepts proposed by Griffith, one may derive a relationship between the crack extension force $\mathscr{G}$ and the stress intensity factor K for various crack modes. For example, for a Griffith crack of length $2a$, centrally located in a plate subjected to a uniformly distributed stress σ at edges far removed from the crack (Fig. 12-5.2), the surfaces of the crack undergo a relative displacement of magnitude $2v$ under a mode I separation (Fig. 12-6.2). For a condition of plane strain,[6] it may be shown that[17]

$$2v = 4(1-\nu^2)\frac{\sigma}{E}\sqrt{a^2 - x^2} \qquad x \leq a \tag{12-6.4}$$

where ν is Poisson's ratio and E is the modulus of elasticity. The problem is to calculate the strain energy released when a crack of half-length a is extended to a half-length $(a + \delta a)$. For constant load σ, the release of potential energy is equal to the release of strain energy under fixed grips as $\delta a \to 0$. Alternatively, we may calculate the change in energy in the plate as a whole, by calculating the work done by the surface forces at the crack tip acting across the length δa when the crack is closed from length $(a + \delta a)$ to length a. In other words, we may employ the principle of virtual work.

In terms of the crack extension force $\mathscr{G}$, the energy change may be expressed in the form

$$\mathscr{G}\,\delta a = \int_a^{a+\delta a} \sigma_{yy} v\, dx \tag{12-6.5}$$

where the plate thickness is taken as unity and where, by Eq. (12-5.12),

$$\sigma_{yy} = \frac{\sigma x}{\sqrt{x^2 - a^2}} \qquad a \leq x \leq a + \delta a \tag{12-6.6}$$

and v is given by Eq. (12-6.4), where we let $a \to a + \delta a$ and $a \leq x \leq a +$

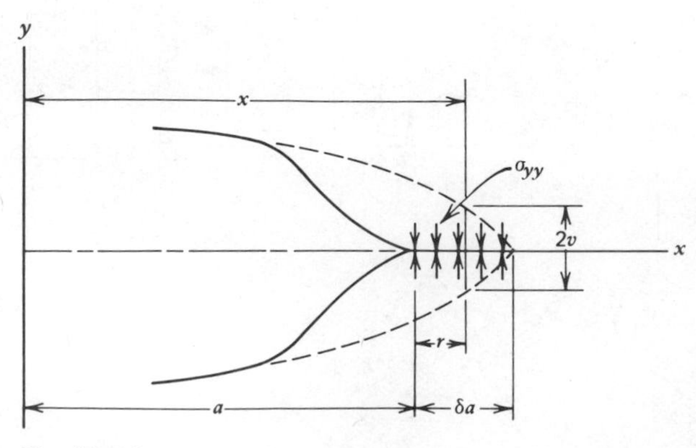

Fig. 12-6.2

δa. Letting $r = x - a$, we may write for sufficiently small r (Fig. 12-6.2)

$$\sigma_{yy} = \frac{\sigma\sqrt{\pi a}}{\sqrt{2\pi r}} = \frac{K_{\mathrm{I}}}{\sqrt{2\pi r}}$$

$$2v = 4(1 - \nu^2)\frac{\sigma}{E}\sqrt{2a}\,\sqrt{\delta a - r} \tag{12-6.7}$$

and, hence, with Eq. (12-6.4) we have

$$\mathscr{G}\,\delta a = 2(1 - \nu^2)\frac{\sigma^2 a}{E}\int_0^{\delta a}\left[\frac{\delta a - r}{r}\right]^{1/2} dr \tag{12-6.8}$$

Integration of Eq. (12-6.8) with the substitution $r = \delta a \sin^2 \omega$ yields for *plane strain*

$$\mathscr{G} = \frac{(1 - \nu^2)\pi a \sigma^2}{E} = (1 - \nu^2)\frac{K_{\mathrm{I}}^2}{E} \tag{12-6.9}$$

where $K_{\mathrm{I}} = \sigma\sqrt{\pi a}$. Similarly, for *plane stress*

$$\mathscr{G} = \frac{K_{\mathrm{I}}^2}{E} \tag{12-6.10}$$

where K_{I} is the stress intensity factor for Mode I opening (Fig. 3-4.1, Table 3-4.2, and Table 12-6.1).

$\mathscr{G}_c$; Critical Value of Crack Extension Force / As noted in Arts. 3-1 and 3-4, under certain conditions of loading, a crack in a structural member may gradually increase in length as the load is increased. This period of gradual increase in crack length may be followed by a rapid (catastrophic) propagation of the crack resulting in complete separation of two parts of the member. In certain fracture mechanics hypotheses, this rapid propagation of the crack is associated with a *critical crack length* a_c. Alternatively, since $\mathscr{G}$, the Griffith crack extension force is related to the crack length a (Eq. 12-6.9), the rapid propagation of the crack may also be associated with $\mathscr{G}_c$, a *critical crack extension force*, defined by

$$\mathscr{G}_c = \frac{(1 - \nu^2)\pi a_c \sigma^2}{E} = (1 - \nu^2)\frac{K_{\mathrm{IC}}^2}{E} \tag{12-6.11}$$

where analogous to $\mathscr{G}_c$, the factor

$$K_{\mathrm{IC}} = \sigma\sqrt{\pi a_c} \tag{12-6.12}$$

is called the *critical stress intensity factor* for Mode I opening of the crack. The factor K_{IC} is also referred to as the *fracture toughness*.[17] Typical values of K_{IC} for several metals are listed in Table 3-4.1 and in Table 12-6.2. These values have been obtained for plane strain conditions according to ASTM Standards, and can be used in the design of the members shown in Tables 3-4.2 and 12-6.1. If the fracture loads for these

Table 12-6.2

K_{IC} Critical Stress Intensity Factor (Fracture Toughness)
(Room Temperature Data)

Material	σ_{ult} MPa	Y MPa	K_{IC} MPa$\sqrt{m}$	*Minimum Values for a, t* mm
		Alloy Steels		
A533B	—	500	175	306.0
2618 Ni Mo V	—	648	106	66.9
V1233 Ni Mo V	—	593	75	40.0
124 K 406 Cr Mo V	—	648	62	22.9
17-7PH	1289	1145	77	11.3
17-4PH	1331	1172	48	4.2
PH 15-7Mo	1600	1413	50	3.1
AISI 4340	1827	1503	59	3.9
		Stainless Steel		
AISI 403	821	690	77	31.1
		Aluminum Alloys		
6061-T651	352	299	29	23.5
2219-T851	454	340	32	22.1
7075-T7351	470	392	31	15.6
7079-T651	569	502	26	6.7
2024-T851	488	444	23	6.7
		Titanium Alloys		
Ti-6Al-4Zr-2Sn-0.5Mo-0.5V	890	836	139	69.1
Ti-6Al-4V-2Sn	852	798	111	48.4
Ti-6.5Al-5Zr-1V	904	858	106	38.2
Ti-6Al-4Sn-1V	889	878	93	28.0
Ti-6Al-6V-2.5Sn	1176	1149	66	8.2

members are to be calculated with reasonable accuracy, it is necessary that the state of stress at the edge of the crack is plane strain over most of the length of the crack edge. This is insured[18] by specifying that the half crack length a (or crack length a when applicable) and thickness t of the cracked member satisfy the relation

$$a, t \geq 2.5\left(\frac{K_{IC}}{Y}\right)^2 \tag{12-6.13}$$

The magnitude of the right side of Eq. (12-6.13) for each metal is listed (in millimeters) in Table 12-6.2. If the half crack length a (or crack length a when applicable) is appreciably less than the value indicated by Eq. (12-6.13), the computed fracture load may be greater than the failure load for another mode of failure (yielding failure, for instance). If the thickness t of the member is small compared to the value given by the right side of Eq. (12-6.13), the state of stress approaches plane stress, appreciable yielding may occur at the crack tip, and the actual fracture load may be as much as several hundred percent greater than the value calculated using K_{IC}.

In addition to the illustrative problem and problem set for Art. 12-6, the reader is referred to the illustrative problem and problem set for Art. 3-4.

EXAMPLE 12-6.1

Brittle Fracture for Combined Tension and Bending

A tool similar to that shown in Fig. E12-6.1 may be used to dig up old road beds before replacing them. Let the tool be made of AISI 4340 steel and heat treated to the properties indicated in Table 12-6.2. The dimensions of the tool are $d = 250$ mm, $2c = 60$ mm, and the width of the rectangular cross section is $t = 25$ mm. Determine the magnitude of the fracture load P for crack length of (a) $a = 5$ mm and (b) $a = 10$ mm.

SOLUTION

Note that both the width t and the crack length a satisfy Eq. (12-6.13); therefore, the tool is assumed to be loaded in the brittle state. At a section through the crack, the tool is subjected to combined axial load (Case 4 of Table 12-6.1) and bending (Case 6 of Table 12-6.1). Since a linear elastic analysis is assumed, the state of stress for combined loading can be obtained by superposition of the two states of stress for the two types of loading. Thus,

$$\sigma f(\lambda) = \sigma_1 f_1(\lambda) + \sigma_2 f_2(\lambda)$$

(a) When $a = 5$ mm, $\lambda = a/2c = 0.083$. From Table 12-6.1 for Case 4 and Case 6, we obtain $f_1(\lambda) = 1.22$ and $f_2(\lambda) = 1.02$.

$$\sigma f(\lambda) = \frac{P}{25(60)} 1.22 + \frac{3(280P)}{2(25)(30)^2} 1.02 = 0.0195P$$

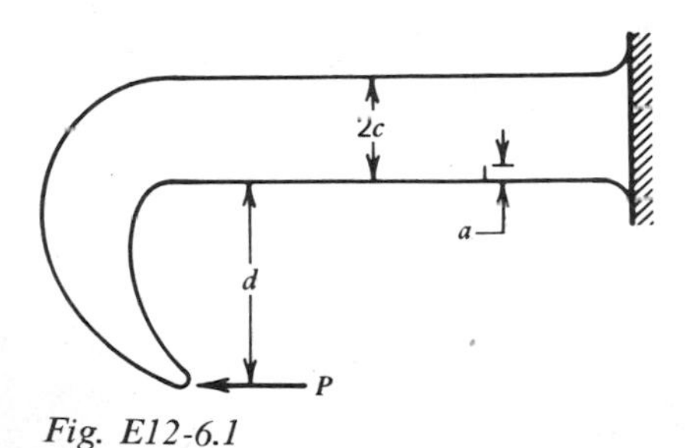

Fig. E12-6.1

This is substituted in the relation

$$K_{IC} = \sigma f(\lambda)\sqrt{\pi a}$$

$$P = \frac{59\sqrt{1000}}{0.0195\sqrt{5\pi}} = 24{,}100 \text{ N} = 24.1 \text{ kN}$$

(b) When $a = 10$ mm, $\lambda = 0.167$

$$\sigma f(\lambda) = \frac{P}{25(60)}1.33 + \frac{3(280P)(1.05)}{2(25)(30)^2} = 0.0205P$$

$$P = \frac{59\sqrt{1000}}{0.0205\sqrt{10\pi}} = 16{,}200 \text{ N} = 16.2 \text{ kN}$$

PROBLEM SET 12-6

1. Solve Example 12-6.1 for the condition that $d = 200$ mm and $2c = 50$ mm.

2. A rectangular section beam has a depth $2c = 150$ mm, a width $t = 25$ mm, and a length $L = 2.00$ m. The beam is loaded as a simply supported beam with a concentrated load P at the center. A notch is machined into the beam on the tension side opposite the point of application of P. The depth of the notch was increased by fatigue loading until $a = 15$ mm. The beam is made of 17-7PH precipitation hardening steel heat treated to give the properties indicated in Table 12-6.2. (a) Determine whether or not plane strain conditions are satisfied for the beam. (b) Determine the fracture load P.

   Ans. (a) Conditions are satisfied
   (b) $P = 65.2$ kN

3. Solve Problem 2 if the beam is made of 2024-T851 aluminum alloy.

4. A closed end cylinder is made of 7079-T651 aluminum alloy. The cylinder has an inside diameter $D = 1000$ mm, a wall thickness $h = 20$ mm, and is subjected to an internal pressure $p = 6.00$ MPa. Determine the length of crack ($2a$) required to cause fracture at this pressure if plane strain conditions are assumed to be satisfied. The inside of the cylinder is covered with a thin layer of rubber to prevent leakage. Determine whether or not conditions are satisfied for a plane strain state of stress.

   Ans. $2a = 19.1$ mm; yes

5. If the crack with length $2a = 19.1$ mm is circumferential instead of longitudinal for the cylinder in Problem 4, determine the internal pressure that will cause fracture.

6. A bar of titanium alloy (Ti-6Al-4Sn-1V) has a rectangular cross section ($t = 30$ mm and $2c = 300$ mm), a length $L = 1$ m, and is heat treated to give the material properties in Table 12-6.2. The bar is subjected to an axial load P whose action line is at the center line of one of the 30 mm edges. If a transverse edge crack at the center of length L has the minimum length required for a plane strain state of stress, determine the magnitude of P to cause brittle fracture.

   Ans. $P = 656$ kN

7. Solve Problem 6 if the action line of load P is at the center of the root of the crack.

REFERENCES

1. C. A. Brabbia and J. J. Connor, *Fundamentals of Finite Element Techniques*, Halsted Press, Wiley, New York, 1974.
2. M. Hetenyi et al., *Handbook of Experimental Stress Analysis*, Wiley, New York, 1950.
3. R. E. Peterson, *Stress Concentration Factors for Design*, Wiley, New York, 1974.
4. J. W. Dally and W. F. Riley, *Experimental Stress Analysis*, McGraw-Hill, New York, 1965.
5. G. N. Savin, *Stress Concentration around Holes*, Pergamon Press, New York, 1961.
6. A. P. Boresi and P. P. Lynn, *Elasticity in Engineering Mechanics*, 2nd Ed., Prentice-Hall, Englewood Cliffs, New Jersey, 1974.
7. C. E. Inglis, "Stresses in the Plate Due to the Presence of Cracks and Sharp Corners," *Transactions of the Institute of Naval Architects*, Vol. 60, London, England, 1913, p. 219.
8. N. I. Muskhelisvili, *Some Basic Problems of the Mathematical Theory of Elasticity*, Wolters-Noordhoff Publishing, Groningen, The Netherlands, 1953.
9. S. Timoshenko and J. Goodier, *Theory of Elasticity*, 3rd Ed., McGraw-Hill, New York, 1970.
10. M. A. Sadowsky and E. Sternberg, "Stress Concentration around a Tri-axial Ellipsoidal Cavity," *Journal of Applied Mechanics*, Vol. 71, 1949, p. 149.
11. H. Neuber, *Kerbspannungslehre*, 2nd Ed., Springer-Verlag, Berlin, 1958.
12. M. M. Frocht, "Photoelastic Studies in Stress Concentration," *Mechanical Engineering*, August 1936, p. 485.
13. E. G. Coker and L. N. G. Filon, *A Treatise on Photoelasticity*, revised by H. T. Jessop, Cambridge University Press, London, 1957, pp. 562–596.

14. R. E. Peterson and A. M. Wahl, "Two- and Three-Dimensional Cases of Stress Concentration, and Comparison with Fatigue Tests," *Transactions American Society of Mechanical Engineers*, Vol. 47, 1925, p. 619; Vol. 59, 1936, p. A15–22.
15. L. S. Jacobsen, "Torsional-Stress Concentrations in Shafts of Circular Cross-Section and Variable Diameter," *Transactions of the American Society of Mechanical Engineers*, Vol. 47, 1925, p. 619.
16. A. A. Griffith and G. I. Taylor, "The Use of Soap Films in Solving Torsional Problems," *Proceedings of the Institute of Mechanical Engineers*, London, October–December 1917, p. 755.
17. J. F. Knott, *Fundamentals of Fracture Mechanics*, Wiley, New York, 1973.
18. M. M. Frocht, *Photoelasticity*, Vols 1 and 2, Wiley, New York, 1948.
19. G. R. Irwin, "Analysis of Stresses and Strains Near the End of a Crack Traversing a Plate," *Journal of Applied Mechanics*, Vol. 24, September 1957, pp. 361–364.

Additional References

1. R. M. Caddell, *Deformation and Fracture of Solids*, Prentice-Hall, Englewood Cliffs, New Jersey, 1980.
2. S. W. Freiman and E. R. Fuller, Jr. (Ed.), *Fracture Mechanics for Ceramics, Rocks, and Concrete*, American Society for Testing Materials, Philadelphia, 1981.
3. R. W. Hertzberg, *Deformation and Fracture of Engineering Materials*, 2nd Ed., Wiley, New York, 1983.

CHAPTER 13

EFFECTIVE STRESS CONCENTRATION FACTORS. APPLICATIONS

As noted in Chapter 12, when the value of the stress concentration factor is based upon tests of the actual material under conditions of use, the stress concentration factor is called the *effective stress concentration factor*. In this chapter, we consider the significance of the stress concentration factor under combined loading conditions for several important practical members and geometries. We also consider the influence of various effects upon the effective stress concentration factor; for example, static loads, repeated loads, residual stresses, large stress gradients, impact loads, and so on.

13-1 STRESS CONCENTRATION FACTORS—COMBINED LOADS

Principle of Superposition/In Art. 12-3, we discussed stress concentrations for several types of notches for simple loading of members made of an isotropic material that is assumed to behave in a linearly elastic manner. Because of the linearity of the response, if these same conditions prevail when such a member is subjected to more complex loading, the loads in some cases may be resolved into simple component parts, for which the results of Chapter 12 hold. Then by means of the principle of superposition, the results may be combined to yield the effect of complex loading.

Infinite Plate with Circular Hole/Consider an infinite plate, with circular hole, subjected to stress $\sigma = \sigma_1$ on two parallel edges far removed from the hole

(Fig. 12-2.1) and to stress $\sigma = \sigma_2$ on the other distant parallel edges. The stress distribution may be derived from Eqs. (12-2.1) by superposition. One need merely set $\sigma = \sigma_1$ and $\theta = 0$ into Eqs. (12-2.1) to obtain stresses due to σ_1. Then set $\sigma = \sigma_2$ and $\theta = \theta + \pi/2$ into Eqs. (12-2.1) to obtain stresses due to σ_2, and add the stresses so obtained to those due to σ_1. Special results are obtained for $\sigma = \sigma_1 = \sigma_2$, the case of uniform tension in all directions (then $\sigma_{\theta\theta(\max)} = 2\sigma$) and for $\sigma = \sigma_1 = -\sigma_2$, the case of pure shear (then $\sigma_{\theta\theta(\max)} = 4\sigma$ for $\theta = \pm\pi/2$). Thus for uniform tension $S_{cc} = 2$ and for uniform shear $S_{cc} = 4$.

Elliptical Hole in Infinite Plate Stressed Uniformly in Directions of Major and Minor Axes of the Hole/Analogous to the circular hole case, the stresses for the state of uniform tension σ on the boundary ($\sigma_{xx} \to \sigma$ for $x \to \infty$ and $\sigma_{yy} \to \sigma$ for $y \to \infty$) may be computed for the elliptical hole. The results are[1]

$$\sigma_{\alpha\alpha} + \sigma_{\beta\beta} = \frac{2\sigma \sinh 2\alpha}{\cosh 2\alpha - \cos 2\beta} \tag{13-1.1}$$

Again since $\sigma_{\alpha\alpha} = 0$ for $\alpha = \alpha_0$ (at the hole),

$$\sigma_{\beta\beta}\big|_{\alpha=\alpha_0} = \frac{2\sigma \sinh 2\alpha_0}{\cosh 2\alpha_0 - \cos 2\beta} \tag{13-1.2}$$

and

$$\sigma_{\beta\beta(\max)} = 2\sigma\left(\frac{a}{b}\right) \tag{13-1.3}$$

which, for $a/b = 1$, becomes equal to 2σ as derived previously for the circular hole.

Pure Shear Parallel to Major and Minor Axes of the Elliptical Hole/Let the infinite plate be subjected to uniform shearing stress τ as shown in Fig. 13-1.1. The stress state due to this case of pure shear parallel to the (x, y)-axes may be found by superposition of the two cases for uniform tension $\sigma(=\tau)$ at $\beta = \pi/4$ and $-\sigma(=-\tau)$ at $\beta = 3\pi/4$; see Figs. 12-2.3 and 12-2.4 and also Eqs. (12-2.7) and (12-2.12). The value of $\sigma_{\beta\beta}$ on the perimeter of the hole ($\alpha = \alpha_0$) may be found in this manner to be

$$\sigma_{\beta\beta}\big|_{\alpha=\alpha_0} = -\frac{2\tau e^{2\alpha_0} \sin 2\beta}{\cosh 2\alpha_0 - \cos 2\beta} \tag{13-1.4}$$

By differentiation of $\sigma_{\beta\beta}$ with respect to β, we may show that the maximum value of $\sigma_{\beta\beta}$ occurs when

$$\tan\beta = -\tanh\alpha_0 = -\frac{b}{a} \tag{13-1.5}$$

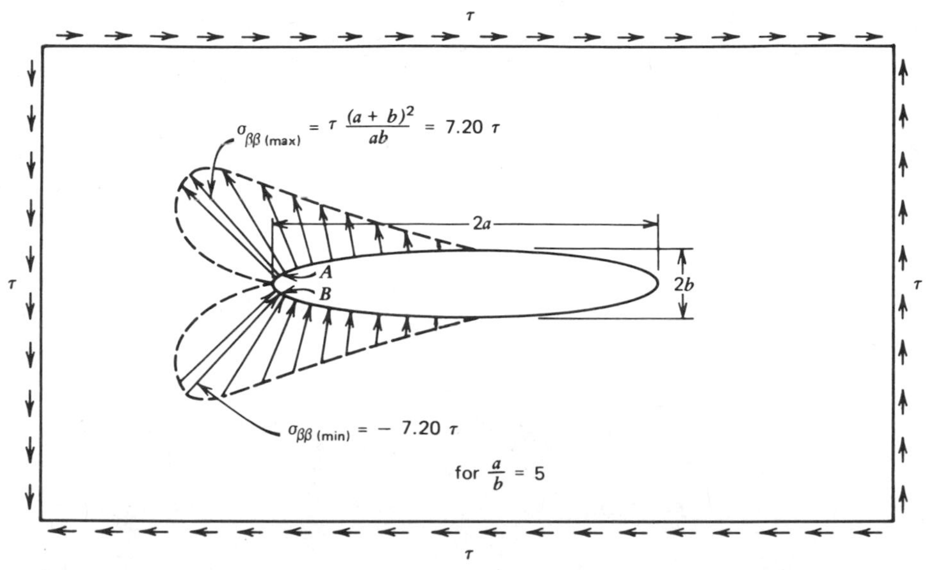

Fig. 13-1.1/Distribution of $\sigma_{\beta\beta}$ around an elliptical hole in an infinite plate loaded in pure shear.

and the maximum value of $\sigma_{\beta\beta}$ is

$$\sigma_{\beta\beta(\max)} = \tau \frac{(\cosh \alpha_0 + \sinh \alpha_0)^2}{\sinh \alpha_0 \cosh \alpha_0} = \tau \frac{(a+b)^2}{ab} \tag{13-1.6}$$

For the case $a/b = 5$, the distribution of $\sigma_{\beta\beta}$ around the hole is given in Fig. 13-1.1, where point A locates the maximum value. Analogously, the minimum value of $\sigma_{\beta\beta}$ is

$$\sigma_{\beta\beta(\min)} = -\tau \frac{(a+b)^2}{ab} \tag{13-1.7}$$

where $\tan \beta = \tanh \alpha_0 = b/a$ (point B in Fig. 13-1.1).

Solutions for the stress distribution around an elliptical hole in a plane isotropic sheet have been obtained for other loadings, for example, pure bending in the plane, as well as for other shapes of holes.[1]

Elliptical Hole in Infinite Plate with Different Loads in Two Perpendicular Directions/Consider an infinite plate with a small elliptical hole (Fig. 13-1.2). Let the plate be subjected to uniformly distributed stresses $\sigma_1 > \sigma_2$ along straight line edges far removed from the hole. Let the major axis of the hole form an angle θ with the edge on which stress σ_1 acts. We wish to compute the maximum value of $\sigma_{\beta\beta}$ at the perimeter of the hole.

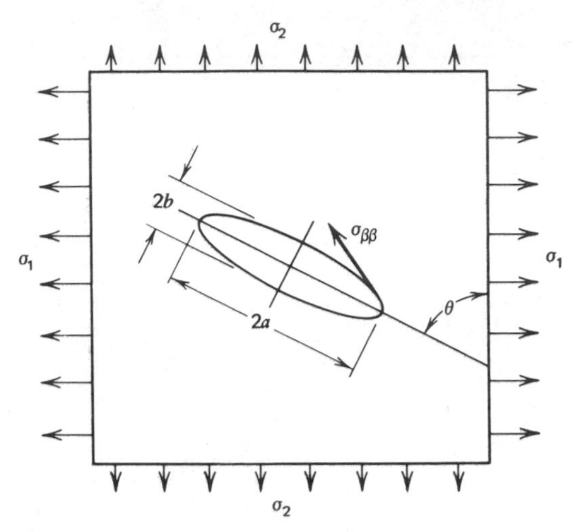

Fig. 13-1.2

The solution to the above problem may be obtained by superposing the loadings of Figs. 12-2.6, 12-2.7, and 13-1.1. By the laws of transformation of stress (see Chapter 1, Eqs. 1-4.19), we compute normal and shearing stresses on planes parallel to the major and minor axes of the ellipse (Fig. 13-1.2), as shown in Fig. 13-1.3. Thus, we obtain

$$\begin{aligned}\sigma_{\mathrm{I}} &= \frac{\sigma_1+\sigma_2}{2}+\frac{\sigma_1-\sigma_2}{2}\cos 2\theta \\ \sigma_{\mathrm{II}} &= \frac{\sigma_1+\sigma_2}{2}-\frac{\sigma_1-\sigma_2}{2}\cos 2\theta \\ \tau_{\mathrm{I,II}} &= \frac{\sigma_1-\sigma_2}{2}\sin 2\theta\end{aligned} \tag{13-1.8}$$

Then the substitutions $\sigma = \sigma_{\mathrm{I}}$ into Eq. (12-2.7), $\sigma = \sigma_{\mathrm{II}}$ into Eq. (12-2.12), and $\tau = \tau_{\mathrm{I,II}}$ into Eq. (13-1.4) and addition of the results yield

$$\begin{aligned}\sigma_{\beta\beta} = \big[&(\sigma_1+\sigma_2)\sinh\alpha_0 + (\sigma_1-\sigma_2)(e^{2\alpha_0}\cos 2\beta - 1)\cos 2\theta \\ &-(\sigma_1-\sigma_2)e^{2\alpha_0}\sin 2\beta\sin 2\theta\big]/(\cosh 2\alpha_0 - \cos 2\beta)\end{aligned} \tag{13-1.9}$$

For a given value of θ, Eq. (13-1.9) gives $\sigma_{\beta\beta}$ as a function of β. Hence, by setting the derivative of $\sigma_{\beta\beta}$ with respect to β equal to zero, we may compute the values of β that give extreme values of $\sigma_{\beta\beta}$. The values of β are solutions of the equation

$$\begin{aligned}1 - \cos 2\beta\cosh 2\alpha_0 - \sin 2\beta\cot 2\theta\sinh 2\alpha_0 \\ = \left(\frac{\sigma_1+\sigma_2}{\sigma_1-\sigma_2}\right)\left(\frac{\sinh 2\alpha_0}{e^{2\alpha_0}}\right)\left(\frac{\sin 2\beta}{\sin 2\theta}\right)\end{aligned} \tag{13-1.10}$$

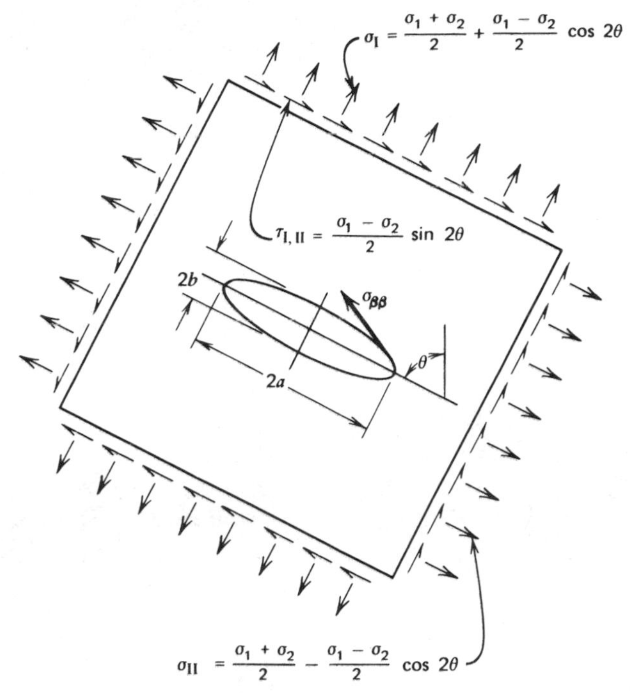

Fig. 13-1.3

In general, Eq. (13-1.10) is satisfied by two values of β, depending upon the quantities σ_1, σ_2, and α_0 ($\tanh \alpha_0 = b/a$), for each value of θ. One value of β is associated with $\sigma_{\beta\beta(\max)}$ and the other with $\sigma_{\beta\beta(\min)}$. Because of symmetry, for given β, the significant values of $\sigma_{\beta\beta}$ may be determined by considering values of θ between 0 and $\pi/2$.

Case 1/$\theta = 0$ or $\theta = \pi/2$. By Eq. (13-1.10), for $\theta = 0$ or $\theta = \pi/2$, we obtain the values $\beta = 0$ and $\beta = \pi/2$. Hence we find for $\theta = 0$; $\beta = 0$:

$$\sigma_{\beta\beta} = \sigma_{\beta\beta(\max)} = \sigma_1\left(1 + 2\frac{a}{b}\right) - \sigma_2 \tag{13-1.11}$$

at the ends of the major axis. For $\theta = \pi/2$; $\beta = 0$:

$$\sigma_{\beta\beta} = \sigma_{\beta\beta(\max)} = -\sigma_1 + \sigma_2\left(1 + 2\frac{a}{b}\right) \tag{13-1.12}$$

at the ends of the major axis. Likewise, we find for $\theta = 0$; $\beta = \pi/2$:

$$\sigma_{\beta\beta} = \sigma_{\beta\beta(\min)} = -\sigma_1 + \sigma_2\left(1 + 2\frac{b}{a}\right) \tag{13-1.13}$$

at the ends of the minor axis. For $\theta = \pi/2$; $\beta = \pi/2$:

$$\sigma_{\beta\beta} = \sigma_{\beta\beta(\min)} = \sigma_1\left(1 + 2\frac{b}{a}\right) - \sigma_2 \tag{13-1.14}$$

at the ends of the minor axis.

Case 2/$0 < \theta < \pi/2$. For a fixed value of β, the extreme values of $\sigma_{\beta\beta}$ occur at values of θ determined by setting the derivative of $\sigma_{\beta\beta}$ (Eq. 13-1.9) with respect to θ equal to zero. Thus, we obtain

$$\tan 2\theta = \frac{e^{2\alpha_0}\sin 2\beta}{1 - e^{2\alpha_0}\cos 2\beta} \tag{13-1.15}$$

Consequently, by Eqs. (13-1.10) and (13-1.15), we find that extreme values of $\sigma_{\beta\beta}$ are obtained when

$$\left(\frac{\sigma_1+\sigma_2}{\sigma_1-\sigma_2}\right)\cos 2\theta = \pm\left[\left(\frac{\sigma_1+\sigma_2}{\sigma_1-\sigma_2}\right)^2 \sinh 2\alpha_0 - \frac{1}{2}(\sinh 2\alpha_0 + \cosh 2\alpha_0)\right] \tag{13-1.16}$$

and

$$e^{2\alpha_0}\cos 2\beta = e^{2\alpha_0}\cosh 2\alpha_0 - 2\left(\frac{\sigma_1+\sigma_2}{\sigma_1-\sigma_2}\right)^2 \sinh^2 2\alpha_0 \tag{13-1.17}$$

provided that

$$\frac{1+\coth 2\alpha_0}{2\coth\alpha_0} \le \left(\frac{\sigma_1+\sigma_2}{\sigma_1-\sigma_2}\right)^2 \le \frac{1+\coth 2\alpha_0}{2\tanh\alpha_0} \tag{13-1.18}$$

where Eq. (13-1.18) follows from Eqs. (13-1.16) and (13-1.17) and the conditions

$$-1 \le \cos 2\theta \le 1, \ -1 \le \cos 2\beta \le 1 \tag{13-1.19}$$

By Eqs. (13-1.16), (13-1.17), and (13-1.9), we find the two values of $\sigma_{\beta\beta}$

$$\sigma_{\beta\beta 1} = -\frac{(\sigma_1-\sigma_2)^2}{2(\sigma_1+\sigma_2)}(1+\coth 2\alpha_0) \tag{13-1.20}$$

$$\sigma_{\beta\beta 2} = \frac{3(\sigma_1-\sigma_2)^2}{2(\sigma_1+\sigma_2)}(1+\coth 2\alpha_0) \tag{13-1.21}$$

Depending upon the sign of the applied stresses, the maximum value of $\sigma_{\beta\beta}$ is given by either the value of $\sigma_{\beta\beta 1}$ or the value of $\sigma_{\beta\beta 2}$ depending upon which is larger. For example, assume that the values of σ_1, σ_2, and α_0 are such that Eq. (13-1.18) is satisfied. Let the elliptical hole be oriented at angle θ (Fig. 13-1.2) given by Eq. (13-1.16). Under these conditions the value of $\sigma_{\beta\beta 2}$ from Eq. (13-1.21), is never greater than the value of $\sigma_{\beta\beta(\max)}$ given by Eqs. (13-1.11) and (13-1.12). However, the stress $\sigma_{\beta\beta 1}$ is a tensile stress when $\sigma_1 + \sigma_2 < 0$. The values of $\sigma_{\beta\beta 1}$ may exceed the maximum tensile stress that can exist for $\theta = 0$ or $\pi/2$. Hence, when σ_1 and σ_2 are both negative (compressive stresses), a tensile stress $\sigma_{\beta\beta 1}$ exists on the perimeter of the elliptical hole. When θ is equal to the value given by Eq. (13-1.16), with the positive sign, $\sigma_{\beta\beta 1}$ is the largest tensile stress that exists for any other value of θ that may be chosen for this state of stress (values of $\sigma_1, \sigma_2, \alpha_0$ that satisfy Eq. 13-1.18). Consequently, the presence of an elliptical hole in a flat plate (even for the case $b/a \approx 0$) may result in a tensile stress on the perimeter of the hole, even when the plate is subjected to negative stresses σ_1 and σ_2 (compression) on its edge (Fig. 13-1.2).

EXAMPLE 13-1.1

Thin Elliptical Hole in Plate

Consider an elliptical hole in a plate with ratio $a/b = 100$ (Fig. 13-1.2). For this large value of a/b, the hole appears as a very narrow slit (crack) in the plate. Let compressive stresses, $\sigma_1 = -20$ MPa and $\sigma_2 = -75$ MPa, be applied to the plate edges. (a) Determine the orientation of the hole (value of θ) for which the tensile stress at the perimeter of the hole is a maximum. (b) Calculate the value of this tensile stress. (c) Calculate the associated value of β (location of the point) for which this tensile stress occurs.

SOLUTION

Since $a/b = 100$, Eq. (12-2.8) indicates that $\coth \alpha_0 = 1/\tanh \alpha_0 = a/b = 100$. Hence, $\alpha_0 = 0.0100$ rad, $\sinh 2\alpha_0 = 0.0200$, $\cosh 2\alpha_0 = 1.000$, $\coth 2\alpha_0 = 50.0$. For these values of σ_1, σ_2, and α_0, Eq. (13-1.18) is satisfied.

(a) The value of θ is given by Eq. (13-1.16). Hence,

$$\cos 2\theta = \left[\frac{-1 - 0.0200 + 2\left(\dfrac{-20-75}{-20+75}\right)^2 (0.020)}{2\left(\dfrac{-20-75}{-20+75}\right)} \right] = 0.2607$$

or

$$\theta = 0.6535 \text{ rad}$$

(b) The maximum value of the tensile stress is given by Eq. (13-1.20). Thus,

$$\sigma_{\beta\beta(\max)} = \sigma_{\beta\beta 1} = -\frac{(-20+75)^2}{2(-20-75)}(1+50) = 812 \text{ MPa tension.}$$

(c) This tensile stress is located on the perimeter of the hole at a value of β given by Eq. (13-1.17)

$$\cos 2\beta = 1 - 2\left(\frac{-20-75}{-20+75}\right)^2 \frac{(0.020)^2}{1.020} = 0.9977$$

or

$$\beta = 0.0342\,\text{rad}$$

This small value of β means that the maximum tensile stress occurs very near the end of the major axis of the elliptical hole (see Figs. 13-1.2 and 12-2.3).

The above computation shows that a slender elliptical hole (long narrow crack) in a plate may result in a high tensile stress concentration even when the applied edge stresses are compressive.

Stress Concentration at Groove in a Circular Shaft / A machine component consists of a circular shaft in which a circumferential circular groove (notch) is cut (Fig. 13-1.4 and Fig. *D*, Table 12-2.3). In practice, the shaft is subjected to an axial force P, a bending moment M, and a twisting moment (torque) T. We wish to compute the maximum principal stress in the cross section of the shaft at the root of the notch. In addition, a shear V may act on the shaft (Fig. *D*, Table 12-2.3). However, this shear has only a small effect on the maximum stress at the root of the notch.[1] Hence, we do not consider its effect.

The maximum principal stress at the root of the notch occurs at point A in Fig. 13-1.4. The stress components at A are σ_{zz} and σ_{zx}. Hence, by Art. 1-4, the maximum principal stress is

$$\sigma_{\max} = \tfrac{1}{2}\sigma_{zz} + \tfrac{1}{2}\sqrt{\sigma_{zz}^2 + 4\sigma_{zx}^2} \tag{13-1.22}$$

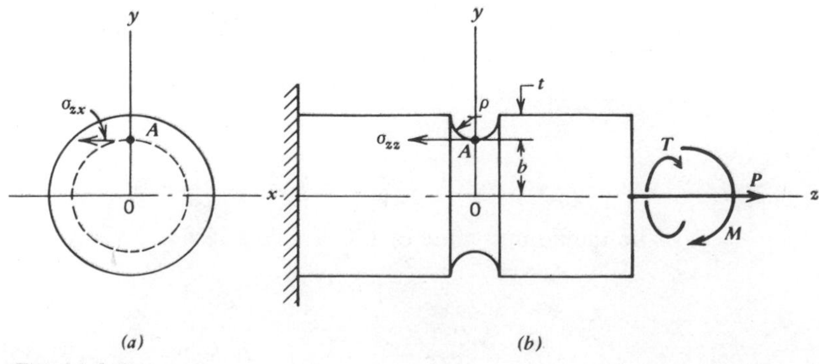

Fig. 13-1.4

The stress component σ_{zz} is produced by the axial load P and bending moment M. Hence,

$$\sigma_{zz} = S_{cc}^{(P)}\frac{P}{A} + S_{cc}^{(M)}\frac{Mc}{I}$$

$$= S_{cc}^{(P)}\frac{P}{\pi b^2} + S_{cc}^{(M)}\frac{4M}{\pi b^3} \tag{13-1.23}$$

where $S_{cc}^{(P)}$ and $S_{cc}^{(M)}$ are the calculated stress concentration factors for axial load and bending moment, respectively. These stress concentration factors are determined from curves 6 and 7 in Fig. 12-2.8. The stress σ_{zx} is given by the relation

$$\sigma_{zx} = S_{cc}^{(T)}\frac{Tc}{J} = S_{cc}^{(T)}\frac{2T}{\pi b^3} \tag{13-1.24}$$

where $S_{cc}^{(T)}$ is the calculated stress concentration factor for torque and is determined from curve 9 of Fig. 12-2.8. For a given set of dimensions of the shaft (Fig. 13-1.4), Eqs. (13-1.22), (13-1.23), and (13-1.24) determine the value of σ_{max}.

PROBLEM SET 13-1

1. In Example 13-1.1, let the hole be circular rather than elliptical. For stresses σ_1, σ_2 as given (a) determine the maximum tensile stress in the plate, and (b) determine the maximum compressive stress in the plate.

2. In Example 13-1.1, let the ratio $b/a = 5$. Determine the orientation of the hole (value of θ) for which the tensile stress at the perimeter of the hole is a maximum, the magnitude of the maximum tensile stress in the plate, and its location.

   Ans. $\sigma_{max} = 57.3$ MPa, $\theta = 0.6406$ rad, $\beta = 0.5837$ rad.

3. A thin-wall cylindrical tank, of diameter D and wall thickness h, is subjected to internal pressure p. A small circular hole exists in the wall of the cylinder. By means of Eqs. (13-1.11) through (13-1.14), derive expressions for the maximum stresses σ_A and σ_B at the hole, on longitudinal and transverse sections of the tank, respectively. Assume that the material remains elastic.

4. A thin-wall cylindrical pressure vessel, of diameter D and wall thickness h, is filled with a fluid whose weight density is ρ (force per unit volume). The fluid is pressurized until the average pressure in the vessel is p (force per unit area). Assume that the difference in pressure between the top and bottom of the tank is small enough to

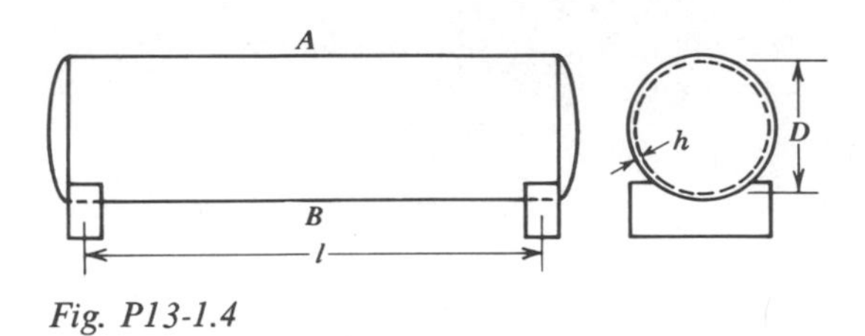

Fig. P13-1.4

be neglected. The vessel is supported near its ends on horizontal supports a distance l apart. Design considerations require that a small circular hole be drilled into the vessel at midspan at either the top (point A) or the bottom (point B) (Fig. P13-1.4).

Show that if the hole is drilled at point A, the maximum stress at the hole due to the fluid and pressure is $\sigma_A = (5pD/4h) + (\rho l^2/8h)$ and if at point B, $\sigma_B = (5pD/4h) - (\rho l^2/8h)$. The weight of the vessel is neglected in estimating the bending stresses at section AB, and the bending stress is assumed to be smaller than the circumferential stress.

5. Let the shaft of Problem 12-2.4 be subjected simultaneously to a bending moment $M = 15.0$ kN · m and a torsional moment $T = 30.0$ kN · m. With the stress concentration factors determined in Problem 12-2.4, compute (a) the maximum principal stress, (b) the maximum shearing stress, and (c) the maximum octahedral shearing stress that occur in the shaft at the root of the groove.

Ans. (a) $\sigma_{max} = 382$ MPa, (b) $\tau_{max} = 236$ MPa, (c) $\tau_{oct} = 205$ MPa

13-2 EFFECTIVE STRESS CONCENTRATION FACTORS

Definition of Effective Stress Concentration Factor / As noted in Chapter 12 and Art. 13-1, the calculated stress concentration factors apply mainly to ideal, elastic material, and they depend mainly on the geometry or form of the abrupt change in section. For these reasons, they are often called *form factors*. However, in applications involving real materials, the significance of a stress concentration factor is not indicated satisfactorily by the calculated value. Rather it is found through experience that the *significant* or *effective stress value* that indicates impending structural damage (failure) of a member depends on the characteristics of the material and on the nature of the load as well as upon the geometry or form of the stress raiser. Consequently, in practice, the *significant (or effective) value* of the stress concentration is obtained by multiplying the nominal stress by a *significant or effective stress concentration factor*,† S_{ce}. Often the nominal stress is computed from an elementary stress formula, such as $\sigma_n = P/A$, $\sigma_n = Mc/I$, etc. Usually, the magnitude of S_{ce} is less than the magnitude S_{cc} of the calculated stress concentration factor for a given stress raiser.

The magnitude of S_{ce} is obtained experimentally in contrast to the calculated value S_{cc}. Ordinarily S_{ce} is obtained by testing two or more samples or sets of specimens of the actual material. One specimen (or set of specimens) is prepared without the presence of the discontinuity or stress raiser, so that the nominal stress is the significant or effective stress. A second specimen (or set of specimens) is prepared with the discontinuity or stress raiser built into the specimens. The second set of specimens is tested in the same manner as the first set. For simple members, such as axial rods, beams, or torsion bars, the stress in each set of specimens is usually calculated by means of elementary formulas.

One may assume that the damage (failure) in the two specimens starts when the stresses in the specimens attain the same value, the loads causing these equal stresses to be unequal. The damaging stress in the specimen with stress raiser is, of course, caused by a lesser load. The significant (or effective) stress concentration factor for the discontinuity or stress raiser may be defined as the ratio of the stress calculated for the load at which structural damage starts in the specimen free of discontinuities to the nominal stress calculated by the same stress formula for the load at which damage starts in the sample containing the stress raiser. Alternatively, one may assume that the significant value of stress that causes damage in a specimen with stress raiser is equal to the nominal value of stress given by ordinary stress equations *plus* some proportion (say $\frac{1}{4}$, $\frac{1}{2}$, $\frac{3}{4}$, etc.) of the *increase* in the calculated stress caused by the stress raiser. This increase in stress is the difference between the nominal value of local stress and the value of the maximum localized stress calculated by means of elasticity theory, or by photoelastic methods, soap-film methods, elastic strain gage methods, etc. Then the nominal stress plus some proportion of the increase in stress caused by the abrupt change of section is considered to be effective in causing damage to the member. For example, if the abrupt change in section is caused by a small hole in the center of a plate subjected to an axial tensile load, and if one half of the difference between the nominal stress $\sigma_n = P/A$ and the maximum theoretical value of the localized stress $\sigma_{max} = 3P/A$ is used, then the significant stress concentration factor would be

† The term strength reduction factor is sometimes used. However, one should note that the strength of the material is not reduced by the stress raiser, but rather the load carrying capacity (strength) of the member is reduced.

$S_{ce} = 1 + \frac{1}{2}(3 - 1) = 2$. Thus the significant (effective) stress σ_e in the member would be considered, for this example, to be $\sigma_e = 2P/A$.

The concept used above can be stated in general form as follows: Let q represent the proportion of the increase in the calculated localized stress σ_c above the nominal stress σ_n to be used for determining the significant stress. The increase in stress at the point of concentration is $\sigma_c - \sigma_n = S_{cc}\sigma_n - \sigma_n = \sigma_n(S_{cc} - 1)$ in which σ_n is the nominal stress given by the elementary stress equation in which the effect of the given discontinuity is neglected. Then, the value of the effective stress is

$$\sigma_e = S_{ce}\sigma_n = \sigma_n + q\sigma_n(S_{cc} - 1)$$

Thus

$$S_{ce} = 1 + q(S_{cc} - 1) \tag{13-2.1}$$

and, hence,

$$q = (S_{ce} - 1)/(S_{cc} - 1) \tag{13-2.2}$$

The ratio q is called the *notch sensitivity index* of the material for the given form of discontinuity and for the given type of loading. For example, in Eq. (13-2.1), if $q = 0$, $S_{ce} = 1$, and the material and member are said to be insensitive to the effects of the stress concentration; whereas if $q = 1$, $S_{ce} = S_{cc}$, and the member is said to be fully sensitive to the effects of the stress concentration. The value of S_{ce} (and hence of q) is determined from tests as described above. It has been found from such tests (Fig. 13-2.1) that the values of S_{ce} and q depend mainly on the ability of the material and the member to make adjustments or accommodations, such as local yielding, which reduce the damaging effects of the localized stress. The ability of the material to make these adjustments or accommodations depends, in turn, on the type of loading applied to

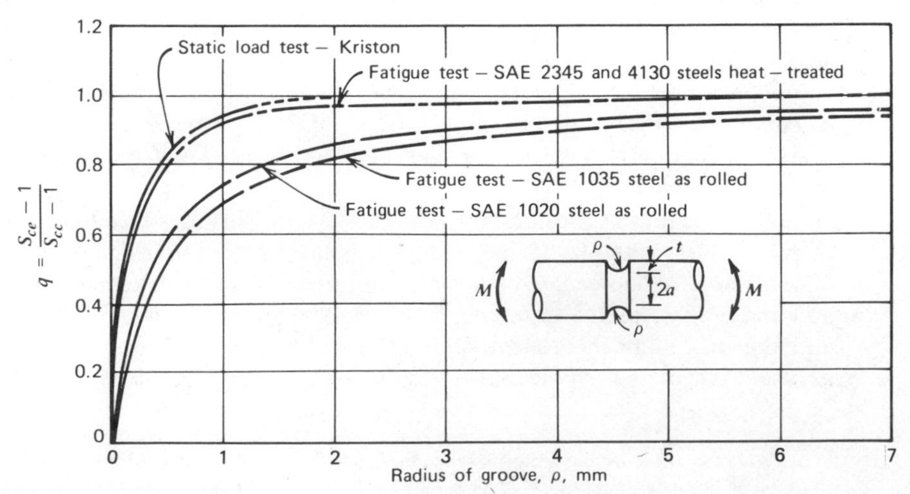

Fig. 13-2.1/The influence of a radius of groove on notch sensitivity index.

the member (whether static, repeated, impact, etc.); on the existence in the member of initial or residual stresses; on the character of the internal structure of the material; on the temperature of the member; on the surface finish at the abrupt change of section; on the stress gradient in the region of the stress concentration, etc. These factors are discussed briefly below.

Static Loads; Ductile Material / At abrupt changes of section in members made of *ductile*† materials (especially metals) and subjected to static loads at ordinary temperatures, the localized stresses at the abrupt change of section are relieved to a large degree by localized yielding of the material that occurs largely, in metals, as slip across intercrystalline planes (see Art. 2-6). Because of this action, the value of q for the conditions specified is very low and lies usually in the range from 0 to 0.1. However, if the use or function of the member is such that the amount of inelastic strain required for this relieving action must be restricted, the value of q may approach 1 (Fig. 13-2.1). If the temperature of a metal member is very low when subjected to static loads, slip in the crystals seem to be reduced and is likely to be less effective in relieving the concentrated stress; hence the value of q may be as much as one half or even greater.

If the metal member is subjected to static load while at an elevated temperature, the mechanism (creep) by which localized yielding occurs may cause the value of q to vary from nearly zero to nearly unity. This situation arises from the fact that the creep of metals may be the result of either one or both of two different inelastic mechanisms, depending on the temperature and stress imposed: (a) Creep may be caused mainly by intercrystalline slip, especially at the lower range of creep temperatures and at relatively high stresses; this type of creep relieves the stress concentration to a large degree ($q = 0$, nearly); or (b) creep may be due to the viscous flow of the unordered (so-called amorphous) grain boundary material, especially at the higher temperatures and lower stresses, and stress concentration is relieved very little by such inelastic deformation ($q = 1$, nearly).

Static Loads; Brittle Material / If the member that contains an abrupt change in cross section is made of a relatively brittle material and is subjected to static loads, q will usually have a value in the range from one half to one, except for certain materials that contain many internal stress raisers inherent in the internal structure of the material such as graphite flakes in gray cast iron. An external stress raiser in the form of an abrupt change in section in such a material as gray cast iron has only a small additional

†See Chapter 2; ductile materials exhibit yield stresses and undergo large plastic strains before fracture.

influence on the strength of the member and, hence, the value of q is relatively small.

Repeated Loads / If a member has an abrupt change in section and is subjected to a load that is repeated many times, the mode of failure is one of progressive localized fracture, even though the material is classed as ductile. Under these conditions, the ability of the material to make adjustments or accommodations by localized yielding is greatly reduced as compared to its ability under static loads. This type of (fatigue) fracture, shown in Fig. 13-2.2, shows little or no evidence of yielding before complete fracture of the member occurs. Thus the value of q for loads repeated a large number of times is relatively large, usually being between one-half and one; the value of unity is approached in general for the harder (heat-treated) metals, and the lower value is approached for

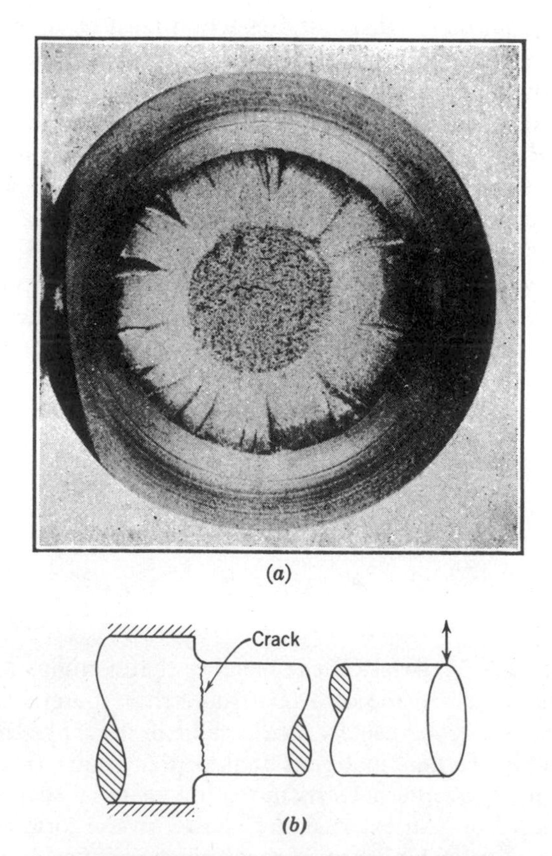

Fig. 13-2.2/Failure by progressive spreading of a crack that starts at region of stress concentration (fatigue fracture).

metals used in their softer condition. Furthermore, the internal structure of metals, especially of steel, has some influence on the value of q. If the pearlitic grain size in steel is very fine, q is likely to be nearly unity but, if the grain size is very coarse, the value of q is decreased.

Residual Stresses/The presence of initial or residual stresses in a member at an abrupt change in section also may influence the value of q. If the member is made of ductile metal and is subjected to static loads at room temperature, localized yielding relieves the effects of residual stresses. Generally, in this case, it is assumed that q is not altered by the residual stresses. On the other hand, if the member is made of brittle material and the residual stresses act along the same directions as the load stresses, the effects of the residual stresses may either add or subtract from the effects of the load stresses, depending on the relative signs of the load stresses and the residuals stresses. Correspondingly, the magnitude of q is increased or decreased. If, however, the member is made of a ductile metal and is subjected to repeated loads, the influence of initial or residual stresses is uncertain. The relatively large inelastic deformation that occurs (in a small volume of the member surrounding the stress concentration) in low cycle fatigue is assumed generally to negate any effect of residual stress on the magnitude of q. However, in the case of high cycle fatigue ($N > 10^6$), inelastic deformation in the region of a stress concentration is ordinarily minimal and residual stresses are assumed generally to alter the magnitude of q; the magnitude of q may either increase or decrease, depending on the sign of the residual stresses.

Very Abrupt Changes in Section. Stress Gradient/Let the change in section of a member be very abrupt; that is, let the hole, fillet, or groove, etc., forming the abrupt change in section have a very small characteristic dimension compared to the dimensions of the section, so that the calculated stress gradient is steep in the region of stress concentration. The value of S_{cc} for such a stress raiser is large, but the value of S_{ce} found from tests of such members, under either static or repeated loads, is usually much smaller than S_{cc}; that is, the value of q is smaller than would be found from tests of members of the same material with less abrupt changes of section. Figure 13-2.1 gives the results of tests of specimens having an abrupt change of section caused by a circumferential groove that show the foregoing facts. In this figure the value of q is plotted as ordinates, and the radius of the groove at the abrupt change of section is plotted as abscissas.

The results of these tests are represented by smooth curves drawn through points (not shown) representing the test data. The data used for each curve were obtained by testing specimens of the same material, the specimens being identical except for the size of the groove radius. The upper curve is for static load tests of specimens of Kriston (one of the plastics), which is a very brittle material. The other curves are for repeated bending-load tests of steels. In these tests, unpublished results of Moore, Jordan, and Morkovin, the values of a/ρ and t/ρ were kept constant, which means that the value of S_{cc} was kept constant (see Neuber's nomograph, Fig. 12-2.8). However the groove radius ρ was varied and all these curves show that when the groove radius approaches very small values, q is quite small, but when the groove radius is relatively large, the value of q approaches unity.

The results of these tests indicate that the damaging effects to a member from notches having small radii at the roots of the notches such as scratches, small holes, grooves, or fillets, or small inclusions, at a section of the member are considerably less than would be indicated by the large values of the theoretical stress at such stress raisers; in other words q (and hence S_{ce}) is relatively small. Much of the available data for the value of S_{ce} and of q have been obtained by conducting repeated load tests of specimens with cross sections of relatively small dimensions containing fillets, grooves, holes, etc., having small radii. These data furnish valuable information for computing significant stresses in a member having such discontinuities within the range of conditions used in the tests, but the values of q are probably unnecessarily small for use in computing S_{ce} by Eq. (13-2.1) for holes, fillets, grooves, etc., whose radii are relatively large.

Significance of Stress Gradient/The question naturally arises as to why the value q for a given material under a given type of loading should depend on the value of the root radius of the notch when it is small, as indicated by the curves of Fig. 13-2.1. Much discussion of this question is found in the technical literature, but no completely satisfactory reason can be given. A possible explanation is as follows: At one or more points on the surface of the member at the root of the notch the stress concentration will have its highest value, but at nearby points in the member in any direction from the root of the notch the values of the stress diminish. For most notches, the highest rate (stress gradient) at which the stress diminishes occurs at points in a cross section of the member at the notch root. Let S be the stress gradient at the root of the notch, that is, S is the slope of a line that is tangent at the root of the notch to the curve of stress distribution on the cross section at the root of the notch. This slope gives the rate at which the stress is diminishing at points just underneath the root of the notch. If S is large, the stress magnitude will diminish rapidly so that the stress at a point just underneath the root of the notch will be only very slightly larger than the value given at this point by the ordinary (nominal) stress equation.

It may be shown that S for notches such as holes, fillets, and grooves is given approximately by the following equation:

$$S = \frac{2.5\sigma_{\max}}{\rho} = \frac{2.5S_{cc}\sigma_n}{\rho} \tag{13-2.3}$$

From Eq. (13-2.3), it is seen that, for a given value of nominal stress σ_n and of S_{cc}, if ρ becomes small, the value of S becomes very large. When ρ is small and S is large, the magnitude of the concentrated stress diminishes so rapidly that only a very thin layer of material at the root of the notch is subjected to the stress concentration. This means that the so-called adjustments or accommodations that take place in the material and that tend to relieve high stresses can take place easier since such a small amount of material is involved. Furthermore, the machining and polishing of a specimen at the root of the notch will frequency result in an increase of the ability of the material in this thin layer (by work hardening) to resist stress. The greater apparent ability of this thin layer of material to resist higher stress plus the fact that the unchanged material (parent material) under this layer is not required to resist the highly concentrated stress also helps explain why q becomes so much smaller as ρ becomes very small.

The foregoing discussion of stress gradient applies mainly to so-called mechanical notches such as holes and fillets rather than to so-called chemical notches such as corrosion pits (see Art. 13-4).

Impact or Energy Loading/If machine parts and structural members are subjected to impact or energy loading, for example, if a member is required to absorb energy delivered to it by a body having a relatively large velocity when it comes in contact with the member, localized stresses have, in general, a large influence in decreasing the load carrying capacity of the member. As discussed in Chapter 2, the energy absorbed per unit volume by a material when stressed within the elastic strength is $\sigma^2/(2E)$; that is, the energy absorbed by a material is proportional to the square of the stress in the material. This means that the small portions of a member where the high localized stresses occur absorb an excessive amount of energy before the main portion of the member can be stressed appreciably and, hence, before the main portion can be made to absorb an appreciable share of the energy delivered to the member. As a result, the small portion where the localized stress occurs is likely to be stressed above the yield stress of the material, and the energy required to be absorbed may be great enough to cause rupture even if the material is relatively ductile; the reader may know that a familiar method of breaking a bar of ductile metal is to file a V-notch in one side of the bar and then to clamp one end of the bar in a vise with the notch close to the face of the vise and strike the bar near the other end a sharp blow with a hammer so that the bar is bent with the notch on the tension side of the bar.

Tests widely used to measure the effects of a notch under impact loads are the Charpy and Izod impact tests. However, neither of these notched-bar single-blow impact tests gives a quantitative value of S_{ce}. These tests are important primarily in determining whether or not a material of known history of manufacture and treatment is substantially the same as similar material that has proved to be satisfactory in service. There is no satisfactory test or method for determining a value of q for stress raisers in members subjected to impact loading.

13-3 EFFECTIVE STRESS CONCENTRATION FACTORS. REPEATED LOADS

The value of the stress concentration factor S_{ce} for a notched member that is subjected to repeated loads (fatigue) is obtained by comparison of data taken from two sets of test specimens. One set of specimens (5 to 10 specimens) is notchfree, the other set is notched. The significant stress in the notchfree specimens is the nominal stress as computed by an elementary stress formula. For the notched specimens, the nominal stress is again computed by the same elementary stress formula as for the notchfree specimens. Both sets of specimens are subjected to the same type of repeated load or fatigue test (say, bending).

It is assumed that the failure (fracture) in each set of specimens for a specified number of cycles N occurs when the stress attains the same value in each set. Since the notch causes a stress concentration, the load required to cause the fracture stress is less for the notched specimens.

To illustrate the method of determining S_{ce} for bending fatigue loading, consider the σ-N diagrams of Fig. 13-3.1. The nominal stress is computed by the equation $\sigma_n = Mc/I$ and plotted as ordinates in Fig. 13-3.1; the abscissa is the number of cycles of bending stress to which the specimen is subjected. For a given value of N, say 300,000 cycles, the value of S_{ce} is computed by taking the ratio of the failure stress of the notchfree specimen to the failure stress of the notched specimen. Thus, for $N = 300{,}000$ cycles, by Fig. 13-3.1, we find

$$S_{ce} = \frac{634.4}{268.9} = 2.36$$

Likewise, for $N = 10^7$ cycles we find

$$S_{ce} = \frac{620.6}{248.2} = 2.50$$

Thus, the value of S_{ce} varies with N. By Fig. 13-3.1, the value of S_{ce} remains relatively constant for $N > 10^7$, since the curves at $N = 10^7$ are changing only very slowly. In fatigue testing, the value of S_{ce} is often

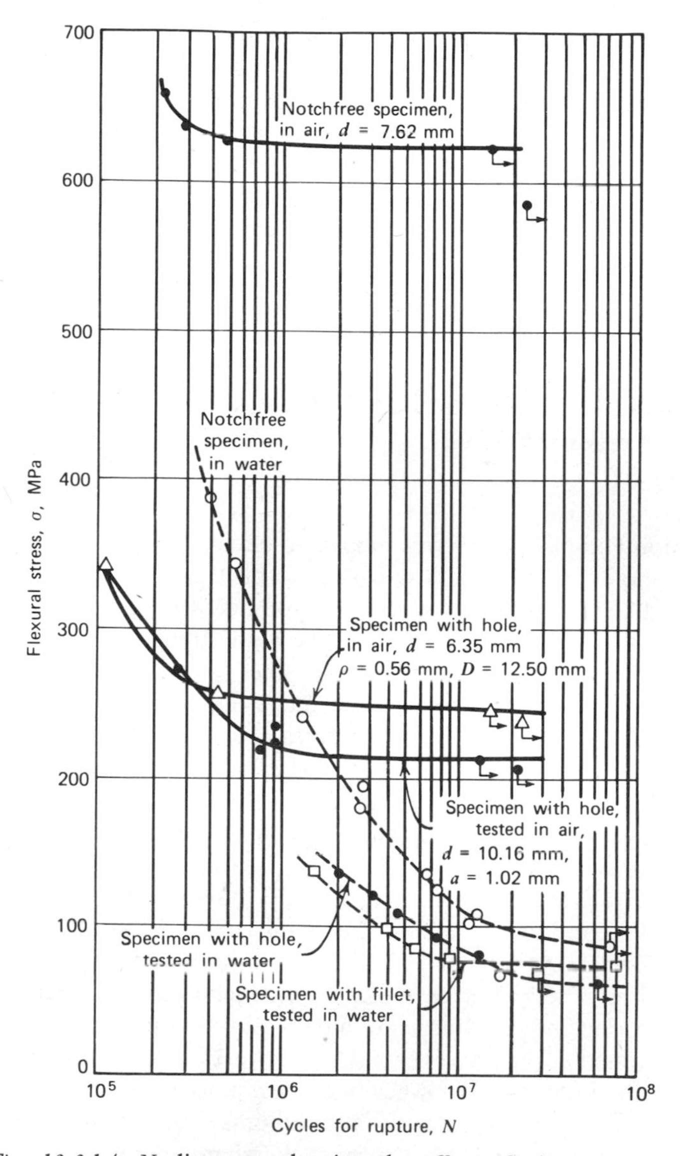

Fig. 13-3.1/σ-N diagrams showing the effect of abrupt changes in cross section and corrosion on the resistance of steel to repeated cycles of completely reversed bending stress. Quenched and tempered SAE 3140 steel was used for all tests. (From Bulletin 293, Engineering Experiment Station, Univ. of Illinois, by T. J. Dolan.)

based upon the endurance limit of the steel (stress at $N = 10^7$).

For the notched specimen with a transverse hole (Fig. 13-3.1), the calculated (elastic) stress concentration factor is $S_{cc} = 3.00$. However, as shown by the results of fatigue tests for these specimens, the significant stress concentration factor is $S_{ce} = 2.5$. Thus, there is a difference between the value of S_{ce} and S_{cc}, the value of S_{ce} being smaller. In the following article, additional conditions that influence the value of S_{ce} are discussed.

13-4 EFFECTIVE STRESS CONCENTRATION FACTORS. OTHER INFLUENCES

Corrosion Fatigue/The severe damaging effects that mechanical notches such as holes and fillets and so-called chemical notches such as corrosion pits are likely to have on the resistance of steel to repeated stress, particularly of alloy steels heat treated to give high strength, are shown in Fig. 13-3.1. The effect of corrosion that takes place while the material is being repeatedly stressed is much more damaging to the fatigue strength of steel than is corrosion that takes place prior to stressing (called stressless corrosion). The main reason for this fact seems to be that the products of the corrosion tend to form a protecting film that excludes the corroding agent from contacting the metal when the protecting film is not subjected to stress. If, however, the rather brittle film is repeatedly stressed in the presence of the corroding agent, it cracks and allows the corroding agent to continue to attack the metal underneath the film. The effect of corrosion on the fatigue strength of steel is shown by the σ-N diagrams in Fig. 13-3.1. For example, the quenched and tempered SAE 3140 steel tested indicated an endurance limit of approximately 620.5 MPa when tested in air (a very mild corrosive medium), and this was reduced to about 68.94 MPa when the specimens were tested in the presence of water; the presence of a small hole caused little further decrease in fatigue strength. Furthermore the shape of the σ-N diagram for stresses above the endurance limit was influenced greatly by the corrosion.

Effect of Range of Stress/In the preceding discussion it was assumed that the member or specimen was subjected to repeated cycles of completely reversed stress, that is, in each stress cycle the stress varied from a given tensile stress to an equal compressive stress. If a specimen in which the stress is concentrated is subjected to repeated cycles of stress in whic' the stress is not completely reversed, it is convenient to consider the cyc. or range of stress to be made up of a steady stress and a completely reve. ed

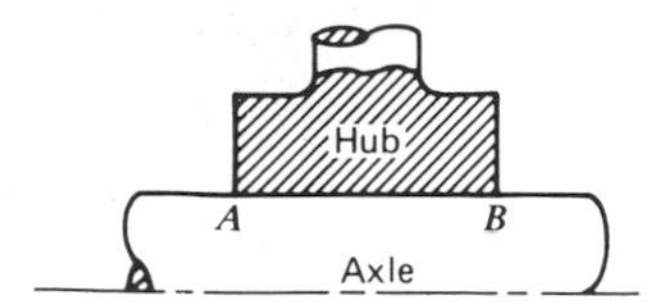

Fig. 13-4.1/Stress concentration occurs at *A* and *B* when wheel is shrunk on axle and axle is subjected to bending.

(alternating) stress superimposed on the steady stress (see Art. 3-5). There is considerable evidence[2] indicating that the damaging effect of the stress concentration in such a repeated cycle of stress is associated only with the completely reversed (alternating) component of the stress cycle and not with the mean stress in the cycle. Thus the stress concentration factor for the particular discontinuity is applied only to the alternating stress component (see Example 13-4.3).

Methods of Reducing Harmful Effects of Stress Concentrations/The problem that frequently arises in engineering is that of reducing the value of a stress concentration below the minimum value that will cause a fatigue fracture to occur or of raising the fatigue strength of the material so that fracture is avoided, rather than that of calculating the effective stress concentration. Some of the ways that have been employed in an attempt to reduce the damaging effects of localized stresses are the following:

1. Reducing the value of the stress concentration by decreasing the abruptness of the change in cross section of the member by use of fillets, etc., either by adding or by removing small amounts of material.
2. Reducing the value of the stress concentration by making the portion of the member in the neighborhood of the stress concentration less stiff; this may be done, for example, by removing material in various ways, as indicated in Figs. 13-4.1 and 13-4.2. It may sometimes be done by substituting a member made of material with a lower modulus of elasticity, such as replacing a steel nut on a steel bolt by a

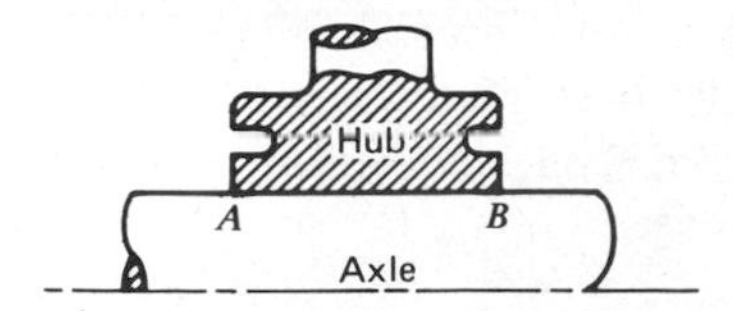

Fig. 13-4.2/Stress concentrations reduced by removing material so that hub at *A* and *B* will deflect more easily.

bronze nut for reducing the stress concentration at the threads of the steel bolt.

3. Increasing the fatigue strength of the material by cold working the portions of the members where the stress concentrations occur; for example, by the cold rolling of fillets and of bearing surfaces on axles, or by the shot blasting or shot peening of surfaces of machine parts. The increased fatigue strength of a member caused by local cold working of the metal at the region of stress concentration in some cases may be due primarily to residual compressive stresses set up in the cold worked metal by the surrounding elastic material as this elastic material attempts to return to its original position when the cold working tool is removed, especially if the repeated cycle of stress is not reversed. Likewise, overstraining of the outer fibers of a beam or of the inner fibers of a thick-walled pressure vessel or pipe may create favorable initial (residual) stresses.
4. Increasing the fatigue strength of the material by alloying and heat treating portions of steel members that resist the high stress, by case hardening, nitriding, flame hardening, etc. In such treatments, however, care must be taken to avoid tensile residual stresses.
5. Reducing the stress concentration by removing surface scratches, tool marks, small laps, and similar stress raisers by creating a smooth surface by polishing.
6. Reducing the stress concentration by the prevention of minute surface corrosion pits by protecting the surface from acid fumes or from moisture through the use of a corrosion-resisting covering, as for example by encasing the member in grease or spraying on a coating of lacquer.

EXAMPLE 13-4.1

Slot in Cantilever Beam

A cantilever beam is made of flat bar of hot-rolled SAE 1020 steel. The beam contains a slot, Fig. *C* of Table 12-2.3, with dimensions $b = 10$ mm, $t = 50$ mm, $\rho = 5$ mm, $h = 25$ mm, and $c = 60$ mm.

Let the beam be subjected at its free end to a large number of completely reversed cycles of bending moment of maximum amplitude M. (a) Compute the significant value of the stress at the top or bottom of the slot (that is, at the root of the notch) in terms of M; make use of Fig. 13-2.1. (b) The maximum utilizable stress for this material under completely reversed cycles of stress is 172 MPa. Compute the allowable moment M based upon a factor of safety of $SF = 4.0$.

SOLUTION

(a) From Fig. 12-2.8, $S_{cc} = 2.8$, and from Fig. 13-2.1 the value of q, the ordinate of the test data for SAE 1020 steel, is $q = 0.94$. Hence, by Eq.

(13-2.1),

$$S_{ce} = 1 + 0.94(2.8 - 1) = 2.70$$

The nominal stress σ_n at the root of the notch (see Table 12-2.3, Fig. C) is

$$\sigma_n = \frac{3Mt}{2h(c^3 - t^3)} = \frac{3M(50)}{2(25)(60^3 - 50^3)} = 32.97 \times 10^{-6}\ M\ (\text{MPa})$$

Hence, the significant (or effective) stress is

$$\sigma_e = S_{ce}\sigma_n = 2.70 \times 33.97 \times 10^{-6} M = 89.0 \times 10^{-6}\ M\ (\text{MPa})$$

(b) The allowable (working) stress is $\sigma_w = \sigma_{max}/SF = 172/4.0 = 43.0$ MPa. Hence, $43.0 = 89.0 \times 10^{-6}\ M$ or $M = 483{,}100\ \text{N} \cdot \text{mm} = 483.1\ \text{N} \cdot \text{m}$.

EXAMPLE 13-4.2

Long Narrow Slot in Cantilever Beam

Let the cantilever beam of Example 13-4.1 be unchanged, except that $\rho = 0.75$ mm. Then the slot approaches a long crack in the bar. Compute the significant stress at the root of the notch.

SOLUTION

(a) From Fig. 12-2.8, we find that $S_{cc} = 6.1$. Figure 13-2.1 shows that for fatigue tests of SAE 1020 steel, $q = 0.69$ when $\rho = 0.75$ mm. Hence,

$$S_{ce} = 1 + 0.69(6.1 - 1) = 4.52$$

and the significant stress is

$$\sigma_e = S_{ce}\sigma_n = 4.52(32.97 \times 10^{-6}\ M) = 149.0 \times 10^{-6}\ M\ (\text{MPa})$$

Comparison with the results of Example 13-4.1 shows that the value of S_{cc} is increased 118 percent, whereas the value of S_{ce} is increased 67 percent, which corresponds to the increase in the significant stress σ_e. These facts indicate that as S_{cc} increases with a decrease in ρ, so does S_{ce} but to a lesser degree.

(b)The allowable stress is $\sigma_w = 43.0$ MPa. Hence $43.0 = 149.0 \times 10^{-6}\ M$ or $M = 288{,}600\ \text{N} \cdot \text{mm} = 288.6\ \text{N} \cdot \text{m}$.

EXAMPLE 13-4.3

Fillet in Bar Subjected to Range of Load

The filleted tension member in Fig. 12-3.1 is made of 2024-T4 aluminum alloy ($E = 72.0$ GPa, $\nu = 0.33$, $\sigma_u = 470$ MPa, $Y = 330$ MPa, and $\sigma_{am} = 190$ MPa for $N = 10^6$ cycles). Perpendicular to the figure, the thickness of the member is 10 mm. The other dimensions are $D = 59$ mm, $d = 50$ mm, and $t = \rho = 3.00$ mm. The member is subjected to a tensile load ranging from $P_{min} = 20.0$ kN to P_{max}. Assuming that $q = 0.95$, determine the magnitude of P_{max} to produce fracture of the tension member in 10^6 cycles. Also determine σ_{max} and σ_{min}.

SOLUTION

The calculated stress concentration factor S_{cc} for the fillet can be read from the curve, fillet $t = \rho$, in Fig. 12-3.1, with $\rho/d = 0.06$. As read from the curve,

$$S_{cc} = 1.90$$

Since $q = 0.95$, Eq. (13-2.1) gives

$$S_{ce} = 1 + 0.95(1.9 - 1) = 1.86$$

Experimental evidence[2] indicates that S_{ce} should be applied only to the alternating part of the stress. Therefore, it is convenient to work with nominal values of the stresses as follows: nominal minimum stress $\sigma_{n(\min)} = P_{min}/A$, nominal maximum stress $\sigma_{n(\max)} = P_{max}/A$, nominal alternating stress σ_{na}, nominal mean stress σ_{nm}, and nominal fatigue strength $\sigma_{nam} = \sigma_{am}/S_{ce}$.

Since 2024-T4 aluminum alloy is a ductile metal, we assume that the Gerber relation, Eq. (3-5.2), is valid when written in terms of nominal stress values.

$$\frac{\sigma_{na}}{\sigma_{nam}} + \left(\frac{\sigma_{nm}}{\sigma_u}\right)^2 = 1 \tag{1}$$

Equation (1) can be interpreted graphically in Fig. 3-5.5; each ordinate to curve AB is reduced in magnitude by the factor $1/S_{ce}$. Nominal stress relations are defined as follows:

$$\sigma_{n(\min)} = \frac{P_{min}}{A} = \frac{20.0 \times 10^3}{50(10)} = 40.0 \text{ MPa} \tag{2}$$

$$\sigma_{nm} = \sigma_{n(\min)} + \sigma_{na} \tag{3}$$

$$\sigma_{n(\max)} = \frac{P_{max}}{A} = \sigma_{n(\min)} + 2\sigma_{na} \tag{4}$$

$$\sigma_{nam} = \frac{\sigma_{am}}{S_{ce}} = \frac{190}{1.86} = 102.2 \text{ MPa} \tag{5}$$

Substitution of Eqs. (3) and (5) into Eq. (1) gives

$$\sigma_{na} = 93.9 \text{ MPa}$$

which when substituted into Eq. (4) gives

$$\sigma_{n(\max)} = 40.0 + 2(93.9) = 227.8 \text{ MPa} = \frac{P_{max}}{A}$$

$$P_{max} = 227.8(50)(10) = 113{,}900 \text{ N} = 113.9 \text{ kN}$$

The assumption, that the effective stress concentration factor should be applied only to the alternating part of the stress, defines the maximum and minimum stresses in the member at the stress concentration as

$$\sigma_{max} = \sigma_{nm} + S_{ce}\sigma_{na} = 133.9 + 1.86(93.9) = 308.6 \text{ MPa}$$
$$\sigma_{min} = \sigma_{nm} - S_{ce}\sigma_{na} = -40.8 \text{ MPa} \tag{6}$$

Since the load cycles between a tensile load of 20.0 kN and a tensile load of 114 kN, the negative sign for σ_{min} may be suspect. The maximum and minimum stresses given by Eqs. (6) give the correct range in stress at the stress concentration for linearly elastic conditions; however, the values given by Eqs. (6) can be only a rough approximation of their true magnitudes. If residual stresses are not present at the stress concentration, linearly elastic analysis gives the maximum stress at the stress concentration as $S_{cc}P_{max}/A = 432.8$ MPa, which exceeds the yield stress of the material. Plasticity theories and experimental evidence indicate that plastic deformation in the region of the stress concentration produces residual stresses that result in a reduction of the mean stress at the stress concentration. The residual stresses at the stress concentration have a sign opposite to the sign of the stresses that caused the inelastic deformation. Thus, the residual stresses decreases the magnitudes of both σ_{max} and σ_{min} so that the values given by Eqs. (6) are realistic.

PROBLEM SET 13-4

1. A flat bar of a relatively brittle material has a fillet as shown in Fig. 12-3.1. The thickness of the bar is 24 mm, $\rho = 3$ mm, $d = 50$ mm, and $D = 68$ mm. The bar is subjected to an axial static tensile load P. If the allowable (working) stress for the material under this condition of loading is $\sigma_w = 14.0$ MPa, compute the maximum allowable load P_{max} for the member. Assume that $q = 0.80$.

2. In Problem 13-1.5, let the twisting moment T and the bending moment M be repeatedly applied through completely reversed cycles. Assume that $q = 0.80$ for the material in the shaft under repeated load. Compute the significant value of the maximum principal stress and maximum shearing stress in the shaft at a cross section through the root of the groove.

   Ans. $\sigma_{max} = 333.8$ MPa; $\tau_{max} = 208.3$ MPa

3. The crank shaft in Fig. P3-5.5 has a diameter $d = 13.0$ mm. Let a small diameter hole be drilled in the crank shaft at a location 50 mm from the load P (measured along the axis of the shaft). Determine the magnitude of the completely reversed load P that can be cycled

10^8 times based on a factor of safety $SF = 1.75$. Assume that $q = 1.00$. Material properties are given in Problem 3-5.5.

4. The load P in Problem 3 is cycled from zero to P_{max}. Determine the magnitude of P_{max} for 10^8 cycles based on a factor of safety $SF = 1.75$. Material properties are given in Problem 3-5.5. Assume that $q = 0.90$ and that the Gerber relation is valid when expressed in terms of nominal stress values (see Eq. (1) of Example 13-4.3). Since the state of stress at the hole is uniaxial, we define S_{cc} for the stress concentration as the ratio of σ_{max} for the crank shaft with the hole to σ_{max} for the crank shaft without the hole.

   Ans. $S_{ce} = 2.98$, $P_{max} = 494.1$ N

5. A crank shaft has a fillet and a minimum diameter of 30 mm. The critical section of the crank shaft is subjected to a bending moment $M = 200P$ and to a torque $T = 180P$ where P is a completely reversed load. The calculated stress concentrations for the fillet are $S_{cc}^{(M)} = 2.50$ for bending and $S_{cc}^{(T)} = 2.00$ for torsion. The shaft is made of a stress relieved cold worked SAE 1060 steel ($E = 200$ GPa, $\nu = 0.29$, $\sigma_u = 810$ MPa, $Y = 620$ MPa, and $\sigma_{am} = 410$ MPa for 10^7 cycles). Determine the completely reversed load P that can be applied 10^7 times based on a factor of safety of $SF = 2.20$. Assume $q = 0.85$.

6. Let the crank shaft in Problem 5 be subjected to a range of load from zero to P_{max}. Determine the completely reversed load P_{max} that can be applied 10^7 times based on a factor of safety $SF = 2.20$. Assume that the Gerber relation is valid when expressed in terms of nominal stress values (see Eq. (1) of Example 13-4.3). Use the maximum shearing stress criterion of failure; therefore, assume that S_{ce} for the fillet is the ratio τ_{max} in the crank shaft with the fillet to τ_{max} in the crank shaft without the fillet.

   Ans. $P_{max} = 1.66$ kN

13-5
EFFECTIVE STRESS CONCENTRATION FACTORS. INELASTIC STRAINS

Consider a flat plate of width l and thickness d, with symmetrically placed edge notches of radius ρ (Fig. 13-5.1a). The elastic stress concentration factor S_{cc} for this case may be found from Fig. 12-2.8 for given values of ρ, t, and a. The tensile stress-strain curve for the material

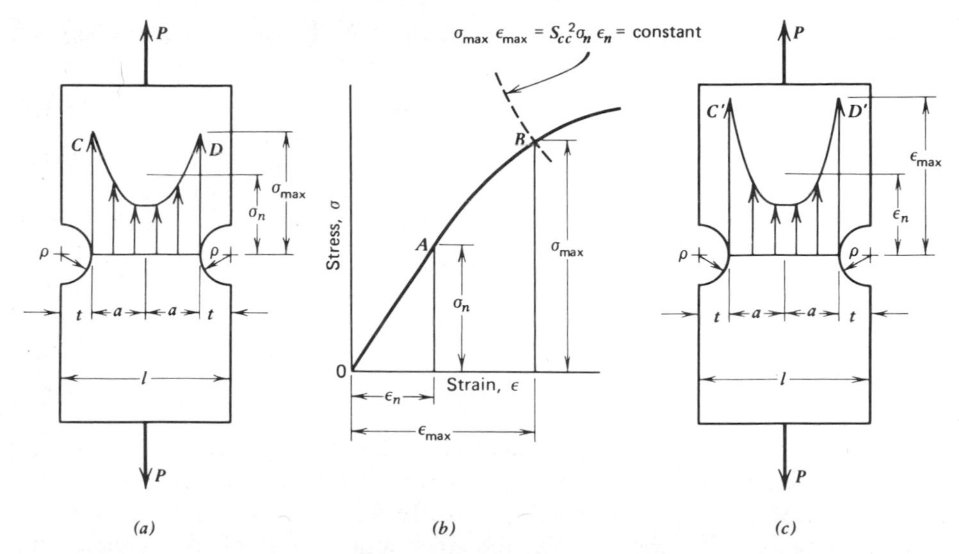

Fig. 13-5.1/(a) Stress distribution. (*b*) Stress-strain curve. (*c*) Strain distribution.

is shown in Fig. 13-5.1*b*. We consider here the problem of determining the maximum stress, σ_{max}, and the maximum strain, ϵ_{max}, at the roots of the edge notches for the case where the axial load P produces inelastic deformation in the material surrounding the notches. Before we present a solution to this problem, we define certain quantities and state a theorem that is employed in obtaining this solution.

In Fig. 13-5.1*a*, we assume that the stress distribution may be represented by the curve CD. The nominal stress on the cross section at the notch is σ_n. One of the quantities we wish to determine is the maximum stress, σ_{max}, where in terms of the significant stress concentration factor S_{ce} and the nominal stress σ_n, we have

$$\sigma_{max} = S_{ce}\sigma_n \tag{13-5.1}$$

Corresponding to the nominal stress σ_n, we have the nominal strain ϵ_n, where (σ_n, ϵ_n) are the coordinates of point A in Fig. 13-5.1*b*. In general, the strain distribution across the specimen and the maximum strain ϵ_{max} in the specimen are not known; see curve $C'D'$ in Fig. 13-5.1*c*. Corresponding to the effective stress concentration factor S_{ce}, we define a strain concentration factor E_{ce} by the relation

$$E_{ce} = \frac{\epsilon_{max}}{\epsilon_n} \tag{13-5.2}$$

We wish to determine the values of both σ_{max} and ϵ_{max}. From curve $0AB$ of Fig. 13-5.1*b*, we note that $(\sigma_{max}, \epsilon_{max})$ are the coordinates of point B. For this purpose, we employ a theorem due to Neuber.[3]

Neuber's Theorem / For relatively sharp notches, the following relation between S_{ce}, S_{cc}, and E_{ce} exists

$$S_{ce} \cdot E_{ce} = S_{cc}^2 \tag{13-5.3}$$

where S_{ce} and E_{ce} are defined by Eqs. (13-5.1) and (13-5.2), respectively, and where S_{cc} is the calculated (theoretical) stress concentration factor.

Equation (13-5.3) holds for σ_{max} above and below the elastic limit of the material. For example, when σ_{max} is below the elastic limit $S_{ce} = S_{cc}$ and also $E_{ce} = S_{cc}$. Hence Eq. (13-5.3) is satisfied identically. When σ_{max} is above the elastic limit, $S_{ce} < S_{cc}$. Hence, by Eq. (13-5.3), $E_{ce} > S_{ce}$. Substitution of Eqs. (13-5.1) and (13-5.2) into Eq. (13-5.3), we find

$$\sigma_{max}\epsilon_{max} = S_{cc}^2\sigma_n\epsilon_n \tag{13-5.4}$$

Equation (13-5.4) may be used to determine the values of σ_{max} and ϵ_{max} (coordinates of point B, Fig. 13-5.1b), since in a typical problem the values of S_{cc}, σ_n, and ϵ_n are usually available; that is, the load, the dimensions of the member, the stress-strain curve of the material, and the value of S_{cc} are known or obtainable. Thus, with these values known, Eq. (13-5.4) may be written

$$\sigma_{max}\epsilon_{max} = \text{constant} \tag{13-5.5}$$

Equation (13-5.5) represents a hyperbola in (σ, ϵ) space (Fig. 13-5.1b). The intersection of this hyperbola with the stress-strain curve occurs at point B (Fig. 13-5.1b). Thus, by plotting Eq. (13-5.5) in Fig. 13-5.1b, we locate point B and, hence, we may read the values of σ_{max} and ϵ_{max} as the coordinates of point B. Then, substitution of σ_{max} and ϵ_{max} values into Eqs. (13-5.1) and (13-5.2) yield values of S_{ce} and E_{ce}, respectively.

EXAMPLE 13-5.1

Application of Neuber's Theorem

Consider a low carbon steel with the stress-strain diagram shown in Fig. E13-5.1a. Let the nominal stress in a notched specimen (Fig. 13-5.1a) be $\sigma_n = 105$ MPa. From Fig. E13-5.1a, we find $\epsilon_n = 0.0005$. Also, let $S_{cc} = 2.43$. By Eq. (13-5.4) we obtain

$$\sigma_{max}\epsilon_{max} = 0.31$$

This curve intersects the stress-strain curve at point B (Fig. E13-5.1a). For point B, we find $\sigma_{max} = 236.3$ MPa and $\epsilon_{max} = 0.0013$. Hence, by Eq. (13-5.1),

$$S_{ce} = \frac{\sigma_{max}}{\sigma_n} = \frac{236.3}{105} = 2.25$$

This value of S_{ce} corresponds to the value of the ordinate of point C, Fig. E13-5.1b, with abscissa value $\sigma_n = 105$ MPa. Proceeding in a similar

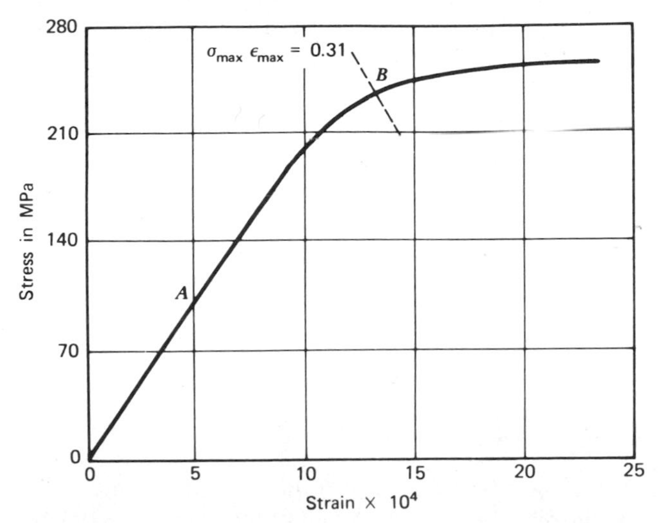

Fig. E13-5.1a/Stress-strain diagram.

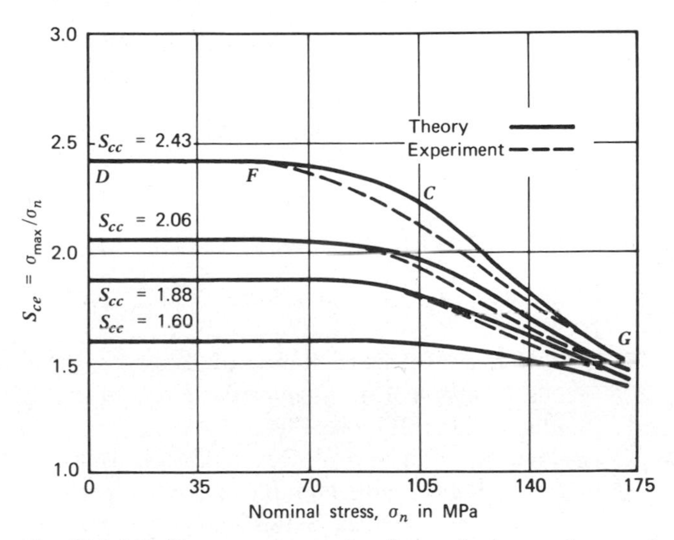

Fig. E13-5.1b/Stress concentration factor for low carbon steel of Fig. E13-5.1*a*. Experimental data from Neuber.

manner, we may plot a continuous curve, FCG, of values of S_{ce} as shown in Fig. E13-5.1*b*. For values of $\sigma_n < 58$ MPa (abscissa of point F), $S_{ce} = \sigma_{max}/\sigma_n = 2.43 = S_{cc}$. For values of $\sigma_n > 58$ MPa, S_{ce} decreases from the value of 2.43 to the value 1.5 at point G. In this region of decreasing value of S_{ce} (from point F to point G), $S_{ce} < S_{cc}$. In this region S_{ce} is the significant (effective) stress concentration factor rather than S_{cc}.

Three other curves for values of S_{cc} equal to 2.06, 1.88, and 1.60, respectively, are also plotted in Fig. E13-5.1*b*, employing the method for $S_{cc} = 2.43$. In addition, experimental values obtained by Neuber[4] are also shown.

PROBLEM SET 13-5

1. In Fig. 13-5.1*a*, let $a = 30$ mm, $t = \rho = 5$ mm and let the thickness of the plate be $d = 12.5$ mm. Let the load $P = 110$ kN. If the stress-strain curve of the material is given by Fig. E13-5.1*a*, determine the stress concentration factor.

2. Let the stress-strain diagram for Problem 1 be flat topped at a stress of 258 MPa. What is the magnitude of the strain at the root of the notch when $S_{ce} = 1.10$?

   Ans. $\epsilon_{max} = 0.0065$

REFERENCES

1. H. Neuber, *Kerbspannungslehre*, 2nd Ed., Springer-Verlag, Berlin, 1958.
2. J. O. Smith, "Effect of Range of Stress on the Fatigue Strength of Metals," *Bulletin* 334, *Engineering Experiment Station*, University of Illinois, Urbana, Illinois, 1942.
3. H. Neuber, "Theory of Stress Concentration for Shear Strained Prismatic Bodies with Nonlinear Stress-Strain Law," *Journal of Applied Mechanics*, Vol. 28, Series E, No. 4, December 1961, pp. 544–550.
4. H. Neuber, "Research on the Distribution of Tension in Notched Construction Parts," WADD Report 60-906, January 1961.

CHAPTER 14

CONTACT STRESSES

14-1 INTRODUCTION

Contact stresses are caused by the pressure of one solid on another over limited areas of contact. Most load-resisting members are designed on the basis of stress in the main body of the member, that is, in portions of the body not affected by the localized stresses at or near a surface of contact between bodies. In other words, most failures (excessive elastic deflection, yielding, and fracture) of members are associated with stresses and strains in portions of the body far removed from the points of application of the loads.

In certain cases, however, the contact stresses created when surfaces of two bodies are pressed together by external loads are the significant stresses; that is, the stresses on or somewhat beneath the surface of the contact are the major cause of failure of one or both of the bodies. For example, contact stresses may be significant at the area (a) between a locomotive wheel and the railroad rail; (b) between a roller or a ball and its race in a roller or ball bearing; (c) between the teeth of a pair of gears in mesh; (d) between the cam and valve tappets of a gasoline engine; etc.

We note that in each of these examples, the members do not necessarily remain in fixed contact. In fact, the contact stresses are often cyclic in nature and are repeated a very large number of times, often resulting in a fatigue failure that starts as a localized fracture (crack) associated with localized stresses; the fact that contact stresses frequently lead to fatigue failure largely explains why these stresses may limit the load carrying capacity of the members in contact and hence may be the significant stresses in the bodies. For example, a railroad rail sometimes fails as a result of "contact stresses"; the failure starts as a localized fracture in the form of a minute transverse crack at a point in the head of the rail somewhat beneath the surface of contact between the rail and the locomotive wheel, and progresses outwardly under the influence of the repeated wheel loads until the entire rail cracks or fractures. This fracture is called a *transverse fissure failure*.

On the other hand, ball bearings and gear teeth sometimes fail as a result of formation of pits (pitting) at the surface of contact. The bottom of such a pit is often located approximately at the point of maximum shearing stress. Steel tappets have been observed to fail by initiation of microscopic cracks at the surface that then spread and cause flaking. Chilled cast-iron tappets have failed by cracks that start beneath the surface, where the shearing stress is highest, and spread to the surface, causing pitting failure.

The principal stresses at or on the contact area between the two curved surfaces that are pressed together are greater than at a point beneath the contact area; whereas the maximum shearing stress is usually greater at a point a small distance beneath the contact surface.

The problem considered here initially is to determine the maximum principal (compressive) and shearing "contact stresses" on and beneath the contact area between two ideal *elastic* bodies having curved surfaces that are pressed together by external loads. Several investigators have attempted to solve this problem. H. Hertz[†] was the first to obtain a satisfactory solution although his solution gives only principal stresses in the contact area.

14-2 THE PROBLEM OF DETERMINING CONTACT STRESSES

Two semicircular disks made of elastic material are pressed together by forces P (Fig. 14-2.1). The two bodies are initially in contact at a single point. Sections of the boundaries of the two bodies at the point of contact are smooth curves before the loads are applied. The principal radii of curvature of the surface of the upper solid at the point of contact are R_1 and R_1'. For the lower solid R_2 and R_2' are the principal radii of curvature, respectively, of the surface of the lower solid at the point of contact. The intersection of the planes in which the radii R_1 and R_2 (or the radii R_1' and R_2') lie form an angle α. Elevation and plan views, respectively, of the two solids are shown in Figs. 14-2.2*a* and *b*. The lines v_1 and v_2, which form the angle α, lie in the plane sections containing the radii R_1 and R_2, respectively. The line of action of load P lies along the axis that passes through the centers of curvature of the solids and through the point of contact. Hence, the line of action of force P is perpendicular to a plane that is tangent to both solids at the point of contact. In other words, it is assumed that there is no tendency for one body to slide with respect to the other and, hence, no friction force is present. The effect of a friction force is discussed in Art. 14-9.

[†] Hertz published a paper in 1881 "On the Contact of Elastic Solids," and in the following year a paper "On the Contact of Rigid Elastic Solids and on Hardness." See reference 1.

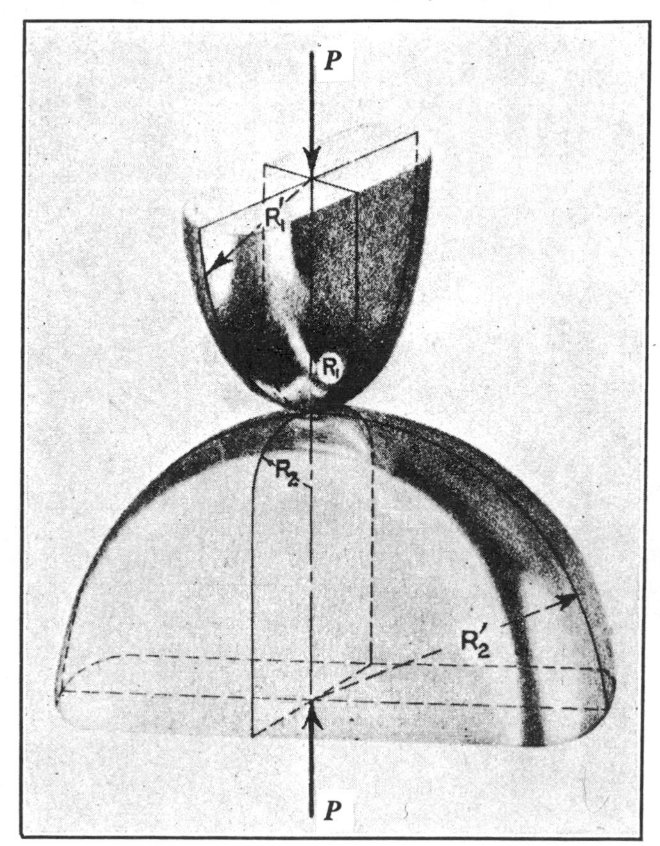

Fig. 14-2.1/Two curved surfaces of different radii pressed against each other.

The effect of the load P is to cause the surface of the solids to be deformed elastically over a region surrounding the initial point of contact, thereby bringing the two bodies into contact over a small area in the neighborhood of the initial point of contact (Fig. 14-2.2*b*). The problem is to determine a relation between the load P and the maximum compressive stress on this small area of contact and to determine the principal stresses at any point in either body on the line of action of the load, designated as the z-axis. The principal stresses σ_{xx}, σ_{yy}, and σ_{zz} acting on a small cube at a point on the z-axis are shown in Fig. 14-2.2*c*. The maximum shearing stress at the point is $\tau_{\max} = \frac{1}{2}(\sigma_{zz} - \sigma_{yy})$ where σ_{zz} and σ_{yy} are the maximum and minimum principal stresses at the point.

The detailed development of the solution of the problem will not be presented here. However, the main assumptions made in the solution are given in order that the limitations on the use of the results may be understood. A brief discussion is given to attempt to explain and justify the assumptions.

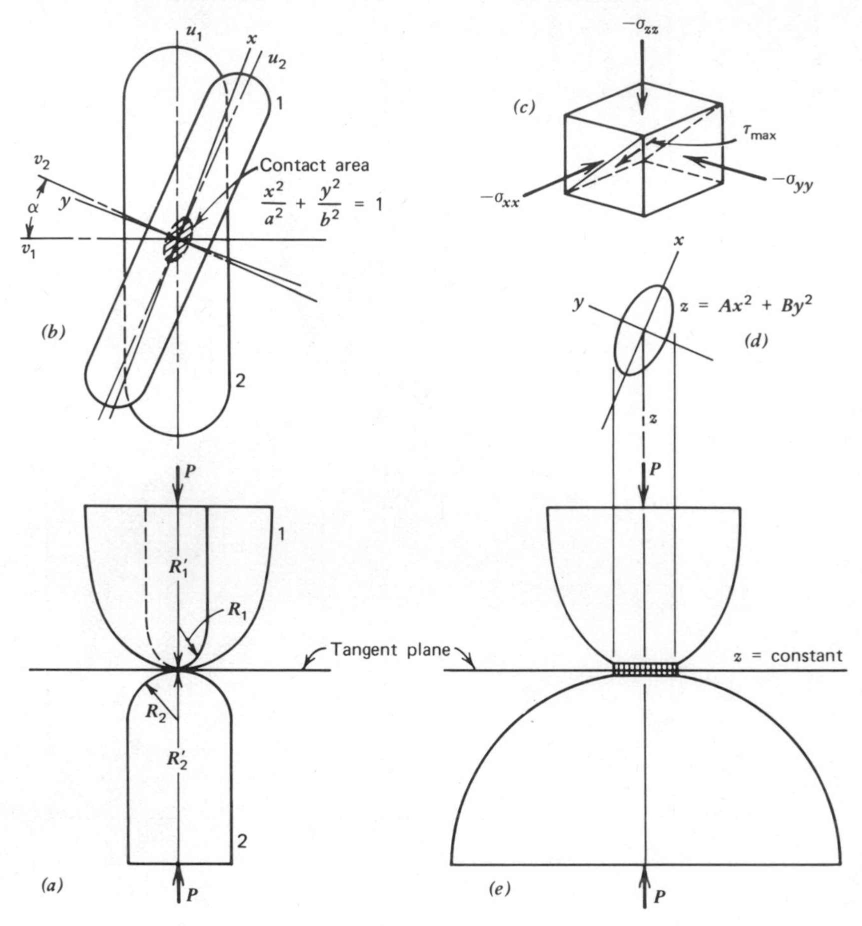

Fig. 14-2.2/Analysis of contact stresses.

14-3 ASSUMPTIONS ON WHICH A SOLUTION FOR CONTACT STRESSES IS BASED

The solution of the problem of the contact stresses in the neighborhood of the point of contact of two bodies is based on the following two assumptions.

(a) Properties of Materials/The material of each body is homogeneous, isotropic, and elastic in accordance with Hooke's law, but the two bodies are not necessarily made of the same material.

(b) Shape of Surfaces near Point of Contact. Before Loading/If two bodies are in contact at a point, there is a common tangent plane to the surfaces at the point of contact. In the solution for contact stresses an expression for the distance between corresponding points on the surfaces near the point of contact is required; corresponding points are points that lie on the surfaces of the bodies and on a line perpendicular to the common tangent plane. Equations that express the distances z from corresponding points to the common tangent plane are needed to determine the deformations of the two bodies near the initial point of contact. In the analysis, an equation that *approximates* the distances z between corresponding points on any two surfaces is used. This equation is

$$z = Ax^2 + By^2 \tag{14-3.1}$$

in which x and y are coordinates with respect to y- and x-axes with origin at the point of contact; these coordinates lie in the tangent plane, and A and B are (positive) constants[1] that depend upon the principal radii of curvature of the surfaces at the point of contact. The derivation of Eq. (14-3.1) is discussed later in this article. Figures 14-2.2*d* and *e* illustrate the fact that the curve representing Eq. (14-3.1) for a constant value of z is an ellipse. This fact will be important in considering the shape of the area of contact between the two bodies.

After Loading/When the loads P are applied to the bodies, their surfaces deform elastically near the point of contact so that a small area of contact is formed. It is assumed that, as this small area of contact forms, points that come into contact are points on the two surfaces that originally were equal distances from the tangent plane. According to Eq. (14-3.1), such equidistant points on the two surfaces lie on an ellipse. Hence the boundary line of the area of contact is assumed to be an ellipse whose equation is

$$\frac{x^2}{a^2} + \frac{y^2}{b^2} = 1 \tag{14-3.2}$$

where x and y are coordinates referred to the same axes as were specified for Eq. (14-3.1). The contact area described by Eq. (14-3.2) is shown in Fig. 14-2.2*b*. Equation (14-3.1) is of sufficient importance to warrant further discussion of its validity, particularly since a method of determining the constants A and B is required in the solution of the problem of finding contact stresses.

Justification of Eq. (14-3.1)/In order to obtain Eq. (14-3.1), an expression is derived first for the perpendicular distance z_1 from the tangent plane to any point on the surface of body 1 near the point of contact, assuming the bodies free from loads and in contact at a point. A portion of body 1 showing the distance z_1 is illustrated in Fig. 14-3.1*a*. Let the points considered lie in the planes of principal radii of curvature. Let u_1 and v_1 be axes in the tangent plane that lie in the planes of principal radii of curvature of body 1. The distance z_1 to point C or D is found as follows. From triangle $0DD'$

$$z_1 = u_1 \tan \tfrac{1}{2}\beta = \tfrac{1}{2}u_1\beta \tag{14-3.3}$$

since the angle β is small. From triangle HKD

$$\tan\beta = \beta = \frac{KD}{HK} = \frac{u_1}{R_1'} \tag{14-3.4}$$

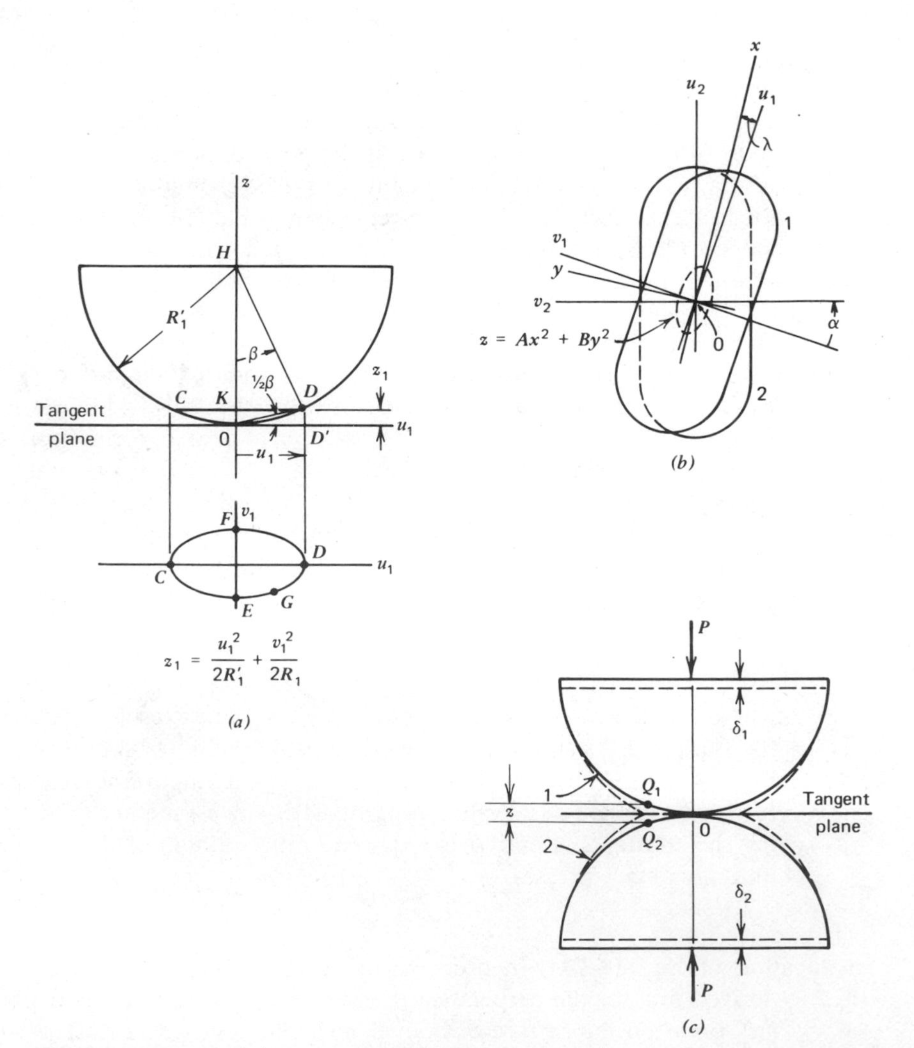

Fig. 14-3.1/Geometry of contact surface.

since the radius R_1' is approximately equal to HK. Substitution of the value of β from Eq. (14-3.4) into Eq. (14-3.3) gives

$$z_1 = \frac{u_1^2}{2R_1'} \tag{14-3.5}$$

In a similar manner, the distance z_1 to the points E and F lying in the plane of radius R_1 is found to be

$$z_1 = \frac{v_1^2}{2R_1} \tag{14-3.6}$$

On the basis of these results, it is assumed that the distance z_1 to any point G not lying in either plane of principal curvature may be approximated by

$$z_1 = \frac{u_1^2}{2R_1'} + \frac{v_1^2}{2R_1} \tag{14-3.7}$$

This assumption seems justified by the fact that Eq. (14-3.7) reduces to Eq. (14-3.6) for $u_1 = 0$, and to Eq. (14-3.5) for $v_1 = 0$. In particular, we note that, if z_1 is constant for all points G, Eq. (14-3.7) is an equation for an ellipse.

Attention is directed now to the second body. The distance z_2 from the tangent plane to any point in the surface of body 2 near the point of contact is obtained in the same way as was z_1 in Eq. (14-3.7). It is

$$z_2 = \frac{u_2^2}{2R_2'} + \frac{v_2^2}{2R_2} \tag{14-3.8}$$

where u_2 and v_2 are coordinates with respect to axes lying in the tangent plane and also in the planes of the principal radii of curvature R_2', and R_2, respectively. The locations of the axes u_1, v_1 and u_2, v_2 are shown in Fig. 14-3.1*b*, which is the same view of the bodies as in Fig. 14-2.2*b*. The axes v_1 and v_2 subtend the angle α that is the angle between the lines v_1 and v_2 of the bodies as shown in Fig. 14-2.2*b*.

The distance z between points on the two surfaces near the point of contact is the numerical sum of z_1 and z_2 given by Eqs. (14-3.7) and (14-3.8). Hence, we find

$$z = z_1 + z_2 = \frac{u_1^2}{2R_1'} + \frac{v_1^2}{2R_1} + \frac{u_2^2}{2R_2'} + \frac{v_2^2}{2R_2} \tag{14-3.9}$$

Equation (14-3.9) may now be transformed into the form of Eq. (14-3.1). The first transformation is the elimination of the coordinates u_2 and v_2 by the relationships

$$\begin{aligned} u_2 &= u_1 \cos\alpha + v_1 \sin\alpha \\ v_2 &= -u_1 \sin\alpha + v_1 \cos\alpha \end{aligned} \tag{14-3.10}$$

When Eqs. (14-3.10) are substituted into Eq. (14-3.9), there results

$$z = A'u_1^2 + 2H'u_1v_1 + B'v_1^2 \tag{14-3.11}$$

where

$$\begin{aligned} 2A' &= \frac{1}{R_1'} + \frac{1}{R_2'}\cos^2\alpha + \frac{1}{R_2}\sin^2\alpha \\ 2H' &= \left[\frac{1}{R_2'} - \frac{1}{R_2}\right]\sin\alpha\cos\alpha \\ 2B' &= \frac{1}{R_1} + \frac{1}{R_2'}\sin^2\alpha + \frac{1}{R_2}\cos^2\alpha \end{aligned} \tag{14-3.12}$$

Equation (14-3.11) is the equation of an ellipse, as shown in Fig. 14-3.1*b*, with center at point 0. To find the equation of the ellipse referred to axes x and y, which coincide with the major and minor axes of the ellipse, the value of the angle λ through which the axes u_1 and v_1 must be rotated in order to eliminate the product term u_1v_1 in Eq. (14-3.11) is required. The transformation is

$$\begin{aligned} u_1 &= x\cos\lambda - y\sin\lambda \\ v_1 &= x\sin\lambda + y\cos\lambda \end{aligned} \tag{14-3.13}$$

If Eqs. (14-3.13) are substituted into Eq. (14-3.11) and the value of the angle λ taken to eliminate the product term u_1v_1, Eq. (14-3.11) becomes

$$z = Ax^2 + By^2 \tag{14-3.14}$$

which is identical in form to Eq. (14-3.1). In the process of making the transformation, it is found that A and B are the roots of a quadratic equation and having the following values:

$$B = \frac{1}{4}\left(\frac{1}{R_1} + \frac{1}{R_2} + \frac{1}{R_1'} + \frac{1}{R_2'}\right) \tag{14-3.15}$$

$$+ \frac{1}{4}\sqrt{\left[\left(\frac{1}{R_1} - \frac{1}{R_1'}\right) + \left(\frac{1}{R_2} - \frac{1}{R_2'}\right)\right]^2 - 4\left(\frac{1}{R_1} - \frac{1}{R_1'}\right)\left(\frac{1}{R_2} - \frac{1}{R_2'}\right)\sin^2\alpha}$$

$$A = \frac{1}{4}\left(\frac{1}{R_1} + \frac{1}{R_2} + \frac{1}{R_1'} + \frac{1}{R_2'}\right) \tag{14-3.16}$$

$$- \frac{1}{4}\sqrt{\left[\left(\frac{1}{R_1} - \frac{1}{R_1'}\right) + \left(\frac{1}{R_2} - \frac{1}{R_2'}\right)\right]^2 - 4\left(\frac{1}{R_1} - \frac{1}{R_1'}\right)\left(\frac{1}{R_2} - \frac{1}{R_2'}\right)\sin^2\alpha}$$

The constants A and B depend on the principal radii of curvature of the two bodies at the point of contact and upon the angle α between the corresponding planes of the principal curvatures.

Note With the definition for radii of curvature in Fig. 14-2.2, it is possible for the ratio B/A to be greater than one or less than one. If $\alpha = 0$ and $B/A > 1$, the *major* axis of the ellipse of contact is in the direction of rolling. If $\alpha = 0$ and $B/A < 1$, the *minor* axis of the ellipse of contact is in the direction of rolling. In this latter case, it is convenient to let R_1 and R_2 be radii of curvature in the plane of rolling so that $B/A > 1$.

Brief Discussion of Solution/It was pointed out earlier in this article that Eq. (14-3.1) is used to estimate the displacement of points on the surfaces of the two bodies that eventually lie within the contact area. In Fig. 14-3.1*c* the solid outline shows the two bodies of Fig. 14-2.1 in contact at one point, before the loads are applied, and the dashed lines show the new positions of the two bodies after the loads P are applied and the two bodies are in contact over a flattened area around the original point of contact 0. The centers of the bodies move toward each other by amounts of δ_1 and δ_2, respectively, which means that the distance between points on the bodies not affected by the local deformation near 0 is decreased by an amount $\delta_1 + \delta_2 = \delta$.

Let w_1 denote the displacement, due to local compression, of point Q_1, Fig. 14-3.1*c*. We take w_1 positive in the direction away from the tangent plane, assumed to remain immovable during local compression. Similarly, let w_2 denote the displacement, due to local compression, of point Q_2, where w_2 is taken positive in the direction away from the tangent plane. These positive directions of w_1 and w_2 conform to the positive directions of displacement in a small loaded region on a part of the boundary of a semi-infinite solid, that is, the positive displacement is directed into the solid. Hence, the distance between two points, such as Q_1 and Q_2 in Fig. 14-3.1, will diminish by $\delta - (w_1 + w_2)$. If, finally, due to the local compression, points Q_1 and Q_2 come inside the surface of contact, we have

$$\delta - (w_1 + w_2) = z_1 + z_2 = z$$

With the expression for z, given by Eq. (14-3.14), we may write

$$w_1 + w_2 = \delta - Ax^2 - By^2 \tag{14-3.17}$$

Equation (14-3.17) has been obtained from geometrical considerations only. To compute the displacements (w_1, w_2), local deformation at the surface of contact must be considered. Under the assumption that the surface of contact is very small compared to the radii of curvatures of the bodies, the solution obtained for semi-infinite bodies subjected to spot loads may be employed to determine $w_1 + w_2$.[2,3] Hertz noted that Eq. (14-3.17) has the same form as that of the Newtonian potential equation for the attraction of a homogeneous mass M in the shape of an ellipsoid upon a unit of mass concentrated at a point P some distance from the ellipsoid. This Newtonian potential function satisfies the same differential equations that are required to be satisfied by the theory of elasticity. The problem is solved by placing into the potential equation the stresses at the contact surface instead of the mass, etc., and the constants are evaluated.[3] The solution is given in terms of elliptic integrals. The results are summarized in the following articles.

14-4 NOTATION AND MEANING OF TERMS

The following notation and interpretations of terms are needed for an understanding of subsequent equations:

P = total force exerted by body 1 on body 2, and vice versa.

E_1, E_2 = tensile (or compressive) moduli of elasticity for bodies 1 and 2.

ν_1, ν_2 = Poisson's ratio for bodies 1 and 2.

a = semimajor axis of ellipse of contact.

b = semiminor axis of ellipse of contact.

$k = b/a = \cos\theta;\ k \leq 1$

$k' = \sqrt{1-k^2} = \sin\theta$

R_1, R_1' = principle values of the radii, respectively, of the surface of body 1 at the point of contact. The plane sections in which R_1, R_1' lie are perpendicular to each other. See Fig. 14-2.1. The signs of R_1 and R_1' are determined as follows. If the center of curvature lies inside the body (that is, if the body surface is convex at the point of contact), the radius is positive. If the center of curvature lies outside the body (that is, if the body surface is concave at the point of contact), the radius is negative. For example, in Fig. 14-2.1, the radii are both positive; in Fig. 14-8.1*c*, R_2 is negative.

R_2, R_2' = same as R_1, R_1', but for body 2.

α = angle between planes of principal curvatures at point of contact (see Fig. 14-2.2*b*).

$k(z/b)$ = relative depth below the surface of contact to a point on the z-axis at which stresses are to be calculated. The reason that the depth is expressed in terms of $k(z/b)$ rather than by z directly is that, in evaluating the integrals obtained in the mathematical solution of the problem, the term $k(z/b)$ can conveniently be replaced by a trigonometric function. Thus

$\cot\phi = k(z/b)$

z_s = depth in either body from surface to point on z-axis at which maximum shearing stress occurs.

In the expressions for the principal stresses, two integrals (called elliptic integrals) are found which involve ϕ [or $k(z/b)$], θ, and k' (that is, b/a). These integrals are denoted as $F(\phi, k')$ and $H(\phi, k')$. Likewise, two integrals involving k' alone, denoted as $K(k')$ and $E(k')$, are required. These elliptic integrals are

$$F(\phi, k') = \int_0^{\phi} \frac{d\theta}{\sqrt{1 - k'^2 \sin^2\theta}}$$

$$H(\phi, k') = \int_0^{\phi} \sqrt{1 - k'^2 \sin^2\theta}\, d\theta$$

$$K(k') = F\left(\frac{\pi}{2}, k'\right) = \int_0^{\pi/2} \frac{d\theta}{\sqrt{1 - k'^2 \sin^2\theta}}$$

$$E(k') = H\left(\frac{\pi}{2}, k'\right) = \int_0^{\pi/2} \sqrt{1 - k'^2 \sin^2\theta}\, d\theta$$

These integrals have been tabulated and are readily available in most mathematical handbooks.

14-5 EXPRESSIONS FOR PRINCIPAL STRESSES

The analysis involving the assumptions and limitations indicated in Art. 14-3 yields the following expressions for the principal stresses σ_{xx}, σ_{yy}, and σ_{zz} at a point on the z-axis; the point is at the distance z from the origin, which lies in the surface of contact of the two elastic bodies; and the stresses act on orthogonal planes perpendicular to the x-, y-, and z-axes, respectively. The solution of this problem is[3]

$$\sigma_{xx} = \left[M(\Omega_x + \nu\Omega'_x)\right]\frac{b}{\Delta} \tag{14-5.1}$$

$$\sigma_{yy} = \left[M(\Omega_y + \nu\Omega'_y)\right]\frac{b}{\Delta} \tag{14-5.2}$$

$$\sigma_{zz} = -\left[\frac{M}{2}\left(\frac{1}{n} - n\right)\right]\frac{b}{\Delta} \tag{14-5.3}$$

in which

$$M = \frac{2k}{k'^2 E(k')} \qquad n = \sqrt{\frac{k^2 + k^2(z/b)^2}{1 + k^2(z/b)^2}}$$

$$\Delta = \frac{1}{A+B}\left(\frac{1-\nu_1^2}{E_1} + \frac{1-\nu_2^2}{E_2}\right) \tag{14-5.3a}$$

where A and B are constants given by Eqs. (14-3.15) and (14-3.16), and where

$$\Omega_x = -\frac{1-n}{2} + k\frac{z}{b}[F(\phi,k') - H(\phi,k')]$$

$$\Omega'_x = -\frac{n}{k^2} + 1 + k\frac{z}{b}\left[\left(\frac{1}{k^2}\right)H(\phi,k') - F(\phi,k')\right]$$

$$\Omega_y = \frac{1}{2n} + \frac{1}{2} - \frac{n}{k^2} + k\frac{z}{b}\left[\frac{1}{k^2}H(\phi,k') - F(\phi,k')\right]$$

$$\Omega'_y = -1 + n + k\frac{z}{b}[F(\phi,k') - H(\phi,k')]$$

We note that the stresses depend upon the variables A, B, k, k', ν_1, ν_2, E_1, E_2, b and z. The first four variables depend only on the shape of the surfaces near the point of contact. Of these four, A and B are found from Eqs. (14-3.15) and (14-3.16), and from Art. 14-4, $k' = \sqrt{1-k^2}$. Therefore, one additional equation is needed for determining the value of k. This equation is

$$\frac{B}{A} = \frac{(1/k^2)E(k') - K(k')}{K(k') - E(k')} \tag{14-5.4}$$

The second group of four variables, ν_1, ν_2, E_1, and E_2, depend only on the physical properties of the two bodies in contact and are found by tests of the material. The variable, b, the semiminor axis of the area of contact, depends upon the eight variables previously listed, but it is important to note that it also depends upon the load P. The equation expressing this fact is

$$b = \sqrt[3]{\frac{3kE(k')}{2\pi}(P\Delta)} = ka \tag{14-5.5}$$

Values of the variable z, which represent the distance of a point *from the surface of contact*, may be chosen. Then the three principal stresses at any point on the z-axis may be obtained.

14-6 METHOD OF COMPUTING CONTACT STRESSES

Principal Stresses / In Art. 14-5 it is noted that the values of A and B must be computed first, and that in Eq. (14-5.4) the ratio B/A determines the value of k (and of k'). It should be remembered (Art. 14-3) that the values of A and B are related to the geometric shape and configuration of the two bodies. Thus, if two cylinders are crossed so that they are in point contact with their longitudinal axes perpendicular, the value of

$B/A = 1$ but, if these cylinders are arranged so that their longitudinal axis are parallel (line contact), $B/A = \infty$. With the values of the four quantities A, B, k, and k' known, the terms in the brackets in Eqs. (14-5.1), (14-5.2), and (14-5.3) can be computed for a value of Poisson's ratio ν. Fortunately the value of ν in these bracket terms has only a small influence on the final values of the stresses. Consequently, we take a value of $\nu = \frac{1}{4}$ to compute these terms. The actual values of ν_1 and ν_2 of the two bodies are used later in computing Δ. Thus, since the terms within the brackets do *not* depend strongly on the elastic constants of the two bodies or on the load P, their magnitudes can be computed and tabulated for use as coefficients in the determination of the ratio b/Δ. For example, let a value of the ratio $B/A = 1.24$ be chosen. From Eq. (14-5.4) it can be shown that $k = 0.866$ and $k' = 0.5$. For specified values of the ratio kz/b, required coefficients can be found for determination of the stresses at depth z below the area of contact. The results of these computations are given in Fig. 14-6.1, in which the coefficients of b/Δ are plotted as abscissas, and the values of kz/b to the point at which the stresses occur are plotted as ordinates. The curves representing σ_{xx}, σ_{yy}, and σ_{zz} show that their largest magnitudes occur when $z = 0$ (at the center of the surface of contact) and that all three stresses decrease as the depth z increases. The principal stress having the greatest magnitude is σ_{zz} and, hence, at each point, $\sigma_{zz} = \sigma_{max}$. In this example, in which $B/A = 1.24$, the value of $\sigma_{max} = 0.67b/\Delta$. The coefficient 0.67 of b/Δ is found at $z = 0$ from the curve σ_{zz}.

Maximum Shearing Stress/The maximum shearing stress at any point is $\frac{1}{2}(\sigma_{max} - \sigma_{min})$. In Fig. 14-6.1 the curves show that the magnitudes of σ_{xx} and σ_{yy} decrease more rapidly than that of σ_{zz} at points just beneath the surface of contact. Because of this fact, the maximum shearing stress at points just beneath the surface of contact increases in magnitude and reaches its maximum value $\frac{1}{2}(\sigma_{zz_s} - \sigma_{yy_s})$ at the depth z_s, as shown by the curve marked τ. In this example in which $B/A = 1.24$, the value of $\tau_{max} = 0.22b/\Delta$, and the depth $kz_s/b = 0.44$, that is, $z_s = 0.44b/0.866 = 0.51b$. The coefficient 0.22 of b/Δ is the ordinate to the τ curve at the depth z_s.

Maximum Octahedral Shearing Stress/The octahedral shearing stress τ_{oct} (see Chapter 1, Eq. 1-4.8) is

$$\tau_{oct} = \frac{1}{3}\sqrt{(\sigma_{xx} - \sigma_{yy})^2 + (\sigma_{yy} - \sigma_{zz})^2 + (\sigma_{zz} - \sigma_{xx})^2}$$

The values of τ_{oct} have been computed by this equation for several points along the z-axis beneath the surface of contact and are plotted as

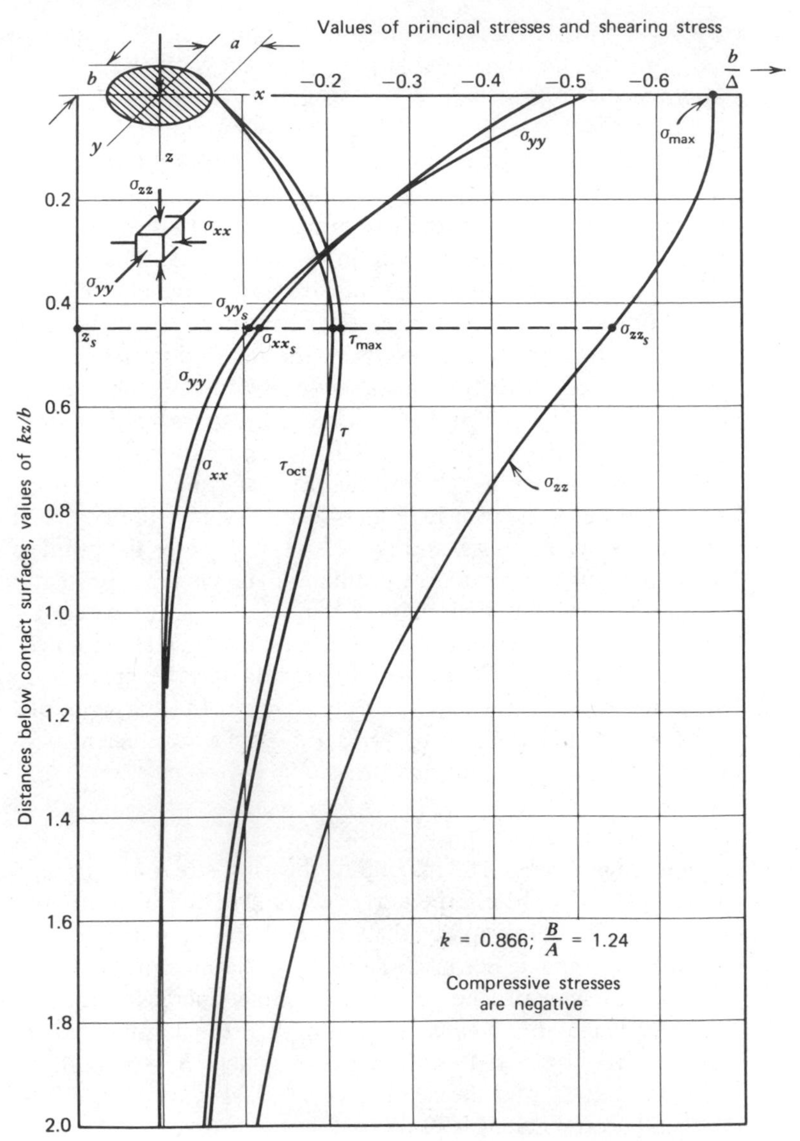

Fig. 14-6.1/Curves showing variation in principal stresses, maximum shearing stress, and octahedral shearing stress with variation in distance below contact surface; $\nu = 0.25$.

ordinates to the curved marked τ_{oct}. In this example, for $B/A = 1.24$, the maximum value of the octahedral shearing stress $\tau_{oct(max)} = 0.21b/\Delta$, and it occurs at the same depth $z_s = 0.51b$ as the maximum shearing stress. The coefficient 0.21 of b/Δ is the ordinate to the τ_{oct} curve at the depth z_s.

Maximum Orthogonal Shearing Stress / As noted above, the maximum shearing stress and maximum octahedral shearing stress occur in the interior of the contacting bodies at points located equidistance from the tangent plane on a line perpendicular to the center of the contacting area. These maximum values are considered to be the significant stresses in certain failure criterion associated with initiation of yielding (Chapter 3). Other shearing stress components that are considered to be significant in the fatigue failure of bearings and other rolling elements (cylinders) in contact are shearing stresses that occur on planes perpendicular and parallel to the tangent plane to the contact area. For example, with reference to Figs. 14-2.2*b* and *d*, orthogonal shearing stress components σ_{xz} and σ_{yz} act on a plane perpendicular to the *z*-axis; they are called orthogonal shearing stresses since they also act on planes that are perpendicular (orthogonal) to the plane of contact (see Art. 1-2). These shearing stress components are zero on the *z*-axis where τ_{max} occurs. We choose the *x*-axis in the direction of rolling and consider only those problems for which the *x*-axis coincides with either the major or minor axes of the contacting ellipse. The maximum orthogonal shearing stress τ_0 is defined as $\sigma_{xz(max)}$, which occurs at points in the interior of the contacting bodies located in the (x, z)-plane equidistant from the *z*-axis at some distance from the contacting surface.

Although τ_0 is always smaller than τ_{max}, τ_0 for a given point in a contacting body changes sign as the rolling element (the contact area) approaches and leaves the region above the point. Therefore, the range of the maximum orthogonal shearing stress is $2\tau_0$, and for most applications this range is greater than the range of the maximum shearing stress τ_{max}. Note that $\sigma_{yz(max)}$ may be greater than τ_0; however, $\sigma_{yz(max)}$ does not change sign during rolling so that the range of σ_{yz} is equal to $\sigma_{yz(max)}$, which is less than $2\tau_0$. The range in shearing stress is considered to be important[4] in studies of fatigue failure due to rolling contact.

The location of the point (perpendicular distance to the point from the tangent plane at the contact area) at which τ_0 occurs, as well as the magnitude of τ_0, is a function of the ellipticity (values of a, b) and orientation of the contact ellipse with respect to the rolling direction.[4] In particular, for toroids under radial loads (Fig. 14-6.2) Fessler and Ollerton[5] have derived expressions for the orthogonal shearing stress components σ_{xz}, σ_{yz} (Fig. 14-6.3). Their results are, with the notations of Arts.

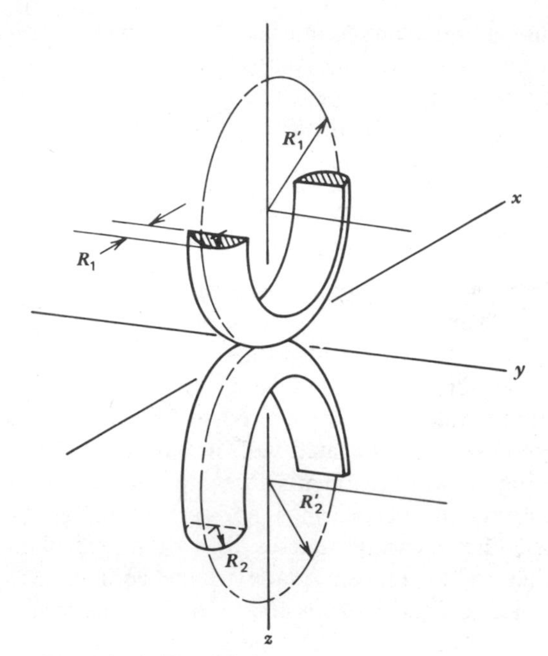

Fig. 14-6.2/Toroids in contact.

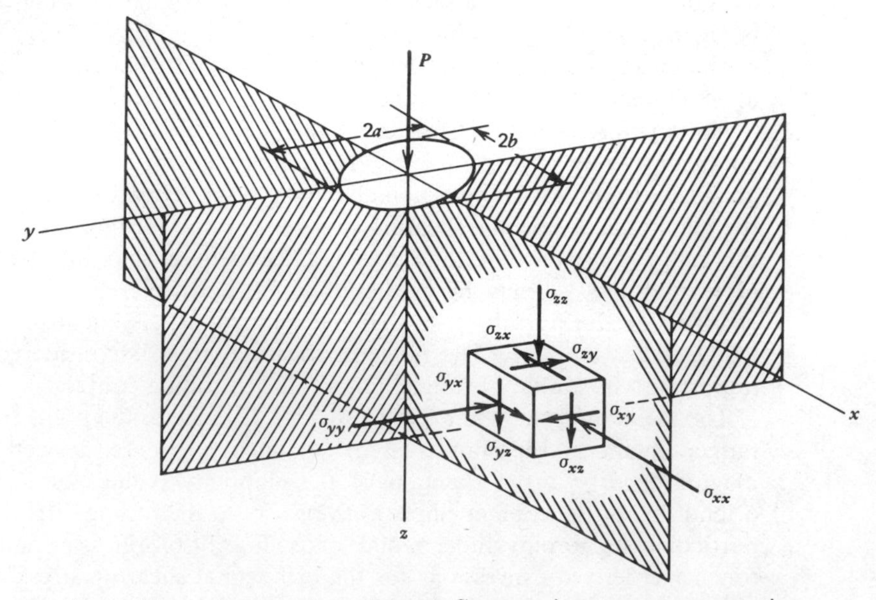

Fig. 14-6.3/State of stress of an element. Compressive stresses are negative.

14-4 and 14-5,

$$\sigma_{xz} = -\frac{3PQ}{2\pi a^2} = -\frac{b}{\Delta}\left[\frac{kQ}{E(k')}\right]$$

$$\sigma_{yz} = -\frac{3PR}{2\pi b^2} = -\frac{b}{\Delta}\left[\frac{R}{kE(k')}\right] \tag{14-6.1}$$

where

$$Q = \frac{\left(\frac{x}{a}\right)\left(\frac{z}{a}\right)^2\left[\left(1+\frac{c^2}{a^2}\right)\frac{c^2}{a^2}\right]^{-3/2}\left(k^2+\frac{c^2}{a^2}\right)^{-1/2}}{\left[\left(\frac{ax}{a^2+c^2}\right)^2+\left(\frac{ay}{b^2+c^2}\right)^2+\left(\frac{az}{c^2}\right)^2\right]} \tag{14-6.2}$$

and

$$R = \frac{\left(\frac{y}{b}\right)\left(\frac{z}{b}\right)^2\left[\left(1-\frac{c^2}{b^2}\right)\frac{c^2}{b^2}\right]^{-3/2}\left(\frac{1}{k^2}+\frac{c^2}{b^2}\right)^{-1/2}}{\left[\left(\frac{bx}{a^2+c^2}\right)^2+\left(\frac{by}{b^2+c^2}\right)^2+\left(\frac{bz}{c}\right)^2\right]} \tag{14-6.3}$$

where c^2 is the positive root of the equation

$$\frac{x^2}{a^2+c^2}+\frac{y^2}{b^2+c^2}+\frac{z^2}{c^2}=1 \tag{14-6.4}$$

The maximum values of σ_{xz} and σ_{yz} occur in the planes of symmetry, $y=0$ and $x=0$, respectively. For the plane of symmetry $y=0$, we get

$$\sigma_{xz} = -\frac{3PQ(y=0)}{2\pi a^2} = -\frac{b}{\Delta}\left[\frac{kQ(y=0)}{E(k')}\right]$$

$$\sigma_{yz} = 0 \tag{14-6.5}$$

and for the plane of symmetry $x=0$, we have

$$\sigma_{xz} = 0$$

$$\sigma_{yz} = -\frac{3PR(x=0)}{2\pi b^2} = -\frac{b}{\Delta}\left[\frac{R(x=0)}{kE(k')}\right] \tag{14-6.6}$$

In particular, we note that along the z-axis ($x=y=0$), $\sigma_{xz}=\sigma_{yz}=0$.

As noted earlier, in rolling contact problems, the perpendicular distance from the tangent plane at the contact area to the point at which the maximum range of τ_0 occurs (as the roller passes over a given region) and the magnitude of τ_0 depend on the ellipticity and the orientation of the contact ellipse with respect to the rolling direction. Consequently, the geometric configuration of the contacting rollers is an important

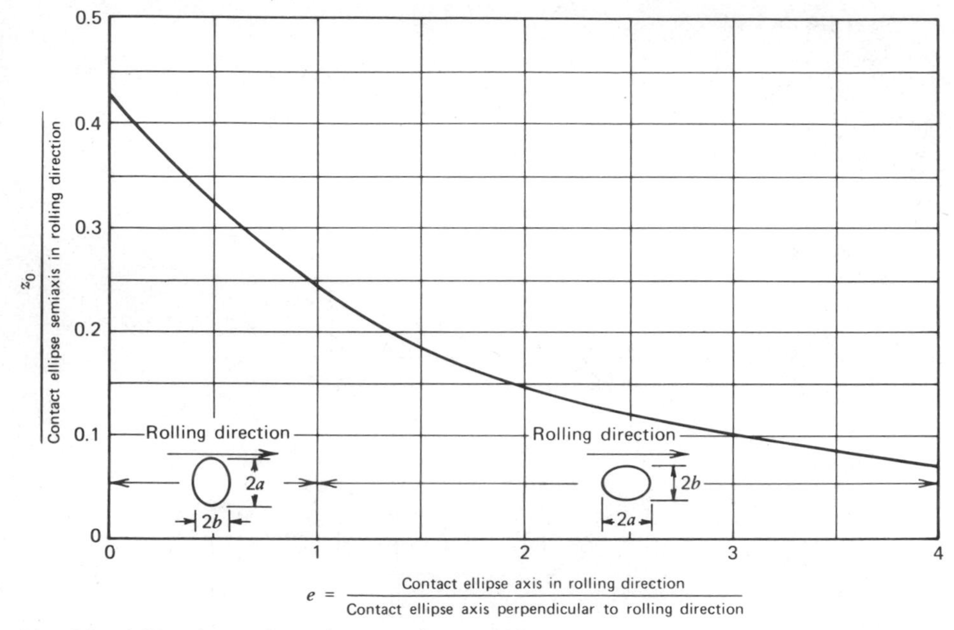

Fig. 14-6.4/Depth z_0 of maximum orthogonal shear stress.

factor. For example, for contacting toroids (Fig. 14-6.2), the depth z_0, beneath the contact surface, at which the maximum orthogonal shearing stress $\tau_0 = \sigma_{zx(\max)}$ occurs is plotted in Fig. 14-6.4; note that the contact ellipse semi-axis in the rolling direction is equal to b for $e < 1$ and is equal to a for $e > 1$. The magnitude of τ_0 is plotted in Fig. 14-6.5.[5]

The range of the maximum orthogonal shearing stress $2\tau_0$ is considered to be important in rolling fatigue problems as long as $2\tau_0$ is greater than the range of the maximum shearing stress $\tau_{\max}$. Fessler and Ollerton[5] found that $2\tau_0 > \tau_{\max}$ for $e < 2.25$ when $\nu = 0.25$ for both bodies and that $2\tau_0 > \tau_{\max}$ for $e < 2.85$ when $\nu = 0.50$ for both bodies. Poisson's ratio for most materials fall between these two values. Furthermore, the contour surfaces for most rolling contact problems result in values of e that are less than 2.25.

Fatigue failure of members under rolling contact depends on the magnitude of $2\tau_0$ for $e < 2.25$; however, the magnitude of the range of the orthogonal shearing stress $2\tau_0$ for fatigue failure for a specified number of cycles is not equal to the range of maximum shearing stress (for the same number of cycles of loading) in specimens (tension-compression or reversed torsion) that are generally used to obtain fatigue properties.[4] The high hydrostatic state of compressive stress in contact problems strengthens the material against fatigue failure. The usual procedure is to obtain the magnitude of τ_0 to initiate fatigue failure for a specified number of cycles for one orientation of the contact ellipse and

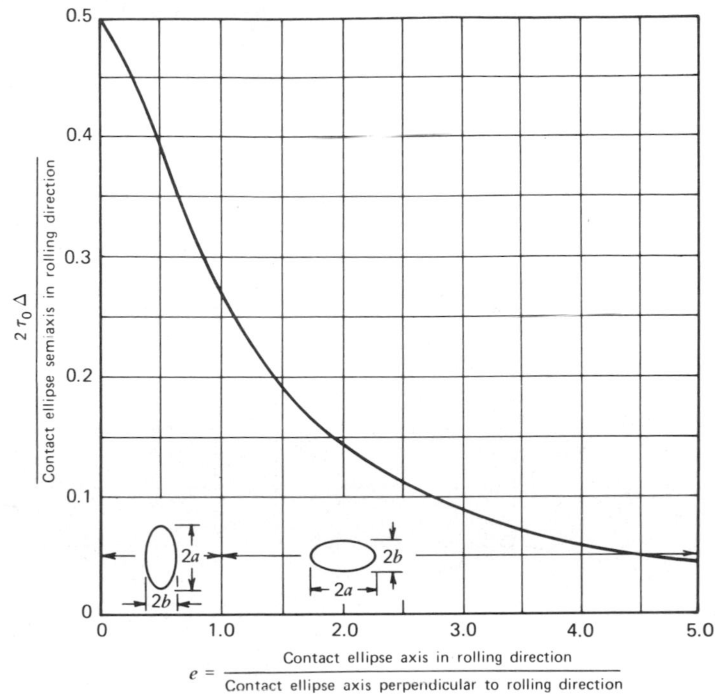

Fig. 14-6.5/Magnitude of maximum orthogonal shearing stress.

one value of b/a, that is, for one value of e, and assume that fatigue failure for another rolling contact stress problem having a different value of e will occur for the same number of cycles when τ_0 has the same magnitude. Since σ_{max} is generally calculated for all contact stress problems, it is not necessary that τ_0 be calculated if the ratio $2\tau_0/\sigma_{max}$ is plotted versus e (Fig. 14-6.6). The curve in Fig. 14-6.6 has been plotted for $\nu = 0.25$. Since σ_{max} depends upon ν while τ_0 does not, the curve in Fig. 14-6.6 is moved slightly for other values of ν. The ratios $2\tau_0/\tau_{max}$ or $2\tau_0/\tau_{oct(max)}$ versus e could also be plotted; however, only one plot is needed and σ_{max} is more often calculated than τ_{max} or $\tau_{oct(max)}$.

To illustrate the use of Fig. 14-6.6 in the analysis of rolling contact stress fatigue problems, let fatigue properties for rolling contact stress be determined using two cylinders that roll together. Since the major axis of the contact ellipse is large compared to the minor axis b, e is assumed to be equal to zero. Let σ_{max1} be the maximum principal stress in the cylinders when fatigue failure of the contact surface occurs after N cycles of load. From Fig. 14-6.6 we read $2\tau_0 = 0.50\sigma_{max1}$. Let the toroids in Fig. 14-6.2 be made of the same material as the cylinders. Let the radii of curvature be such that $e = 1.3$; for this value of e we read $2\tau_0 = 0.40\sigma_{max2}$ from Fig. 14-6.6. The magnitude of σ_{max2} such that fatigue failure of the toroids occurs after the same number of cycles N as the two cylinders is

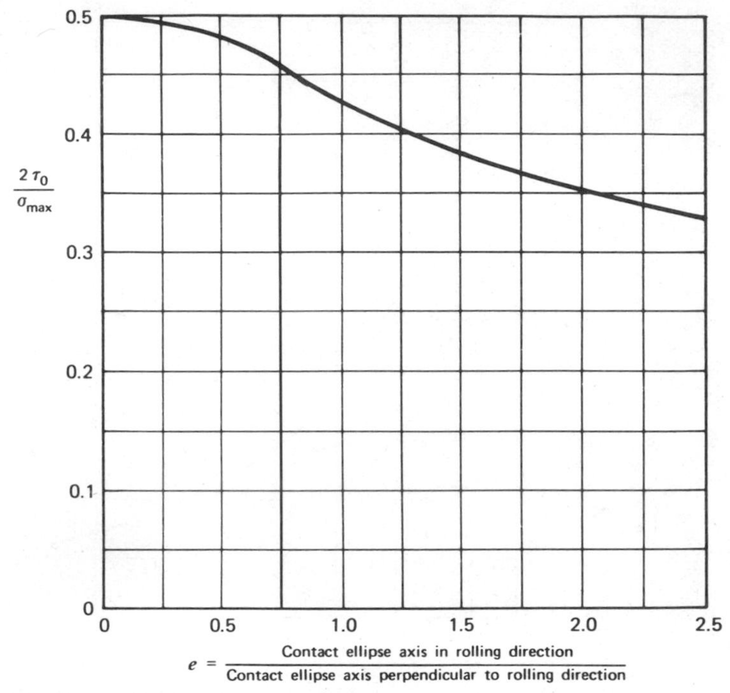

Fig. 14-6.6/Ratio of range of maximum orthogonal shearing stress to σ_{max} ($\nu = 0.25$).

obtained by setting the two values of $2\tau_0$ equal. Thus, we obtain $\sigma_{max2} = 1.25\sigma_{max1}$. This result is based on the assumption that the fatigue strength of the material is the same for both types of loading. Moyer and Morrow[4] indicate that the fatigue strength is not the same because of the size effect. Because of the larger volume of material under stress for the cylinders than for the toroids, the fatigue strength for the toroids should be larger than for the cylinders. Thus, the magnitude of $2\tau_0$ for the toroids should be larger than for the cylinders in order to produce fatigue failure in the same number of cycles. Our result that $\sigma_{max2} = 1.25\sigma_{max1}$ is therefore conservative. If fatigue properties for materials are obtained for $e = 0$, the use of Fig. 14-6.6 without correction for size effect predicts conservative results.

Curves for Computing Stresses for Any Value of B/A /The above example in which $B/A = 1.24$ ($k = 0.866$) shows that for a value of B/A or k a set of curves may be drawn representing the values of the principal stresses σ_{xx}, σ_{yy}, and σ_{zz} along the z-axis at small distances z beneath the surface of contact. These curves may be used to find the magnitude and location of the maximum shearing stress and the maximum octahedral shearing stress. Curves may also be constructed for a wide range of values of the ratio B/A. For each value of B/A the maximum values of stresses may be found from the equations

$$\sigma_{\max} = -c_\sigma\left(\frac{b}{\Delta}\right)$$
$$\tau_{\max} = c_\tau\left(\frac{b}{\Delta}\right) \qquad (14\text{-}6.7)$$
$$\tau_{\text{oct(max)}} = c_G\left(\frac{b}{\Delta}\right)$$

where the values of the coefficients c_σ, c_τ, and c_G may be read from the curves shown in Fig. 14-6.7. In the example in which $B/A = 1.24$ the values of the coefficients were given as $c_\sigma = 0.67$, $c_\tau = 0.22$, and $c_G = 0.21$. In Figs. 14-6.7 and 14-6.8, values of these coefficients for use in Eq. (14-6.7) are given as ordinates to the curves marked c_σ, c_τ, and c_G for a range of values B/A from 1 to 10,000. The values of k that are required in computing the semimajor axis a and the semiminor axis b of the area of contact are given as ordinates to the curved marked k. The value of b that may be computed by using Eq. (14-5.5) is found as follows. Equation

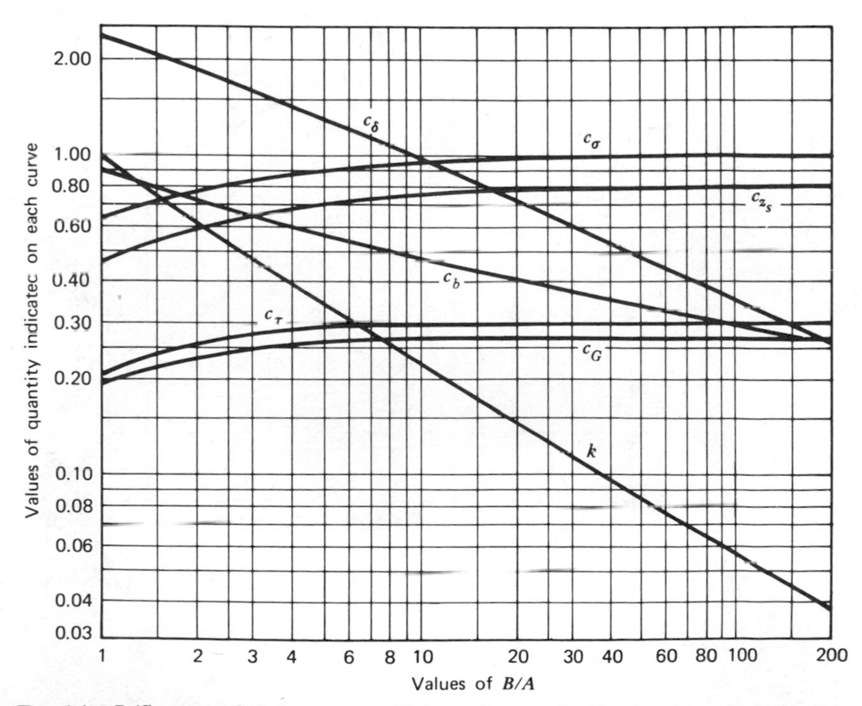

Fig. 14-6.7/Stress and deflection coefficients for two bodies in contact at a point.

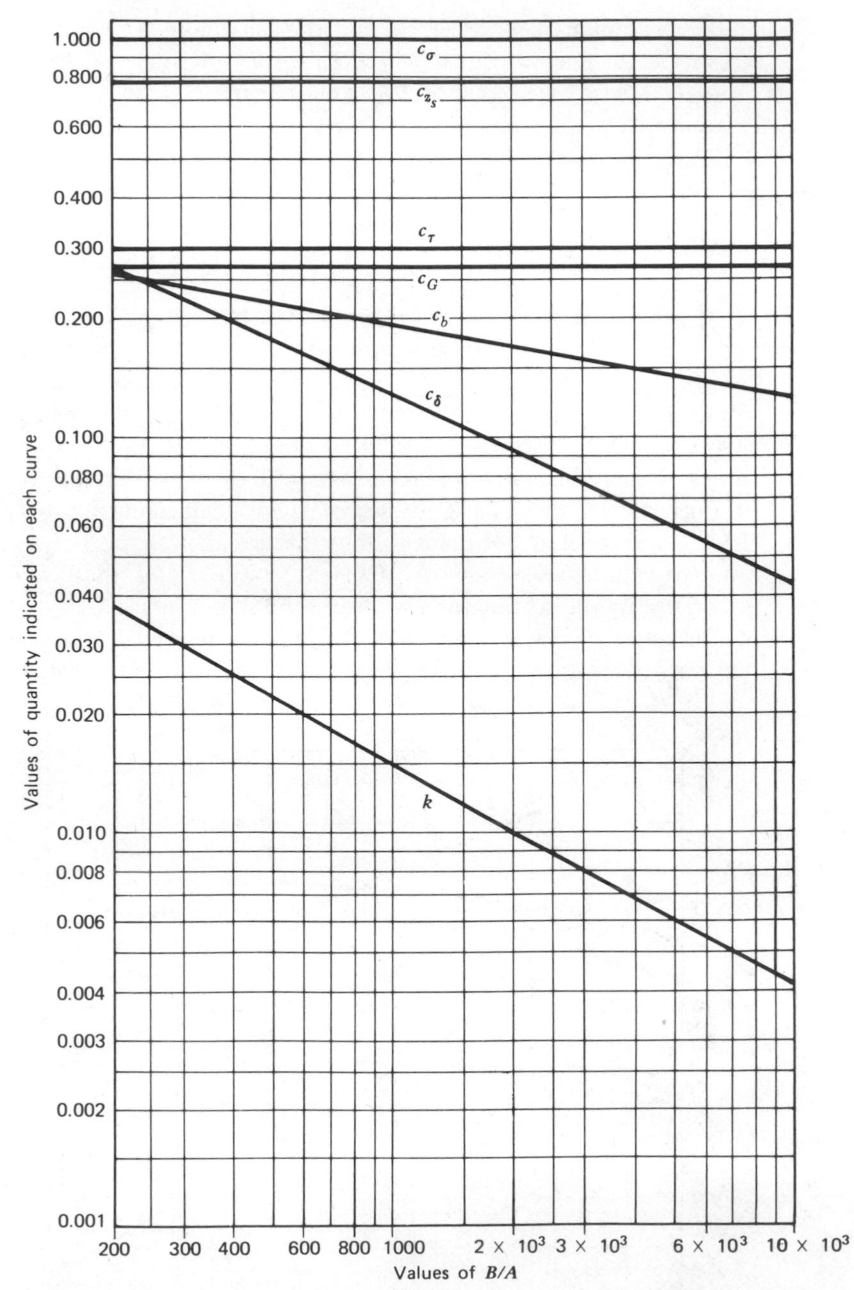

Fig. 14-6.8/Stress and deflection coefficients for two bodies in contact at a point.

(14-5.5) is rewritten as

$$b = c_b \sqrt[3]{P\Delta} \tag{14-6.8}$$

in which $c_b = \sqrt[3]{3kE(k')/2\pi}$. The values of k (and k') as found from the curve marked k are used to compute the value of the coefficient c_b. These values of c_b are given as ordinates to the curve marked c_b. The length of the semimajor axis is $a = b/k$. The depth z_s below the surface of contact to the location on the z-axis of the point at which the maximum stresses τ_{max} and $\tau_{oct(max)}$ occur is

$$z_s = c_{zs} b \tag{14-6.9}$$

The coefficient c_{zs} is plotted in Figs. 14-6.7 and 14-6.8. The examples following Art. 14-7 illustrate the use of Figs. 14-6.7 and 14-6.8.

14-7 DEFLECTION OF BODIES IN POINT CONTACT

An expression for the distance $\delta = \delta_1 + \delta_2$ through which two bodies in contact at a point move toward each other when acted upon by a load P (Fig. 14-3.1*c*) is given by Eq. (14-3.17). The deflection δ is sometimes called the *approach* because it expresses the sum of the "deflections" of the two bodies as they approach each other. The expression for the value of δ is given by the following equation due to Hertz[1]

$$\delta = \frac{3kPK(k')}{2\pi}\left(\frac{A+B}{b/\Delta}\right) \tag{14-7.1}$$

where P is the load, $K(k')$ is the complete elliptic integral described in Art. 14-4, and A, B, k, Δ, and b are defined in Arts. 14-4 and 14-5. In order to achieve a convenient use of Eq. (14-7.1), the substitution of

$$c_\delta = \frac{3kK(k')}{2} \tag{14-7.2}$$

is made to obtain

$$\delta = c_\delta \frac{P}{\pi}\left(\frac{A+B}{b/\Delta}\right) \tag{14-7.3}$$

In Eq. (14-7.2), the value c_δ depends only upon k (and k'), and since from Eq. (14-5.4) there is a value of B/A corresponding to each value of k, there is a value of c_δ corresponding to each value of B/A. In Figs. 14-6.7 and 14-6.8 values of this coefficient have been computed by Eq. (14-7.2) and plotted as ordinates for the curve marked c_δ. Equation (14-7.3) gives approximate results since the elastic strains in the two bodies away from the contact region are neglected.

EXAMPLE 14-7.1

Contact Stresses between Two Semicircular Disks

Let the two semicircular disks in Fig. 14-2.1 be made of steel ($E_1 = E_2 =$ 200 GPa and $\nu_1 = \nu_2 = 0.29$). The radii of curvature of the two surfaces at the point of contact are $R_1 = 60$ mm, $R_1' = 130$ mm, $R_2 = 80$ mm, and $R_2' = 200$ mm. The angle α between the planes of minimum curvature is $\pi/3$ rad. If the load $P = 4.50$ kN, determine the maximum principal stress, the maximum shearing stress, and the maximum octahedral shearing stress in the disks, and state the location of the point where each of these stresses occur. Determine the distance that the two disks move toward each other because of load P.

SOLUTION

All stress and displacement calculations require first that values be obtained for B, A, and Δ; these are given by Eqs. (14-3.15), (14-3.16), and (14-5.3a).

$$B = \frac{1}{4}\left[\frac{1}{60} + \frac{1}{80} + \frac{1}{130} + \frac{1}{200}\right] + \frac{1}{4}\left\{\left[\frac{1}{60} - \frac{1}{130} + \frac{1}{80} - \frac{1}{200}\right]^2 - 4\left[\frac{1}{60} - \frac{1}{130}\right]\left[\frac{1}{80} - \frac{1}{200}\right]\sin^2\frac{\pi}{3}\right\}^{1/2}$$

$$= 0.01255 \text{ mm}^{-1}$$

$$A = \frac{1}{4}\left[\frac{1}{60} + \frac{1}{80} + \frac{1}{130} + \frac{1}{200}\right] - \frac{1}{4}\left\{\left[\frac{1}{60} - \frac{1}{130} + \frac{1}{80} - \frac{1}{200}\right]^2 - 4\left[\frac{1}{60} - \frac{1}{130}\right]\left[\frac{1}{80} - \frac{1}{200}\right]\sin^2\frac{\pi}{3}\right\}^{1/2}$$

$$= 0.00838 \text{ mm}^{-1}$$

$$\Delta = \frac{2(1-\nu^2)}{(A+B)E} = \frac{2\left[1-(0.29)^2\right]}{(0.01255 + 0.00838)(200 \times 10^3)}$$

$$= 438 \times 10^{-6} \text{ mm}^3/\text{N}$$

$$\frac{B}{A} = 1.50$$

The coefficients needed to calculate b, σ_{max}, τ_{max}, $\tau_{oct(max)}$, z_s, and the deflection are read from Fig. 14-6.7 for the ratio $B/A = 1.50$. Hence, by Fig. 14-6.7 we obtain the following coefficients: $c_b = 0.77$, $c_\sigma = 0.72$, $c_\tau = 0.24$, $c_G = 0.22$, $c_{z_s} = 0.53$, and $c_\delta = 2.10$. For known values of c_b, P, and Δ, Eq. (14-6.8) gives

$$b = c_b \sqrt[3]{P\Delta} = 0.77\sqrt[3]{(4.5 \times 10^3)(438 \times 10^{-6})} = 0.965 \text{ mm}$$

from which

$$\frac{b}{\Delta} = \frac{0.965}{438 \times 10^{-6}} = 2203 \text{ MPa}$$

Values of σ_{max}, τ_{max}, $\tau_{oct(max)}$, z_s, and δ are obtained by substituting known values of the coefficients into Eqs. (14-6.7), (14-6.9), and (14-7.3).

$$\sigma_{max} = -c_\sigma \frac{b}{\Delta} = -0.72(2203) = -1586 \text{ MPa}$$

$$\tau_{max} = c_\tau \frac{b}{\Delta} = 0.24(2203) = 529 \text{ MPa}$$

$$\tau_{oct(max)} = c_G \frac{b}{\Delta} = 0.22(2203) = 485 \text{ MPa}$$

$$z_s = c_{z_s} b = 0.53(0.965) = 0.51 \text{ mm}$$

$$\delta = c_\delta \frac{P}{\pi}\left[\frac{A+B}{b/\Delta}\right] = \frac{2.10(4.5 \times 10^3)}{\pi}\left[\frac{0.01255 + 0.00838}{2203}\right]$$

$$= 0.029 \text{ mm}$$

The maximum Hertz stress $\sigma_{max} = -1586$ MPa occurs at the contact surface under the load. The maximum shearing stress and maximum octahedral shearing stress occurs at the depth $z_s = 0.51$ mm below the surface of contact; yielding would be expected to be initiated at this depth if the load is increased to the magnitude required to produce yielding.

EXAMPLE 14-7.2

Contact Stresses in a Steel Ball Bearing

A steel ball bearing consisting of an inner race, an outer race, and 12 balls is shown in Fig. E14-7.2 ($E = 200$ GPa, $\nu = 0.29$, and $Y = 1600$ MPa). A rated load of $P_0 = 4.2$ kN is given in a manufacturer's handbook for this bearing when operated at 3000 rpm. An empirical relation[6] is used to determine the load P on the topmost ball that bears the largest portion of the load; $P = 5P_0/n = 1.75$ kN in which n is the number of balls. (a) At the region of contact between the inner race and the topmost ball, determine the maximum principal stress, the maximum shearing stress, the maximum octahedral shearing stress, the dimensions of the area of contact, the maximum orthogonal shearing stress, and the distance from the point of contact to the point where these stresses occur.

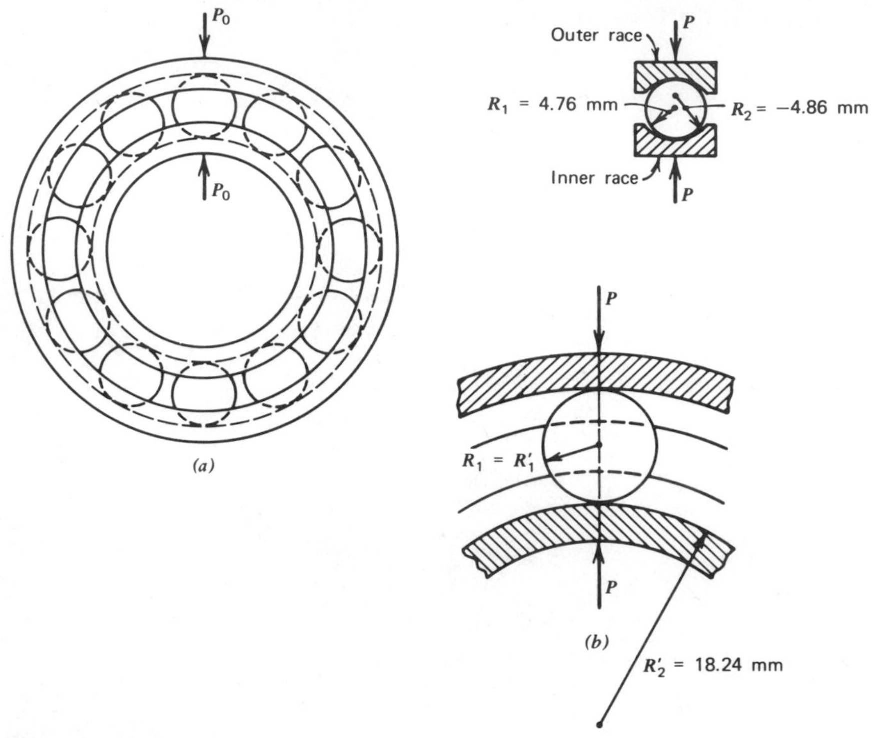

Fig. E14-7.2/Contact load in ball bearing.

(b) What is the factor of safety against initiation of yielding based on the octahedral shearing stress criterion of failure?

SOLUTION

(a) Let the ball be designated as body 1 and the inner race as body 2 so that $R_1 = R_1' = 4.76$ mm, $R_2 = -4.86$ mm, and $R_2' = 18.24$ mm. We substitute these values in Eqs. (14-3.15) and (14-3.16) to obtain values for A and B. The following results are obtained.

$$B = 0.13245 \text{ mm}^{-1}; \qquad A = 0.00216 \text{ mm}^{-1}; \qquad B/A = 61.3$$

$$\Delta = \frac{2}{A+B}\,\frac{1-\nu^2}{E} = \frac{2(1-0.29^2)}{(0.13246+0.00216)(200\times 10^3)}$$

$$= 68.0 \times 10^{-6} \text{ mm}^3/\text{N}$$

By Fig. 14-6.7, with $B/A = 61.3$, we obtain the following values for the coefficients: $c_b = 0.32$, $k = 0.075$, $c_\sigma = 1.00$, $c_\tau = 0.30$, $c_G = 0.27$, and

$c_{z_s} = 0.78$. Hence,

$$b = c_b \sqrt[3]{P\Delta} = 0.32\sqrt[3]{1.75 \times 10^3(68.0 \times 10^{-6})} = 0.1574 \text{ mm}$$

$$a = \frac{b}{k} = \frac{0.1574}{0.075} = 2.099 \text{ mm}$$

$$\frac{b}{\Delta} = \frac{0.1574}{68.0 \times 10^{-6}} = 2315 \text{ MPa}$$

$$\sigma_{\max} = -c_\sigma \frac{b}{\Delta} = -1.00(2315) = -2315 \text{ MPa}$$

$$\tau_{\max} = c_\tau \frac{b}{\Delta} = 0.30(2315) = 695 \text{ MPa}$$

$$\tau_{\text{oct(max)}} = c_G \frac{b}{\Delta} = 0.27(2315) = 625 \text{ MPa}$$

The maximum principal stress occurs under the load at the contact tangent plane. The maximum shearing stress and maximum octahedral shearing stress is located at a distance

$$z_s = c_{zs} b = 0.78(0.1574) = 0.123 \text{ mm}$$

from the contact tangent plane directly under the load. The magnitude of the maximum orthogonal shearing stress is obtained from Fig. 14-6.5. Since b is in the direction of rolling, $e = b/a = k = 0.075$. From Fig. 14-6.5, we obtain

$$\tau_0 = \frac{0.486b}{2\Delta} = \frac{0.486(0.1574)}{2(68.0 \times 10^{-6})} = 556 \text{ MPa}$$

The location of the maximum orthogonal shearing stress from the contact tangent plane is obtained from Fig. 14-6.4.

$$z_0 = 0.41b = 0.41(0.1574) = 0.065 \text{ mm}$$

(b) Since contact stresses are not linearly related to load P, the safety factor is not equal to the ratio of the maximum octahedral shearing stress in the specimen used to obtain material properties and the maximum octahedral shearing stress in the ball bearing. The magnitude of the yield load P_Y for a single ball may be obtained from the relation

$$(\tau_{\text{oct}})_Y = \frac{\sqrt{2}}{3} Y = \frac{\sqrt{2}}{3} 1600 = c_G \frac{b}{\Delta} = c_G c_b \frac{\sqrt[3]{P_Y \Delta}}{\Delta}$$

$$= 0.27(0.32) \frac{\sqrt[3]{P_Y(68.0 \times 10^{-6})}}{68.0 \times 10^{-6}}$$

from which

$$P_Y = 3076 \text{ N}$$

The safety factor SF is equal to the ratio of P_Y to P

$$SF = \frac{P_Y}{P} = \frac{3076}{1750} = 1.76$$

Significance of Stresses/In the preceding examples the magnitudes of the maximum principal stresses are quite large in comparison with values of this stress usually found in direct tension, bending, and torsion. In these problems, as in all contact stress problems, the three principal stresses at the point of maximum values are all compressive stresses. As a result, the maximum shearing stress and the maximum octahedral shearing stress are always less than one half the maximum principal stress; we recall that for a state of uniaxial stress (one principal stress) the maximum shearing stress is one half the principal stress. In fact by a comparison of the values of c_σ, c_τ, and c_G for various values of B/A in Figs. 14-6.7 and 14-6.8, we see that, when $B/A = 1$, $c_\tau = 0.32 c_\sigma$, and $c_G = 0.30 c_\sigma$ and, when $B/A = 100$ or larger, $c_\tau = 0.30 c_\sigma$, and $c_G = 0.27 c_\sigma$. Thus τ_{max} and $\tau_{oct(max)}$ are always slightly smaller than one third of the maximum principal stress σ_{max}. This fact is of special importance if the maximum shearing stress or the octahedral shearing stress is considered to be the cause of structural damage (failure) of the member; for, if the shearing stresses are relatively small in comparison to the maximum principal stress, very high principal stresses can occur. However, the maximum utilizable values of the maximum shearing stress or maximum octahedral shearing stress are not easily determined, because in many problems involving two bodies under pressure at a small area of contact, such as occurs in rolling bearings, there are additional factors that affect the behavior of the material, for example, sliding friction, the effect of a lubricant, the effect of repeated loads, the effect of variation in the metal near the surface of contact such as that due to case hardening, and the effects of metallurgical changes that often occur in such parts as the races of ball bearings due to the repeated stressing.

PROBLEM SET 14-7

1. A steel railway car wheel may be considered to be approximately a cylinder with a radius of 440 mm. The wheel rolls on a steel rail whose top surface may be considered to be approximately another cylinder with a radius of 330 mm. For the steel wheel and the steel rail, $E = 200$ GPa, $\nu = 0.29$, and $Y = 880$ MPa. If the wheel load is 110 kN, determine σ_{max}, τ_{max}, $\tau_{oct(max)}$, $2\tau_0$, and the factor of safety against initiation of yielding based upon the maximum shearing stress criterion.

2. Determine the vertical displacement of the center of the wheel in Problem 1 due to the deflections in the region of contact.

   Ans. $\delta = 0.116$ mm

3. Compute in terms of P (Newtons) the maximum principal stress, the maximum shearing stress, and the maximum octahedral shearing stress in two steel balls ($E = 200$ GPa and $\nu = 0.29$) 200 mm in diameter pressed together by a force P.

4. Solve Problem 3 for the condition that a single steel ball is pressed against a thick flat steel plate.

   Ans. $\sigma_{max} = -61\sqrt[3]{P}$ (MPa), $\tau_{max} = 20\sqrt[3]{P}$ (MPa), $\tau_{oct(max)} = 18\sqrt[3]{P}$ (MPa)

5. Solve Problem 3 for the condition that a single steel ball is pressed against the inside of a thick spherical steel race of inner radius 200 mm.

6. A feed roll (a device used to surface finish steel shafts) consists of two circular cylindrical steel rollers each 200 mm in diameter and arranged so that their longitudinal axes are parallel. A steel shaft (60 mm in diameter) is fed between the rollers in such a manner that its longitudinal axis is perpendicular to that of the rollers. The total load P between the shaft and the rollers is 4.5 kN. Determine the values of the maximum principal stress and the maximum shearing stress in the shaft. Determine the distance from the plane of contact to the point of maximum shearing stress, $E = 200$ GPa and $\nu = 0.29$.

   Ans. $\sigma_{max} = -1589$ MPa, $\tau_{max} = 517$ MPa, $z_s = 0.515$ mm

7. The longitudinal axes of the two feed rollers in Problem 6 are rotated in parallel planes until they form an angle of $\pi/6$ radians. The steel shaft is then fed between the two rollers at an angle of $\pi/12$ radians with respect to each of the rollers; again $P = 4.5$ kN. Determine the maximum principal stress, the maximum shearing stress and, the distance from the plane of contact to the maximum shearing stress.

8. A cast iron push rod ($E = 117$ GPa, $\nu = 0.20$) in a valve gear is operated by a steel cam ($E = 200$ GPa, $\nu = 0.29$) (Fig. P14-7.8). The cam is cylindrical in shape and has a radius of curvature of 5.00 mm at its tip. The surface of the push rod that contacts the cam is spherical in shape with a radius of curvature 4.00 m so that the rod and cam are in point contact. If the allowable maximum principal

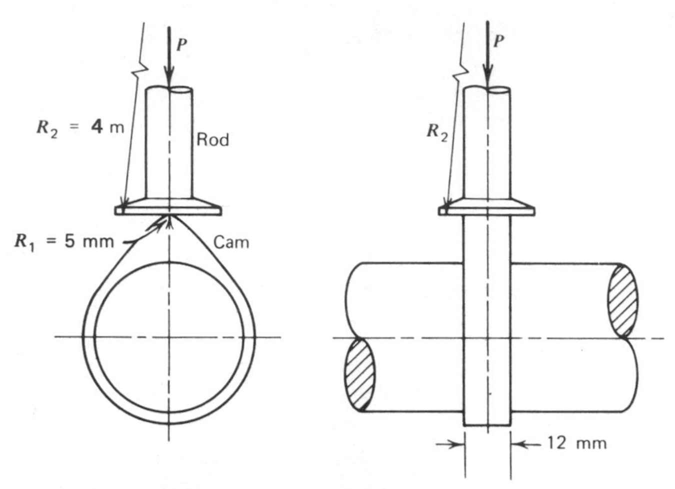

Fig. P14-7.8/Contact load in valve tappet.

stress for cast iron is -1400 MPa, determine the maximum load P that may act on the rod.

Ans. $P = 5.58$ kN

9. A fatigue testing machine to determine fatigue life under rolling contact consists of a steel toroid (body 2) rolling on a steel cylinder (body 1), where $R_1 = 32$ mm, $R_1' = \infty$, $R_2 = 32$ mm, and $R_2' = 20$ mm. For steel, $E = 200$ GPa, $\nu = 0.29$, and the Rockwell C hardness is 64.0. (a) Determine an expression for σ_{max} in terms of P. (b) Fatigue test results indicate that fatigue failure occurs at approximately $N = 10^9$ cycles with $\sigma_{max} = -2758$ MPa. Determine the applied load P (Newtons). Since $\alpha = 0$ and R_1 and R_2 lie in the x-z-plane (see Fig. 14-6.2), b (the minor semiaxis of the contact eclipse) is in the direction of rolling.

10. In the fatigue testing machine of Problem 9, the same cylinder is used, but the toroid is replaced by a second toroid, where $R_1 = \infty$, $R_1' = 32$ mm, $R_2 = 12.8$ mm, and $R_2' = 32$ mm. For the same steel properties as in Problem 9, (a) determine σ_{max} for fatigue failure at approximately $N = 10^9$ cycles. Neglect size effects and assume that fatigue failure is governed by the maximum range, $2\tau_0$, of the orthogonal shearing stress. (b) Determine the required load P. Since $\alpha = 0$ and R_1' and R_2' lie in the x-z-plane (see Fig. 14-6.2), a (the major semi-axis of the ellipse) lies in the direction of rolling.

Ans. $P = 2.60$ kN

11. A hard steel ball ($E = 200$ GPa, $\nu = 0.29$) of diameter 50 mm is pressed against a thick aluminum plate ($E = 72.0$ GPa, $\nu = 0.33$, and $Y = 450$ MPa). Determine the magnitude of load P_Y required to initiate yield in the aluminum plate according to the maximum octahedral shearing stress criterion of failure.

12. For a safety factor $SF = 1.75$, (a) recalculate the required load in Problem 11. (b) For this load, determine the displacement (approach) δ of the ball relative to the plate.

    Ans. $P = 196$ N, $\delta = 0.006$ mm

14-8
STRESS FOR TWO BODIES IN CONTACT OVER NARROW RECTANGULAR AREA (LINE CONTACT). LOADS NORMAL TO CONTACT AREA

If two cylindrical surfaces are in contact, the contact region is approximately along a straight line element before loads are applied. Figure 14-8.1 illustrates contact between two circular cylinders, the line of contact being perpendicular to the paper. Figure 14-8.1*b* also shows a line contact of a circular cylinder resting upon a plane. Figure 14-8.1*c* shows a line contact of a small circular cylinder resting inside a larger hollow cylinder. In these cases, the radii R_1' and R_2', which lie in a plane perpendicular to the paper, are each infinitely large so that $1/R_1'$ and $1/R_2'$ each vanish identically and the angle $\alpha = 0$ (Fig. 14-3.1*b*). Therefore, from Eqs. (14-3.15) and (14-3.16) the expressions for B and A are

$$B = \frac{1}{2}\left(\frac{1}{R_1} + \frac{1}{R_2}\right) \qquad A = 0 \qquad \frac{B}{A} = \infty$$

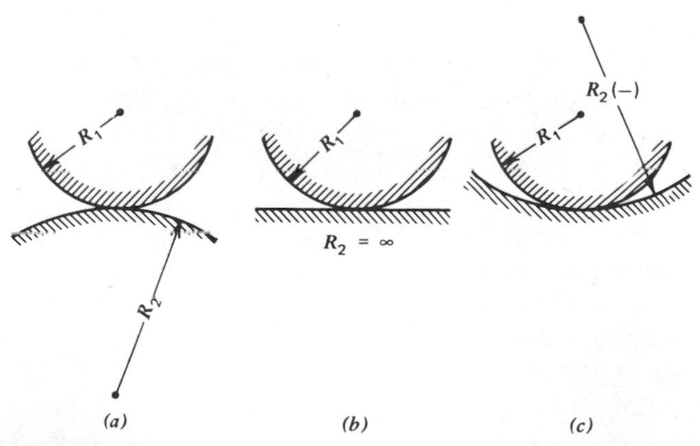

Fig. 14-8.1/Line contact between cylindrical bodies.

where R_1 and R_2 are the radii of curvature of the cylindrical surfaces (Fig. 14-8.1). Note that R_2 is negative in Fig. 14-8.1*c*. Hence the value of the ratio B/A is infinitely large, and from Eq. (14-5.4) the corresponding value of k approaches zero. But k is the ratio of the semiminor axis b of the area of contact to semimajor axis a, (Eq. 14-5.5) and, therefore, a must be infinitely large, which is the case of contact along a line between two bodies. The area of contact, when a distributed load of w Newtons per unit length is applied, is a long narrow rectangle of width $2b$ in the x-direction and length $2a$ in the y-direction. When $k=0$, Eqs. (14-5.1), (14-5.2), and (14-5.3) for the stresses at points on the z-axis at various distances z/b below the contact surface do not involve elliptic functions. In this case,

$$\sigma_{yy} = -2\nu\left[\sqrt{1+\left(\frac{z}{b}\right)^2} - \frac{z}{b}\right]\frac{b}{\Delta} \tag{14-8.1}$$

$$\sigma_{xx} = -\left[\frac{\left(\sqrt{1+(z/b)^2} - z/b\right)^2}{\sqrt{1+(z/b)^2}}\right]\frac{b}{\Delta} \tag{14-8.2}$$

$$\sigma_{zz} = -\left[\frac{1}{\sqrt{1+(z/b)^2}}\right]\frac{b}{\Delta} \tag{14-8.3}$$

The value of b from Eq. (14-5.5) for the limiting case in which $k=0$ is

$$b = \sqrt{2w\Delta/\pi} \tag{14-8.4}$$

in which w is the load per unit length of the contact area. The value of Δ is (Eq. 14-5.3a)

$$\Delta = \frac{1}{(1/2R_1)+(1/2R_2)}\left(\frac{1-\nu_1^2}{E_1} + \frac{1-\nu_2^2}{E_2}\right) \tag{14-8.5}$$

where R_1 and R_2 are the radii of curvature of the cylindrical surfaces as shown in Fig. 14-8.1. The values of the stresses at a point on the line of contact are obtained from Eqs. (14-8.1), (14-8.2), and (14-8.3) by setting $z=0$.

Maximum Principal Stresses: $k=0$ / It is seen from Eqs. (14-8.1), (14-8.2), and (14-8.3) that the principal stresses σ_{xx}, σ_{yy}, and σ_{zz} have their maximum numerical value when $z/b=0$, that is, at the surface of contact. These stresses are

$$\sigma_{xx} = -\frac{b}{\Delta} \qquad \sigma_{yy} = -2\nu\left(\frac{b}{\Delta}\right) \qquad \sigma_{zz} = -\frac{b}{\Delta} \tag{14-8.6}$$

Maximum Shearing Stress: $k = 0$/The shearing stress at any point on the z-axis is $\tau = \frac{1}{2}(\sigma_{xx} - \sigma_{zz})$. If the expressions for σ_{xx} and σ_{zz} from Eq. (14-8.2) and (14-8.3) are substituted in this equation for τ and the first derivative of τ with respect to z is equated to zero, the value of z (or z/b) found from the resulting equation is the distance below the contact surface at which the greatest value τ_{max} of the shearing stress occurs. The value thus found is $z_s/b = 0.7861$. At this point, the principal stresses are, from Eqs. (14-8.1), (14-8.2), and (14-8.3),

$$\sigma_{xx} = -0.1856\left(\frac{b}{\Delta}\right); \qquad \sigma_{yy} = -0.9718\frac{\nu b}{\Delta}; \qquad \sigma_{zz} = -0.7861\left(\frac{b}{\Delta}\right) \tag{14-8.7}$$

Hence,

$$\tau_{max} = \frac{1}{2}(\sigma_{xx} - \sigma_{zz}) = 0.300\left(\frac{b}{\Delta}\right) \tag{14-8.8}$$

At the depth $z_s/b = 0.7861$, σ_{xx} is smaller than σ_{yy} for values of ν greater than about 0.19.

Maximum Octahedral Shearing Stress: $k = 0$/The maximum octahedral shearing stress occurs at the same point as the maximum shear and is found by substituting the values of σ_{xx}, σ_{yy}, and σ_{zz} from Eq. (14-8.7) into Eq. (1-4.8). The result is

$$\tau_{oct(max)} = 0.27\frac{b}{\Delta} \tag{14-8.9}$$

We note that the coefficients for determining the quantities σ_{max}, τ_{max}, $\tau_{oct(max)}$ and z_s as obtained from Figs. 14 6.7 and 14 6.8 for values of B/A greater than about 50 are 1.00, 0.30, 0.27, and 0.78, respectively, and these are the same coefficients found for the case of line contact between two bodies. This fact means that when the ratio B/A is about 50 or larger the area of contact between the two bodies is very nearly a long narrow rectangle.

PROBLEM SET 14-8

1. A fatigue testing machine rolls together two identical steel disks ($E = 200$ GPa, $\nu = 0.29$), with radii 40 mm and thickness $h = 20$ mm. In terms of the applied load P, determine σ_{max}, τ_{max}, $\tau_{oct(max)}$, and τ_0.

2. Test data for the disks of Problem 1 indicate that fatigue failure occurs at approximately 10^8 cycles of load for $\sigma_{max} = -1380$ MPa. (a) Determine the corresponding value of load P. (b) For a fatigue

failure at 10^8 cycles and a factor of safety $SF = 2.50$, determine the value of σ_{max}.

Ans. (a) $P = 21.9$ kN, (b) $\sigma_{max} = -872$ MPa

3. The rail in Problem 14-7.1 wears in service until the top of the rail is flat with a width $h = 100$ mm. (a) For other conditions given in Problem 14-7.1 remaining constant, determine the values of σ_{max} and τ_{max}. (b) Using the maximum shearing stress criterion of failure, determine the safety factor SF against initiation of yield.

4. A cylindrical steel roller, with diameter 30 mm, is used as a follower on a steel cam. The surface of the cam at the contact region is cylindrical with radius of curvature 6 mm. Under no load, the follower and cam are in line contact over a length of 15 mm. For a value $\sigma_{max} = -1000$ MPa, determine the corresponding applied load P, ($E = 200$ GPa, $\nu = 0.29$).

   Ans. $P = 1.85$ kN

14-9 STRESSES FOR TWO BODIES IN LINE CONTACT. LOADS NORMAL AND TANGENT TO CONTACT AREA

In the preceding articles, the contact stresses in two elastic bodies held in contact by forces normal to the area of contact have been found. Frequently, the normal force is accompanied by a tangential (frictional) force in the contact area such as occurs when the teeth of spur gears come into contact or when a shaft rotates in a bearing. The frictional force that results from the sliding contact lies in the plane of the area of contact in a direction perpendicular to the normal force. The presence of frictional force causes the maximum values of the contact stresses in the two elastic bodies to become substantially larger than those produced by a normal force acting alone. Furthermore, the presence of a frictional force combined with a normal force causes certain changes in the nature of the stresses. For example, when a normal force acts alone, the three principal stresses are compressive stresses at every point in the body near the contact area, and this fact makes it difficult to understand how a crack can form and progressively spread to cause a separation type of failure such as occurs in pitting failures of some bearing surfaces. But, when a frictional force is introduced, two of the three principal stresses are changed into tensile stresses in the region immediately behind the frictional force (see Figs. 14-9.2*b* and 14-9.2*c*). If the coefficient of friction for the two surfaces of contact is sufficiently large, these tensile stresses are relatively large. However, if these tensile stresses are nominally small, as they probably are on well-lubricated surfaces, their values may be raised by stress concentration that results from surface irregularities or from small microscopic cracks that usually exist in the surfaces of real materials. These tensile stresses, when considered in conjunction with the many other factors involved, such as wear, nonhomogeneity of the material, and type of lubrication, help in explaining why a crack may develop and progressively spread in the surface of contact of such parts as gear teeth, roller bearings, etc.

The addition of a frictional force to a normal force on the contact surface also causes a change in the shearing stresses in the region of the contact surface. One important change is that the location of the point at which the maximum shearing stress occurs moves from beneath the surface of contact toward the contact area. In fact, when the coefficient of friction is $\frac{1}{10}$ or greater, this point is located in the contact surface. The foregoing remarks also apply to the maximum octahedral shearing stress.

The facts described above may be illustrated for an elastic cylindrical roller pressed against the plane surface of another elastic body.

Roller on Plane / Let Fig. 14-9.1a represent the cross section of a long roller of elastic material that rests upon a flat surface of a thick, solid elastic body.

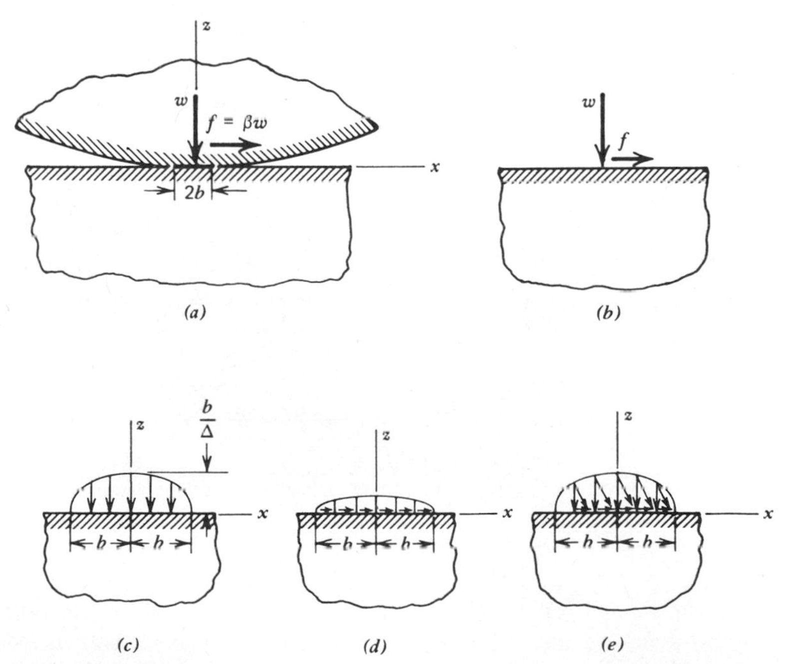

Fig. 14-9.1/Tangential (shearing) forces in addition to normal forces on the contact area.

The roller is subjected to a distributed load of w Newtons per unit length, which presses it against the body over a long narrow area of contact whose width is $2b$. A lateral distributed load of f Newtons per unit length causes the roller to slide on the body. If the coefficient of sliding friction is designated as β, then $f = \beta w$. In Fig. 14-9.1b, a part of the solid body is shown with the distributed loads w and f acting on the contact area. In Fig. 14-9.1c, which is an enlarged view of the part near the contact area, the ordinates to the ellipse show the distribution of normal stresses over this area, and the maximum stress is $\sigma_{zz} = -b/\Delta$ (Eq. 14-8.6). R. D. Mindlin[7] has found that when sliding occurs, the shearing stresses on the contact area due to the frictional force f are distributed as ordinates to an ellipse as shown in Fig. 14-9.1d, and the maximum shearing stress σ_{zx} at the center is $\sigma_{zx} = \beta(b/\Delta)$. Figure 14-9.1$e$ shows the distribution of the combined normal and friction stresses on the contact surface. C. K. Liu[8] has derived the equations for the stresses σ_{zz}, σ_{xx}, σ_{yy}, and σ_{zx} at any point in the body. These equations are

$$
\begin{aligned}
\sigma_{zz} &= -\frac{b}{\pi\Delta}\left[z(b\phi_1 - x\phi_2) + \beta z^2\phi_2\right] \\
\sigma_{xx} &= -\frac{b}{\pi\Delta}\left\{z\left(\frac{b^2+2z^2+2x^2}{b}\phi_1 - \frac{2\pi}{b} - 3x\phi_2\right)\right. \\
&\quad \left. + \beta\left[(2x^2 - 2b^2 - 3z^2)\phi_2 + \frac{2\pi x}{b} + 2(b^2 - x^2 - z^2)\frac{x}{b}\phi_1\right]\right\} \\
\sigma_{yy} &= -\frac{2\nu b}{\pi\Delta}\left\{z\left(\frac{b^2+x^2+z^2}{b}\phi_1 - \frac{\pi}{b} - 2x\phi_2\right)\right. \\
&\quad \left. + \beta\left[(x^2 - b^2 - z^2)\phi_2 + \frac{\pi x}{b} + (b^2 - x^2 - z^2)\frac{x}{b}\phi_1\right]\right\} \\
\sigma_{zx} &= -\frac{b}{\pi\Delta}\left\{z^2\phi_2 + \beta\left[(b^2 + 2x^2 + 2z^2)\frac{z}{b}\phi_1 - 2\pi\frac{z}{b} - 3xz\phi_2\right]\right\}
\end{aligned}
\tag{14-9.1}
$$

where ϕ_1, ϕ_2 are

$$\phi_1 = \frac{\pi(M+N)}{MN\sqrt{2MN + 2x^2 + 2z^2 - 2b^2}}$$

$$\phi_2 = \frac{\pi(M-N)}{MN\sqrt{2MN + 2x^2 + 2z^2 - 2b^2}}$$

where $M = \sqrt{(b+x)^2 + z^2}$ and $N = \sqrt{(b-x)^2 + z^2}$. The values of stress as given by Eq. (14-9.1) do not depend on y because it is assumed that either a state of plane strain or of plane stress exists relative to the (x, z) plane.

Principal Stresses / In Eq. (14-9.1) σ_{yy} is a principal stress, say σ_3, but σ_{zz} and σ_{xx} are not principal stresses because of the presence of the shearing stress σ_{zx} that acts on these planes. Let the other two principal stresses at any point be designated by σ_1 and σ_2. These two stresses may be found from plane theory (Art. 1-4) with the values of σ_{zz}, σ_{xx}, and σ_{zx} for the point. The principal stresses σ_1, σ_2, and σ_3 for points on the surface† and at a distance $z = b/4$ underneath the surface have been computed by this theory for a value of friction coefficient of $\frac{1}{3}$, and their values have been plotted in Figs. 14-9.2*a*, *b*, and *c*. Each principal stress has its maximum value in the surface of the body at a distance of about $0.3b$ from the center of the area of contact in the direction of the frictional force. These maximum values, all of which occur at the same point, are $\sigma_1 = -1.4b/\Delta$, $\sigma_2 = -0.72b/\Delta$, and $\sigma_3 = -0.53b/\Delta$. These values may be compared with $\sigma_1 = -b/\Delta$, $\sigma_2 = -b/\Delta$, and $\sigma_3 = -0.5b/\Delta$, as found from Eq. (14-8.6) for the normal distributed load w only. This comparison shows that the frictional force corresponding to a coefficient of friction of $\frac{1}{3}$ increases the maximum principal stress by 40 percent. Furthermore, the curves in Fig. 14-9.2 show that the principal stresses σ_2 and σ_3 are tension stresses near the edge of the contact area opposite to the direction of the frictional force. The largest magnitudes of these stresses are $\frac{2}{3}(b/\Delta)$ and $\frac{1}{6}(b/\Delta)$, respectively, but these values are sometimes quite large. The presence of the tensile stresses in the surface aids in understanding the occurrence of fatigue failure by pitting, etc., of bearing surfaces subjected to repeated loads.

Maximum Shearing Stress / From the values of maximum and minimum principal stresses at a point in the surface of contact, the maximum shearing stress at the point on the surface is found to be

$$\tau_{\max} = \frac{1}{2}\left(-\frac{1.4b}{\Delta} + \frac{0.53b}{\Delta}\right) = -0.43\left(\frac{b}{\Delta}\right) \qquad (14\text{-}9.2)$$

To determine whether or not this value of the shearing stress is the maximum value occurring in the body, it is necessary to compute the maximum shearing stress at all other points, and especially at points inside the body under the contact area, since in all previous results presented in this chapter the maximum shearing stress was found to be a subsurface shearing stress. The values of shearing stress at points on the surface and at points below the surface a distance of $z = b/4$ (where the maximum subsurface shear occurs) have been computed by making use of the principal stresses in Fig. 14-9.2 and are represented as ordinates to the curves in Figs. 14-9.3*a*, *b*, and *c*. There are three extreme values of

†A special method of evaluating Eq. (14-9.1) may be used when solving for the stresses on the surface where $z = 0$. (See reference 8.)

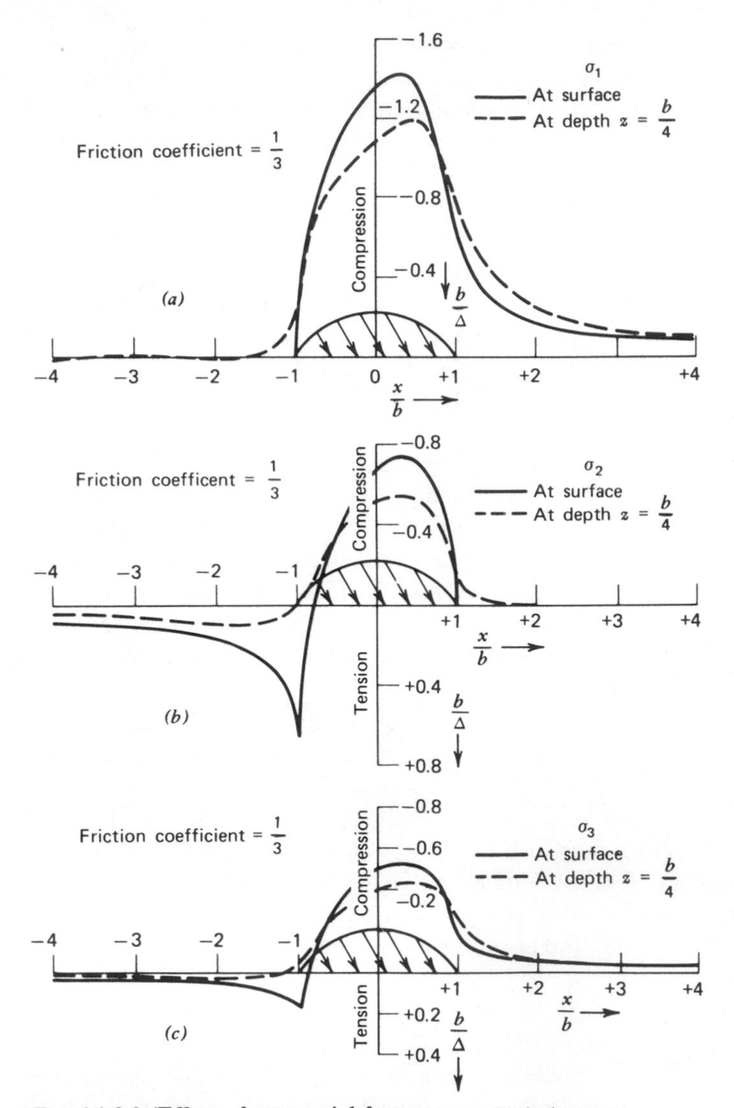

Fig. 14-9.2/Effect of tangential force on contact stresses.

shearing stresses at each point; they are

$$
\begin{aligned}
\tau_1 &= \tfrac{1}{2}(\sigma_1 - \sigma_3) \\
\tau_2 &= \tfrac{1}{2}(\sigma_1 - \sigma_2) \qquad (14\text{-}9.3) \\
\tau_3 &= \tfrac{1}{2}(\sigma_2 - \sigma_3)
\end{aligned}
$$

From Figs. 14-9.3*a* and *c*, we see that the ordinates to the curves representing τ_1 and τ_3 at a depth $z = b/4$ underneath the surface are everywhere smaller than at the surface. This fact is true of the curves for

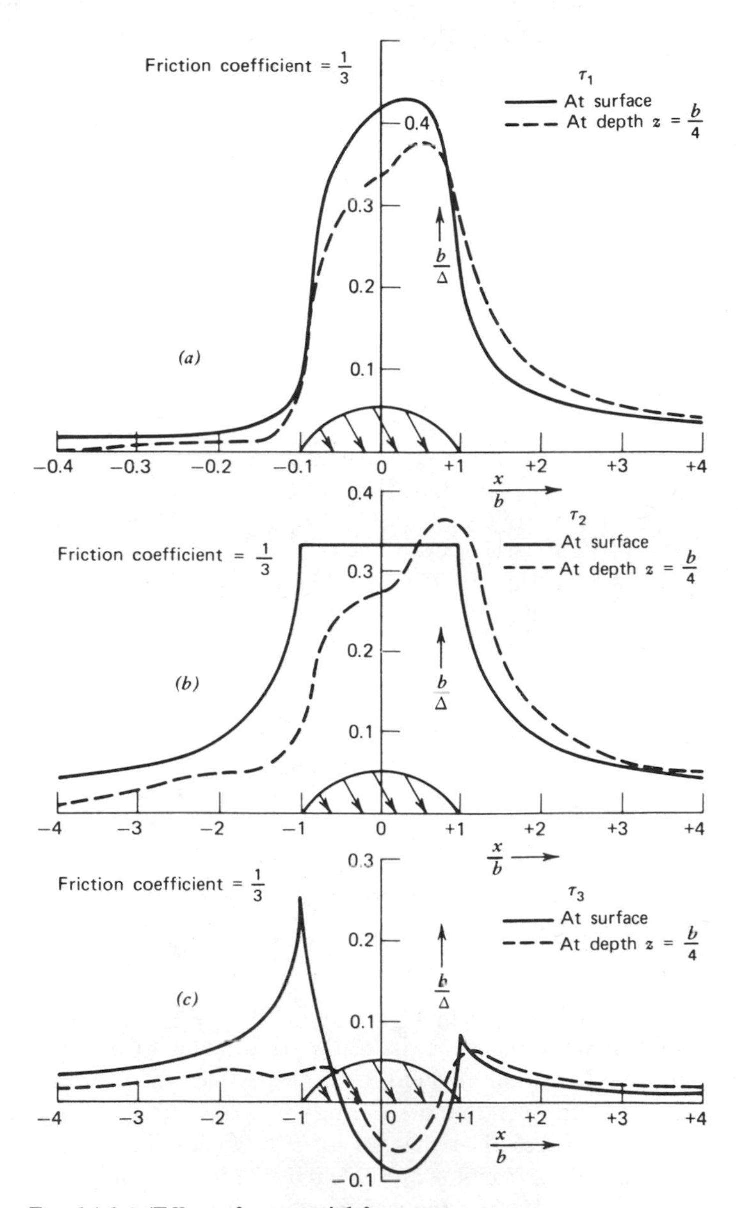

Fig. 14-9.3/Effect of tangential force on contact stresses.

these values at all depths. However, in Fig. 14-9.3*b*, the curve for τ_2 at $z = b/4$ rises above the curve representing values of τ_2 at the surface. Such curves for values of τ_2 have been plotted for several different depths, and it is found that the largest value of τ_2 is $0.36b/\Delta$. This value occurs at a depth of about $b/4$ below the surface. Therefore, the value of $\tau_1 = 0.43b/\Delta$ as given by Eq. (14-9.3) is the maximum shearing stress,

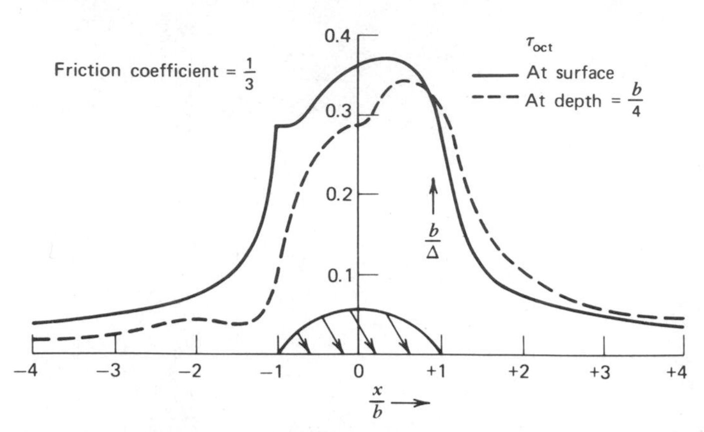

Fig. 14-9.4/Effect of tangential force on contact stresses.

and it occurs at a point in the contact area about $0.3b$ from the center of the area. In Eq. (14-9.3) the maximum value of τ_2, which always occurs below the surface, does not exceed τ_1 until the coefficient of friction has a value less than $\frac{1}{10}$.

Maximum Octahedral Shearing Stress/In Fig. 14-9.4 the ordinates to the curves represent the values of the octahedral shearing stresses τ_{oct} that have been computed at each point from Eq. (1-4.8) by substitution of values of the principal stresses obtained from Fig. 14-9.2. The maximum value is $\tau_{oct(max)} = 0.37b/\Delta$, and this value occurs in the contact area at the same point that the maximum principal stress and maximum shearing stresses occur (Figs. 14-9.2 and 14-9.3).

Effect of Magnitude of Friction Coefficient/The magnitude of the coefficient of friction determines the size of the frictional distributed load f for a given value of w and, therefore, of the values of the maximum principal stresses, the maximum shearing stresses, and the maximum octahedral shearing stress. The changes in the maximum contact stresses with the coefficient of friction are given by Table 14-9.1. The increases in the maximum values of the tensile and compressive principal stresses caused by the frictional distributed load are very nearly proportional to the increases in the friction coefficient. For small values of the friction coefficient, the values of shearing stress are increased only slightly by an increase in the friction coefficient, whereas there is a small decrease in octahedral shear up to a friction coefficient of $\frac{1}{6}$. For the case of a disk in rolling contact with a cylinder and for $\beta = \frac{1}{3}$, Mitsuda[9] has noted that the range of maximum octahedral shear stress is 26 percent larger than the range of maximum orthogonal shear stress.

Table 14-9.1

Values of Contact Stresses between Two Long Cylindrical Bodies Sliding against Each Other While in Line Contact (Normal and Friction Forces)

Coefficient of Friction	*0*	$\frac{1}{12}$	$\frac{1}{9}$	$\frac{1}{6}$	$\frac{1}{3}$
Kind of Stress and Its Location	*Values of Stress in Terms of b/Δ Corresponding to the Above Friction Coefficients*				
Maximum tensile principal stress that occurs in surface at $x = -b$	0	$\frac{2}{12}\frac{b}{\Delta}$	$\frac{2}{9}\frac{b}{\Delta}$	$\frac{2}{6}\frac{b}{\Delta}$	$\frac{2}{3}\frac{b}{\Delta}$
Maximum compressive principal stress that occurs in the surface between $x=0$ and $x=0.3b$	$-\frac{b}{\Delta}$	$-1.09\frac{b}{\Delta}$	$-1.13\frac{b}{\Delta}$	$-1.19\frac{b}{\Delta}$	$-1.40\frac{b}{\Delta}$
Maximum shearing stress†	$0.300\frac{b}{\Delta}$	$0.308\frac{b}{\Delta}$	$0.310\frac{b}{\Delta}$	$0.339\frac{b}{\Delta}$	$0.435\frac{b}{\Delta}$
Maximum octahedral shearing stress†	$0.272\frac{b}{\Delta}$	$0.265\frac{b}{\Delta}$	$0.255\frac{b}{\Delta}$	$0.277\frac{b}{\Delta}$	$0.368\frac{b}{\Delta}$

† Note that these stresses occur at the surface when the friction coefficient is $\frac{1}{10}$ or larger.

Range of Shearing Stress for One Load Cycle/The *magnitude* of the maximum shearing stress or maximum octahedral shearing stress serves to indicate if yielding has taken place or to determine the factor of safety against impending yielding. However, in the case of fatigue loading, the *range* of shearing stress on a given plane in a given direction in that plane is more commonly used to indicate the severity of a given loading. In the absence of friction, the *range* of the orthogonal shearing stress for two cylinders in rolling contact is $2\tau_0 = 0.50b/\Delta(\tau_0 = \sigma_{zx(\max)})$; this range is greater than either the corresponding range of maximum shearing stress $(0.30b/\Delta)$ or the corresponding range of octahedral shearing stress $(0.27b/\Delta)$.

The presence of friction has little influence on the range of the orthogonal shearing stress. However, the ranges of shearing stress on certain planes increases with the coefficient of friction; the maximum range is as large as $0.67b/\Delta$ for $\beta = \frac{1}{3}$. Furthermore, this range occurs for points at the free surface of the contacting bodies where fatigue failures are more likely to be initiated.

For rolling cylinders in contact, Smith and Liu[8] have determined the principal state of stress for a point in the surface of one of the rolling cylinders for the special case that the coefficient of friction was $\beta = \frac{1}{3}$. The principal stresses are indicated in Table 14-9.2 for a volume element located at a point 0 for several locations of point 0 relative to the contact surface. For each location of point 0, the shearing stress and the octahedral shearing stress attain maximum values on certain planes passing through point 0.

The largest values of these maxima (τ_{max} and $\tau_{oct(max)}$) occur for the volume element located at point 0 in Fig. (*D*) of Table 14-9.2. Smith and Liu noted the two sets of planes on which the largest values of τ_{max} and

Table 14-9.2

Principal Stresses at Fixed Point 0 as Contact Surface Moves Relative to 0

	Position of Contact Surface Relative to Fixed Point 0	*Direction of Principle Stresses of Fixed Point 0*
(*A*)	z; 0; x; b, b; $10b$	$\sigma_3 = 0$; $\sigma_1 \cong 0$, $\sigma_2 \cong 0$
(*B*)	0; x; b	$\sigma_3 = 0$; $\sigma_1 = -0.67\frac{b}{\Delta}$, $\sigma_2 = -0.17\frac{b}{\Delta}$
(*C*)	0; x; $0.55b$	28°; $\sigma_2 = -69\frac{b}{\Delta}$; $\sigma_1 = -1.33\frac{b}{\Delta}$; $\sigma_3 = -0.51\frac{b}{\Delta}$
(*D*)	0; x; $0.306b$	36°; $\sigma_2 \doteq -0.72\frac{b}{\Delta}$; $\sigma_1 = -1.39\frac{b}{\Delta}$; $\sigma_3 = -0.53\frac{b}{\Delta}$
(*E*)	0; x	45; $\sigma_2 = -0.67\frac{b}{\Delta}$; $\sigma_1 = -1.33\frac{b}{\Delta}$; $\sigma_3 = -0.50\frac{b}{\Delta}$
(*F*)	0; x; $-0.50b$	60°; $\sigma_3 = -0.36\frac{b}{\Delta}$; $\sigma_1 = -1.04\frac{b}{\Delta}$; $\sigma_2 = -0.37\frac{b}{\Delta}$
(*G*)	0; x; $-0.83b$	73°; $\sigma_3 = 0.05\frac{b}{\Delta}$; $\sigma_1 = -0.61\frac{b}{\Delta}$; $\sigma_2 = -0.14\frac{b}{\Delta}$
(*H*)	0; x; $-b$	$\sigma_3 = 0$; $\sigma_1 = 0.67\frac{b}{\Delta}$, $\sigma_2 = 0.17\frac{b}{\Delta}$
(*I*)	0; x; $-10b$	$\sigma_3 = 0$; $\sigma_1 = 0$, $\sigma_2 = 0$

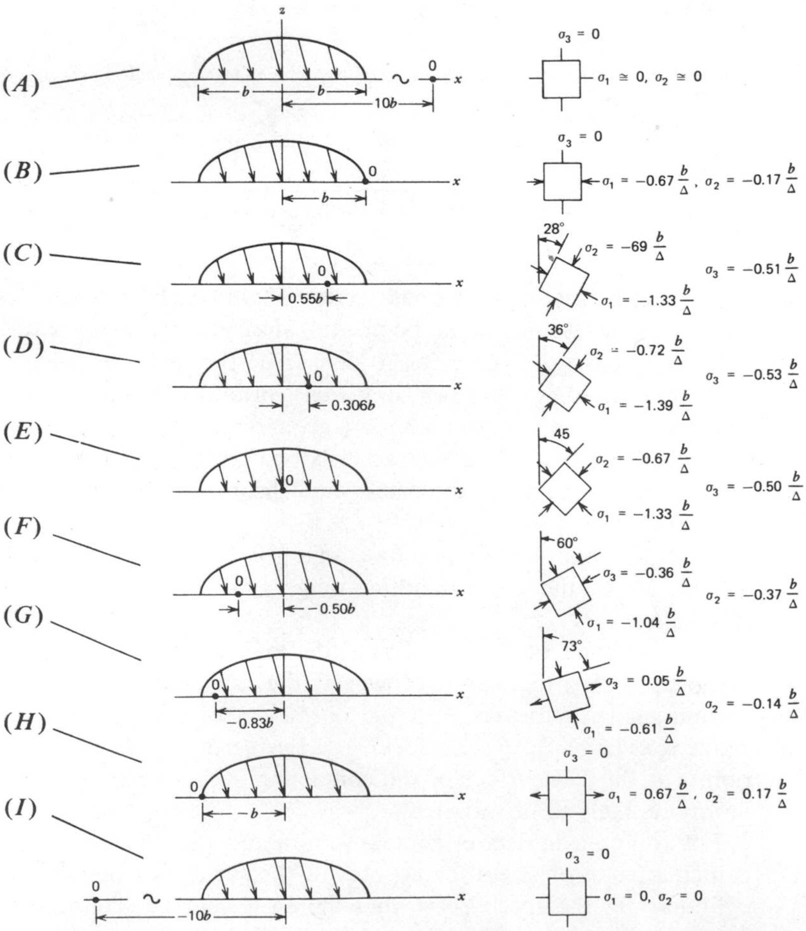

$\tau_{oct(max)}$ occur and determined the magnitude and sense of the shearing stress acting on these planes for various locations of point 0. They also defined the maximum range of shearing stress on either set of planes to be the magnitude of the maximum diameter of the shearing stress envelope. (These maximum ranges were found to be $0.53b/\Delta$ and $0.63b/\Delta$.)[8]

Note that in Figs. (B) of Table 14-9.2 the maximum shearing stress for these two states of stress occur on the same planes and they are opposite in sign; the range of shearing stress for these planes is $0.67b/\Delta$. Only four sets of the infinite number of planes through point 0 (including planes perpendicular to the x- and z-axes on which $\tau_0 = \sigma_{zx(max)}$ occur) were investigated; therefore, the range of shearing stress ($0.67b/\Delta$) may not be the largest that occurs. However, the results do indicate that a tangential component of force at the contact surface increases the probability of fatigue failure particularly if the coefficient of friction approaches a value of $\frac{1}{3}$.

EXAMPLE 14-9.1

Contact Stress in Cylinders with Friction

The fatigue testing machine described in Problem 14-8.1 has two identical steel disks ($E = 200$ GPa and $\nu = 0.29$) rolling together. The identical disks have a radius of curvature of 40 mm and a width $h = 20$ mm. For rolling without friction, a load $P = 24.1$ kN produces the following stresses: $\sigma_{max} = 1445$ MPa, $\tau_{max} = 433$ MPa, and $\tau_{oct(max)} = 361$ MPa. Let the cylinder be subjected to a load $P = 24.1$ kN and be rotated at slightly different speeds so that the roller surfaces slide across each other. If the coefficient of sliding friction is $\frac{1}{9}$, determine σ_{max}(tension), σ_{max}(compression), τ_{max}, and $\tau_{oct(max)}$.

SOLUTION

From Table 14-9.1 the value of the stresses are found as follows:

$$\sigma_{max}(\text{tension}) = \frac{2}{9}\frac{b}{\Delta}$$

$$\sigma_{max}(\text{compression}) = -1.13\frac{b}{\Delta}$$

$$\tau_{max} = 0.310\frac{b}{\Delta}$$

$$\tau_{oct(max)} = 0.255\frac{b}{\Delta}$$

The magnitude of Δ and b are given by Eqs. (14-8.5) and (14-8.4).

$$\Delta = 2R\left(\frac{1-\nu^2}{E}\right) = \frac{2(40)(1-0.29^2)}{200 \times 10^3} = 0.0003664$$

$$b = \sqrt{\frac{2P\Delta}{h\pi}} = \sqrt{\frac{2(24.1 \times 10^3)(0.0003664)}{20\pi}} = 0.5301 \text{ mm}$$

$$\frac{b}{\Delta} = 1447 \text{ MPa}$$

Therefore, we have the following results:

$$\sigma_{max}(\text{tension}) = \frac{2}{9}(1447) = 322 \text{ MPa}$$

$$\sigma_{max}(\text{compression}) = -1.13(1447) = -1635 \text{ MPa}$$

$$\tau_{max} = 0.310(1447) = 449 \text{ MPa}$$

$$\tau_{oct(max)} = 0.255(1447) = 369 \text{ MPa}$$

The friction force (coefficient of sliding friction is $\frac{1}{9}$) increases the maximum compression stress by 13.1 percent, increases the maximum shearing stress by 3.7 percent, and increases the maximum octahedral shearing stress by 2.2 percent.

PROBLEM SET 14-9

1. Two cylindrical steel rollers ($E = 200$ MPa and $\nu = 0.29$) each 80 mm in diameter and 150 mm long are mounted on parallel shafts and are loaded by a force $P = 80$ kN. The two cylinders are rotated at slightly different speeds so that the roller surfaces slide across each other. If the coefficient of sliding friction is $\beta = \frac{1}{3}$, determine the maximum compressive principal stress, the maximum shearing stress, and the maximum octahedral shearing stress.

2. The two cylinders in Problem 1 are hardened. It is found that fatigue failures occur in the cylinders after 10^9 cycles for $\sigma_{max} = -1500$ MPa when $\beta = 0$. (a) Determine the load P that can be applied to the cylinders to cause fatigue failure after 10^9 cycles ($\beta = 0$). (b) Determine the load P that can be applied to the cylinders to cause fatigue failure at approximately 10^9 cycles for $\beta = \frac{1}{3}$. Assume that σ_{max} to cause fatigue failure is inversely proportional to the range of shearing stress on a given plane and that the maximum range of shearing stress is $0.67b/\Delta$ for $\beta = \frac{1}{3}$.

   Ans. (a) $P = 194$ kN, (b) $P = 55.2$ kN

REFERENCES

1. H. Hertz, *Gesammetlte Werke*, Vol. 1, Lepzig, 1895. For an English translation, see H. Hertz, *Miscellaneous Papers*, Macmillan, New York, 1896.
2. S. Timoshenko and J. Goodier, *Theory of Elasticity*, 3rd Ed., McGraw-Hill, New York, 1970.
3. H. R. Thomas and V. A. Hoersch, "Stresses Due to the Pressure of One Elastic Solid on Another," *Bulletin* 212, Engineering Experiment Station, University of Illinois, Urbana, June 15, 1930.
4. G. J. Moyar and JoDean Morrow, "Surface Failure of Bearings and Other Roller Elements," *Bulletin* 468, Engineering Experiment Station, University of Illinois, Urbana, 1964.
5. H. Fessler and E. Ollerton, "Contact Stresses in Toroids under Radial Loads," *British Journal of Applied Physics*, Vol. 8, No. 10, October 1957, p. 387.
6. R. K. Allen, *Rolling Bearings*, Pitman, London, 1945.
7. R. D. Mindlin, "Compliance of Elastic Bodies in Contact," *Journal of Applied Mechanics*, Vol. 16, No. 3, September 1949, p. 259.
8. J. O. Smith and C. K. Liu, "Stresses due to Tangential and Normal Loads on an Elastic Solid with Applications to Some Contact Stress Problems," *Journal of Applied Mechanics*, Vol. 20, No. 2, June 1953, p. 157.
9. T. Mitsuda, "An Investigation of Pitting and Shelling Failure in Rolling Contact," *Ph. D. Thesis*, Department of Theoretical and Applied Mechanics, University of Illinois, Urbana, 1965, p. 16 and Table 1.

CHAPTER 15

FINITE ELEMENT METHODS

15-1 INTRODUCTION

Finite element methods, in a narrow sense, may be considered as modifications of the Rayleigh-Ritz method in the calculus of variations in that trial functions or approximating functions are assumed for the unknown functions to be determined. The constants appearing in the trial functions, in turn, are evaluated by some method, say, the Rayleigh-Ritz method, which requires that the potential energy of the system be stationary. Ordinarily, in finite element methods, the physical body or the region of interest is subdivided into an assemblage of a finite number of individual, one-, two-, or three-dimensional subregions or elements interconnected at a number of specified nodal points. To each of these subregions or elements, trial functions for the unknown quantities of interest are defined in a piecewise manner over the assemblage. The constants appearing in the trial functions represent approximate values of the exact solution at the nodes of the elements. Various techniques may be employed to determine the constants; for example, the Rayleigh-Ritz method,[1] the partition method,[2] collocation, least squares,[3] or, in general, the method of weighted residuals.[4,4a] In particular, in structural mechanics, the Rayleigh-Ritz method is employed to develop the finite element approximation of structures.

As noted above, the essence of finite element methods lies in the idealization of a physical continuous medium by an assemblage of discrete elements and localized trial functions. These methods make it possible for analysts to model a continuous medium with its governing differential equations by a corresponding discretized system with much simpler algebraic (matrix) equations. (For a philosophical discussion on the discrete methods of structural mechanics, see reference 5.) Finite element methods have been employed widely in structural problems, as well as nonstructural problems of heat transfer, fluid mechanics, and two- and three-dimensional studies in continuum mechanics. In the following, we direct our attention mainly to finite element methods in solid mechanics.

The idea of representing an elastic continuum by an assemblage of a number of component parts has been used by structural engineers for the analysis of continuous beams and rigid frames.[6] An extension of this approach to two- or three-dimensional elasticity problems is the so-called *lumped-parameter* method where multidimensional elastic bodies are modeled by combinations of members or components, such as elastic bars, strings, dashpots, point masses, and so on.[7,8] The lumped-parameter method requires considerable ingenuity and experience, particularly in accurately modeling a body by springs, bars, masses, etc.

A generalization of the discrete method of structural mechanics, the finite element method,[9,9a] employs two- or three-dimensional continuous discrete elements, whose elastic force displacement relations are obtained from energy considerations. This approach eliminates the modeling of the body by assemblages of idealized component parts as in the lumped-parameter method. Furthermore, it leads to the representation of elastic behavior of a physical continuous medium in a fairly routine manner. The formal matrix formulation of the finite element method was propounded by Argyris,[10] but the basic foundations of the method can be traced back to the work of Courant,[11] Synge,[12] and Turner et al.[13] The use of matrix formulations in the development of the method (although theoretically unnecessary) renders an elegantly clear representation of the discretized system, and facilitates automatic computation by high speed electronic computers.

In the application of finite element methods to solid elastic bodies, after replacement of the physical solid by an assemblage of discrete elements, it is necessary to evaluate element properties (stiffness or flexibility) by the use of trial functions, which are approximations of the unknown quantities to be determined. For the formulation of element force-displacement relations, two major distinct approaches have been employed; namely, the force method and the displacement method. In the force (or flexibility) method, stress fields are approximated by trial functions, and the principle of minimum complementary energy (Art. 4-2) is employed in the formulation. In the displacement (or stiffness) method, the displacement fields are approximated by trial functions, and the principle of stationary potential energy (Art. 4-1) is used in the formulation. Since the displacement method affords a simpler development and greater ease in computer programming for most highly complex structures,[14] only that method will be treated here. Nevertheless, certain advantages of the force method in some problems should not be overlooked. Besides these two major approaches, two alternative methods based on modified energy principle considerations are noteworthy. These are mixed methods derived on the basis of Reissner's variational principle[15] and hybrid methods constructed from modified energy principles.[16]

One of the most critical features of the displacement method is the choice of trial displacement fields, since this choice affects the convergence of the method. Convergence of the sequence of approximate solutions, as the region of interest is further subdivided, is a matter of great practical and theoretical importance. In general, successful displacement trial functions must satisfy interelement compatibility, and include provisions for a state of constant strain, and for kinematic rigid body modes. When these criteria are met, the element is said to be a conforming element, and the approximate solutions converge monotonically to the corresponding exact solutions under certain sequences of mesh subdivisions idealizing the physical continuum. The selected trial functions are then expressed *in terms of element nodal displacements which are regarded as unknown quantities to be determined by the method.*

With trial displacements for an element specified, the state of strain within that element is uniquely defined. In turn, with the state of strain defined, the state of stress throughout the element may be calculated by appropriate stress-strain relations for each element. This circumstance reflects the capability of the finite element method to deal with problems where the elastic material properties vary from one element to another. Thus, the method provides a powerful means of treating anisotropic and nonhomogeneous three-dimensional elasticity problems or two-dimensional (plane) problems with variable thickness perpendicular to the plane. Also, the method is well suited to the incremental method of stress analysis in nonlinear and elastoplastic problems.

With the states of stress and strain of an element specified in terms of element nodal displacements, the element potential energy is expressed in terms of the unknown element nodal displacements; subsequently, the extremization of the potential energy with respect to these parameters leads to the formulation of the element force-displacement relations. Physically, the extremal requirement of the element potential energy places each element in a state of overall equilibrium by determining a system of fictitious forces concentrated at the nodes of the element. This system of forces balances the boundary stresses and any distributed loads acting on the element. However, local equilibrium within each element and on element boundaries ordinarily is not ensured.

Finally, all the element force-displacement equations are assembled or superimposed, corresponding to the manner in which they are interconnected, to form the stiffness equations for the entire assemblage of elements. These equations are called the structural stiffness equations. Since the conditions of overall equilibrium of each element have been satisfied, the construction of the structural stiffness equations further ensures the fulfillment of structural nodal equilibrium. Alternatively, the procedure may be considered as a means in which the unknown displacements (nodal displacements) are determined in a sense that the actual total structural potential energy is approximately minimized.

These operations are common to all solid mechanics (elasticity) problems and can be easily standardized in matrix notation and programmed for computers as a finite element displacement method.

Summary. Basic Steps in the Derivation of Element Characteristics by the Stiffness Method / Preliminary to the derivation of the element stiffness relation, one must identify the nature of the problem; namely, is it a beam problem, a two-dimensional (plane) problem, an axially symmetric problem, a plate (bending) problem, a shell problem, or a general three-dimensional problem? However, for any type of problem, the basic stiffness method as related to the development of element relations may be subdivided into seven general steps as follows:

Step 1 Choose a suitable element coordinate system and number the nodes of the basic element.

Step 2 Choose the displacement (trial or approximating) functions that define the displacement components at points in the element.

Step 3 Express the displacement within the element in terms of the displacement of the nodes of the element.

Step 4 Relate the strains in the element to the element displacement and, hence, by Step 3, to the nodal displacements of the element.

Step 5 Relate stresses in the element to the strains in the element for elastic material (or for inelastic material, if appropriate).

Step 6 Represent the stresses in the element by statically equivalent forces acting at the nodes of the element. Hence, derive an expression for the element stiffness matrix using the results of Steps 4 and 5.

Step 7 Express the stresses in the element in terms of element nodal displacements, using the results of Steps 4 and 5, to obtain the stress-displacement matrix for the element.

Once the general element relations have been obtained, the effects of each element are assembled (added together properly) to obtain a relation between the nodal displacements (vector) and the structural stiffness matrix. This equation is then solved to obtain the nodal displacements of the structure (elements). Once these nodal displacements have been calculated, the element stress components (Step 5) and the element strain components (Step 4) may be computed.

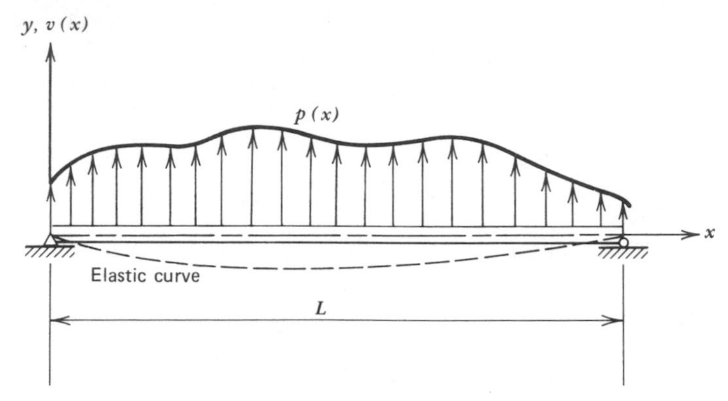

Fig. 15-2.1/Elastic beam.

15-2
FINITE ELEMENT METHODS FOR BEAMS†

As an example of the displacement finite element method, let us consider an elementary beam problem. In this example, the formulation of the required properties of the finite element and the solution procedures is illustrated in matrix form.

By the theory of straight beams,[17] the elastic curve of a beam (Fig. 15-2.1) is governed by the differential equation

$$EI\frac{d^4v}{dx^4} = p(x) \tag{15-2.1}$$

and the bending moment M of the beam is defined by

$$M(x) = EI\frac{d^2v}{dx^2} \tag{15-2.2}$$

In Eqs. (15-2.1) and (15-2.2), E is Young's modulis, I is the moment of inertia of the beam cross section, $v(x)$ is the displacement of the elastic curve, and $p(x)$ is the distributed load taken positive in the sense of the positive y-axis (Fig. 15-2.1). In beam theory, the term d^2v/dx^2 is an approximation for the curvature.

For the finite element analysis of beams, consider a prismatic cantilever beam with uniformly distributed load $p(x) = -p_0$, where p_0 is a constant. Let the beam be idealized as an assemblage of n beam elements of length h (Fig. 15-2.2). The ith beam element with axial coordinates axes x and y, with two element nodes (1,2) and four nodal displacements is shown in Fig. 15-2.3; where (v_1, v_2) and (θ_1, θ_2) denote the

† Much of the material presented in this and subsequent articles is based upon a set of notes written by Paul P. Lynn, Boulder, CO and one of the authors (A. P. Boresi).

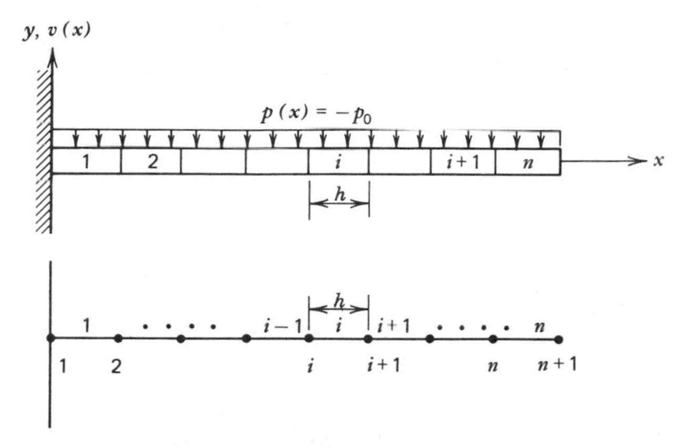

Fig. 15-2.2/Finite element idealization of a cantilever beam.

inplane nodal end displacements and nodal end rotations of the ith element (see Step 1, Art 15-1). The four nodal displacements of the ith element may be represented in matrix form as

$$\{\delta\}_i = \begin{Bmatrix} \delta_1 \\ \delta_2 \\ \delta_3 \\ \delta_4 \end{Bmatrix} = \begin{Bmatrix} v_1 \\ \theta_1 \\ v_2 \\ \theta_2 \end{Bmatrix}; \qquad \delta_1 = v_1,\ \delta_2 = \theta_1,\ \delta_3 = v_2,\ \delta_4 = \theta_2 \tag{15-2.3}$$

where for simplicity we have denoted element nodal displacements by δ and where the subscript i denotes the ith beam element.

We note that the beam element has four degrees of freedom (one displacement and one rotation at each of two nodes). Hence, our displacement (trial) function $v_i^*(x)$ which approximates $v(x)$ [see Eqs. (15-2.1) and (15-2.2)] in the element must contain four unknown coeffi-

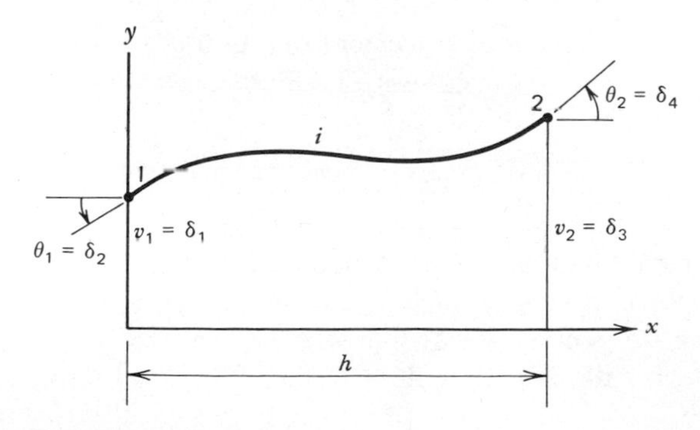

Fig. 15-2.3/Beam element i.

cients (or generalized coordinates), say, $\alpha_1, \alpha_2, \alpha_3, \alpha_4$. Thus, representing $v_i^*(x)$ in the simplest polynomial form in x, we write (Step 2, Art. 15-1)

$$v_i^*(x) = \alpha_1 + \alpha_2 x + \alpha_3 x^2 + \alpha_4 x^3 \tag{15-2.4a}$$

or, in matrix form,

$$v_i^*(x) = [N]_i \{\alpha\}_i \tag{15-2.4b}$$

in which $[N]_i = [1 \quad x \quad x^2 \quad x^3]$ and $\{\alpha\}_i = [\alpha_1 \quad \alpha_2 \quad \alpha_3 \quad \alpha_4]^T$, where T denotes transpose. The trial function $v_i^*(x)$ is a cubic polynomial in x. By means of Eq. (15-2.4), we can express the nodal displacements $(\delta_1, \delta_2, \delta_3, \delta_4)$, Eq. (15-2.3), in terms of $(\alpha_1, \alpha_2, \alpha_3, \alpha_4)$.

Thus, substitution of the element nodal coordinates (Fig. 15-2.3) into Eq. (15-2.4) yields, with Eq. (15-2.3) (since $v_i^*(0) = \delta_1$, $dv_i^*(0)/dx = \theta_1 = \delta_2$, at $x = 0$, with similar relations for $x = h$; Fig. 15-2.3)

$$\begin{Bmatrix} \delta_1 \\ \delta_2 \\ \delta_3 \\ \delta_4 \end{Bmatrix} = \begin{bmatrix} 1 & 0 & 0 & 0 \\ 0 & 1 & 0 & 0 \\ 1 & h & h^2 & h^3 \\ 0 & 1 & 2h & 3h^2 \end{bmatrix} \begin{Bmatrix} \alpha_1 \\ \alpha_2 \\ \alpha_3 \\ \alpha_4 \end{Bmatrix}$$

Denoting the 4 by 4 matrix in the above equation by $[A]_i$, we may write

$$\{\delta\}_i = [A]_i \{\alpha\}_i \tag{15-2.5}$$

By matrix inversion of $[A]_i$, we obtain

$$\{\alpha\}_i = [A]_i^{-1} \{\delta\}_i \tag{15-2.6}$$

where

$$[A]_i^{-1} = \frac{1}{h^3} \begin{bmatrix} h^3 & 0 & 0 & 0 \\ 0 & h^3 & 0 & 0 \\ -3h & -2h^2 & 3h & -h^2 \\ 2 & h & -2 & h \end{bmatrix} \tag{15-2.6a}$$

Thus, the element displacement function $v_i^*(x)$ (or in matrix form $\{v^*\}_i$) is uniquely determined by $\{\delta\}_i$ (Step 3, Art. 15-1); that is (by Eqs. 15-2.4 and 15-2.6),

$$v_i^*(x) = \{v^*\}_i = [N]_i [A]_i^{-1} \{\delta\}_i \tag{15-2.7}$$

From Eq. (15-2.7) we can calculate the appropriate strain quantity (Step 4, Art. 15-1) which, in beam theory, is the curvature of the elastic curve. In general, we denote strain quantities by the symbol ϵ. Thus, we have, for the ith element, in terms of the trial function, $v_i^*(x)$,

$$\{\epsilon\}_i = \frac{d^2 v_i^*}{dx^2} \tag{15-2.8}$$

In terms of nodal displacements, Eqs. (15-2.8) and (15-2.6) yield (since $d^2v_i^*/dx^2 = 2\alpha_3 + 6\alpha_4 x$)

$$\{\epsilon\}_i = [B]_i[A]_i^{-1}\{\delta\}_i \tag{15-2.9}$$

where the matrix $[B]_i$ is defined as

$$[B]_i = [0 \quad 0 \quad 2 \quad 6x] \tag{15-2.9a}$$

The bending moment of the beam is directly related to the curvature of the elastic curve (see Eq. 15-2.2). Considering the bending moment M represented as a generalized stress quantity, $\{\sigma\}_i$, with Eqs. (15-2.2), (15-2.8), and (15-2.9), we may express the stress-strain relation of the ith element as a linear relation between $\{\sigma\}_i$ and $\{\epsilon\}_i$ (Step 5, Art, 15-1); namely,

$$\{\sigma\}_i = [D]_i\{\epsilon\}_i = [D]_i[B]_i[A]_i^{-1}\{\delta\}_i \tag{15-2.10}$$

in which $[D]_i = EI$ is called the elasticity matrix connecting the generalized element stress and strain quantities.

We now employ the principle of virtual work for elastic bodies (see Art. 4-1; also reference 18) to derive the element force-displacement relations (stiffness equation) (see Step 6, Art. 15-1). Accordingly, we introduce a set of virtual nodal displacements $\{\bar{\delta}\}_i$, Art. 4-1. Under this virtual displacement, the nodal displacements $\{\delta\}_i \to \{\delta\}_i + \{\bar{\delta}\}_i$ and the strain components $\{\epsilon\}_i \to \{\epsilon\}_i + \{\bar{\epsilon}\}_i$, where the bar denotes virtual changes. Hence, the internal virtual work $\delta\overline{U}_i$ is

$$\delta\overline{U}_i = \int_0^h \{\bar{\epsilon}\}_i^T\{\sigma\}_i\, dx$$

where

$$\{\bar{\epsilon}\}_i = [B]_i[A]_i^{-1}\{\bar{\delta}\}_i \quad \text{and} \quad \{\sigma\}_i = [D]_i[B]_i[A]_i^{-1}\{\delta\}_i.$$

Taking the transpose of $\{\bar{\epsilon}\}_i$, we find†

$$\delta\overline{U}_i = \int_0^h \{\bar{\delta}\}_i^T[A^{-1}]_i^T[B]_i^T[D]_i[B]_i[A^{-1}]_i\{\delta\}_i\, dx$$

$$= \{\bar{\delta}\}_i^T[A^{-1}]_i^T\left[\int_0^h [B]_i^T[D]_i[B]_i\, dx\right][A^{-1}]_i\{\delta\}_i \tag{15-2.11}$$

The external virtual work $\delta\overline{W}_i$ is given by

$$\delta\overline{W}_i = \int_0^h \{v\}_i^T\{p(x)\}_i\, dx$$

where $\{\bar{v}\}_i = [N]_i[A^{-1}]_i\{\bar{\delta}\}_i$ and $\{p(x)\}_i = \{-p_0\}_i$. Hence, $\delta\overline{W}_i$ becomes

$$\delta\overline{W}_i = \{\bar{\delta}\}_i^T[A^{-1}]_i^T\left\{\int_0^h [N]_i^T\{-p_0\}_i\right\} dx \tag{15-2.12}$$

†The transpose of the matrix $\{\bar{\epsilon}\}_i$ is $\{\bar{\epsilon}\}_i^T = \{\bar{\delta}\}_i^T[A^{-1}]_i^T[B]_i^T$. In Eq. (15-2.11) the constant matrices $\{\bar{\delta}\}_i$, $[A^{-1}]_i$ and $\{\delta\}_i$ are taken outside of the integral sign, since they do not depend on x.

On equating the internal energy and external virtual work (see Art. 4-1) and cancelling the term $\{\bar{\delta}\}_i^T$, we arrive at the element stiffness equation (element equilibrium equation)

$$[A^{-1}]_i^T\left[\int_0^h [B]_i^T[D]_i[B]_i\,dx\right][A^{-1}]_i\{\delta\}_i$$
$$=[A^{-1}]_i^T\left\{\int_0^h [N]_i^T\{-p_0\}_i\right\}dx$$

which may be written as

$$[K]_i\{\delta\}_i=\{F\}_i \tag{15-2.13}$$

where the element stiffness matrix is

$$[K]_i=[A^{-1}]_i^T\left[\int_0^h [B]_i^T[D]_i[B]_i\,dx\right][A^{-1}]_i=[A^{-1}]_i^T[\bar{K}]_i[A^{-1}]_i \tag{15-2.14}$$

where

$$[\bar{K}]_i=\int_0^h [B]_i^T[D]_i[B]_i\,dx \tag{15-2.14a}$$

is called the generalized element stiffness matrix, and the element force (or load) matrix is

$$\{F\}_i=[A^{-1}]_i^T\left\{\int_0^h [N]_i^T\{-p_0\}_i\right\}dx=[A^{-1}]_i^T\{\bar{P}\}_i \tag{15-2.15}$$

where

$$\{\bar{P}\}_i=\left\{\int_0^h [N]_i^T\{-p_0\}_i\right\}dx \tag{15-2.15a}$$

is called the generalized surface load matrix.

To obtain explicit expressions of the matrices $[K]_i$ and $\{F\}_i$, we must perform the matrix multiplications and integrations indicated in Eqs. (15-2.14) and (15-2.15). Thus, we find with Eq. (15-2.6a)

$$[K]_i=\frac{EI}{h^3}\begin{bmatrix} 12 & 6h & -12 & 6h \\ 6h & 4h^2 & -6h & 2h^2 \\ -12 & -6h & 12 & -6h \\ 6h & 2h^2 & -6h & 4h^2 \end{bmatrix} \tag{15-2.16}$$

and

$$\{F\}_i=\frac{p_0}{h^3}[-h^4/2 \quad -h^5/12 \quad -h^4/2 \quad h^5/12]^T \tag{15-2.17}$$

In this example, our aim is to present the finite element method in the most direct manner, and to clarify the understanding of the method. However, the use of so-called isoparametric finite elements is more effective than our use of nodal (generalized) displacements. Conse-

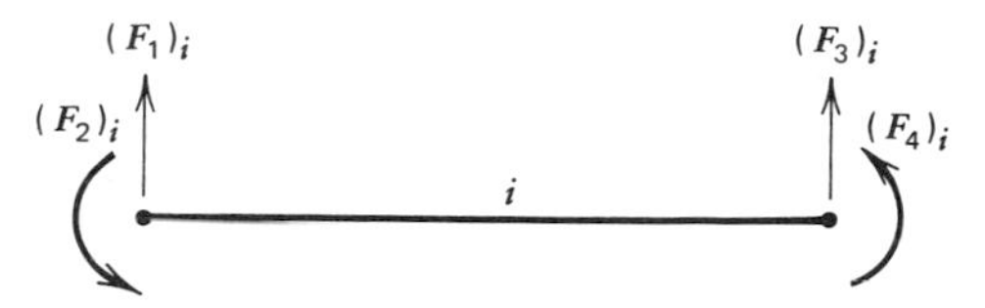

Fig. 15-2.4/Element nodal forces.

quently, to evaluate the element stiffness matrix $[K]_i$, Eq. (15-2.14), and the element force matrix $\{F\}_i$, Eq. (15-2.15), we must evaluate the matrix $[A^{-1}]$, Eq. (15-2.6) which, from an efficient computational viewpoint, is undesirable. The isoparametric finite element formulation achieves the relationship between element displacements at any point and the element nodal displacements directly through the use of *interpolation functions* (*shape functions*), a procedure which does not require the evaluation of $[A^{-1}]$. Rather, the element matrices corresponding to the required degrees of freedom are obtained directly.[19] The element nodal force matrix $\{F\}_i$ may be considered as a set of fictitious nodal forces $(F_1)_i$, $(F_2)_i$, $(F_3)_i$, and $(F_4)_i$ acting on the ith beam element with positive senses as shown in Fig. 15-2.4. We note that $(F_1)_i$,$(F_3)_i$ are transverse forces and $(F_2)_i$ and $(F_4)_i$ are moments. From the definition of the element stiffness matrix, the element $(K_{rs})_i$ of the matrix $[K]_i$ represents physically the nodal force element $(F_r)_i$ produced by a unit nodal displacement $(\delta_s)_i$.

In order to formulate the force-displacement relations (or the so-called structural stiffness equation) for the entire beam, it is necessary to combine the individual element stiffness and force matrices into the structural (total) stiffness equation of the beam. For this purpose, we number the structural nodes (**1, 2, . . . , i, . . . , n + 1**) in boldface letters as shown in Fig. 15-2.5, and assign structural nodal displacements

Force	F_1	F_3			F_{2i-1}	F_{2i+1}	F_{2i+3}	· · ·	F_{2n+1}
Moment	F_2	F_4			F_{2i}	F_{2i+2}	F_{2i+4}	· · ·	F_{2n+2}
Element Number	1	2			i	$i+1$			
Structural Node	**1**	**2**	**3**	· · ·	**i**	**i + 1**	**i + 2**	· · ·	**n + 1**
Deflection	δ_1	δ_3			δ_{2i-1}	δ_{2i+1}	δ_{2i+3}	· · ·	δ_{2n+1}
Rotation	δ_2	δ_4			δ_{2i}	δ_{2i+2}	δ_{2i+4}	· · ·	δ_{2n+2}

Fig. 15-2.5/Structural numbering systems.

$$\delta_{2i-1} = (\delta_1)_i \qquad (\delta_3)_i = \delta_{2i+1} = (\delta_1)_{i+1} \qquad \delta_{2i+3} = (\delta_3)_{i+1}$$
$$\delta_{2i} = (\delta_2)_i \qquad (\delta_4)_i = \delta_{2i+2} = (\delta_2)_{i+1} \qquad \delta_{2i+4} = (\delta_4)_{i+1}$$

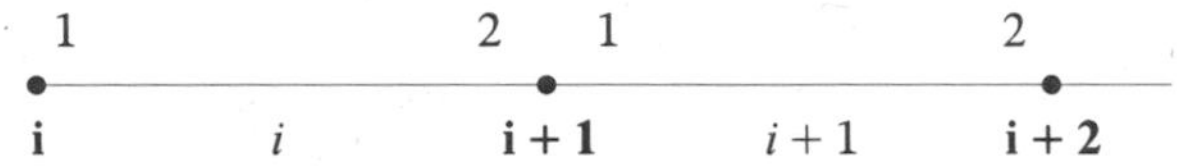

1 2 1 2

i *i* **i + 1** *i* + 1 **i + 2**

Fig. 15-2.6/Assemblage of i and $i+1$ elements.

$(\delta_1, \delta_2, \ldots, \delta_{2i-1}, \delta_{2i}, \ldots, \delta_{2n+2})$ and the corresponding nodal forces $(F_1, F_2, \ldots, F_{2i-1}, F_{2i}, \ldots, F_{2n+2})$. For example, two nodal displacements (δ_1, δ_2) and a pair of corresponding nodal forces (F_1, F_2) are assigned at the node **1**.

Consider now the assemblage of two adjacent beam elements (i) and $(i+1)$ that lie to the right of nodes (i) and $(i+1)$, respectively, as shown in Fig. 15-2.6. Continuity of displacement and slope at the end points of adjacent beam elements of the assembled system requires that the element nodal displacement and the structural nodal displacement assigned at a common node must be identical. For example, at node **(i + 1)** the element nodal displacements of the ith element $(\delta_3)_i$ and $(\delta_4)_i$ and of the $(i+1)$th element $(\delta_1)_{i+1}$ and $(\delta_2)_{i+1}$ are equal to the structural nodal displacements δ_{2i+1} and δ_{2i+2}, respectively. This requirement is shown schematically in Fig. 15-2.6.

We may calculate the total structural forces which act on the node **(i + 1)**, by adding the contributions of element nodal forces from elements i and $(i+1)$. Accordingly, from the element stiffness equations for the elements i and $(i+1)$ [see Eqs. (15-2.13), (15-2.16), (15-2.17)], we obtain

$$(F_3)_i = -\frac{p_0 h}{2} = (K_{31})_i(\delta_1)_i + (K_{32})_i(\delta_2)_i + (K_{33})_i(\delta_3)_i + (K_{34})_i(\delta_4)_i$$

$$(F_4)_i = \frac{p_0 h^2}{12} = (K_{41})_i(\delta_1)_i + (K_{42})_i(\delta_2)_i + (K_{43})_i(\delta_3)_i + (K_{44})_i(\delta_4)_i$$

$$(F_1)_{i+1} = -\frac{p_0 h}{2} = (K_{11})_{i+1}(\delta_1)_{i+1} + (K_{12})_{i+1}(\delta_2)_{i+1} + (K_{13})_{i+1}(\delta_3)_{i+1} + (K_{14})_{i+1}(\delta_4)_{i+1}$$

$$(F_2)_{i+1} = -\frac{p_0 h^2}{12} = (K_{21})_{i+1}(\delta_1)_{i+1} + (K_{22})_{i+1}(\delta_2)_{i+1} + (K_{23})_{i+1}(\delta_3)_{i+1} + (K_{24})_{i+1}(\delta_4)_{i+1} \qquad (15\text{-}2.18)$$

Converting all the element nodal displacements to their corresponding structural nodal displacements in Eq. (15-2.18), we obtain

$$(F_3)_i = -\frac{p_0 h}{2} = (K_{31})_i \delta_{2i-1} + (K_{32})_i \delta_{2i} + (K_{33})_i \delta_{2i+1} + (K_{34})_i \delta_{2i+2}$$

$$(F_4)_i = \frac{p_0 h^2}{12} = (K_{41})_i \delta_{2i-1} + (K_{42})_i \delta_{2i} + (K_{43})_i \delta_{2i+1} + (K_{44})_i \delta_{2i+2}$$

$$(F_1)_{i+1} = -\frac{p_0 h}{2} = (K_{11})_{i+1} \delta_{2i+1} + (K_{12})_{i+1} \delta_{2i+2} + (K_{13})_{i+1} \delta_{2i+3} + (K_{14})_{i+1} \delta_{2i+4}$$

$$(F_2)_{i+1} = -\frac{p_0 h^2}{12} = (K_{21})_{i+1} \delta_{2i+1} + (K_{22})_{i+1} \delta_{2i+2} + (K_{23})_{i+1} \delta_{2i+3} + (K_{24})_{i+1} \delta_{2i+4} \qquad (15\text{-}2.19)$$

where $(F_3)_i$ and $(F_1)_{i+1}$ are transverse forces and $(F_4)_i$ and $(F_2)_{i+1}$ are moments acting at node $(\mathbf{i+1})$. Therefore, resultant force and moment F_{2i+1} and F_{2i+2} acting at node $(\mathbf{i+1})$ are, respectively,

$$F_{2i+1} = (F_3)_i + (F_1)_{i+1}$$

and

$$F_{2i+2} = (F_4)_i + (F_2)_{i+1}$$

By Eq. (15-2.19), we may write F_{2i+1} and F_{2i+2} in the form

$$F_{2i+1} = -p_0 h = (K_{31})_i \delta_{2i-1} + (K_{32})_i \delta_{2i} + \{(K_{33})_i + (K_{11})_{i+1}\} \delta_{2i+1} + \{(K_{34})_i + (K_{12})_{i+1}\} \delta_{2i+2} + (K_{13})_{i+1} \delta_{2i+3} + (K_{14})_{i+1} \delta_{2i+4}$$

$$F_{2i+2} = 0 = (K_{41})_i \delta_{2i-1} + (K_{42})_i \delta_{2i} + \{(K_{43})_i + (K_{21})_{i+1}\} \delta_{2i+1} + \{(K_{44})_i + (K_{22})_{i+1}\} \delta_{2i+2} + (K_{23})_{i+1} \delta_{2i+3} + (K_{24})_{i+1} \delta_{2i+4} \qquad (15\text{-}2.20)$$

Equation (15-2.20) represents the stiffness equation of the structural node $(\mathbf{i+1})$. It may be considered also as the equilibrium condition for the node. This fact is schematically illustrated in Fig. 15-2.7.

The form of Eq. (15-2.20) suggests that the stiffness equation for the assemblage of i and $i+1$ elements can be obtained simply by the superposition of the element stiffness and force matrices of two adjacent elements. This superposition is indicated schematically in Fig. 15-2.8, where the shaded areas indicate regions of coupling between individual elements; that is, in the shaded areas the terms are obtained by additions of the superposed elements of the ith and $(i+1)$th element matrices. Assembling all the elements in this manner to form the structural

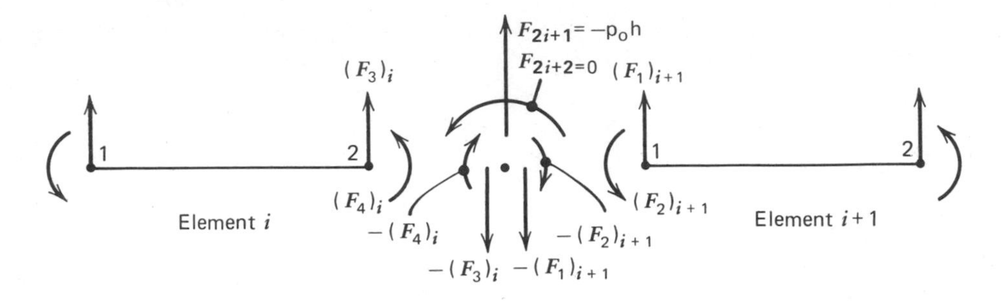

Fig. 15-2.7/Forces acting on $i+1$ node.

stiffness equation of the beam, we obtain the matrix equation

$$\{F\} = [S]\{\delta\} \tag{15-2.21}$$

Equation (15-2.21) may be written in the explicit form of Equation (15-2.22).

Since the beam is clamped at node **1**, the boundary conditions are $\delta_1 = \delta_2 = 0$. Thus, the corresponding nodal forces F_1 and F_2 are the unknown reactive forces at the fixed support. The other nodal forces are calculated according to the superposition rule described earlier.

If we consider (δ_1, δ_2) as the constrained nodal displacements $\{\delta_c\}$ and $(\delta_3, \delta_4, \ldots, \delta_{2n+2})$ as the free nodal displacements $\{\delta_f\}$, we can partition the $\{F\}$ and $[S]$ matrices accordingly. Hence, we may write Eq. (15-2.22) in the form

$$\left\{\begin{array}{c} F_c \\ \hline F_f \end{array}\right\} = \left[\begin{array}{c|c} S_{cc} & S_{cf} \\ \hline S_{fc} & S_{ff} \end{array}\right] \left\{\begin{array}{c} \delta_c \\ \hline \delta_f \end{array}\right\}$$

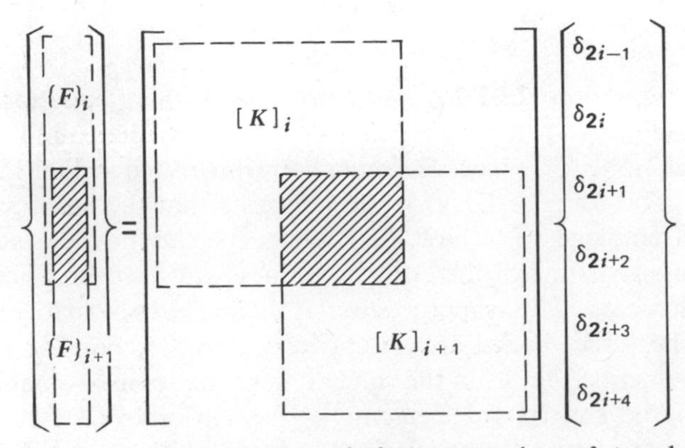

Fig. 15-2.8/Superposition of element matrices of i and $i+1$ beam elements.

$$
\begin{Bmatrix} F_1 \\ F_2 \\ F_3 \\ F_4 \\ F_5 \\ F_6 \\ F_7 \\ \cdot \\ \cdot \\ F_{2n+1} \\ F_{2n+2} \end{Bmatrix} = \frac{EI}{h^3} \begin{bmatrix}
12 & 6h & -12 & 6h & 0 & 0 & \cdot & \cdot & \cdot & \cdot & \cdot \\
6h & 4h^2 & -6h & 2h^2 & 0 & 0 & \cdot & \cdot & \cdot & \cdot & \cdot \\
-12 & -6h & 24 & 0 & -12 & 6h & \cdot & \cdot & \cdot & \cdot & \cdot \\
6h & 2h^2 & 0 & 8h^2 & -6h & 2h^2 & 0 & \cdot & \cdot & \cdot & \cdot \\
0 & 0 & -12 & -6h & 24 & 0 & -12 & 6h & 0 & \cdot & \cdot \\
0 & 0 & 6h & 2h^2 & 0 & 8h^2 & -6h & 2h^2 & 0 & \cdot & \cdot \\
0 & 0 & 0 & 0 & -12 & -6h & 24 & 0 & \cdot & \cdot & \cdot \\
\cdot & \cdot & \cdot & \cdot & 6h & 2h^2 & 0 & 8h^2 & \cdot & \cdot & \cdot \\
\cdot & \cdot & \cdot & \cdot & \cdot & \cdot & \cdot & \cdot & \cdot & \cdot & \cdot \\
\cdot & \cdot & \cdot & \cdot & \cdot & \cdot & \cdot & \cdot & \cdot & \cdot & \cdot \\
\cdot & \cdot & \cdot & \cdot & \cdot & \cdot & \cdot & \cdot & \cdot & \cdot & \cdot
\end{bmatrix} \begin{Bmatrix} \delta_1 \\ \delta_2 \\ \delta_3 \\ \delta_4 \\ \delta_5 \\ \delta_6 \\ \delta_7 \\ \cdot \\ \cdot \\ \delta_{2n+1} \\ \delta_{2n+2} \end{Bmatrix} \tag{15-2.22}
$$

Expansion of this equation then yields

$$\{F_c\} = [S_{cc}]\{\delta_c\} + [S_{cf}]\{\delta_f\} \tag{15-2.23}$$

and

$$\{F_f\} = [S_{fc}]\{\delta_c\} + [S_{ff}]\{\delta_f\} \tag{15-2.24}$$

Since $\{\delta_c\} = 0$, (alternatively, $\{\delta_c\}$ may be prescribed nonzero boundary displacements), we find from Eq. (15-2.24) that

$$\{\delta_f\} = [S_{ff}]^{-1}\{F_f\} \tag{15-2.25}$$

Substitution of Eq. (15-2.25) into Eq. (15-2.23) yields, with $\{\delta_c\} = 0$

$$\{F_c\} = [S_{cf}][S_{ff}]^{-1}\{F_f\} \tag{15-2.26}$$

For the numerical solutions of Eqs. (15-2.25) and (15-2.26) we must rely on high speed electronic digital computers. Indeed, the entire finite element procedure is usually programmed for automatic computation. When all the nodal displacements are known, the state of element stress (bending moment) can be calculated by Eq. (15-2.10).

For this particular example, the solution of Eq. (15-2.25) gives the free end deflection of the cantilever as

$$\delta_{2n+1} = -\frac{p_0 L^4}{8EI}$$

and the corresponding free end rotation

$$\delta_{2n+2} = -\frac{p_0 L^3}{6EI}$$

where L is the total length of the beam.

In this problem, δ_{2n+1} and δ_{2n+2} are the same as the exact solutions of beam theory. This result is due to the fact that the trial function $v_i^*(x)$ employed here is the solution of the homogeneous beam equation

$$\frac{d^4 v(x)}{dx^4} = 0$$

Ordinarily, for two- or three-dimensional finite elements, there is no such simple exact displacement function. Hence, use of *approximate* trial functions is required. As a result, the method yields, in general, approximate solutions which may be improved by increasing the number of properly chosen element subdivisions, provided proper trial functions are employed. Consequently, the choice of trial functions and the convergence of the approximate solutions to the true solutions require careful study (see reference 19, Chapter 4).

15-3 PLANE PROBLEMS OF ELASTICITY

The basic equations of plane stress problems are derived in Chapters 1 and 2. In this article, we present in greater detail the finite element

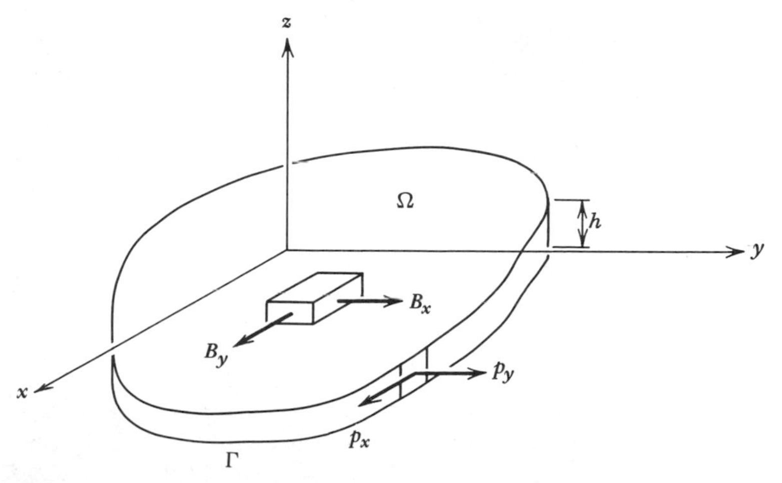

Fig. 15-3.1/A two-dimensional elasticity problem.

method of analysis of this important problem. The first successful application of this approach was in the field of aeronautical structural analysis.[13] The practical value of the method has been clearly demonstrated by numerous works.[20,21]

Consider a plane region Ω of thickness h, under the action of body forces B_x, B_y, and tractions (p_x, p_y) on the boundary Γ. Two classical plane problems (plane strain and plane stress and/or generalized plane stress) will be treated.† Mathematically, these problems are equivalent. However, the material constants enter in a slightly different way (see Table 15-3.1).

In the finite element methods, different shaped elements may be employed to approximate the region Ω (Fig. 15-3.1). In the following analysis, we will employ a triangular element and a rectangular element (Fig. 15-3.2). It may be shown that the following trial functions satisfy appropriate convergence criteria.

For the triangular element:‡

$$\begin{aligned} u &= \alpha_1 + \alpha_2 x + \alpha_3 y \\ v &= \alpha_4 + \alpha_5 x + \alpha_6 y \end{aligned} \tag{15-3.1}$$

For the rectangular element:

$$\begin{aligned} u &= \alpha_1 + \alpha_2 x + \alpha_3 y + \alpha_4 xy \\ v &= \alpha_5 + \alpha_6 x + \alpha_7 y + \alpha_8 xy \end{aligned} \tag{15-3.2}$$

† Chapters 1 and 2. See also reference 22.

‡ For simplicity, the superscript asterisk, which indicates the trial displacement functions, will be dropped.

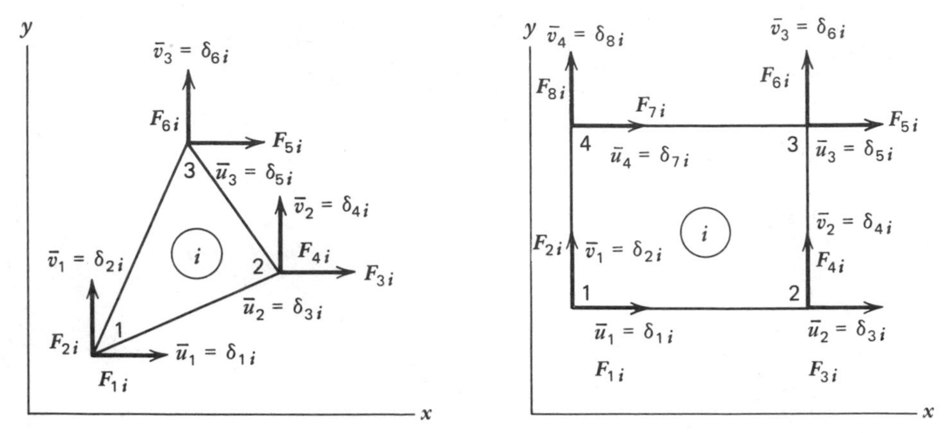

Fig. 15-3.2/Triangular and rectangular elements.

In each case, two nodal displacements $(\bar{u}, \bar{v})$ are assigned at each corner node of the elements (Fig. 15-3.2).

Other types of elements or element shapes have been used[23] but the elements shown in Fig. 15-3.2 represent the simplest types and have yielded reasonable results in practical applications.[9] With triangular elements, one may approximate closely any given boundary of a problem. Furthermore, the triangular element may be readily used to form nonuniform nets or subdivisions. However, for problems with rectangular domains, rectangular elements may prove to be advantageous. In the following, we treat the detailed formulation of these two types of elements.

Triangular Finite Elements/Consider a typical element i (Fig. 15-3.3) with nodes 1, 2, 3, numbered in counterclockwise order and with node 1 at the origin of (x, y)-coordinates.

The (x, y)-coordinates are called element *local coordinates* and $(\mathbf{x}, \mathbf{y})$ are called structural or *global coordinates*. Use of element local coordinates facilitates the formulation of element properties. One may express element properties in terms of the global coordinates by means of a coordinate transformation. Therefore, without loss of generality, we let the (x, y)-coordinates be parallel to the $(\mathbf{x}, \mathbf{y})$ coordinates.†

From Eq. (15-3.1) we write the trial functions in the matrix form

$$\begin{Bmatrix} u \\ v \end{Bmatrix}_i = [N]_i \{\alpha\}_i \tag{15-3.3}$$

†Note that nodal displacements and forces remain invariant under coordinate translation.

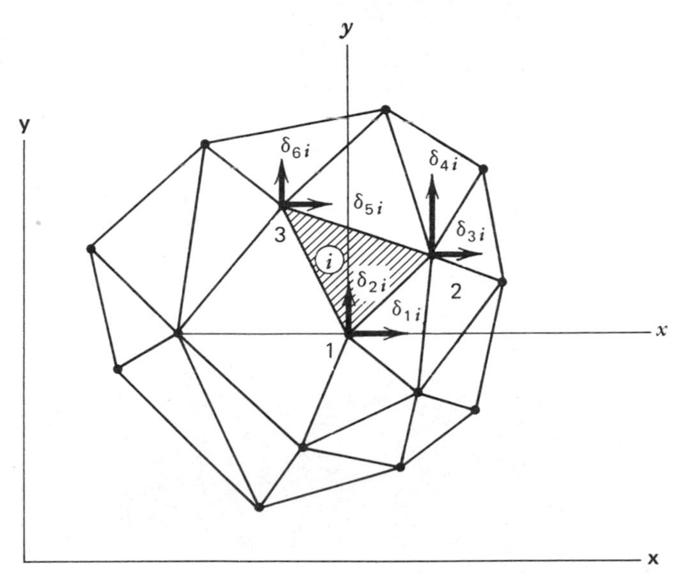

Fig. 15-3.3/Triangular element i.

where the matrix

$$[N]_i = \begin{bmatrix} 1 & x & y & 0 & 0 & 0 \\ 0 & 0 & 0 & 1 & x & y \end{bmatrix}$$

and

$$\{\alpha\}_i^T = [\alpha_1 \quad \alpha_2 \quad \alpha_3 \quad \alpha_4 \quad \alpha_5 \quad \alpha_6]$$

Substitution of the (x, y) nodal coordinates of nodes 1, 2, 3 into Eq. (15-3.3) gives

$$\{\delta\}_i = [A]_i\{\alpha\}_i \tag{15-3.4}$$

where the element nodal displacement $\{\delta\}_i$ is defined by

$$\{\delta\}_i^T = [\bar{u}_{1i} \quad \bar{v}_{1i} \quad \bar{u}_{2i} \quad \bar{v}_{2i} \quad \bar{u}_{3i} \quad \bar{v}_{3i}]$$

and matrix $[A]_i$ is the six by six square matrix defined in terms of nodal coordinates $(x_1, y_1, x_2, y_2, x_3, y_3)$

$$[A]_i = \begin{bmatrix} 1 & 0 & 0 & 0 & 0 & 0 \\ 0 & 0 & 0 & 1 & 0 & 0 \\ 1 & x_2 & y_2 & 0 & 0 & 0 \\ 0 & 0 & 0 & 1 & x_2 & y_2 \\ 1 & x_3 & y_3 & 0 & 0 & 0 \\ 0 & 0 & 0 & 1 & x_3 & y_3 \end{bmatrix} \tag{15-3.5}$$

Inverting the matrix $[A]_i$, we find

$$\{\alpha\}_i = [A]_i^{-1}\{\delta\}_i \tag{15-3.5a}$$

where[†]

$$[A]_i^{-1} = \frac{1}{\Delta}\begin{bmatrix} \Delta & 0 & 0 & 0 & 0 & 0 \\ y_2 - y_3 & 0 & y_3 & 0 & -y_2 & 0 \\ x_3 - x_2 & 0 & -x_3 & 0 & x_2 & 0 \\ 0 & \Delta & 0 & 0 & 0 & 0 \\ 0 & y_2 - y_3 & 0 & y_3 & 0 & -y_2 \\ 0 & x_3 - x_2 & 0 & -x_3 & 0 & x_2 \end{bmatrix} \tag{15-3.6}$$

where $\Delta = x_2 y_3 - x_3 y_2$ represents twice the area of the triangular element. By the definition of strain (Chapter 2), the element strain $\{\epsilon\}_i$ is

$$\{\epsilon\}_i = \begin{Bmatrix} \epsilon_x \\ \epsilon_y \\ \gamma_{xy} \end{Bmatrix}^{\ddagger} = \begin{Bmatrix} u_x \\ v_y \\ u_y + v_x \end{Bmatrix}$$

Thus, in matrix form, we have

$$\{\epsilon\}_i = [B]_i \{\alpha\}_i \tag{15-3.7}$$

where matrix $[B]_i$ is given by

$$[B]_i = \begin{bmatrix} 0 & 1 & 0 & 0 & 0 & 0 \\ 0 & 0 & 0 & 0 & 0 & 1 \\ 0 & 0 & 1 & 0 & 1 & 0 \end{bmatrix}$$

Having established the strain quantities, we express the corresponding stresses through the linear elastic stress-strain relations. Symbolically, we have

$$\{\sigma\}_i = [D]_i \{\epsilon\}_i = [D]_i [B]_i \{\alpha\}_i \tag{15-3.8}$$

where $\{\sigma\}_i$ is the column *stress matrix* defined by

$$\{\sigma\}_i^T = [\sigma_x \quad \sigma_y \quad \tau_{xy}]$$

and the elasticity matrix $[D]_i$ is (see Chapter 2)

$$[D]_i = \mu \begin{bmatrix} 1 & D_{12} & 0 \\ D_{12} & 1 & 0 \\ 0 & 0 & D_{33} \end{bmatrix}$$

where μ, D_{12}, and D_{33} are defined in Table (15-3.1)

[†] It is not necessary to invert the $[A]_i$ matrix explicitly (algebraically). The inversion may be performed implicitly within an electronic computer during the processes of calculating the element stiffness matrix.

[‡] In this chapter we use the stress (and associated strain) notations of row II, Table 1-3.1, page 11.

Table 15-3.1

	Plane Strain	*Plain Stress*
μ	$\dfrac{E(1-\nu)}{(1+\nu)(1-2\nu)}$	$\dfrac{E}{1-\nu^2}$
D_{12}	$\dfrac{\nu}{1-\nu}$	ν
D_{33}	$\dfrac{(1-2\nu)}{2(1-\nu)}$	$\dfrac{1-\nu}{2}$

Substitution of the column matrix $\{\alpha\}_i = [A]_i^{-1}\{\delta\}_i$ into Eq. (15-3.8) yields the stress matrix in terms of the element nodal displacement quantities. Thus,

$$\begin{Bmatrix} \sigma_x \\ \sigma_y \\ \tau_{xy} \end{Bmatrix}_i = \begin{bmatrix} 0 & \mu & 0 & 0 & 0 & \mu D_{12} \\ 0 & \mu D_{12} & 0 & 0 & 0 & \mu \\ 0 & 0 & \mu D_{33} & 0 & \mu D_{33} & 0 \end{bmatrix} [A]_i^{-1}\{\delta\}_i \tag{15-3.9}$$

Since $[A]_i^{-1}$ is constant, Eq. (15-3.9) indicates that a state of constant stress exists within the element. When $\{\delta\}_i$ is known, the stress quantities may be evaluated by means of Eq. (15-3.9).

For the plane region, the generalized element stiffness matrix $[\bar{K}]_i$ is given by the area integral (see Eq. 15-2.14a)

$$[\bar{K}]_i = h \iint_{A_i} [B]_i^T [D]_i [B]_i \, dA_i \tag{15-3.10}$$

where A_i is the area of the ith element and h is the thickness of the region. On carrying out the matrix multiplication and integrating over A_i, we arrive at the matrix $[\bar{K}]_i$ in the form

$$[\bar{K}]_i = \frac{h\,\Delta\mu}{2} \begin{bmatrix} 0 & 0 & 0 & 0 & 0 & 0 \\ 0 & 1 & 0 & 0 & 0 & D_{12} \\ 0 & 0 & D_{33} & 0 & D_{33} & 0 \\ 0 & 0 & 0 & 0 & 0 & 0 \\ 0 & 0 & D_{33} & 0 & D_{33} & 0 \\ 0 & D_{12} & 0 & 0 & 0 & 1 \end{bmatrix} \tag{15-3.11}$$

We note that the matrix $[\bar{K}]_i$ is symmetric and its principal diagonal elements are positive. Unfortunately, there is no direct way to make an *equilibrium check* on each column of the matrix $[\bar{K}]_i$ as there is for the element stiffness matrix $[K]_i$ (see Art. 15-2). Nevertheless, an alternative method of checking does exist.

Analogous to Eq. (15-2.13), the equilibrium conditions of the element are

$$\{F\}_i = [K]_i\{\delta\}_i \tag{15-3.12}$$

Analogous to Eqs. (15-2.13) and (15-2.14), and with Eq. (15-3.4), we may rewrite Eq. (15-3.12) in the form

$$\{F\}_i = [A^{-1}]_i^T[\bar{K}]_i[A]_i^{-1}\{\delta\}_i = [A^{-1}]_i^T[\bar{K}]_i\{\alpha\}_i \tag{15-3.13}$$

where we have used the relation (see Eq. 15-2.14)

$$[K]_i = [A^{-1}]_i^T[\bar{K}]_i[A]_i^{-1} \tag{15-3.14}$$

Then, an equilibrium check on the forces in the x and y directions may be performed on the columns of the matrix $[[A^{-1}]_i^T[\bar{K}]_i]$ (see Art. 15-2). Multiplication of $[A^{-1}]_i^T$ and $[\bar{K}]_i$ yields the matrix

$$[A^{-1}]_i^T[\bar{K}]_i = \frac{h\mu}{2}\begin{bmatrix} 0 & (y_2 - y_3) & D_{33}(x_3 - x_2) & 0 & D_{33}(x_3 - x_2) & D_{12}(y_2 - y_3) \\ 0 & D_{12}(x_3 - x_2) & D_{33}(y_2 - y_3) & 0 & D_{33}(y_2 - y_3) & (x_3 - x_2) \\ 0 & y_3 & -D_{33}x_3 & 0 & -D_{33}x_3 & D_{12}y_3 \\ 0 & -D_{12}x_3 & D_{33}y_3 & 0 & D_{33}y_3 & -x_3 \\ 0 & -y_2 & D_{33}x_2 & 0 & D_{33}x_2 & -D_{12}y_2 \\ 0 & D_{12}x_2 & -D_{33}y_2 & 0 & -D_{33}y_2 & x_2 \end{bmatrix} \tag{15-3.15}$$

We note that each column of this matrix satisfies the conditions

$$\sum F_x = \text{row}(1) + \text{row}(3) + \text{row}(5) = 0$$
$$\sum F_y = \text{row}(2) + \text{row}(4) + \text{row}(6) = 0 \tag{15-3.16}$$

The zeros in the first and fourth columns of the matrix represent nodal forces induced by unit values of α_1 and α_4 which correspond to rigid body·translations in the x- and y-directions, respectively.

By Eq. (15-3.14) we may then compute the element stiffness matrix $[K]_i$. In practice, the evaluation of the matrix $[K]_i$ is seldom performed explicitly; instead, the calculation is usually performed automatically by computing machines for individual finite elements.† However, when the algebraic inversion of the matrix $[A]_i$ is relatively simple, it is desirable to calculate the element stiffness matrix in its explicit form in order to reduce costly computer time. Furthermore, it is instructive to calculate the element stiffness matrix explicitly.

Accordingly, from Eqs. (15-3.6), (15-3.14) and (15-3.15), we obtain Eq. (15-3.17). Again the symmetry, the positive diagonal elements, and the column equilibrium check of the $[K]_i$ matrix should be noted.

Consider next the *element load matrix* (nodal force matrix) associated with load vectors

$$\{X\}_i = \begin{Bmatrix} B_x \\ B_y \end{Bmatrix} \quad \text{and} \quad \{p\}_i = \begin{Bmatrix} p_x \\ p_y \end{Bmatrix} \tag{15-3.18}$$

† For the automatic generation of the element stiffness matrices, see references 24 and 25.

$$
[K]_i = \frac{h\mu}{2\Delta}
\left[\begin{array}{c|c|c|c|c|c}
(y_2 - y_3)^2 + D_{33}(x_3 - x_2)^2 & D_{12}(x_3 - x_2)(y_2 - y_3) + D_{33}(x_3 - x_2)(y_2 - y_3) & y_3(y_2 - y_3) - D_{33}x_3(x_3 - x_2) & D_{33}y_3(x_3 - x_2) - D_{12}x_3(y_2 - y_3) & -y_2(y_2 - y_3) + D_{33}x_2(x_3 - x_2) & -D_{33}y_2(x_3 - x_2) + D_{12}x_2(y_2 - y_3) \\
\hline
D_{12}(x_3 - x_2)(y_2 - y_3) + D_{33}(y_2 - y_3)(x_3 - x_2) & D_{33}(y_2 - y_3)^2 + (x_3 - x_2)^2 & D_{12}y_3(x_3 - x_2) - D_{33}x_3(y_2 - y_3) & D_{33}y_3(y_2 - y_3) - x_3(x_3 - x_2) & -D_{12}y_2(x_3 - x_2) + D_{33}x_2(y_2 - y_3) & -D_{33}y_2(y_2 - y_3) + x_2(x_3 - x_2) \\
\hline
y_3(y_2 - y_3) - D_{33}x_3(x_3 - x_2) & -D_{33}x_3(y_2 - y_3) + D_{12}y_3(x_3 - x_2) & y_3^2 + D_{33}x_3^2 & -(D_{12} + D_{33})x_3y_3 & -y_3y_2 - D_{33}x_3x_2 & D_{33}x_3y_2 + D_{12}x_2y_3 \\
\hline
-D_{12}x_3(y_2 - y_3) + D_{33}y_3(x_3 - x_2) & D_{33}y_3(y_2 - y_3) - x_3(x_3 - x_2) & -(D_{12} + D_{33})x_3y_3 & D_{33}y_3^2 + x_3^2 & D_{12}x_3y_2 + D_{33}x_2y_3 & -D_{33}y_3y_2 - x_3x_2 \\
\hline
-y_2(y_2 - y_3) + D_{33}x_2(x_3 - x_2) & D_{33}x_2(y_2 - y_3) - D_{12}y_2(x_3 - x_2) & -y_2y_3 - D_{33}x_2x_3 & D_{33}x_2y_3 + D_{12}x_3y_2 & y_2^2 + D_{33}x_2^2 & -(D_{12} + D_{33})x_2y_2 \\
\hline
D_{12}x_2(y_2 - y_3) - D_{33}y_2(x_3 - x_2) & -D_{33}y_2(y_2 - y_3) + x_2(x_3 - x_2) & D_{12}x_2y_3 + D_{33}x_3y_2 & -D_{33}y_2y_3 - x_2x_3 & -(D_{12} + D_{33})x_2y_2 & D_{33}y_2^2 + x_2^2
\end{array}\right]
\qquad (15\text{-}3.17)
$$

Analogous to Eqs. (15-2.15) and (15-2.15a), we define generalized element body force and surface load matrices as

$$\{\overline{Q}\}_i = h \iint_{A_i} [N]_i^T \{X\}_i \, dA_i \tag{15-3.19}$$

$$\{\overline{P}\}_i = h \int_{\Gamma} [N]_i^T \{p\}_i \, dS_i \tag{15-3.20}$$

where, analogous to Eq. (15-3.14), the element body force and element surface load matrices are

$$\{Q\}_i = [A^{-1}]_i^T [\overline{Q}]_i \tag{15-3.19a}$$

$$\{P\}_i = [A^{-1}]_i^T [\overline{P}]_i \tag{15-3.20a}$$

We first evaluate the generalized element body forces. The integrand of Eq. (15-3.19) is

$$[N]_i^T \{X\}_i = \begin{bmatrix} B_x & xB_x & yB_x & B_y & xB_y & yB_y \end{bmatrix}^T \tag{15-3.21}$$

Assuming (B_x, B_y) to be constant within the element and integrating Eq. (15-3.21) over A_i, we obtain the element body force matrix

$$\{\overline{Q}\}_i = \frac{h\Delta}{2} \begin{bmatrix} B_x & B_x I_2 & B_x I_3 & B_y & B_y I_2 & B_y I_3 \end{bmatrix}^T \tag{15-3.22}$$

where (see Appendix 15A):

$$\begin{aligned} I_1 &= \Delta/2 = \text{the area of the triangular element} \\ I_2 &= (x_1 + x_2 + x_3)/3 \\ I_3 &= (y_1 + y_2 + y_3)/3 \end{aligned} \tag{15-3.23}$$

Substitution of Eq. (15-3.22) into Eq. (15-3.19a) yields the element body force matrix

$$\{Q\}_i = [A^{-1}]_i^T \{\overline{Q}\}_i = \frac{h}{2} \begin{Bmatrix} [\Delta + I_2(y_2 - y_3) + I_3(x_3 - x_2)] B_x \\ [\Delta + I_2(y_2 - y_3) + I_3(x_3 - x_2)] B_y \\ (y_3 I_2 - x_3 I_3) B_x \\ (y_3 I_2 - x_3 I_3) B_y \\ (-y_2 I_2 + x_2 I_3) B_x \\ (-y_2 I_2 + x_2 I_3) B_y \end{Bmatrix} \tag{15-3.24}$$

As checks on the correctness of $\{Q\}_i$, we have the conditions

$$\text{Row }(1) + \text{row }(3) + \text{row }(5) = \frac{h\Delta}{2} B_x$$

$$\text{Row }(2) + \text{row }(4) + \text{row }(6) = \frac{h\Delta}{2} B_y$$

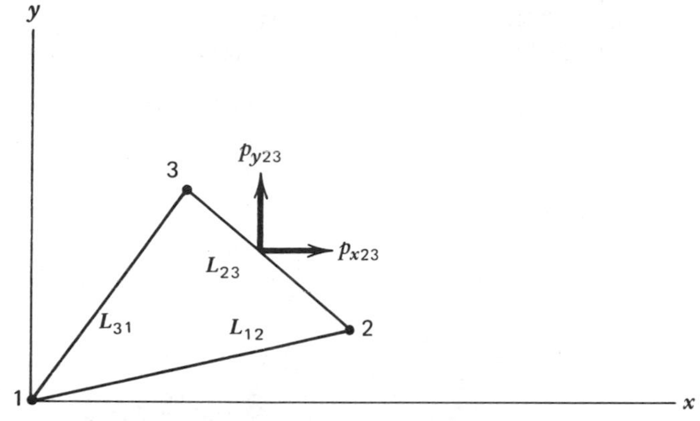

Fig. 15-3.4/A triangular element with boundary tractions.

Consider now the surface tractions (stresses) p_{x23} and p_{y23} acting on the L_{23} edge of the element (Fig. 15-3.4). The equation of the L_{23} line is $y = a_{23}x + b_{23}$, where $a_{23} = (y_3 - y_2)/(x_3 - x_2)$ and $b_{23} = (y_2x_3 - y_3x_2)/(x_3 - x_2)$. The equations of lines L_{31} and L_{12} may be obtained by proper cyclic index permutation ($1 \to 2 \to 3 \to 1$).

Evaluating the integrand of Eq. (15-3.20), we find

$$[N]_i^T\{p\}_i = [p_{x23} \quad xp_{x23} \quad yp_{x23} \quad p_{y23} \quad xp_{y23} \quad yp_{y23}]^T \tag{15-3.25}$$

In view of the approximation of constant stress quantities within the element, it is reasonable to assume constant boundary tractions acting on the sides of the element. Consequently, we regard p_{x23} and p_{y23} as constants along the L_{23} segment. Integrating Eq. (15-3.25) with respect to

$$dS = \frac{L_{23}}{x_3 - x_2} dx$$

between the limits x_2, x_3, we obtain the element surface load matrix $\{\bar{P}\}_i$ due to surface tractions p_{x23} and p_{y23} on L_{23}.

$$\{\bar{P}\}_i = hL_{23}\begin{Bmatrix} p_{x23} \\ \frac{1}{2}p_{x23}(x_3 + x_2) \\ \frac{1}{2}p_{x23}(y_3 + y_2) \\ p_{y23} \\ \frac{1}{2}p_{y23}(x_3 + x_2) \\ \frac{1}{2}p_{y23}(y_3 + y_2) \end{Bmatrix}$$

Summation of this equation, with cyclic permutations of indices ($1 \to 2 \to 3 \to 1$), gives the $\{\bar{P}\}_i$ matrix for all three edges of the element under boundary tractions. Denoting the special summation described above as $\sum_c$, and performing the matrix multiplication $[A^{-1}]_i^T\{\bar{P}\}_i$, by Eq. (15-3.20a) we find the element *surface force* or *load matrix* to be

$$\{P\}_i = \frac{h}{\Delta}\begin{Bmatrix} \sum_c [\Delta + \frac{1}{2}(y_2 - y_3)(x_2 + x_3) + \frac{1}{2}(x_3 - x_2)(y_3 + y_2)]\, L_{23} p_{x23} \\ \sum_c [\Delta + \frac{1}{2}(y_2 - y_3)(x_2 + x_3) + \frac{1}{2}(x_3 - x_2)(y_3 + y_2)]\, L_{23} p_{y23} \\ \sum_c \frac{1}{2}[y_3(x_3 + x_2) - x_3(y_3 + y_2)]\, p_{x23} L_{23} \\ \sum_c \frac{1}{2}[y_3(x_3 + x_2) - x_3(y_3 + y_2)]\, p_{y23} L_{23} \\ \sum_c \frac{1}{2}[-y_2(x_3 + x_2) + x_2(y_3 + y_2)]\, p_{x23} L_{23} \\ \sum_c \frac{1}{2}[-y_2(x_3 + x_2) + x_2(y_3 + y_2)]\, p_{y23} L_{23} \end{Bmatrix} \tag{15-3.26}$$

Checks on Eq. (15-3.26) exist in the form

$$\begin{aligned} \text{Row (1)} + \text{row (3)} + \text{row (5)} &= \sum_c p_{x23} L_{23} h \\ \text{Row (2)} + \text{row (4)} + \text{row (6)} &= \sum_c p_{y23} L_{23} h \end{aligned} \tag{15-3.27}$$

Finally, the addition of $\{Q\}_i$ and $\{P\}_i$ yields the desired *element force* or *load matrix* $\{F\}_i$.[†] The element properties $\{K\}_i$ and $\{F\}_i$ are then evaluated for each element, that is, for $i = 1, 2, 3, \ldots, n$.

Assemblage of the Structural Stiffness Matrix/To solve a plane elasticity problem, it is necessary to combine the individual element stiffness matrices $[K]_i$ and the individual load matrices $\{F\}_i$ to form the *structural stiffness matrix* $[S]$ and the *structural load matrix* $\{F\}$, respectively. The process of assemblage is most conveniently carried out within an electronic computer. To explain this process, we introduce two sets of node numbering systems. We let numerals in boldface refer to the structural system, and numerals in lightface refer to the element system (Fig. 15-3.5). Thus, for the idealization of the plane problem shown in Fig. 15-3.5, we have fifteen structural nodes ($\mathbf{n_s}$) and three element nodes (n_e). The structural nodal displacement numbering system ($\delta_{\mathbf{1}}, \delta_{\mathbf{2}}, \delta_{\mathbf{3}}, \ldots, \delta_{\mathbf{2n_s}}$) and the element nodal displacement numbering system ($\delta_{1i}, \delta_{2i}, \ldots, \delta_{6i}$) are also indicated in Fig. 15-3.5.

[†] Instead of forming the nodal load matrix $\{F\}_i$ as described here, the external loads may be simply lumped and distributed to the element nodes according to a certain arithmetic rule. For example ($p_{x23} L_{23}$) may be divided into two equal parts and assigned to the nodes 2 and 3, respectively. However, in general, this technique is a cruder approximation than the method described here.

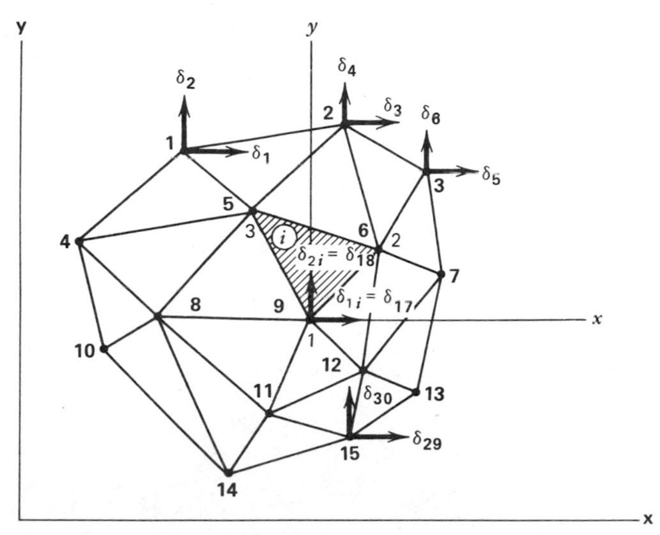

Fig. 15-3.5/Finite element idealization of a plane problem.

Accordingly, for the subdivision under consideration, we have 6 element nodal displacements for each element and 30 structural nodal displacements. If the net of subdivision is regular, the numbering can be so arranged that a mathematical relationship exists between element node numbers and structural node numbers. However, generally such a relationship is not readily obtainable. Hence, ordinarily the relation between n_e and $\mathbf{n}_s$ for each element is obtained manually from a figure (Fig. 15-3.5) and fed into a computer as input data. For example, in the ith element, we have a correspondence between (1, 2, 3) and (**9**, **6**, **5**).

The element nodal displacements and the structural nodal displacements for the ith element are shown in Fig. 15-3.6. According to the element and structural numbering systems, we find

$$\left[\delta_{1i} \quad \delta_{2i} \quad \delta_{3i} \quad \delta_{4i} \quad \delta_{5i} \quad \delta_{6i}\right]^T \leftrightarrow \left[\delta_{\mathbf{17}} \quad \delta_{\mathbf{18}} \quad \delta_{\mathbf{11}} \quad \delta_{\mathbf{12}} \quad \delta_{\mathbf{9}} \quad \delta_{\mathbf{10}}\right]^T \tag{15-3.28}$$

Correspondingly, we have the following relation between element loads $\{F\}_i$ and structural loads $\{F\}$:

$$\lfloor F_{1i} \quad F_{2i} \quad F_{3i} \quad F_{4i} \quad F_{5i} \quad F_{6i}\rfloor^T \leftrightarrow \lfloor F_{\mathbf{17}} \quad F_{\mathbf{18}} \quad F_{\mathbf{11}} \quad F_{\mathbf{12}} \quad F_{\mathbf{9}} \quad F_{\mathbf{10}}\rfloor^T \tag{15-3.29}$$

The double-headed arrows in Eqs. (15-3.28) and (15-3.29) indicate the correspondence of the quantities on the left to the quantities on the right. Generally, the ith element loads make up only a part of the total structural loads; for example, the structural load $F_{\mathbf{17}}$ may include contributions from elements adjacent to the ith element.

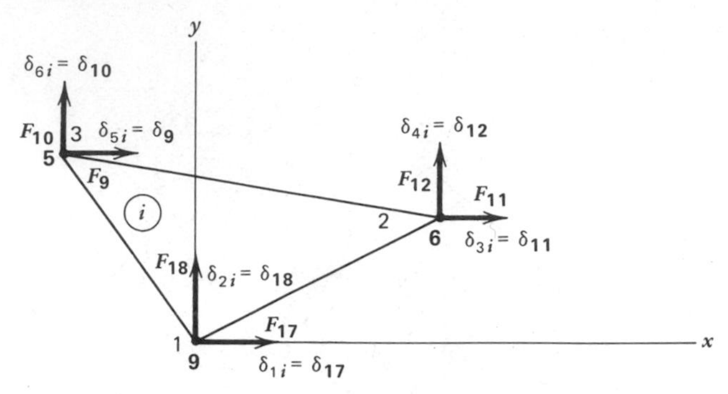

Fig. 15-3.6/Element and structural numbering systems.

The matrices $\{F\}$ and $[S]$ are related by the equation

$$\begin{Bmatrix} F_1 \\ F_2 \\ \vdots \\ F_{17} \\ \vdots \\ F_{30} \end{Bmatrix} = \begin{bmatrix} S_{1,1,} & S_{1,2} & \cdots & S_{1,17} & \cdots & S_{1,30} & \\ S_{2,1} & S_{2,2} & \cdots & S_{2,17} & \cdots & S_{2,30} & \\ \vdots & & \vdots & & & \vdots & \vdots \\ S_{17,1} & S_{17,2} & \cdots & S_{17,17} & \cdots & S_{17,30} & \\ \vdots & & \vdots & & & \vdots & \vdots \\ S_{30,1} & S_{30,2} & \cdots & S_{30,17} & \cdots & S_{30,30} & \end{bmatrix} \begin{Bmatrix} \delta_1 \\ \delta_2 \\ \vdots \\ \delta_{17} \\ \vdots \\ \delta_{30} \end{Bmatrix} \tag{15-3.30}$$

where $\{F\}$ is the structural load matrix, $[S\}$ is the structural stiffness matrix, and $\{\delta\}$ is the structural nodal matrix. To obtain the matrices $\{F\}$, $[S]$, and $\{\delta\}$, we must superimpose matrices $\{F\}_i$, $[K]_i$ and $\{\delta\}_i$, properly.

As an example, let us consider the elements F_{17}, F_{18}, $S_{17,17}$ and $S_{17,18}$ of matrices $\{F\}$ and $[S]$ of Eq. (15-3.30). We isolate the elements surrounding node **9** as shown in Fig. 15-3.7. The elements are numbered $i, i+1, \ldots, i+4$. The following correspondence between the element and structural nodal displacement exists:

Elements	*Correspondence between Nodal Displacements*		
	Element		*Structural*
i	$\delta_{1i}, \quad \delta_{2i}$	$=$	δ_{17}, δ_{18}
$i+1$	$\delta_{3(i+1)}, \delta_{4(i+1)}$	$=$	δ_{17}, δ_{18}
$i+2$	$\delta_{3(i+2)}, \delta_{4(i+2)}$	$=$	δ_{17}, δ_{18}
$i+3$	$\delta_{5(i+3)}, \delta_{6(i+3)}$	$=$	δ_{17}, δ_{18}
$i+4$	$\delta_{5(i+4)}, \delta_{6(i+4)}$	$=$	δ_{17}, δ_{18}

Hence, by adding all the element nodal loads or forces from the elements

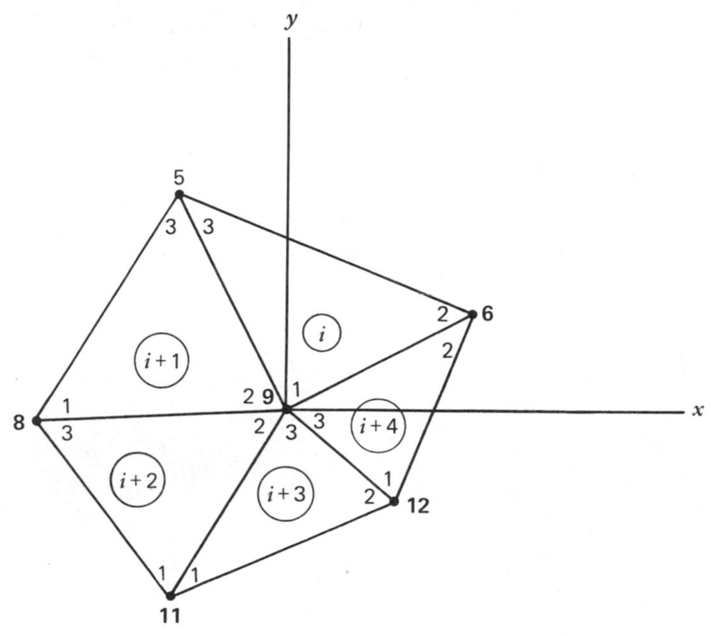

Fig. 15-3.7/Elements surrounding node **9**.

surrounding node **9**, we have

$$F_{17} = (F_1)_i + (F_3)_{i+1} + (F_3)_{i+2} + (F_5)_{i+3} + (F_5)_{i+4}$$
$$F_{18} = (F_2)_i + (F_4)_{i+1} + (F_4)_{i+2} + (F_6)_{i+3} + (F_6)_{i+4} \qquad (15\text{-}3.31)$$

To form $S_{17,17}$ (and then $S_{17,18}$), we let δ_{17} (and then δ_{18}) = 1 and set all the other nodal displacements to zero. Next we consider the structural nodal force F_{17} induced by the above set of nodal displacements. By definition, these are $S_{17,17}$ (or $S_{17,18}$); Eq. (15-3.30). Thus, by summing up the contribution of the force to F_{17} from every element which is joined to node **9**, we obtain

$$S_{17,17} = (k_{11})_i + (k_{33})_{i+1} + (k_{33})_{i+2} + (k_{55})_{i+3} + (k_{55})_{i+4}$$
$$S_{17,18} = (k_{12})_i + (k_{34})_{i+1} + (k_{34})_{i+2} + (k_{56})_{i+3} + (k_{56})_{i+4}$$
$$(15\text{-}3.32)$$

If the nodal force contributions to F_{18} are summed, we find

$$S_{18,18} = (k_{22})_i + (k_{44})_{i+1} + (k_{44})_{i+2} + (k_{66})_{i+3} + (k_{66})_{i+4}$$
$$S_{18,17} = (k_{21})_i + (k_{43})_{i+1} + (k_{43})_{i+2} + (k_{65})_{i+3} + (k_{65})_{i+4}$$
$$(15\text{-}3.33)$$

Since $k_{ij} = k_{ji}$ for each element, we note that $S_{17,18} = S_{18,17}$. In other words, the structural stiffness matrix $[S]$ is symmetric. Accordingly, proceeding from one structural node to another in this manner, the matrices $\{F\}$ and $[S]$ can be constructed.

In computer programs, however, it is often more convenient to assemble matrices $\{F\}$ and $[S]$ element-wise, since the element properties are computed element by element. Let us consider inserting the properties of the ith element into proper positions in matrices $\{F\}$ and $[S]$. By means of Eqs. (15-3.29) and (15-3.31), $\{F\}_i$ is added into the left-hand side of Eq. (15-3.30). For the insertion of the first row of matrix $[K]_i$ into matrix $[S]$, by Eqs. (15-3.28) and (15-3.29), we note the following relations:

$$
\begin{aligned}
(k_{11})_i &\leftrightarrow S_{\mathbf{17,17}} \\
(k_{12})_i &\leftrightarrow S_{\mathbf{17,18}} \\
(k_{13})_i &\leftrightarrow S_{\mathbf{17,11}} \\
(k_{14})_i &\leftrightarrow S_{\mathbf{17,12}} \\
(k_{15})_i &\leftrightarrow S_{\mathbf{17,9}} \\
(k_{16})_i &\leftrightarrow S_{\mathbf{17,10}}
\end{aligned}
\tag{15-3.34}
$$

where we have employed the conditions $k_{ij} = k_{ji}$ and $S_{\mathbf{i,j}} = S_{\mathbf{j,i}}$.

Hence, elements in the first row of matrix $[K]_i$, $(k_{1j})_i$ $(j = 1, 2, \ldots, 6)$, are to be inserted into appropriate locations of matrix $[S]$ as indicated by Eq. (15-3.34). Similar superposition may be performed for the other rows of matrix $[K]_i$. With the superposition being carried out termwise for all the elements $(i = 1, 2, \ldots, n)$ by means of an automatic computer routine, we may generate the structural matrices $[S]$ and $\{F\}$. Thus, formulation of matrices $[S]$ and $\{F\}$ proceeds automatically, and the process leads to the structural equation

$$\{F\} = [S]\{\delta\} \tag{15-3.35}$$

If at certain nodes the nodal displacements $\{\delta_c\}$ are prescribed or constrained as a result of given boundary conditions, appropriate rows and columns of matrices $[S]$, $\{F\}$, and $\{\delta\}$ may be deleted. This process may be accomplished either during or after formulation of matrices $[S]$, $\{F\}$, and $\{\delta\}$. The system of equations (Eq. 15-3.35) so reduced may then be solved for the unknown nodal displacements $\{\delta_f\}$ from the equation $\{\delta\} = \{\delta_c\} + \{\delta_f\}$. Back substitution of the nodal displacements into Eq. (15-3.9) yields the stress field.

Interpretation of Element Boundary Stresses/When the true stress distribution $\sigma_\alpha(x, y)$ is approximated element-wise by a set of constant element stress fields σ_α^*, as for the triangular element presented above, discontinuities of the approximate stress field appear along the element boundaries. Consequently, an interpretation of the approximate element boundary stress is desirable.

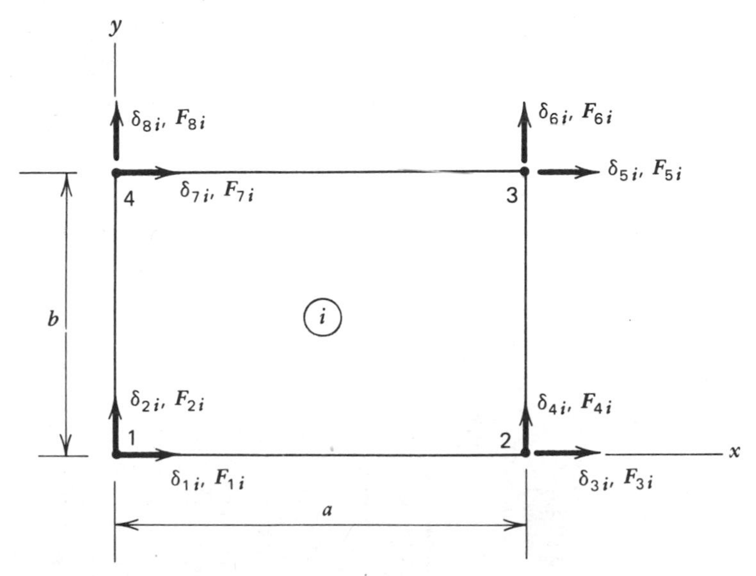

Fig. 15-3.8/A rectangular element.

We described the process of assembling the individual element stiffness matrices into the total structural stiffness matrix equation. Once the nodal point displacements have been determined, the element stresses may be computed by means of Eq. (15-3.9). Assuming that the elements are complete and compatible,[19] it may be shown that the element displacements are continuous at element boundaries. However, the continuity of element displacements does not mean that the element stresses are continuous across element boundaries. The element stresses are related to the element strains (see Eqs. 15-3.7 and 15-3.8); and, for the triangular element considered here, are constant within each element. Consequently, the stresses in adjacent elements may differ greatly if large finite elements are used. The stress differences between elements decrease as smaller finite elements are used; that is, as the model is refined. In practice, the stresses at the boundary between elements are often taken to be the average of the element stresses.†

Rectangular Finite Elements/With the selection of a proper set of trial displacement fields for the rectangular finite element, the remainder of the finite element formulation follows closely that for triangular elements.

A rectangular element with corner nodes (1, 2, 3, 4) ordered counterclockwise and with node 1 at the origin of (x, y)-element local coordi-

†A different method of stress averaging process is described in reference 26. A rather intuitive stress averaging technique may be found in reference 1.

nates is shown in Fig. 15-3.8. Again, we let the element local coordinates be parallel to the global coordinates. The numbering systems for column matrices $\{\delta\}_i$ and $\{F\}_i$ are also shown in Fig. 15-3.8.

Let the trial functions be taken in the form (see Art. 4-6, additional reference 6)†

$$\begin{aligned} u &= \alpha_1 + \alpha_2 x + \alpha_3 y + \alpha_4 xy \\ v &= \alpha_5 + \alpha_6 x + \alpha_7 y + \alpha_8 xy \end{aligned} \qquad (15\text{-}3.2)$$

Then matrix $[A]_i$ becomes [see Eq. (15-3.4)]

$$[A]_i = \begin{bmatrix} 1 & 0 & 0 & 0 & 0 & 0 & 0 & 0 \\ 0 & 0 & 0 & 0 & 1 & 0 & 0 & 0 \\ 1 & a & 0 & 0 & 0 & 0 & 0 & 0 \\ 0 & 0 & 0 & 0 & 1 & a & 0 & 0 \\ 1 & a & b & ab & 0 & 0 & 0 & 0 \\ 0 & 0 & 0 & 0 & 1 & a & b & ab \\ 1 & 0 & b & 0 & 0 & 0 & 0 & 0 \\ 0 & 0 & 0 & 0 & 1 & 0 & b & 0 \end{bmatrix}$$

since $x_2 = x_3 = a$, $y_3 = y_4 = b$, and $y_2 = x_4 = 0$, where (a, b) denote the lengths of the element sides (see Fig. 15-3.8). The inverse of matrix $[A]_i$ is

$$[A]_i^{-1} = \frac{1}{ab}\begin{bmatrix} ab & 0 & 0 & 0 & 0 & 0 & 0 & 0 \\ -b & 0 & b & 0 & 0 & 0 & 0 & 0 \\ -a & 0 & 0 & 0 & 0 & 0 & a & 0 \\ 1 & 0 & -1 & 0 & 1 & 0 & -1 & 0 \\ 0 & ab & 0 & 0 & 0 & 0 & 0 & 0 \\ 0 & -b & 0 & b & 0 & 0 & 0 & 0 \\ 0 & -a & 0 & 0 & 0 & 0 & 0 & a \\ 0 & 1 & 0 & -1 & 0 & 1 & 0 & -1 \end{bmatrix} \qquad (15\text{-}3.36)$$

By definition of the strain matrix $\{\epsilon\}_i = [B]_i\{\alpha\}_i$, we find matrix $[B]_i$. Thus

$$[B]_i = \begin{bmatrix} 0 & 1 & 0 & y & 0 & 0 & 0 & 0 \\ 0 & 0 & 0 & 0 & 0 & 0 & 1 & x \\ 0 & 0 & 1 & x & 0 & 1 & 0 & y \end{bmatrix} \qquad (15\text{-}3.37)$$

The elasticity matrix $[D]_i$ again is defined by Eq. (15-3.8). Forming the

†The stiffness matrix of an improved form of this element is given in Art. 7-7 of this reference. In addition questions of incompatibility are discussed and isoparametric formulations are presented. These questions are beyond the scope of this text. The treatment of the rectangular finite element as presented here are intended to be heuristic.

matrix $[D]_i[B]_i$, we write

$$\begin{Bmatrix} \sigma_x \\ \sigma_y \\ \tau_{xy} \end{Bmatrix} = \begin{bmatrix} 0 & \mu & 0 & \mu y & 0 & 0 & \mu D_{12} & \mu D_{12} x \\ 0 & \mu D_{12} & 0 & \mu D_{12} y & 0 & 0 & \mu & \mu x \\ 0 & 0 & \mu D_{33} & \mu D_{33} x & 0 & \mu D_{33} & 0 & \mu D_{33} y \end{bmatrix} [A]_i^{-1}\{\delta\}_i \tag{15-3.38}$$

Equations (15-3.38) indicates that the stress distribution over the rectangular element is linear in x and y.

Having the $[D]_i$ and $[B]_i$ matrices, we evaluate the generalized element stiffness matrix $[\bar{K}]_i$ according to Eq. (15-3.10). First, we compute the integrand of Eq. (15-3.10) as

$$[B]_i^T[D]_i[B]_i = \mu \begin{bmatrix} 0 & 0 & 0 & 0 & 0 & 0 & 0 & 0 \\ 0 & 1 & 0 & y & 0 & 0 & D_{12} & D_{12}x \\ 0 & 0 & D_{33} & D_{33}x & 0 & D_{33} & 0 & D_{33}x \\ 0 & y & D_{33}x & y^2 + D_{33}x^2 & 0 & D_{33}x & D_{12}y & (D_{12}+D_{33})xy \\ 0 & 0 & 0 & 0 & 0 & 0 & 0 & 0 \\ 0 & 0 & D_{33} & D_{33}x & 0 & D_{33} & 0 & D_{33}y \\ 0 & D_{12} & 0 & D_{12}y & 0 & 0 & 1 & x \\ 0 & D_{12}x & D_{33}y & (D_{12}+D_{33})xy & 0 & D_{33}y & x & x^2 + D_{33}y^2 \end{bmatrix} \tag{15-3.39}$$

Next, we integrated each term in the matrix over the area of the element. Integration formulas for the rectangular element are listed in Appendix (15A-2). Integration yields the matrix $[\bar{K}]_i$ in the form

$$[\bar{K}]_i = \mu hab \begin{bmatrix} 0 & 0 & 0 & 0 & 0 & 0 & 0 & 0 \\ 0 & 1 & 0 & \frac{b}{2} & 0 & 0 & D_{12} & \frac{D_{12}a}{2} \\ 0 & 0 & D_{33} & \frac{D_{33}a}{2} & 0 & D_{33} & 0 & \frac{D_{33}b}{2} \\ 0 & \frac{b}{2} & \frac{D_{33}a}{2} & \frac{b^2 + D_{33}a^2}{3} & 0 & \frac{D_{33}a}{2} & \frac{D_{12}b}{2} & \frac{(D_{12}+D_{33})ab}{4} \\ 0 & 0 & 0 & 0 & 0 & 0 & 0 & 0 \\ 0 & 0 & D_{33} & \frac{D_{33}a}{2} & 0 & D_{33} & 0 & \frac{D_{33}b}{2} \\ 0 & D_{12} & 0 & \frac{D_{12}b}{2} & 0 & 0 & 1 & \frac{a}{2} \\ 0 & \frac{D_{12}a}{2} & \frac{D_{33}b}{2} & \frac{(D_{12}+D_{33})ab}{4} & 0 & \frac{D_{33}b}{2} & \frac{a}{2} & \frac{a^2 + D_{33}b^2}{3} \end{bmatrix} \tag{15-3.40}$$

Finally, performing the multiplication $[A^{-1}]_i^T[\bar{K}]_i[A]_i^{-1}$, we obtain the element stiffness matrix $[K]_i$ in the form that follows, Eq. (15-3.41).

$$
[K]_i = \frac{h}{ab}\begin{bmatrix}
\frac{b^2 + a^2 D_{33}}{3} & \frac{ab(D_{12}+D_{33})}{4} & -\frac{b^2}{3}+\frac{a^2 D_{33}}{6} & \frac{ab(D_{12}-D_{33})}{4} & -\frac{b^2+a^2 D_{33}}{6} & -\frac{ab(D_{12}+D_{33})}{4} & \frac{b^2}{6}-\frac{a^2 D_{33}}{3} & \frac{ab(-D_{12}+D_{33})}{4} \\
\frac{ab(D_{12}+D_{33})}{4} & \frac{a^2+b^2 D_{33}}{3} & \frac{ab(-D_{12}+D_{33})}{4} & \frac{a^2}{6}-\frac{b^2 D_{33}}{3} & -\frac{ab(D_{12}+D_{33})}{4} & -\frac{a^2+b^2 D_{33}}{6} & \frac{ab(D_{12}-D_{33})}{4} & -\frac{a^2}{3}+\frac{b^2 D_{33}}{6} \\
-\frac{b^2}{3}+\frac{a^2 D_{33}}{6} & \frac{ab(-D_{12}+D_{33})}{4} & \frac{b^2+a^2 D_{33}}{3} & -\frac{ab(D_{12}+D_{33})}{4} & \frac{b^2}{6}-\frac{a^2 D_{33}}{3} & \frac{ab(D_{12}-D_{33})}{4} & -\frac{b^2+a^2 D_{33}}{6} & \frac{ab(D_{12}+D_{33})}{4} \\
\frac{ab(D_{12}-D_{33})}{4} & \frac{a^2}{6}-\frac{b^2 D_{33}}{3} & -\frac{ab(D_{12}+D_{33})}{4} & \frac{a^2+b^2 D_{33}}{3} & \frac{ab(-D_{12}+D_{33})}{4} & -\frac{a^2}{3}+\frac{b^2 D_{33}}{6} & \frac{ab(D_{12}+D_{33})}{4} & -\frac{a^2+b^2 D_{33}}{6} \\
-\frac{b^2+a^2 D_{33}}{6} & -\frac{ab(D_{12}+D_{33})}{4} & \frac{b^2}{6}-\frac{a^2 D_{33}}{3} & \frac{ab(-D_{12}+D_{33})}{4} & \frac{b^2+a^2 D_{33}}{3} & \frac{ab(D_{12}+D_{33})}{4} & -\frac{b^2}{3}+\frac{a^2 D_{33}}{6} & \frac{ab(D_{12}-D_{33})}{4} \\
-\frac{ab(D_{12}+D_{33})}{4} & -\frac{a^2+b^2 D_{33}}{6} & \frac{ab(D_{12}-D_{33})}{4} & -\frac{a^2}{3}+\frac{b^2 D_{33}}{6} & \frac{ab(D_{12}+D_{33})}{4} & \frac{a^2+b^2 D_{33}}{3} & \frac{ab(-D_{12}+D_{33})}{4} & \frac{a^2}{6}-\frac{b^2 D_{33}}{3} \\
\frac{b^2}{6}-\frac{a^2 D_{33}}{3} & \frac{ab(D_{12}-D_{33})}{4} & -\frac{b^2+a^2 D_{33}}{6} & \frac{ab(D_{12}+D_{33})}{4} & -\frac{b^2}{3}+\frac{a^2 D_{33}}{6} & \frac{ab(-D_{12}+D_{33})}{4} & \frac{b^2+a^2 D_{33}}{3} & -\frac{ab(D_{12}+D_{33})}{4} \\
\frac{ab(-D_{12}+D_{33})}{4} & -\frac{a^2}{3}+\frac{b^2 D_{33}}{6} & \frac{ab(D_{12}+D_{33})}{4} & -\frac{a^2+b^2 D_{33}}{6} & \frac{ab(D_{12}-D_{33})}{4} & \frac{a^2}{6}-\frac{b^2 D_{33}}{3} & -\frac{ab(D_{12}+D_{33})}{4} & \frac{a^2+b^2 D_{33}}{3}
\end{bmatrix}
\tag{15-3.41}
$$

The matrix $[K]_i$ is symmetrical, and its diagonal elements are positive. An equilibrium check on the rectangular element stiffness matrix may be performed as in the case of the triangular element.

To compute the element load matrix, we first observed that the matrix $[N]_i$ is [see Eqs. (15-3.2) and (15-3.3)]

$$[N]_i = \begin{bmatrix} 1 & x & y & xy & 0 & 0 & 0 & 0 \\ 0 & 0 & 0 & 0 & 1 & x & y & xy \end{bmatrix} \tag{15-3.42}$$

Consider the case for constant body forces (B_x, B_y) and surface tractions (p_x, p_y) Fig. 15-3.9. By means of matrices $[N]_i$, $[A]_i$, $\{X\}_i$, and $\{p\}_i$ we find the element load matrices $\{Q\}_i$ and $\{P\}_i$. They are

$$\{Q\}_i = \frac{abh}{4}\begin{Bmatrix} B_x \\ B_y \\ B_x \\ B_y \\ B_x \\ B_y \\ B_x \\ B_y \end{Bmatrix} \quad \text{and} \quad \{P\}_i = \frac{h}{2}\begin{Bmatrix} ap_{x12} + bp_{x41} \\ ap_{y12} + bp_{y41} \\ ap_{x12} + bp_{x23} \\ ap_{y12} + bp_{y23} \\ ap_{x23} + bp_{x23} \\ ap_{y34} + bp_{y23} \\ ap_{x34} + bp_{x41} \\ ap_{y34} + bp_{y41} \end{Bmatrix} \tag{15-3.43}$$

A more accurate approximation to applied external loads may be obtained by using linear or quadratic approximations of the load. For example, let us consider a linear variation of the boundary stress component p_x which varies from p_{x2} to p_{x3} along the edge 2–3 of the element (Fig. 15-3.10). The corresponding element nodal load matrix is

$$\{p\}_i = h\begin{Bmatrix} 0 \\ 0 \\ \left(\frac{b}{2}\right)p_{x2} + \frac{b}{6}(p_{x3} - p_{x2}) \\ 0 \\ \left(\frac{b}{2}\right)p_{x2} + \frac{b}{3}(p_{x3} - p_{x2}) \\ 0 \\ 0 \\ 0 \end{Bmatrix} \tag{15-3.44}$$

Equation (15-3.44) shows that one half of the load (bp_{x2}) goes to F_{3i} and F_{5i}, and the load $(b/2)(p_{x3} - p_{x2})$ is distributed to F_{3i} and F_{5i}, in the ratio of 1 to 2. Hence, in this case, the conversion of external load into nodal loads by means of energy considerations is equivalent to a simple static distribution of the external load.

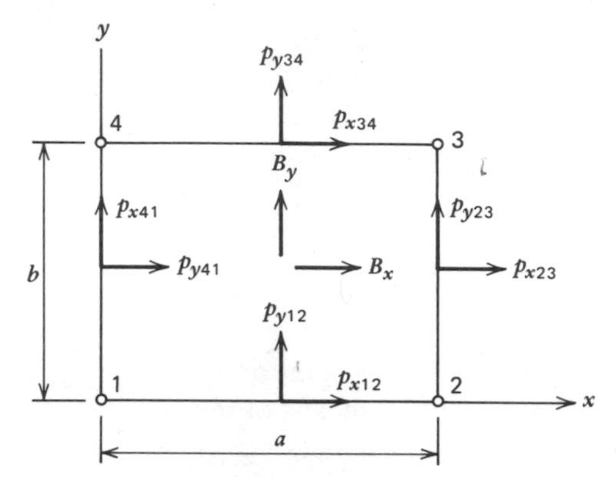

Fig. 15-3.9/External loads on a rectangular element.

Coordinate Transformations/It has been noted that the finite element properties are best formulated in terms of conveniently selected element local coordinates. If the chosen element local coordinates coincide in direction with the global coordinates of the problem, then all element properties remain unchanged under a coordinate transformation. However, for certain problems, it may be more convenient to assign nodal displacement and load quantities in terms of global coordinates which do not agree in direction with the individual element local coordinates. Consider, for example, the circular disk shown in Fig. 15-3.11. Polar coordinates $(\mathbf{x}, \mathbf{y})$ are naturally suitable for global coordinates. Referring to Fig. 15-3.11, let the ith element have local coordinates (x, y). At each node of the element, the global coordinates are designated as $(\mathbf{x}_1, \mathbf{y}_1)$, $(\mathbf{x}_2, \mathbf{y}_2)$, and

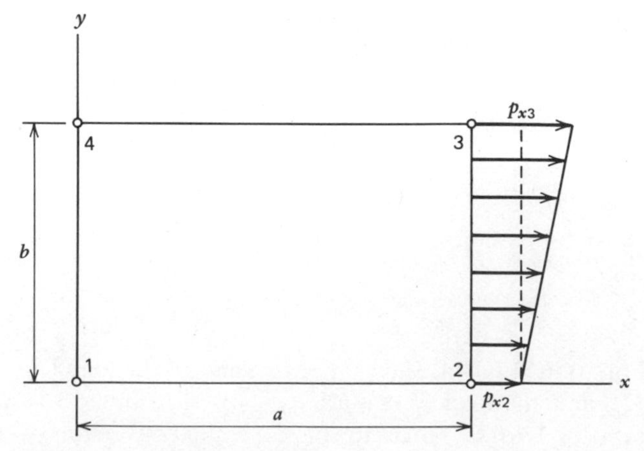

Fig. 15-3.10/Linear approximation of the boundary traction.

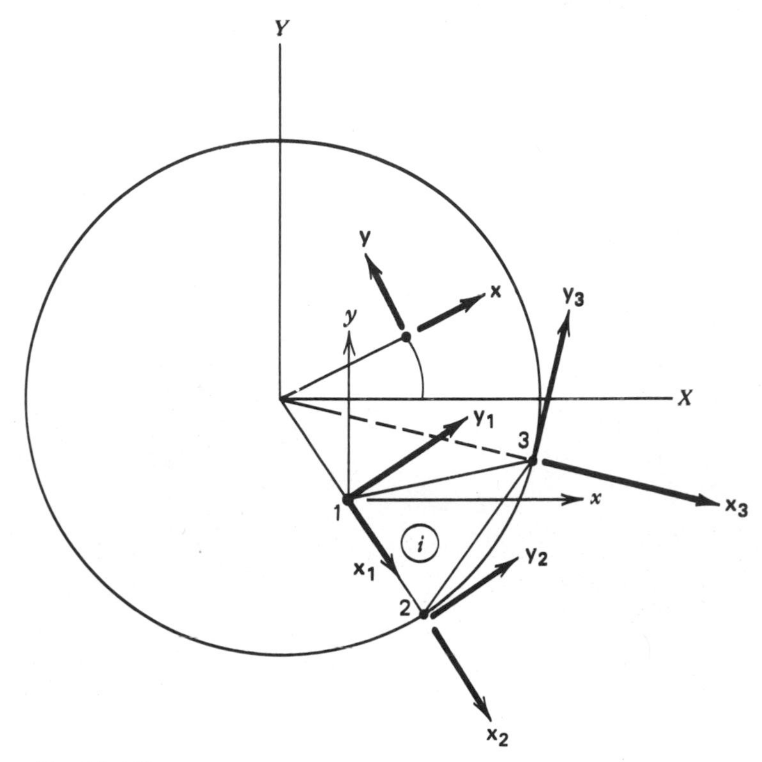

Fig. 15-3.11

$(\mathbf{x}_3, \mathbf{y}_3)$. To formulate the finite element method in terms of global coordinates, we require the transformation of the element properties from (x, y) element coordinates to the global system $(\mathbf{x}, \mathbf{y})$.

We first consider the element node 3 of element i, as shown in Fig. 5-3.12. The nodal quantities $(\bar{u}_3', \bar{v}_3')$ and (F_{x3}', F_{y3}') with respect to the local coordinates must be transformed into components $(\bar{u}_3, \bar{v}_3)$ and (F_{x3}, F_{y3}) in terms of global coordinates $(\mathbf{x}_3, \mathbf{y}_3)$, where primed quantities refer to local coordinates (x, y) and nonprimed quantities refer to global coordinates $(\mathbf{x}_3, \mathbf{y}_3)$.

By a coordinate transformation, we have

$$\begin{Bmatrix} \bar{u}_3' \\ \bar{v}_3' \end{Bmatrix} = [T_3] \begin{Bmatrix} u_3 \\ \bar{v}_3 \end{Bmatrix} \quad \text{and} \quad \begin{Bmatrix} F_{x3}' \\ F_{y3}' \end{Bmatrix} = [T_3] \begin{Bmatrix} F_{x3} \\ F_{y3} \end{Bmatrix} \qquad (15\text{-}3.45)$$

where

$$[T_3] = \begin{bmatrix} \cos\phi_3 & \sin\phi_3 \\ -\sin\phi_3 & \cos\phi_3 \end{bmatrix} \qquad (15\text{-}3.45\text{a})$$

and where ϕ_3 is the angle between axes x and $\mathbf{x}_3$. The transformation

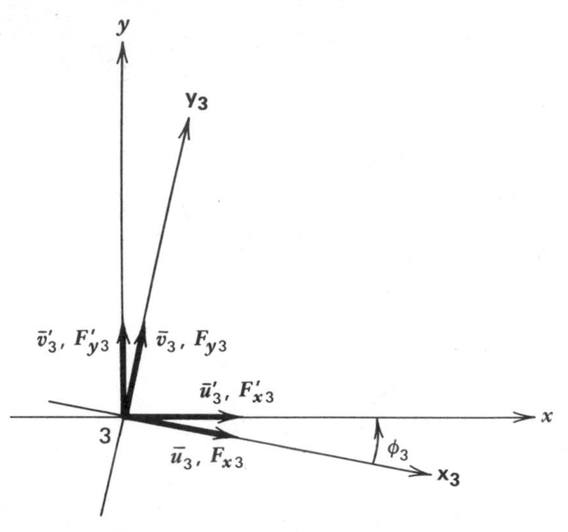

Fig. 15-3.12

matrix $[T_3]$ is called the orthogonal transformation matrix. The inverse of $[T_3]$ is equal to its transpose; that is, $[T_3]^{-1} = [T_3]^T$. We may prove this result as follows: by definition of the coordinate transformation,

$$\begin{Bmatrix} x \\ y \end{Bmatrix} = [T_3]\begin{Bmatrix} \mathbf{x}_3 \\ \mathbf{y}_3 \end{Bmatrix}$$

Also, on the basis of orthogonal rotations of axes, we have

$$[x \quad y]\begin{Bmatrix} x \\ y \end{Bmatrix} = [\mathbf{x}_3 \quad \mathbf{y}_3]\begin{Bmatrix} \mathbf{x}_3 \\ \mathbf{y}_3 \end{Bmatrix}$$

Accordingly, we may write

$$[x \quad y]\begin{Bmatrix} x \\ y \end{Bmatrix} = [\mathbf{x}_3 \quad \mathbf{y}_3][T_3]^T\begin{Bmatrix} x \\ y \end{Bmatrix} = [\mathbf{x}_3 \quad \mathbf{y}_3][T_3]^T[T_3]\begin{Bmatrix} \mathbf{x}_3 \\ \mathbf{y}_3 \end{Bmatrix}$$

Hence, $[T_3]^T[T_3] = [I]$, where $[I]$ is the unit identity matrix. Thus, $[T_3]^T = [T_3]^{-1}$. Similar transformations may be formulated at nodes 1 and 2 with orthogonal transformation matrices $[T_1]$ and $[T_2]$ defined in the sense of Eq. (15-3.45a).

Accordingly, the element nodal quantities are transformed into global nodal quantities by the relations

$$\{\delta'\}_i = [T]_i\{\delta\}_i \quad \text{and} \quad \{F'\}_i = [T]_i\{F\}_i \qquad (15\text{-}3.46)$$

with $[T]_i$ defined as

$$[T]_i = \begin{bmatrix} T_1 & 0 & 0 \\ 0 & T_2 & 0 \\ 0 & 0 & T_3 \end{bmatrix} \qquad (15\text{-}3.47)$$

Hence, the element stiffness equation, $\{F'\}_i = [K']_i\{\delta'\}_i$, in local coordinates, may be written as

$$[T]_i\{F\}_i = [K']_i[T]_i\{\delta\}_i$$

Solving for $\{F\}_i$, with $[T]_i^{-1} = [T]_i^T$ [Eq. (15-3.47)], we find

$$\{F\}_i = [T]_i^T[K']_i[T]_i\{\delta\}_i = [K]_i\{\delta\}_i \qquad (15\text{-}3.48)$$

Therefore, the element stiffness matrix $[K]_i$ in terms of global coordinates is

$$[K]_i = [T]_i^T[K']_i[T]_i \qquad (15\text{-}3.49)$$

The corresponding nodal load matrix can be transformed in a similar manner by the use of Eq. (15-3.46). Thus, we obtain

$$\{F\}_i = [T]_i^T\{F'\}_i \qquad (15\text{-}3.50)$$

Consequently, the transformation of element properties from element local coordinates to global coordinates can be performed implicitly by means of Eqs. (15-3.49) and (15-3.50) and digital computer computation on matrices.

Second Order Triangular Element / As has been pointed out earlier, the linear triangular element leads to a constant element stress field. Hence, at times it might not be sufficiently accurate in regions of high stress gradients. Furthermore, near the boundary of an elastic body the stress averaging technique fails, and some kind of extrapolation has to be used to determine the stresses at the boundary. This shortcoming may be rectified in several ways. One method is simply to subdivide the regions near the high stress gradients and boundaries into finer mesh subdivisions. Another method is to employ so-called *refined elements*.†

The *refined element* method essentially employs terms beyond first order x and y terms in the polynomial interpolations of the element displacements. Consequently, quadratic, or higher degree element stress fields may be achieved. As a result, the number of element degrees of freedom may be increased, and discontinuities in the approximating stress fields along the element interfaces may be reduced.

Here we present a *second order triangular in-plane element* where *complete* quadratic polynomials in x and y are used to represent u and v element displacements. This refined element has been extensively investigated, and its behavior in two-dimensional stress analysis has been well established.[23]

By selecting element local x- and y-axes parallel to global structural coordinate axes, we define the second order triangular element displace-

† The use of *refined elements* was proposed by Fraeijs deVeubeke, reference 27.

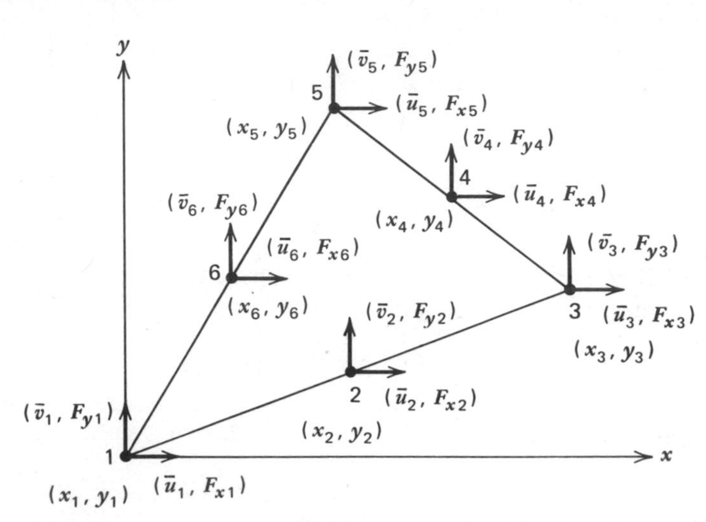

Fig. 15-3.13/Element nodal displacements and forces for second order in-plane triangular elements.

ments as

$$\begin{Bmatrix} u \\ v \end{Bmatrix}_i = \begin{bmatrix} 1 & x & y & x^2 & xy & y^2 & 0 & 0 & 0 & 0 & 0 & 0 \\ 0 & 0 & 0 & 0 & 0 & 0 & 1 & x & y & x^2 & xy & y^2 \end{bmatrix} \{\alpha\}_i$$

$$= [N]_i \{\alpha\}_i \tag{15-3.51}$$

where $\{\alpha\}_i = \begin{bmatrix} \alpha_1 & \alpha_2 & \alpha_3 & \alpha_4 & \alpha_5 & \alpha_6 & \alpha_7 \\ \alpha_8 & \alpha_9 & \alpha_{10} & \alpha_{11} & \alpha_{12} \end{bmatrix}^T$. In order to accommodate the additional six generalized coordinates α_i, we introduce so-called *midpoint nodes* at each side of the triangle as shown in Fig. 15-3.13 where element node 1 is placed at the origin of the element local coordinates, and the remaining element nodes are numbered consecutively counterclockwise around the element.

Assignment of u and v displacements at each element node yields a total of 12 nodal displacements for the ith element.†

$$\{\delta\}_i = [\bar{u}_1 \quad \bar{v}_1 \quad \bar{u}_2 \quad \bar{v}_2 \quad \bar{u}_3 \quad \bar{v}_3 \quad \bar{u}_4 \quad \bar{v}_4 \quad \bar{u}_5 \quad \bar{v}_5 \quad \bar{u}_6 \quad \bar{v}_6]^T$$

Corresponding to these element nodal displacements, we assign element nodal forces as shown in Fig. 15-3.13. The element nodal force matrix is designated as

$$\{F\}_i = \begin{bmatrix} F_{x1} & F_{y1} & F_{x2} & F_{y2} & F_{x3} & F_{y3} & F_{x4} & F_{y4} \\ F_{x5} & F_{y5} & F_{x6} & F_{y6} \end{bmatrix}^T$$

† Instead of local x and y Cartesian coordinates, so-called *triangular coordinates* may be used advantageously for higher order refined elements. For detailed use of this coordinate system, see reference 23.

The matrix $[A]_i$ which relates $\{\delta\}_i$ and $\{\alpha\}_i$ is then

$$[A]_i = \begin{bmatrix} 1 & 0 & 0 & 0 & 0 & 0 & 0 & 0 & 0 & 0 & 0 & 0 \\ 0 & 0 & 0 & 0 & 0 & 0 & 1 & 0 & 0 & 0 & 0 & 0 \\ 1 & x_2 & y_2 & x_2^2 & x_2 y_2 & y_2^2 & 0 & 0 & 0 & 0 & 0 & 0 \\ 0 & 0 & 0 & 0 & 0 & 0 & 1 & x_2 & y_2 & x_2^2 & x_2 y_2 & y_2^2 \\ 1 & x_3 & y_3 & x_3^2 & x_3 y_3 & y_3^2 & 0 & 0 & 0 & 0 & 0 & 0 \\ 0 & 0 & 0 & 0 & 0 & 0 & 1 & x_3 & y_3 & x_3^2 & x_3 y_3 & y_3^2 \\ 1 & x_4 & y_4 & x_4^2 & x_4 y_4 & y_4^2 & 0 & 0 & 0 & 0 & 0 & 0 \\ 0 & 0 & 0 & 0 & 0 & 0 & 1 & x_4 & y_4 & x_4^2 & x_4 y_4 & y_4^2 \\ 1 & x_5 & y_5 & x_5^2 & x_5 y_5 & y_5^2 & 0 & 0 & 0 & 0 & 0 & 0 \\ 0 & 0 & 0 & 0 & 0 & 0 & 1 & x_5 & y_5 & x_5^2 & x_5 y_5 & y_5^2 \\ 1 & x_6 & y_6 & x_6^2 & x_6 y_6 & y_6^2 & 0 & 0 & 0 & 0 & 0 & 0 \\ 0 & 0 & 0 & 0 & 0 & 0 & 1 & x_6 & y_6 & x_6^2 & x_6 y_6 & y_6^2 \end{bmatrix} \tag{15-3.52}$$

In view of the size of matrix $[A]_i$, its inversion will be left for a computer. The matrix $[B]_i$ which defines the element strains (Eq. 15-3.7) is then

$$[B]_i = \begin{bmatrix} 0 & 1 & 0 & 2x & y & 0 & 0 & 0 & 0 & 0 & 0 & 0 \\ 0 & 0 & 0 & 0 & 0 & 0 & 0 & 0 & 1 & 0 & x & 2y \\ 0 & 0 & 1 & 0 & x & 2y & 0 & 1 & 0 & 2x & y & 0 \end{bmatrix} \tag{15-3.53}$$

Equation (15-3.53) shows that the element strains vary linearly over the element, with corresponding linear stress variations. It follows from the definition of the element stress matrix (Eq. 15-3.8) that

$$\{\sigma\}_i = [D]_i [B]_i [A]_i^{-1} \{\delta\}_i \tag{15-3.54}$$

where the elasticity matrix $[D]_i$ is defined by Eq. (15-3.8) and the product of $[D]_i[B]_i$ is

$$[D]_i[B]_i = \begin{bmatrix} 0 & \mu & 0 & 2\mu x & \mu y & 0 \\ 0 & \mu D_{12} & 0 & 2\mu D_{12} x & \mu D_{12} y & 0 \\ 0 & 0 & \mu D_{33} & 0 & \mu D_{33} x & 2\mu D_{33} y \end{bmatrix.$$
$$\begin{matrix} 0 & 0 & \mu D_{12} & 0 & \mu D_{12} x & 2\mu D_{12} y \\ 0 & 0 & \mu & 0 & \mu x & 2\mu y \\ 0 & \mu D_{33} & 0 & 2\mu D_{33} x & \mu D_{33} y & 0 \end{bmatrix} \tag{15-3.55}$$

With the aid of matrices $[D]_i$ and $[B]_i$, we then construct the element generalized stiffness matrix $[K]_i$ (Eq. 15-3.10). Finally, we obtain Equa-

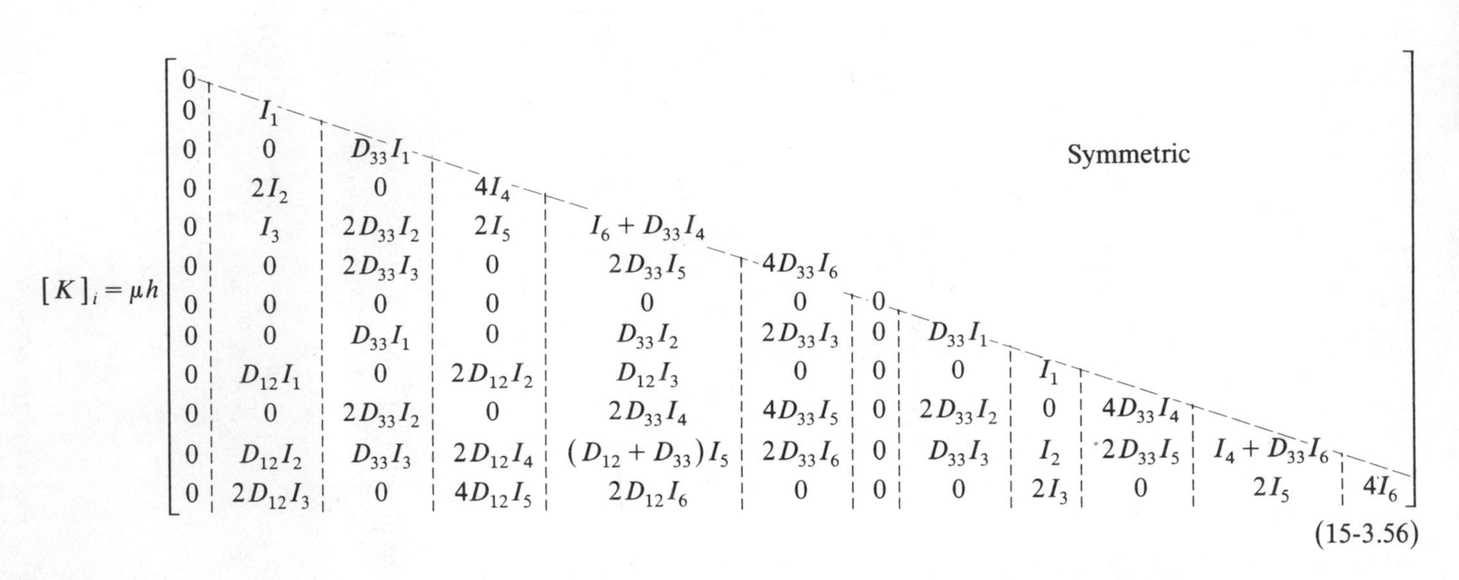

$$
[K]_i = \mu h \begin{bmatrix}
0 & & & & & & & & & & & \\
0 & I_1 & & & & & & & & & & \\
0 & 0 & D_{33} I_1 & & & & & \text{Symmetric} & & & & \\
0 & 2I_2 & 0 & 4I_4 & & & & & & & & \\
0 & I_3 & 2D_{33} I_2 & 2I_5 & I_6 + D_{33} I_4 & & & & & & & \\
0 & 0 & 2D_{33} I_3 & 0 & 2D_{33} I_5 & 4D_{33} I_6 & & & & & & \\
0 & 0 & 0 & 0 & 0 & 0 & 0 & & & & & \\
0 & 0 & D_{33} I_1 & 0 & D_{33} I_2 & 2D_{33} I_3 & 0 & D_{33} I_1 & & & & \\
0 & D_{12} I_1 & 0 & 2D_{12} I_2 & D_{12} I_3 & 0 & 0 & 0 & I_1 & & & \\
0 & 0 & 2D_{33} I_2 & 0 & 2D_{33} I_4 & 4D_{33} I_5 & 0 & 2D_{33} I_2 & 0 & 4D_{33} I_4 & & \\
0 & D_{12} I_2 & D_{33} I_3 & 2D_{12} I_4 & (D_{12} + D_{33}) I_5 & 2D_{33} I_6 & 0 & D_{33} I_3 & I_2 & 2D_{33} I_5 & I_4 + D_{33} I_6 & \\
0 & 2D_{12} I_3 & 0 & 4D_{12} I_5 & 2D_{12} I_6 & 0 & 0 & 0 & 2I_3 & 0 & 2I_5 & 4I_6
\end{bmatrix} \tag{15-3.56}
$$

tion (15-3.56) in which the following notations are introduced:

$$I_1 = \iint_{A_i} dA_i = \tfrac{1}{2}[x_1(y_3 - y_5) + x_3(y_5 - y_1) + x_5(y_1 - y_3)] = \Delta/2$$

$$I_2 = \iint_{A_i} x\,dA_i = \frac{\Delta}{6}(x_1 + x_3 + x_5)$$

$$I_3 = \iint_{A_i} y\,dA_i = \frac{\Delta}{6}(y_1 + y_3 + y_5)$$

$$I_4 = \iint_{A_i} x^2\,dA_i = \frac{\Delta}{24}\left[(x_1 + x_3 + x_5)^2 + (x_1^2 + x_3^2 + x_5^2)\right]$$

$$I_5 = \iint_{A_i} xy\,dA_i = \frac{\Delta}{24}[(x_1 + x_3 + x_5)(y_1 + y_3 + y_5) + (x_1y_1 + x_3y_3 + x_5y_5)]$$

$$I_6 = \iint_{A_i} y^2\,dA_i = \frac{\Delta}{24}\left[(y_1 + y_3 + y_5)^2 + (y_1^2 + y_3^2 + y_5^2)\right]$$

Once again, we have to rely on a computer to perform the triple matrix multiplication which defines the matrix $[K]_i$ (Eq. 15-3.14) and produce the required element stiffness matrix $[K]_i$.

The generalized element load matrices $\{\overline{Q}\}_i$, $\{\overline{P}\}_i$ can be formulated by means of Eqs. (15-3.19) and (15-3.20). For illustrative purpose, we consider here the conversion of element boundary traction force along the edge 3-4-5 into its equivalent element nodal forces (Fig. 15-3.14). Since the element stresses vary linearly, it is sufficient to approximate the boundary traction piecewise-linearly.

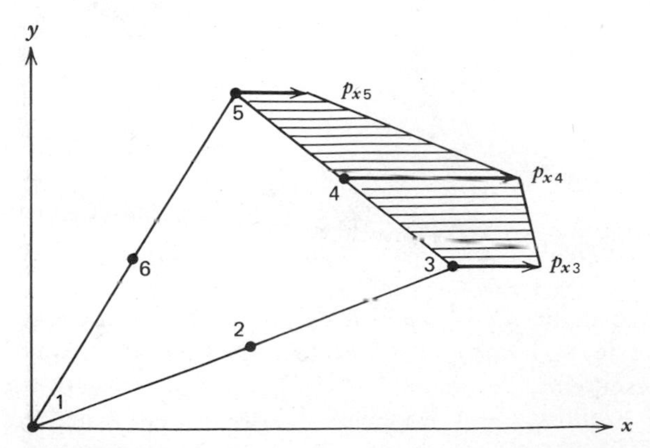

Fig. 15-3.14/Piecewise linear distribution of p_x along 3-4-5 edge.

Let the ith element be subjected to the boundary traction along its 3-4-5 edge. At the element nodes 3, 4, and 5, let the x-component of the load intensities p_{x35} be specified as p_{x3}, p_{x4}, and p_{x5}, respectively (Fig. 15-3.14). Note that the double subscripts 35 on p_x are used to indicate the line segment on which p_x acts. Hence, the linear approximations of the p_{x35} force are

$$p_{x34} = \frac{x_4 p_{x3} - x_3 p_{x4}}{x_4 - x_3} + \frac{p_{x4} - p_{x3}}{x_4 - x_3} x$$

or

$$p_{x34} = \frac{y_4 p_{x3} - y_3 p_{x4}}{y_4 - y_3} + \frac{p_{x4} - p_{x3}}{y_4 - y_3} y$$

between nodes 3 and 4, and

$$p_{x45} = \frac{x_5 p_{x4} - x_4 p_{x5}}{x_5 - x_4} + \frac{p_{x5} - p_{x4}}{x_5 - x_4} x$$

or

$$p_{x45} = \frac{y_5 p_{x4} - y_4 p_{x5}}{y_5 - y_4} + \frac{p_{x5} - p_{x4}}{y_5 - y_4} y$$

for segment 4-5. Similar expressions may be obtained for the y component of the boundary traction p_{y35} along the same edge.

Next, we calculate the integrand of Eq. (15-3.20) $[N]_i^T\{p\}_i$, which results in

$$[N]_i^T\{p\}_i = h\begin{Bmatrix} p_{x35} \\ xp_{x35} \\ yp_{x35} \\ x^2p_{x35} \\ xyp_{x35} \\ y^2p_{x35} \\ p_{y35} \\ xp_{y35} \\ yp_{y35} \\ x^2p_{y35} \\ xyp_{y35} \\ y^2p_{y35} \end{Bmatrix}$$

Integration of $[N]_i^T\{p\}_i$ along the 3-4-5 edge yields the load matrix $\{\overline{P}\}_i$ due to p_{x35} and p_{y35}. Due to rather lengthy expression of $\{\overline{P}\}_i$, it is presented in Appendix 15A-3. Premultiplication of the matrix $[A^{-1}]_i^T$ on $\{\overline{P}\}_i$ subsequently produces the desired $\{P\}_i$ matrix.

In general, two or more boundaries of the element may be loaded. Then, similar calculations must be repeated for all boundaries to obtain

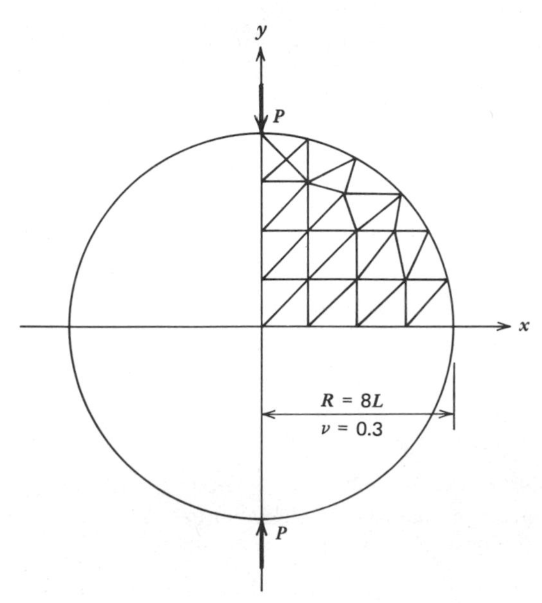

Fig. 15-3.15/Diametrically loaded circular disk (L is unit of length).

the element load matrix. When concentrated loads act, it is generally desirable to provide nodal points under each load such that direct conversions into element nodal forces are possible.

A Plane Circular Disk Subjected to Diametrically Directed Forces *P*/Consider, as a numerical example, a circular disk under a pair of diametrically applied loads. The problem is novel in a sense that it has a finite domain and an analytical solution;† hence, it can be used to check the accuracy of the finite element approximate solutions.

The nature of the problem calls for the use of in-plane triangular elements. Both the linear triangular element (LTE) and the second order triangular element (STE) are employed in this numerical example. Figure 15-3.15 shows a 30 triangular elements idealization of a quarter of the disk, with the radius R taken as 8 units of length (L). Due to the double symmetries about the x- and y-axes, it is only necessary to consider a quarter of the disk. By symmetry, the u-nodal displacements along the y-axis and the v-nodal displacements along the x-axis are zero, where u, v denote (x, y) displacement components, respectively.

The next step is to apply the theoretical formulations of the triangular element properties developed in the previous articles. This is accom-

† Reference 22, p. 318.

plished by developing appropriate computer programs. One of the advantages of the finite element method is that these computer programs can be standardized in a small number of computational steps, as follows:

1. Input the structural and element data. This includes the structural and element node numbers, the total unknown and constrained nodal displacements, and element material properties, and so on.
2. Compute element stiffness $[K]_i$ and element load $\{F\}_i$ matrices for each element.
3. Compute, element by element, the $[K]_i$ and $\{F\}_i$ matrices, and superimpose them properly to form the structural stiffness equation, $[S]\{\delta\} = \{F\}$.
4. Numerically solve the structural stiffness equation to obtain the structural nodal displacements, $\{\delta\}$, and, hence, the element nodal displacements $\{\delta\}_i$.
5. Compute the element stress and strain fields with the computed element nodal displacements $\{\delta\}_i$.
6. Print the calculated results, either in tabular or graphical forms.

In view of the wide availability of finite element computer programs,† complete FORTRAN programs are not presented here. However, from the authors' personal computer program output, we present the σ_y distribution along the x-axis as shown in Fig. 15-3.16. In this figure, two LTE solutions (30 elements and 109 elements) and one STE solution (30 elements) are plotted along with the analytical solution.[22] The number of degrees of freedom (NDF) for each of the three finite element solutions and the required computational time (CP time) on a CDC 6400 computer are given in Table 15-3.2.

From Fig. 15-3.16, we note that the 30-element STE has an advantage over the 30-element LTE in accuracy and in capability of producing

†See references 14 and 19. There exist many general purpose computer programs. Also, there exist several more specialized programs for less comprehensive purposes and for teaching purposes. For example, in the general purpose category, well-known programs are NASTRAN (*NA*sa *STR*uctural *AN*alysis, U.S. National Aeronautics and Space Administration), STRUDL (*STRU*ctural *D*esign *L*anguage, Massachusetts Institute of Technology), and SAP (*S*tructural *A*nalysis *P*rogram, E. L. Wilson, University of California, Berkeley). In the more limited purpose programs, individual minicomputer vendors often have their own software available. Alternatively, such programs are available from other sources. For example, CAL (*C*omputer *A*nalysis *L*anguage for Static and Dynamic Analysis of Structural Systems) is a matrix interpretive language and a small capacity structural analysis program, and it contains 3 to 8 node plane stress isoparametric elements. It is written in FORTRAN IV and has run on the CDC 6400, but it is designed to run on minicomputers (16 bit words) with a minimum of modification. CAL may be obtained from National Information Service, Earthquake Engineering, Computer Program Applications, Earthquake Engineering Research Center, University of California, Berkeley, California 94720.

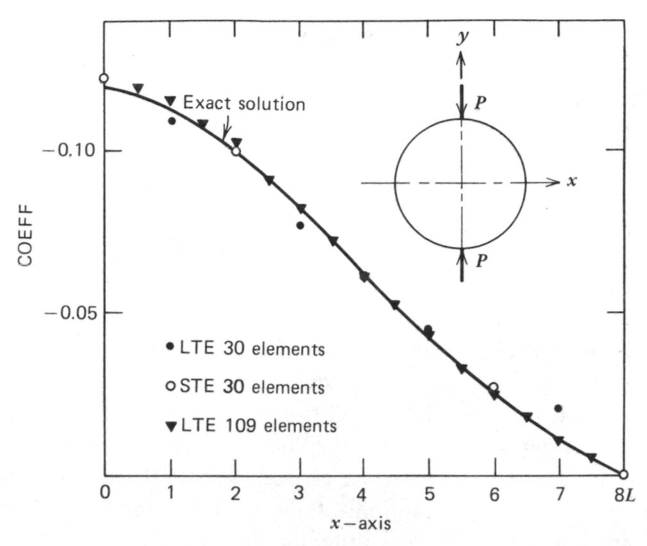

Fig. 15-3.16/Distribution of normal stress (σ_y) along the x-axis (L is unit of length); σ_y = COEFF (P/L^2).

stress values at the boundaries. However, there is very little difference in the accuracy of the 30-element STE and 109-element LTE solutions (109-element mesh subdivision is not shown here). The greater accuracy of the 30-element STE is due to its larger number of nodal unknowns (Table 15-3.2). As a result, it requires considerably more computational time. On the other hand, the 109-element LTE yields results comparable to those of the 30-element STE, while the required CP time is about one-third of the STE. This result is due to the explicit formulations of matrices $[A]_i^{-1}$ and $[K]_i$ in the LTE, and the more complex element-wise machine inversion of the matrix $[A]_i^{-1}$ and the calculation of the matrix $[K]_i$ in the STE. In the calculation of stress quantities, a similar drawback exists for STE. However, when a high degree of accuracy is required in the neighborhood of boundaries or near regions of stress concentrations, the use of STE may be desirable. Often, the type of problem may dictate the choice of the element. For example, in variable

Table 15-3.2

Number of Degrees of Freedom (NDF) and Computational Time (CP)

Type of Finite Element	*NDF*	*CDC 6400 CP Time (sec)*
30-element LTE	36	6.59
30-element STE	132	30.4
109-element LTE	121	9.79

thickness or nonhomogeneous in-plane problems, we may want to employ the larger number of mesh subdivisions of the LTE to achieve better approximations of the geometrical or material aspects of the problems. Accordingly, in general, the advantages and disadvantages of these two in-plane finite elements (LTE and STE) are governed by the problem objectives. The choice of the element is often determined by the desired accuracy and the kind of problem being treated.

Cantilever Beam with Parabolic End Load / As another example, let us consider a rectangular cantilever beam under a parabolically distributed end load P. In the classical elasticity solution of this problem,[22] the distributions of the bending stress σ_x and the shearing stress τ_{xy} of the beam are linear and parabolic, respectively. Hence, we study the finite element approximate solutions and compare them to the elementary elasticity solution.

In Fig. 15-3.17, we subdivide the cantilever beam into a triangular mesh (designated by 4 by 12 mesh). The length of the beam is taken as 24 units of length (L) which is three times the depth of the beam. The structural node numbers shown in the figure are for the LTE. The nodes are numbered along the depth of the beam to produce the smallest possible *band-width* of the structural stiffness matrix $[S]$. The band-width is defined as the maximum width of the band of nonzero elements in the $[S]$ matrix occupying the area along the principal diagonal as shown in Fig. 15-3.18. Taking advantage of the strongly banded and symmetric nature of the $[S]$ matrix, we need only use the elements within this band in the computer solution of the structural stiffness equation. Conse-

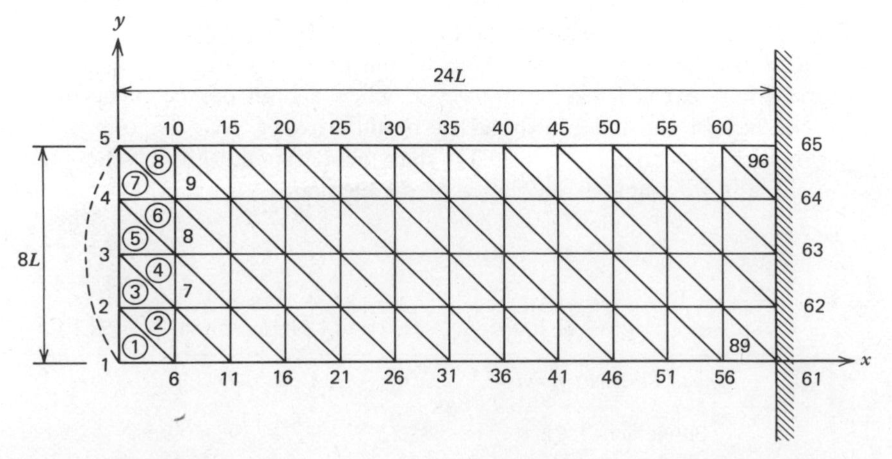

Fig. 15-3.17 / Cantilever beam with parabolic end load P, (4×12) linear triangular finite element idealization.

quently, the smaller the band-width, the less computer storage space required.

To demonstrate the band-width of matrix $[S]$ resulting from the mesh subdivision shown in Fig. 15-3.17, we consider the first element (element numbers are in circles). For this element, we have structural nodes 1, 6, and 2, with corresponding structural nodal displacements (1,2), (11,12), and (3,4), respectively. This suggests that the range of the structural nodal displacements associated with this element is from 1 to 12. Accordingly, referring to Fig. 15-3.18, we see that the maximum band-width in this case is 12. Note that a much larger band-width could result, should we number the structural nodes lengthwise along the beam. Hence, we must be careful in the selection of node numbers for efficient computation.

Except for input data, the computational steps remain essentially unchanged from those for the previous disk problem. However, special programs may be developed to take advantage of the regular mesh subdivision. For instance, a data generating subprogram can be developed to reduce the amount of input data, and the computation of element stiffness matrices may be limited to two typical elements; that is, elements 1 and 2. However, such a computer time-saving feature is applicable mainly to problems modeled by rectangular elements.

From the LTE and STE finite element solutions based on the 4 by 12 mesh, the bending stress σ_x and the shearing stress τ_{xy} distributions at the midspan and at the fixed end of the beam are given in Figs. 15-3.19 through 15-3.22. The approximate finite element solutions compare

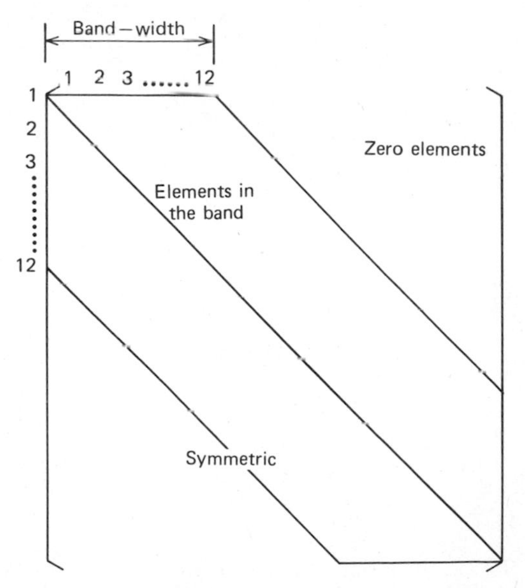

Fig. 15-3.18/Structural stiffness matrix $[S]$.

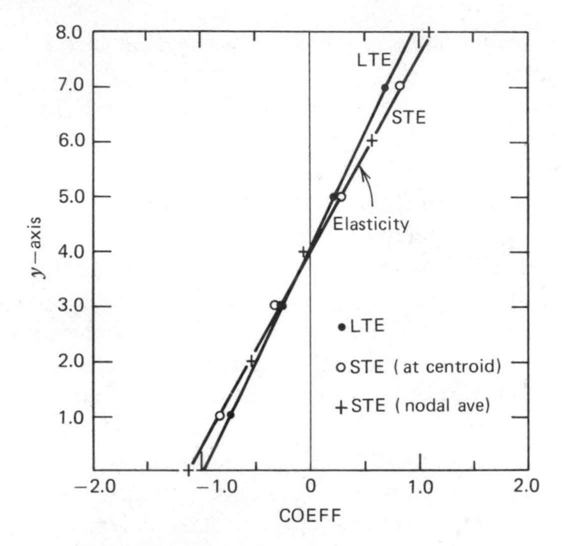

Fig. 15-3.19/Distribution of normal stress (σ_x) at midspan (4×12 mesh size); $\sigma_x = \text{COEFF}\ (P/L^2)$.

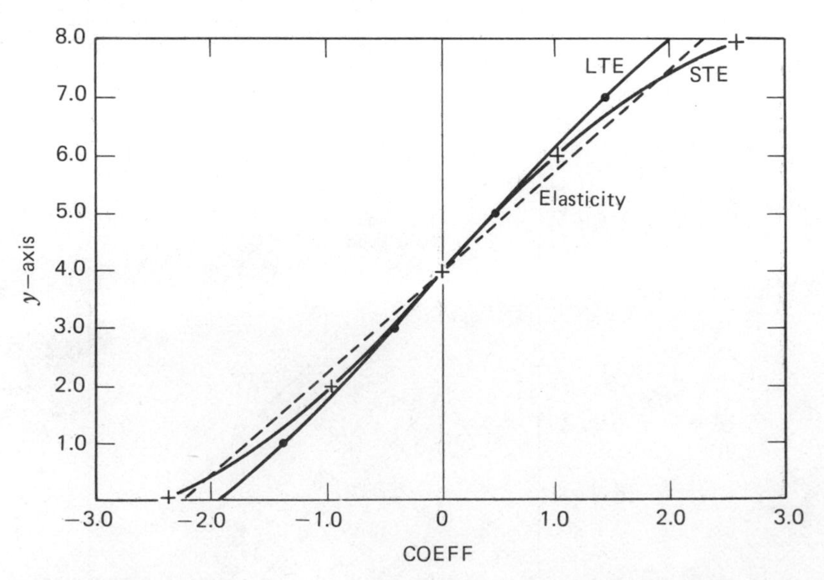

Fig. 15-3.20/Distribution of normal stress (σ_x) of beam (4×12 mesh size); $\sigma_x = \text{COEFF}\ (P/L^2)$.

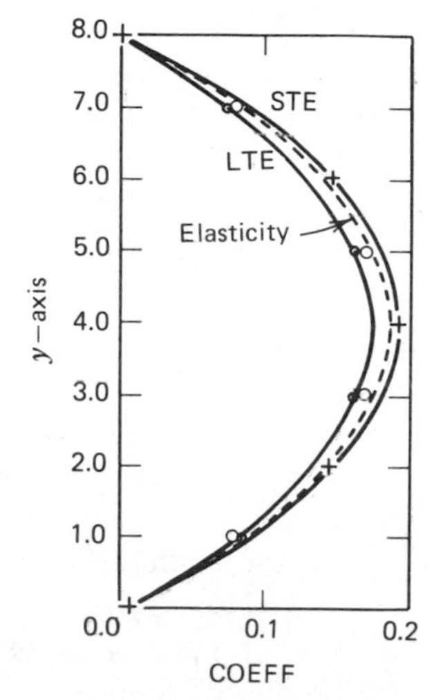

Fig. 15-3.21/Distribution of shearing stress (τ_{xy}) (4×12 mesh size); $\tau_{xy} =$ COEFF (P/L^2).

favorably with the elementary elasticity solution at the midspan of the beam. However, near the fixed end, both the bending stress and the shearing stress predicted by the finite element solutions differ widely from the corresponding stresses given by the elementary elasticity solution. The discrepancies, however, are not due principally to the inaccuracy of the finite element solution. Rather, they are the results of the inability of the elementary elasticity solution to satisfy the fixed end

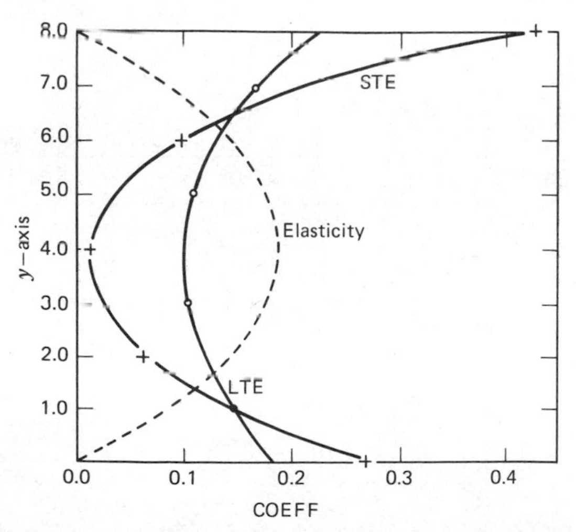

Fig. 15-3.22/Distribution of shearing stress (τ_{xy}) at fixed end (4×12 mesh size); $\tau_{xy} =$ COEFF (P/L^2).

boundary conditions.[22,28] The unsymmetrical shearing stress distribution about the beam neutral axis at the fixed end (Fig. 15-3.22) may be attributed to the nonsymmetric triangulation of the mesh subdivision. From these figures, we can observe once again the improved accuracy of the refined STE over that of the LTE. The ability of the STE of producing the stress values at the boundary is apparent in Figs. 15-3.20 and 15-3.22.

To illustrate the nature of the convergence of the finite element solutions, three different mesh subdivisions (Fig. 15-3.23) are considered. The improvement on the stresses from the 4 by 12 mesh to 8 by 24 mesh are presented in Tables 15-3.3 to 15-3.6. These tables show that reasonable accuracy may be obtained with the 4 by 12 mesh.

It has been shown that the displacement finite element method essentially minimizes the potential energy of a given elastic system.[19] Furthermore, when the sequence of meshes is taken such that the nth mesh contains all the nodal points of the previous meshes, the convergence of the approximate potential energies is monotonic decreasing. Since the

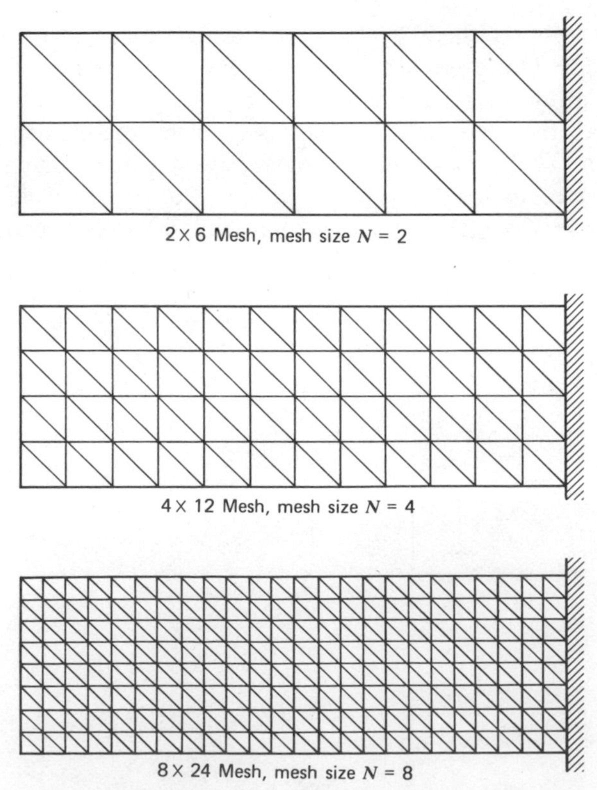

Fig. 15-3.23/Triangular mesh subdivisions of cantilever beam.

Table 15-3.3

Bending Stress Distribution at Midspan $\sigma_x = \text{COEFF}(P/L^2)^\dagger$

y	σ_x *(Elasticity)*	σ_x *(4×12 LTE)*	σ_x *(4×12 STE)*	σ_x *(8×24 LTE)*	σ_x *(8×24 STE)*
8.0	1.12500		1.1154		1.1239
7.5	0.98437			0.9351	
7.0	0.84375	0.6934	0.8229		0.8424
6.5	0.70312			0.6662	
6.0	0.56250		0.5613		0.5622
5.5	0.42187			0.3973	
5.0	0.28125	0.2165	0.2692		0.2821
4.5	0.14062			0.1284	
4.0	0.00000		−0.0955		0.0019
3.5	−0.14062			−0.1406	
3.0	−0.28125	−0.2602	−0.2845		−0.2782
2.5	−0.42187			−0.4095	
2.0	−0.56250		−0.5461		−0.5584
1.5	−0.70312			−0.6785	
1.0	−0.84375	−0.7369	−0.8383		−0.8385
0.5	−0.98437			−0.9475	
0.0	−1.12500		−1.1028		−1.1860

† L is the unit of length, P is end load, and COEFF is a factor which when multiplied by P/L^2 gives σ_x. The tabulated values are values of COEFF.

Table 15-3.4

Shearing Stress Distribution at Midspan $\tau_{xy} = \text{COEFF}\ (P/L^2)^\dagger$

y	τ_{xy} *(Elasticity)*	τ_{xy} *(4×12 LTE)*	τ_{xy} *(4×12 STE)*	τ_{xy} *(8×24 LTE)*	τ_{xy} *(8×24 STE)*
8.0	0.000000		0.0080		0.0020
7.5	0.043945			0.0454	
7.0	0.082031	0.0705	0.0769		0.0836
6.5	0.114258			0.1127	
6.0	0.140625		0.1459		0.1420
5.5	0.161133			0.1576	
5.0	0.175781	0.1616	0.1692		0.1770
4.5	0.184570			0.1799	
4.0	0.187500		0.1922		0.1887
3.5	0.184570			0.1798	
3.0	0.175781	0.1605	0.1692		0.1770
2.5	0.161133			0.1573	
2.0	0.140625		0.1456		0.1420
1.5	0.114258			0.1125	
1.0	0.082031	0.0806	0.0769		0.0836
0.5	0.043945			0.0453	
0.0	0.000000		0.0080		0.0020

† L is the unit of length, P is end load, and COEFF is a factor which when multiplied by P/L^2 gives τ_{xy}. The tabulated values are values of COEFF.

Table 15-3.5

Bending Stress Distribution at Fixed End $\sigma_x = \text{COEFF}(P/L^2)^{\dagger}$

y	σ_x *(Elasticity)*	σ_x *(4×12 LTE)*	σ_x *(4×12 STE)*	σ_x *(8×24 LTE)*	σ_x *(8×24 STE)*
8.0	2.25000		2.5077		2.8438
7.5	1.96875			1.8899	
7.0	1.68750	1.3697			1.5395
6.5	1.40625			1.2573	
6.0	1.12500		0.9605		0.9753
5.5	0.84375			0.7314	
5.0	0.56250	0.4481			0.4775
4.5	0.28125			0.2456	
4.0	0.00000		−0.0083		−0.0029
3.5	−0.28125			−0.2281	
3.0	−0.56250	−0.4140			−0.4842
2.5	−0.84375			−0.7112	
2.0	−1.12500		−0.9736		−0.9841
1.5	−1.40625			−1.2386	
1.0	−1.68750	−1.4036			−1.5477
0.5	−1.96875			−1.9462	
0.0	−2.25000		−2.3590		−2.6341

$^{\dagger}L$ is the unit of length, P is end load, and COEFF is a factor which when multiplied by P/L^2 gives σ_x. The tabulated values are values of COEFF.

Table 15-3.6

Shearing Shear Distribution at Fixed End $\tau_{xy} = \text{COEFF}(P/L^2)^{\dagger}$

y	τ_{xy} *(Elasticity)*	τ_{xy} *(4×12 LTE)*	τ_{xy} *(4×12 STE)*	τ_{xy} *(8×24 LTE)*	τ_{xy} *(8×24 STE)*
8.0	0.000000		0.4332		0.5826
7.5	0.043945			0.2328	
7.0	0.082031	0.1618			0.2095
6.5	0.114258			0.1393	
6.0	0.140625		0.0985		0.0661
5.5	0.161133			0.0861	
5.0	0.175781	0.1073			0.0152
4.5	0.184570			0.0623	
4.0	0.187500		0.0115		0.0017
3.5	0.184570			0.0617	
3.0	0.175781	0.1039			0.0151
2.5	0.161133			0.0831	
2.0	0.140625		0.0626		0.0617
1.5	0.114258			0.1317	
1.0	0.082031	0.1427			0.1698
0.5	0.043945			0.2106	
0.0	0.000000		0.2680		0.4000

$^{\dagger}L$ is the unit of length, P is end load, and COEFF is a factor which when multiplied by P/L^2 gives τ_{xy}. The tabulated values are values of COEFF.

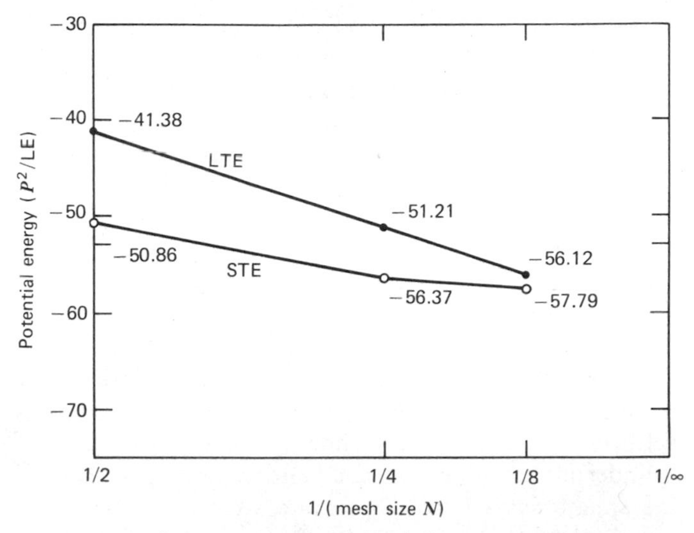

Fig. 15-3.24/Convergence of cantilever beam potential energy.

sequence of meshes used here meets these requirements, we may expect monotonic convergence of the beam potential energy with increasing number of elements. The monotonic decreasing approximate beam potential energies are shown in Fig. 15-3.24 where the abscissa is taken as the reciprocal of the mesh size N (N is defined in Fig. 15-3.23). In this scale, we get an indication of what the exact potential energy might be, since the right-hand side of the scale corresponds to the infinitely fine mesh subdivision. It may also be seen that with the refined STE, the use of the 8 by 24 mesh results in only 2.5 percent improvement on the potential energy over that of the 4 by 12 mesh. On the other hand, the number of unknown nodal displacements is 1632 for the 8 by 24 mesh and 432 for the 4 by 12 mesh. Compared to the large degree of refinement in the mesh subdivision, the corresponding improvement on the energy is very small, and continued refinement of the mesh is not warranted. Accordingly, for most practical purposes, the exact beam potential energy may be approximated by $\pi = -57.79p^2/(\mathrm{LE})$.

15-4 AXIALLY SYMMETRICAL STATES OF STRESS

In Chapter 11, we considered briefly the important class of problems known as *torsionless axially symmetrical states*. This class of problems may be defined relative to cylindrical coordinates (r, θ, z) (Fig. 15-4.1) by the conditions that the corresponding displacement components (u, v, w)

are such that v vanishes and (u, w) are independent of polar angle θ (Art. 11-1). Thus, $u = u(r, z)$, $w = w(r, z)$, and $v = 0$. It then follows by the strain-displacement relations (Chapter 1) and the stress-strain relations (Chapter 2) that the stress components $\tau_{\theta z} = \tau_{r\theta} = 0$, and that the stress components σ_r, σ_θ, σ_z, τ_{rz} are functions of coordinates (r, z) only. Physically, we classify this type of behavior as *torsionless symmetry* relative to the axis z. In contrast to torsionally symmetric problems, there also exist axially symmetric problems for which $u = w = 0$ and $v = v(r, z) \neq 0$. For example, in Art. 15-5, we treat the case of torsion of bars with circular cross section, the radius of which varies with the axial coordinate z (see also Chapter 5).

In the present article, we treat the case of torsionless symmetry only. Examples of such problems include the disk under constant angular velocity,[22] the half-space under circular spot loading,[29] the circular cylinder under uniform internal and external pressures (Chapter 11) and the sphere subject to temperature which is a function only of radial coordinate r.[22] In general, obtaining exact elasticity solutions of general axisymmetrical problems is a formidable problem, the number of such available solutions being quite small. Accordingly, in this article, we consider approximate solutions to the elasticity problem by finite element methods. This approach is particularly adaptable to complex engineering problems, and has been applied with success in the design of complicated aerospace structural systems.[30,31]

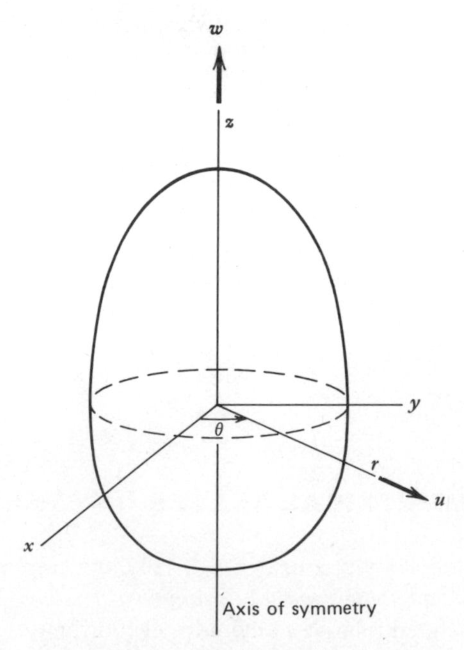

Fig. 15-4.1/Solid of revolution: cylindrical coordinates.

As applied to axisymmetric problems, finite element methods seek solutions for the radial and axial displacement components (u, w) of a body. In this sense, the stress analysis resembles closely that of the two-dimensional elasticity problems in (x, y) (see Chapter 2 and Art. 15-3). However, axisymmetrical problems are further characterized by the existence of a third extensional strain (ϵ_θ) induced by radial displacements. (See Eqs. 11-1.2.)

Consider a solid of revolution, relative to cylindrical coordinates (r, θ, z), Fig. 15-4.1. Let the solid be subdivided into toroidal elements. A typical toroidal element is shown in Fig. 15-4.2. The cross section of the element is triangular in shape. To the triangular cross section of the finite element, assign nodal points 1, 2, and 3 in counterclockwise order. Since the body element and the deformation is axisymmetric about axis z, the points 1, 2, 3 define circles 1, 2, 3 along which nodal quantities remain constant. Accordingly, we may consider the ith element to be a plane triangle as shown in Fig. 15-4.3.

By assigning (u, w) displacement components at each nodal point of the element, we obtain a total of six nodal displacements for the ith element

$$\{\delta\}_i = [\bar{u}_1 \quad \bar{w}_1 \quad \bar{u}_2 \quad \bar{w}_2 \quad \bar{u}_3 \quad \bar{w}_3]^T \tag{15-4.1}$$

The corresponding element nodal forces are

$$\{F\}_i = [F_{r1} \quad F_{z1} \quad F_{r2} \quad F_{z2} \quad F_{r3} \quad F_{z3}]^T \tag{15-4.2}$$

Positive senses of these quantities are defined in Fig. 15-4.3

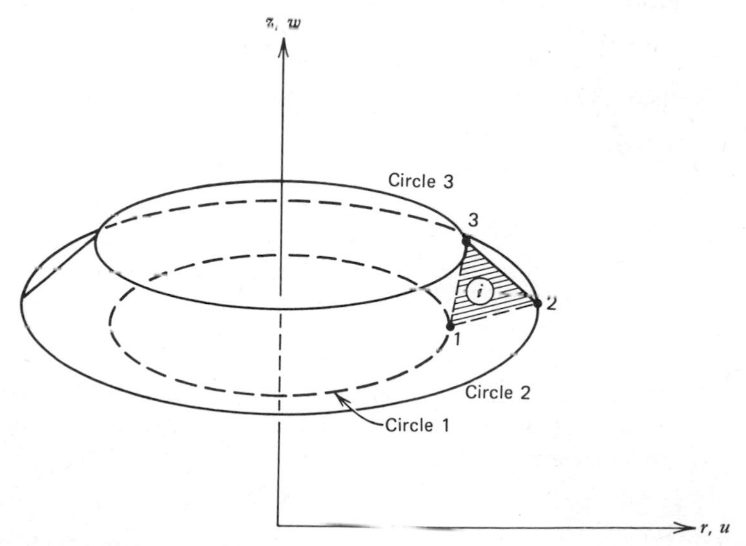

Fig. 15-4.2/Toroidal element.

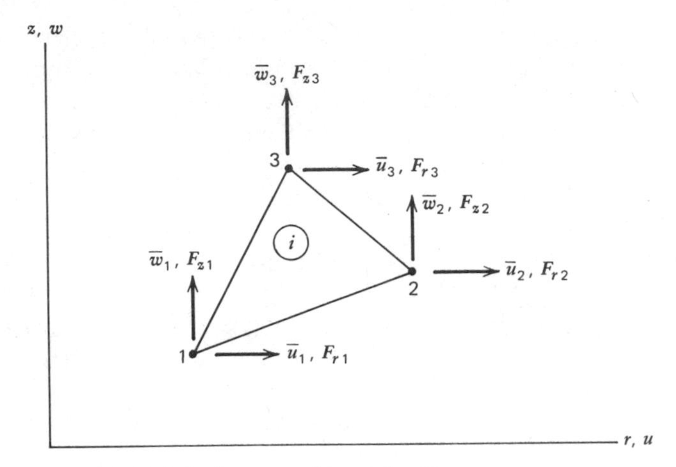

Fig. 15-4.3/i th element: plane triangle.

In view of the convergence criteria for conforming finite element displacement fields,[19] we choose the displacement field in the simplest linear form. Thus,

$$\begin{Bmatrix} u \\ w \end{Bmatrix}_i = [N]_i\{\alpha\}_i \tag{15-4.3}$$

where the matrix $[N]_i$ is defined as

$$[N]_i = \begin{bmatrix} 1 & r & z & 0 & 0 & 0 \\ 0 & 0 & 0 & 1 & r & z \end{bmatrix}$$

The above displacement fields satisfy required convergence criteria.[19] Substituting nodal coordinates of the element, (r_1, z_1) (r_2, z_2), (r_3, z_3), into Eq. (15-4.3), we obtain

$$\{\delta\}_i = [A]_i\{\alpha\}_i \tag{15-4.4}$$

where the matrix $[A]_i$ is given by

$$[A]_i = \begin{bmatrix} 1 & r_1 & z_1 & 0 & 0 & 0 \\ 0 & 0 & 0 & 1 & r_1 & z_1 \\ 1 & r_2 & z_2 & 0 & 0 & 0 \\ 0 & 0 & 0 & 1 & r_2 & z_2 \\ 1 & r_3 & z_3 & 0 & 0 & 0 \\ 0 & 0 & 0 & 1 & r_3 & z_3 \end{bmatrix}$$

$\{\delta\}_i$ is defined by Eq. (15-4.1) and $\{\alpha\}_i$ is the element generalized coordinates matrix. Inverting matrix $[A]_i$, we obtain for the element generalized coordinates

$$\{\alpha\}_i = [A]_i^{-1}\{\delta\}_i \tag{15-4.5}$$

where the inverse of $[A]_i$ is given by

$$
[A]_i^{-1} = \frac{1}{\Delta}\begin{bmatrix}
z_3r_2 - z_2r_3 & 0 & z_1r_3 - z_3r_1 & 0 & z_2r_1 - z_1r_2 & 0 \\
z_2 - z_3 & 0 & z_3 - z_1 & 0 & z_1 - z_2 & 0 \\
r_3 - r_2 & 0 & r_1 - r_3 & 0 & r_2 - r_1 & 0 \\
0 & z_3r_2 - z_2r_3 & 0 & z_1r_3 - z_3r_1 & 0 & z_2r_1 - z_1r_2 \\
0 & z_2 - z_3 & 0 & z_3 - z_1 & 0 & z_1 - z_2 \\
0 & r_3 - r_2 & 0 & r_1 - r_3 & 0 & r_2 - r_1
\end{bmatrix}
$$

and where $\Delta = (z_1 - z_2)r_3 + (z_2 - z_3)r_1 + (z_3 - z_1)r_2$ is equal to twice the area of the element triangle (1, 2, 3) (Fig. 15-4.3).

For axisymmetrical deformation of solids, the strain components are defined by the relations (see Eqs. 11-1.2 and 1-9.6).

$$
\epsilon_r = u_r \qquad \epsilon_\theta = u/r \qquad \epsilon_z = w_z \qquad \gamma_{rz} = u_z + w_r \tag{15-4.6}
$$

where subscripts (r, z) on u, w denote partial derivatives. The stress-strain relations for an isotropic material are (Eqs. 2-4.5)

$$
\begin{aligned}
\sigma_r &= \lambda e + 2G\epsilon_r \\
\sigma_\theta &= \lambda e + 2G\epsilon_\theta \\
\sigma_z &= \lambda e + 2G\epsilon_z \\
\tau_{rz} &= G\gamma_{rz}
\end{aligned} \tag{15-4.7}
$$

where λ and G are the Lamé elastic coefficients, and $e = \epsilon_r + \epsilon_\theta + \epsilon_z$ is the volumetric strain.

Substitution of the assumed element displacement fields (Eq. 15-4.3) into the definition of strain components (Eq. 15-4.6) yields

$$
\begin{Bmatrix} \epsilon_r \\ \epsilon_\theta \\ \epsilon_z \\ \gamma_{rz} \end{Bmatrix} = \begin{bmatrix}
0 & 1 & 0 & 0 & 0 & 0 \\
\frac{1}{r} & 1 & \frac{z}{r} & 0 & 0 & 0 \\
0 & 0 & 0 & 0 & 0 & 1 \\
0 & 0 & 1 & 0 & 1 & 0
\end{bmatrix} \begin{Bmatrix} \alpha_1 \\ \alpha_2 \\ \alpha_3 \\ \alpha_4 \\ \alpha_5 \\ \alpha_6 \end{Bmatrix} \tag{15-4.8}
$$

or in matrix notation

$$
\{\epsilon\}_i = [B]_i \{\alpha\}_i
$$

By means of Eq. (15-4.7), we have for the stress matrix $\{\sigma\}_i$

$$
\begin{Bmatrix} \sigma_r \\ \sigma_\theta \\ \sigma_z \\ \tau_{rz} \end{Bmatrix} = \lambda \begin{bmatrix}
1+\mu & 1 & 1 & 0 \\
1 & 1+\mu & 1 & 0 \\
1 & 1 & 1+\mu & 0 \\
0 & 0 & 0 & \frac{\mu}{2}
\end{bmatrix} \begin{Bmatrix} \epsilon_r \\ \epsilon_\theta \\ \epsilon_z \\ \gamma_{rz} \end{Bmatrix} \tag{15-4.9}
$$

or in matrix notation

$$
\{\sigma\}_i = [D]_i \{\epsilon\}_i
$$

where $\mu = (1 - 2\nu)/\nu$. The representation of stresses within the element can be obtained by elimination of $\{\epsilon\}_i$ from Eq. (15-4.9) by means of Eq. (15-4.8). Thus, we find

$$\begin{Bmatrix} \sigma_r \\ \sigma_\theta \\ \sigma_z \\ \tau_{rz} \end{Bmatrix} = \lambda \begin{bmatrix} \frac{1}{r} & 2+\mu & \frac{z}{r} & 0 & 0 & 1 \\ \frac{1+\mu}{r} & 2+\mu & \frac{(1+\mu)z}{r} & 0 & 0 & 1 \\ \frac{1}{r} & 2 & \frac{z}{r} & 0 & 0 & 1+\mu \\ 0 & 0 & \frac{\mu}{2} & 0 & \frac{\mu}{2} & 0 \end{bmatrix} [A]_i^{-1}\{\delta\}_i \tag{15-4.10}$$

By the theory of the finite element displacement method, the generalized element stiffness matrix of the torsionless axially symmetric problem may be written in the form (see Eqs. 15-2.14 and 15-3.10)

$$[\bar{K}]_i = 2\pi \iint_A [B]_i^T [D]_i [B]_i r\,dr\,dz \tag{15-4.11}$$

where the integration is over the area A of the element triangle in the r, z-plane (Fig. 15-4.3). After matrix multiplications are performed, the integrand of Eq. (15-4.11) takes the form

$$r[B]_i^T[D]_i[B]_i = \lambda \begin{bmatrix} \frac{1+\mu}{r} & 2+\mu & \frac{(1+\mu)z}{r} & 0 & 0 & 1 \\ 2+\mu & r(4+2\mu) & (2+\mu)z & 0 & 0 & 2r \\ \frac{z(1+\mu)}{r} & (2+\mu)z & \frac{(1+\mu)z^2}{r} + \frac{\mu r}{2} & 0 & \frac{\mu r}{2} & z \\ 0 & 0 & 0 & 0 & 0 & 0 \\ 0 & 0 & \frac{\mu r}{2} & 0 & \frac{\mu r}{2} & 0 \\ 0 & 2r & z & 0 & 0 & (1+\mu)r \end{bmatrix} \tag{15-4.12}$$

Then, integrating Eq. (15-4.11), we obtain

$$[\bar{K}]_i = 2\pi\lambda \begin{bmatrix} (1+\mu)I_4 & (2+\mu)I_1 & (1+\mu)I_5 & 0 & 0 & I_1 \\ (2+\mu)I_1 & (4+2\mu)I_2 & (2+\mu)I_3 & 0 & 0 & 2I_2 \\ (1+\mu)I_5 & (2+\mu)I_3 & (1+\mu)I_6 + \frac{\mu}{2}I_2 & 0 & \frac{\mu}{2}I_2 & I_3 \\ 0 & 0 & 0 & 0 & 0 & 0 \\ 0 & 0 & \frac{\mu}{2}I_2 & 0 & \frac{\mu}{2}I_2 & 0 \\ I_1 & 2I_2 & I_3 & 0 & 0 & (1+\mu)I_2 \end{bmatrix} \tag{15-4.13}$$

where the following notations for the integrals have been used:

$$I_1 = \iint_A dr\,dz \qquad I_2 = \iint_A r\,dr\,dz \qquad I_3 = \iint_A z\,dr\,dz \tag{15-4.13a}$$

$$I_4 = \iint_A \frac{1}{r}\,dr\,dz \qquad I_5 = \iint_A \frac{z}{r}\,dr\,dz \qquad I_6 = \iint_A \frac{z^2}{r}\,dr\,dz$$

The first three integrals have been treated in the finite element formulation of plane strain and stress problems. They are (see Appendix 15A-1)

$$I_1 = \frac{\Delta}{2}$$

$$I_2 = \frac{\Delta}{6}(r_1 + r_2 + r_3) \tag{15-4.14}$$

$$I_3 = \frac{\Delta}{6}(z_1 + z_2 + z_3)$$

The integrands of (I_4, I_5, I_6) contain the term $1/r$ and, hence, become infinite (singular) if r_1, r_2, or r_3 vanish. Numerical integration techniques have been proposed to avoid this difficult by evaluating the integrands at selected points within the element through the use of proper weight factors.[31,32] However, in the present formulation we employ an exact procedure and, thus, avoid numerical errors due to integration weighting factors.

We first write equations of lines L_{12}, L_{23}, and L_{31} bounding the element triangle (Fig. 15-4.4). For line L_{23}, we have

$$z = a_{23} r + b_{23} \tag{15-4.15}$$

where a_{23} and b_{23} are defined by

$$a_{23} = (z_3 - z_2)/(r_3 - r_2) \qquad b_{23} = (z_2 r_3 - z_3 r_2)/(r_3 - r_2)$$

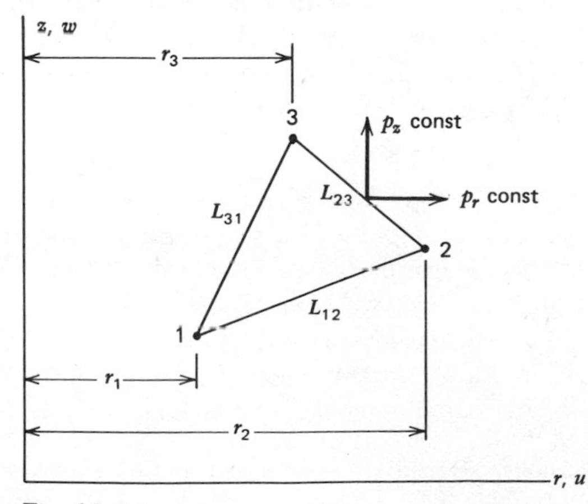

Fig. 15-4.4

The equations for the other lines, L_{31} and L_{12}, are obtained by cyclic permutation ($1 \to 2 \to 3 \to 1$) of the indices in Eq. (15-4.15).

On performing the integrations with respect to z, we find from Eq. (15-4.13a) that

$$I_4 = \int_{r_1}^{r_3} \frac{1}{r}(a_{31}r + b_{31})\,dr + \int_{r_3}^{r_2} \frac{1}{r}(a_{23}r + b_{23})\,dr - \int_{r_1}^{r_2} \frac{1}{r}(a_{12}r + b_{12})\,dr$$

$$I_5 = \int_{r_1}^{r_3} \frac{1}{2r}(a_{31}r + b_{31})^2\,dr + \int_{r_3}^{r_2} \frac{1}{2r}(a_{23}r + b_{23})^2\,dr - \int_{r_1}^{r_2} \frac{1}{2r}(a_{12}r + b_{12})^2\,dr$$

$$I_6 = \int_{r_1}^{r_3} \frac{1}{3r}(a_{31}r + b_{31})^3\,dr + \int_{r_3}^{r_2} \frac{1}{3r}(a_{23}r + b_{23})^3\,dr - \int_{r_1}^{r_2} \frac{1}{3r}(a_{12}r + b_{12})^3\,dr \qquad (15\text{-}4.16)$$

Integration of Eq. (15-4.16) yields†

$$I_4 = \sum_c \left[(a_{12} - a_{31})r_1 + (b_{12} - b_{31})\ln r_1\right]$$

$$I_5 = \sum_c \left[\tfrac{1}{4}(a_{12}^2 - a_{31}^2)r_1^2 + (a_{12}b_{12} - a_{31}b_{31})r_1 + \tfrac{1}{2}(b_{12}^2 - b_{31}^2)\ln r_1\right]$$

$$I_6 = \sum_c \left[\tfrac{1}{9}(a_{12}^3 - a_{31}^3)r_1^3 + \tfrac{1}{2}(a_{12}^2 b_{12} - a_{31}^2 b_{31})r_1^2 + (a_{12}b_{12}^2 - a_{31}b_{31}^2)r_1 + \tfrac{1}{3}(b_{12}^3 - b_{31}^3)\ln r_1\right] \qquad (15\text{-}4.17)$$

These expressions for I_4, I_5, I_6 are generally valid when r_1, r_2, and r_3 for the element are distinct and not equal to zero. However, Eq. (15-4.17) is not valid for certain orientations of the element. For example, when $r_1 = 0$ or $r_2 = r_3$, the integration formulas are undefined (see Eqs. 15-4.15 and 15-4.16). In order to overcome this difficulty, we treat such special cases separately.

Case I. When any two of the values r_1, r_2, r_3 for an element are equal, but none are zero, a pair of the subscripted a and b values defined in Eq. (15-4.15) becomes infinite [see Eq. (15-4.15)]. For example, if $r_2 = r_3$ and (r_1, r_2) are not zero, we have $a_{23} \to \infty$ and $b_{23} \to \infty$. This singularity, however, can be easily removed by consideration of the definition of the integrals in Eq. (15-4.16) where the integrations from r_3 to r_2 may be dropped. The same objective may be conveniently achieved by redefining

† The notation $\sum_c$ is employed to indicate summation by cyclic permutations ($1 \to 2 \to 3 \to 1$) of the indices.

a_{23} and b_{23}. Thus, we take

$$\begin{aligned} a_{23} &= a_{23} \quad \text{and} \quad b_{23} = b_{23}, \quad \text{for } r_2 \neq r_3 \\ a_{23} &= b_{23} = 0, \quad \text{for } r_2 = r_3 \end{aligned} \tag{15-4.18}$$

Accordingly, in this case, all the subscripted a and b values in Eq. (15-4.17) are redefined in the sense of Eq. (15-4.18).

Case II. If any one of the values r_1, r_2, r_3 becomes zero, the logarithmic term which contains that r value tends to order $(-\infty)$ (see Eq. 15-4.17). However, the limit of the product of the logarithmic term and its coefficient can be shown to exist, and is always equal to zero, by means of L'Hopital's rule. Therefore, in this case, we simply delete the logarithmic term whenever its argument (r) vanishes.

Case III. When two of the radii, say r_1 and r_3, are zero, the limits of all the integrals in Eq. (15-4.17) fail to exist. In this case, we have from Eq. (15-4.17), by setting $r_1 = r_3$.

$$I_4 = b_{12} \ln r_1 + (b_{23} - b_{12}) \ln r_2 - b_{23} \ln r_3 \tag{15-4.19}$$

Thus, the limits of $(b_{12} \ln r_1)$ and $(b_{23} \ln r_3)$, as $r_1 \to 0$ and $r_3 \to 0$, are undefined. Similar singularities exist in the other integrals (I_5, I_6). Through physical considerations in this particular case, we may once again, however, remove the difficulty. Since for $r_1 = r_3 = 0$ the corresponding element nodal radial displacements $\bar{u}_1$ and $\bar{u}_3$ are also equal to zero (Fig. 15-4.4), we find $\alpha_1 = \alpha_3 = 0$. Consequently, the assumed element displacement fields then take on the form

$$\begin{aligned} u &= \alpha_2 r \\ w &= \alpha_4 + \alpha_5 r + \alpha_6 z \end{aligned} \tag{15-4.20}$$

With Eq. (15-4.20), we now rederive the matrices $[A]_i$, $[B]_i$, and $[\bar{K}]_i$. Including the rows and columns corresponding to α_1 $(=0)$ and α_3 $(=0)$ in the formulation, we have the matrix $[A]_i^{-1}$

$$[A]_i^{-1} = [A_0]_i^{-1}$$

$$= \frac{1}{\Delta} \begin{bmatrix} 0 & 0 & 0 & 0 & 0 & 0 \\ \frac{(z_2 - z_3) r_1}{r_1 + r_2 + r_3} & 0 & \frac{(z_3 - z_1) r_2}{r_1 + r_2 + r_3} & 0 & \frac{(z_1 - z_2) r_3}{r_1 + r_2 + r_3} & 0 \\ 0 & 0 & 0 & 0 & 0 & 0 \\ 0 & z_3 r_2 - z_2 r_3 & 0 & z_1 r_3 - z_3 r_1 & 0 & z_2 r_1 - z_1 r_2 \\ 0 & z_2 - z_3 & 0 & z_3 - z_1 & 0 & z_1 - z_2 \\ 0 & r_3 - r_2 & 0 & r_1 - r_3 & 0 & r_2 - r_1 \end{bmatrix} \tag{15-4.21}$$

and the matrix $[\bar{K}]_i$

$$[\bar{K}]_i = [\bar{K}_0]_i = 2\pi\lambda \begin{bmatrix} 0 & 0 & 0 & 0 & 0 & 0 \\ 0 & (4+2\mu)I_2 & 0 & 0 & 0 & 2I_2 \\ 0 & 0 & 0 & 0 & 0 & 0 \\ 0 & 0 & 0 & 0 & 0 & 0 \\ 0 & 0 & 0 & 0 & \frac{\mu}{2}I_2 & 0 \\ 0 & 2I_2 & 0 & 0 & 0 & (1+\mu)I_2 \end{bmatrix} \tag{15-4.22}$$

In the second row of Eq. (15-4.21), the factors $r_1/(r_1+r_2+r_3)$, $r_2/(r_1+r_2+r_3)$, and $r_3/(r_1+r_2+r_3)$ are introduced so that when two of the radii vanish, one nonzero element remains in the row. The absence of the (I_4, I_5, I_6) integrals in Eq. (15-4.22) implies that the singularity has been removed. Accordingly, for Case III, we set $I_4 = I_5 = I_6 = 0$, and redefine the $[A]_i^{-1}$ and $[\bar{K}]_i$ matrices as $[A_0]_i^{-1}$ and $[\bar{K}_0]_i$, respectively.

The element stiffness matrix $[K]_i$ in terms of the generalized element stiffness matrix may be obtained as [see Eqs. (15-2.14) and (15-3.14)]

$$[K]_i = [A^{-1}]_i^T[\bar{K}]_i[A]_i^{-1}$$

Because of the complexity of $[\bar{K}]_i$, a direct algebraic evaluation of the $[K]_i$ matrix is not attempted [see Eq. (15-3.17)]. Instead, the matrix multiplication is left to be performed by the computer during the process of evaluating element properties.

Next we consider the formulation of element load matrices. Basically there are two types of loads; that is, body forces and surface pressure forces. Let the body force $\{X\}_i$ and surface force $\{p\}_i$ be expressed in terms of their (r, z) components (B_r, B_z) and (p_r, p_z). Thus,

$$\{X\}_i = \begin{Bmatrix} B_r \\ B_z \end{Bmatrix} \quad \text{and} \quad \{p\}_i = \begin{Bmatrix} p_r \\ p_z \end{Bmatrix} \tag{15-4.23}$$

According to the definition of the generalized element load matrices (Eqs. 15-3.19 and 15-3.20), we have, for the axially symmetric case,

$$\{Q\}_i = 2\pi \iint_A [N]_i^T \{X\}_i r\,dr\,dz \tag{15-4.24}$$

for the body force, and

$$\{\bar{P}\}_i = 2\pi \int_S [N]_i^T \{p\}_i r\,ds \tag{15-4.25}$$

for the surface pressure load. In Eq. (15-4.25), the integration with respect to θ has been performed, and the line integration is taken along the line segment upon which $\{p\}_i$ acts.

For axisymmetrical surface pressure loads, Eq. (15-4.25) must be evaluated for those elements subject to the pressure. When it is assumed that the surface pressure may be approximated as constant on L_{23} (see Fig. 15-4.4), we obtain by Eq. (15-4.25) the result

$$\{\bar{P}\}_i = 2\pi \begin{Bmatrix} \frac{1}{2}p_r(r_3 + r_2)L_{23} \\ \frac{1}{3}p_r(r_3^2 + r_3 r_2 + r_2^2)L_{23} \\ p_r\left[\frac{a_{23}}{3}(r_3^2 + r_3 r_2 + r_2^2) + \frac{b_{23}}{2}(r_3 + r_2)\right]L_{23} \\ \frac{1}{2}p_z(r_3 + r_2)L_{23} \\ \frac{1}{3}p_z(r_3^2 + r_3 r_2 + r_2^2)L_{23} \\ p_z\left[\frac{a_{23}}{3}(r_3^2 + r_3 r_2 + r_2^2) + \frac{b_{23}}{2}(r_3 + r_2)\right]L_{23} \end{Bmatrix} \tag{15-4.26}$$

Premultiplication of Eq. (15-4.26) by $[A^{-1}]_i^T$ then gives the corresponding element load matrix $\{P\}_i$. If more than one side of the element is acted on by pressure, the $\{\bar{P}\}_i$ matrix must be modified accordingly. The calculation of $\{\bar{P}\}_i$ for element pressure loads approximated in a linear manner is left for the reader.

Thick-Walled Cylinder under Internal Pressure / One of the classical exact elasticity solutions in the class of axisymmetric problems is Lamé's solution of a thick-walled cylinder with internal pressure (Chapter 11). Let the inner and outer radii of the cylinder be a and b, respectively, and let the internal pressure be denoted by p_i. A slice of the cylinder idealized by a 2 by 8 mesh is shown in Fig. 15-4.5, where the radii are taken as $a = 10$ and $b = 18$ units of length. The boundary condition for the idealized system may be physically represented by discrete rollers at nodes along

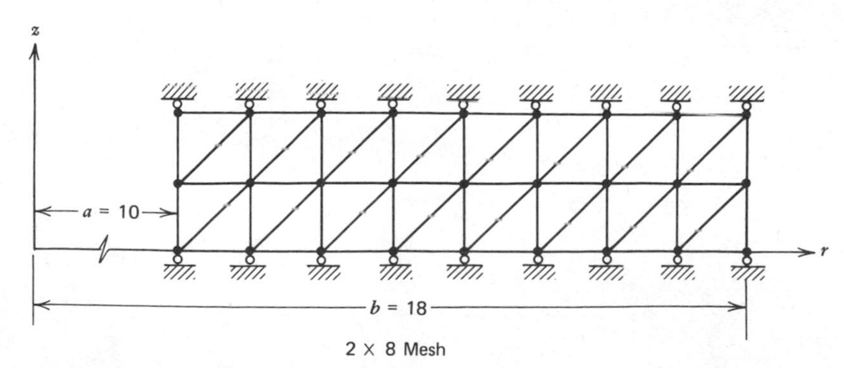

Fig. 15-4.5/Cylinder section: 2 × 8 mesh.

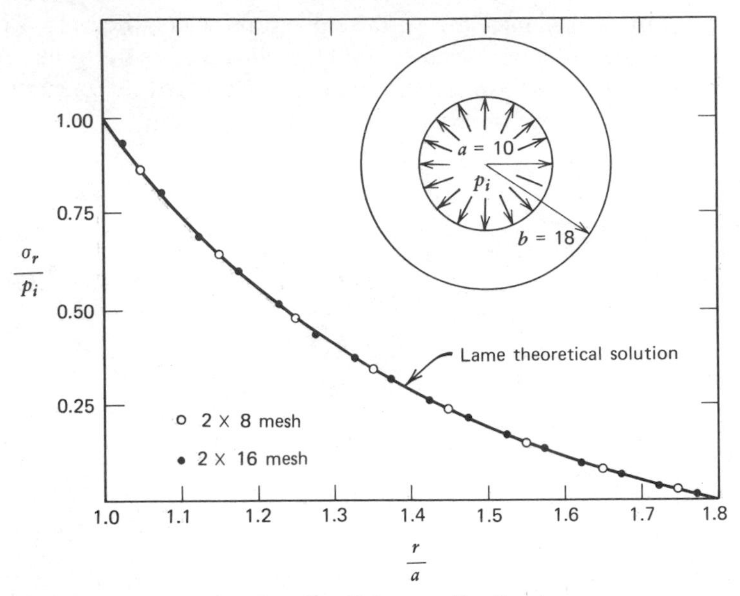

Fig. 15-4.6/Nondimensional radial stress distribution.

the top and bottom faces of the slice (Fig. 15-4.5); that is, at these nodes, the w nodal displacements (along the axis z of the cylinder) are to be constrained to vanish.

The finite element solutions are plotted in Figs. 15-4.6 and 15-4.7; namely, nondimensional radial (σ_r) and hoop (σ_θ) stress components as a function of the radial coordinate of the cylinder. Also shown are the Lamé solutions (Art. 11-2). These numerical finite element values are obtained for two mesh subdivisions; namely, a 2 by 8 mesh and a 2 by 16

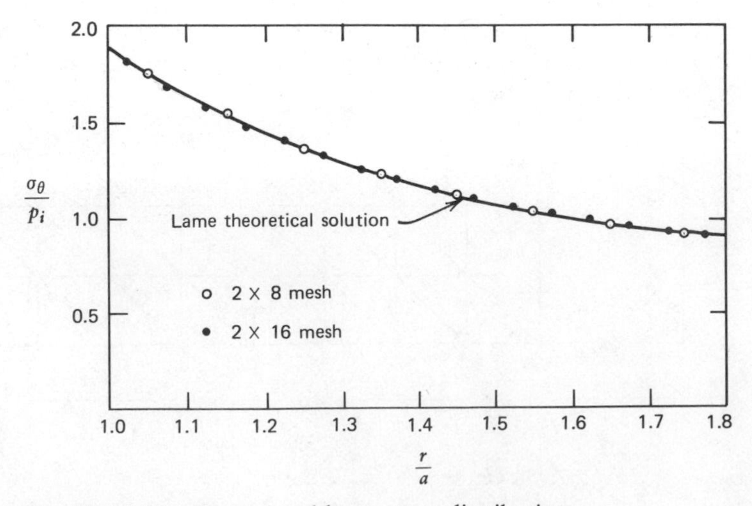

Fig. 15-4.7/Nondimensional hoop stress distribution.

mesh. Excellent agreement between the Lamée solutions and approximate solutions is obtained.

Cylindrical Bar with Central Spheroidal Cavity in Tension / Cast steel members are very often used in many machine components. Due to the manufacturing processes, such a member may contain small imperfections (holes). Under loading, these small cavities act as stress raisers and lead to appreciable stress concentrations in the member (Chapter 12). The effect of these small cavities on the endurance limit of the cast steel members subjected to stress reversals is of considerable practical interest.

To investigate local stresses around a spheroidal cavity, we consider a centrally located spheroidal cavity of a cylindrical bar under uniform tension of intensity p. For the finite element analysis, let the diameter and length of the cylindrical bar be both 10 units of length, and let the average of two spheroidal axes, a and b, be of 2 units of length. Figure 15-4.8 shows the coordinate system and the orientation of the a- and b-axes together with a parameter $R = a/b$ defining the shape of the spheroid. For example, when $R = 1.0$, we have a spherical cavity; and for $R = 2.0$, the cavity becomes a flat spheroid resembling a flat penny crack. We consider several cases for R values ranging from 0.5 to 2.0.

Because of high stress gradients around the cavity, we split the domain of interest roughly into two parts. One is the outer domain, which is away

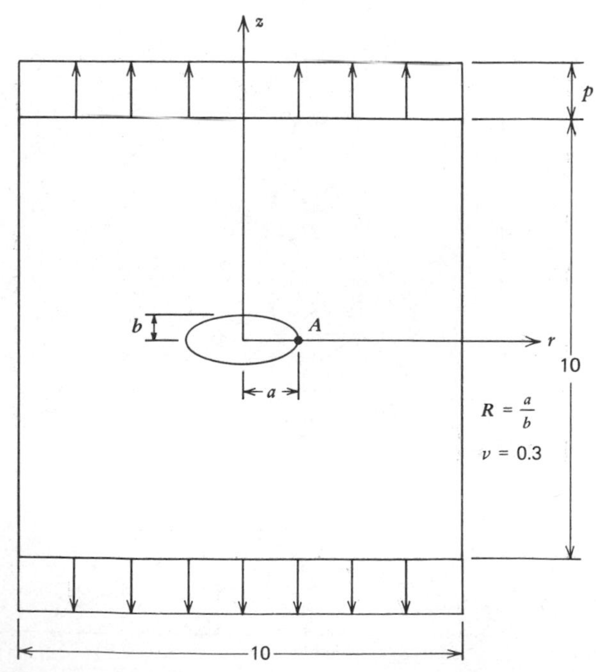

Fig. 15-4.8/Cylindrical bar with a spheroidal cavity.

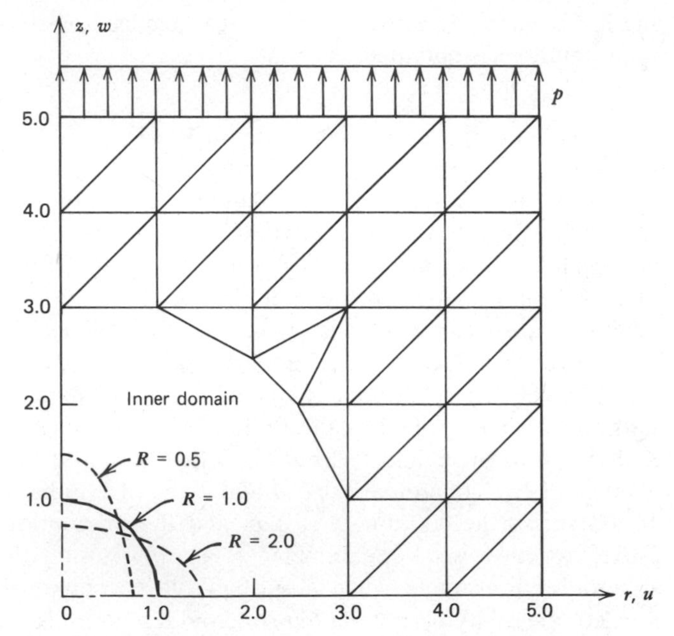

Fig. 15-4.9/Outer domain mesh subdivision, $r \geq 3.0$.

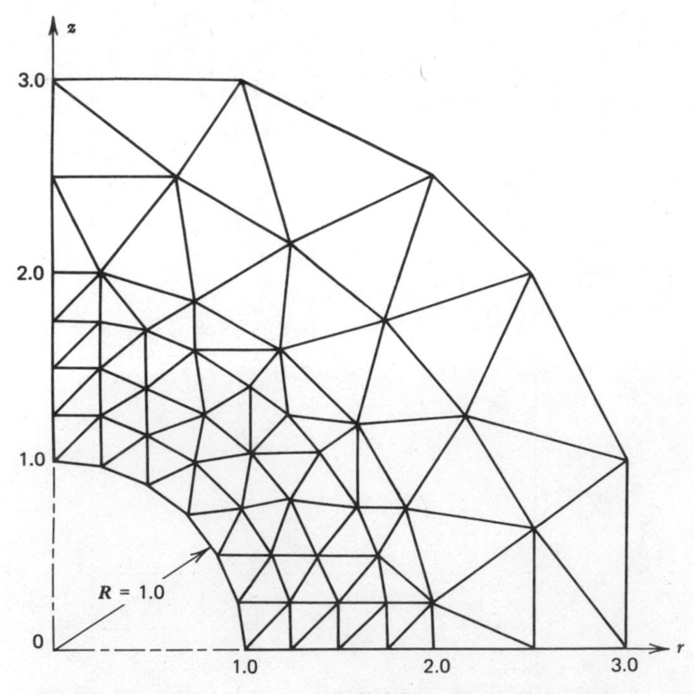

Fig. 15-4.10/Inner domain subdivision, $R = 1.0$; $r \leq 3.0$.

from the cavity, for which a coarse mesh may be used; the other is the inner domain near the region of the cavity for which a finer mesh is used. For this analysis, the elements and nodal points of the outer domain subdivision (Fig. 15-4.9) remain unchanged for different values of R, while the mesh of the inner domain (Fig. 15-4.10) is adjusted to accommodate different cavities. The mesh subdivision of Fig. 15-4.10 is for the spherical cavity with $R = 1.0$.

The distribution of stress component σ_z along the r axis is illustrated for $R = 0.5$, $R = 1.0$, and $R = 2.0$ in Fig. 15-4.11. Since the finite element solutions fail to give values at the edge of the cavity (Point A in Fig. 15-4.8), graphical extrapolation is used to obtain stresses at A. The maximum stress $(\sigma_z)_{\max}$ for the case of spherical cavity ($R = 1.0$) is determined to be 2.05 times the uniform tension p applied at the end of the bar which is very close to the theoretical value of $2.045p$ obtained by Goodier.[33]

If we define a stress concentration factor K as the ratio of the stress σ_z at A to the uniform tension p, we can plot a curve relating K and R values (Fig. 15-4.12). It is noteworthy that the entire stress distributions

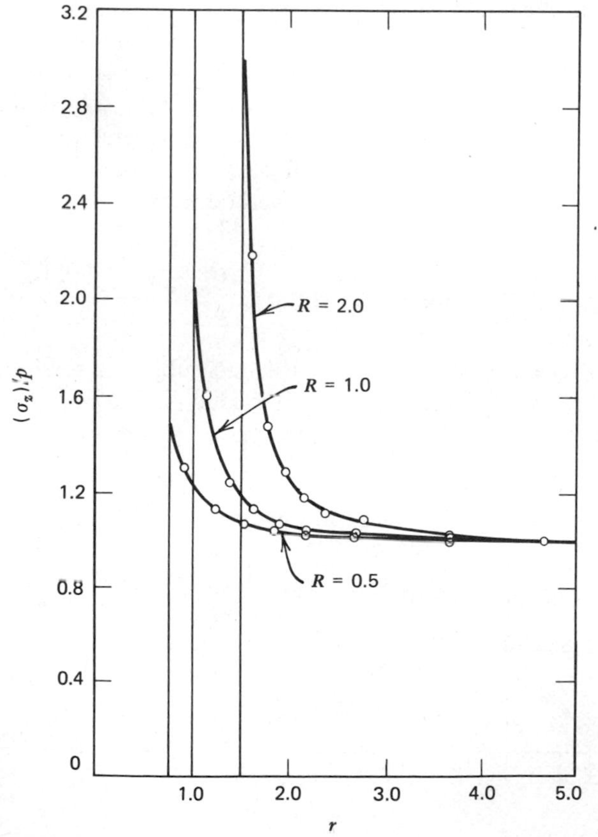

Fig. 15-4.11/Stress component σ_z at point A.

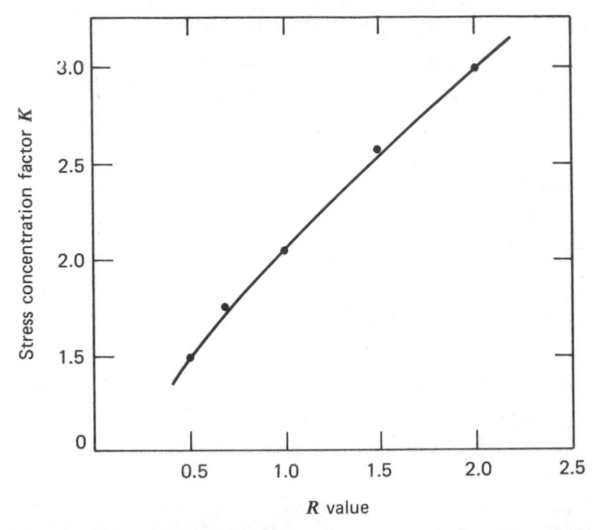

Fig. 15-4.12/Variation of stress concentration factor K.

and deformations of the bar (which are not shown) are also available from the analysis.

Other interesting problems, such as a cylindrical bar with a ring-shaped cavity or with a circumferential groove, a large sphere under uniform tension with a spheroidal cavity, etc., can be solved approximately by the finite element method (Fig. 15-4.13). The versatility and generality of the finite element method, as illustrated by these examples, are attractive features of the finite element method.[1]

15-5 TORSION OF PRISMATIC BARS

Although problems of prismatic bars subject to torsional forces fall in the category of three-dimensional elasticity, the Saint-Venant semi-inverse method (Art. 5-2) enables us to construct two-dimensional torsional

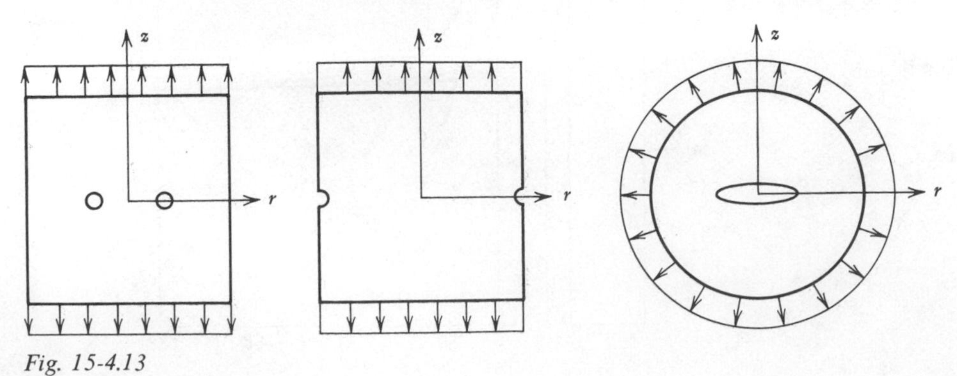

Fig. 15-4.13

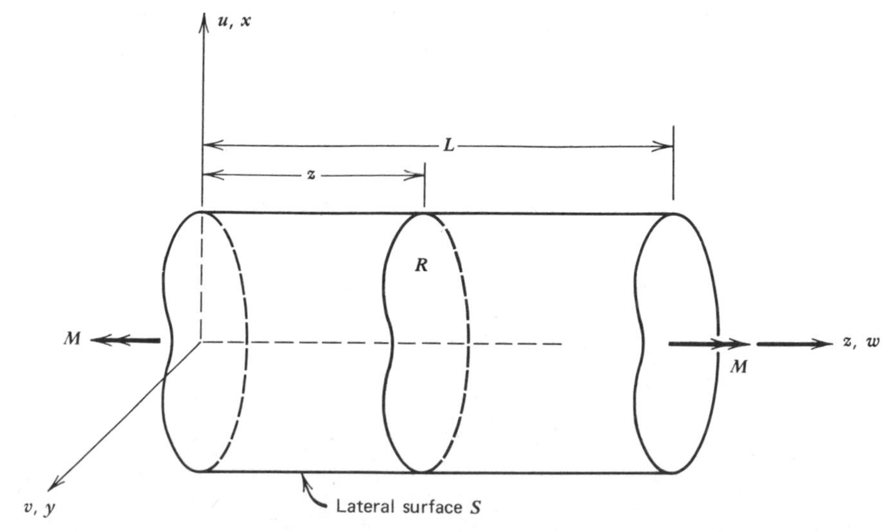

Fig. 15-5.1

finite element models. Applications of finite element methods to the Saint-Venant torsion problem have been given by Herrmann[34] and by Zienkiewicz.[1] A more general formulation of the problem is presented in this article. The procedure may be regarded as a *semi-inverse finite element method*. As an application of the technique, we consider torsion of a shaft with circular cross section, the diameter of which varies with axial coordinate.

Saint-Venant Torsion Problem / Let us consider first a prismatic bar with straight line generators parallel to the z-axis. We neglect body forces and allow no external forces on the lateral surface (Fig. 15-5.1). A twisting moment M is applied to each end of the bar.

After Saint-Venant, we assume the displacement components (Art. 5-2)

$$u = -\theta zy, \qquad v = \theta zx, \qquad w = \theta\psi(x, y) \tag{15-5.1}$$

where $\psi(x, y)$ denotes the *warping function* of the bar cross section, and θ denotes the angle of twist per unit length of the bar. Equation (15-5.1) satisfies the equations of elasticity, provided ψ satisfies the equation, $\Delta^2\psi = 0$.[22]

The *trial warping function* $\psi(x, y)$ may be taken as† a linear function of (x, y); namely,

$$\psi(x, y) = \alpha_1 + \alpha_2 x + \alpha_3 y \tag{15-5.2}$$

Thus, the simplest set of conforming trial displacement functions which

† The asterisk that indicates an approximate quantity will be dropped for convenience.

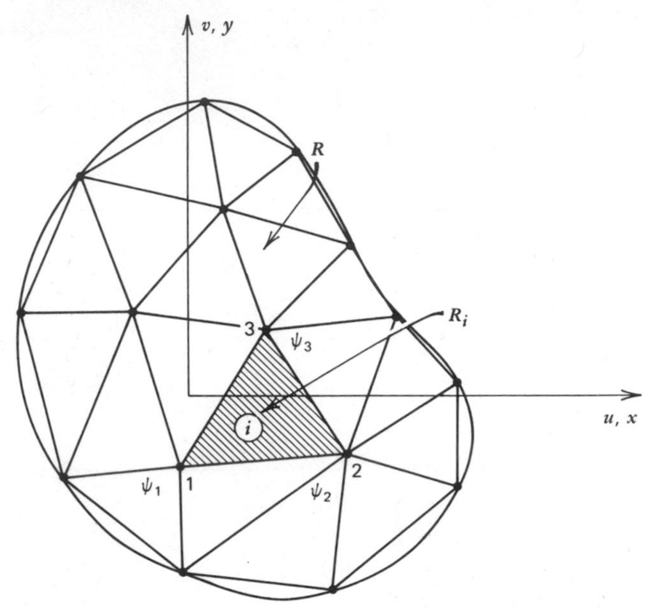

Fig. 15-5.2

satisfy appropriate convergence criteria for this problem is

$$u = -\theta yz, \qquad v = \theta xz, \qquad w = \theta(\alpha_1 + \alpha_2 x + \alpha_3 y) \tag{15-5.3}$$

Let us subdivide the bar cross section R into triangular elements, Fig. 15-5.2. Three nodal *warping displacement quantities* (ψ_1, ψ_2, ψ_3) are assigned to the three vertices of each element (ith element). These nodal quantities are regarded as the element nodal displacements; that is,

$$\{\delta\}_i = [\psi_1 \quad \psi_2 \quad \psi_3]^T \tag{15-5.4}$$

With the aid of Eq. (15-5.2), we find

$$\{\delta\}_i = [A]_i\{\alpha\}_i \quad \text{or} \quad \{\alpha\}_i = [A]_i^{-1}\{\delta\}_i \tag{15-5.5}$$

where

$$\{\alpha\}_i = [\alpha_1 \quad \alpha_2 \quad \alpha_3]^T,$$

$$[A]_i = \begin{bmatrix} 1 & x_1 & y_1 \\ 1 & x_2 & y_2 \\ 1 & x_3 & y_3 \end{bmatrix},$$

$$[A]_i^{-1} = \frac{1}{\Delta}\begin{bmatrix} x_2y_3 - x_3y_2 & x_3y_1 - x_1y_3 & x_1y_2 - x_2y_1 \\ y_2 - y_3 & y_3 - y_1 & y_1 - y_2 \\ x_3 - x_2 & x_1 - x_3 & x_2 - x_1 \end{bmatrix}$$

where (x_j, y_j), $j = 1, 2, 3$, are coordinates of the element vertices, and $\Delta = x_1(y_3 - y_1) + x_2(y_3 - y_1) + x_3(y_1 - y_2)$ is equal to twice the area of the element triangle.

By Eqs. (5-2.2), the nonzero strain components of the problem are

$$\{\epsilon\}_i = \begin{Bmatrix} \gamma_{xz} \\ \gamma_{yz} \end{Bmatrix} = \theta \begin{Bmatrix} \dfrac{\partial \psi}{\partial x} - y \\ \dfrac{\partial \psi}{\partial y} + x \end{Bmatrix} = \theta \begin{Bmatrix} \alpha_2 - y \\ \alpha_3 + x \end{Bmatrix} \qquad (15\text{-}5.6)$$

Equation (15-5.6) may be written as the sum of two matrices in the form

$$\{\epsilon\}_i = \theta \begin{Bmatrix} \alpha_2 \\ \alpha_3 \end{Bmatrix} + \theta \begin{Bmatrix} -y \\ x \end{Bmatrix}$$

Thus, in finite element displacement notation, we may write

$$\{\epsilon\}_i = \theta\big[[B]_i\{\alpha\}_i + \{B^0\}_i\big] = \theta\big[[B]_i[A]_i^{-1}\{\delta\}_i + \{B^0\}_i\big] \qquad (15\text{-}5.7)$$

where

$$[B]_i = \begin{bmatrix} 0 & 1 & 0 \\ 0 & 0 & 1 \end{bmatrix} \quad \text{and} \quad \{B^0\}_i = \begin{Bmatrix} -y \\ x \end{Bmatrix}$$

For an isotropic material, we have the element stress column matrix

$$\{\sigma\}_i = \begin{Bmatrix} \tau_{xz} \\ \tau_{yz} \end{Bmatrix} = G\{\epsilon\}_i$$

or

$$\{\sigma\}_i = G\theta\big[[B]_i\{\alpha\}_i + \{B^0\}_i\big] = G\theta\big[[B]_i[A]_i^{-1}\{\delta\}_i + \{B^0\}_i\big] \qquad (15\text{-}5.8)$$

Written out in detail, Eq. (15-5.8) becomes

$$\{\sigma\}_i = \frac{G\theta}{\Delta}\begin{bmatrix} y_2 - y_3 & y_3 - y_1 & y_1 - y_2 \\ x_3 - x_2 & x_1 - x_3 & x_2 - x_1 \end{bmatrix}\{\delta\}_i + G\theta\begin{Bmatrix} -y \\ x \end{Bmatrix} \qquad (15\text{-}5.9)$$

When a displacement finite element is applied in a semi-inverse method such as that employed in the Saint-Venant torsion problem, the element stiffness and load matrices often lose direct physical meaning. For example, in the present case, we are led rather to an *element warping stiffness matrix* and an *equivalent element load matrix*. These element properties may be formulated for the torsion problem as follows:

First, consider the element strain energy

$$U_i = \frac{L}{2} \iint \{\epsilon\}_i^T \{\sigma\}_i \, dR$$

By Eqs. (15-5.7) and (15-5.8), we find

$$U_i = \frac{G\theta^2 L}{2} \iint_{R_i} \left[\{\delta\}_i^T [A^{-1}]_i^T [B]_i^T + \{B^0\}_i^T \right] \times \left[[B]_i [A]_i^{-1} \{\delta\}_i + \{B^0\}_i \right] dR$$

Since matrix multiplication is distributive, on expanding the integrand of the U_i expression, we obtain

$$U_i = \frac{G\theta^2 L}{2} \iint_{R_i} \left[\{\delta\}_i^T [A^{-1}]_i^T [B]_i^T [B]_i [A]_i^{-1} \{\delta\}_i + 2\{\delta\}_i^T [A^{-1}]_i^T [B]_i^T \{B^0\}_i + \{B^0\}_i^T \{B^0\}_i \right] dR \tag{15-5.10}$$

where the identity

$$\{B^0\}_i^T [B]_i [A]_i^{-1} \{\delta\}_i = \{\delta\}_i^T [A^{-1}]_i^T [B]_i^T \{B^0\}_i$$

has been used.

The total strain energy U is now obtained by taking the sum of U_i over all the elements $(1, 2, \ldots, n)$. Hence,

$$U = \sum_{i=1}^{n} U_i$$

Accordingly, the total potential energy of the problem becomes

$$\pi = U - M\theta L = \pi\left[(\psi_1, \psi_2, \psi_3)_i, \theta \right] \tag{15-5.11}$$

Hence, requiring that the total potential energy be stationary[18]; that is, $\delta\pi = 0$, we have

$$\frac{\partial \pi}{\partial \{\delta\}} = \sum_{i=1}^{n} \frac{\partial \pi_i}{\partial \{\delta\}_i} = 0 \tag{15-5.12}$$

and

$$\frac{\partial \pi}{\partial \theta} = 0 \tag{15-5.13}$$

Equation (15-5.12) yields

$$\sum_{i=1}^{n} \left[[A^{-1}]_i^T [\bar{K}^w]_i [A]_i^{-1} \{\delta\}_i - [A^{-1}]_i^T \{\bar{F}^w\}_i \right] = 0 \tag{15-5.14}$$

where we have introduced the generalized element warping stiffness and element load matrices, $[\bar{K}^w]_i$, $\{\bar{F}^w\}_i$, defined by

$$[\bar{K}^w]_i = L \iint_{R_i} [B]_i^T [B]_i \, dR, \qquad \{\bar{F}^w\}_i = -L \iint_{R_i} [B]_i^T \{B^0\}_i \, dR \tag{15-5.15}$$

The superscript w indicates the element properties associated with warping of the prismatic bar cross section. Evaluation of Eq. (15-5.15) yields

$$[\bar{K}^w]_i = \frac{L\Delta}{2}\begin{bmatrix} 0 & 0 & 0 \\ 0 & 1 & 0 \\ 0 & 0 & 1 \end{bmatrix} \quad \text{and} \quad \{\bar{F}^w\}_i = \frac{L\Delta}{2}\begin{Bmatrix} 0 \\ \bar{y} \\ -\bar{x} \end{Bmatrix} \tag{15-5.16}$$

where $\bar{x} = (x_1 + x_2 + x_3)/3$ and $\bar{y} = (y_1 + y_2 + y_3)/3$. The corresponding *element warping stiffness matrix* $[K^w]_i$ and *equivalent element load matrix* $\{F^w\}_i$ are, respectively

$$[K^w]_i = [A^{-1}]_i^T [\bar{K}^w]_i [A]_i^{-1}$$

$$= \frac{L}{2\Delta}\left[\begin{array}{c|c|c} (y_2-y_3)^2 & (y_2-y_3)(y_3-y_1) & (y_2-y_3)(y_1-y_2) \\ +(x_3-x_2)^2 & +(x_3-x_2)(x_1-x_3) & +(x_3-x_2)(x_2-x_1) \\ \hline (y_3-y_1)(y_2-y_3) & (y_3-y_1)^2 & (y_3-y_1)(y_1-y_2) \\ +(x_1-x_3)(x_3-x_2) & +(x_1-x_3)^2 & +(x_1-x_3)(x_2-x_1) \\ \hline (y_1-y_2)(y_2-y_3) & (y_1-y_2)(y_3-y_1) & (y_1-y_2)^2 \\ +(x_2-x_1)(x_3-x_2) & +(x_2-x_1)(x_1-x_3) & +(x_2-x_1)^2 \end{array}\right] \tag{15-5.17}$$

$$\{F^w\}_i = [A^{-1}]_i^T \{\bar{F}^w\}_i = \frac{L}{2}\begin{Bmatrix} \bar{y}(y_2-y_3) - \bar{x}(x_3-x_2) \\ \bar{y}(y_3-y_1) - \bar{x}(x_1-x_3) \\ \bar{y}(y_1-y_2) - \bar{x}(x_2-x_1) \end{Bmatrix} \tag{15-5.18}$$

The matrix $[K^w]_i$ possesses properties of symmetry and positive definiteness. Furthermore, the sum of each column of matrix $[K^w]_i$ vanishes, the condition indicating over all element equilibrium conditions.

On denoting the *structural warping stiffness matrix* and *equivalent structural load matrices* as $[S^w]$ and $\{F^w\}$, respectively, Eq. (15-5.14) reduces to the *structural warping stiffness equation*

$$\{F^w\} = [S^w]\{\delta\} \tag{15-5.19}$$

The assemblage of matrices $[S^w]$ and $\{F^w\}$ from the corresponding element matrices follows that employed in Art. 15-4.

Reconsider, now, Eq. (15-5.13)

$$G\theta \sum_{i=1}^{n} \left[\{\delta\}_i^T [K^w]_i \{\delta\}_i - 2\{\delta\}_i^T \{F^w\}_i + L \iint_{R_i} \{B^0\}_i^T \{B^0\}_i \, dR \right] - ML = 0$$

Regrouping this equation, we have

$$G\theta \sum_{i=1}^{n} \Bigg[\{\delta\}_i^T [[K^w]_i \{\delta\}_i - \{F^w\}_i] - \{\delta\}_i^T \{F^w\}_i + L \iint_{R_i} \{B^0\}_i^T \{B^0\}_i \, dR \Bigg] - ML = 0$$

In view of the element warping stiffness equation, $[K^w]_i \{\delta\}_i - \{F^w\}_i = 0$, the above equation further reduces to

$$G\theta \sum_{i=1}^{n} \Bigg[-\{\delta\}_i^T \{F^w\}_i + L \iint_{R_i} \{B^0\}_i^T \{B^0\}_i \, dR \Bigg] = ML \tag{15-5.20}$$

By Eq. (15-5.20) the angle of twist θ per unit length of the bar may be written in the form

$$\theta = \frac{M}{C} \tag{15-5.21}$$

We note that in the solution of Eq. (15-5.19) for the unknown structural warping nodal displacements $\{\delta\}$, there is no need to specify nodal boundary displacements. The formulation is specified completely by the minimum potential energy principle. There are no essential displacement boundary conditions associated with the problem. However, to eliminate rigid body translation of the bar, we must prescribe a value of ψ (say zero) at some point in R.

The method presented above is quite general, as it is applicable to both *singly* and *multiconnected* regions of R. Moreover, the method can be employed in the development of more *refined* torsional finite elements.† If reentrant corners are present in the bar cross section, large stress concentrations occur in the vicinity of the corners.[35] This effect can be visualized by considering the Prandtl soap film analogy (Art. 5-4), since steep slopes of the soap film are required to accommodate the rapidly changing boundary curvature along the reentrant corners. Accordingly, in finite element numerical analysis, finer element subdivision near the vicinity of reentrant corners is recommended, if accurate representation of stress concentration effects is to be obtained.

Torsion of Bars of Varying Circular Cross Section / As another application of the *semi-inverse displacement finite element method* in torsion of bars, consider the torsion of a bar with circular section and radius which varies

† For a refined element, we may use a complete quadratic trial function for $\psi(x, y)$ together with additional mid-point nodes on the sides of the triangular element.

along the axis of the bar. In practice, this type of problem arises when one requires gradual or abrupt changes in the cross section of a torsion member. In particular, stress concentrations arising in the transition region between sections of two different diameters are of considerable practical interest.

Consider cylindrical coordinates (r, θ, z) with the z-axis coinciding with the axis of the bar (Fig. 15-5.3). Let the corresponding displacement components be (u, v, w). The semi-inverse solution of the problem, which is due to Mitchell† assumes that during twist the only nonzero displacement component is v; that is, $u = w = 0$. Furthermore, because of axial symmetry, the displacement v is independent of θ.

Accordingly, the nonzero strain components are

$$\gamma_{r\theta} = \frac{\partial v}{\partial r} - \frac{v}{r}, \qquad \gamma_{\theta z} = \frac{\partial v}{\partial z}$$

Introducing a function $\psi(r, z) = v/r$, which denotes the angle of rotation of an elemental tube of radius r within the shaft as indicated in Fig. 15-5.3, we have‡

$$v = r\psi(r, z) \tag{15-5.22}$$

and

$$\gamma_{r\theta} = r\frac{\partial \psi}{\partial r}, \qquad \gamma_{\theta z} = r\frac{\partial \psi}{\partial z} \tag{15-5.23}$$

The function $\psi(r, z)$ is called a *twisting function* of the bar.

By a derivation analogous to that for the convergence criteria of the Saint-Venant torsion finite element, we may show that for a trial function ψ, the continuity requirements are the continuity of ψ and mean convergence of the first partial derivatives of ψ. Accordingly, we consider a triangular element on the r-z-plane (ith element) as shown in Fig. 15-5.4, together with its trial displacement field

$$\psi(r, z) = \alpha_1 + \alpha_2 r + \alpha_3 z \tag{15-5.24}$$

The element nodal displacements are the three ψ values at the vertices of

†For justification of the semi-inverse method, see reference 28, pp. 325–327. The theory is due to J. H. Mitchell, reference 36.

‡The equations of equilibrium reduce to the single equation [see Eqs. (1-5.6), (1-9.6), and (2-4.9)]

$$\frac{\partial^2 v}{\partial r^2} + \frac{1}{r}\frac{\partial v}{\partial r} - \frac{v}{r^2} + \frac{\partial^2 v}{\partial z^2} = 0,$$

which may be equivalently expressed in the form

$$\frac{\partial}{\partial r}\left[r^3 \frac{\partial}{\partial r}\left(\frac{v}{r}\right)\right] + \frac{\partial}{\partial z}\left[r^3 \frac{\partial}{\partial z}\left(\frac{v}{r}\right)\right] = 0$$

This equation suggests the use of the function $v/r = \psi(r, z)$.

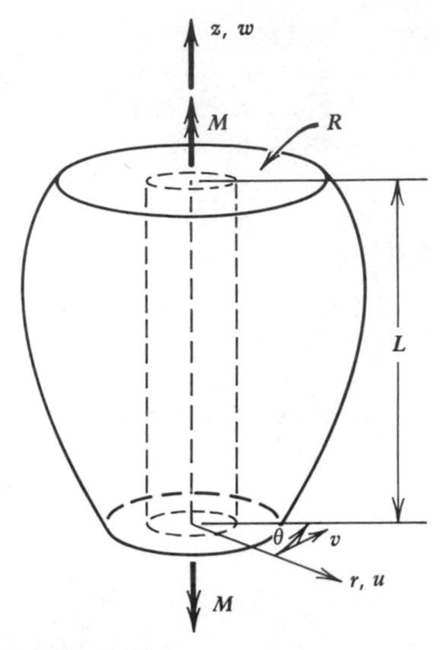

Fig. 15-5.3

the triangle. Thus, by the finite element formulation, we have

$$\{\delta\}_i = [A]_i\{\alpha\}_i \quad \text{and} \quad \{\alpha\}_i = [A]_i^{-1}\{\delta\}_i \qquad (15\text{-}5.25)$$

where $\{\delta\}_i = [\psi_1 \quad \psi_2 \quad \psi_3]^T$, $\{\alpha\}_i = [\alpha_1 \quad \alpha_2 \quad \alpha_3]^T$, and matrices $[A]_i$ and $[A]_i^{-1}$ are similar to those defined in Eq. (15-5.5) with (x_j, y_j) replaced by (r_j, z_j). Equation (15-5.23) gives the strain matrix

$$\{\epsilon\}_i = \begin{Bmatrix} \gamma_{r\theta} \\ \gamma_{\theta z} \end{Bmatrix} = [B]_i\{\alpha\}_i = [B]_i[A]_i^{-1}\{\delta\}_i \qquad (15\text{-}5.26)$$

where

$$[B]_i = \begin{bmatrix} 0 & r & 0 \\ 0 & 0 & r \end{bmatrix}$$

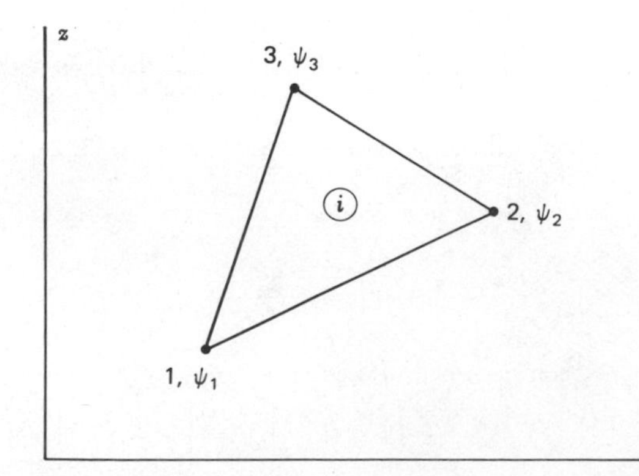

Fig. 15-5.4

The corresponding stress matrix for isotropic material is

$$\{\sigma\}_i = \begin{Bmatrix} \tau_{r\theta} \\ \tau_{\theta z} \end{Bmatrix} = G[B]_i\{\alpha\}_i = G[B]_i[A]_i^{-1}\{\delta\}_i \qquad (15\text{-}5.27)$$

and the element strain energy U_i is

$$U_i = \tfrac{1}{2}\{\delta\}_i^T[A^{-1}]_i^T[\bar{K}^t]_i[A]_i^{-1}\{\delta\}_i$$

where we have introduced the generalized element twisting stiffness matrix

$$[\bar{K}^t]_i = 2\pi G \iint_{A_i} [B]_i^T[B]_i r\,dr\,dz \qquad (15\text{-}5.28)$$

The integration in Eq. (15-5.28) is over the area of the triangle A_i. The superscript t denotes the quantity is associated with the twisting function. Carrying out the matrix multiplication and the integration of Eq. (15-5.28), we obtain explicitly

$$[\bar{K}^t]_i = 2\pi G \Delta I \begin{bmatrix} 0 & 0 & 0 \\ 0 & 1 & 0 \\ 0 & 0 & 1 \end{bmatrix} \qquad (15\text{-}5.29)$$

where $I = \frac{1}{20}[(r_1+r_2+r_3)(r_1^2+r_2^2+r_3^2)+r_1r_2r_3]$ and $\Delta = 2A_i$ (see Appendix 15A-1). From the definition $[K^t]=[A^{-1}]_i^T[\bar{K}^t]_i[A]_i^{-1}$, we obtain the *element twisting stiffness matrix*

$$[K^t]_i = \frac{2\pi G I}{\Delta}\left[\begin{array}{c|c|c} (z_2-z_3)^2 & (z_2-z_3)(z_3-z_1) & (z_2-z_3)(z_1-z_2) \\ +(r_3-r_2)^2 & +(r_3-r_2)(r_1-r_3) & +(r_3-r_2)(r_2-r_1) \\ \hline (z_3-z_1)(z_2-z_3) & (z_3-z_1)^2 & (z_3-z_1)(z_1-z_2) \\ +(r_1-r_3)(r_3-r_2) & +(r_1-r_3)^2 & +(r_1-r_3)(r_2-r_1) \\ \hline (z_1-z_2)(z_2-z_3) & (z_1-z_2)(z_3-z_1) & (z_1-z_2)^2 \\ +(r_2-r_1)(r_3-r_2) & +(r_2-r_1)(r_1-r_3) & +(r_2-r_1)^2 \end{array}\right] \qquad (15\text{-}5.30)$$

Note that except for the factor $2\pi GI/\Delta$, the $[K^t]_i$ matrix is similar to the $[K^w]_i$ matrix in Eq. (15-5.17).

Consider next the potential energy of external loads. Assuming the bar is fixed at $z=0$, (i.e., $\psi=0$ at $z=0$) and that τ is distributed linearly over the end $z=L$ (Fig. 15-5.5), we have for the potential energy of the end moment M

$$W = -\iint_R \{v\}^T\{\tau\} r\,dr\,d\theta \qquad (15\text{-}5.31)$$

where R is the area of the end cross section at $z=L$ and $v=r\psi$.

If the end shear τ is expressed in terms of the applied end moment M, as in the case of a prismatic circular bar, we have (see Fig. 15-5.5)

$$\tau = \frac{2M}{\pi a^4} r$$

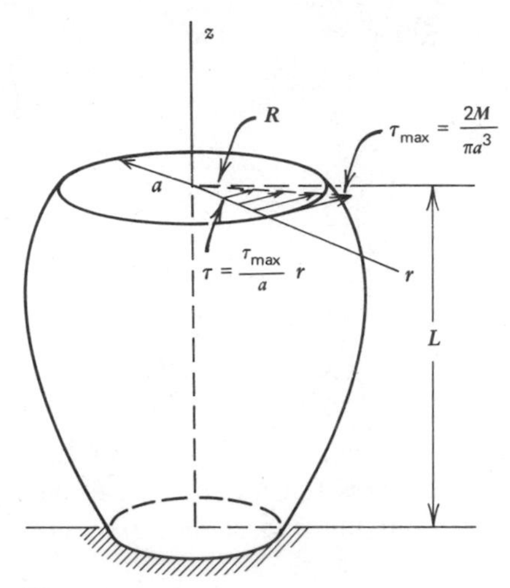

Fig. 15-5.5

Defining the applied end shear to be $\tau = (2Mr)/(\pi a^4)$ for elements which subdivide the end radius a, and the applied stress to be $\tau = 0$ for all the other elements, we may write

$$W = -\sum_{i=1}^{n} \frac{4M}{a^4} \int_{c_i} \{\alpha\}_i^T [N]_i^T r^3 \, dr = -\sum_{i=1}^{n} \frac{4M}{a^4} \{\delta\}_i^T [A^{-1}]_i^T \int_{c_i} [N]_i^T r^3 \, dr \tag{15-5.32}$$

in which $[N]_i = [1 \quad r \quad z]$, and the line integral is along the ith element edge c_i which coincides with the end radius a (Fig. 15-5.6)

Let the *generalized element twisting load* and *element twisting load matrices* be respectively

$$\{\bar{F}^t\}_i = \frac{4M}{a^4} \int_{c_i} [N]_i^T r^3 \, dr; \qquad \{F^t\}_i = [A^{-1}]_i^T \{\bar{F}^t\}_i$$

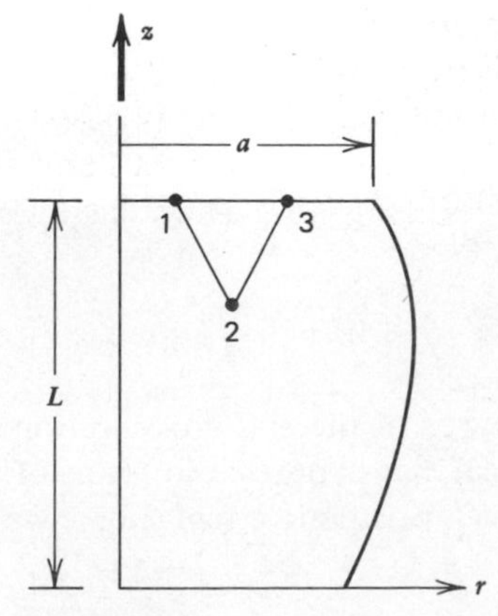

Fig. 15-5.6

If the ith element under consideration is located at the end of the bar as shown in Fig. 15-5.6, we have

$$\{\bar{F}^t\}_i = \frac{4M}{a^4}\begin{Bmatrix} \frac{1}{4}(r_3^4 - r_1^4) \\ \frac{1}{5}(r_3^5 - r_1^5) \\ \frac{L}{4}(r_3^4 - r_1^4) \end{Bmatrix} \tag{15-5.33}$$

$$\{F^t\}_i = \frac{4M}{a^4\Delta}$$

$$\times \begin{Bmatrix} \frac{1}{4}(r_3^4 - r_1^4)[(r_2z_3 - r_3z_2) + L(r_3 - r_2)] + \frac{1}{5}(r_3^5 - r_1^5)(z_2 - z_3) \\ \frac{1}{4}(r_3^4 - r_1^4)[(r_3z_1 - r_1z_3) + L(r_1 - r_3)] + \frac{1}{5}(r_3^5 - r_1^5)(z_3 - z_1) \\ \frac{1}{4}(r_3^4 - r_1^4)[(r_1z_2 - r_2z_1) + L(r_2 - r_1)] + \frac{1}{5}(r_3^5 - r_1^5)(z_1 - z_2) \end{Bmatrix} \tag{15-5.34}$$

From minimum potential energy considerations, we obtain

$$\sum_{i=1}^{n} [[K^t]_i\{\delta\}_i - \{F^t\}] = 0$$

or in the form of the *structural twisting stiffness equation*

$$\{F^t\} = [S^t]\{\delta\} \tag{15-5.35}$$

Since we have assumed $\psi = 0$ at $z = 0$, the nodal displacements at $z = 0$ must be set equal to zero in solving Eq. (15-5.35). The finite element model presented here applies equally well for hollow circular bars with variable cross-sectional area.

Refined Torsion Elements/By means of Saint-Venant's torsion problem, we illustrate briefly two typical refined elements. One is a second order element similar to the one we employed earlier in Art. 15-3; the other is a quadrilateral element which is composed of four linear triangular torsion elements.

The formulation of the second order triangular torsion element closely follows that of the first order element. Hence, we list only the essential results here. By taking the warping function as

$$\psi(x, y) = \alpha_1 + \alpha_2 x + \alpha_3 y + \alpha_4 x^2 + \alpha_5 xy + \alpha_6 y^2 \tag{15-5.36}$$

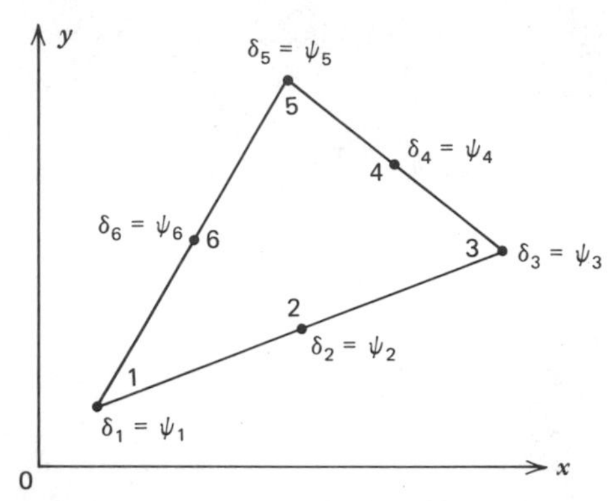

Fig. 15-5.7/Second order triangular torsion element.

and introducing the midpoint nodes as shown in Fig. 15-5.7, we have

$$[A]_i = \begin{bmatrix} 1 & x_1 & y_1 & x_1^2 & x_1y_1 & y_1^2 \\ 1 & x_2 & y_2 & x_2^2 & x_2y_2 & y_2^2 \\ 1 & x_3 & y_3 & x_3^2 & x_3y_3 & y_3^2 \\ 1 & x_4 & y_4 & x_4^2 & x_4y_4 & y_4^2 \\ 1 & x_5 & y_5 & x_5^2 & x_5y_5 & y_5^2 \\ 1 & x_6 & y_6 & x_6^2 & x_6y_6 & y_6^2 \end{bmatrix} \tag{15-5.37}$$

$$[B]_i = \begin{bmatrix} 0 & 1 & 0 & 2x & y & 0 \\ 0 & 0 & 1 & 0 & x & 2y \end{bmatrix} \quad \text{and} \quad \{B^0\}_i = \begin{Bmatrix} -y \\ x \end{Bmatrix} \tag{15-5.38}$$

Accordingly, the stress matrix is

$$\{\sigma\}_i = G\theta[B]_i[A]_i^{-1}\{\delta\}_i + \{B^0\}_i G\theta \tag{15-5.39}$$

where $\{\delta\}_i = [\psi_1\ \psi_2 \cdots \psi_6]^T$. Forming the generalized element warping stiffness matrix, we obtain

$$[\bar{K}^w]_i = L\iint_R [B]_i^T\, dR$$

Integration yields

$$[\bar{K}^w]_i = \frac{L\Delta}{2}\begin{bmatrix} 0 & 0 & 0 & 0 & 0 & 0 \\ 0 & 1 & 0 & 2\bar{x} & \bar{y} & 0 \\ 0 & 0 & 1 & 0 & \bar{x} & 2\bar{y} \\ 0 & 2\bar{x} & 0 & 3\bar{x}^2 + \frac{1}{3}X & \frac{3}{2}\overline{xy} + \frac{1}{6}Z & 0 \\ 0 & \bar{y} & \bar{x} & \frac{3}{2}\overline{xy} + \frac{1}{6}Z & \begin{bmatrix} \frac{3}{4}(\bar{x}^2 + \bar{y}^2) \\ + \frac{1}{12}(X+Y) \end{bmatrix} & \frac{3}{2}\overline{xy} + \frac{1}{6}Z \\ 0 & 0 & 2\bar{y} & 0 & \frac{3}{2}\overline{xy} + \frac{1}{6}Z & 3\bar{y}^2 + \frac{1}{3}Y \end{bmatrix} \tag{15-5.40}$$

in which $\bar{x}=\frac{1}{3}(x_1+x_3+x_5)$, $\bar{y}=\frac{1}{3}(y_1+y_3+y_5)$, $X=x_1^2+x_3^2+x_5^2, Y=y_1^2+y_3^2+y_5^2$, and $Z=x_1y_1+x_3y_3+x_5y_5$. The corresponding generalized element load matrix is

$$\{\bar{F}^w\}_i=-L\iint_R [B]_i^T\{B^0\}_i\,dR=\frac{L\Delta}{2}\begin{Bmatrix} 0 \\ \bar{y} \\ -\bar{x} \\ \frac{3}{2}\bar{x}\bar{y}+\frac{1}{6}Z \\ \frac{3}{4}(\bar{y}^2-\bar{x}^2)+\frac{1}{12}(Y-X) \\ -\frac{3}{2}\bar{x}\bar{y}-\frac{1}{6}Z \end{Bmatrix} \tag{15-5.41}$$

The relations $[K^w]_i=[A^{-1}]_i^T[\bar{K}^w]_i[A]_i^{-1}$ and $\{F^w\}_i=[A^{-1}]_i^T\{\bar{F}^w\}_i$ then define the $[K^w]_i$ and $\{F^w\}_i$ matrices.

Consider next a quadrilateral element consisting of four linear torsion elements as shown in Fig. 15-5.8. Basically, this element has five degrees of nodal freedom, and its behavior does not differ from that of the four elements composing it. However, the main motivation of developing this element lies in reducing the size of the structural matrix $[S^w]$. Since, in this case, we can eliminate the nodal unknown ψ associated with the inner node (ψ_5 at node 5), the quadrilateral element nodal freedom reduces to four ($\psi_1, \psi_2, \psi_3, \psi_4$), and, therefore, the rank of its stiffness matrix becomes 4.

To illustrate the method, let us consider the total stiffness equation of the four interconnected elements which form the quadrilateral (Fig. 15-5.8)

$$\begin{Bmatrix} F_1 \\ F_2 \\ F_3 \\ F_4 \\ F_5 \end{Bmatrix}=\begin{bmatrix} k_{11} & k_{12} & k_{13} & k_{14} & k_{15} \\ k_{21} & k_{22} & k_{23} & k_{24} & k_{25} \\ k_{31} & k_{32} & k_{33} & k_{34} & k_{35} \\ k_{41} & k_{42} & k_{43} & k_{44} & k_{45} \\ k_{51} & k_{52} & k_{53} & k_{54} & k_{55} \end{bmatrix}\begin{Bmatrix} \delta_1 \\ \delta_2 \\ \delta_3 \\ \delta_4 \\ \delta_5 \end{Bmatrix} \tag{15-5.42}$$

Note that for convenience the superscript w denoting the quantities associated with the warping of the bar cross section is dropped. Equation (15-5.42) may be partitioned to give

$$\begin{Bmatrix} F_{1-4} \\ \hline F_5 \end{Bmatrix}=\left[\begin{array}{c|c} k_{(1-4)(1-4)} & k_{(1-4)5} \\ \hline k_{5(1-4)} & k_{55} \end{array}\right]\begin{Bmatrix} \delta_{(1-4)} \\ \hline \delta_5 \end{Bmatrix} \tag{15-5.43}$$

Expanding Eq. (15-5.43), we have

$$\{F_{(1-4)}\}=[k_{(1-4)(1-4)}]\{\delta_{(1-4)}\}+[k_{(1-4)5}]\{\delta_5\}$$

$$\{F_5\}=[k_{5(1-4)}]\{\delta_{(1-4)}\}+[k_{55}]\{\delta_5\}$$

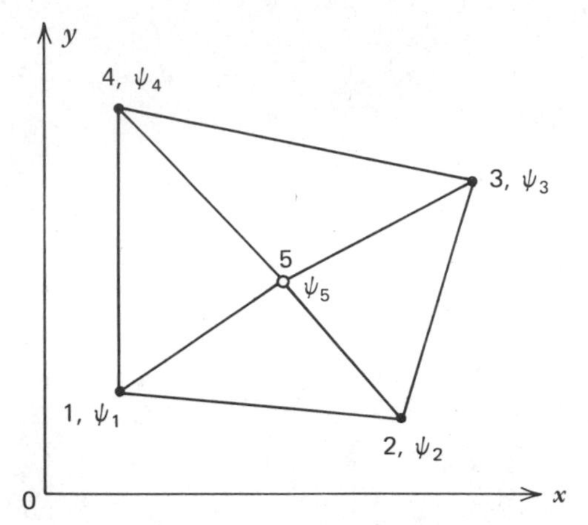

Fig. 15-5.8/Quadrilateral element.

On solving for $\{\delta_5\}$ from the second equation and substituting it into the first equation, we find

$$\left\{F_{(1-4)} - k_{(1-4)5}k_{55}^{-1}F_5\right\} = \left[k_{(1-4)(1-4)} - k_{(1-4)5}k_{55}^{-1}k_{5(1-4)}\right]\left\{\delta_{(1-4)}\right\} \tag{15-5.44}$$

Since $\{\delta_{(1-4)}\}_i = [\psi_1\ \psi_2\ \psi_3\ \psi_4]^T$ represents the four nodal displacements of the quadrilateral, we conclude that Eq. (15-5.44) is indeed the quadrilateral element stiffness equation.

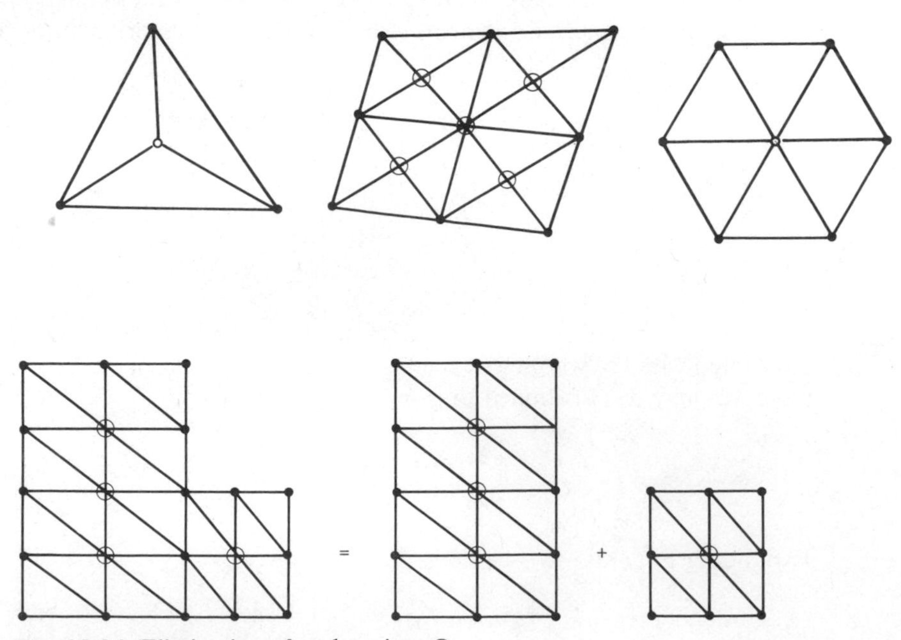

Fig. 15-5.9/Elimination of node points O.

Accordingly, for this quadrilateral torsion element, the element load matrix is

$$\{\hat{F}^w\}_i = \{F_{(1-4)} - k_{(1-4)5}k_{55}^{-1}F_5\} \tag{15-5.45}$$

and the element warping stiffness matrix is

$$[\hat{K}^w]_i = [k_{(1-4)(1-4)} - k_{(1-4)5}k_{55}^{-1}k_{5(1-4)}] \tag{15-5.46}$$

Equations (15-5.45) and (15-5.46) are called the modified element load and the modified stiffness matrices, respectively.

The described technique is not limited to forming a quadrilateral element. The method is quite general and can be applied to any substructure which is a part of an overall finite element idealization. This point is schematically illustrated in Fig. 15-5.9 where nodal displacements associated with nodes designated by the small circles can be eliminated.

Torsion of Square Prismatic Bars / As a numerical example, let us consider a square prismatic bar in torsion, Fig. 15-5.10. The analytical elasticity solution is (see Art. 5-3) $M = k_1 G\theta(2a)^4$ with $k_1 = 0.1406$, and $\tau_{max} = M/(k_2(2a)^3)$ with $k_2 = 0.208$, where k_1 and k_2 are factors defining the torsional rigidity and the maximum shear respectively.

For the finite element analysis, we subdivide a quarter of the cross section into mesh sizes defined in Fig. 15-5.11. In Table 15-5.1, the finite element solutions for k_1 and k_2 are compared with those obtained from the analytical elasticity solution.

It is noteworthy that for the same mesh size both linear triangular and quadrilateral torsion elements require the same storage space for the structural stiffness equation, while the improved accuracy of the latter is obvious. However, this improvement is obtained at the expense of additional computational time. The best indication of the overall

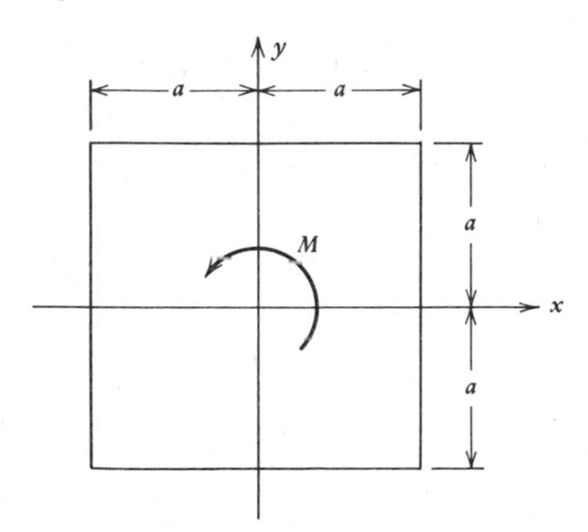

Fig. 15-5.10

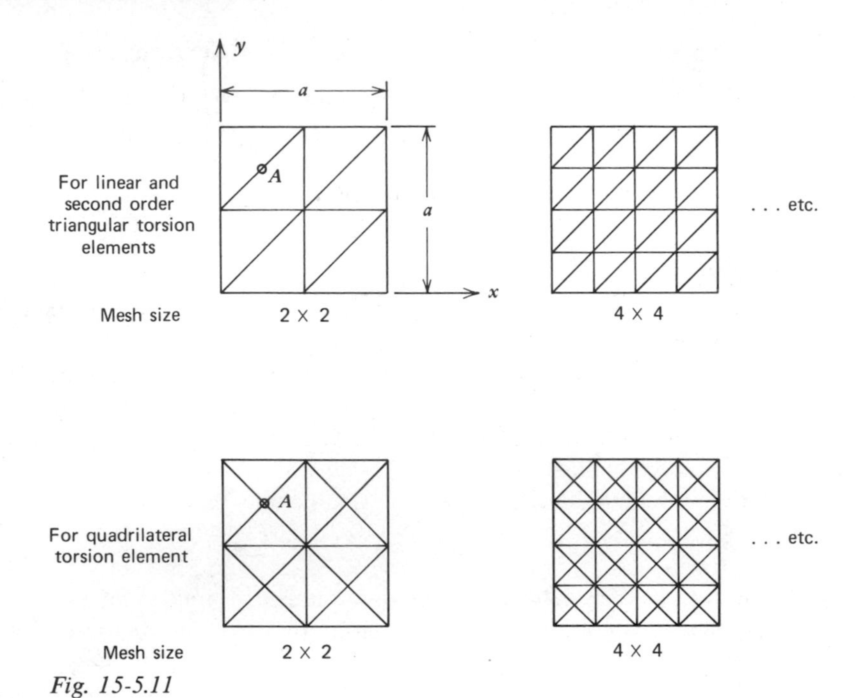

Fig. 15-5.11

Table 15-5.1

List of k_1 and k_2 Values†

Mesh Size		*Linear Triangular Torsion Element*	*Second Order Triangular Torsion Element*	*Quadrilateral Torsion Element*
2×2	k_1	0.15104	0.14110	0.14881
	k_2	0.34177	0.32732	0.33096
4×4	k_1	0.14363	0.14060	0.14278
	k_2	0.25994	0.25537	0.23077
8×8	k_1	0.14138	0.14058	0.14114
	k_2	0.23053	0.22941	0.22978
16×16	k_1	0.14078	—	0.14072
	k_2	0.21852	—	0.20543
Analytical solution	k_1	0.1406	0.1406	0.1406
	k_2	0.208	0.208	0.208

† The average maximum shear at point A (Fig. 15-5.11) is used to calculate k_2.

accuracy of the finite element solutions is convergence of the approximate k_1 values. Analytically, the monotonic converging upper bound nature of the approximate k_1 values can be established.

To show this, consider the total potential energy (Eq. 15-5.11)

$$\pi = U - M\theta L \tag{a}$$

By the principle of conservation of energy, we note that $U = \frac{1}{2}M\theta L$; hence

$$\pi = -\tfrac{1}{2}M\theta L \tag{b}$$

Eliminating θ by means of Eq. (15-5.21), we have

$$\pi = -\frac{1}{2}\frac{M^2}{C}L \tag{c}$$

where C is the torsional constant and, in this case, $C = k_1 G(2a)^4$. Since finite element solutions minimize the total potential energy π, we must have

$$\pi_{\text{approx.}} \geq \pi_{\text{exact}} \tag{d}$$

Equations (c) and (d) imply that

$$C_{\text{approx.}} \geq C_{\text{exact}}$$

and, therefore, we have

$$k_{1\ \text{approx.}} \geq k_{1\ \text{exact}} \tag{e}$$

Since the sequence of mesh sizes (Fig. 15-5.11) conforms to the requirements of the monotonic convergence[19] of $\tau_{\text{approx.}}$, we conclude that $k_{1\ \text{approx.}}$ also converges monotonically. Indeed, plots of the k_1 values against mesh sizes show the desired convergence (Fig. 15-5.12). From Fig. 15-5.12, the improved accuracy of the second order triangular torsion element is apparent.

As a second example in the torsion of square bars, we briefly present numerical results of torsion of a square bar made of two different materials as shown in Fig. 15-5.13. The 10 by 10 mesh subdivision for a quarter of the section is shown in Fig. 15-5.14. By means of the linear triangular torsion element, three different cases are investigated. They are

Case 1. $G_1/G_2 = 1.0$

Case 2. $G_1/G_2 = 3.0$

Case 3. $G_1/G_2 = 1/3$

where G_1 and G_2 are shearing rigidities of the material 1 and 2, and G_2 is taken as $G_2 = G$.

Figures 15-5.15 and 15-5.16 are stress contour plots of shearing stress components τ_{xz} and τ_{yz} throughout the quarter section. It may be seen

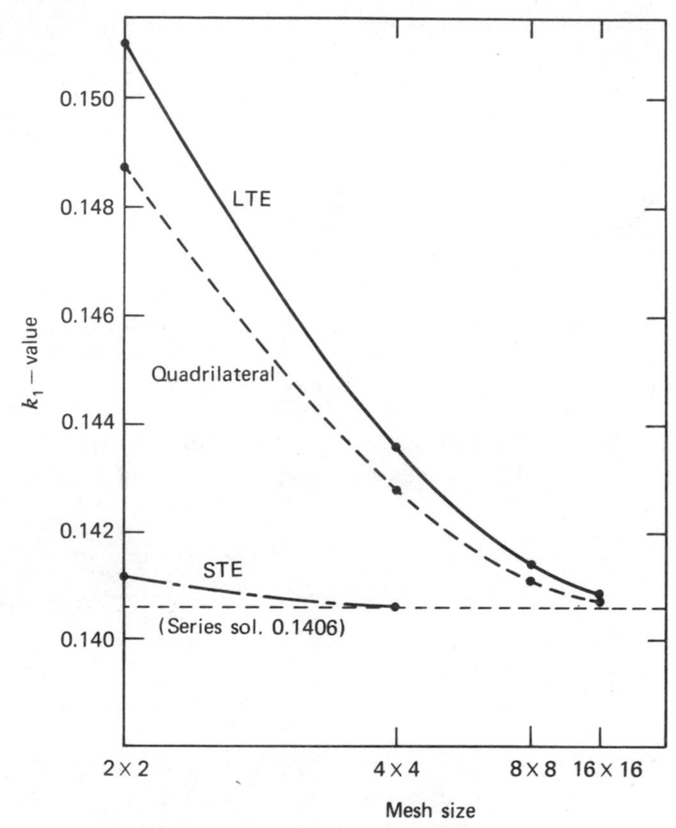

Fig. 15-5.12/Convergence of approximate values of k_1.

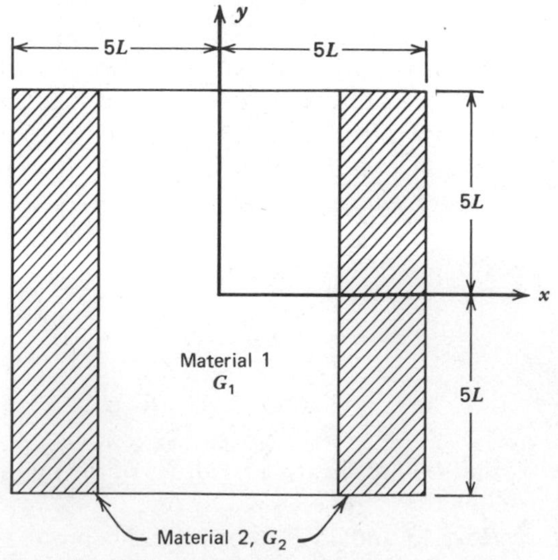

Fig. 15-5.13/A square bar with two materials.

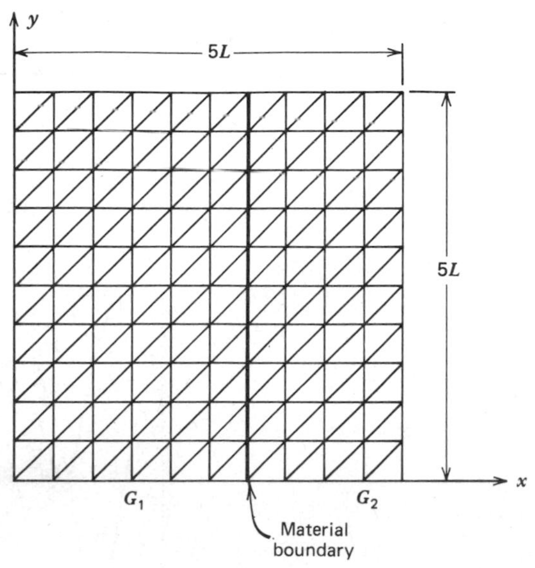

Fig. 15-5.14/Quarter of square cross section: 10×10 mesh.

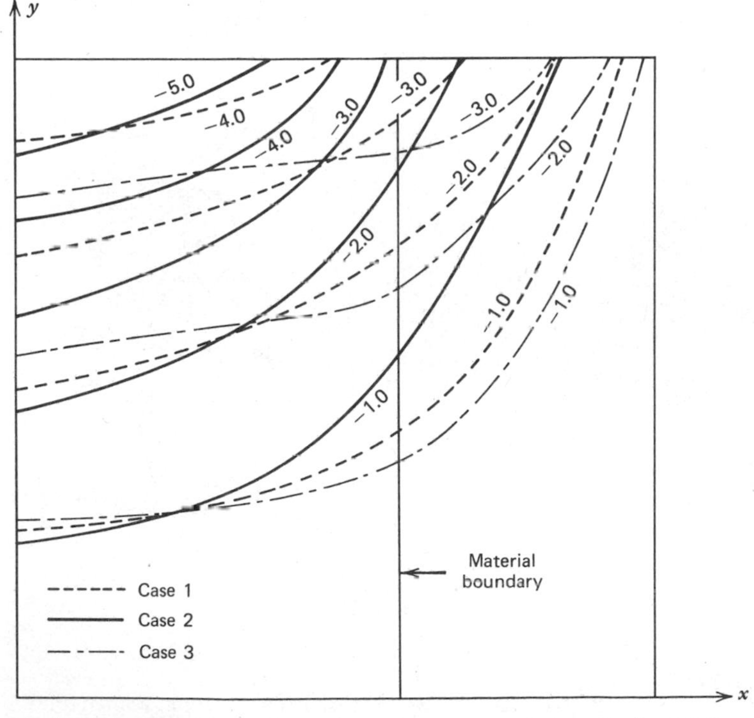

Fig. 15-5.15/COEFF contours for τ_{xz} ($\tau_{xz} = \text{COEFF} \times 10^{-3}\ M/L^3$).

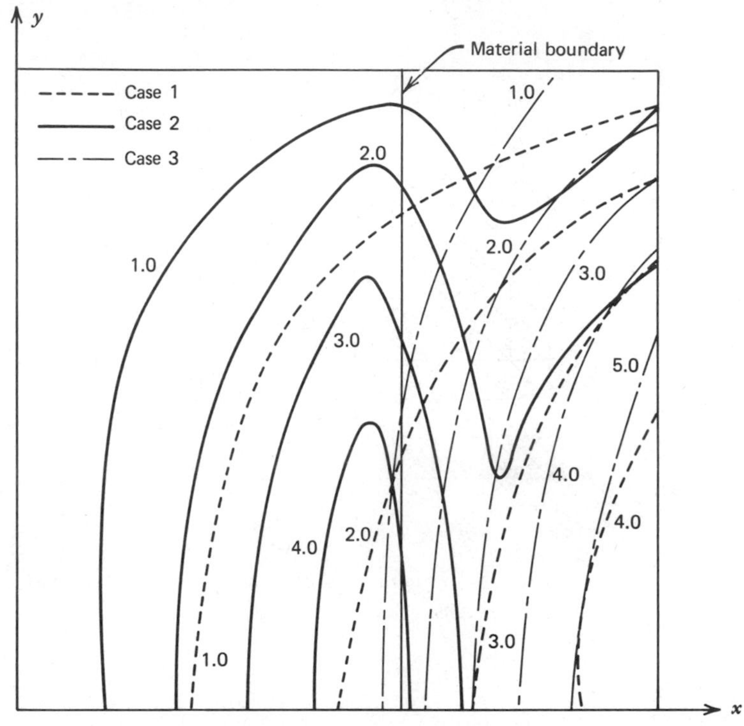

Fig. 15-5.16/COEFF contours for τ_{yz} ($\tau_{yz} = \text{COEFF} \times 10^{-3}\ M/L^3$).

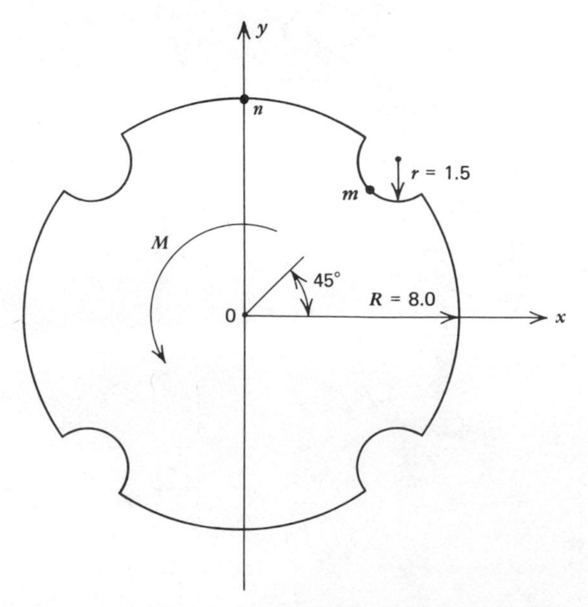

Fig. 15-5.17/Splined shaft with four semicircular keyways.

that the maximum shear stress for Case 2 occurs in the upper left-hand corner; whereas the maximum shear stress for Case 3 occurs in the lower right-hand corner. From these results, it is apparent that the behavior of a nonhomogeneous bar in torsion is quite different from that of a homogeneous bar under otherwise similar conditions. This example demonstrates the versatility of finite element methods in dealing with nonhomogeneous material problems of elasticity.

Splined Shafts / A splined shaft is a circular prismatic bar with several semicircular keyway cuts. An application of the quadrilateral element in torsion of such a shaft with four semicircular keyways as shown in Fig. 15-5.17 is presented.

In this example, the radii of the shaft and keyways are taken as $R = 8.0$ and $r = 1.5$ units of length, respectively. Since the shaft is symmetric with respect to the x- and y-axes, we need consider only a quarter section of the shaft. The section is then idealized by 84 quadrilateral elements. The boundary conditions are furnished by constraining all the nodal displacements along the x- and y-axes, since they are axes of symmetry where the w displacement vanishes.

From the numerical analysis, we obtain the torsional constant (torsional rigidity) of the splined shaft to be $C = 0.781\ (G(\pi R^4)/2)$. Since the quantity in parentheses is the torsional constant of a circular shaft of

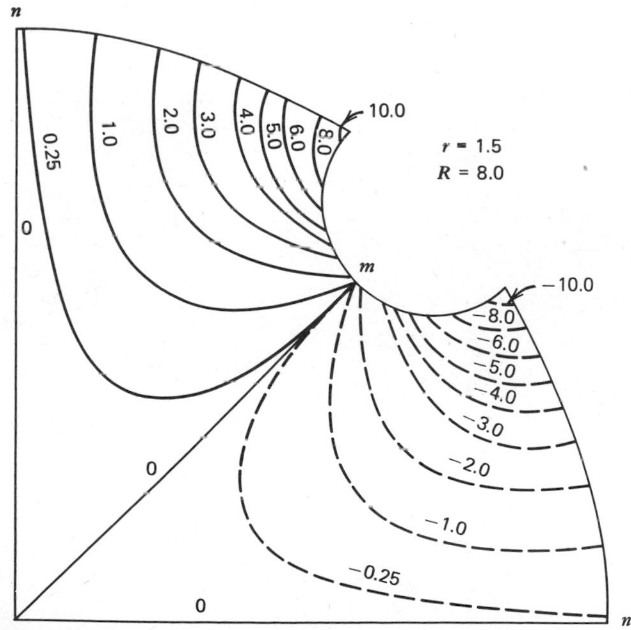

Fig. 15-5.18/Warping displacement COEFF contours; displacement $w = (\text{COEFF})(M/C)$.

radius R without keyways, the factor 0.781 represents a reduction in the torsional rigidity of the splined shaft. At the root of the keyway (point m) the shearing stress becomes maximum. The ratio of this stress τ_m to the stress at midway between grooves τ_n (Fig. 15-5.17) is numerically determined to be $\tau_m/\tau_n = 1.8$.

From the nodal displacements obtained by the finite element analysis, a complete picture of the deformed shape can be constructed. To illustrate this, the warping of the shaft cross section is plotted in Fig. 15-5.18, where the coefficients must be multiplied by M/C to yield the w displacement. The nodal line extending from point 0 to point m indicates that an octant of the cross section could have been used for this analysis instead of the quadrant. However, this nodal line serves as a check in the validity of the numerical analysis. Quite often such sketches of deformed elastic bodies provide visual or intuitive means of checking the correctness of computer programs or the input data for the stress analyst.

Torsional Stresses in Shafts Having Fillets/As an application of the element for torsion of bars of varying circular section, let us consider torsion of a circular shaft with a small radius fillet as shown in Fig. 15-5.19. Because of its considerable practical interest, this problem has been studied by many different methods. We show how the finite element solution compares with other solutions.

In this study, we shall consider principally the variation of the stress concentration factor k with respect to the fillet radius r. For this purpose, we let $D/d = 1.50$ (see Fig. 15-5.19), and subdivide the region of the shaft roughly into 120 linear triangular elements with finer mesh

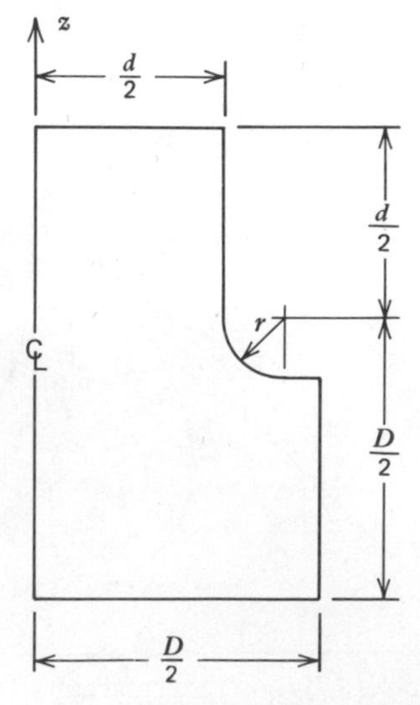

Fig. 15-5.19/Circular shaft with a small fillet.

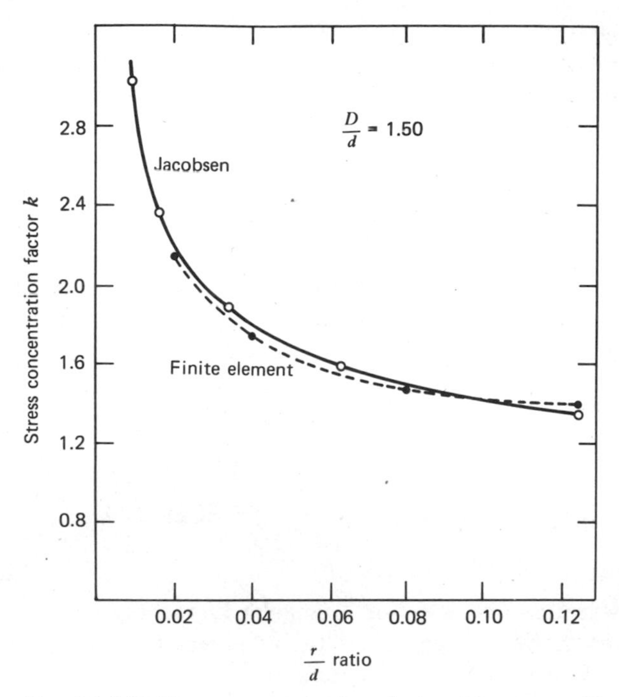

Fig. 15-5.20/Stress concentration factor k versus r/d ratio.

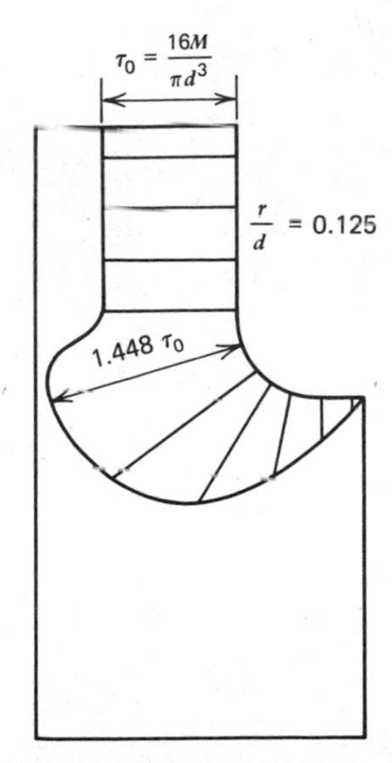

Fig. 15-5.21/Shearing distribution along the boundary.

near the fillet. The finite element analyses were done for four different values of r/d; namely, $r/d=0.02$, 0.04, 0.08, and 0.125. By defining the stress concentration factor k as the ratio between the peak stress τ_{max} in the fillet to the nominal computed torsional stress in the smaller shaft τ_0, we obtain four numerical values of k. They are plotted in Fig. 15-5.20 together with Jacobsen's results.† Figure 15-5.20 presents evidence that the two methods are in good agreement. The shearing stress distribution along the boundary of the filleted member for the case $r/d=0.125$ is shown in Fig. 15-5.21. It may be seen that the maximum stress occurs at a point located near the point of tangency of the fillet with the outline of the smaller part. This result agrees with that found by Willers in 1907.[38]

APPENDIX 15A

INTEGRATION FORMULAS FOR FINITE ELEMENTS

15A-1. Integration Formulas for Triangular Elements / Let a triangle (1, 2, 3) with area A be defined in the (x, y) plane. In terms of coordinates of the vertices, (x_1, y_1), (x_2, y_2) and (x_3, y_3) we have the following integrals:

$$\iint_A dx\,dy = \tfrac{1}{2}[x_1(y_2-y_3)+x_2(y_3-y_1)+x_3(y_1-y_2)] = A$$

$$\iint_A x\,dx\,dy = \frac{A}{3}[x_1+x_2+x_3]$$

$$\iint_A y\,dx\,dy = \frac{A}{3}[y_1+y_2+y_3]$$

$$\iint_A x^2\,dx\,dy = \frac{A}{12}\left[x_1+x_2+x_3)^2+(x_1^2+x_2^2+x_3^2)\right]$$

$$\iint_A y^2\,dx\,dy = \frac{A}{12}\left[(y_1+y_2+y_3)^2+(y_1^2+y_2^2+y_3^2)\right]$$

$$\iint_A xy\,dx\,dy = \frac{A}{12}[(x_1+x_2+x_3)(y_1+y_2+y_3)+(x_1y_1+x_2y_2+x_3y_3)]$$

$$\iint_A x^3\,dx\,dy = \frac{A}{10}\left[(x_1+x_2+x_3)(x_1^2+x_2^2+x_3^2)+x_1x_2x_3\right]$$

$$\iint_A y^3\,dx\,dy = \frac{A}{10}\left[(y_1+y_2+y_3)(y_1^2+y_2^2+y_3^2)+y_1y_2y_3\right]$$

†An electrical analogy was successfully employed by L. S. Jacobsen in obtaining stress concentration factors; see reference 37.

$$\iint_A x^4\,dx\,dy = \frac{A}{15}\left[(x_1 + x_2 + x_3)\left(x_1^3 + x_2^3 + x_3^3 + x_1x_2x_3\right) + x_1^2x_2^2 + x_2^2x_3^2 + x_3^2x_1^2\right]$$

$$\iint_A y^4\,dx\,dy = \frac{A}{15}\left[(y_1 + y_2 + y_3)\left(y_1^3 + y_2^3 + y_3^3 + y_1y_2y_3\right) + y_1^2y_2^2 + y_2^2y_3^2 + y_3^2y_1^2\right]$$

$$\iint_A x^2y\,dx\,dy = \frac{A}{30}\left[(y_1 + y_2 + y_3)(x_1 + x_2 + x_3)^2 + y_1\left(2x_1^2 - x_2x_3\right) + y_2\left(2x_2^2 - x_3x_1\right) + y_3\left(2x_3^2 - x_1x_2\right)\right]$$

$$\iint_A xy^2\,dx\,dy = \frac{A}{30}\left[(x_1 + x_2 + x_3)(y_1 + y_2 + y_3)^2 + x_1\left(2y_1^2 - y_2y_3\right) + x_2\left(2y_2^2 - y_3y_1\right) + x_3\left(2y_3^2 - y_1y_2\right)\right]$$

$$\begin{aligned}\iint_A x^3y\,dx\,dy = \frac{A}{60}\Big\{&(y_1 + y_2 + y_3)\left[(x_1 + x_2 + x_3)\left(x_1^2 + x_2^2 + x_3^2\right) + 2x_1x_2x_3\right] + (x_1y_1 + x_2y_2 + x_3y_3)\left(x_1^2 + x_2^2 + x_3^2\right)\\ &+ 2(x_1 + x_2 + x_3)\left(x_1^2y_1 + x_2^2y_2 + x_3^2y_3\right)\Big\}\end{aligned}$$

$$\begin{aligned}\iint_A xy^3\,dx\,dy = \frac{A}{60}\Big\{&(x_1 + x_2 + x_3)\left[(y_1 + y_2 + y_3)\left(y_1^2 + y_2^2 + y_3^2\right) + 2y_1y_2y_3\right] + (x_1y_1 + x_2y_2 + x_3y_3)\left(y_1^2 + y_2^2 + y_3^2\right)\\ &2(y_1 + y_2 + y_3)\left(x_1y_1^2 + x_2y_2^2 + x_3y_3^2\right)\Big\}\end{aligned}$$

$$\begin{aligned}\iint_A x^2y^2\,dx\,dy = \frac{A}{180}\Big\{&\left[(x_1 + x_2 + x_3)(y_1 + y_2 + y_3)\right]^2\\ &+ \left(x_1^2 + x_2^2 + x_3^2\right)\left(y_1^2 + y_2^2 + y_3^2\right)\\ &+ 2\left(x_1^2y_1^2 + x_2^2y_2^2 + x_3^2y_3^2\right)\\ &+ 4(x_1 + x_2 + x_3)\left(x_1y_1^2 + x_2y_2^2 + x_3y_3^2\right)\\ &+ 4(y_1 + y_2 + y_3)\left(x_1^2y_1 + x_2^2y_2 + x_3^2y_3\right)\\ &+ 2y_1y_2x_1(x_2 + x_3) + 2y_2y_3x_2(x_3 + x_1)\\ &+ 2y_3y_1x_3(x_1 + x_2)\Big\}\end{aligned}$$

15A-2. Integration Formulas for Rectangular Elements/For the rectangular element shown in Fig. 15-3.8, we have the following integration formulas:

$$\iint_A dx\,dy = ab \qquad \iint_A x\,dx\,dy = \tfrac{1}{2}(a^2b) \qquad \iint_A y\,dx\,dy = \tfrac{1}{2}(ab^2)$$

$$\iint_A x^2\,dx\,dy = \tfrac{1}{3}(a^3b) \qquad \iint_A y^2\,dx\,dy = \tfrac{1}{3}(ab^3) \qquad \iint_A xy\,dx\,dy = \tfrac{1}{4}(a^2b^2)$$

$$\iint_A x^3\,dx\,dy = \tfrac{1}{4}(a^4b) \qquad \iint_A y^3\,dx\,dy = \tfrac{1}{4}(ab^4) \qquad \iint_A x^4\,dx\,dy = \tfrac{1}{5}(a^5b)$$

$$\iint_A y^4\,dx\,dy = \tfrac{1}{5}(ab^5) \qquad \iint_A x^2y\,dx\,dy = \tfrac{1}{6}(a^3b^2) \qquad \iint_A xy^2\,dx\,dy = \tfrac{1}{6}(a^2b^3)$$

$$\iint_A x^3y\,dx\,dy = \tfrac{1}{8}(a^4b^2) \qquad \iint_A xy^3\,dx\,dy = \tfrac{1}{8}(a^2b^4) \qquad \iint_A x^2y^2\,dx\,dy = \tfrac{1}{9}(a^3b^3)$$

15A-3. The $\{\overline{P}\}_i$ Load Matrix for STE/There are 12 elements in the $\{\overline{P}\}_i$ matrix due to the element boundary traction along the length L_{35}.

$$(1) = \frac{hL_{35}}{4}(p_{x3} + 2p_{x4} + p_{x5})$$

$$(2) = \frac{hL_{35}}{12}[(2x_3 + x_4)p_{x3} + (x_3 + 4x_4 + x_5)p_{x4} + (x_4 + 2x_5)p_{x5}]$$

$$(3) = \frac{hL_{35}}{12}[(2y_3 + y_4)p_{x3} + (y_3 + 4y_4 + y_5)p_{x4} + (y_4 + 2y_5)p_{x5}]$$

$$(4) = \frac{hL_{35}}{24}[(3x_3^2 + 2x_3x_4 + x_4^2)p_{x3} + (x_3^2 + 2x_3x_4 + 6x_4^2 + 2x_4x_5 + x_5^2)p_{x4} + (x_4^2 + 2x_4x_5 + 3x_5^2)p_{x5}]$$

$$(5) = \frac{hL_{35}}{24}[(3x_3y_3 + x_3y_4 + x_4y_3 + x_4y_4)p_{x3} + (x_3y_3 + x_3y_4 + x_4y_3 + 6x_4y_4 + x_4y_5 + x_5y_4 + x_5y_5)p_{x4} + (x_4y_4 + x_4y_5 + x_5y_4 + 3x_5y_5)p_{x5}]$$

$$(6) = \frac{hL_{35}}{24}[(3y_3^2 + 2y_3y_4 + y_4^2)p_{x3} + (y_3^2 + 2y_3y_4 + 6y_4^2 + 2y_4y_5 + y_5^2)p_{x4} + (y_4^2 + 2y_4y_5 + 3y_5^2)p_{x5}]$$

$$(7) = \frac{hL_{35}}{4}(p_{y3} + 2p_{y4} + p_{y5})$$

$$(8) = \frac{hL_{35}}{12}[(2x_3 + x_4)p_{y3} + (x_3 + 4x_4 + x_5)p_{y4} + (x_4 + 2x_5)p_{y5}]$$

$$(9) = \frac{hL_{35}}{12}\left[(2y_3 + y_4)p_{y3} + (y_3 + 4y_4 + y_5)p_{y4} + (y_4 + 2y_5)p_{y5}\right]$$

$$(10) = \frac{hL_{35}}{24}\left[(3x_3^2 + 2x_3x_4 + x_4^2)p_{y3} + (x_3^2 + 2x_3x_4 + 6x_4^2 + 2x_4x_5 + x_5^2)p_{y4} + (x_4^2 + 2x_4x_5 + 3x_5^2)p_{y5}\right]$$

$$(11) = \frac{hL_{35}}{24}\left[(3x_3y_3 + x_3y_4 + x_4y_3 + x_4y_4)p_{y3} + (x_3y_3 + x_3y_4 + x_4y_3 + 6x_4y_4 + x_4y_5 + x_5y_4 + x_5y_5)p_{y4} + (x_4y_4 + x_4y_5 + x_5y_4 + 3x_5y_5)p_{y5}\right]$$

$$(12) = \frac{hL_{35}}{24}\left[(3y_3^2 + 2y_3y_4 + y_4^2)p_{y3} + (y_3^2 + 2y_3y_4 + 6y_4^2 + 2y_4y_5 + y_5^2)p_{y4} + (y_4^2 + 2y_4y_5 + 3y_5^2)p_{y5}\right]$$

REFERENCES

1. O. C. Zienkiewicz, *The Finite Element Method*, 3rd Ed., McGraw-Hill, New York, 1977.
2. H. L. Langhaar and S. C. Chu, "Piecewise Polynomials and the Partition Method for Ordinary Differential Equations," *Proceedings, 4th Southeastern Conference on Theoretical and Applied Mechanics*, 1968.
3. S. H. Crandall, *Engineering Analysis*, McGraw-Hill, New York, 1956.
4. B. A. Finlayson and L. E. Scriven, "The Method of Weighted Residuals—A Review," *Journal of Applied Mechanics Reviews*, Vol. 19, No. 9, September 1966.

4a. B. A. Finlayson, *The Method of Weighted Residuals and Variational Principles with Application in Fluid Mechanics, Heat and Mass Transfer*, Academic Press, New York, 1972.

5. J. H. Argyris and P. C. Patton, "Computer Oriented Research in a University Milieu," *Journal of Applied Mechanics Reviews*, Vol. 19, No. 12, December, 1966.
6. Yuan-Yu Hsien, *Elementary Theory of Structures*, 2nd Ed., Prentice-Hall, Englewood Cliffs, New Jersey, 1982.
7. N. M. Newmark, "Numerical Methods of Analysis of Bars, Plates, and Elastic Bodies," Chapter 9, in *Numerical Methods of Analysis in Engineering*, Macmillan, New York, 1949.
8. A. H. S. Ang, "Numerical Approach for Wave Motions in Nonlinear Solid Media," *Proceedings of the Conference on Matrix Methods in Structural Mechanics*, AFFDL-TR-66-80, Wright-Patterson Air Force Base, Ohio, October 1965.

9. R. W. Clough, "The Finite Element Method in Structural Mechanics," Chapter 7, in *Stress Analysis*, Wiley, New York, 1965.
9a. L. J. Segerlind, *Applied Finite Element Analysis*, Wiley, New York, 1976.
10. J. H. Argyris and S. Kelsey, *Energy Theorems and Structural Analysis*, Butterworths, London, 1960.
11. R. Courant, "Variational Methods for the Solution of Problems of Equilibrium and Vibrations," *Bulletin*, *American Mathematical Society*, Vol. 49, 1943.
12. J. L. Synge, *The Hypercircle in Mathematical Physics*, Cambridge University Press, Cambridge, 1957.
13. M. J. Turner, R. W. Cough, H. C. Martin, and L. J. Topp, "Stiffness and Deflection Analysis of Complex Structures," *Journal of the Aeronautical Sciences*, Vol. 23, No. 9, 1956, pp. 805–823.
14. K.-J. Bathe and E. L. Wilson, *Numerical Methods in Finite Element Analysis*, Prentice-Hall, Englewood Cliffs, New Jersey, 1976, Chapter 3.
15. L. R. Herrmann, "A Bending Analysis for Plates," *Proceedings of the Conference on Matrix Methods in Structural Mechanics*, AFFDL-TR-66-80, Wright-Patterson Air Force Base, Ohio, October 1965.
16. T. H. H. Pian and P. Tong, "Basis of Finite Element Methods for Solid Continua," *International Journal for Numerical Methods in Engineering*, Vol. 1, No. 1, January 1969.
17. E. P. Popov, *Mechanics of Materials*, 2nd Ed., Prentice-Hall, Englewood Cliffs, New Jersey, 1978, Chapter 11.
18. H. L. Langhaar, *Energy Methods in Applied Mechanics*, Wiley, New York, 1962, Chapter 1.
19. K.-J. Bathe, *Finite Element Procedures in Engineering Analysis*, Prentice-Hall, Englewood Cliffs, New Jersey, 1982, Chapter 5.
20. R. W. Clough, "The Finite Element Method in Plane Stress Analysis," *Proceedings of the 2nd ASCE Conference on Electronic Computation*, Pittsburgh, September 1960.
21. J. H. Argyris, S. Kelsey, and H. Kamel, "Matrix Methods of Structural Analysis—A Precis of Recent Developments," in *Matrix Methods of Structural Analysis*, B. F. deVeubeke, Ed., Macmillan, New York, 1964.
22. A. P. Boresi and P. P. Lynn, *Elasticity in Engineering Mechanics*, Prentice-Hall, Englewood Cliffs, New Jersey, 1974, Chapter 4.
23. C. A. Felippa, "Refined Finite Element Analysis of Linear and Nonlinear Two-Dimensional Structures," Report No. 66-22, Structure and Materials Research Department of Civil Engineering, University of California, Berkeley, October 1966.
24. R. W. Luft, J. M. Roesset, and Jerome J. Connor, "Automatic Generation of Finite Element Matrices," *Fifth Conference on Electronic Computation*, August 31, 1970, American Society of Civil Engineers, New York.

25. R. H. Gunderson and A. Cetiner, "A Stiffness Matrix Generator," *Fifth Conference on Electronic Computation*, August 31, 1970, American Society of Civil Engineers, New York.
26. M. J. Turner, H. C. Martin, and R. C. Weikel, "Further Development and Applications of the Stiffness Method," in *Matrix Methods of Structural Analysis*, B. F. deVeubeke, Ed., Macmillan, New York, 1964, pp. 203–266.
27. F. deVeubeke, "Displacement and Equilibrium Models in Finite Element Method," in *Stress Analysis*, O. C. Zienkiewicz and G. S. Holister, Eds., Wiley, New York, 1965, Chapter 9.
28. S. P. Timoshenko and J. N. Goodier, *Theory of Elasticity*, 3rd Ed., McGraw-Hill, New York, 1970, Arts. 21 and 24.
29. A. E. H. Love, *The Mathematical Theory of Elasticity*, 4th Ed., Dover, New York, 1944, Chapter VIII.
30. R. W. Clough and Y. Rashid, "Finite Element Analysis of Axisymmetric Solids," *Journal of Engineering Mechanics, ASCE*, Vol. 18, February 1965, pp. 71–85.
31. E. L. Wilson, "Structural Analysis of Axisymmetric Solids," *Journal of the American Institute of Aeronautics and Astronomy*, Vol. 3, No. 12, Dec. 1965, pp. 2269–2274.
32. J. L. Meek and G. Carey, "Axisymmetric Solution of Elastic-Plastic Problems by Finite Element Methods," Bulletin No. 11, Department of Civil Engineering, University of Queensland, Australia, 1969.
33. J. N. Goodier, "Concentration of Stress around Spherical and Cylindrical Inclusions and Flaws," *Journal of Applied Mathematics, Trans. ASME*, Vol. 55, 1933, p. A-39.
34. L. R. Herrmann, "Elastic Torsional Analysis of Irregular Shapes," *Journal of the Engineering Mechanics Division, ASCE*, Vol. 91, No. EM6, December 1965, pp. 11–19.
35. J. H. Tuth, "Torsional Stress Concentration in Angle and Square Tube Fillets," *Journal of Applied Mechanics*, Vol. 17, No. 4, 1950, pp. 388–390.
36. J. H. Mitchell, "Torsion of a Rod of Varying Cross Section," *Proceedings of the London Mathematical Society*, Vol. 31, 1899, pp. 140–141.
37. L. S. Jacobsen, "Torsional Stress in Shafts Having Grooves or Fillets," *Transactions of the ASME*, Vol. 57, 1935, pp. A154–155.
38. F. A. Willers, "Die Torsion eines Rotationskorpers um seine Achse," *Zeitung fur Mathematik und Physik*, Vol. 55, 1970, p. 225.

Additional References

1. K. J. Bathe, *Finite Element Procedures in Engineering Analysis*, Prentice-Hall, Englewood Cliffs, New Jersey, 1982.

2. K. J. Bathe, and E. L. Wilson, *Numerical Methods in Finite Element Analysis*, Prentice-Hall, Englewood Cliffs, New Jersey, 1976.
3. E. B. Becker, G. F. Carey, and J. T. Oden, *Finite Elements*, Prentice-Hall, Englewood Cliffs, New Jersey, 1981.
4. Y. K. Cheung, *Finite Strip Method in Structural Analysis*, Pergamon Press, New York, 1976.
5. Y. K. Cheung, and M. F. Yeo, *A Practical Introduction to Finite Element Analysis*, Pitman, London, 1979.
6. R. D. Cook, *Concepts and Applications of Finite Element Analysis*, 2nd Ed., Wiley, New York, 1981.
7. C. S. Desai, *Elementary Finite Element Method*, Prentice-Hall, Englewood Cliffs, New Jersey, 1979.
8. G. Dhatt, G. Touzot, and G. Cantin (translator), *The Finite Element Method Displayed*, Wiley, New York, 1984.
9. R. H. Gallagher, *Finite Element Analysis Fundamentals*, Prentice-Hall, Englewood Cliffs, New Jersey, 1975.
10. B. Irons, and N. Shrive, *Finite Element Primer*, Wiley, New York, 1983.
11. K. H. Huebner, *The Finite Element Method*, Wiley, New York, 1975.
12. D. H. Norrie, and G. deVries, *The Finite Element Method—Fundamentals and Applications*, Academic Press, New York, 1973.
13. J. T. Oden, *Finite Elements of Nonlinear Continua*, McGraw-Hill, New York, 1972.
14. K. C. Rockey, H. R. Evans, D. W. Griffiths, and D. A. Nethercot, *The Finite Element Method*, 2nd Ed., Wiley, New York, 1975, 1983.
15. M. F. Rubinstein, *Structural Systems—Statics, Dynamics and Stability*, Prentice-Hall, Englewood Cliffs, New Jersey, 1970.
16. L. J. Segerlind, *Applied Finite Element Analysis*, Wiley, New York, 1976.
17. I. M. Smith, *Programming the Finite Element Method*, Wiley, New York, 1982.
18. G. Strang, and G. J. Fix, *An Analysis of the Finite Element Method*, Prentice-Hall, Englewood Cliffs, New Jersey, 1973.
19. W. Weaver, Jr., and P. R. Johnston, *Finite Elements for Structural Analysis*, Prentice-Hall, Englewood Cliffs, New Jersey, 1984.
20. J. R. Whiteman, *A Bibliography for Finite Elements*, Academic Press, New York, 1975.

APPENDIX

SECOND MOMENT (MOMENT OF INERTIA) OF A PLANE AREA

A-1 MOMENTS OF INERTIA OF A PLANE AREA

The derivation of load-stress formulas for torsion members and for beams may require solutions of one or more of the following integrals:

$$I_x = \int y^2 \, dA \tag{A-1.1}$$

$$I_y = \int x^2 \, dA \tag{A-1.2}$$

$$J = \int r^2 \, dA \tag{A-1.3}$$

$$I_{xy} = \int xy \, dA \tag{A-1.4}$$

where dA is an element of the plane area A lying in the (x, y)-plane in Fig. A-1.1. Area A represents the cross-sectional area of a member subjected to bending and or torsional loads.

The integrals in Eqs. (A-1.1), (A-1.2), and (A-1.3) are commonly called moments of inertia of the area A because of the similarity with integrals that define the moment of inertia of bodies in the field of dynamics. Since an area cannot have an inertia, moment of inertia of an area is a misnomer. We use the term because of common usage.

The integral represented by Eq. (A-1.4) is called the product of inertia. Its sign can be negative. The moment of inertia and product of inertia is given the symbol I if the axes about which the moments are taken lie in the plane of the area (see Eqs. A-1.1, A-1.2, and A-1.4). When the axis about which the moment is taken is perpendicular to the area (see Eq.

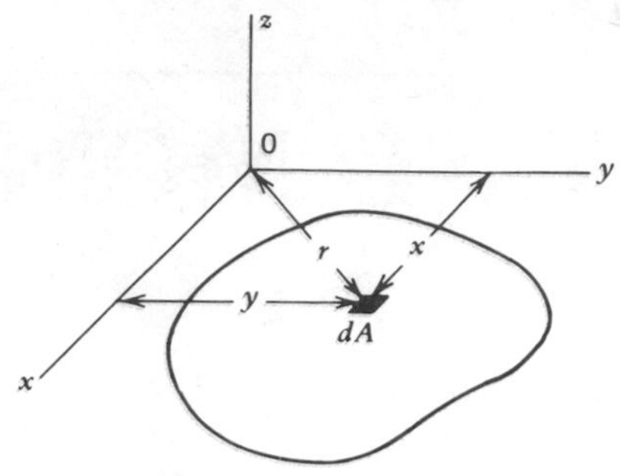

Fig. A-1.1

A-1.3), the moment of inertia is given the symbol J and is called the polar moment of inertia of the area.

A-2 PARALLEL AXIS THEOREM

In the application of Eqs. (A-1.1), (A-1.2), (A-1.3), and (A-1.4) to engineering problems, it is convenient to know these integrals for coordinate axes at the centroid of area A. The values of the integrals for a few cross sections are listed in Table A-2.1. Often, practical members have cross sections that are composed of two or more of simple cross sections (Table A-2.1). Moments of inertia for composite areas are obtained by application of the parallel axis theorem.

Let it be required to obtain moments of inertia for area A in Fig. A-2.1 for coordinate axes (x', y', z'). Area A lies in the (x', y')-plane. First, locate coordinate axes (x, y, z) with axes parallel, respectively, to the (x', y', z')-axes and with the origin 0 at the centroid of A. Let the distances of the centroid 0 from the axes (x', y') be $(\bar{x}, \bar{y})$. Then, $\bar{r} = \sqrt{\bar{x}^2 + \bar{y}^2}$ is the distance between the z'-axis and the z-axis. Using Eqs.

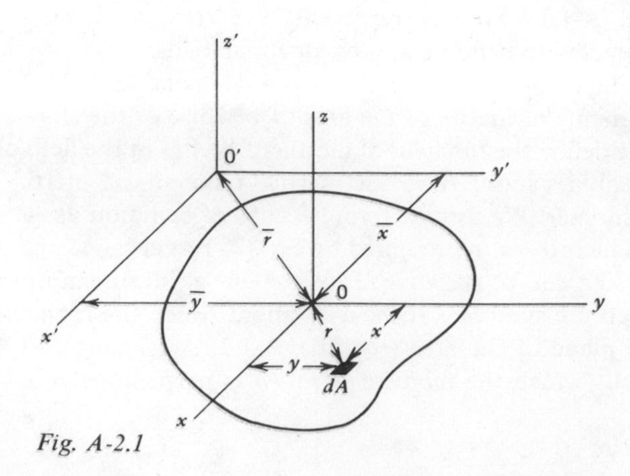

Fig. A-2.1

Table A-2.1

Moments of Inertia of Common Plane Areas

Shape	Formulas
Rectangle	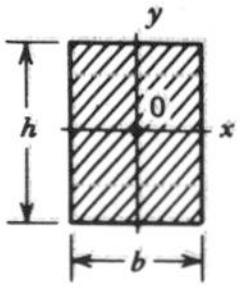 $I_x = bh^3/12$ $I_y = hb^3/12$ $J_0 = (bh^3 + hb^3)/12$ $I_{xy} = 0$
Right Triangle	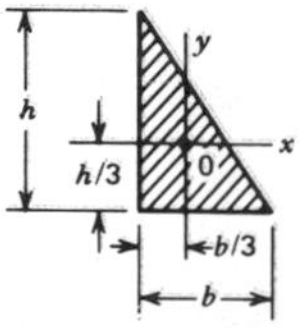 $I_x = bh^3/36$ $I_y = hb^3/36$ $J_0 = (bh^3 + hb^3)/36$ $I_{xy} = -b^2h^2/72$
Circle	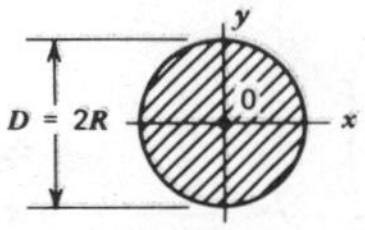 $I_x = \pi D^4/64 = \pi R^4/4$ $I_y = \pi D^4/64 = \pi R^4/4$ $J_0 = \pi D^4/32 = \pi R^4/2$ $I_{xy} = 0$
Ellipse	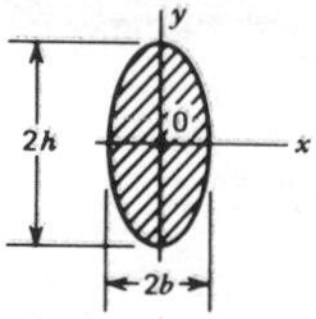 $I_x = \pi bh^3/4$ $I_y = \pi hb^3/4$ $J_0 = \pi bh(h^2 + b^2)/4$ $I_{xy} = 0$
Semicircle	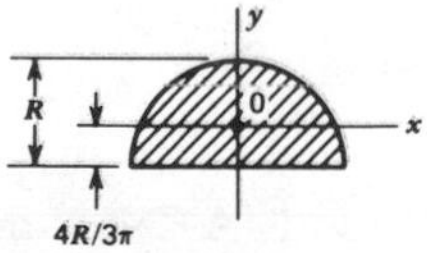 $I_x = \pi R^4(1/8 - 8/9\pi^2)$ $I_y = \pi R^4/8$ $J_0 = \pi R^4(1/4 - 8/9\pi^2)$ $I_{xy} = 0$
Semi-ellipse	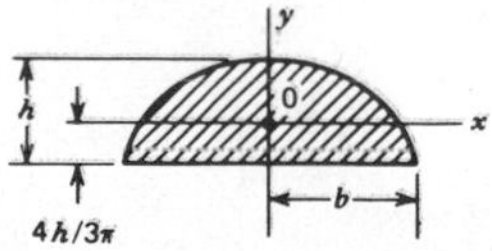 $I_x = \pi bh^3(1/8 - 8/9\pi^2)$ $I_y = \pi hb^3/8$ $J_0 = \pi bh(h^2/8 - 8h^2/9\pi^2 + b^2/8)$ $I_{xy} = 0$

(A-1.1), (A-1.2), (A-1.3), and (A-1.4), we obtain

$$I_{x'} = \int (y + \bar{y})^2 \, dA = I_x + A\bar{y}^2 \tag{A-2.1}$$

$$I_{y'} = \int (x + \bar{x})^2 \, dA = I_y + A\bar{x}^2 \tag{A-2.2}$$

$$J_{0'} = \int [(x + \bar{x})^2 + (y + \bar{y})^2] \, dA = J_0 + A\bar{r}^2 \tag{A-2.3}$$

$$I_{x'y'} = \int (x + \bar{x})(y + \bar{y}) \, dA = I_{xy} + A\overline{xy} \tag{A-2.4}$$

where integrals $\int y\,dA$ and $\int x\,dA$ are zero since the first moment of an area with respect to an axis through the centroid of the area vanishes. Equations (A-2.1) through (A-2.4) represent parallel axes formulas for moments of inertia for an area. They may be employed to obtain the moments of inertia of composite areas.

EXAMPLE A-2.1

Moments of Inertia for Z-bar

A Z-bar has the cross section shown in Fig. EA-2.1. Determine I_x, I_y, and I_{xy} for the centroidal axes (x, y) shown.

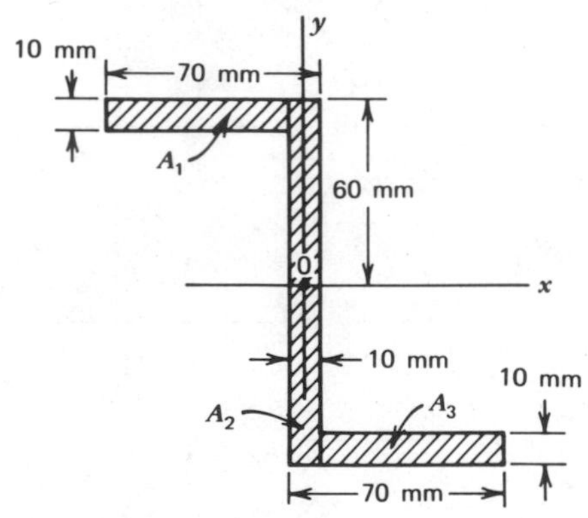

Fig. EA-2.1

SOLUTION

The area is divided into three rectangular areas A_1, A_2, and A_3 (Fig. EA-2.1). Using Eqs. (A-2.1), (A-2.2), and (A-2.4), and Table A-2.1, we obtain

$$I_x = \frac{60(10)^3}{12} + 60(10)(55)^2 + \frac{10(120)^3}{12} + 120(10)(0)^2$$

$$+ \frac{60(10)^3}{12} + 60(10)(-55)^2 = 5.08 \times 10^6 \text{ mm}^4$$

$$I_y = \frac{10(60)^3}{12} + 60(10)(-35)^2 + \frac{120(10)^3}{12} + 120(10)(0)^2$$

$$+ \frac{10(60)^3}{12} + 60(10)(35)^2 = 1.84 \times 10^6 \text{ mm}^4$$

$$I_{xy} = 60(10)(-35)(55) + 120(10)(0)(0) + 60(10)(35)(-55)$$

$$= -2.31 \times 10^6 \text{ mm}^2$$

PROBLEM SET A-2

1. Derive the expressions for I_x and I_{xy} for the right triangle in Table A-2.1.

2. Derive the expression for I_x for the semiellipse in Table A-2.1.

3. Determine I_x, I_y, and I_{xy} for the centroidal axes for the cross-sectional area shown in Fig. PA-2.3.

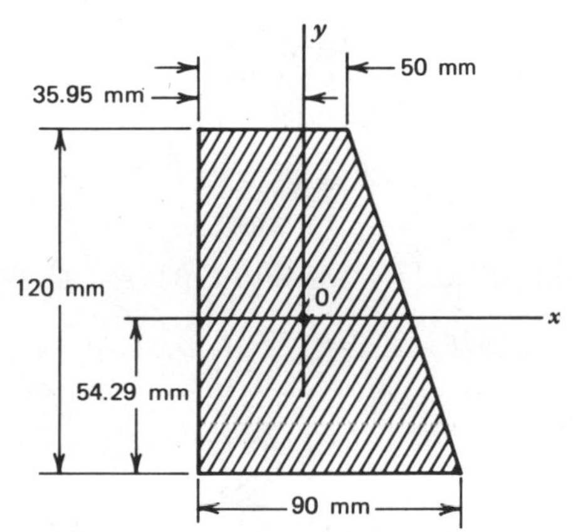

Fig. PA-2.3

Ans. $I_x = 9.806 \times 10^6\ \text{mm}^4$, $I_y = 3.982 \times 10^6\ \text{mm}^4$,
$I_{xy} = -1.634 \times 10^6\ \text{mm}^4$

4. Determine I_x, I_y, and I_{xy} for the centroidal axes for the cross sectional area shown in Fig. PA-2.4.

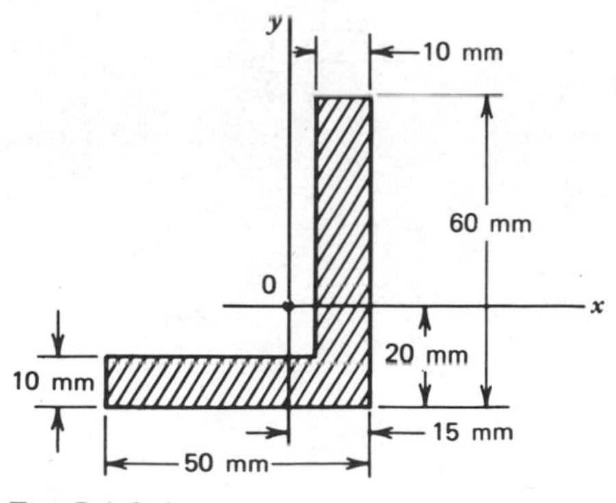

Fig. PA-2.4

Ans. $I_x = 333.3 \times 10^3\ \text{mm}^4$, $I_y = 208.3 \times 10^3\ \text{mm}^4$, $I_{xy} = 150.0 \times 10^3\ \text{mm}^4$

A-3
TRANSFORMATION EQUATIONS FOR MOMENTS AND PRODUCTS OF INERTIA

Let I_x, I_y, and I_{xy} be known moments and product of inertia for area A (Fig. A-3.1) for (x, y) rectangular axes that lie in the plane of the area. Consider the (X, Y) coordinate axes that have the same origin and same plane as the (x, y)-axes. We wish to derive transformation equations by which I_X, I_Y, and I_{XY} are obtained in terms of I_x, I_y, I_{xy}, and θ, the angle through which the (x, y)-axes must be rotated to coincide with the (X, Y)-axes; θ is positive in the counterclockwise sense. Consider an element of area dA at X and Y coordinates given by the relations

$$\begin{aligned} X &= x\cos\theta + y\sin\theta \\ Y &= y\cos\theta - x\sin\theta \end{aligned} \tag{A-3.1}$$

Substitution of Eqs. (A-3.1) into Eqs. (A-1.1), (A-1.2), and (A-1.4) gives

$$I_X = \int (y\cos\theta - x\sin\theta)^2\, dA = I_x\cos^2\theta + I_y\sin^2\theta - 2I_{xy}\sin\theta\cos\theta$$

$$I_Y = \int (x\cos\theta + y\sin\theta)^2\, dA = I_x\sin^2\theta + I_y\cos^2\theta + 2I_{xy}\sin\theta\cos\theta$$

$$\begin{aligned} I_{XY} &= \int (x\cos\theta + y\sin\theta)(y\cos\theta - x\sin\theta)\, dA \\ &= (I_x - I_y)\sin\theta\cos\theta + I_{xy}(\cos^2\theta - \sin^2\theta) \end{aligned} \tag{A-3.2}$$

With double angle identities, Eq. (A-3.2) can be written in the form

$$\begin{aligned} I_X &= \frac{I_x + I_y}{2} + \frac{I_x - I_y}{2}\cos 2\theta - I_{xy}\sin 2\theta \\ I_Y &= \frac{I_x + I_y}{2} - \frac{I_x - I_y}{2}\cos 2\theta + I_{xy}\sin 2\theta \\ I_{XY} &= \frac{I_x - I_y}{2}\sin 2\theta + I_{xy}\cos 2\theta \end{aligned} \tag{A-3.3}$$

Note the similarity between the transformation equations for moments

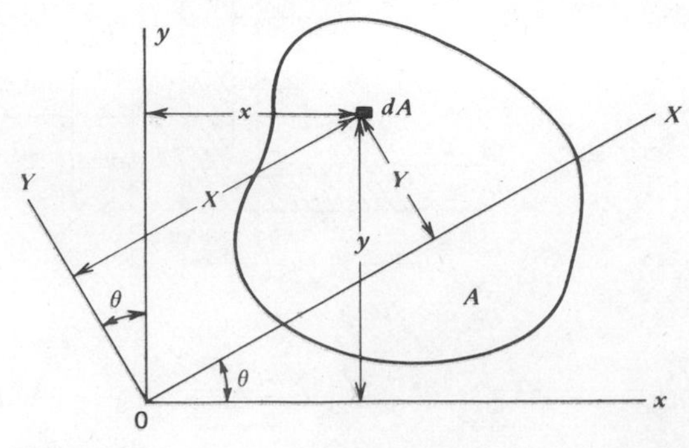

Fig. A-3.1

and products of inertia and the transformation equations of stress given by Eq. (1-4.19). Like stress components and strain components, moments and products of inertia transform according to the rule for second-order symmetric tensors.

Principal Axes of Inertia/There are two values of θ for which $I_{XY}=0$. To determine these values, let $I_{XY}=0$. Then, the third of Eqs. (A-3.3) yields

$$\tan 2\theta = -\frac{2I_{xy}}{I_x - I_y} \tag{A-3.4}$$

The two values of θ given by Eq. (A-3.4) locate two positions of axes (X, Y) that represent the principal axes of inertia for a given cross sectional area. In the discussion that follows, we assume for definitiveness that I_x is greater than I_y. Then, the maximum moment of inertia, which we take to be I_X and which is associated with the X-axis, will occur for the X-axis located at the smallest of the two values of θ from the x-axis; the direction is counterclockwise for a positive value of θ and clockwise for a negative value of θ. Since $I_x - I_y > 0$, if we substitute the value of θ given by Eq. (A-3.4) into the first and second of Eqs. (A-3.3), we find that

$$\begin{aligned} I_X &= \frac{I_x + I_y}{2} + \sqrt{\left[\frac{I_x - I_y}{2}\right]^2 + I_{xy}^2} \\ I_Y &= \frac{I_x + I_y}{2} - \sqrt{\left[\frac{I_x - I_y}{2}\right]^2 + I_{xy}^2} \end{aligned} \tag{A-3.5}$$

as the principal moments of inertia for the cross-sectional area A.

EXAMPLE A-3.1

Principal Axes for Z-Bar

Locate the principal axes and determine the principal moments of inertia I_X and I_Y for the Z-bar whose dimensions are specified in Fig. EA-2.1.

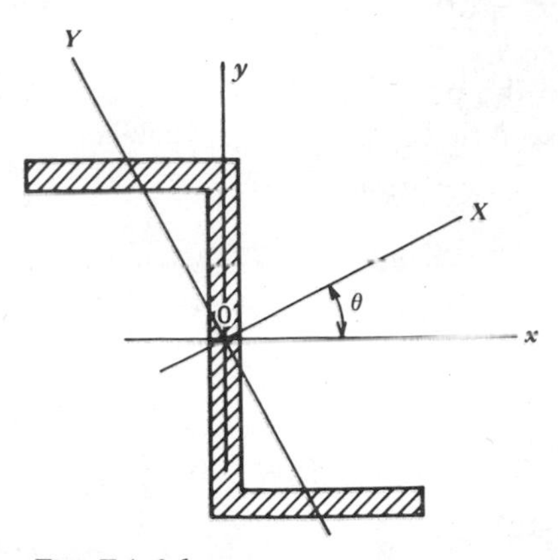

Fig. EA-3.1

SOLUTION

Since $I_x = 5.08 \times 10^6$ mm^4, $I_y = 1.84 \times 10^6$ mm^4, and $I_{xy} = -2.31 \times 10^6$ mm^4, the principal values for the moments of inertia are given by Eqs. (A-3.5).

$$I_X = \frac{5.08 \times 10^6 + 1.84 \times 10^6}{2} + \sqrt{\left[\frac{5.08 \times 10^6 - 1.84 \times 10^6}{2}\right]^2 + (-2.31 \times 10^6)^2}$$
$$= 6.281 \times 10^6 \text{ mm}^4$$

$$I_Y = \frac{5.08 \times 10^6 + 1.84 \times 10^6}{2} - \sqrt{\left[\frac{5.08 \times 10^6 - 1.84 \times 10^6}{2}\right] + (-2.31 \times 10^6)^2}$$
$$= 0.639 \times 10^6 \text{ mm}^4$$

The location of the X-axis is given by Eq. (A-3.4). Thus,

$$\tan 2\theta = \frac{-2(-2.31 \times 10^6)}{5.08 \times 10^6 - 1.84 \times 10^6} = 1.4259$$
$$\theta = 0.4796 \text{ rad}$$

Hence, the X-axis is located at 0.4796 rad, measured counterclockwise from the x-axis, as shown in Fig. EA-3.1.

PROBLEM SET A-3.1

1. Locate principal axes (X, Y) and determine I_X and I_Y for the cross-sectional area in Problem A-2.3.

   Ans. $I_X = 10.233 \times 10^6$ mm^4, $I_Y = 3.555 \times 10^6$ mm^4, $\theta = 0.2557$ rad

2. Locate the principal axes (X, Y) and determine I_X and I_Y for the cross sectional area in Problem A-2.4.

   Ans. $I_X = 433.3 \times 10^3$ mm^4, $I_Y = 108.3 \times 10^3$ mm^4, $\theta = -0.5880$ rad

INDEX